Information-theoretic Cryptography

This book offers a mathematical foundation for modern cryptography. It is primarily intended as an introduction for graduate students. Readers should have a basic knowledge of probability theory, but familiarity with computational complexity is not required. Starting from Shannon's classic result on secret key cryptography, fundamental topics of cryptography, such as the secret key agreement, authentication, secret sharing, and secure computation, are covered. Particular attention is drawn to how correlated randomness can be used to construct cryptographic primitives. To evaluate the efficiency of such constructions, information-theoretic tools, such as smooth min/max entropies and information spectrum, are developed. The broad coverage means the book will be useful to experts as well as students in cryptography as a reference for information-theoretic concepts and tools.

Himanshu Tyagi is an associate professor in the Department of Electrical Communication Engineering at the Indian Institute of Science, Bangalore. A specialist in information theory, his current research focuses on blockchains for crowdsourced networks, distributed statistics under information constraints, and privacy for federated learning data pipelines. He has served as an associate editor for the journal *IEEE Transactions on Information Theory* and was awarded the Indian National Science Academy (INSA) Young Scientist medal for 2020.

Shun Watanabe is an associate professor in the Department of Computer and Information Sciences at the Tokyo University of Agriculture and Technology. His research focuses on the intersection of information theory and cryptography. He is a senior member of the Institute of Electrical and Electronics Engineers (IEEE) and served as an associate editor of the journal *IEEE Transactions on Information Theory* from 2016 to 2020.

Information-theoretic Cryptography

HIMANSHU TYAGI
Indian Institute of Science

SHUN WATANABE
Tokyo University of Agriculture and Technology

Shaftesbury Road, Cambridge CB2 8EA, United Kingdom

One Liberty Plaza, 20th Floor, New York, NY 10006, USA

477 Williamstown Road, Port Melbourne, VIC 3207, Australia

314–321, 3rd Floor, Plot 3, Splendor Forum, Jasola District Centre, New Delhi – 110025, India

103 Penang Road, #05–06/07, Visioncrest Commercial, Singapore 238467

Cambridge University Press is part of Cambridge University Press & Assessment, a department of the University of Cambridge.

We share the University's mission to contribute to society through the pursuit of education, learning and research at the highest international levels of excellence.

www.cambridge.org
Information on this title: www.cambridge.org/9781108484336

DOI: 10.1017/9781108670203

First published 2023

A catalogue record for this publication is available from the British Library

ISBN 978-1-108-48433-6 Hardback

Dedicated to our families, students, and teachers

Contents

Preface

Cryptography is the science underlying all online payment systems and more recent blockchain and crypto engineering waves. Despite its popularity and its importance, it is quite difficult for most people outside the theoretical cryptography community to verify the security of basic cryptographic primitives. The reason for this is that cryptography has evolved rather rapidly and using a language which is not easily accessible by outsiders. Even for people working in related theoretical areas such as information theory or quantum physics, it is not easy to follow the ideas and details.

When the authors, both working on information-theoretic cryptography, started to explore all the details of results they had known for a while, they were surprised to find how scattered the information was and how many subtle points had to be kept in mind. In fact, we attended a crypto conference back in 2014 and found that the problems studied are closely related to those studied in information theory and will be of interest to information theorists and communication engineers. Yet, most talks would have been inaccessible to many regular attendees of information theory conferences. Even in books, most presentations target a computer science audience, assuming familiarity with complexity theory abstractions and ways of thinking.

This is an obstacle for many. Admittedly, most exciting developments in practical cryptography have relied on computationally secure primitives, which can only be treated formally using these abstractions. Nonetheless, one would expect that information-theoretically secure cryptography could be understood without overly subscribing to these abstractions. Unfortunately, these details are scattered across technical papers and online lecture notes. This book is an attempt to collate this information in one place, in a form that is accessible to anyone with basic knowledge of probability.

So, we present the outcome of our effort. It is a book focusing purely on information-theoretic cryptography – all the primitives presented are information-theoretically secure. This theory, where the availability of certain correlated random variables replaces computational assumptions, is as old as cryptography and has developed in parallel, intertwined with computational cryptography. We believe we have only partially succeeded in our effort. It became clear to us why certain formalism is unavoidable and why complete proofs would be too long to write, even in a book. Nonetheless, we have tried to bring together all these insights and we have tried to offer a self-contained book. The writing took us two years more than we originally planned (in part owing to the pandemic and related distractions).

This book is dedicated to all the theoretical cryptographers who have carefully articulated some of the most tedious and verbose of formal arguments in human history. It is unbelievable what has been achieved in a short span of 30–40 years!

The authors would like to thank the following colleagues for their comments and feedback: Te Sun Han, Mitsugu Iwamoto, Navin Kashyap, Ryutaroh Matsumoto, Anderson Nascimento, Manoj Prabhakaran, Vinod Prabhakaran, Vincent Tan, and Hirosuke Yamamoto. The authors would like to especially thank Prakash Narayan for constant encouragement for writing this book, a project which could have stalled at any stage without some encouragement.

The authors would like to thank the students who took their courses for proofreading the draft of the book and spotting many typos.

Finally, the authors are grateful to Julie Lancashire, an editor at Cambridge University Press, for providing an opportunity to write the book. The editing team at Cambridge University Press were very helpful and patiently kept up with the changes to the plan that we made along the way.

1 Introduction

1.1 Information Theory and Cryptography

Information theory is a close cousin of probability theory. While probability allows us to model what it means to accurately know or estimate an unknown, information theory allows us to exactly capture the amount of uncertainty. The foremost exponent of this notion is Shannon's entropy $H(X)$ for a random variable X. Without knowing anything else about X, there is an "uncertainty" of $H(X)$ about X. When we know a correlated random variable Y, this uncertainty reduces to $H(X \mid Y)$, the conditional Shannon entropy of X given Y. Shannon defined this reduction in uncertainty $I(X \wedge Y) := H(X) - H(X \mid Y)$ as the measure of information revealed by Y about X. Over the years, Shannon theory has evolved to provide a comprehensive justification for these measures being appropriate measures of information. These measures of information are now gainfully and regularly applied across areas such as biology, control, economics, machine learning, and statistics.

Cryptography theory, the science of maintaining secrets and honesty in protocols, adopted these notions at the outset. Indeed, it was Shannon himself who wrote the first paper to mathematically model secure encryption, and he naturally adopted his notions of information in modeling security. Encryption requires us to send a message M over a public communication channel in such a manner that only the legitimate receiver, and no one else, gets to know M from the communication C. Shannon considered $I(M \wedge C)$ as the measure of information about M leaked by the communication.[1] Thus, at the outset information theory provided a way to measure secrecy in cryptography. The two theories were joined at birth!

But a major development took place in cryptography in the late 1970s. Diffie and Hellman invented an interesting key exchange scheme which was not information-theoretically secure, but was secure in practice. Specifically, their scheme relied on the fact that discrete exponentiation is easy to compute, but (computationally) very difficult to invert. This insight led to the quick development of many fascinating and practical cryptography protocols, all seemingly

[1] To be precise, Shannon's original paper did not consider partial information leakage and did not talk about $I(M \wedge C)$, but the notion was clear and was picked up in subsequent works.

difficult to break in practice but clearly not secure in the information-theoretic sense of Shannon. This was the birth of the field of computational cryptography.

In another remarkable insight, Goldwasser and Micali formulated the notion of semantic security for formally analyzing computational security. This new formulation related security of encryption to the ability to test certain hypotheses about messages by looking at the communication. Over the years this idea evolved, in particular to handle the challenge of formalizing security for an adversary that can deviate from the protocol. The modern framework defines security in terms of the difference in the ability of the adversary in an ideal (secure) system and a system under attack. If every adversarial behavior for a protocol can be "simulated" in the ideal system, the protocol is deemed to be secure.

This modern formulation is very flexible: it can be applied to both information-theoretic and computational settings. In the computational setting, the adversary is restricted to using polynomial-time algorithms; in the information-theoretic setting, there is no computational restriction on the adversary. It is a subtle point, but there is no strict hierarchy between the two notions. They should be viewed as two different assumption classes under which one can analyze the security of various cryptographic primitives.

Specifically, computational cryptography assumes the availability of certain computational primitives such as one-way functions which are easy to compute but computationally hard to invert. Using such primitives, we design cryptographic protocols that remain secure as long as the adversary is computationally restricted to using polynomial-time algorithms. On the other hand, *information-theoretic cryptography* seeks to establish cryptographic protocols that are information-theoretically secure. Often this requires additional resources; for instance, encryption is possible only when the parties share secret keys and two-party secure computation requires the availability of nontrivial correlated observations (such as oblivious transfer).

This book is a comprehensive presentation of information-theoretically secure cryptographic primitives, with emphasis on formal security analysis.

1.2 Overview of Covered Topics

As mentioned in the previous section, a systematic study of cryptography with security analysis in mind was initiated in Shannon's landmark paper. The focus of Shannon's paper was enabling secure communication between legitimate parties over a public communication channel that may not be secure. This is an important problem which cryptography has solved since then, and has enabled secure banking and communication over the Internet. Among other things, Shannon's focus in the paper was the secret key encryption system described in Figure 1.1. In this system, Party 1 (sender) sends a secret message to Party 2 (receiver). To secure their message, the parties encrypt and decrypt the message using a shared secret key (a sequence of random bits). One of Shannon's contributions is

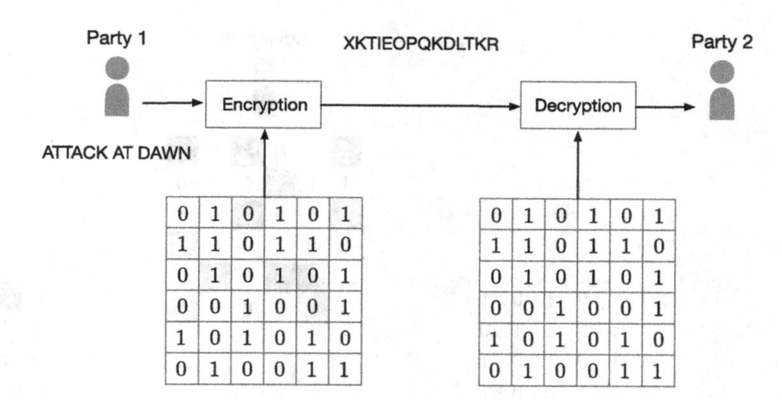

Figure 1.1 A description of secret key encryption. We depict a common anecdotal motivation where a General wants to command their officer to "attack at dawn" over an insecure channel.

to formally describe the notion of security for this secret key encryption system. In particular, Shannon defined the notion of *perfect secrecy* which ensures that an eavesdropper observing all the communication sent over the channel cannot glean any information about the message the sender wants to send to the receiver. The main theoretical result Shannon established was the following: in order to attain perfect secrecy, the length of the shared secret key must be as long as the length of the message. In Chapter 3, we will cover the definition of perfect secrecy and Shannon's result, as well as some other relevant concepts in secret key encryption. We also define notions of approximate secrecy as a relaxation to the perfect secrecy requirement, which lays the foundation for security analysis of modern cryptographic schemes.

As suggested by Shannon's pessimistic result, one of the most important problems in cryptography is how to share a secret key among the legitimate parties. In most current technology, the secret key is exchanged by using so-called public key cryptography, which guarantees security against a computationally bounded adversary. However, in certain applications, it is desirable to have a method to share a secret key even against an adversary who has unlimited computational power. In fact, utilizing certain physical phenomena, methods to share a secret key have been proposed, such as quantum key distribution, or key generation using a wireless communication signal. In Chapter 10, we will cover the data processing part of those key agreement methods, termed the secret key agreement problem.[2] When a secret key is shared using a certain physical carrier, the secret key observed by the receiver is disturbed by noise; furthermore, a part of the key may be leaked to the adversary; see Figure. 1.2. Thus, we have to correct

[2] Technically speaking, in quantum key distribution, we need to consider the density operator instead of random variables so that we can take care of an adversary who may have quantum memory to store eavesdropped signals, which is beyond the scope of this book. However, apart from mathematical difficulties arising from analysis of the density operator, most of the cryptographic concepts necessary for quantum key distribution are covered in this book.

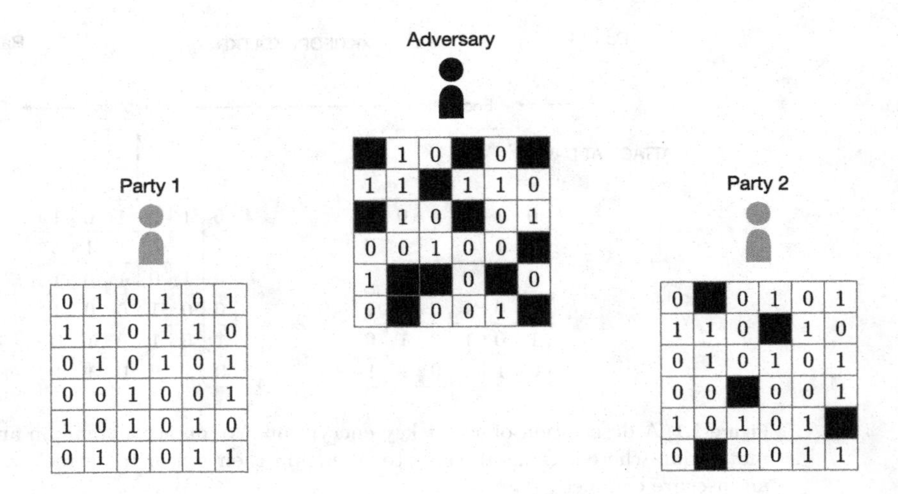

Figure 1.2 A description of the situation in secret key agreement.

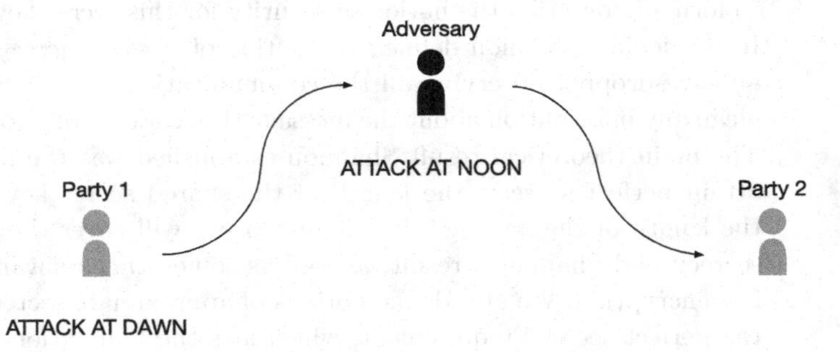

Figure 1.3 A description of authentication. An adversary intercepts the legitimate "attack at dawn" message and replaces it with a fake "attack at noon" message.

the discrepancy between the keys observed by the parties; also, we have to eliminate the information leaked to the adversary. The data processing handling the former problem is termed *information reconciliation*, and it will be covered in Chapter 6. On the other hand, the data processing handling the latter problem is termed *privacy amplification*, and it will be covered in Chapter 7. Furthermore, in Chapter 4, we will cover a cryptographic tool termed the universal hash family, which is used for information reconciliation and privacy amplification.

Along with secret key encryption, the second important problem for enabling secure communication over a public channel is that of *authentication*. In this problem, we would like to prevent an adversary from forging or substituting a transmitted message; see Figure. 1.3. This topic will be covered in Chapter 8 after the key tool, the *strong universal hash family*, is presented in Chapter 4. One of the major results in the authentication problem is that a secure authentication scheme can be realized by using a secret key of length that is of logarithmic order of the message length. This is in contrast to Shannon's result

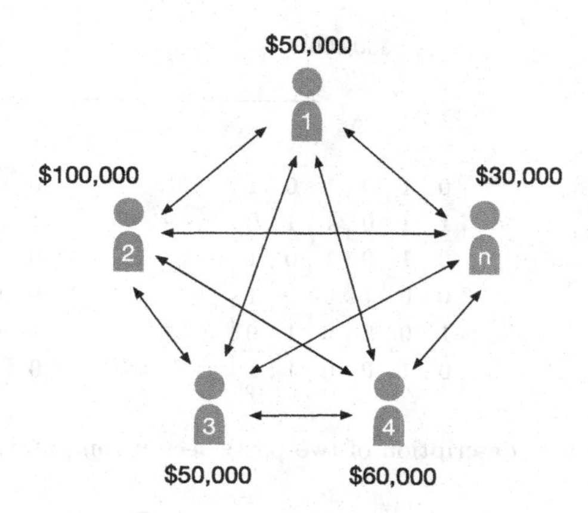

Figure 1.4 A description of multiparty secure computation.

for secure encryption, and suggests that authentication has a much milder secret key requirement in comparison to encryption.

In the second part of the book, we move beyond secure message transmission to more modern problems in cryptography, the ones that are fueling the blockchain and web3.0 applications. Specifically, we present the concepts of secure computation in Chapters 12–19. In secure computation, two or more parties with their own private data seek to execute a program on their collective data and ascertain the outcome. However, the parties do not want to reveal their data to each other beyond the minimal revealed by the outcome. For an anecdotal example, the parties may represent employees who want to compute the average of their salaries without revealing their individual salaries to others; see Figure. 1.4. An important question in this problem is how many dishonest parties we can tolerate in order to realize secure computation. It turns out that, depending on the power of the adversary (whether the adversary can modify the protocol or not), we can realize secure computation when honest parties form a majority or a "supermajority."

For the two-party setting, one dishonest party means a dishonest majority, and we cannot realize secure computation for any nontrivial function. For instance, the parties cannot compare who is wealthier without leaking the value of their salaries to each other. However, if we assume that the parties have some additional resources at their disposal – for instance, they may have access to correlated observations – then it is possible to realize secure computation; see Figure 1.5. In contrast to the reliable communication system in which noise is always troublesome, it turns out that noise can be used as a resource to realize certain tasks in cryptography; this is covered in Chapter 13 and Chapter 14.

When parties in a peer-to-peer network do not trust each other, it is difficult to get consensus among the parties. This is one of the major problems that must be addressed when we consider more than two parties. An important example is the *broadcast* problem where one party wants to send the same message to multiple

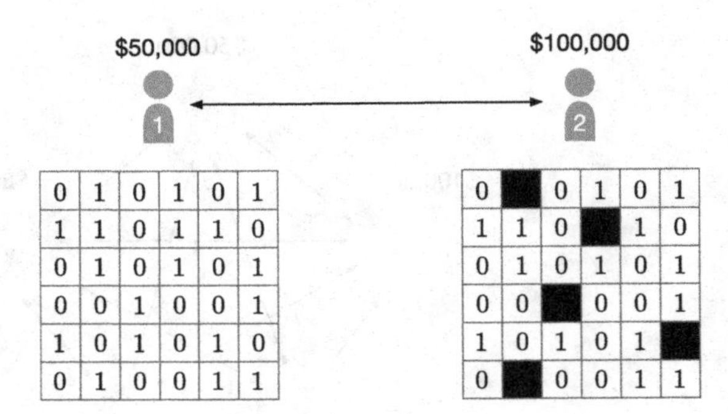

Figure 1.5 A description of two-party secure computation using noisy correlation.

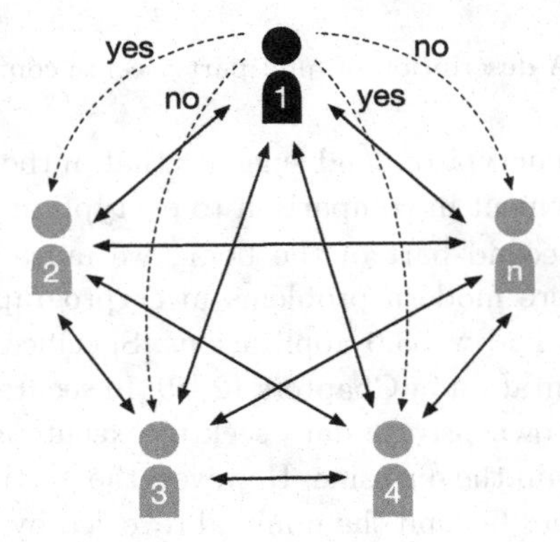

Figure 1.6 A description of broadcast.

parties. But the first party can cheat and send different messages to different parties; see Figure 1.6. In order to detect such an attack, other parties need additional communication to confirm that the message from Party 1 is consistent. Broadcast is a fundamental primitive for multiparty secure computation and is covered in Chapter 18. A key result of this problem is that, if honest parties in the network form a "supermajority," it is possible to realize a secure broadcast. The main secure function computation results are given in Chapters 12, ,17, and 19: two-party secure computing with passive adversary is in Chapter 12, with active adversary in Chapter 17, and multiparty secure computing in Chapter 19.

1.3 Overview of the Technical Framework

Our technical agenda in this book is two-fold. First, we want to lay foundations for a "resource theory" of cryptography, explaining how the availability of different kinds of correlation makes different cryptographic primitives feasible even under information-theoretic security. Second, we want to bring out through examples how various notions of information-theoretic security have evolved in cryptography, culminating in the composable security framework for an active adversary. We elaborate on each of these items below.

1.3.1 A Resource Theory for Cryptography

Right at the outset, Shannon established a result showing that information-theoretically secure encryption requires the legitimate parties to share a secret key that is as large as the length of the messages they want to transmit. For authenticating a message (showing that the message is indeed sent by a legitimate user), parties need to share secret keys of logarithmic length in the message. Thus, if one has secret shared randomness, secret keys, then both authentication and encryption are possible. In fact, this secret shared randomness can itself be extracted from any nontrivial correlated observations (of appropriate amount) that the parties share. So, we can update the observation above and claim that any correlated observations which can give secret keys of appropriate size are a sufficient resource for both encryption and authentication, two primitives which alone account for most cryptographic applications used by industry today.

Interestingly, this shared secret key resource is not sufficient for two-party secure computing (beyond that for trivial functions). We will see that we need a more interesting correlation for secure computing, where either party has some part of the correlation left to itself. A prototypical example is oblivious transfer where one party gets two random bits (K_0, K_1) and the second gets a random bit B and K_B. In this correlation, both parties have some randomness that is not available to the other, and yet there is reasonable correlation between their observations. For multiparty secure computing, we actually do not need any additional resources beyond shared secret keys if a sufficiently large fraction of parties is honest. But the availability of primitives such as digital signatures, Byzantine agreement, and secure broadcast allow more efficient implementation. Indeed, recent progress in blockchains uses these primitives to enable very complex secure multiparty consensus.

An important point to note here is that enabling different cryptographic primitives requires different amounts of resources. For example, to extract a secret key of length equal to the message (which is needed for secure encryption), we need a sufficient amount of a specific type of correlation. The understanding of such tight results is not very mature, but there is a lot of clarity in some settings. For instance, we know exactly how long a secret key can be extracted using independent copies of given correlated random variables. However, such results are not well understood for general secure computing problems. For instance, we do not know what is the most efficient manner of using oblivious transfer to securely compute a given function.

We provide tools to tackle such questions as well. There are two parts to such results. The first part is a scheme which efficiently converts one type of resource into another. As an example, the *leftover hash lemma* shows that a random hash of length ℓ applied to a random variable X will produce an output that is uniform and independent of another random variable Y as long as ℓ is less than roughly the conditional minimum entropy of X given Y. The second part is the so-called converse result in information theory, an impossibility result showing that nothing better will be possible. Most of our converse results in this book rely on a very general result relating the cryptographic task to the difficulty of a statistical problem involving testing the correlation in the resource.

When implementing secure computing of a function using simpler primitives, we view the function using its Boolean circuit. The problem then reduces to computing each binary gate, say NAND, securely, but without revealing intermediate outputs. For this latter requirement, we need a scheme called *secret sharing* which allows multiple parties to get parts of a secret in such a manner that they can reconstruct it only when they come together. It turns out that this secret sharing can be implemented without any additional resources. Thus, the only additional resource needed is a secure implementation of gates such as NAND.

There is subtlety when handling an active adversary. The protocol above requires each party to complete some computations locally, and an active adversary can modify these computations to its advantage. To overcome this difficulty, we introduce a very interesting primitive called *zero-knowledge proofs*, which allow one party to establish that it has indeed completed the desired calculation without revealing the input or output. Almost the same setup extends to multiple parties as well. In fact, as mentioned earlier, if there are sufficiently many honest parties, we can compute any function securely. At a high level, this works out because now there are multiple honest players holding shares of inputs or outputs, and they can have a majority over dishonest players. One extra primitive needed here is *verifiable secret sharing*, which we build using interesting polynomial constructions over finite fields.

1.3.2 Formal Security Definitions

We have already outlined earlier in the chapter how initial security definitions were formulated using information theory. If we want to ensure that a communication C does not reveal any information about a message M, we will require that the mutual information $I(M \wedge C)$ is small. Similarly, with some more care, we can capture the low leakage requirements for secure computing using different mutual information quantities. Note that here we already assume a distribution on M, or on inputs in secure computing. This distribution need not be fixed, and we can consider the worst-case distribution. It is an interesting philosophical exercise to consider what this distribution represents, but we will simply take this distribution as the prior knowledge the adversary has about the unknown.

Instead of using mutual information quantities, an alternative approach (which is perhaps more popular in cryptography) is to define security operationally. For instance, indistinguishable security requires that for any two messages of the adversary's choice, by looking at the communication the probability of error for the adversary to find out which message was sent remains close to 1/2 (which corresponds to a random guess). These operational notions can be shown to be roughly equivalent to notions using information quantities. It is important to note that this equivalence is valid for the information-theoretic security setting, but for the computational setting there is no closed form information quantity available. An advantage of having these closed form quantities is that when analyzing security for specific protocols, these quantities can be broken down into smaller components using so-called "chain rules," allowing simpler analysis. Of course, even the operational notions can be broken down into such smaller components using so-called "hybrid arguments." We will see all this in the book.

Both the notions of security mentioned above are great for curtailing leaked information, namely information revealed beyond what is expected. This suffices for handling a passive adversary who does not deviate from the protocol, but may try to get information it is not entitled to. However, these notions are not sufficient for handling an active adversary who may not even follow the protocol. A major achievement of cryptography theory is to have a sound method for analyzing security in the presence of an active adversary. To formalize security in this setting, the first thing to realize is that we must moderate our expectations: some attacks are unavoidable. For instance, if a party modifies its input to the protocol, there is nothing much that we can do about it. To formalize this, we can at the outset think of an ideal protocol to which parties give inputs and receive outputs as expected, allowing for the unavoidable (admissible) attacks. Security can then be defined as how different can information extracted by an adversary be when using an ideal protocol versus the protocol that we are analyzing.

This formulation is profound, but tedious to use in practice. In fact, most proofs in the literature omit many details, and perhaps writing all the details will make them very long. It is a fascinating story that the most basic result in two-party secure computation, namely the fact that the availability of oblivious transfer allows one to compute any function securely, has been proved several times over a span of roughly 30 years, and each time for a more powerful adversary. We believe this is because these arguments are so delicate. In this book, we make an attempt to collect many of these examples in one place, slowly building concepts to handle security analysis for an active adversary.

1.4 Possible Course Plans

With a few exceptions, the dependency of chapters in this book is depicted in Figure 1.7. Even though later chapters tend to depend on earlier chapters, dependencies are partial. For instance, the information-theoretic tools provided in Chapters 5–7 are not necessary to understand most parts of Chapters 15–19; knowledge of the secret key agreement in Chapter 10 is only necessary to derive

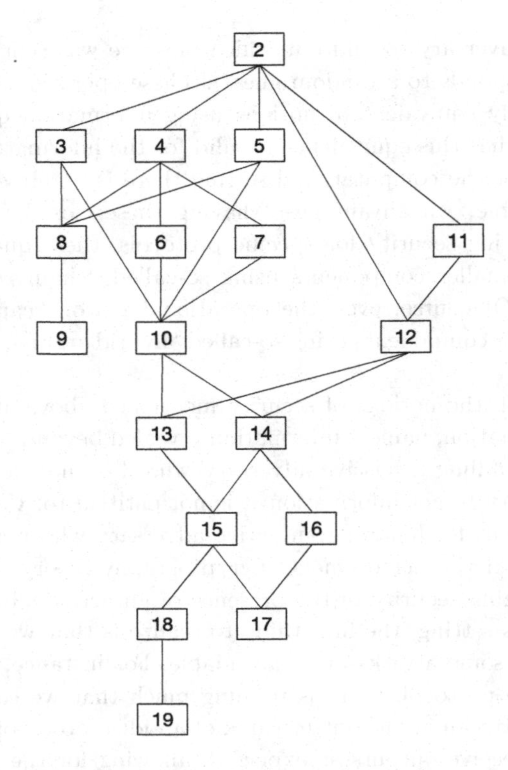

Figure 1.7 Dependence graph of the book; the numbers refer to the corresponding chapters.

the impossibility results of oblivious transfer and bit commitment in Chapter 13 and Chapter 14. Keeping in mind these partial dependencies, we suggest the following three possible course plans.

Basic Cryptography
Starting with Chapter 2, this course will cover two basic problems of cryptography: encryption (Chapter 3) and authentication (Chapter 8). Also, as preparatory tools for authentication, we will cover the basic concepts of universal hash family (Chapter 4) and hypothesis testing (Chapter 5, but Sections 5.5 and 5.6 may be skipped). Then, we will introduce computationally secure encryption and authentication (Chapter 9), in order to highlight how information-theoretic cryptography is related to computationally secure cryptography. If time permits, secret sharing (Chapter 11) may be covered as well since it does not require any other prerequisites from earlier chapters.

Cryptography from Correlated Randomness
This is an advanced course which will start in the same manner as the previous basic course, but will focus on the role of correlated randomness in en-

abling cryptographic primitives. In addition to the topics above, we will cover secret key agreement (Chapter 10), oblivious transfer (Chapter 13), and bit commitment (Chapter 14). As preparatory tools for these problems, we will cover the universal hash family (Chapter 4), hypothesis testing (Chapter 5), information reconciliation (Chapter 6), and random number generation (Chapter 9). In order to motivate the oblivious transfer problem, we can cover some material from Chapter 12. This course will be suitable for graduate students working in cryptography, information theory, or quantum information theory. It highlights the main theme of this book: the use of information-theoretic tools in cryptography.

Secure Computation
This course will cover basic results on secure computation. Starting with Chapter 2, we will first cover secret sharing (Chapter 11), which is a basic tool for secure computation. Then, we will proceed to the two-party secure computation problem (Chapter 12). We will move then to the oblivious transfer (Chapter 13) and the bit commitment (Chapter 14) problems, but will not cover constructions as that would require prerequisites from earlier chapters. After that, we will cover some selected topics from Chapters 15–19. There are two possible paths. In the first one, we will highlight the completeness of oblivious transfer (Chapter 17) after covering the notions of active adversary, composability (Chapter 15), and zero-knowledge proofs (Chapter 16). In the second one, we will highlight the honest majority/supermajority threshold of multiparty secure computation (Chapter 19). This requires as prerequisites the notions of active adversary and composability (Chapter 15, if the active adversary is discussed) and broadcast (Chapter 18, which may be omitted if we assume existence of the broadcast). Selection of topics depends on the preference of the instructor; however, multiparty secure computation (Chapter 19) may be easier to digest compared to the completeness of oblivious transfer (Chapter 17).

Each chapter is supplemented by several problems. Some of these are just exercises to confirm results provided in the chapter or to fill omitted steps of proofs. Others are meant to be pointers to interesting results that are beyond the scope of the book. For some selected problems (mainly for those that are used in later chapters), answers are provided in the Appendix.

1.5 References and Additional Reading

The topic of cryptography is as classic as it is popular. There are already several excellent textbooks that provide a review of different aspects of this vast area. We review some of them below. For a historical perspective of cryptography and information theory, see [310, 313].

Most existing textbooks on cryptography are based on computational complexity theory. A thorough exposition can be found in the two volume textbook by Goldreich [141, 142]. A more introductory but rigorous treatment can be found in the textbook by Katz and Lindell [187]. In addition to these general textbooks, there are textbooks specializing in specific topics, such as two-party secure computation by Hazay and Lindell [166] or digital signatures by Katz [185].

Another important topic of cryptography is algebraic construction and practical implementation of cryptographic primitives. Textbooks covering these topics include those by Blahut [34] and Stinson [318]. In particular, the former is written by an information theorist with engineers in mind.

Some topics of information-theoretic cryptography, such as encryption or the secret key agreement, are treated in information theory textbooks. For instance, see the textbooks by Csiszár and Körner [88] and El Gamal and Kim [131]. A popular application of information theoretic-cryptography is physical layer security; a good resource on this topic is the textbook by Bloch and Barros [37]. In a different flavor, the book by Cramer, Damgård, and Nielsen contains a thorough treatment of information-theoretically secure multiparty computation and secret sharing [77].

Recent development of information-theoretic cryptography is closely tied to quantum information science. The classic textbook by Nielsen and Chuang broadly covers quantum computation, quantum cryptography, and quantum information theory [261]. For a more thorough treatment of quantum information theory, see the textbooks by Hayashi [161] and Wilde [348]. These books also treat problems related to converting one type of randomness to another.

Part I

External Adversary: Encryption, Authentication, Secret Key

2 Basic Information Theory

We begin with a short primer on basic notions of information theory. We review the various measures of information and their properties, without discussing their operational meanings.

At a high level, we can categorize information measures into three categories: measures of uncertainty, measures of mutual information, and measures of statistical distances. Quantities such as Shannon entropy fall in the first category, and provide a quantitative measure of uncertainty in a random variable. In addition to Shannon's classic notion of entropy, we also need a related notion of entropy due to Rényi. These quantities will be used throughout this book, but will be central to our treatment of randomness extraction (namely, the problem of generating random and independent coin tosses using a biased coin, correlated with another random variable).

The measures of mutual information capture the information revealed by a random variable about another random variable. As opposed to statistical notions such as mean squared error, which capture the notion of estimating or failing to estimate, mutual information allows us to quantify partial information and gives a mathematical equivalent of the heuristic phrase: "X gives *a bit of information* about Y." For us, Shannon's mutual information and a related quantity (using total variation distance) will be used to measure the "information leaked" to an adversary. These quantities are central to the theme of this book.

Finally, we need the notions of distances between probability distributions. Statistical inference entails determining the properties of the distribution generating a random variable X, by observing X. The closer two distributions P and Q are, the more difficult it is to distinguish whether X is generated by P or by Q. Information theory provides a battery of measures for "distances" between two probability distributions. In fact, our notions of information-theoretic security will be defined using these distances.

In addition, we present the properties of these quantities and inequalities relating them. To keep things interesting, we have tried to present "less well-known" proofs of these inequalities – even a reader familiar with these bounds may find something interesting in this chapter. For instance, we prove Fano's inequality using data processing inequality, we prove continuity of entropy using Fano's inequality, we provide multiple variational formulae for information quantities, and we provide two different proofs of Pinsker's inequality.

We start by covering the essential notions of probability.

2.1 Very Basic Probability

Since our focus will be on discrete random variables, the reader only needs to be familiar with very basic probability theory. We review the main concepts and notations in this section.

Let Ω be a set of finite cardinality. A discrete *probability distribution* P on Ω can be described using its *probability mass function* (pmf) $p\colon \Omega \to [0,1]$ satisfying $\sum_{\omega\in\Omega} p(\omega) = 1$. We can think of Ω as the underlying "probability space" with "probability measure" P. Since we restrict our attention to discrete random variables throughout this book, we will specify the probability distribution P using the corresponding pmf. In particular, with an abuse of notation, we use $\mathrm{P}(\omega)$ to denote the probability of the event $\{\omega\}$ under P.

For a probability distribution P on Ω, we denote by $\mathtt{supp}(\mathrm{P})$ its *support set* given by

$$\mathtt{supp}(\mathrm{P}) = \{\omega \in \Omega : \mathrm{P}(\omega) > 0\}.$$

A *discrete random variable* X is a mapping $X\colon \Omega \to \mathcal{X}$ where $\mathcal{X}$ is a set of finite cardinality. Without loss of generality, we will assume that the mapping X is onto, namely $\mathcal{X} = \mathtt{range}(X)$. It is not necessary that the probability space (Ω, P) is discrete; but it suffices for our purpose. We can associate with a random variable X a distribution P_X, which is the distribution "induced" on the output $\mathcal{X}$ by X. Since X is a discrete random variable, P_X is a discrete probability distribution given by

$$\mathrm{P}_X(x) = \Pr(X = x) = \sum_{\omega\in\Omega: X(\omega)=x} \mathrm{P}(\omega);$$

often, we will simply say $\mathrm{P} = \mathrm{P}_X$ is a probability distribution on $\mathcal{X}$, without referring to the underlying probability space Ω. Throughout the book, unless otherwise stated, when we say random variable, we refer only to discrete random variables.

Notation for random variables: In probability, we often choose the set $\mathcal{X}$ as $\mathbb{R}$, the set of real numbers. But in this book we will often treat random binary vectors or messages as random variables. Thus, we will allow for an arbitrary finite set $\mathcal{X}$. We will say that "X is a random variable taking values in the set $\mathcal{X}$," for any finite set $\mathcal{X}$. We denote the probability distribution of X by P_X and, for every realization $x \in \mathcal{X}$, denote by $\mathrm{P}_X(x)$ the probability $\Pr(X = x)$. Throughout the book, we denote the random variables by capital letters such as X, Y, Z, etc., realizations by the corresponding small letters such as x, y, z, etc., and the corresponding range-sets by calligraphic letters such as $\mathcal{X}, \mathcal{Y}, \mathcal{Z}$, etc.

Associated with a random variable X taking values in $\mathcal{X} \subset \mathbb{R}$, there are two fundamental quantities: its *expected value* $\mathbb{E}[X]$ given by

$$\mathbb{E}[X] = \sum_{x \in \mathcal{X}} x \mathrm{P}_X(x),$$

and its *variance* $\mathbb{V}[X]$ given by

$$\mathbb{V}[X] = \mathbb{E}[(X - \mathbb{E}[X])^2].$$

An important property of expectation is its *linearity*. Namely,

$$\mathbb{E}[X + Y] = \mathbb{E}[X] + \mathbb{E}[Y].$$

Before proceeding, denoting by $\mathbf{1}[\mathcal{S}]$ the *indicator function* for the set $\mathcal{S}$, we note a useful fact about the binary random variable $\mathbf{1}[X \in \mathcal{A}]$:

$$\mathbb{E}[\mathbf{1}[X \in \mathcal{A}]] = \Pr(X \in \mathcal{A}),$$

for any subset $\mathcal{A}$ of $\mathcal{X}$. Heuristically, the expected value of X serves as an estimate for X and the variance of X serves as an estimate for the error in the estimate. We now provide two simple inequalities that formalize this heuristic.

THEOREM 2.1 (Markov's inequality) *For a random variable X taking values in $[0, \infty)$ and any $t > 0$, we have*

$$\Pr(X \geq t) \leq \frac{\mathbb{E}[X]}{t}.$$

Proof We note that any random variable X can be expressed as

$$X = X\mathbf{1}[X \geq t] + X\mathbf{1}[X < t].$$

Then,

$$\begin{aligned}
\mathbb{E}[X] &= \mathbb{E}[X\mathbf{1}[X \geq t]] + \mathbb{E}[X\mathbf{1}[X < t]] \\
&\geq \mathbb{E}[X\mathbf{1}[X \geq t]] \\
&\geq \mathbb{E}[t\mathbf{1}[X \geq t]] \\
&= t\Pr(X \geq t),
\end{aligned}$$

where the first inequality uses the fact that X is nonnegative and the second inequality uses the fact that $X\mathbf{1}[X \geq t]$ is either 0 or exceeds t with probability 1. □

Applying Markov's inequality to $|X - \mathbb{E}[X]|^2$, we obtain the following bound.

THEOREM 2.2 (Chebyshev's inequality) *For a random variable X taking values in $\mathbb{R}$ and any $t > 0$, we have*

$$\Pr(|X - \mathbb{E}[X]| \geq t) \leq \frac{\mathbb{V}[X]}{t^2}.$$

The previous bound says that the random variable X lies within the interval $[\mathbb{E}[X] - \sqrt{\mathbb{V}[X]/\varepsilon}, \mathbb{E}[X] + \sqrt{\mathbb{V}[X]/\varepsilon}]$ with probability exceeding $1-\varepsilon$. Such an interval is called the $(1-\varepsilon)$-confidence interval. In fact, the $\sqrt{1/\varepsilon}$ dependence of the accuracy on the probability of error can often be improved to $\sqrt{\ln 1/\varepsilon}$ using a "Chernoff bound." We will present such bounds later in the book.

When we want to consider multiple random variables X and Y taking values in $\mathcal{X}$ and $\mathcal{Y}$, respectively, we consider their joint distribution P_{XY} specified by the joint probability distribution $\mathrm{P}_{XY}(x,y)$. For discrete random variables, we can define the *conditional distribution* $\mathrm{P}_{X|Y}$ using the conditional probabilities given by[1]

$$\mathrm{P}_{X|Y}(x|y) := \frac{\mathrm{P}_{XY}(x,y)}{\mathrm{P}_{Y}(y)}$$

for every $x \in \mathcal{X}$ and $y \in \mathcal{Y}$ such that $\mathrm{P}_Y(y) > 0$; if $\mathrm{P}_Y(y) = 0$, we can define $\mathrm{P}_{X|Y}(x|y)$ to be an arbitrary distribution on $\mathcal{X}$. A particular quantity of interest is the *conditional expectation* $\mathbb{E}[X|Y]$ of X given Y, which is a random variable that is a function of Y, defined as

$$\mathbb{E}[X|Y](y) := \sum_{x \in \mathcal{X}} x \mathrm{P}_{X|Y}(x|y), \quad \forall y \in \mathcal{Y}.$$

Often, we use the alternative notation $\mathbb{E}[X|Y=y]$ for $\mathbb{E}[X|Y](y)$. Note that $\mathbb{E}[X|Y]$ denotes the random variable obtained when y is replaced with the random Y with distribution P_Y.

In information theory, it is customary to use the terminology of a *channel* in place of conditional distributions. Simply speaking, a channel is a randomized mapping. For our use, we will define this mapping using the conditional distribution it induces between the output and the input. Formally, we have the definition below.

DEFINITION 2.3 (Channels) For finite alphabets $\mathcal{X}$ and $\mathcal{Y}$, a channel W with input alphabet $\mathcal{X}$ and output alphabet $\mathcal{Y}$ is given by probabilities $W(y|x)$, $y \in \mathcal{Y}, x \in \mathcal{X}$, where $W(y|x)$ denotes the probability of observing y when the input is x.

Often, we abuse the notation and use $W\colon \mathcal{X} \to \mathcal{Y}$ to represent the channel. Also, for a channel $W\colon \mathcal{X} \to \mathcal{Y}$ and input distribution P on $\mathcal{X}$, we denote by $\mathrm{P} \circ W$ the distribution induced on the output Y of the channel when the input X has distribution X. Namely,

$$(\mathrm{P} \circ W)(y) = \sum_{x \in \mathcal{X}} \mathrm{P}(x) W(y|x).$$

Furthermore, we denote by $\mathrm{P} \times W$ the joint distribution of (X, Y).

[1] In probability theory, conditional probability densities are technically difficult to define and require several conditions on the underlying probability space. However, since we restrict our attention to discrete random variables, the conditional pmf serves this purpose for us.

2.2 The Law of Large Numbers

Chebyshev's inequality tells us that a good estimate for a random variable X is its expected value $\mathbb{E}[X]$, up to an accuracy of roughly $\pm\sqrt{\mathbb{V}[X]}$. In fact, this estimate is pretty sharp asymptotically when X is a sum of independent random variables.

Specifically, let $X_1, \ldots, X_n$ be *independent and identically distributed* (i.i.d.) random variables, that is, they are independent and have the same (marginal) distribution, say, P_X. An important fact to note about independent random variables is that their variance is *additive*. Indeed, for independent random variables X and Y,

$$\begin{aligned}
\mathbb{V}[X+Y] &= \mathbb{E}\big[(X+Y)^2\big] - \mathbb{E}[X+Y]^2 \\
&= \mathbb{E}\big[X^2\big] + \mathbb{E}\big[Y^2\big] + 2\mathbb{E}[XY] - \big(\mathbb{E}[X]^2 + \mathbb{E}[Y]^2 + 2\mathbb{E}[X]\,\mathbb{E}[Y]\big) \\
&= \mathbb{V}[X] + \mathbb{V}[Y] + 2(\mathbb{E}[XY] - \mathbb{E}[X]\,\mathbb{E}[Y]) \\
&= \mathbb{V}[X] + \mathbb{V}[Y],
\end{aligned}$$

where in the final identity we used the observation that $\mathbb{E}[XY] = \mathbb{E}[X]\,\mathbb{E}[Y]$ for independent random variables.

In fact, X and Y such that $\mathbb{E}[XY] = \mathbb{E}[X]\,\mathbb{E}[Y]$ are called *uncorrelated*, and this is the only property we need to get additivity of variance. Note that in general, random variables may be uncorrelated but not independent.[2]

Returning to i.i.d. random variables $X_1, X_2, \ldots, X_n$ with common distribution P_X, by the linearity of expectation we have

$$\mathbb{E}\left[\sum_{i=1}^n X_i\right] = \sum_{i=1}^n \mathbb{E}[X_i] = n\mathbb{E}[X],$$

and since the variance is additive for independent random variables,

$$\mathbb{V}\left[\sum_{i=1}^n X_i\right] = \sum_{i=1}^n \mathbb{V}[X_i] = n\mathbb{V}[X].$$

Therefore, by Chebyshev's inequality,

$$\Pr\left(\left|\sum_{i=1}^n X_i - n\mathbb{E}[X]\right| \geq t\right) \leq \frac{n\mathbb{V}[X]}{t^2},$$

or equivalently,

$$\Pr\left(\left|\sum_{i=1}^n X_i - n\mathbb{E}[X]\right| \geq \sqrt{\frac{n\mathbb{V}[X]}{\varepsilon}}\right) \leq \varepsilon$$

[2] For example, consider the random variable X taking values $\{-1, 0, 1\}$ with equal probabilities and $Y = 1 - |X|$. For these random variables, $\mathbb{E}[X] = 0 = \mathbb{E}[X]\,\mathbb{E}[Y]$ and $\mathbb{E}[XY] = 0$ since $Y = 0$ whenever $X \neq 0$. But clearly X and Y are not independent.

for every $\varepsilon \in (0,1)$. Thus, with large probability, $\frac{1}{n}\sum_{i=1}^{n} X_i$ is roughly within $\pm\sqrt{\mathbb{V}[X]/n}$ of its expected value $\mathbb{E}[X]$. We have proved the *weak law of large numbers*.

THEOREM 2.4 (Weak law of large numbers) *Let $X_1, X_2, \ldots, X_n$ be i.i.d. with common distribution P_X over a finite set $\mathcal{X} \subset \mathbb{R}$. For every $\delta > 0$ and $\varepsilon \in (0,1)$, we have*

$$\Pr\left(\left|\frac{1}{n}\sum_{i=1}^{n} X_i - \mathbb{E}[X]\right| > \delta\right) \le \varepsilon$$

for every n sufficiently large.

Alternatively, we can express the previous result as follows: for every $\delta > 0$,

$$\lim_{n\to\infty} \Pr\left(\left|\frac{1}{n}\sum_{i=1}^{n} X_i - \mathbb{E}[X]\right| > \delta\right) = 0.$$

In fact, a stronger version of the result above holds – we can exchange the limit and the probability. This result is a bit technical, and it says that for all "sample paths" $\{X_i\}_{i=1}^{\infty}$ the sample average $\frac{1}{n}\sum_{i=1}^{n}$ converges to $\mathbb{E}[X]$ as n goes to infinity. We state this result without proof.

THEOREM 2.5 (Strong law of large numbers) *Let $X_1, X_2, \ldots, X_n$ be i.i.d. with common distribution P_X over a finite set $\mathcal{X} \subset \mathbb{R}$. Then,*

$$\Pr\left(\lim_{n\to\infty} \frac{1}{n}\sum_{i=1}^{n} X_i = \mathbb{E}[X]\right) = 1.$$

In summary, in this section we have learnt that the average of a large number of independent random variables can be approximated, rather accurately, by its expected value.

2.3 Convex and Concave Functions

Convex and concave functions play an important role in information theory. Informally speaking, convex and concave functions, respectively, are those whose graphs look like a "cup" and "cap." In particular, a function f is concave if $-f$ is convex. We provide the formal definition and a key property below.

The domain of a convex function must be a *convex set*, which we define first. For simplicity, we restrict ourselves to convex sets that are subsets of $\mathbb{R}^d$.

DEFINITION 2.6 (Convex set) For a natural number $d \in \mathbb{N}$, a set $\mathcal{S} \subset \mathbb{R}^d$ is a convex set if for any two points s_1 and s_2 in $\mathcal{S}$ and any $\theta \in [0,1]$, we must have $\theta s_1 + (1-\theta)s_2 \in \mathcal{S}$. Namely, if two points belong to $\mathcal{S}$, then all the points in the straight line joining these two points also belong to $\mathcal{S}$.

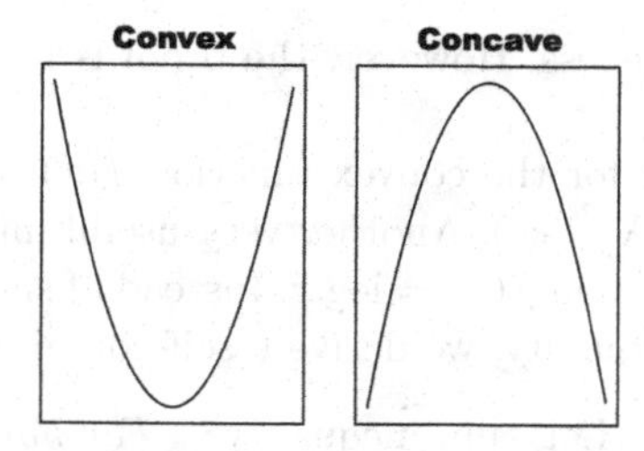

Figure 2.1 Depiction of convex and concave functions.

DEFINITION 2.7 (Convex and concave functions) For $d \in \mathbb{N}$, let $\mathcal{S} \subset \mathbb{R}^d$ be a convex set. Then, a function $f: \mathcal{S} \to \mathbb{R}$ is a convex function if for every $\theta \in [0, 1]$ and every pair of points $s_1, s_2 \in \mathcal{S}$, we have

$$f(\theta s_1 + (1 - \theta)s_2) \leq \theta f(s_1) + (1 - \theta)f(s_2), \tag{2.1}$$

and it is concave if

$$f(\theta s_1 + (1 - \theta)s_2) \geq \theta f(s_1) + (1 - \theta)f(s_2). \tag{2.2}$$

In particular, when strict inequality holds in (2.1) (respectively (2.2)) for every $\theta \in (0, 1)$ and $s_1 \neq s_2$, then the function is a strict convex function (respectively strict concave function).

Simply speaking, a function is convex (respectively concave) if the value of the function at the (weighted) average of two points is less than (respectively more than) the average of the values at the point. Note that linear functions are both convex and concave, but not strict convex nor strict concave. Examples of strict convex functions include e^x, e^{-x}, x^2, etc. and examples of strict concave functions include $\ln x$ (natural logarithm), $\sqrt{x}$, etc. We depict the shape of convex and concave functions in Figure 2.1.

It is clear from the definition of convex functions that, for a random variable X taking values in a finite set $\mathcal{X}$ and a convex function f, we must have $\mathbb{E}[f(X)] \geq f(\mathbb{E}[X])$. This is the powerful Jensen inequality.

LEMMA 2.8 (Jensen's inequality) *Let $f: \mathbb{R}^n \to \mathbb{R}$ be a convex function and X a random variable taking values in $\mathbb{R}^n$. Then,*

$$f(\mathbb{E}[X]) \leq \mathbb{E}[f(X)]. \tag{2.3}$$

Similarly, if f is a concave function, then

$$f(\mathbb{E}[X]) \geq \mathbb{E}[f(X)]. \tag{2.4}$$

In particular, when f is strict convex (respectively strict concave), then strict inequality holds in (2.3) (respectively (2.4)) unless $X = \mathbb{E}[X]$ with probability 1.

In fact, this inequality holds for more general random variables than those considered in this book – there is no need for the assumption of finiteness or even discreteness. However, the proof is technical and beyond the scope of this book.

Note that for the convex function $f(x) = x^2$, this inequality implies $\mathbb{V}[X] = \mathbb{E}[X^2] - \mathbb{E}[X]^2 \geq 0$. Another very useful implication of this inequality is for the concave function $f(x) = \log x$. Instead of showing that $\log x$ is concave and using Jensen's inequality, we derive a self-contained inequality which is very handy.

LEMMA 2.9 (Log-sum inequality) *For nonnegative numbers* $\{(a_i, b_i)\}_{i=1}^n$,

$$\sum_{i=1}^n a_i \log \frac{a_i}{b_i} \geq \sum_{i=1}^n a_i \log \frac{\sum_{i=1}^n a_i}{\sum_{i=1}^n b_i},$$

with equality[3] *if and only if* $a_i = b_i$ *for all* i.

Proof The inequality is trivial if all a_i are 0, or if there exists an i such that $b_i = 0$ but $a_i \neq 0$. Otherwise, by rearranging the terms, it suffices to show that

$$\sum_{i=1}^n a_i \log \frac{a_i / \sum_{j=1}^n a_j}{b_i / \sum_{j=1}^n b_j} \geq 0,$$

which holds if and only if

$$\sum_{i=1}^n a_i' \log \frac{a_i'}{b_i'} \geq 0,$$

where $a_i' = a_i / \sum_{j=1}^n a_j$ and $b_i' = b_i / \sum_{j=1}^n b_j$. Note that the previous inequality is simply our original inequality for the case when $\sum_{i=1}^n a_i = \sum_{i=1}^n b_i = 1$. Thus, without loss of generality we can assume that a_i and b_i, $1 \leq i \leq n$, both constitute pmfs. Then, since $\ln x \leq x - 1$, $\log x \leq (x-1)\log e$, applying this inequality for $x = a_i/b_i$ we get

$$\sum_{i=1}^n a_i \log \frac{b_i}{a_i} \leq \log e \sum_{i=1}^n a_i \left(\frac{b_i}{a_i} - 1 \right) = 0,$$

which establishes the inequality.

Equality can hold only if equality holds for every instance of $\ln x \leq x - 1$ used in the proof, which happens only if $x = 1$. Thus, equality holds only if $a_i = b_i$ for every $i \in \{1, \ldots, n\}$. □

2.4 Total Variation Distance

Suppose we observe a sample X taking values in a discrete set $\mathcal{X}$. How difficult is it to determine if X is generated from a distribution P or Q? We need a precise quantitative handle on this difficulty for this book. Indeed, the formal notion of information-theoretic security that we define in this book will rely on difficulty in solving such problems. This problem is one of the fundamental problems in

[3] We follow the convention that $0 \log(0/0) = 0 \log 0 = 0$ throughout the book.

statistics – the *binary hypothesis testing* problem. Later in the book, we will revisit this problem in greater detail. In this section, we only present a quick version that will help us motivate an important notion.

Formally, we can apply a channel $T\colon \mathcal{X} \to \{0,1\}$ to make this decision, where the outputs 0 and 1 indicate P and Q, respectively. Such a decision rule is called a (hypothesis) *test*. For this channel denoting the test, let $T(1|x) = 1 - T(0|x)$ denote the probability with which T declares 1 for input $x \in \mathcal{X}$. When X was generated using P, the test T makes an error if its output is 1, which happens with probability $\sum_{x\in\mathcal{X}} \mathrm{P}(x)T(1|x)$. Similarly, when X was generated using Q, the probability of making an error is $\sum_{x\in\mathcal{X}} \mathrm{Q}(x)T(0|x)$.

One simple notion of performance of the test T is the average of these two errors. It corresponds to the probability of error in T's output when the input distribution for X is chosen to be P or Q with equal probability. Specifically, the average probability of error $P_{\mathtt{err}}(T)$ is given by

$$P_{\mathtt{err}}(T|\mathrm{P},\mathrm{Q}) := \frac{1}{2}\left[\sum_{x\in\mathcal{X}} \mathrm{P}(x)T(1|x) + \sum_{x\in\mathcal{X}} \mathrm{Q}(x)T(0|x)\right],$$

and a measure of difficulty of the hypothesis testing problem described above is the minimum average probability of error

$$P_{\mathtt{err}}(\mathrm{P},\mathrm{Q}) = \min_T P_{\mathtt{err}}(T|\mathrm{P},\mathrm{Q}),$$

where the minimum is over all channels $T\colon \mathcal{X} \to \{0,1\}$.

In fact, the minimizing T is easy to find (it is sometimes called the *Bayes optimal test* or simply the *Bayes test*).

LEMMA 2.10 (Bayes test) *For distributions* P *and* Q *on a discrete set* $\mathcal{X}$, $P_{\mathtt{err}}(\mathrm{P},\mathrm{Q})$ *is attained by the deterministic test which outputs* 0 *for* $x \in \mathcal{X}$ *such that* $\mathrm{P}(x) \ge \mathrm{Q}(x)$ *and* 1 *otherwise, i.e.,*

$$T^*(0|x) = \begin{cases} 1, & \text{if } \mathrm{P}(x) \ge \mathrm{Q}(x), \\ 0, & \text{otherwise.} \end{cases}$$

Furthermore, the minimum average probability of error is given by

$$P_{\mathtt{err}}(\mathrm{P},\mathrm{Q}) = \frac{1}{2}\left[1 - \sum_{x\in\mathcal{X}:\mathrm{P}(x)\ge\mathrm{Q}(x)} (\mathrm{P}(x) - \mathrm{Q}(x))\right].$$

Proof It is easy to verify that

$$P_{\mathtt{err}}(T^*|\mathrm{P},\mathrm{Q}) = \frac{1}{2}\left[1 - \sum_{x\in\mathcal{X}:\mathrm{P}(x)\ge\mathrm{Q}(x)} (\mathrm{P}(x) - \mathrm{Q}(x))\right].$$

Thus, it suffices to show that the right-hand side of the expression above is less than $P_{\mathtt{err}}(T|\mathrm{P},\mathrm{Q})$ for every test T. To this end, note that

$$\begin{aligned}
2P_{\mathrm{err}}(T|\mathrm{P},\mathrm{Q}) &= \sum_{x\in\mathcal{X}} \mathrm{P}(x)T(1|x) + \sum_{x\in\mathcal{X}} \mathrm{Q}(x)T(0|x)\\
&= \sum_{x\in\mathcal{X}} \mathrm{P}(x)T(1|x) + \sum_{x\in\mathcal{X}} \mathrm{Q}(x)(1-T(1|x))\\
&= \sum_{x\in\mathcal{X}} (\mathrm{P}(x)-\mathrm{Q}(x))T(1|x) + \sum_{x\in\mathcal{X}} \mathrm{Q}(x)\\
&\geq \sum_{x\in\mathcal{X}:\mathrm{P}(x)<\mathrm{Q}(x)} (\mathrm{P}(x)-\mathrm{Q}(x)) + 1\\
&= \sum_{x\in\mathcal{X}} (\mathrm{P}(x)-\mathrm{Q}(x))T^*(1|x) + 1\\
&= 1 - \sum_{x\in\mathcal{X}:\mathrm{P}(x)\geq\mathrm{Q}(x)} (\mathrm{P}(x)-\mathrm{Q}(x)),
\end{aligned}$$

where the inequality holds since $T(1|x)$ lies in $[0,1]$ and we have dropped only positive terms from the preceding expression. □

The optimal test T^* that emerges from the previous result is a natural one: declare P when you observe x which has higher probability of occurrence under P than under Q. Thus, the difficulty of resolving the hypothesis testing problem above is determined by the quantity $\sum_{x\in\mathcal{X}:\mathrm{P}(x)\geq\mathrm{Q}(x)} \mathrm{P}(x)-\mathrm{Q}(x)$.

We note that the optimal test T^* is a deterministic one. Further, note that a deterministic test can be characterized by the subset $\mathcal{A} = \{x \in \mathcal{X} : T(0|x) = 1\}$ of $\mathcal{X}$, and its probability of error is given by

$$P_{\mathrm{err}}(T|\mathrm{P},\mathrm{Q}) = \frac{1}{2}\left[\mathrm{P}(\mathcal{A}^c) + \mathrm{Q}(\mathcal{A})\right] = \frac{1}{2}(1 - (\mathrm{P}(\mathcal{A}) - \mathrm{Q}(\mathcal{A}))).$$

By comparing with the optimal probability of error for a deterministic test (attained by T^*), we get

$$\sum_{x\in\mathcal{X}:\mathrm{P}(x)>\mathrm{Q}(x)} \mathrm{P}(x)-\mathrm{Q}(x) = \max_{\mathcal{A}\subset\mathcal{X}} \mathrm{P}(\mathcal{A}) - \mathrm{Q}(\mathcal{A}),$$

where the maximum is attained by the set $\mathcal{A}^* = \{x \in \mathcal{X} : \mathrm{P}(x) > \mathrm{Q}(x)\}$ (corresponding to T^*).[4]

We would like to treat $\max_{\mathcal{A}\subset\mathcal{X}} \mathrm{P}(\mathcal{A})-\mathrm{Q}(\mathcal{A})$ as a notion of "distance" between the distributions P and Q; our little result above already tells us that the closer P and Q are in this distance, the harder it is to tell them apart using statistical tests. In fact, this quantity is rather well suited to being termed a distance. The next result shows an equivalent form for the same quantity that better justifies its role as a distance.

LEMMA 2.11 *For two distributions* P *and* Q *on a finite set* $\mathcal{X}$, *we have*

$$\max_{\mathcal{A}\subset\mathcal{X}} \mathrm{P}(\mathcal{A}) - \mathrm{Q}(\mathcal{A}) = \frac{1}{2}\sum_{x\in\mathcal{X}} |\mathrm{P}(x)-\mathrm{Q}(x)|.$$

We leave the proof as an (interesting) exercise (see Problem 2.1).

[4] Alternatively, we can include x with $\mathrm{P}(x) = \mathrm{Q}(x)$ into $\mathcal{A}^*$.

The expression on the right-hand side of the previous result is (up to normalization) the ℓ_1 distance between finite-dimensional vectors $(\mathrm{P}(x), x \in \mathcal{X})$ and $(\mathrm{Q}(x), x \in \mathcal{X})$. We have obtained the following important notion of distance between P and Q.

DEFINITION 2.12 (Total variation distance) For discrete distributions P and Q on $\mathcal{X}$, the *total variation distance* $d_{\mathtt{var}}(\mathrm{P}, \mathrm{Q})$ between P and Q is given by[5]

$$d_{\mathtt{var}}(\mathrm{P}, \mathrm{Q}) := \frac{1}{2} \sum_{x \in \mathcal{X}} |\mathrm{P}(x) - \mathrm{Q}(x)| = \max_{\mathcal{A} \subset \mathcal{X}} \mathrm{P}(\mathcal{A}) - \mathrm{Q}(\mathcal{A}).$$

By the foregoing discussion and the definition above, we have

$$P_{\mathtt{err}}(\mathrm{P}, \mathrm{Q}) = \frac{1}{2} \left(1 - d_{\mathtt{var}}(\mathrm{P}, \mathrm{Q})\right). \tag{2.5}$$

THEOREM 2.13 (Properties of total variation distance) *For distributions* P, Q, *and* R *on a finite set* $\mathcal{X}$, *the following hold.*

1. (Nonnegativity) $d_{\mathtt{var}}(\mathrm{P}, \mathrm{Q}) \geq 0$ *with equality if and only if* $\mathrm{P} = \mathrm{Q}$.
2. (Triangular inequality) $d_{\mathtt{var}}(\mathrm{P}, \mathrm{Q}) \leq d_{\mathtt{var}}(\mathrm{P}, \mathrm{R}) + d_{\mathtt{var}}(\mathrm{R}, \mathrm{Q})$.
3. (Normalization) $d_{\mathtt{var}}(\mathrm{P}, \mathrm{Q}) \leq 1$ *with equality if and only if* $\mathtt{supp}(\mathrm{P}) \cap \mathtt{supp}(\mathrm{Q}) = \emptyset$.

Proof The first two properties are easy to check; we only show the third one. For that, we note that $\mathrm{P}(\mathcal{A}) - \mathrm{Q}(\mathcal{A}) \leq 1$ for every $\mathcal{A} \subset \mathcal{X}$, whereby $d_{\mathtt{var}}(\mathrm{P}, \mathrm{Q}) \leq 1$. Further, denoting $\mathcal{A}^* = \{x : \mathrm{P}(x) > \mathrm{Q}(x)\}$, we have $d_{\mathtt{var}}(\mathrm{P}, \mathrm{Q}) = \mathrm{P}(\mathcal{A}^*) - \mathrm{Q}(\mathcal{A}^*)$ whereby $d_{\mathtt{var}}(\mathrm{P}, \mathrm{Q}) = 1$ holds if and only if

$$\mathrm{P}(\mathcal{A}^*) - \mathrm{Q}(\mathcal{A}^*) = 1.$$

This in turn is possible if and only if $\mathrm{P}(\mathcal{A}^*) = 1$ and $\mathrm{Q}(\mathcal{A}^*) = 0$, which is the same as $\mathcal{A}^* = \mathtt{supp}(\mathrm{P})$ and $\mathtt{supp}(\mathrm{Q}) \subset \mathcal{A}^c$, completing the proof. □

Next, we note a property which must be satisfied by any reasonable measure of distance between distributions – the *data processing inequality.*

LEMMA 2.14 (Data processing inequality for total variation distance) *Let* P_X *and* Q_X *be distributions on* $\mathcal{X}$ *and* $T\colon \mathcal{X} \to \mathcal{Y}$ *be a channel. Denote by* P_Y *and* Q_Y *the distribution of the output of* T *when the input distribution is* P_X *and* P_Y, *respectively. Then,*

$$d_{\mathtt{var}}(\mathrm{P}_Y, \mathrm{Q}_Y) \leq d_{\mathtt{var}}(\mathrm{P}_X, \mathrm{Q}_X).$$

Proof Define the conditional distribution $W(y|x) := \Pr(T(X) = y \mid X = x)$, $x \in \mathcal{X}, y \in \mathcal{Y}$. Then, the distributions P_Y and Q_Y are given by

$$\mathrm{P}_Y(y) = \sum_{x \in \mathcal{X}} \mathrm{P}_X(x) W(y|x), \quad \mathrm{Q}_Y(y) = \sum_{x \in \mathcal{X}} \mathrm{Q}_X(x) W(y|x).$$

[5] Other names used for the distance are *variational distance* and *statistical distance.*

Thus, we get

$$\begin{aligned}
d_{\mathrm{var}}(\mathrm{P}_Y, \mathrm{Q}_Y) &= \frac{1}{2}\sum_{y\in\mathcal{Y}} |\mathrm{P}_Y(y) - \mathrm{Q}_Y(y)| \\
&= \frac{1}{2}\sum_{y\in\mathcal{Y}} \Big|\sum_{x\in\mathcal{X}} W(y|x)(\mathrm{P}_X(x) - \mathrm{Q}_X(x))\Big| \\
&\le \frac{1}{2}\sum_{y\in\mathcal{Y}}\sum_{x\in\mathcal{X}} W(y|x)|\mathrm{P}_X(x) - \mathrm{Q}_X(x)| \\
&= d_{\mathrm{var}}(\mathrm{P}_X, \mathrm{Q}_X),
\end{aligned}$$

which completes the proof. □

We can see the property above directly from the connection between hypothesis testing and total variation distance seen earlier. Indeed, we note that $P_{\mathrm{err}}(\mathrm{P}_Y, \mathrm{Q}_Y) \ge P_{\mathrm{err}}(\mathrm{P}_X, \mathrm{Q}_X)$ since the optimal test for P_Y versus Q_Y can be used as a test for P_X versus Q_X as well, by first transforming the observation X to $Y = T(X)$. By (2.5), this yields the data processing inequality above.

We close this section with a very useful property, which will be used heavily in formal security analysis of different protocols later in the book.

LEMMA 2.15 (Chain rule for total variation distance) *For two distributions* P_{XY} *and* Q_{XY} *on* $\mathcal{X}\times\mathcal{Y}$, *we have*

$$d_{\mathrm{var}}(\mathrm{P}_{XY}, \mathrm{Q}_{XY}) \le d_{\mathrm{var}}(\mathrm{P}_X, \mathrm{Q}_X) + \mathbb{E}_{\mathrm{P}_X}\left[d_{\mathrm{var}}(\mathrm{P}_{Y|X}, \mathrm{Q}_{Y|X})\right],$$

and further,

$$d_{\mathrm{var}}(\mathrm{P}_{X_1\cdots X_n}, \mathrm{Q}_{X_1\cdots X_n}) \le \sum_{i=1}^n \mathbb{E}_{\mathrm{P}_{X^{i-1}}}\left[d_{\mathrm{var}}(\mathrm{P}_{X_i|X^{i-1}}, \mathrm{Q}_{X_i|X^{i-1}})\right],$$

where X^i *abbreviates the random variable* $(X_1, \ldots, X_i)$.

Proof The proof simply uses the triangular inequality for total variation distance. Specifically, we have

$$d_{\mathrm{var}}(\mathrm{P}_{XY}, \mathrm{Q}_{XY}) \le d_{\mathrm{var}}(\mathrm{P}_{XY}, \mathrm{P}_X\mathrm{Q}_{Y|X}) + d_{\mathrm{var}}(\mathrm{Q}_{XY}, \mathrm{P}_X\mathrm{Q}_{Y|X}).$$

For the first term on the right-hand side, we have

$$\begin{aligned}
d_{\mathrm{var}}(\mathrm{P}_X\mathrm{P}_{Y|X}, \mathrm{P}_X\mathrm{Q}_{Y|X}) &= \frac{1}{2}\sum_{x\in\mathcal{X}} \mathrm{P}_X(x)\sum_{y\in\mathcal{Y}} |\mathrm{P}_{Y|X}(y|x) - \mathrm{Q}_{Y|X}(y|x)| \\
&= \mathbb{E}_{\mathrm{P}_X}\left[d_{\mathrm{var}}(\mathrm{Q}_{Y|X}, \mathrm{P}_{Y|X})\right],
\end{aligned}$$

and further, for the second term,

$$\begin{aligned}
d_{\mathrm{var}}(\mathrm{Q}_X\mathrm{Q}_{Y|X}, \mathrm{P}_X\mathrm{Q}_{Y|X}) &= \frac{1}{2}\sum_{x\in\mathcal{X}, y\in\mathcal{Y}} |\mathrm{Q}_X(x)\mathrm{Q}_{Y|X}(y|x) - \mathrm{P}_X(x)\,\mathrm{Q}_{Y|X}(y|x)| \\
&= \frac{1}{2}\sum_{x\in\mathcal{X}}\sum_{y\in\mathcal{Y}} \mathrm{Q}_{Y|X}(y|x)|\mathrm{P}_X(x) - \mathrm{Q}_X(x)| \\
&= d_{\mathrm{var}}(\mathrm{P}_X, \mathrm{Q}_X).
\end{aligned}$$

The claim follows upon combining these expressions with the previous bound; the general proof for $n \geq 2$ follows by applying this inequality repeatedly. □

In the proof above, we noted that

$$d_{\mathtt{var}}(\mathrm{P}_{XY}, \mathrm{P}_X \mathrm{Q}_{Y|X}) = \mathbb{E}_{\mathrm{P}_X}\left[d_{\mathtt{var}}(\mathrm{P}_{Y|X}, \mathrm{Q}_{Y|X})\right],$$

a useful expression of independent interest. Further, for product distributions P_{X^n} and Q_{X^n} (corresponding to independent random variables), the previous result implies the *subadditivity property*

$$d_{\mathtt{var}}(\mathrm{P}_{X^n}, \mathrm{Q}_{X^n}) \leq \sum_{i=1}^n d_{\mathtt{var}}(\mathrm{P}_{X_i}, \mathrm{Q}_{X_i}).$$

2.5 Kullback–Leibler Divergence

In this book, we use total variation distance to define our notion of security. However, there is a close cousin of this notion of distance that enjoys great popularity in information theory – the *Kullback–Leibler divergence.*

DEFINITION 2.16 (KL divergence) For two discrete distributions P and Q on $\mathcal{X}$, the Kullback–Leibler (KL) divergence $D(\mathrm{P}\|\mathrm{Q})$ between P and Q is given by

$$D(\mathrm{P}\|\mathrm{Q}) = \begin{cases} \sum_{x\in\mathcal{X}} \mathrm{P}(x) \log \frac{\mathrm{P}(x)}{\mathrm{Q}(x)}, & \mathtt{supp}(\mathrm{P}) \subset \mathtt{supp}(\mathrm{Q}), \\ \infty, & \mathtt{supp}(\mathrm{P}) \not\subset \mathtt{supp}(\mathrm{Q}) \end{cases}$$

where log is to the base 2. This convention will be followed throughout – *all our logarithms are to the base 2, unless otherwise stated.*

KL divergence is not a metric, but is a very useful notion of "distance" between distributions. Without giving it an "operational meaning" at the outset, we simply note some of its useful properties in this chapter.

THEOREM 2.17 (Properties of KL divergence) *For distributions P and Q on a finite set $\mathcal{X}$, the following hold.*

1. (Nonnegativity) $D(\mathrm{P}\|\mathrm{Q}) \geq 0$ *with equality if and only if* $\mathrm{P} = \mathrm{Q}$.
2. (Convexity) $D(\mathrm{P}\|\mathrm{Q})$ *is a convex function of the pair* (P, Q) *(over the set of pairs of distributions on $\mathcal{X}$).*
3. (Data processing inequality) *For a channel $T \colon \mathcal{X} \to \mathcal{Y}$, denote by P_Y and Q_Y the distribution of the output of T when the input distribution is P_X and Q_X, respectively. Then, $D(\mathrm{P}_Y\|\mathrm{Q}_Y) \leq D(\mathrm{P}_X\|\mathrm{Q}_X)$.*

Proof The proofs of all these properties use the log-sum inequality (see Lemma 2.9). In fact, the first property is equivalent to the log-sum inequality with vectors $(\mathrm{P}(x), x \in \mathcal{X})$ and $(\mathrm{Q}(x), x \in \mathcal{X})$ in the role of $(a_1, \ldots, a_k)$ and $(b_1, \ldots, b_k)$, respectively.

For the second property, consider pairs $(\mathrm{P}_1, \mathrm{Q}_1)$ and $(\mathrm{P}_2, \mathrm{Q}_2)$ of distributions on $\mathcal{X}$. Further, for $\theta \in [0,1]$, consider the pair $(\mathrm{P}_\theta, \mathrm{Q}_\theta)$ of distributions on $\mathcal{X}$ given by $\mathrm{P}_\theta := \theta \mathrm{P}_1 + (1-\theta)\mathrm{P}_2$ and $\mathrm{Q}_\theta := \theta \mathrm{Q}_1 + (1-\theta)\mathrm{Q}_2$. Then, we have

$$
\begin{aligned}
D(\mathrm{P}_\theta \| \mathrm{Q}_\theta) &= \sum_x \mathrm{P}_\theta(x) \log \frac{\mathrm{P}_\theta(x)}{\mathrm{Q}_\theta(x)} \\
&= \sum_x (\theta \mathrm{P}_1(x) + (1-\theta)\mathrm{P}_2(x)) \log \frac{\theta \mathrm{P}_1(x) + (1-\theta)\mathrm{P}_2(x)}{\theta \mathrm{Q}_1(x) + (1-\theta)\mathrm{Q}_2(x)} \\
&\le \sum_x \theta \mathrm{P}_1(x) \log \frac{\theta \mathrm{P}_1(x)}{\theta \mathrm{Q}_1(x)} + (1-\theta)\mathrm{P}_2(x) \log \frac{(1-\theta)\mathrm{P}_2(x)}{(1-\theta)\mathrm{Q}_2(x)} \\
&= \theta D(\mathrm{P}_1 \| \mathrm{Q}_1) + (1-\theta) D(\mathrm{P}_2 \| \mathrm{Q}_2),
\end{aligned}
$$

where the inequality is by the log-sum inequality.

Finally, for the data processing inequality, with $W(y|x) := \Pr(T(x) = y \mid X = x)$, we get

$$
\begin{aligned}
D(\mathrm{P}_Y \| \mathrm{Q}_Y) &= \sum_{y \in \mathcal{Y}} \mathrm{P}_Y(y) \log \frac{\mathrm{P}_Y(y)}{\mathrm{Q}_Y(y)} \\
&= \sum_{y \in \mathcal{Y}} \sum_{x \in \mathcal{X}} \mathrm{P}_X(x) W(y|x) \log \frac{\sum_{x \in \mathcal{X}} \mathrm{P}_X(x) W(y|x)}{\sum_{x \in \mathcal{X}} \mathrm{Q}_X(x) W(y|x)} \\
&\le \sum_{y \in \mathcal{Y}} \sum_{x \in \mathcal{X}} \mathrm{P}_X(x) W(y|x) \log \frac{\mathrm{P}_X(x) W(y|x)}{\mathrm{Q}_X(x) W(y|x)} \\
&= D(\mathrm{P}_X \| \mathrm{Q}_X),
\end{aligned}
$$

where the inequality uses the log-sum inequality and our convention that $0 \log(0/0) = 0$. □

The convexity of $D(\mathrm{P}\|\mathrm{Q})$ in the pair (P, Q) is a very useful property. In particular, it implies that $D(\mathrm{P}\|\mathrm{Q})$ is convex in P for a fixed Q and convex in Q for a fixed P. Also, later in the book, we will see a connection between KL divergence and probability of error for binary hypothesis testing, and the data processing inequality for KL divergence has similar interpretation to that for total variation distance – adding noise ("data processing") gets distributions closer and makes it harder to distinguish them.

We close this section with a chain rule for KL divergence.

LEMMA 2.18 (Chain rule for KL divergence) *For distributions $\mathrm{P}_{X_1 \cdots X_n}$ and $\mathrm{Q}_{X_1 \cdots X_n}$ on a discrete set $\mathcal{X}_1 \times \mathcal{X}_2 \times \cdots \times \mathcal{X}_n$, we have*

$$
D(\mathrm{P}_{X_1 \cdots X_n} \| \mathrm{Q}_{X_1 \cdots X_n}) = \sum_{i=1}^n \mathbb{E}_{\mathrm{P}_{X^{i-1}}} \left[D(\mathrm{P}_{X_i | X^{i-1}} \| \mathrm{Q}_{X_i | X^{i-1}}) \right],
$$

where X^i abbreviates the random variable $(X_1, \ldots, X_i)$.

Proof It suffices to show the result for $n = 2$. We have

$$\begin{aligned} D(\mathrm{P}_{XY}\|\mathrm{Q}_{XY}) &= \mathbb{E}_{\mathrm{P}_{XY}}\left[\log\frac{\mathrm{P}_{XY}(X,Y)}{\mathrm{Q}_{XY}(X,Y)}\right] \\ &= \mathbb{E}_{\mathrm{P}_{XY}}\left[\log\frac{\mathrm{P}_X(X)}{\mathrm{Q}_X(X)} + \log\frac{\mathrm{P}_{Y|X}(Y|X)}{\mathrm{Q}_{Y|X}(Y|X)}\right] \\ &= \mathbb{E}_{\mathrm{P}_X}\left[\log\frac{\mathrm{P}_X(X)}{\mathrm{Q}_X(X)}\right] + \mathbb{E}_{\mathrm{P}_{XY}}\left[\log\frac{\mathrm{P}_{Y|X}(Y|X)}{\mathrm{Q}_{Y|X}(Y|X)}\right] \\ &= D(\mathrm{P}_X\|\mathrm{Q}_X) + \mathbb{E}_{\mathrm{P}_X}\left[D(\mathrm{P}_{Y|X}\|\mathrm{Q}_{Y|X})\right]; \end{aligned}$$

the proof for general n is obtained by applying this identity recursively. □

We note that, unlike the chain rule for total variation distance, the chain rule above holds with equality. In particular, KL divergence is seen to be *additive* for product distributions; namely, for $\mathrm{P}_{X_1\cdots X_n} = \prod_{i=1}^n \mathrm{P}_{X_i}$ and $\mathrm{Q}_{X_1\cdots X_n} = \prod_{i=1}^n \mathrm{Q}_{X_i}$, we have

$$D(\mathrm{P}_{X_1\cdots X_n}\|\mathrm{Q}_{X_1\cdots X_n}) = \sum_{t=1}^{n} D(\mathrm{P}_{X_i}\|\mathrm{Q}_{X_i}).$$

2.6 Shannon Entropy

Probabilistic modeling allows us to capture uncertainty in our knowledge of a quantity (modeled as a random variable). But to build a formal theory for security, we need to quantify what it means to have partial knowledge of a random variable – to have a "bit" of knowledge about a random variable X. Such a quantification of uncertainty is provided by the information-theoretic notion of *Shannon entropy.*

DEFINITION 2.19 (Shannon entropy) For a random variable X, the Shannon entropy of X is given by[6]

$$H(\mathrm{P}_X) := \sum_{x\in\mathcal{X}} \mathrm{P}_X(x)\log\frac{1}{\mathrm{P}_X(x)},$$

where log is to the base 2. For brevity, we often abbreviate $H(\mathrm{P}_X)$ as $H(X)$. However, the reader should keep in mind that H is a function of the distribution P_X of X.

We have not drawn this quantity out of the hat and proposed it as a measure of uncertainty. There is a rich theory supporting the role of entropy as a measure of uncertainty or a measure of randomness. However, the details are beyond the scope of our book. In fact, a heuristic justification for entropy as a measure of randomness comes from the following observation: denoting by $\mathrm{P}_{\mathtt{unif}}$ the uniform distribution on $\mathcal{X}$, we have

$$H(\mathrm{P}) = \log|\mathcal{X}| - D(\mathrm{P}\|\mathrm{P}_{\mathtt{unif}}). \tag{2.6}$$

[6] We follow the convention $0\log 0 = 0$.

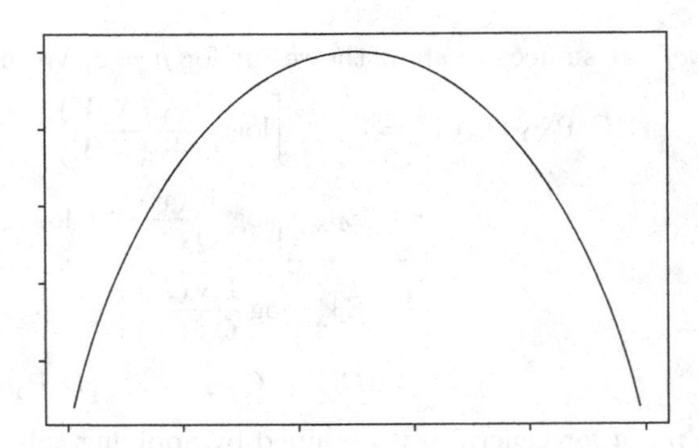

Figure 2.2 The binary entropy function $h(p)$.

That is, Shannon entropy $H(\mathrm{P})$ is a measure of how far P is from a uniform distribution. Heuristically, the uniform distribution is the "most random" distribution on $\mathcal{X}$, and therefore, Shannon entropy is indeed a measure of uncertainty or randomness.

We present the properties of Shannon entropy, which can be derived readily from the properties of KL divergence and (2.6).

THEOREM 2.20 (Shape of the Shannon entropy functional) *Consider a probability distribution* P *on a finite set* $\mathcal{X}$. *Then, the following properties hold.*

1. (Nonnegativity) $H(\mathrm{P}) \geq 0$ *with equality if and only if* $|\mathrm{supp}(\mathrm{P})| = 1$, *namely* P *is the distribution of a constant random variable.*
2. (Boundedness) $H(\mathrm{P}) \leq \log |\mathcal{X}|$ *with equality if and only if* P *is the uniform distribution on* $\mathcal{X}$.
3. (Concavity) $H(\mathrm{P})$ *is a concave function of* P, *namely for every* $\theta \in [0, 1]$ *and two probability distributions* Q_1 *and* Q_2 *on* $\mathcal{X}$,

$$H(\theta \mathrm{Q}_1 + (1 - \theta)\mathrm{Q}_2) \geq \theta H(\mathrm{Q}_1) + (1 - \theta) H(\mathrm{Q}_2).$$

Proof The nonnegativity property is easy to see: each term in the expression for entropy is nonnegative, whereby $H(\mathrm{P})$ is 0 if and only if each term in the sum is 0. This can only happen if $\mathrm{P}(x) = 1$ for one $x \in \mathcal{X}$ and $\mathrm{P}(x) = 0$ for the rest.

The boundedness property follows from (2.6) using the nonnegativity of KL divergence. Further, the concavity of Shannon entropy also follows from (2.6) using the convexity of $D(\mathrm{P} \| \mathrm{P}_{\mathtt{unif}})$ in P. □

For the special case of binary random variables ($\mathcal{X} = \{0, 1\}$), Shannon entropy $H(\mathrm{P})$ depends only on $p := \mathrm{P}(1)$ and is denoted using the function $h \colon [0, 1] \to [0, 1]$, termed the *binary entropy function*. That is, $h(p) := p \log \frac{1}{p} + (1 - p) \log \frac{1}{(1-p)}$, $p \in [0, 1]$. We depict $h(p)$ in Figure 2.2.

Next, we seek a notion of residual uncertainty, the uncertainty remaining in X when a correlated random variable Y is revealed. Such a notion is given by the *conditional Shannon entropy*, or simply, conditional entropy, defined below.

DEFINITION 2.21 (Conditional Shannon entropy) For discrete random variables X and Y, the conditional Shannon entropy $H(X|Y)$ is given by $\mathbb{E}_{\mathrm{P}_Y}\left[H(\mathrm{P}_{X|Y})\right]$.

Using the expression for Shannon entropy, it is easy to check that

$$H(Y|X) = \sum_{x,y} \mathrm{P}_{XY}(x,y) \log \frac{1}{\mathrm{P}_{Y|X}(y|x)} = \mathbb{E}\left[-\log \mathrm{P}_{Y|X}(Y|X)\right].$$

Note that the random variable in the expression on the right-hand side is $-\log \mathrm{P}_{Y|X}(Y|X)$, and we take expectation over $(X,Y) \sim \mathrm{P}_{XY}$.

We present now a chain rule for Shannon entropy, which allows us to divide the *joint entropy* $H(X_1,\ldots,X_n)$ of random variables $(X_1,\ldots,X_n)$ into "smaller" components.

LEMMA 2.22 (Chain rule for entropy) *For discrete random variables* $(X_1,\ldots,X_n)$ *and* Y, *we have*

$$H(X_1,\ldots,X_n \mid Y) = \sum_{i=1}^{n} H(X_i|X^{i-1},Y). \tag{2.7}$$

Proof We can derive this using the chain rule for KL divergence and (2.6). Alternatively, we can see it directly as follows. First consider the case $n=2$ and Y is a constant. We have

$$\begin{aligned} H(X_2|X_1) &= \mathbb{E}\left[-\log \mathrm{P}_{X_2|X_1}(X_2|X_1)\right] \\ &= \mathbb{E}[-\log \mathrm{P}_{X_1X_2}(X_1,X_2)] + \mathbb{E}[\log \mathrm{P}_{X_1}(X_1)] \\ &= H(X_1,X_2) - H(X_1), \end{aligned}$$

where we used the Bayes rule in the second identity. The result for general n, but with Y constant, is obtained by applying this result recursively. The more general result when Y is not constant follows from (2.7) with constant Y upon noting that $H(X_1,\ldots,X_n|Y) = H(X_1,\ldots,X_n,Y) - H(Y)$. □

We close this section by commenting on the notation $H(Y|X)$, which is, admittedly, a bit informal. It will be more appropriate to view conditional entropy as a function of (P,W) where P is the distribution of X and W is the channel with X as the input and Y as the output. In particular, we use the notation $H(W|\mathrm{P})$ to denote $H(Y|X)$. Note that

$$H(W|\mathrm{P}) = \sum_{x\in\mathcal{X}} \mathrm{P}(x) \sum_{y\in\mathcal{Y}} W(y|x) \log \frac{1}{W(y|x)},$$

and that $H(W|\mathrm{P})$ is a linear function of P and a concave function of W.

Finally, we note the following important consequence of concavity of $H(P)$.

LEMMA 2.23 (Conditioning reducing entropy) *For a probability distribution* P *on a finite set* $\mathcal{X}$ *and a channel* $W: \mathcal{X} \to \mathcal{Y}$, *we have* $H(W|\mathrm{P}) \le H(\mathrm{P} \circ W)$. *(In our alternative notation,* $H(Y|X) \le H(Y)$ *for all random variables* (X, Y).*)*

Proof Since $H(\cdot)$ is a concave function, we have

$$H(W|\mathrm{P}) = \sum_x \mathrm{P}(x) H(W_x) \le H\Big(\sum_x \mathrm{P}(x) W_x\Big) = H(\mathrm{P} \circ W),$$

where we abbreviate the output distribution $W(\cdot|x)$ as W_x. □

2.7 Mutual Information

How much information does a random variable Y reveal about another random variable X? A basic postulate of information theory is that "Information is reduction in Uncertainty." We already saw how to measure uncertainty: $H(X)$ is the uncertainty in a random variable X and $H(X|Y)$ is the uncertainty remaining in X once Y is revealed. Thus, a measure of information provided by the postulate above is $H(X) - H(X|Y)$. This fundamental measure of information is called the *mutual information*.

DEFINITION 2.24 (Mutual information) Given a joint distribution P_{XY}, the mutual information between X and Y is given by $I(X \wedge Y) = H(X) - H(X|Y)$. We will use an alternative definition where we represent the mutual information as a function of the input distribution $\mathrm{P} = \mathrm{P}_X$ and the channel $W = \mathrm{P}_{Y|X}$. Namely, we represent mutual information $I(X \wedge Y)$ as $I(\mathrm{P}, W)$.

Before proceeding, we note several alternative expressions for mutual information.

LEMMA 2.25 (Alternative expressions for mutual information) *For discrete random variables* (X, Y), *the following quantities are equal to* $I(X \wedge Y)$:

1. $H(Y) - H(Y|X)$;
2. $H(X) + H(Y) - H(X, Y)$;
3. $H(X, Y) - H(X|Y) - H(Y|X)$;
4. $D(\mathrm{P}_{XY} \| \mathrm{P}_X \times \mathrm{P}_Y)$.

Proof The equality of $H(X) - H(X|Y)$ with expressions in 1–3 follows upon noting that $H(X|Y) = H(X, Y) - H(Y)$ and $H(Y|X) = H(X, Y) - H(X)$. The equality with $D(\mathrm{P}_{XY} \| \mathrm{P}_X \times \mathrm{P}_Y)$ follows since $H(X, Y) = \mathbb{E}_{\mathrm{P}_{XY}}[-\log \mathrm{P}_{XY}(X, Y)]$, $H(X) = \mathbb{E}_{\mathrm{P}_{XY}}[-\log \mathrm{P}_X(X)]$, and $H(Y) = \mathbb{E}_{\mathrm{P}_{XY}}[-\log \mathrm{P}_Y(Y)]$, whereby

$$H(X) + H(Y) - H(X, Y) = \mathbb{E}_{\mathrm{P}_{XY}}\left[\log \frac{\mathrm{P}_{XY}(X, Y)}{\mathrm{P}_X(X)\,\mathrm{P}_Y(Y)}\right] = D(\mathrm{P}_{XY} \| \mathrm{P}_X \times \mathrm{P}_Y).$$

□

The simple mathematical expressions above lead to a rather remarkable theory. First, we observe that the information revealed by Y about X, $I(X \wedge Y)$, coincides with the information revealed by X about Y, $I(Y \wedge X)$. Further, $I(X \wedge Y) = D(\mathrm{P}_{XY} \| \mathrm{P}_X \times \mathrm{P}_Y)$ allows us to interpret mutual information as a measure of how "dependent" X and Y are. In particular, $I(X \wedge Y) = 0$ if and only if X and Y are independent, and equivalently, $H(X) = H(X|Y)$ if and only if X and Y are independent. Also, since KL divergence is nonnegative, it follows that conditioning reduces entropy, a fact we already saw using concavity of entropy.

Next, we define *conditional mutual information* of X and Y given Z as

$$I(X \wedge Y|Z) := H(X|Z) - H(X|Y, Z).$$

It is easy to verify that

$$I(X \wedge Y|Z) = \mathbb{E}_{\mathrm{P}_Z}\left[D(\mathrm{P}_{XY|Z} \| \mathrm{P}_{X|Z} \times \mathrm{P}_{Y|Z})\right],$$

whereby $I(X \wedge Y|Z) = 0$ if and only if X and Y are independent given Z. Further, we can use the chain rule for KL divergence or Shannon entropy to obtain the following chain rule for mutual information.

LEMMA 2.26 (Chain rule for mutual information) *For discrete random variables* $(X_1, \ldots, X_n, Y)$, *we have*

$$I(X_1, X_2, \ldots, X_n \wedge Y) = \sum_{i=1}^{n} I(X_i \wedge Y \mid X^{i-1}),$$

where $X^{i-1} = (X_1, \ldots, X_{i-1})$ *for* $1 < i \leq n$ *and* X^0 *is a constant random variable.*

Finally, we present a data processing inequality for mutual information which is, in fact, a consequence of the data processing inequality for KL divergence. To present this inequality, we need to introduce the notion of a Markov chain.

DEFINITION 2.27 (Markov chains) Random variables X, Y, Z form a Markov chain if X and Z are independent when conditioned on Y, i.e., when $I(X \wedge Z|Y) = 0$ or, equivalently, $\mathrm{P}_{XYZ} = \mathrm{P}_{X|Y}\mathrm{P}_{Z|Y}\mathrm{P}_Y$. This definition extends naturally to multiple random variables: $X_1, \ldots, X_n$ form a Markov chain if $X^{i-1} = (X_1, \ldots, X_{i-1})$, X_i, and $X_{i+1}^n = (X_{i+1}, \ldots, X_n)$ form a Markov chain for every $1 \leq i \leq n$. We use the notation $X_1 \multimap X_2 \multimap \cdots \multimap X_n$ to indicate that $X_1, \ldots, X_n$ form a Markov chain.[7]

A specific example is when $Z = f(Y)$ for some function f. In this case, for every X we have $X \multimap Y \multimap Z$. Heuristically, if $X \multimap Y \multimap Z$ holds, then Z can contain no more information about X than Y. The following result establishes this bound formally.

LEMMA 2.28 (Data processing inequality for mutual information) *If* $X \multimap Y \multimap Z$, *then* $I(X \wedge Z) \leq I(X \wedge Y)$. *Equivalently,* $H(X|Y) \leq H(X|Z)$.

[7] That is, there is no more information about X_{i+1}^n in X^i than that contained in X_i.

Proof Instead of taking recourse to the data processing inequality for KL divergence, we present an alternative proof. We have

$$\begin{aligned} I(X \wedge Z) &= I(X \wedge Y, Z) - I(X \wedge Y|Z) \\ &\leq I(X \wedge Y, Z) \\ &= I(X \wedge Y) + I(X \wedge Z|Y) \\ &= I(X \wedge Y), \end{aligned}$$

where the inequality holds since conditional mutual information is nonnegative and the final identity holds since $X \multimap Y \multimap Z$. □

2.8 Fano's Inequality

We now prove Fano's inequality – a lower bound for the probability of error for guessing a random variable X using another random variable Y. In particular, Fano's inequality provides a lower bound for the probability of error in terms of the mutual information $I(X \wedge Y)$ between X and Y. Alternatively, we can view Fano's inequality as an upper bound for the conditional entropy $H(X|Y)$, in terms of the probability of error. It is a very useful inequality and is applied widely in information theory, statistics, communications, and other related fields. In fact, the proof of Fano's inequality we present uses nothing more than the data processing inequality.

THEOREM 2.29 (Fano's inequality) *For discrete random variables X and Y, consider $\hat{X}$ such that $X \multimap Y \multimap \hat{X}$.*[8] *Then, we have*

$$H(X|Y) \leq \Pr(\hat{X} \neq X) \log(|\mathcal{X}| - 1) + h\big(\Pr(\hat{X} \neq X)\big),$$

where $h(t) = -t \log t - (1-t) \log(1-t)$ is the binary entropy function.

Proof Instead of P_{XY}, consider the distribution Q_{XY} given by

$$\mathrm{Q}_{XY}(x, y) = \frac{1}{|\mathcal{X}|} \mathrm{P}_Y(y), \quad x \in \mathcal{X}, y \in \mathcal{Y},$$

namely, the distribution when X is uniform and independent of Y. We can treat the estimate $\hat{X}$ as a randomized function of Y, expressed using the channel $\mathrm{P}_{\hat{X}|Y}$. Let $\mathrm{Q}_{XY\hat{X}}$ be given by

$$\mathrm{Q}_{XY\hat{X}}(x, y, \hat{x}) = \mathrm{Q}_{XY}(x, y) \mathrm{P}_{\hat{X}|Y}(\hat{x}|y), \quad \forall x, \hat{x} \in \mathcal{X}, y \in \mathcal{Y}.$$

We note that the probability of correctness of the estimate $\hat{X}$ under Q_{XY} is the same as that of a "random guess." Indeed, we have

$$\mathrm{Q}_{XY\hat{X}}(X = \hat{X}) = \sum_{x \in \mathcal{X}} \sum_{y \in \mathcal{Y}} \mathrm{Q}_{XY}(x, y) \mathrm{P}_{\hat{X}|Y}(x|y)$$

[8] We can view $\hat{X}$ as an "estimate" of X formed from Y.

$$= \frac{1}{|\mathcal{X}|} \sum_{x \in \mathcal{X}} \sum_{y \in \mathcal{Y}} \mathrm{P}_Y(y) \, \mathrm{P}_{\hat{X}|Y}(x|y)$$

$$= \frac{1}{|\mathcal{X}|}. \qquad (2.8)$$

The main idea behind our proof is the following. For any distribution P_{XY}, the difference between the performance of the estimator $\hat{X}$ under P_{XY} and Q_{XY}, the independent distribution, is bounded by the "distance" between these distributions. We formalize this using the data processing inequality.

Formally, consider the channel $W\colon \mathcal{X} \times \mathcal{Y} \times \mathcal{X} \to \{0,1\}$ given by $W(1|x,y,\hat{x}) = \mathbf{1}[x = \hat{x}]$. Then, by the data processing inequality we get

$$\begin{aligned} D(\mathrm{P}_{XY\hat{X}} \circ W \| \mathrm{Q}_{XY\hat{X}} \circ W) &\leq D(\mathrm{P}_{XY\hat{X}} \| \mathrm{Q}_{XY\hat{X}}) \\ &= D(\mathrm{P}_{XY} \| \mathrm{Q}_{XY}) \\ &= \log |\mathcal{X}| - H(X|Y), \end{aligned}$$

where we used the chain rule for KL divergence in the first identity and the second identity can be verified by a direct calculation (see Problem 2.2) . Further, denoting $p = \mathrm{P}_{XY\hat{X}}(X = \hat{X})$ and $q = \mathrm{Q}_{XY\hat{X}}(X = \hat{X})$, we get

$$\begin{aligned} D(\mathrm{P}_{XY\hat{X}} \circ W \| \mathrm{Q}_{XY\hat{X}} \circ W) &= p \log \frac{p}{q} + (1-p) \log \frac{1-p}{1-q} \\ &= \log |\mathcal{X}| - (1-p)\log(|\mathcal{X}| - 1) - h(1-p), \end{aligned}$$

where in the second identity we used the expression for probability of correctness q under $\mathrm{Q}_{XY\hat{X}}$ computed in (2.8). Upon combining this expression with the previous bound, we get

$$H(X|Y) \leq (1-p)\log(|\mathcal{X}| - 1) + h(1-p),$$

which completes the proof since $\mathrm{P}_{XY\hat{X}}(X \neq \hat{X}) = 1 - p$. □

When X is uniform, it follows from Fano's inequality that

$$\Pr(X \neq \hat{X}) \geq 1 - \frac{I(X \wedge Y) + 1}{\log |\mathcal{X}|},$$

an inequality used popularly for deriving lower bounds in statistics.

2.9 Maximal Coupling Lemma

Earlier in Section 2.4, we saw that $d_{\mathrm{var}}(\mathrm{P},\mathrm{Q}) = \max_{\mathcal{A} \subset \mathcal{X}} (\mathrm{P}(\mathcal{A}) - \mathrm{Q}(\mathcal{A}))$, a formula that expresses total variation distance as an optimization problem. Such expressions are sometimes called "variational formulae," leading to the alternative name *variational distance* for total variation distance. In fact, we saw in Lemma 2.10 an operational interpretation of this formula, where we can associate with $\mathcal{A}$ the hypothesis test which declares P when $X \in \mathcal{A}$. In this section, we will see yet another variational formula for total variation distance, which is equally important and interesting. This one also gives another operational meaning to the total variation distance, in the context of *optimal transportation cost*.

As a motivation, consider the following optimal transportation problem. A commodity is stored across multiple warehouses labeled by elements of $\mathcal{X}$, with warehouse $x \in \mathcal{X}$ having a fraction $\mathrm{P}(x)$ of it. We need to transfer this commodity to multiple destinations, labeled again by the elements of $\mathcal{X}$, with destination $y \in \mathcal{X}$ seeking $\mathrm{Q}(y)$ fraction of it. Towards that, we assign a fraction $W(y|x)$ of commodity at $x \in \mathcal{X}$ to be shipped to $y \in \mathcal{X}$. Suppose that we incur a cost of 1 when we send a "unit" of the commodity from source x to a destination y that differs from x, and no cost when sending from x to x. What is the minimum possible cost that we can incur?

More precisely, we can represent the input fractions using a random variable X with probability distribution P and the output fractions using a random variable Y with probability distribution Q. While the marginal distributions of X and Y are fixed, our assignment W defines a joint distribution for X and Y. Such a joint distribution is called a *coupling* of distributions P and Q.

DEFINITION 2.30 (Coupling) For two probability distributions P and Q on $\mathcal{X}$, a coupling of P and Q is a joint distribution P_{XY} such that $\mathrm{P}_X = \mathrm{P}$ and $\mathrm{P}_Y = \mathrm{Q}$. The set of all couplings of P and Q is denoted by $\pi(\mathrm{P}, \mathrm{Q})$. Note that $\pi(\mathrm{P}, \mathrm{Q})$ contains $\mathrm{P} \times \mathrm{Q}$ and is, therefore, nonempty

We now express the previous optimal transportation problem using this notion of couplings. An assignment W coincides with a coupling P_{XY} of P and Q, and the cost incurred by this assignment is given by

$$C(X, Y) := \mathbb{E}_{\mathrm{P}_{XY}}[\mathbf{1}[X \neq Y]] = \Pr(X \neq Y).$$

Thus, in the optimal transport problem specified above, the goal is to find the minimum cost[9]

$$C^*(\mathrm{P}, \mathrm{Q}) = \min_{\mathrm{P}_{XY} \in \pi(\mathrm{P}, \mathrm{Q})} C(X, Y) = \min_{\mathrm{P}_{XY} \in \pi(\mathrm{P}, \mathrm{Q})} \Pr(X \neq Y).$$

Interestingly, $C^*(\mathrm{P}, \mathrm{Q})$ coincides with $d_{\mathrm{var}}(\mathrm{P}, \mathrm{Q})$.

LEMMA 2.31 (Maximal coupling lemma) *For probability distributions* P *and* Q *on* $\mathcal{X}$, *we have*

$$d_{\mathrm{var}}(\mathrm{P}, \mathrm{Q}) = C^*(\mathrm{P}, \mathrm{Q}).$$

Proof Consider a coupling $\mathrm{P}_{XY} \in \pi(\mathrm{P}, \mathrm{Q})$. Then, for any x, we have

$$\begin{aligned} \mathrm{P}(x) &= \Pr(X = x, Y \neq X) + \Pr(X = x, Y = X) \\ &\leq \Pr(X = x, Y \neq X) + \Pr(Y = x) \\ &= \Pr(X = x, Y \neq X) + \mathrm{Q}(x), \end{aligned}$$

[9] We will see soon that there is a coupling that attains the minimum, justifying the use of min instead of inf in the definition.

whereby

$$P(x) - Q(x) \le \Pr(X = x, Y \neq X), \quad \forall x \in \mathcal{X}.$$

Summing over x such that $P(x) \ge Q(x)$, we get

$$d_{\mathtt{var}}(P, Q) \le \sum_{x : P(x) \ge Q(x)} \Pr(X = x, Y \neq X) \le \Pr(Y \neq X).$$

Since this bound holds for every coupling $P_{XY} \in \pi(P, Q)$, we obtain

$$d_{\mathtt{var}}(P, Q) \le C^*(P, Q).$$

For the other direction, let U be a binary random variable, with $\Pr(U = 0) = \sum_{x \in \mathcal{X}} \min\{P(x), Q(x)\}$. Noting that

$$\begin{aligned} 1 - \sum_{x \in \mathcal{X}} \min\{P(x), Q(x)\} &= \sum_{x \in \mathcal{X}} (P(x) - \min\{P(x), Q(x)\}) \\ &= \sum_{x \in \mathcal{X} : P(x) \ge Q(x)} (P(x) - Q(x)) \\ &= d_{\mathtt{var}}(P, Q), \end{aligned}$$

i.e., U is the Bernoulli random variable with parameter $d_{\mathtt{var}}(P, Q)$. Conditioned on $U = 0$, we sample $X = x, Y = y$ with probability

$$\Pr(X = x, Y = y | U = 0) = \min\{P(x), Q(x)\} \mathbf{1}[x = y] / (1 - d_{\mathtt{var}}(P, Q)).$$

Conditioned on $U = 1$, we sample $X = x$ and $Y = y$ with probability

$$\begin{aligned} &\Pr(X = x, Y = y | U = 1) \\ &\quad = \frac{(P(x) - Q(x))(Q(y) - P(y))}{d_{\mathtt{var}}(P, Q)^2} \mathbf{1}[P(x) \ge Q(x), Q(y) \ge P(y)]. \end{aligned}$$

Thus, $X = Y$ if and only if $U = 0$, whereby

$$\Pr(X \neq Y) = \Pr(U = 1) = d_{\mathtt{var}}(P, Q).$$

It only remains to verify that $P_{XY} \in \pi(P, Q)$. Indeed, note that

$$\begin{aligned} \sum_{y \in \mathcal{Y}} P_{XY}(x, y) &= \Pr(U = 0) \sum_{y \in \mathcal{X}} \Pr(X = x, Y = y | U = 0) \\ &\quad + \Pr(U = 1) \sum_{y \in \mathcal{X}} \Pr(X = x, Y = y | U = 1) \\ &= \min\{P(x), Q(x)\} \sum_{y \in \mathcal{X}} \mathbf{1}[y = x] \\ &\quad + (P(x) - Q(x)) \mathbf{1}[P(x) \ge Q(x)] \sum_{y \in \mathcal{X} :\, Q(y) \ge P(y)} \frac{(Q(y) - P(y))}{d_{\mathtt{var}}(P, Q)} \\ &= \min\{P(x), Q(x)\} + (P(x) - Q(x)) \mathbf{1}[P(x) \ge Q(x)] \\ &= P(x) \end{aligned}$$

for every $x \in \mathcal{X}$, and similarly, $\sum_{x\in\mathcal{X}} \mathrm{P}_{XY}(x, y) = \mathrm{Q}(y)$, which shows that $C^*(\mathrm{P}, \mathrm{Q}) \le d_{\mathtt{var}}(\mathrm{P}, \mathrm{Q})$ and completes the proof. □

2.10 A Variational Formula for KL Divergence

We have seen two variational formulae for total variation distance. In fact, a very useful variational formula can be given for KL divergence as well.

LEMMA 2.32 (A variational formula for KL divergence) *For probability distributions* P *and* Q *on a set* $\mathcal{X}$ *such that* $\mathrm{supp}(\mathrm{P}) \subset \mathrm{supp}(\mathrm{Q})$, *we have*

$$D(\mathrm{P}\|\mathrm{Q}) = \max_{R} \sum_{x\in\mathcal{X}} \mathrm{P}(x) \log \frac{R(x)}{\mathrm{Q}(x)},$$

where the max *is over all probability distributions* R *on* $\mathcal{X}$ *such that* $\mathrm{supp}(\mathrm{P}) \subset \mathrm{supp}(R)$. *The* max *is attained by* $R = \mathrm{P}$.

Proof Using the expression for KL divergence, we have

$$\begin{aligned} D(\mathrm{P}\|\mathrm{Q}) &= \sum_{x\in\mathcal{X}} \mathrm{P}(x) \log \frac{\mathrm{P}(x)}{\mathrm{Q}(x)} \\ &= \sum_{x} \mathrm{P}(x) \log \frac{R(x)}{\mathrm{Q}(x)} + D(\mathrm{P}\|R) \\ &\ge \sum_{x} \mathrm{P}(x) \log \frac{R(x)}{\mathrm{Q}(x)}, \end{aligned}$$

with equality if and only if $\mathrm{P} = R$. □

In fact, a similar formula can be attained by restricting R to a smaller family of probability distributions containing P. A particular family of interest is the "exponentially tilted family" of probability distributions given by $R_f(x) = \mathrm{Q}(x)2^{f(X)}/\mathbb{E}_{\mathrm{Q}}\left[2^{f(X)}\right]$, where $f\colon \mathcal{X} \to \mathbb{R}$, which gives the following alternative variational formula.

LEMMA 2.33 *For probability distributions* P *and* Q *on a set* $\mathcal{X}$ *such that* $\mathrm{supp}(\mathrm{P}) \subset \mathrm{supp}(\mathrm{Q})$, *we have*

$$D(\mathrm{P}\|\mathrm{Q}) = \max_{f} \mathbb{E}_{\mathrm{P}}[f(X)] - \log \mathbb{E}_{\mathrm{Q}}\left[2^{f(X)}\right],$$

where the maximum is over all functions $f\colon \mathcal{X} \to \mathbb{R}$ *and is attained by* $f(x) = \log(P(x)/Q(x))$.

2.11 Continuity of Entropy

Next, we present a bound that relates $H(\mathrm{P}) - H(\mathrm{Q})$ to $d_{\mathtt{var}}(\mathrm{P}, \mathrm{Q})$. We will show that $|H(\mathrm{P}) - H(\mathrm{Q})|$ is roughly $\mathcal{O}(d_{\mathtt{var}}(\mathrm{P}, \mathrm{Q}) \log 1/d_{\mathtt{var}}(\mathrm{P}, \mathrm{Q}))$ with the constant depending on $\log|\mathcal{X}|$. As a consequence, we find that entropy is continuous in P, for distributions P supported on a finite set $\mathcal{X}$.

THEOREM 2.34 (Continuity of entropy) *For probability distributions* P *and* Q *on* $\mathcal{X}$, *we have*

$$|H(\mathrm{P}) - H(\mathrm{Q})| \le d_{\mathtt{var}}(\mathrm{P},\mathrm{Q}) \log(|\mathcal{X}| - 1) + h(d_{\mathtt{var}}(\mathrm{P},\mathrm{Q})),$$

where $h(\cdot)$ *denotes the binary entropy function.*

Proof Consider a coupling P_{XY} of P and Q ($\mathrm{P}_X = \mathrm{P}$ and $\mathrm{P}_Y = \mathrm{Q}$). Then, $H(X) = H(\mathrm{P})$ and $H(Y) = H(\mathrm{Q})$, whereby

$$\begin{aligned} |H(\mathrm{P}) - H(\mathrm{Q})| &= |H(X) - H(Y)| = |H(X|Y) - H(Y|X)| \\ &\le \max\{H(X|Y), H(Y|X)\}. \end{aligned}$$

By Fano's inequality, we have

$$\max\{H(X|Y), H(Y|X)\} \le \Pr(X \neq Y) \log(|\mathcal{X}| - 1) + h(\Pr(X \neq Y)).$$

The bound above holds for every coupling. Therefore, choosing the coupling that attains the lower bound of $d_{\mathtt{var}}(\mathrm{P},\mathrm{Q})$ in the maximal coupling lemma (Lemma 2.31), we get

$$\max\{H(X|Y), H(Y|X)\} \le d_{\mathtt{var}}(P,Q) \log(|\mathcal{X}| - 1) + h(d_{\mathtt{var}}(P,Q)).$$ □

2.12 Hoeffding's Inequality

Earlier, in Section 2.2, we saw that a sum of i.i.d. random variables $S_n = \sum_{i=1}^n X_i$ takes values close to $n\mathbb{E}[X_1]$ with high probability as n increases. Specifically, in proving the weak law of large numbers, we used Chebyshev's inequality for S_n. We take a short detour in this section and present a "concentration inequality" which often gives better estimates for $|S_n - n\mathbb{E}[X_1]|$ than Chebyshev's inequality. This is a specific instance of the *Chernoff bound* and applies to bounded random variables – it is called *Hoeffding's inequality.* The reason for presenting this bound here is twofold: first, indeed, we use Hoeffding's inequality for our analysis in this book; and second, we will use Hoeffding's lemma to prove Pinsker's inequality in the next section.

Consider a random variable X taking values in a finite set $\mathcal{X} \subset \mathbb{R}$ and such that $\mathbb{E}[X] = 0$. By Markov's inequality applied to the random variable $e^{\lambda X}$, where $\lambda > 0$, we get

$$\begin{aligned} \Pr(X > t) &= \Pr(\lambda X > \lambda t) \\ &= \Pr(e^{\lambda X} > e^{\lambda t}) \\ &\le \mathbb{E}\left[e^{\lambda(X-t)}\right] \end{aligned}$$

for every $t \in \mathbb{R}$ and every $\lambda > 0$. This very simple bound is, in fact, very powerful, and is called the *Chernoff bound.*

Of particular interest are *sub-Gaussian* random variables, namely random variables which have similar tail probabilities $\Pr(X > t)$ to Gaussian random variables. It is a standard fact that, for a Gaussian random variable G with zero mean and unit variance, $\Pr(G > t) \le e^{-\frac{t^2}{2}}$. Roughly speaking, sub-Gaussian random variables are those which have similarly decaying tail probabilities. Using the Chernoff bound given above, we can convert this requirement of quadratically exponential decay of tail probabilities to that for the *log-moment generating function* $\psi_X(\lambda) := \ln \mathbb{E}[e^{\lambda X}]$, $\lambda \in \mathbb{R}$.

Formally, we have the following definition.

DEFINITION 2.35 (Sub-Gaussian random variables) A random variable X is sub-Gaussian with variance parameter σ^2 if $\mathbb{E}[X] = 0$ and for every $\lambda \in \mathbb{R}$ we have

$$\ln \mathbb{E}[e^{\lambda X}] \le \frac{\lambda^2 \sigma^2}{2}.$$

Recall that the log-moment generating function of the standard Gaussian random variable is given by $\psi_G(\lambda) = \lambda^2/2$. Thus, the definition above requires that X has log-moment generating function dominated by that of a Gaussian random variable with variance σ^2.

Using the Chernoff bound provided above, we get a Gaussian-like tail bound for sub-Gaussian random variables.

LEMMA 2.36 (Sub-Gaussian tails) *For a sub-Gaussian random variable X with variance parameter σ^2, we have for every $t > 0$ that*

$$\Pr(X > t) \le e^{-\frac{t^2}{2\sigma^2}}. \tag{2.9}$$

Proof Since X is sub-Gaussian with variance parameter σ^2, using the Chernoff bound, for every $\lambda > 0$ we have

$$\Pr(X > t) \le e^{-\lambda t} \mathbb{E}[e^{\lambda X}] \le e^{-\lambda t + \lambda^2 \sigma^2/2},$$

which upon minimizing the right-hand side over $\lambda > 0$ gives the desired bound (2.9), which is attained by $\lambda = \frac{t}{\sigma^2}$. □

Next, we make the important observation that the sum of independent sub-Gaussian random variables is sub-Gaussian too. This, when combined with the previous observations, gives a concentration bound for sums of sub-Gaussian random variables.

LEMMA 2.37 (Sum of sub-Gaussian random variables) *Let $X_1, \ldots, X_n$ be independent and sub-Gaussian random variables with variance parameters $\sigma_1^2, \ldots, \sigma_n^2$, respectively. Then, for every $t > 0$, we have*

$$\Pr\left(\sum_{i=1}^{n} X_i \ge t \right) > e^{-\frac{t^2}{2\sum_{i=1}^{n} \sigma_i^2}}.$$

Proof First, we note that for every $\lambda > 0$ we have

$$\mathbb{E}\left[e^{\lambda \sum_{i=1}^n X_i}\right] = \prod_{i=1}^n \mathbb{E}\left[e^{\lambda X_i}\right] \leq e^{\lambda^2 \sum_{i=1}^n \sigma_i^2/2},$$

where we used the independence of the X_i for the identity and the fact that they are sub-Gaussian for the inequality. Thus, $\sum_{i=1}^n X_i$ is sub-Gaussian with variable parameter $\sum_{i=1}^n \sigma_i^2$, and the claim follows from Lemma 2.36. □

Finally, we come to the Hoeffding inequality, which provides a concentration bound for bounded random variable. The main technical component of the proof is to show that a bounded random variable is sub-Gaussian. We show this first.

LEMMA 2.38 (Hoeffding's lemma) *For a random variable X taking finitely many values in the interval $[a, b]$ and such that $\mathbb{E}[X] = 0$, we have*

$$\ln \mathbb{E}\left[e^{\lambda X}\right] \leq \frac{(b-a)^2 \lambda^2}{8}.$$

Proof As a preparation for the proof, we first note that for any random variable Y taking values in $[a, b]$, we have

$$\mathbb{V}[Y] \leq \frac{(b-a)^2}{4}.$$

The first observation we make is that $\mathbb{V}[Y] = \min_{\theta \in [a,b]} \mathbb{E}\left[(Y-\theta)^2\right]$. This can be verified[10] by simply optimizing over θ. It follows that

$$\mathbb{V}[Y] \leq \min_{\theta \in [a,b]} \max\{(\theta - a)^2, (b-\theta)^2\} = \frac{(a-b)^2}{4}.$$

Next, we note that the function $\psi(\lambda) = \ln \mathbb{E}\left[e^{\lambda X}\right]$ is a twice continuously differentiable function over $\lambda \in \mathbb{R}$ for a discrete and finite random variable X. Thus, by Taylor's approximation,

$$\psi(\lambda) \leq \psi(0) + \psi'(0)\lambda + \max_{c \in (0,\lambda]} \psi''(c)\frac{\lambda^2}{2}.$$

A simple calculation shows that $\psi'(\lambda) = \mathbb{E}\left[X e^{\lambda X}\right] / \mathbb{E}\left[e^{\lambda X}\right]$, and it follows that

$$\psi(0) = \psi'(0) = 0.$$

Also, differentiating once again, we get

$$\psi''(\lambda) = \frac{\mathbb{E}\left[X^2 e^{\lambda X}\right]}{\mathbb{E}\left[e^{\lambda X}\right]} - \left(\frac{\mathbb{E}\left[X e^{\lambda X}\right]}{\mathbb{E}\left[e^{\lambda X}\right]}\right)^2.$$

Denoting by $\mathrm{Q}(x)$ the probability distribution $\mathrm{Q}(x) = \mathrm{P}_X(x)\, e^{\lambda x} / \mathbb{E}\left[e^{\lambda X}\right]$, we note that $\psi''(\lambda)$ is the variance of X under Q. Thus, by our observation earlier,

[10] We only consider discrete random variables, wherein the proofs are technically straightforward.

$\psi''(\lambda) \le (b-a)^2/4$. Combining these bounds with the Taylor approximation, we get

$$\psi(\lambda) \le \frac{(b-a)^2\lambda^2}{8},$$

which completes the proof. □

Thus, a zero-mean random variable taking values in $[a, b]$ is sub-Gaussian with variance parameter $(b-a)^2/4$. We obtain Hoeffding's inequality as a consequence of this fact and Lemma 2.37. We state this final form for random variables which need not be zero-mean. This can be done simply by noting that if $X \in [a, b]$, then even $X - \mathbb{E}[X]$ takes values in an interval of length $b - a$.

THEOREM 2.39 (Hoeffding's inequality) *Consider discrete, independent random variables $X_1, \ldots, X_n$ that take values in the interval $[a_i, b_i]$, $1 \le i \le n$. Then, for every $t > 0$, we have*

$$\Pr\left(\sum_{i=1}^{n}(X_i - \mathbb{E}[X_i]) > t\right) \le e^{-\frac{2t^2}{\sum_{i=1}^{n}(b_i-a_i)^2}}.$$

2.13 Pinsker's Inequality

Returning to the discussion on information-theoretic quantities, we now establish a relation between $d_{\mathtt{var}}(\mathrm{P}, \mathrm{Q})$ and $D(\mathrm{P}\|\mathrm{Q})$. Roughly, we show that $d_{\mathtt{var}}(\mathrm{P}, \mathrm{Q})$ is less than $\sqrt{D(\mathrm{P}\|\mathrm{Q})}$.

THEOREM 2.40 (Pinsker's inequality) *For probability distributions* P *and* Q *on $\mathcal{X}$, we have*

$$d_{\mathtt{var}}(\mathrm{P}, \mathrm{Q})^2 \le \frac{\ln 2}{2} D(\mathrm{P}\|\mathrm{Q}).$$

Proof We obtain Pinsker's inequality as a consequence of the variational formula for KL divergence given in Lemma 2.33 and Hoeffding's lemma (Lemma 2.38). Consider the set $\mathcal{A}$ such that $d_{\mathtt{var}}(\mathrm{P}, \mathrm{Q}) = P(\mathcal{A}) - Q(\mathcal{A})$, and let $f_\lambda(x) = \lambda(\mathbf{1}[\{x \in \mathcal{A}\}] - \mathrm{Q}(\mathcal{A}))$. Then, it is easy to see that $\mathbb{E}_{\mathrm{P}}[f_\lambda(X)] = \lambda d_{\mathtt{var}}(\mathrm{P}, \mathrm{Q})$ and $\mathbb{E}_{\mathrm{Q}}[f_\lambda(X)] = 0$. Using this specific choice of $f = f_\lambda$ in the variation formula for KL divergence given in Lemma 2.33, we get

$$D(\mathrm{P}\|\mathrm{Q}) \ge \lambda d_{\mathtt{var}}(\mathrm{P}, \mathrm{Q}) - \log \mathbb{E}_{\mathrm{Q}}\left[2^{f_\lambda(X)}\right].$$

Note that the random variable $\mathbf{1}[x \in \mathcal{A}] - \mathrm{Q}(\mathcal{A})$ is zero-mean under Q and takes values between $-\mathrm{Q}(\mathcal{A})$ and $1 - \mathrm{Q}(\mathcal{A})$. Thus, by Hoeffding's lemma,

$$\log \mathbb{E}_{\mathrm{Q}}\left[2^{f_\lambda(X)}\right] = \frac{1}{\ln 2} \ln \mathbb{E}_{\mathrm{Q}}\left[e^{(\ln 2) f_\lambda(X)}\right] \le \frac{(\ln 2)\,\lambda^2}{8}.$$

Upon combining the two bounds above, we obtain

$$D(\mathrm{P}\|\mathrm{Q}) \ge \lambda d_{\mathtt{var}}(\mathrm{P}, \mathrm{Q}) - \frac{(\ln 2)\,\lambda^2}{8}, \quad \forall \lambda > 0,$$

which on maximizing the right-hand side over $\lambda > 0$ yields the claimed inequality. □

2.14 Rényi Entropy

In addition to Shannon entropy, in this book, we rely on another related measure of uncertainty and randomness: the Rényi entropy. Below we review some basic properties of this quantity.

DEFINITION 2.41 (Rényi entropy) For a probability distribution P on $\mathcal{X}$ and $\alpha \geq 0, \alpha \neq 1$, the *Rényi entropy* of order α, denoted by $H_\alpha(\mathrm{P})$, is defined as

$$H_\alpha(\mathrm{P}) := \frac{1}{1-\alpha} \log \sum_{x \in \mathcal{X}} \mathrm{P}(x)^\alpha.$$

As for Shannon entropy, we use the notation $H_\alpha(X)$ and $H_\alpha(\mathrm{P})$ to denote the Rényi entropy of order α for a random variable X with distribution P.

THEOREM 2.42 (Properties of $H_\alpha(\mathrm{P})$) *For a probability distribution* P *on a finite set* $\mathcal{X}$ *and* $\alpha > 0, \alpha \neq 1$, *the following properties hold.*

1. $0 \leq H_\alpha(\mathrm{P}) \leq \log |\mathcal{X}|$.
2. $H_\alpha(\mathrm{P})$ *is a nonincreasing function of* α.
3. $\lim_{\alpha \to \infty} H_\alpha(\mathrm{P}) = \min_{x \in \mathcal{X}} - \log \mathrm{P}(x)$.
4. $\lim_{\alpha \to 1} H_\alpha(\mathrm{P}) = H(\mathrm{P})$.

Proof of 1. For $0 < \alpha < 1$ and $\alpha > 1$, respectively, we note that $\sum_{x \in \mathcal{X}} \mathrm{P}(x)^\alpha > 1$ and $\sum_{x \in \mathcal{X}} \mathrm{P}(x)^\alpha \leq 1$. Therefore, $H_\alpha(\mathrm{P}) \geq 0$.

Next, for $\alpha \in (0, 1)$, note that

$$\begin{aligned}\sum_x \mathrm{P}(x)^\alpha &= \mathbb{E}_{\mathrm{P}}\left[\left(\frac{1}{\mathrm{P}(X)}\right)^{1-\alpha}\right] \\ &\leq |\mathcal{X}|^{1-\alpha},\end{aligned}$$

where the inequality uses Jensen's inequality applied to the concave function $t^{1-\alpha}$ for $\alpha \in (0, 1)$ and $t > 0$. Thus, $H_\alpha(\mathrm{P}) \leq \log |\mathcal{X}|$ for $\alpha \in (0, 1)$. The proof for $\alpha > 1$ can be completed similarly by noting that $t^{1-\alpha}$ is a convex function for $\alpha > 1$ and $t > 0$.

Proof of 2. Consider the function $f(\alpha) = H_\alpha(\mathrm{P})$ for $\alpha > 0$ and $\alpha \neq 1$. Then, denoting $\mathrm{P}_\alpha(x) = \mathrm{P}(x)^\alpha / \sum_{x'} \mathrm{P}(x')^\alpha$, we can verify that

$$f'(\alpha) = -\frac{1}{(1-\alpha)^2} \sum_{x \in \mathcal{X}} \mathrm{P}_\alpha(x) \log \frac{\mathrm{P}_\alpha(x)}{\mathrm{P}(x)} = -\frac{1}{(1-\alpha)^2} D(\mathrm{P}_\alpha \| \mathrm{P}) \leq 0$$

whereby f is a nonincreasing function.

Claims 3 and 4 can be verified by directly computing the limits. □

From Claim 4, we can regard the Shannon entropy as the Rényi entropy of order $\alpha = 1$. The Rényi entropy for $\alpha = 0$ is $H_0(\mathrm{P}) = \log |\mathtt{supp}(\mathrm{P})|$, and it is referred to as the max-entropy, denoted $H_{\max}(\mathrm{P})$. On the other hand, the Rényi entropy for $\alpha \to \infty$ is referred to as the min-entropy, denoted $H_{\min}(\mathrm{P})$. We will see operational meanings of the max-entropy and the min-entropy in Chapter 6 and Chapter 7, respectively.

2.15 References and Additional Reading

The content of this chapter concerns many basic quantities in information theory. Many of these appeared in Shannon's seminal work [304] and the relevance of some of them in the context of security appeared in [305]. However, quantities such as total variation distance and Kullback–Leibler divergence are statistical in origin and did not directly appear in Shannon's original work. A good reference for their early use is Kullback's book [207] and references therein. But these notions are now classic and can be accessed best through standard textbooks for information theory such as [75, 88, 151]. In our presentation, some of the proofs are new and not available in these textbooks. In particular, our proof of Fano's inequality based on data processing inequality is a folklore and underlies many generalizations of Fano's inequality; our presentation is closest to that in [150]. Our discussion on the variational formula for Kullback–Leibler divergence, Hoeffding's inequality, and proof of Pinsker's inequality using these tools is based on the presentation in the excellent textbook [40] on concentration inequalities. Our bound for continuity of Shannon entropy using the maximal coupling lemma is from [10, 362] (references we found in [88, Problem 3.10]). Rényi entropy was introduced in [291] and has emerged as an important tool for single-shot results in information theory and information-theoretic cryptography.

Problems

2.1 For two distributions P and Q on a finite set $\mathcal{X}$, prove the following equivalent forms of the total variation distance for discrete distributions P and Q:

$$\begin{aligned} d_{\mathtt{var}}(\mathrm{P}, \mathrm{Q}) &= \sup_{A \subset \mathcal{X}} \mathrm{P}(A) - \mathrm{Q}(A) \\ &= \sup_{A \subset \mathcal{X}} |\mathrm{P}(A) - \mathrm{Q}(A)| \\ &= \sum_{x \in \mathcal{X} : \mathrm{P}(x) \geq \mathrm{Q}(x)} \mathrm{P}(x) - \mathrm{Q}(x) \\ &= \frac{1}{2} \sum_{x \in \mathcal{X}} |\mathrm{P}(x) - \mathrm{Q}(x)|. \end{aligned}$$

2.2 For distributions P_{XY} and $\mathrm{Q}_{XY} = \mathrm{P}_{\mathtt{unif}} \times \mathrm{P}_Y$ on $\mathcal{X} \times \mathcal{Y}$, where $\mathrm{P}_{\mathtt{unif}}$ denotes the uniform distribution on $\mathcal{X}$, show that

$$D(\mathrm{P}_{XY}\|\mathrm{Q}_{XY}) = \log|\mathcal{X}| - H(X \mid Y).$$

The quantity on the left-hand side was used to define a security index in [90].

2.3 Show the following inequalities for entropies (see [228] for other such inequalities and their application in combinatorics):

1. $H(X_1, X_2, X_3) \leq \frac{1}{2}[H(X_1, X_2) + H(X_2, X_3) + H(X_3, X_1)]$,
2. $H(X_1, X_2, X_3) \geq \frac{1}{2}[H(X_1, X_2|X_3) + H(X_2, X_3|X_1) + H(X_3, X_1|X_2)]$.

2.4 In this problem we outline an alternative proof of Fano's inequality. Consider random variables X, Y, and $\hat{X}$ satisfying the Markov relation $X \multimap Y \multimap \hat{X}$ and such that X and $\hat{X}$ both take values in the same finite set $\mathcal{X}$. Denote by E the random variable $\mathbb{1}\left[\hat{X} \neq X\right]$. Show that $H(X \mid \hat{X}) \leq \Pr(E = 1)\log(|\mathcal{X}| - 1) + H(E)$ and conclude that

$$H(X \mid Y) \leq \Pr\left(X \neq \hat{X}\right)\log(|\mathcal{X}| - 1) + h\left(\Pr\left(X \neq \hat{X}\right)\right).$$

2.5 In this problem we outline an alternative proof of Pinsker's inequality. Using the data processing inequality for KL divergence, show that Pinsker's inequality holds if and only if it holds for distributions on $\mathcal{X}$ with $|\mathcal{X}| = 2$. Further, show that Pinsker's inequality for binary alphabet holds, that is, show that for every $p, q \in (0, 1)$ we have

$$p\log\frac{p}{q} + (1-p)\log\frac{(1-p)}{(1-q)} \geq \frac{2}{\ln 2}\cdot(p-q)^2.$$

2.6 Let $(X_1, \ldots, X_n) \in \{0,1\}^n$ be distributed uniformly over all binary sequences with less than np ones, with $0 \leq p \leq 1/2$. Show that $\Pr(X_i = 1) \leq h(p)$ for all $1 \leq i \leq n$, and that $H(X_1, \ldots, X_n) \leq \sum_{i=1}^n H(X_i)$. Conclude that for every $t \leq np$,

$$\sum_{i=0}^{t}\binom{n}{i} \leq 2^{nh(p)}.$$

2.7 We now use Problem 2.6 and Pinsker's inequality to derive a Hoeffding-type bound. Consider i.i.d. random variables $X_1, \ldots, X_n$ with common distribution $\mathtt{Bernoulli}(p)$, $0 \leq p \leq 1/2$.

1. For any sequence $\mathbf{x} \in \{0,1\}^n$ with $n\theta$ ones, show that

$$\Pr(X^n = \mathbf{x}) = 2^{-nD_2(\theta\|p) - nh(\theta)},$$

where $D_2(q\|p)$ denotes the KL divergence between Bernoulli distributions with parameters q and p.

2. Use Problem 2.6 to conclude that $\Pr(\sum_{i=1}^n X_i = n\theta) \leq 2^{-nD_2(\theta\|p)}$ and then Pinsker's inequality to conclude that for all $\theta > p$,

$$\Pr\left(\sum_{i=1}^n X_i > n\theta\right) \leq ne^{-2n(p-\theta)^2}.$$

2.8 Establish the following variational formula for a discrete distribution P over real numbers. For all $\lambda > 0$,

$$\log \mathbb{E}_P\left[2^{\lambda(X-\mathbb{E}_P[X])}\right] = \max_{Q:\mathtt{supp}(Q)\subset\ \mathtt{supp}(P)} \lambda(\mathbb{E}_Q[X] - \mathbb{E}_P[X]) - D(Q\|P).$$

Show that if $|X| \leq 1$ with probability 1 under P, then $\mathbb{E}_Q[X] - \mathbb{E}_P[X] \leq 2d(P,Q)$. Finally, conclude that if $P(|X| \leq 1) = 1$, then

$$\log \mathbb{E}_P\left[2^{\lambda(X-\mathbb{E}_P[X])}\right] \leq \frac{2\lambda^2}{\ln 2}.$$

2.9 For two pmfs P and Q on a finite set $\mathcal{X}$ and $0 < \theta < 1$, define

$$d_\theta(P,Q) := \max_A P(A) - \frac{1-\theta}{\theta} Q(A).$$

Suppose that $B \sim \mathtt{Ber}(\theta)$ is generated. If $B = 1$, then a sample X from P is generated. If $B = 0$, then a sample X from Q is generated. Consider the minimum probability of error in estimating B from X given by $P_e^*(\theta) = \min_{f:\mathcal{X}\to\{0,1\}} \Pr(B \neq f(X))$. Show that

$$P_e^*(\theta) = \theta(1 - d_\theta(P,Q))$$

and, further, that

$$d_\theta(P,Q) = \frac{1}{2\theta}\sum_x |\theta P(x) - (1-\theta)Q(x)| + \frac{2\theta - 1}{2\theta}.$$

2.10 Show that the Rényi entropy $H_\alpha(\mathrm{P})$ for a distribution P on a finite cardinality set $\mathcal{X}$ is a concave function of P.

3 Secret Keys and Encryption

In this chapter, we consider one of the most basic problems in cryptography, namely the problem of secure communication. In this problem, two legitimate parties seek to exchange a message securely, over a communication channel that is compromised. Specifically, an eavesdropper can read all the messages sent over the communication channel, and the parties do not want to reveal their message to the eavesdropper. To that end, the parties "encrypt" the message using a "secret key" which they share, and send the encrypted message over the communication channel. Analyzing the security of this scheme requires a clear definition of what it means for a scheme to be secure and what constitutes a secret key. We present these definitions in this chapter, and further, present several variants and relaxations. At the end, we will converge on the notion of indistinguishable security that is roughly equivalent to all the definitions considered. Once the definition of security is set, we study how efficiently a secret key can be used to send messages securely – how many bits of secret key are needed to send a one-bit message securely. We provide answers to these questions below.

3.1 Secure Communication

We consider the communication system described in Figure 3.1. A sender would like to send a message M to a receiver. The communication channel between the sender and the receiver is not secure, i.e., any communication over the channel may be observed by an eavesdropper. To prevent the eavesdropper from learning the message, the sender encrypts the message into a ciphertext C by using a *secret key* K shared with the receiver. This secret key is a random variable K that is independent of the message M and is shared between the sender and the receiver upfront, before the communication starts. Then, upon receiving the ciphertext, the receiver reproduces the message using the secret key.

Formally, an *encryption scheme* is described by a pair of mappings $f\colon \mathcal{M}\times\mathcal{K} \to \mathcal{C}$ and $g\colon \mathcal{C} \times \mathcal{K} \to \mathcal{M}$, where $\mathcal{M}$ is the message alphabet, $\mathcal{K}$ is the key alphabet, and $\mathcal{C}$ is the ciphertext alphabet. For simplicity, we only consider encryption schemes that reproduce the message perfectly at the receiver, i.e., $g(f(m,k),k) = m$ for every $m \in \mathcal{M}$ and $k \in \mathcal{K}$.

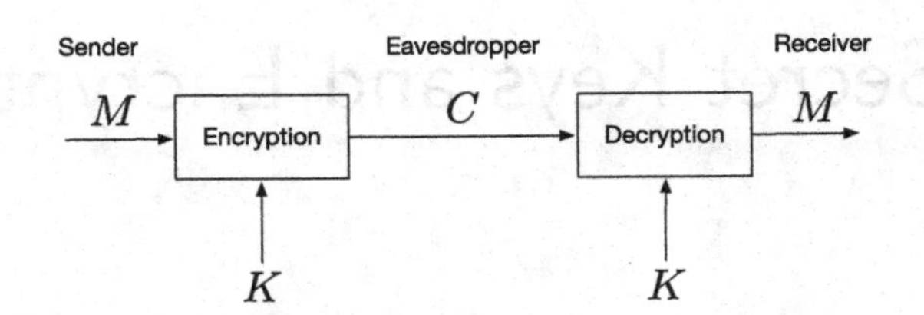

Figure 3.1 The secure communication system.

For a given message M and secret key K, the eavesdropper tries to predict the message from the ciphertext $C = f(M, K)$. If the a posteriori distribution $\mathrm{P}_{M|C}$ of message M given ciphertext C is the same as the a priori distribution P_M of the message, then there is, in essence, no information leaked about the message M by C. Then, the best thing the eavesdropper can do is ignore C and guess the message without observing the ciphertext.

This is our ultimate goal, and an encryption scheme satisfying this condition is termed perfectly secure.

DEFINITION 3.1 (Perfect security) An encryption scheme is *perfectly secure* if, for any distribution P_M of the message,

$$\mathrm{P}_{M|C}(m|c) = \mathrm{P}_M(m)\,,\ \forall (m,c) \in \mathcal{M} \times \mathcal{C}; \tag{3.1}$$

or equivalently,

$$\mathrm{P}_{C|M}(c|m) = \mathrm{P}_C(c)\,,\ \forall (m,c) \in \mathcal{M} \times \mathcal{C}. \tag{3.2}$$

From the definition of mutual information, we readily have the following equivalent form of perfect security (Problem 3.1).

PROPOSITION 3.2 *An encryption scheme is perfectly secure if and only if, for every distribution* P_M *of the message,*

$$I(M \wedge C) = 0.$$

To illustrate an application of the definition above, we consider the example of a substitution cipher.

EXAMPLE 3.3 For the alphabet $\mathcal{X} = \{a, b, c\}$, consider the set $\{\sigma_1, \ldots, \sigma_6\}$ of all permutations on $\mathcal{X}$, and let K be the key uniformly distributed on $\mathcal{K} = \{1, \ldots, 6\}$. First, suppose that only one symbol from $\mathcal{X}$ is transmitted, i.e., $\mathcal{M} = \mathcal{X}$. For a given message $m \in \mathcal{M}$ and key $k \in \mathcal{K}$, the substitution cipher encrypts the message into the ciphertext $f(m, k) = \sigma_k(m)$. Since the key K is uniformly distributed on $\mathcal{K}$, the ciphertext $\sigma_K(m)$ is uniformly distributed on $\mathcal{X}$, irrespective of the message m, i.e., $\mathrm{P}_{C|M}(c|m) = 1/3$ for every (m, c), which implies $\mathrm{P}_C(c) = 1/3$ for any distribution P_M of the message. Thus, (3.2) is satisfied, and the "one symbol substitution cipher" described above is perfectly secure.

Next, suppose that two symbols from $\mathcal{X}$ are to be transmitted, i.e., $\mathcal{M} = \mathcal{X}^2$. For a given message $m = (x_1, x_2) \in \mathcal{X}^2$ and key $k \in \mathcal{K}$, the message is encrypted into the ciphertext $f(m, k) = (\sigma_k(x_1), \sigma_k(x_2))$. For a message M uniformly distributed on $\mathcal{M}$, let $C = f(M, K)$ be the resulting ciphertext. Since σ_k is a permutation, if the ciphertext $c = (y_1, y_2)$ satisfies $y_1 = y_2$, then the original message $m = (x_1, x_2)$ must satisfy $x_1 = x_2$ as well. By noting this property, when $y_1 = y_2$, the posterior distribution for M given C is $\mathrm{P}_{M|C}(x_1, x_2|y_1, y_2) = 1/3$ if $x_1 = x_2$ and 0 otherwise; on the other hand, the prior distribution is $\mathrm{P}_M(x_1, x_2) = 1/9$ for every $(x_1, x_2) \in \mathcal{M}$. Thus, the substitution cipher is not perfectly secure if it is used for transmission of more than one symbol.

Remark 3.1 In Definition 3.1, perfect security is defined by taking the worst case over message distributions. For some fixed distributions, unsatisfactory crypto schemes may satisfy perfect security. For instance, the substitution cipher used for two symbols in Example 3.3 satisfies (3.1) (and (3.2)) if we only consider message distributions on the support $\{(x_1, x_2) \in \mathcal{X}^2 : x_1 \neq x_2\}$.

Example 3.3 suggests that the substitution cipher is perfectly secure if independent keys are used for every message.[1] By generalizing this observation, we can construct a perfectly secure encryption scheme known as the *one-time pad.* For simplicity of notation, we consider messages on the binary alphabet of length n, i.e., $\mathcal{M} = \{0, 1\}^n$. Suppose that the secret key K is uniformly distributed on $\mathcal{K} = \{0, 1\}^n$. Then, for a given message $m = (m_1, \ldots, m_n)$ and key $k = (k_1, \ldots, k_n)$, the one-time pad encrypts the message into

$$c = (c_1, \ldots, c_n) = (m_1 \oplus k_1, \ldots, m_n \oplus k_n). \tag{3.3}$$

By a slight abuse of notation, we denote $c = m \oplus k$ if there is no confusion from the context.

THEOREM 3.4 *When the secret key K is uniformly distributed on $\{0, 1\}^n$, the one-time pad encryption scheme of (3.3) is perfectly secure.*

Proof For any ciphertext $c \in \{0, 1\}^n$ and message $m \in \{0, 1\}^n$, since the key K is uniformly distributed on $\{0, 1\}^n$, we have

$$\begin{aligned}\mathrm{P}_{C|M}(c|m) &= \Pr(m \oplus K = c)\\ &= \Pr(K = m \oplus c)\\ &= \frac{1}{2^n},\end{aligned}$$

which also implies $\mathrm{P}_C(c) = 1/2^n$ for any message distribution P_M. Thus, (3.2) is satisfied, and the one-time pad is perfectly secure. □

Even though the one-time pad is a simple and perfectly secure crypto scheme, it uses an n-bit secret key to transmit an n-bit message, which is not efficient in

[1] In fact, there is no need to use general permutations, and it suffices to use "cyclic permutations."

terms of key consumption. A reader may wonder whether we can come up with a perfectly secure scheme with a shorter key length by making the encryption scheme more complicated. Unfortunately, that is not possible. The following "impossibility theorem" claims that the one-time pad is optimal in the sense of key consumption.

THEOREM 3.5 (Impossibility theorem) *For any perfectly secure encryption scheme with a message alphabet $\mathcal{M}$ using secret key K, it must hold that*

$$H(K) \geq \log |\mathcal{M}|.$$

Proof We prove the following for an arbitrary message distribution P_M:

$$H(K) \geq H(M). \tag{3.4}$$

Then, by taking the uniform distribution on $\mathcal{M}$, we have the claim of the theorem. In fact, (3.4) follows from basic manipulations of Shannon entropy as follows:

$$\begin{aligned} H(M) &= H(M|C) &(3.5)\\ &= H(M|C) - H(M|K,C)\\ &= H(M|C) - \big[H(M,K|C) - H(K|C)\big]\\ &= H(K|C) - H(K|M,C)\\ &\leq H(K|C)\\ &\leq H(K). &(3.6) \end{aligned}$$

We leave it as an exercise to justify each manipulation (Problem 3.2). □

Remark 3.2 (Perfect secret keys) It is clear from the discussion above that a secret key is a necessary resource for enabling information-theoretically secure encryption. To keep the treatment simple, we will often consider an *n-bit perfect secret key*, which is simply a K distributed uniformly over $\{0,1\}^n$. This accounts for n units of "resource" consumed.

3.2 Approximate Security

When restricting to perfect secret keys, in order to transmit an n-bit message in a perfectly secure manner, an n-bit perfect secret key is necessary and sufficient. In practice, it is more convenient to consider an approximately secure scheme by allowing small deviation from perfect security. Admitting this deviation allows us to consider an imperfect secret key as a resource, which is what one can expect to implement in practice anyway. Motivated by these considerations, in the next few sections, we discuss the approximate security of a secret key, and the security of the one-time pad using an approximately secure key. Finally, we discuss an important tool for analyzing the security of complicated protocols: *composability*. At a high level, a composability result allows us to separate the security analysis of an encryption scheme and the security of a secret key.

3.2.1 Indistinguishable and Semantic Security

In order to motivate the next security definition, let us consider the following hypothetical experiment for a given encryption scheme (f, g). The sender chooses one of the two messages $m_0, m_1 \in \mathcal{M}$ at random, and sends the encrypted ciphertext $f(m_b, K)$ to the eavesdropper. If the eavesdropper can distinguish which message was sent, then the security of the encryption scheme is suspicious. Conversely, for any pair (m_0, m_1) of messages, if the eavesdropper cannot distinguish which message was sent in the experiment above, then we can say that the encryption scheme is secure.

Formally, this experiment can be described as follows. Let $\phi\colon \mathcal{C} \to \{0,1\}$ be the eavesdropper's guessing function. Then, the maximum probability with which the eavesdropper can correctly distinguish the messages chosen by the sender is given by

$$\mathrm{P}_{\mathtt{dist}}(m_0, m_1) := \max_{\phi:\mathcal{C}\to\{0,1\}} \sum_{b\in\{0,1\}} \frac{1}{2} \Pr\left(\phi(f(m_b, K)) = b\right). \tag{3.7}$$

Note that without observing the ciphertext the eavesdropper can correctly guess the message with probability at most 1/2. Using variational distance introduced in Section 2.4 (see Lemma 2.10 in particular), the distinguishing probability $\mathrm{P}_{\mathtt{dist}}(m_0, m_1)$ can be described as follows.

LEMMA 3.6 *For a given encryption scheme* (f, g)*, the distinguishing probability in* (3.7) *satisfies*

$$\mathrm{P}_{\mathtt{dist}}(m_0, m_1) = \frac{1}{2}\left[1 + d_{\mathtt{var}}(\mathrm{P}_{C|M=m_0}, \mathrm{P}_{C|M=m_1})\right].$$

Proof Let $C_b = f(m_b, K)$ for $b \in \{0,1\}$. For an arbitrary guessing function $\phi\colon \mathcal{C} \to \{0,1\}$, the probability of correct guessing is given by

$$\begin{aligned}
&\frac{1}{2}\Pr\left(\phi(C_0) = 0\right) + \frac{1}{2}\Pr\left(\phi(C_1) = 1\right) \\
&\quad = \frac{1}{2}\left[1 + \Pr\left(\phi(C_0) = 0\right) - \Pr\left(\phi(C_1) = 0\right)\right] \\
&\quad = \frac{1}{2}\left[1 + \Pr\left(C_0 \in \phi^{-1}(0)\right) - \Pr\left(C_1 \in \phi^{-1}(0)\right)\right] \\
&\quad \le \frac{1}{2}\left[1 + \max_{\mathcal{A}\subseteq\mathcal{C}}\left(\Pr\left(C_0 \in \mathcal{A}\right) - \Pr\left(C_1 \in \mathcal{A}\right)\right)\right] \\
&\quad = \frac{1}{2}\left[1 + d_{\mathtt{var}}(\mathrm{P}_{C|M=m_0}, \mathrm{P}_{C|M=m_1})\right],
\end{aligned}$$

where the previous identity follows from Lemma 2.11. Furthermore, the equality in the inequality is attained for ϕ such that $\phi(c) = 0$ if and only if $\mathrm{P}_{C|M=m_0}(c) \ge \mathrm{P}_{C|M=m_1}(c)$. ☐

Lemma 3.6 claims that the increase in the distinguishing probability over the random guess is characterized by the variational distance between the ciphertexts corresponding to those messages. Motivated by this observation, we introduce the notion of indistinguishable secure encryption scheme as follows.

DEFINITION 3.7 (Indistinguishable security) For a given $0 \le \delta < 1$, an encryption scheme is *δ-indistinguishable secure* if

$$d_{\mathrm{var}}\left(\mathrm{P}_{C|M=m_0}, \mathrm{P}_{C|M=m_1}\right) \le \delta, \ \forall m_0 \ne m_1 \in \mathcal{M}.$$

Indistinguishable security quantifies how difficult it will be for the eavesdropper to distinguish a pair of messages.

But we may be interested in an alternative notion of security, where we seek to forbid the eavesdropper from learning the value of a function of the message. Consider the following example for motivation.

EXAMPLE 3.8 In order to save 1 bit secret key from the one-time pad described in (3.3), let us consider the following crypto scheme: for a given message $m = (m_1, \ldots, m_n) \in \{0,1\}^n$ and key $k = (k_1, \ldots, k_{n-1}) \in \{0,1\}^{n-1}$, set the ciphertext $c = (c_1, \ldots, c_n)$ as $c_i = m_i \oplus k_i$ for $1 \le i \le n-1$ and

$$c_n = m_n \oplus \bigoplus_{i=1}^{n-1} k_i.$$

Note that the eavesdropper can compute $\oplus_{i=1}^n c_i = \oplus_{i=1}^n m_i$ to find the parity of the message. Thus, this scheme falls short of our new requirement of security where we seek to forbid the eavesdropper from ascertaining any function of the message.

Formally, the notion of security below requires that the eavesdropper cannot get any information about any function of the message.

DEFINITION 3.9 (Semantic security) For $0 \le \delta < 1$, an encryption scheme is *δ-semantic secure* if there exists a random variable U such that, for every distribution P_M of the message, for every function $\psi \colon \mathcal{M} \to \mathcal{T}$ taking values in a finite cardinality set $\mathcal{T}$, and every estimator $\hat{\psi} \colon \mathcal{C} \to \mathcal{T}$, the following holds:[2]

$$\Pr\left(\hat{\psi}(C) = \psi(M)\right) - \Pr\left(\hat{\psi}(U) = \psi(M)\right) \le \delta. \tag{3.8}$$

Note that the second term on the left-hand side of (3.8) is the probability that the eavesdropper correctly guesses $\psi(M)$ using local randomness U instead of the ciphertext C. Thus, the left-hand side of (3.8) quantifies the improvement in the probability of correctly guessing $\psi(M)$ when the ciphertext is available, in comparison to a "random guess."

But is semantic security a strictly more stringent requirement in comparison to indistinguishable security? For the encryption scheme in Example 3.8, we can verify that it is not δ-semantic secure for any $0 \le \delta < 1/2$ (Problem 3.5). Interestingly, we can even verify that the scheme is not δ-indistinguishable secure

[2] We emphasize that the random variable U is independent of (M, C), and the choice of U does not depend on the distribution P_M nor functions $\psi, \hat{\psi}$.

for any $0 \le \delta < 1$ (Problem 3.4); in other words, this unsatisfactory scheme can be excluded by the indistinguishable security requirement.

In fact, indistinguishable security is equivalent to semantic security, which is the content of the result below.

PROPOSITION 3.10 (Relation between indistinguishable and semantic security) *If an encryption scheme is δ-indistinguishable secure, then it is δ-semantic secure. On the other hand, if a crypto scheme is δ-semantic secure, it is 2δ-indistinguishable secure.*

Proof First, suppose that the scheme is δ-indistinguishable secure. For arbitrarily fixed message $m_1 \in \mathcal{M}$, let U be the random variable distributed according to $\mathrm{P}_{C|M=m_1}$. Then, for an arbitrary message M and functions ψ and $\hat{\psi}$, denoting $C_m = f(m, K)$, we have

$$\begin{aligned}
&\Pr\big(\hat{\psi}(C) = \psi(M)\big) - \Pr\big(\hat{\psi}(U) = \psi(M)\big) \\
&\quad = \Pr\big(\hat{\psi}(C) = \psi(M)\big) - \Pr\big(\hat{\psi}(C_{m_1}) = \psi(M)\big) \\
&\quad = \sum_{m \in \mathcal{M}} \mathrm{P}_M(m)\big[\Pr\big(\hat{\psi}(C_m) = \psi(m)\big) - \Pr\big(\hat{\psi}(C_{m_1}) = \psi(m)\big)\big] \\
&\quad \le \Pr\big(\hat{\psi}(C_{m_0}) = \psi(m_0)\big) - \Pr\big(\hat{\psi}(C_{m_1}) = \psi(m_0)\big)
\end{aligned}$$

for $m_0 \in \mathcal{M}$ that maximizes the right-hand side above. By setting $\mathcal{A} = \{c \in \mathcal{C} : \hat{\psi}(c) = \psi(m_0)\}$, we have

$$\begin{aligned}
\Pr\big(\hat{\psi}(C_{m_0}) = \psi(m_0)\big) &- \Pr\big(\hat{\psi}(C_{m_1}) = \psi(m_0)\big) \\
&= \Pr\big(C_{m_0} \in \mathcal{A}\big) - \Pr\big(C_{m_1} \in \mathcal{A}\big) \\
&\le \max_{\mathcal{B} \subseteq \mathcal{C}} \big[\Pr\big(C_{m_0} \in \mathcal{B}\big) - \Pr\big(C_{m_1} \in \mathcal{B}\big)\big] \\
&= d_{\mathtt{var}}(\mathrm{P}_{C|M=m_0}, \mathrm{P}_{C|M=m_1}) \\
&\le \delta,
\end{aligned}$$

where the last identity follows from Lemma 2.11 and the last inequality holds since the scheme is δ-indistinguishable secure. Thus, indistinguishable security implies semantic security too.

Next, suppose that the scheme is δ-semantic secure. For a pair of messages (m_0, m_1) such that $m_0 \neq m_1$, consider the message $M = m_B$, where B is the uniform random variable on $\{0,1\}$. Consider a function $\psi\colon \mathcal{M} \to \{0,1\}$ such that $\psi(m_b) = b$, $b \in \{0,1\}$, and arbitrary for other inputs $m \notin \{m_0, m_1\}$. Note that

$$\Pr\big(\hat{\psi}(U) = \psi(M)\big) = \frac{1}{2}\Pr\big(\hat{\psi}(U) = 0\big) + \frac{1}{2}\Pr\big(\hat{\psi}(U) = 1\big) = \frac{1}{2}$$

for any $\hat{\psi}$. Then, if the scheme is δ-semantic secure, for every $\hat{\psi}$, we have

$$\frac{1}{2} + \delta \ge \Pr\big(\hat{\psi}(C) = \psi(M)\big) = \frac{1}{2}\Pr\big(\hat{\psi}(C_0) = 0\big) + \frac{1}{2}\Pr\big(\hat{\psi}(C_1) = 1\big),$$

where $C_b = f(m_b, K)$. It follows from Lemma 3.6 (see also Lemma 2.10) that

$$d_{\mathtt{var}}(\mathrm{P}_{C|M=m_0}, \mathrm{P}_{C|M=m_1}) \le 2\delta.$$

Since $m_0 \neq m_1$ are arbitrary, we have proved that it is 2δ-indistinguishable secure. □

As a consequence of Proposition 3.10, we can use the indistinguishable security criterion without worrying that the value of some function of the message might be leaked to the eavesdropper.

3.2.2 Mutual Information and Total Variation Security

The notions of security we saw above start by motivating an operational notion of "information leakage," and require the leakage to be limited. A more direct notion of security can simply require that the information about the message M contained in the communication C seen by the eavesdropper is limited, as measured by a metric for measuring information. A widely used metric for measuring the leaked information is Shannon's mutual information, which leads to the definition of security below.

DEFINITION 3.11 (Mutual information security) For a given $\delta \ge 0$, an encryption scheme is δ-mutual information (δ-MI) secure if[3]

$$\max_{\mathrm{P}_M} I(M \wedge C) \le \delta,$$

where the maximum is over all distributions P_M for the message.

But, once again, this new notion of security can be related to the notion of indistinguishable security we saw earlier. This is the content of the result below.

PROPOSITION 3.12 (Relation between indistinguishable and MI security) *If an encryption scheme is δ-MI secure, then it is $2\sqrt{(2\ln 2)\delta}$-indistinguishable secure. On the other hand, if an encryption scheme is δ-indistinguishable secure, then it is δ'-MI secure with $\delta' = \delta \log(|\mathcal{M}| - 1) + h(\min\{\delta, 1/2\})$.*

Proof For an arbitrarily fixed P_M, we can write

$$I(M \wedge C) = D(\mathrm{P}_{MC} \| \mathrm{P}_M \times \mathrm{P}_C).$$

Then, by Pinsker's inequality (cf. Theorem 2.40), if the scheme is δ-MI secure, we have

$$\mathbb{E}_{\mathrm{P}_M}\left[d_{\mathtt{var}}(\mathrm{P}_{C|M}, \mathrm{P}_C)\right] = d_{\mathtt{var}}(\mathrm{P}_{MC}, \mathrm{P}_M \times \mathrm{P}_C) \le \sqrt{\frac{\ln 2}{2}\delta}.$$

[3] Since the distribution of the message often represents the belief of the eavesdropper about the message, which takes into account prior information about the message, we consider the worst-case input distribution in the definition of security.

In particular, this holds for the message M that is randomly chosen from an arbitrary but fixed pair of messages (m_0, m_1) where $m_0 \neq m_1$. Thus, by the triangular inequality, we have

$$\begin{aligned} d_{\mathtt{var}}(\mathrm{P}_{C|M=m_0}, \mathrm{P}_{C|M=m_1}) &\le d_{\mathtt{var}}(\mathrm{P}_{C|M=m_0}, \mathrm{P}_C) + d_{\mathtt{var}}(\mathrm{P}_{C|M=m_1}, \mathrm{P}_C) \\ &\le \sqrt{(2\ln 2)\delta}, \end{aligned}$$

which proves the first claim.

To prove the second claim, for an arbitrary distribution P_M, if the scheme is δ-indistinguishable secure, we have

$$\begin{aligned} d_{\mathtt{var}}(\mathrm{P}_{MC}, \mathrm{P}_M \times \mathrm{P}_C) &= \sum_{m \in \mathcal{M}} \mathrm{P}_M(m)\, d_{\mathtt{var}}(\mathrm{P}_{C|M=m}, \mathrm{P}_C) \\ &= \sum_{m \in \mathcal{M}} \mathrm{P}_M(m)\, d_{\mathtt{var}}\Big(\mathrm{P}_{C|M=m}, \sum_{m' \in \mathcal{M}} \mathrm{P}_M(m')\, \mathrm{P}_{C|M=m'}\Big) \\ &\le \sum_{m,m' \in \mathcal{M}} \mathrm{P}_M(m)\, \mathrm{P}_M(m')\, d_{\mathtt{var}}(\mathrm{P}_{C|M=m}, \mathrm{P}_{C|M=m'}) \\ &\le \delta, \end{aligned} \tag{3.9}$$

where we used the triangular inequality.

Thus, by the continuity bound for the entropy function (cf. Theorem 2.34), we obtain

$$\begin{aligned} I(M \wedge C) &= H(M) - H(M|C) \\ &= \sum_{c \in \mathcal{C}} \mathrm{P}_C(c)\, [H(M) - H(M|C=c)] \\ &\le \sum_{c \in \mathcal{C}} \mathrm{P}_C(c)\, \big[d_{\mathtt{var}}(\mathrm{P}_M, \mathrm{P}_{M|C=c}) \log(|\mathcal{M}| - 1) \\ &\qquad\qquad + h(d_{\mathtt{var}}(\mathrm{P}_M, \mathrm{P}_{M|C=c}))\big] \\ &\le d_{\mathtt{var}}(\mathrm{P}_{MC}, \mathrm{P}_M \times \mathrm{P}_C) \log(|\mathcal{M}| - 1) \\ &\qquad\qquad + h(d_{\mathtt{var}}(\mathrm{P}_{MC}, \mathrm{P}_M \times \mathrm{P}_C)) \\ &\le \delta \log(|\mathcal{M}| - 1) + h(\min\{\delta, 1/2\}), \end{aligned}$$

where we used the concavity of the binary entropy function in the second-last inequality. □

By Proposition 3.12, if an encryption scheme is δ-indistinguishable secure for δ much smaller than the message length $\log |\mathcal{M}|$, then it is also δ'-MI secure for a satisfactory δ'. A typical choice of δ is very small compared to the message length, whereby we can say, informally, that "indistinguishable security implies MI security."

Similar to MI security, we can use the total variation distance to measure the leakage to get the following alternative notion of security.

DEFINITION 3.13 (Total variation security) For a given $\delta \geq 0$, an encryption scheme is δ-total variation (δ-TV) secure if

$$\max_{P_M} d_{\mathtt{var}}(P_{MC}, P_M \times P_C) \leq \delta,$$

where the maximum is over all distributions P_M for the message.

As in Proposition 3.12, we can relate TV security to indistinguishable security as well.

PROPOSITION 3.14 (Relation between indistinguishable security and TV security) *If an encryption scheme is δ-TV secure, then it is 2δ-indistinguishable secure. On the other hand, if an encryption scheme is δ-indistinguishable secure, then it is δ-TV secure.*

We leave the proof as an exercise; see Problem 3.6.

Thus, indistinguishable security is roughly equivalent to both MI and TV security. Furthermore, indistinguishable security has a computational analog which will be discussed in Chapter 9, and it is closely related to simulation based security to be discussed in Chapters 12–19 on secure computing. Motivated by these observations, we focus on indistinguishable security in the rest of this chapter.

3.2.3 Impossibility Theorem for Approximate Security

To end this section, we extend the impossibility theorem for perfect security (Theorem 3.5) to approximate security.

THEOREM 3.15 *If an encryption scheme (f, g) is δ-indistinguishable secure, then it holds that*

$$\delta \geq 1 - \frac{|\mathcal{K}|}{|\mathcal{M}|},$$

or equivalently,

$$\log |\mathcal{K}| \geq \log |\mathcal{M}| - \log(1/(1-\delta)).$$

Proof By Proposition 3.14, if the scheme is δ-indistinguishable secure, we have

$$\delta \geq d_{\mathtt{var}}(P_{MC}, P_M \times P_C)$$

for any distribution P_M. Set P_M to be the uniform distribution on $\mathcal{M}$. For a given (arbitrary) encryption scheme (f, g) satisfying $g(f(m,k),k) = m$, let

$$\mathcal{A} = \{(m,c) \in \mathcal{M} \times \mathcal{C} : \exists k \in \mathcal{K} \text{ such that } g(c,k) = m\},$$

namely, $\mathcal{A}$ is the set of messages and ciphertexts that are consistent for some values of keys.

Then, since we are restricting to error-free encryption schemes whereby $M = g(C, K)$, we have

$$P_{MC}(\mathcal{A}) = 1. \tag{3.10}$$

On the other hand,

$$\begin{aligned}
\mathrm{P}_M \times \mathrm{P}_C(\mathcal{A}) &= \sum_{m\in\mathcal{M}}\sum_{c\in\mathcal{C}} \frac{1}{|\mathcal{M}|}\mathrm{P}_C(c)\,\mathbf{1}[\exists k \in \mathcal{K} \text{ such that } g(c,k)=m] \\
&\le \sum_{k\in\mathcal{K}}\sum_{m\in\mathcal{M}}\sum_{c\in\mathcal{C}} \frac{1}{|\mathcal{M}|}\mathrm{P}_C(c)\,\mathbf{1}[g(c,k)=m] \\
&= \sum_{k\in\mathcal{K}}\sum_{c\in\mathcal{C}} \frac{1}{|\mathcal{M}|}\mathrm{P}_C(c) \\
&= \frac{|\mathcal{K}|}{|\mathcal{M}|}, \qquad (3.11)
\end{aligned}$$

where the second-last identity follows from the fact that $\mathbf{1}[g(c,k)=m]=1$ for exactly one $m \in \mathcal{M}$. By using Lemma 2.11, together with (3.10) and (3.11) we obtain

$$\begin{aligned}
d_{\mathrm{var}}(\mathrm{P}_{MC}, \mathrm{P}_M \times \mathrm{P}_C) &\ge \mathrm{P}_{MC}(\mathcal{A}) - \mathrm{P}_M \times \mathrm{P}_C(\mathcal{A}) \\
&\ge 1 - \frac{|\mathcal{K}|}{|\mathcal{M}|},
\end{aligned}$$

which completes the proof. □

Theorem 3.15 claims that, even if the perfect security requirement is relaxed to approximate security, the length of the secret key must be almost as large as the length of the message.

3.3 Approximately Secure Keys

When seeking approximate security of encryption schemes, it is no longer necessary to use perfect secret keys. One can make do with approximately secure keys, defined below.

We present this definition keeping in mind the formulation we will see later in Chapter 10, where we will study the generation of secret keys using public discussion. In such cases, the eavesdropper may have side-information Z that might be correlated to the agreed secret key K. Thus, we need to argue that the key K is independent of the side-information Z and is uniformly distributed on the range $\mathcal{K}$. Our notion of approximately secure keys seeks to relax this notion.

In order to motivate the security definition, let us consider a hypothetical experiment similar to the one discussed at the beginning of Section 3.2.1. Suppose that there is a key agreement protocol in which the sender and the receiver agree on K and the eavesdropper observes Z. Let random variable B be distributed uniformly on $\{0,1\}$, and let U be a random variable that is independent of Z and is uniformly distributed on $\mathcal{K}$. In addition to the side-information Z, the eavesdropper is provided with $\hat{K}$, where

$$\hat{K} = \begin{cases} K, & \text{if } B = 0, \\ U, & \text{if } B = 1. \end{cases}$$

In our hypothetical experiment, the eavesdropper seeks to ascertain B using $(\hat{K}, Z)$. Towards that, the eavesdropper outputs an estimate $\phi(\hat{K}, Z)$ of B. Then, the optimal probability that the eavesdropper correctly distinguishes between $B = 0$ and $B = 1$ is given by

$$\mathrm{P}_{\mathtt{dist}}(\mathrm{P}_{KZ}) := \max_{\phi:\mathcal{K}\times\mathcal{Z}\to\{0,1\}} \Pr\big(\phi(\hat{K}, Z) = B\big). \tag{3.12}$$

Note that, without observing $(\hat{K}, Z)$, the eavesdropper can correctly guess B with probability at most $1/2$.

In fact, in a similar manner to Lemma 3.6 (see, also, Theorem 2.4), we can characterize $\mathrm{P}_{\mathtt{dist}}(\mathrm{P}_{KZ})$ in terms of total variation distance (the proof is left as an exercise Problem 3.8).

LEMMA 3.16 *For a given secret key K with the eavesdropper's side-information Z, the distinguishing probability $\mathrm{P}_{\mathtt{dist}}$ satisfies*

$$\mathrm{P}_{\mathtt{dist}}(\mathrm{P}_{KZ}) = \frac{1}{2}\big[1 + d_{\mathtt{var}}(\mathrm{P}_{KZ}, \mathrm{P}_U \times \mathrm{P}_Z)\big],$$

where P_U is the uniform distribution on $\mathcal{K}$.

Thus, the improvement of the correctly distinguishing probability over the random guess of B is given by the total variation distance between the distribution P_{KZ} of the actual key and the distribution $\mathrm{P}_U \times \mathrm{P}_Z$ of the ideal key. Motivated by this observation, we define the security of a shared key as follows.

DEFINITION 3.17 For a given $0 \le \delta < 1$, a secret key K on $\mathcal{K}$ constitutes a δ-secure key for side-information Z if

$$d_{\mathtt{var}}(\mathrm{P}_{KZ}, \mathrm{P}_U \times \mathrm{P}_Z) \le \delta,$$

where P_U is the uniform distribution on $\mathcal{K}$.

Having defined the notion of approximately secure keys, we now exhibit how approximately secure keys lead to approximately secure encryption schemes. We show this first for the specific example of a one-time pad, before establishing a general "composition theorem" to establish a general principle.

3.3.1 One-time Pad Using an Approximately Secure Key

Consider the one-time pad encryption scheme seen earlier, with the perfectly secure key replaced with an approximately secure key. Specifically, for a given secret key K on $\mathcal{K} = \{0, 1\}^n$ with the eavesdropper's side-information Z, the sender encrypts a message $m = (m_1, \ldots, m_n)$ using (3.3). The result below establishes approximate security of this scheme against an eavesdropper who observes $C = m \oplus K$ and Z; the notion of indistinguishable security can be extended to the present case by replacing C in Definition 3.7 with (C, Z).

THEOREM 3.18 *For a secret key K that is a δ-secure key for side-information Z, the one-time pad using K is 2δ-indistinguishable secure (against an eavesdropper observing (C, Z)).*

Proof For any $m_0 \neq m_1$, by the triangular inequality, we have

$$\begin{aligned} & d_{\mathtt{var}}(\mathrm{P}_{CZ|M=m_0}, \mathrm{P}_{CZ|M=m_1}) \\ & \quad \leq d_{\mathtt{var}}(\mathrm{P}_{CZ|M=m_0}, \mathrm{P}_U \times \mathrm{P}_Z) + d_{\mathtt{var}}(\mathrm{P}_{CZ|M=m_1}, \mathrm{P}_U \times \mathrm{P}_Z). \end{aligned} \tag{3.13}$$

Then, by noting that $\mathrm{P}_{CZ|M}(c, z|m_b) = \mathrm{P}_{KZ}(c \oplus m_b, z)$ for $b \in \{0, 1\}$, which uses the independence of (K, Z) and M, we have

$$\begin{aligned} d_{\mathtt{var}}(\mathrm{P}_{CZ|M=m_b}, \mathrm{P}_U \times \mathrm{P}_Z) &= \frac{1}{2} \sum_{c \in \{0,1\}^n} \sum_{z \in \mathcal{Z}} \left| \mathrm{P}_{CZ|M}(c, z|m_b) - \mathrm{P}_U(c)\,\mathrm{P}_Z(z) \right| \\ &= \frac{1}{2} \sum_{c \in \{0,1\}^n} \sum_{z \in \mathcal{Z}} \left| \mathrm{P}_{KZ}(c \oplus m_b, z) - \frac{1}{2^n} \mathrm{P}_Z(z) \right| \\ &= \frac{1}{2} \sum_{k \in \{0,1\}^n} \sum_{z \in \mathcal{Z}} \left| \mathrm{P}_{KZ}(k, z) - \frac{1}{2^n} \mathrm{P}_Z(z) \right| \\ &= d_{\mathtt{var}}(\mathrm{P}_{KZ}, \mathrm{P}_U \times \mathrm{P}_Z) \\ &\leq \delta, \end{aligned}$$

which together with (3.13) imply the claim of the theorem. □

3.3.2 Composable Security

Theorem 3.18 claims that a δ-secure key can be used for the one-time pad encryption scheme, if we are willing to settle for approximate security of the resulting encryption scheme. But how about other encryption schemes?

Typically, a crypto system is composed of smaller modules such as the encryption scheme, key management module, etc. From a system designer's perspective, it is desirable that each module can be constructed without worrying about its role in the larger system. For instance, when we construct a secret key agreement protocol, we should not worry about where the secret key will be used and yet ensure security of any system that uses the agreed secret key. A notion of security which ensures a "graceful" degradation of security when multiple modules are composed in a system is called *composable security.* In fact, the security of secret keys in Definition 3.17 satisfies the following notion of composability.

THEOREM 3.19 *Let K be a δ-secure key for side-information Z. Further, let (f, g) be a δ'-indistinguishable secure encryption scheme when $f\colon \mathcal{M} \times \mathcal{K} \to \mathcal{C}$ is used with a perfect key. Then, the encryption scheme when f is used with the key K is $(2\delta + \delta')$-indistinguishable secure.*

Proof For a message $m \in \mathcal{M}$, let

$$\begin{aligned} \mathrm{Q}_{CZ|M}(c, z|m) &:= \Pr\left((f(m,U), Z) = (c,z)\right) \\ &= \Pr\left(f(m,U) = c\right)\mathrm{P}_Z(z), \end{aligned} \tag{3.14}$$

where U, denoting a perfect secret key, is the random variable that is independent of Z and uniformly distributed on $\mathcal{K}$. By noting this and by using the triangular inequality twice, for any $m_0 \neq m_1$, we have

$$\begin{aligned} &d_{\mathtt{var}}\left(\mathrm{P}_{CZ|M=m_0}, \mathrm{P}_{CZ|M=m_1}\right) \\ &\le d_{\mathtt{var}}\left(\mathrm{Q}_{CZ|M=m_0}, \mathrm{Q}_{CZ|M=m_1}\right) \\ &\quad + d_{\mathtt{var}}\left(\mathrm{P}_{CZ|M=m_0}, \mathrm{Q}_{CZ|M=m_0}\right) + d_{\mathtt{var}}\left(\mathrm{P}_{CZ|M=m_1}, \mathrm{Q}_{CZ|M=m_1}\right) \\ &= d_{\mathtt{var}}\left(\mathrm{Q}_{C|M=m_0}, \mathrm{Q}_{C|M=m_1}\right) \\ &\quad + d_{\mathtt{var}}\left(\mathrm{P}_{CZ|M=m_0}, \mathrm{Q}_{CZ|M=m_0}\right) + d_{\mathtt{var}}\left(\mathrm{P}_{CZ|M=m_1}, \mathrm{Q}_{CZ|M=m_1}\right) \\ &\le \delta' + d_{\mathtt{var}}\left(\mathrm{P}_{CZ|M=m_0}, \mathrm{Q}_{CZ|M=m_0}\right) + d_{\mathtt{var}}\left(\mathrm{P}_{CZ|M=m_1}, \mathrm{Q}_{CZ|M=m_1}\right), \end{aligned} \tag{3.15}$$

where we used (3.14) in the identity and the δ'-indistinguishable security of the encryption scheme with perfect key in the second inequality.

Next, by the data processing inequality of the variational distance (cf. Lemma 2.14), for $b \in \{0, 1\}$, we have

$$\begin{aligned} d_{\mathtt{var}}\left(\mathrm{P}_{CZ|M=m_b}, \mathrm{Q}_{CZ|M=m_b}\right) &\le d_{\mathtt{var}}\left(\mathrm{P}_{KZ}, \mathrm{P}_U \times \mathrm{P}_Z\right) \\ &\le \delta, \end{aligned}$$

which together with (3.15) implies the claim of the theorem. □

Note that Theorem 3.18 can be derived as a corollary of Theorem 3.4 and Theorem 3.19. Also, note that the proof of Theorem 3.18 uses the structure of the one-time pad scheme while the proof of Theorem 3.19 does not rely on any structural assumptions about the underlying encryption scheme. The latter type of argument is sometimes refereed to as a *black-box reduction* while the former type of argument is refereed to as a *non-black-box reduction*.

In modern cryptography, a standard approach is to build a complex crypto system using composably secure modules. In fact, the δ-secure key in the sense of Definition 3.17 can be securely used in any other cryptographic schemes, such as the authentication to be discussed in Chapter 8. Formalizing composable security requires an abstract language to describe unspecified cryptographic schemes, which we do not pursue in this chapter. We will come back to the topic of composable security in Chapter 15 in the context of secure computation.

3.4 Security and Reliability

When the sender and the receiver share a secret key by running a secret key agreement protocol, the key K_1 obtained by the sender and the key K_2 obtained by the receiver may disagree with small probability. In this section, we shall extend the notion of approximately secure key to incorporate such a disagreement.

DEFINITION 3.20 For a given $0 \le \varepsilon, \delta < 1$, a pair (K_1, K_2) of secret keys on $\mathcal{K}$ constitute an (ε, δ)-secure key for side-information Z if

$$\Pr\left(K_1 \neq K_2\right) \le \varepsilon$$

and

$$d_{\mathtt{var}}\left(\mathrm{P}_{K_1 Z}, \mathrm{P}_U \times \mathrm{P}_Z\right) \le \delta,$$

where P_U is the uniform distribution on $\mathcal{K}$.

Even though the (ε, δ)-security in Definition 3.20 is natural in the context of encryption schemes for transmitting a message from the sender to the receiver, it has a drawback that the definition is not symmetric with respect to the two legitimate parties. Moreover, since it assumes that an encrypted communication is from the sender to the receiver rather than in the opposite direction or a mixture of both directions, it is not appropriate for composable security. For these reasons, the following alternative definition may be better suited in some cases.

DEFINITION 3.21 For a given $0 \le \varepsilon < 1$, a pair (K_1, K_2) of secret keys on $\mathcal{K}$ is an ε-secure key for side-information Z if

$$d_{\mathtt{var}}\left(\mathrm{P}_{K_1 K_2 Z}, \mathrm{P}_{UU} \times \mathrm{P}_Z\right) \le \varepsilon,$$

where

$$\mathrm{P}_{UU}(k_1, k_2) := \frac{1}{|\mathcal{K}|}\mathbf{1}[k_1 = k_2]$$

is the uniform distribution on the diagonal of $\mathcal{K} \times \mathcal{K}$.

In fact, the two definitions are almost equivalent.

PROPOSITION 3.22 *If (K_1, K_2) is an (ε, δ)-secure key for side-information Z in the sense of Definition 3.20, then it is a $(\varepsilon + \delta)$-secure key in the sense of Definition 3.21. On the other hand, if (K_1, K_2) is an ε-secure key for side-information Z in the sense of Definition 3.21, then it is an $(\varepsilon, \varepsilon)$-secure key in the sense of Definition 3.20.*

Proof First, suppose that (K_1, K_2, Z) is (ε, δ)-secure in the sense of Definition 3.20. By the triangular inequality, we have

$$\begin{aligned} & d_{\mathtt{var}}\left(\mathrm{P}_{K_1 K_2 Z}, \mathrm{P}_{UU} \times \mathrm{P}_Z\right) \\ & \quad \le d_{\mathtt{var}}\left(\mathrm{P}_{K_1 K_2 Z}, \mathrm{P}_{K_1 K_1 Z}\right) + d_{\mathtt{var}}\left(\mathrm{P}_{K_1 K_1 Z}, \mathrm{P}_{UU} \times \mathrm{P}_Z\right), \end{aligned} \tag{3.16}$$

where

$$\mathrm{P}_{K_1 K_1 Z}(k_1, k_2, z) = \mathrm{P}_{K_1 Z}(k_1, z)\, \mathbf{1}[k_1 = k_2].$$

Note that the variational distance can be written as (cf. Lemma 2.11)

$$d_{\mathtt{var}}(P,Q) = P(\{x : P(x) > Q(x)\}) - Q(\{x : P(x) > Q(x)\}) \tag{3.17}$$
$$= \max_{\mathcal{A}\subset\mathcal{X}} \left[P(\mathcal{A}) - Q(\mathcal{A})\right]. \tag{3.18}$$

Note also that $\mathrm{P}_{K_1K_2Z}(k_1,k_2,z) > \mathrm{P}_{K_1K_1Z}(k_1,k_2,z) = 0$ for $k_1 \neq k_2$ with $\mathrm{P}_{K_1K_2Z}(k_1,k_2,z) > 0$ and $\mathrm{P}_{K_1K_2Z}(k_1,k_2,z) \le \mathrm{P}_{K_1K_1Z}(k_1,k_2,z)$ for $k_1 = k_2$. Thus, by applying (3.17), the first term of (3.16) can be bounded as

$$d_{\mathtt{var}}(\mathrm{P}_{K_1K_2Z}, \mathrm{P}_{K_1K_1Z}) = \Pr\left(K_1 \neq K_2\right) \le \varepsilon.$$

On the other hand, the second term of (3.16) can be bounded as

$$d_{\mathtt{var}}(\mathrm{P}_{K_1K_1Z}, \mathrm{P}_{UU}\times\mathrm{P}_Z) = d_{\mathtt{var}}(\mathrm{P}_{K_1Z}, \mathrm{P}_U\times\mathrm{P}_Z) \le \delta,$$

whereby (K_1, K_2, Z) is $(\varepsilon+\delta)$-secure in the sense of Definition 3.21.

Next, suppose that (K_1, K_2, Z) is ε-secure in the sense of Definition 3.21. By applying (3.18), we have

$$\begin{aligned}
\varepsilon &\ge d_{\mathtt{var}}(\mathrm{P}_{K_1K_2Z}, \mathrm{P}_{UU}\times\mathrm{P}_Z) \\
&\ge \mathrm{P}_{K_1K_2Z}(\{(k_1,k_2,z) : k_1 \neq k_2) - \mathrm{P}_{UU}\times\mathrm{P}_Z(\{(k_1,k_2,z) : k_1 \neq k_2\}) \\
&= \Pr\left(K_1 \neq K_2\right).
\end{aligned}$$

On the other hand, by the monotonicity of the variational distance (cf. Lemma 2.14), we have

$$d_{\mathtt{var}}(\mathrm{P}_{K_1Z}, \mathrm{P}_U\times\mathrm{P}_Z) \le d_{\mathtt{var}}(\mathrm{P}_{K_1K_2Z}, \mathrm{P}_{UU}\times\mathrm{P}_Z) \le \varepsilon.$$

Thus, (K_1, K_2, Z) is $(\varepsilon, \varepsilon)$-secure in the sense of Definition 3.20. □

Finally, we state a composition theorem for using a secret key (K_1, K_2) in an encryption scheme. A secret key (K_1, K_2) with K_1 observed by the sender and K_2 observed by the receiver is used in an encryption scheme (f, g), where the sender transmits $C = f(m, K_1)$ for a message $m \in \mathcal{M}$ and the receiver decrypts C as $\hat{M} = g(C, K_2) = g(f(m, K_1), K_2)$. Such an encryption scheme using (K_1, K_2) is *ε-reliable* if

$$\Pr\left(\hat{M} \neq m\right) \le \varepsilon$$

for every message $m \in \mathcal{M}$.

We have the following composition theorem when using an (ε, δ)-secure key with an encryption scheme; we leave the proof as an exercise in Problem 3.9.

THEOREM 3.23 *Let (K_1, K_2) be an (ε, δ)-secure key with side-information Z. Further, let (f, g), when used with a perfect key, be an ε'-reliable and δ'-indistinguishable secure encryption scheme. Then, (f, g) using (K_1, K_2) is $(\varepsilon + \varepsilon')$-reliable and $(2\delta + \delta')$-indistinguishable secure.*

3.5 References and Additional Reading

The concept of perfect security was introduced by Shannon in his seminal paper, and the impossibility theorem (Theorem 3.5) was also proved there [305]. Shannon's work laid the foundation of information-theoretic cryptography, and many researchers have extended his results in various directions. For an exposition of information-theoretic cryptography up to the 1990s, see [235, 236, 238, 351, 356].

The concept of semantic security was first introduced in the context of computational security [147], and it was also shown that computational indistinguishable security implies computational semantic security. For a detailed exposition on computational security, see textbooks [141, 187]. Some basic results on computationally secure encryption schemes will be presented in Chapter 9 of this book as well.

In the context of information-theoretic security, the relation between semantic security and indistinguishable security was clarified in [22, 104, 179, 180] (see also [258] for an exposition).

Various extensions of Shannon's impossibility theorem to approximate security have been derived in the literature (cf. [101, 165, 180, 198, 249, 276]). In particular, the bound in Theorem 3.15 can be found in [101, 276].

Theorem 3.15 relies on the fact that any pair of messages cannot be distinguished. For the relaxed requirement that any message of sufficient entropy cannot be predicted, the notion of entropic security, it is known that the required length of secret key can be reduced from Shannon's impossibility result [104, 295]; for more recent advances on entropic security, see also [223].

Another direction of relaxation is to consider the secrecy of data that is correlated with the message, instead of the secrecy of the message itself. In this case, it is known that the required length of the secret key can be reduced to the mutual information between the data and the message [357].

When the receiver does not need to reproduce the sender's data fully, i.e., when distortion is allowed, the required length of secret key can be reduced as well. The combination of the rate-distortion theory and secure communication has been studied in the literature; for instance, see [300, 301, 358].

In our exposition, we have focused on approximate security using total variation distance since it is widely used in cryptography and is compatible with the notion of simulation based security that we will encounter in Chapter 15. However, various other security criteria have been proposed in the literature; see [51, 176, 181, 223]. Heuristically, a natural security definition should be defined as the gain obtained from the cipher-text in the probability of guessing (a function of) the message. Semantic security is defined as an additive gain, and total variation distance captures the behavior of additive gain well. On the other hand, we can define security by a multiplicative gain [176], and other quantities, such as Rényi information measures or maximal correlation, can suitably capture the behavior of multiplicative gain.

Problems

3.1 Prove Proposition 3.2.

3.2 Provide justification for each manipulation in steps (3.5) to (3.6).

3.3 For a perfectly secure encryption scheme with secret key K, message M, and communication C, prove that $H(K|C) \geq H(M)$ holds for any distribution P_M.

3.4 Verify that the encryption scheme in Example 3.8 is not δ-indistinguishable secure for any $0 \leq \delta < 1$.

HINT Consider a pair of messages with different parities.

3.5 Verify that the encryption scheme in Example 3.8 is not δ-semantic secure for any $0 \leq \delta < 1/2$.

HINT For the message M distributed uniformly on $\{0,1\}^n$, verify that the second term on the left-hand side of (3.8) is 1/2 while the first term can be 1 for an appropriately chosen estimator.

3.6 Prove Proposition 3.14.

HINT For each pair $m_0 \neq m_1$, consider the message M uniformly distributed on $\{m_0, m_1\}$ and use the triangular inequality.

3.7 When a one-time pad encryption scheme is used twice with the same key, prove that it is not δ-indistinguishable secure for any $0 \leq \delta < 1$.

HINT Consider the pair of messages $m_0 = (m, m) \in \{0,1\}^{2n}$ and $m_1 = (m, m') \in \{0,1\}^{2n}$ for some $m \neq m' \in \{0,1\}^n$.

3.8 Prove Lemma 3.16.

3.9 Prove Theorem 3.23.

3.10 When an encryption scheme (f, g) satisfies $|\mathcal{C}| = |\mathcal{M}|$ and $g(f(m,k),k) = m$ for every $m \in \mathcal{M}$ and $k \in \mathcal{K}$, prove the following claims [179].

1. The matrix $|\mathcal{M}| \times |\mathcal{M}|$ with (c, m)th entry given by $P_{C|M}(c|m) = \Pr(f(m,K) = c)$ for distribution P_{CM} induced by the encryption scheme is doubly stochastic.[4]
2. Conversely, for a given doubly stochastic matrix with entries $P_{C|M}(c|m)$, there exists an encryption scheme (f, g) that induces the distribution $P_{C|M}(c|m)$ for some secret key K.

HINT The assumption on (f, g) implies that $f(\cdot, k)$ is a bijection from $\mathcal{M}$ to $\mathcal{C}$ for each $k \in \mathcal{K}$; use the Birkhoff–von Neumann theorem which states that a doubly stochastic matrix can be expressed as a convex combination of permutation matrices.

[4] A matrix is termed doubly stochastic if all of its entries are nonnegative and each of its rows and columns sums to 1.

3.11 By Problem 3.10, there exists an encryption scheme corresponding to the following doubly stochastic matrix:

$$\mathrm{P}_{C|M} = \begin{bmatrix} N^{-1}+\delta & N^{-1}-\delta & \cdots & N^{-1}+\delta & N^{-1}-\delta \\ N^{-1}-\delta & N^{-1}+\delta & \cdots & N^{-1}-\delta & N^{-1}+\delta \\ N^{-1} & N^{-1} & \cdots & N^{-1} & N^{-1} \\ \vdots & \vdots & \ddots & \vdots & \vdots \\ N^{-1} & N^{-1} & \cdots & N^{-1} & N^{-1} \end{bmatrix},$$

where $N = |\mathcal{M}| = |\mathcal{C}|$. For this encryption scheme, verify the following claims [179].

1. It is 2δ-indistinguishable secure.
2. For the uniform distribution P_M, the a posteriori distribution $\mathrm{P}_{M|C}$ is the transpose of $\mathrm{P}_{C|M}$.
3. For the uniform distribution P_M,

$$d_{\mathtt{var}}(\mathrm{P}_{M|C=c}, \mathrm{P}_M) = \begin{cases} N\delta/2, & \text{if } c = 1, 2, \\ 0, & \text{else.} \end{cases}$$

Remark 3.3 This example shows that δ-indistinguishable security for small δ does not imply the following security for small δ': $d_{\mathtt{var}}(\mathrm{P}_{M|C=c}, \mathrm{P}_M) \le \delta'$ for every P_M and $c \in \mathcal{C}$.

4 Universal Hash Families

In this chapter, we introduce a family of hash functions, termed the *universal hash family*, and its variants. A "hash function" is a function that scrambles and shrinks a given input sequence. Depending on the application, we require a hash function to have various properties. For instance, a cryptographic hash function, such as the popular hash function called the Secure Hash Algorithm (SHA), requires that it is computationally difficult to find two different inputs having the same function output value.

Typically, a cryptographic hash function used in practice is deterministic. On the other hand, the universal hash family considered in this chapter is a randomized function – for a given input, we apply a randomly chosen function from a family of functions. Usually, the randomness of choosing a function is supplied by a random key shared among the parties, and a universal hash is categorized as a keyed hash function. On the other hand, the above-mentioned cryptographic hash function is categorized as an unkeyed hash function.

We present basic properties of universal hash families which will be used to obtain the keyed hash functions needed at various places throughout this book.

4.1 A Short Primer on Finite Fields

To begin with, we digress from the main topic of this chapter and briefly review the basics of finite fields, which will be used throughout this chapter as well as later in the book. For resources on more details on the topic, see the references in Section 4.7. Readers familiar with finite fields may skip this section and directly go to Section 4.2.

Roughly speaking, a field is a set in which we can conduct the four basic arithmetic operations of addition, subtraction, multiplication, and division. For instance, the set of rational numbers $\mathbb{Q}$, the set of real numbers $\mathbb{R}$, and the set of complex numbers $\mathbb{C}$ are fields. Finite fields are fields having a finite number of elements; the number of elements in a field is called its *order* or, sometimes, its *size*. Instead of providing a formal definition of the field, which we expect the reader to review from a more standard reference on the topic, we illustrate the properties of finite fields with simple examples.

EXAMPLE 4.1 (The field $\mathbb{F}_2$ of binary operations) Let $\mathbb{F}_2 = \{0, 1\}$. For two elements $x, y \in \mathbb{F}_2$, addition $x + y$ is defined as $x \oplus y$ and multiplication $x \cdot y$ is defined as the logical AND operation $x \wedge y$. Then, $\mathbb{F}_2$ equipped with two operations $(+, \cdot)$ constitutes the finite field of order (size) 2.

EXAMPLE 4.2 (The prime field $\mathbb{F}_p$) For a prime number p, $\mathbb{F}_p = \{0, 1, \ldots, p-1\}$ equipped with addition and multiplication operations modulo p constitutes the finite field of order p. For any nonzero element $a \in \mathbb{F}_p$, the existence of the multiplicative inverse of a can be guaranteed as follows. Since the greatest common divisor (gcd) of a and p is 1, by the extended Euclidean algorithm, we can identify integers u and v satisfying $ua + vp = 1$. Then, $u \mod p$ is the multiplicative inverse of a in $\mathbb{F}_p$.

Just as we define *polynomials* over $\mathbb{R}$, we can define a polynomial over a field $\mathbb{F}_p$ as a mapping $f\colon \mathbb{F}_p \to \mathbb{F}_p$ given by

$$f(x) = \sum_{i=0}^{n} a_i x^i,$$

where the coefficients are $a_0, \ldots, a_n \in \mathbb{F}_p$. The largest i such that $a_i \neq 0$ is termed the *degree* of the polynomial $f(x)$ and is denoted by $\deg(f)$.

For given polynomials $f(x) = \sum_{i=0}^{n} a_i x^i$ and $g(x) = \sum_{i=0}^{n} b_i x^i$ on $\mathbb{F}_p$, the addition $f(x)+g(x)$ and multiplication $f(x)g(x)$ are defined in the same manner as those for polynomials on $\mathbb{R}$, i.e.,[1]

$$f(x) + g(x) = \sum_{i=0}^{n} (a_i + b_i) x^i,$$

$$f(x)g(x) = \sum_{k=0}^{2n} \Bigg(\sum_{\substack{i,j=0:\\ i+j=k}}^{n} a_i b_j \Bigg) x^k.$$

For a given polynomial $f(x)$ on $\mathbb{F}_p$, an element $a \in \mathbb{F}_p$ satisfying $f(a) = 0$ is termed a *root* of $f(x)$. When $f(x)$ has a root a, it can be factorized as $f(x) = (x - a)g(x)$ for some polynomial $g(x)$ on $\mathbb{F}_p$ with $\deg(g) = \deg(f) - 1$. In general, a polynomial may not have any root belonging to $\mathbb{F}_p$. Also, a polynomial of degree n has at the most n roots.[2] This latter fact will be frequently used throughout the book.

DEFINITION 4.3 (Irreducible polynomials) A polynomial $f(x)$ over a field $\mathbb{F}$ is termed *reducible* if it can be factorized as $f(x) = g(x)h(x)$ for polynomials f and g over $\mathbb{F}$ satisfying $\deg(g) > 0$ and $\deg(h) > 0$. When $f(x)$ is not reducible, it is termed irreducible.

[1] Even though both $f(x)$ and $g(x)$ have the same number of terms $n+1$ for simplicity of notation here, there is no loss in generality since we can append terms with coefficient 0 to make the number of terms of the two polynomials the same.

[2] When $f(x)$ is factorized as $f(x) = (x - a)^m g(x)$, a is counted as m roots of $f(x)$, and m is termed the multiplicity.

For instance, the polynomial $f(x) = x^2 + 1$ over $\mathbb{F}_2$ is reducible since it can be factorized as $x^2+1 = (x+1)(x+1)$ on $\mathbb{F}_2$. On the other hand, $f(x) = x^2+x+1$ over $\mathbb{F}_2$ is irreducible. An irreducible polynomial over a field $\mathbb{F}$ does not have any root in $\mathbb{F}$, but the converse need not be true since it may be factorized into polynomials of degree larger than 1.

When a polynomial $f(x)$ over $\mathbb{F}$ is irreducible, it does not have any root belonging to $\mathbb{F}$, i.e., $f(a) \neq 0$ for every $a \in \mathbb{F}$. However, the same polynomial can have roots when considered as a polynomial over an *extension field* of $\mathbb{F}$. In fact, this can be used to construct larger fields. For instance, we can extend the set $\mathbb{R}$ of real numbers to the set $\mathbb{C}$ of complex numbers by introducing an imaginary unit, which is a root of the irreducible polynomial $f(x) = x^2 + 1$ over $\mathbb{R}$. In a similar manner, we can extend the finite field $\mathbb{F}_p$ by introducing roots of an irreducible polynomial over $\mathbb{F}_p$. The example below illustrates how we can extend $\mathbb{F}_2$ to obtain a field of order 4.

EXAMPLE 4.4 (Extending $\mathbb{F}_2$ to $\mathbb{F}_4$) Denote by α a root of the irreducible polynomial $f(x) = x^2 + x + 1$ over $\mathbb{F}_2$, namely a fictitious element for which $\alpha^2 + \alpha + 1 = 0$. Since f is irreducible over $\mathbb{F}_2$, $\alpha \notin \mathbb{F}_2$. Define

$$\mathbb{F}_4 := \{0, 1, \alpha, 1+\alpha\},$$

i.e., $\mathbb{F}_4$ comprises elements of the form $a_0 + a_1\alpha$ where a_0 and a_1 are from $\mathbb{F}_2$. For two elements $a = a_0 + a_1\alpha$ and $b = b_0 + b_1\alpha$, we can define addition using $a+b = (a_0+b_0)+(a_1+b_1)\alpha$, where $a_i + b_i$ is in $\mathbb{F}_2$. For multiplication, we can treat a and b as polynomials and write $a \cdot b = a_0 \cdot b_0 + (a_0 \cdot b_1 + a_1 \cdot b_0)\alpha + a_1 \cdot b_1\alpha^2$, where $a_i \cdot b_j$ is multiplication in $\mathbb{F}_2$. But now we have an "α^2" term which needs to be eliminated to get back an element from $\mathbb{F}_4$. We do this using the relation $\alpha^2 = -\alpha - 1 = \alpha + 1$ which holds since $f(\alpha) = 0$ by our assumption. Namely, we define

$$a \cdot b = (a_0 \cdot b_0 - a_1 \cdot b_1) + (a_0 \cdot b_1 + a_1 \cdot b_0 - a_1 \cdot b_1)\alpha.$$

More generally, by introducing a root $\alpha \in \mathbb{F}_p$ of an irreducible polynomial $f(x)$ of degree k, we can construct a finite field of order p^k as follows:

$$\mathbb{F}_{p^k} = \left\{ \sum_{i=0}^{k-1} a_i\alpha^i : a_i \in \mathbb{F}_p, \forall i \in \{0, 1, \ldots, k-1\} \right\}.$$

The addition of two elements $\sum_{i=0}^{k-1} a_i\alpha^i$ and $\sum_{i=0}^{k-1} b_i\alpha^i$ is given by $\sum_{i=0}^{k-1}(a_i + b_i)\alpha^i$, where $a_i + b_i$ is addition in $\mathbb{F}_p$. For multiplication, we exploit the fact that $f(\alpha) = 0$. Specifically, $(\sum_{i=0}^{k-1} a_i\alpha^i) \cdot (\sum_{i=0}^{k-1} b_i\alpha^i)$ can be expressed as $\sum_{i=0}^{k-1} c_i\alpha^i$ by relating the terms α^i with $i \geq k$ to those with $i < k$ using $f(\alpha) = 0$.

Note that in the example for $\mathbb{F}_4$ above, we have $(1+\alpha) \cdot \alpha = \alpha^2 + \alpha = f(\alpha) + 1 = 1$. It follows that α is the (multiplicative) inverse of $(1+\alpha)$. In general, the existence of multiplicative inverses can be guaranteed as follows. First, we notice that the elements of $\mathbb{F}_{p^k}$ can be represented as polynomials over $\mathbb{F}_p$ with degree less than k, i.e.,

$$\mathbb{F}_{p^k} = \{g(\alpha) : g(x) \text{ is a polynomial over } \mathbb{F}_p \text{ with } \deg(g) < k\}.$$

For any nonzero polynomial $g(x)$ with $\deg(g) < k$, since $f(x)$ is irreducible, the gcd of $g(x)$ and $f(x)$ is a constant polynomial. Thus, by the polynomial version of the extended Euclidean algorithm, we can identify polynomials $u(x)$ and $v(x)$ satisfying $u(x)g(x) + v(x)f(x) = 1$. Then, with $u'(x) := u(x) \mod f(x)$, $u'(\alpha)$ is the multiplicative inverse of $g(\alpha)$.

Note that the elements of the finite field $\mathbb{F}_{p^k}$ can be identified with elements in the product set $\mathbb{F}_p^k$. Further, in a similar manner as above, for $q = p^k$, we can also extend $\mathbb{F}_q$ to the field $\mathbb{F}_{q^s}$ by introducing a root of an irreducible polynomial $f(x)$ of degree s over $\mathbb{F}_q$. It should be noted that $\mathbb{F}_{q^s}$ and $\mathbb{F}_{p^{ks}}$ are *algebraically isomorphic*. In fact, an important result for finite fields is that all finite fields of the same order are algebraically isomorphic.

We conclude this section by recalling a result from linear algebra, but this time for linear mappings over a finite field. The set $\mathbb{F}^k$ can be viewed as a k-dimensional vector space over a finite field $\mathbb{F}$ comprising vectors $x = (x_1, \ldots, x_k)$ with $x_i \in \mathbb{F}$. Note that for $x \in \mathbb{F}^k$ and $a \in \mathbb{F}$, $ax := (a \cdot x_1, a \cdot x_2, \ldots, a \cdot x_k)$ is in $\mathbb{F}^k$, and further, for $y \in \mathbb{F}^k$ and $b \in \mathbb{F}$, then $ax + by \in \mathbb{F}^k$. A linear mapping $\phi\colon \mathbb{F}^k \to \mathbb{F}^\ell$ can be represented by a $k \times \ell$ matrix Φ consisting of entries from $\mathbb{F}$ and such that $\phi(x) = \Phi x$, where Φx is the multiplication of matrix Φ and vector x defined using operations from $\mathbb{F}$.

All the basic notions and results of linear algebra extend to the finite field setup. The notion of basis of a vector space can be defined using linearly independent vectors with coefficients coming from $\mathbb{F}$, which in turn can be used to define the rank of a matrix. We note that the range $\phi(\mathbb{F}^k)$ is a vector space over $\mathbb{F}$ too. The *rank* $\mathrm{rank}(\phi)$ of a linear mapping $\phi\colon \mathbb{F}^k \to \mathbb{F}^\ell$ is defined as the dimension of the vector space $\phi(\mathbb{F}^k)$, which equals the rank of the matrix Φ representing ϕ. The following result is the counterpart of the rank-nullity theorem applied to finite fields.

THEOREM 4.5 (Rank-nullity) *Consider a finite field $\mathbb{F}$. For a linear mapping $\phi\colon \mathbb{F}^k \to \mathbb{F}^\ell$ with $\mathrm{rank}(\phi) = r$, the kernel $\phi^{-1}(0)$ of ϕ is a $(k-r)$-dimensional vector space over $\mathbb{F}$ whereby $|\phi^{-1}(0)| = |\mathbb{F}|^{k-r}$.*

As a corollary, since $\phi(x) = y = \phi(x')$ implies that $\phi(x - x') = 0$, we obtain the following.

COROLLARY 4.6 *Consider a finite field $\mathbb{F}$. For a linear mapping $\phi\colon \mathbb{F}^k \to \mathbb{F}^\ell$ with $\mathrm{rank}(\phi) = r$, $|\phi^{-1}(y)| = |\mathbb{F}|^{k-r}$ for every $y \in \mathbb{F}^\ell$.*

4.2 Universal Hash Family: Basic Definition

Before we start describing universal hashing, we take a step back and revisit its nonuniversal variant. Perhaps the first application of hashing appears in databases. A *hash table* is a data structure implementation where data entries are partitioned into *bins*, and the hash value of an entry is the index of its bin. The goal of such an arrangement is to ensure that the comparison of data points can be facilitated by comparison of their hash values alone, which are much smaller in number than data values. This requirement is captured by the probability of collision of hash values, as illustrated by the simple example below.

EXAMPLE 4.7 Suppose that data values comprise n-bit vectors $\{0,1\}^n$, and we seek to hash them to m-bits, namely partition the data values into 2^m bins for $m \ll n$. A simple strategy for doing so is to retain the first m bits as the hash, i.e., $f(x_1, \ldots, x_n) = (x_1, \ldots, x_m)$; each bin in this scheme consists of 2^{n-m} data values. To evaluate the probability of hash collision, namely the probability that hash values of two different data values coincide, assume that data values X and X' are distributed independently and uniformly over $\{0,1\}^n$. For this distribution, the probability of hash collision equals $\Pr(f(X) = f(X')) = 2^{-(n-m)}$.

The probability of hash collision attained by the previous example for uniformly distributed data values is small enough for many applications. But what about the probability of hash collision for other distributions? For instance, if the data value is distributed so that the first m bits are fixed with probability one, the probability of hash collision is one as well. A universal hash ensures that the probability of collision remains small for every distribution on data. Our first definition below captures this basic requirement. Unlike the hash construction in the example above, universal hash constructions are often randomized and the probability of hash collision is analyzed for the worst-case input pair. We denote the set of data values by $\mathcal{X}$ and the set of hash values by $\mathcal{Y}$. Most practical implementations will operate with both $\mathcal{X}$ and $\mathcal{Y}$ corresponding to fixed length binary sequences. However, the general form we adopt will help with presentation.

DEFINITION 4.8 (Universal hash family) A class $\mathcal{F}$ of functions from $\mathcal{X}$ to $\mathcal{Y}$ is called a universal hash family (UHF) if for any pair of distinct elements x, x' in $\mathcal{X}$ and for F generated uniformly from $\mathcal{F}$,

$$\Pr\left(F(x) = F(x')\right) \le \frac{1}{|\mathcal{Y}|}. \tag{4.1}$$

Skeptical readers might be worried about the existence of such families. We set their mind at ease right away by pointing out that the class of all mappings from $\mathcal{X}$ to $\mathcal{Y}$ is indeed a UHF (the proof is left as an exercise, see Problem 4.1).

A hash table can be implemented using a UHF by computing a hash value $F(x)$ for a given data value x and F generated uniformly over $\mathcal{F}$. But this will require us to generate a uniform random variable over a set of cardinality $|\mathcal{F}|$. For instance, $|\mathcal{X}| \log |\mathcal{Y}|$ bits of randomness will be needed to specify a function from the UHF described above, i.e., the set of all mappings from $\mathcal{X}$ to $\mathcal{Y}$. This is unacceptable for most applications. We seek to design a UHF which, for a given input alphabet size (in bits) $n = \log |\mathcal{X}|$ and output alphabet size $k = \log |\mathcal{Y}|$, has $\log |\mathcal{F}|$ as small as possible. The construction we describe next has $n = 2k$ and $\log |\mathcal{F}| = k$, which is much better than the aforementioned naive construction for the $n = \mathcal{O}(k)$ regime.

CONSTRUCTION 4.1 (Linear mappings over a finite field) For $q = 2^k$, let $\mathbb{F}_q$ denote the finite field of order q. We represent $\mathcal{X} = \mathbb{F}_q \times \mathbb{F}_q$, $\mathcal{Y} = \mathbb{F}_q$, and associate with each $u \in \mathbb{F}_q$ a mapping $f_u : \mathcal{X} \to \mathcal{Y}$ given by

$$f_u(x_1, x_2) = x_2 \cdot u + x_1 \qquad \text{for all } x_1, x_2 \in \mathbb{F}_q.$$

Our desired UHF is given by the family $\mathcal{F} = \{f_u : u \in \mathbb{F}_q\}$. Clearly, $|\mathcal{X}| = 2^{2k}$, $|\mathcal{Y}| = 2^k$, and $|\mathcal{F}| = 2^k$.

LEMMA 4.9 *The family $\mathcal{F}$ given in Construction 4.1 constitutes a UHF.*

Proof We need to verify (4.1). Let $x = (x_1, x_2)$ and $x' = (x_1', x_2')$. Then, $f_u(x) = f_u(x')$ can hold only if

$$(x_2 - x_2') \cdot u = (x_1' - x_1).$$

If $x_2 = x_2'$, the previous identity can hold only if $x_1 = x_1'$ holds as well. Otherwise, the identity can hold only if $u = (x_2 - x_2')^{-1} \cdot (x_1' - x_1)$. Thus, for $x \neq x'$, $f_u(x) = f_u(x')$ can hold for at most one value of u, which yields (4.1). □

Note that Construction 4.1 can be applied for any q equal to the power of a prime number. But we do not gain much from such an extension for two reasons. First, most practical implementations represent $\mathcal{X}$ in bits. Second, and more importantly, the efficiency of these constructions is measured in terms of two parameters: the *compression ratio*[3] given by $\frac{\log |\mathcal{X}|}{\log |\mathcal{Y}|}$ and the *randomness cost* given by $\frac{\log |\mathcal{F}|}{\log |\mathcal{Y}|}$. We seek UHFs with lower randomness costs; in some applications, a higher compression ratio is preferable, but not necessarily so. Going forward in this chapter, we will benchmark all schemes in terms of these parameters.

Construction 4.1 with $q = 2^k$ has compression ratio of 2 and randomness cost of 1. Using q that are powers of primes other than 2 in Construction 4.1 offers no additional gains in compression ratios or randomness costs.

By using more sophisticated observations from linear algebra, we have a construction which has a more flexible choice of parameters as follows.

CONSTRUCTION 4.2 (Linear mappings over a vector space) With $q = 2^k$, $s \geq t$, and $\mathbb{F}_q$ the finite field of order q, let $\mathcal{X} = \mathbb{F}_{q^s}$ and $\mathcal{Y} = \mathbb{F}_{q^t}$. Note that $\mathbb{F}_{q^s}$ and $\mathbb{F}_{q^t}$ can be regarded as s- and t-dimensional vector spaces over $\mathbb{F}_q$ as well as extension fields of orders q^s and q^t, respectively.

Consider the vector space interpretation $\mathbb{F}_{q^s}$, and let $\phi\colon \mathbb{F}_{q^s} \to \mathbb{F}_{q^t}$ be a linear mapping of rank t (full rank); see Section 4.1 for the notion of rank of a linear mapping. A simple example of such a mapping ϕ is a projection of the s-dimensional input to its first t coordinates.

Using this mapping ϕ, we can associate with each $u \in \mathbb{F}_{q^s}$ a linear mapping $f_u\colon \mathcal{X} \to \mathcal{Y}$ given by

$$f_u(x) = \phi(x \cdot u),$$

[3] The term compression ratio is motivated by the hash table application where the hash value represents a compressed version of the data value.

where $x \cdot u$ denotes multiplication in the extension field interpretation of $\mathbb{F}_{q^s}$. Note that the mapping $u \mapsto x \cdot u$ is one-to-one for a nonzero x.

Our next UHF $\mathcal{F}$ is given by $\mathcal{F} = \{f_u : u \in \mathbb{F}_{q^s}\}$.

LEMMA 4.10 *The family $\mathcal{F}$ given in Construction 4.2 constitutes a UHF with compression ratio and randomness cost both equal to (s/t).*

Proof To verify the UHF property (4.1), note that for $x \neq x'$ the identity $f_u(x) = f_u(x')$ holds if and only if $\phi(x \cdot u) = \phi(x' \cdot u)$, which can occur only if $(x - x') \cdot u \in \phi^{-1}(0)$ since ϕ is a linear mapping. Using the rank-nullity theorem for finite fields (Theorem 4.5), $|\phi^{-1}(0)| = q^{s-t}$, since $\text{rank}(\phi) = t$. Therefore, when $x' \neq x$, there are exactly q^{s-t} values of u given by the set $(x-x')^{-1} \cdot \phi^{-1}(0)$ for which the equality above can hold. Since $|\mathcal{X}| = q^s$, $|\mathcal{Y}| = q^t$, and $|\mathcal{F}| = q^s$, the fraction of mappings for which $f_u(x) = f_u(x')$ for $x \neq x'$ equals $q^{s-t}/|\mathcal{F}| = 1/q^t = 1/|\mathcal{Y}|$, which yields (4.1). In fact, (4.1) holds with equality for every $x \neq x'$. Also, both the compression ratio and randomness cost for this UHF are (s/t). □

This new construction allows us to obtain arbitrarily high compression ratio, but for a higher randomness cost. In fact, this tradeoff is fundamental and a scheme with compression ratio r must have randomness cost at least $r - 1$; see Problem 4.3.

4.3 Strong Universal Hash Families

We motivated a UHF as a tool for obtaining compression in hash tables. But how is this tool related to cryptographic applications? This connection is best understood by observing additional structure in the naive construction of Example 4.7. For uniformly distributed data values X, the output $f(X)$ is distributed uniformly as well. A universal variant of this property will require $F(x)$ to be uniformly distributed for each x, namely that the statistics of the output $F(x)$ remain the same for every x. Thus, an observer of $F(x)$ gets no information about x.

For our applications, this property alone will not suffice. We still need low collision probabilities to ensure that different inputs will produce independent hashes, namely for $x' \neq x$, we need $F(x)$ and $F(x')$ to behave roughly as two independent and uniformly distributed random variables.

Both these properties are captured by the following definition of a *strong UHF*.

DEFINITION 4.11 (Strong universal hash family) A class of functions $\mathcal{F}$ from $\mathcal{X}$ to $\mathcal{Y}$ is called a strong UHF if the following properties hold. For F distributed uniformly over $\mathcal{F}$ and every $x \in \mathcal{X}$,

$$\Pr\left(F(x) = y\right) = \frac{1}{|\mathcal{Y}|} \quad \text{for all } y \in \mathcal{Y}, \tag{4.2}$$

and for every $x' \neq x$ and $y, y' \in \mathcal{Y}$,

$$\Pr\left(F(x) = y, F(x') = y'\right) \leq \frac{1}{|\mathcal{Y}|^2}. \tag{4.3}$$

Note that for a strong UHF $\mathcal{F}$ and $x \neq x'$,

$$\Pr\left(F(x) = F(x')\right) = \sum_{y \in \mathcal{Y}} \Pr\left(F(x) = F(x') = y\right) \leq \frac{1}{|\mathcal{Y}|},$$

where the inequality is by (4.3). Thus, a strong UHF is indeed a UHF.

Remark 4.1 (On imposing uniformity condition (4.2)) There are various definitions for strong UHFs in the literature, and some of them do not impose the additional uniformity condition (4.2). For the purpose of this book, we primarily apply strong UHFs to design message authentication codes in Chapter 8. Using strong UHFs that satisfy the uniformity condition (4.2) allows us to exhibit different levels of security under different types of attacks (see Remark 8.2). In any case, most current constructions of strong UHFs do satisfy this condition.

For both Construction 4.1 and Construction 4.2, neither the uniformity requirement (4.2) nor the collision probability requirement (4.3) is satisfied. Thus, the UHFs of Construction 4.1 and Construction 4.2 are not strong UHFs. In order to satisfy the more severe requirements in (4.2) and (4.3), we modify Construction 4.2 as follows.

CONSTRUCTION 4.3 (Strong UHF from linear mappings) Using the same notation as Construction 4.2, we set $\mathcal{X} = \mathbb{F}_{q^s}$ and $\mathcal{Y} = \mathbb{F}_{q^t}$. For $u \in \mathbb{F}_{q^s}$ and $v \in \mathbb{F}_{q^t}$, define the mapping $f_{u,v} \colon \mathcal{X} \to \mathcal{Y}$ as

$$f_{u,v}(x) = \phi(x \cdot u) + v.$$

The family $\mathcal{F}$ is given by the set of mappings $\{f_{u,v} : u \in \mathbb{F}_{q^s}, v \in \mathbb{F}_{q^t}\}$.

LEMMA 4.12 *The family $\mathcal{F}$ given in Construction 4.3 constitutes a strong UHF with compression ratio (s/t) and randomness cost $(s/t) + 1$.*

Proof For property (4.2), let U and V be distributed uniformly over $\mathbb{F}_{q^s}$ and $\mathbb{F}_{q^t}$, respectively, and let F be distributed uniformly over $\mathcal{F}$. Then, using independence of U and V and the fact that V is uniform, we get

$$\begin{aligned}\Pr(F(x) = y | U = u) &= \Pr(\phi(u \cdot x) + V = y | U = u) \\ &= \Pr(V = y - \phi(u \cdot x)) \\ &= \frac{1}{q^t},\end{aligned}$$

which establishes (4.2) upon taking expectation over U on both sides.

For property (4.3), for $x \neq x'$ and $y, y' \in \mathcal{Y}$, the identities $f_{u,v}(x) = y$ and $f_{u,v}(x') = y'$ can hold only if u belongs to $(x - x')^{-1} \cdot \phi^{-1}(y - y')$, which is a set of cardinality q^{s-t} by Corollary 4.6. Also, for each fixed u, the identity $\phi(x \cdot u) + v = y$ holds for a unique v, and for a $u \in (x - x')^{-1} \cdot \phi^{-1}(y - y')$, the same v must satisfy $\phi(x' \cdot u) + v = y'$ as well for $f_{u,v}(x') = y'$ to hold. Thus, $f_{u,v}(x) = y$ and $f_{u,v}(x') = y'$ can hold for q^{s-t} values of pairs (u, v) whereby $\Pr(F(x) = y, F(x') = y') = q^{s-t}/q^{s+t} = q^{-2t} = 1/|\mathcal{Y}|^2$, establishing (4.3).

Finally, it can easily be checked that the compression ratio for this scheme is (s/t) and the randomness cost is $(s/t) + 1$. □

As pointed out at the end of the previous section, the least randomness cost for a UHF with compression ratio r is $r - 1$. Thus, the family $\mathcal{F}$ of Construction 4.3 has almost optimal randomness cost for a UHF with compression ratio (s/t), while offering the additional structure of a strong UHF (see also Problem 4.4). On the other hand, the UHF of Construction 4.1 has a lower randomness cost for a compression ratio of 2. Later in the chapter, we will combine these two constructions to obtain a family which retains the advantages of both.

But before we proceed, let us look at some use-cases for these constructions.

4.4 Example Applications

We now outline some applications of UHFs and strong UHFs. We only provide a high-level description of these applications, relegating more detailed discussions to later chapters focusing on specific applications.

4.4.1 Compression

We have already seen the application of UHFs in implementing hash tables where each data value x is stored as its hash $F(x)$ in a table. This hash table can be used to enable several basic database operations. For a simple example, suppose that we want to see if a data value x' lies in the database. We can simply compute its hash $F(x')$ and see if it belongs to the hash table. If no match is found, then clearly the data value is not present in the database. On the other hand, if a match is found, then with probability $1 - 1/|\mathcal{Y}|$ the data value is indeed present in the database. Since hash values can be represented by only $\log |\mathcal{Y}|$ bits, as opposed to $\log |\mathcal{X}|$ bits required by the original data values, the search over hash values is much quicker than a search over the original data values. In effect, we have obtained a compression of the original database with compression ratio of the UHF capturing the gain.

Note that the probabilistic guarantee above is for one search, and by union bound, the probability of hash collision increases linearly with the number of searches. In practice, we can follow a hash-match with a search over elements with that hash to verify a match. This in turn can be implemented by another independent hash. Such procedures are bread-and-butter for database theory and even cache design, but are beyond the scope of this book.

In a different direction, information-theoretic compression studied in source coding and communication complexity considers the task of describing a random variable X to a receiver with knowledge of the generating distribution P_X.

The interesting case is when the sender itself does not know P_X. In this case, the sender (or the compressor) can simply use a UHF with $|\mathcal{Y}| = 2^k$ and sends the hash $F(X)$. Note that since the distribution of X is unknown to the sender, the universality of our hash family is crucial here. The receiver (or the decompressor) uses the knowledge of P_X and forms a guess list of size 2^ℓ of likely values of X. It computes $F(x')$ for each value of x' in the guess list and declares the value that matches the received hash $F(X)$. A simple union bound together with (4.1) can be used to show that the decoder declares the correct value of X with probability at least $1 - 2^{-(k-\ell)}$; the details are left as an exercise in Problem 4.5.

4.4.2 Random Number Generation

Random number generation is a basic requirement in many cryptographic applications and even beyond. The requirement here is to generate a uniform random variable using a perhaps nonuniform source of randomness. Specifically, given a random variable X, we seek a mapping f such that $f(X)$ is distributed uniformly. Property (4.2) tells us that the output $F(x)$ of a uniformly chosen mapping from a strong UHF is distributed uniformly for every x, and thereby also for a random X. But this is not what we are looking for: here we started with a uniformly distributed F and generated another uniform random variable. To ensure that the generated randomness is indeed extracted from X, we require that $F(X)$ is independent of F and is distributed uniformly. Perhaps one can heuristically be convinced that a strong UHF can yield such a mapping. Surprisingly, as we shall see in Chapter 7, a UHF itself yields such a random mapping where $F(X)$ is almost independent of F and is distributed almost uniformly. Thus, a UHF can be used for randomness extraction, a tool with a plethora of applications in theory and practice.

To get a high-level idea of how a UHF is useful for random number generation, let us consider the following simple example. Let X be the uniform random variable on an unknown support $\mathcal{E} \subseteq \{0,1\}^n$ with $|\mathcal{E}| = 2^r$. If the support is known, we can easily generate r uniformly random bits by a one-to-one mapping from $\mathcal{E}$ to $\{0,1\}^r$. If we do not know the support $\mathcal{E}$, we can use a UHF from $\{0,1\}^n$ to $\{0,1\}^k$ instead. In Chapter 7, we shall show that $F(X)$ is statistically close to uniform distribution. For now, to build heuristics, we show closeness to uniform random variable in another sense, namely that of collision probability.[4]

For a discrete random variable X, the *collision probability* $p_{\mathtt{col}}(X)$ is defined as the probability that two independent copies of X coincide, which in turn is given by

$$p_{\mathtt{col}}(X) := \sum_x \mathrm{P}_X(x)^2 .$$

[4] Closeness to uniformity evaluated by collision probability is similar in essence to that evaluated by entropy; see Problem 4.7.

Note that for the uniform distribution on $\{0,1\}^k$, the collision probability is 2^{-k}. In fact, for any random variable supported on $\{0,1\}^k$, the collision probability is at least 2^{-k}, a lower bound that is attained by the uniform distribution (see Problem 4.7). We evaluate the uniformity of $F(X)$ by comparing its collision probability with 2^{-k}.

For X with $p_{\mathtt{col}}(X) = 2^{-r}$ and the output $F(X)$ of a mapping F chosen uniformly from a UHF, the expected collision probability is

$$\begin{aligned}
&\mathbb{E}_F\left[p_{\mathtt{col}}(F(X))\right] \\
&\quad = \Pr\left(F(X_1) = F(X_2)\right) \\
&\quad = \Pr\left(X_1 = X_2\right) + \Pr\left(X_1 \neq X_2\right)\Pr\left(F(X_1) = F(X_2) \middle| X_1 \neq X_2\right) \\
&\quad \leq p_{\mathtt{col}}(X) + (1 - p_{\mathtt{col}}(X))2^{-k} \\
&\quad \leq 2^{-k}\left(1 + 2^{-(r-k)}\right),
\end{aligned}$$

where the first inequality follows from the UHF property (4.1). Thus, if the length k of the extracted random number is smaller than the "initial randomness" r bits, then the multiplicative "gap from uniformity" is small for $F(X)$.

4.4.3 Authentication

Message authentication code (MAC) is a basic primitive of cryptography; we will explore this in a later chapter and build on it throughout the book. In fact, the strong UHF plays an important role in message authentication schemes. In a message authentication scheme, we transmit a tag t computed from a message m along with the message itself so that if an adversary modifies this message along the way, the receiver can determine that the message has been modified. Also, an adversary cannot fabricate a new message and a tag consistent with it. We shall formalize these security requirements in Chapter 8, but for now we just provide a heuristic explanation of how a strong UHF can play a role in enabling an authentication scheme.

Assume that the sender and the receiver can generate random mapping F distributed uniformly over a strong UHF using shared randomness, namely a random variable available to both of them but not to the adversary. The sender can add $t = F(m)$ as the tag for message m. On receiving (m, t), the receiver can accept the message if $t = F(m)$. If an adversary has modified the message m to $m' \neq m$ during transmission, property (4.3) will ensure that the probability $F(m') = t$ is small. On the other hand, if an adversary (without the knowledge of F) sends (m, t) of its own, by property (4.2), the probability $\Pr(F(m) = t)$ that this message m will be accepted by the receiver is small.

4.5 Almost Universal Hash Families

As pointed out earlier, the construction of UHFs and strong UHFs that we have presented achieves almost optimal tradeoff between compression ratio and randomness cost. But for the cryptographic applications above, the requirement of randomness cost of the same order as compression ratio may not be acceptable. We move beyond this seeming dead-end by noticing that requirements (4.1) and (4.3) are related to the probability of error or independence requirements in applications. In many cases, we can relax these requirements and allow different quantities on the right-hand sides of (4.1) and (4.3). This leads to the definition of an *almost UHF* and a *strong almost UHF* which can serve the same purpose as a UHF and strong UHF in applications, but lends itself to more efficient constructions.

DEFINITION 4.13 (Almost UHF and strong almost UHF) A class $\mathcal{F}$ of functions from $\mathcal{X}$ to $\mathcal{Y}$ is called a δ-almost UHF (δ-UHF) if for any pair of distinct elements x, x' in $\mathcal{X}$ and for F generated uniformly from $\mathcal{F}$,

$$\Pr\left(F(x) = F(x')\right) \le \delta. \tag{4.4}$$

It is called a strong δ-almost UHF (strong δ-UHF) if it satisfies (4.2) and

$$\Pr\left(F(x) = y, F(x') = y'\right) \le \frac{\delta}{|\mathcal{Y}|} \tag{4.5}$$

for every $x' \ne x$ and $y, y' \in \mathcal{Y}$.

Note that a δ-UHF and strong δ-UHF correspond to a UHF and strong UHF, respectively, for $\delta = 1/|\mathcal{Y}|$. The main motivation for defining this general notion is that we can obtain more efficient constructions for larger values of δ. We illustrate this gain by providing an extension of Construction 4.1 that can attain a higher compression ratio at the same randomness cost (for a δ greater than $1/|\mathcal{Y}|$).

CONSTRUCTION 4.4 (A randomness efficient almost UHF) Following the same notation as Construction 4.1, for an integer $1 \le r \le q$, let $\mathcal{X} = \mathbb{F}_q^r$ and $\mathcal{Y} = \mathbb{F}_q$. For each $u \in \mathbb{F}_q$, denote by $f_u \colon \mathcal{X} \to \mathcal{Y}$ the mapping

$$f_u(x) = \sum_{i=0}^{r-1} x_i \cdot u^i,$$

where $x = (x_0, \ldots, x_{r-1})$ with each $x_i \in \mathbb{F}_q$, namely the polynomial of degree $r-1$ in u with coefficients given by x. The family $\mathcal{F}$ is given by the set of mappings $\{f_u : u \in \mathbb{F}_q\}$. Note that Construction 4.1 corresponds to $r = 2$.

LEMMA 4.14 *The family $\mathcal{F}$ given in Construction 4.4 constitutes a δ-UHF for $\delta = (r-1)/q$ with compression ratio r and randomness cost* 1.

Proof For $x' \ne x$, the identity $f_u(x) = f_u(x')$ holds if and only if

$$\sum_{i=0}^{r-1} (x_i - x_i') \cdot u^i = 0.$$

Since the left-hand side above is a (nontrivial) polynomial of degree at most $r-1$, it can have at most $r-1$ roots. Thus, there are at most $r-1$ values of u for which

$f_u(x) = f_u(x')$ can hold, which implies $\Pr(F(x) = F(x')) \leq (r-1)/q$. Further, the compression ratio r and randomness cost are 1 since $|\mathcal{X}| = q^r$, $|\mathcal{Y}| = q$, and $|\mathcal{F}| = q$. □

Thus, by relaxing δ by a factor of $r - 1$, we were able to gain a factor r in compression ratio. In fact, by setting $q \leftarrow 2^{s+t}$ and $r \leftarrow 2^s + 1$, we can get a 2^{-t}-UHF with compression ratio of $2^s + 1$ and a constant randomness cost; note that the compression ratio is exponential in parameter s. However, Construction 4.4 still does not satisfy (4.3) (see Problem 4.6). In general, we note that, while the polynomial construction has a lower randomness cost than the extended field based Construction 4.3, the latter satisfies (4.3) and constitutes a strong UHF. The next result provides a tool to combine these two constructions to retain the good features of both.

LEMMA 4.15 (Composition lemma) *Let $\mathcal{F}_1$ be a δ_1-UHF consisting of mappings from $\mathcal{X} \to \mathcal{Y}_1$ and $\mathcal{F}_2$ be a strong δ_2-UHF consisting of mappings from $\mathcal{Y}_1 \to \mathcal{Y}_2$. Then, the family $\mathcal{F}$ consisting of composition mappings $f = f_2 \circ f_1$ given by $f(x) = f_2(f_1(x))$, $f_1 \in \mathcal{F}_1$, $f_2 \in \mathcal{F}_2$, constitutes a strong $(\delta_1 + \delta_2)$-UHF.*

Proof Let F_1 and F_2 be distributed uniformly over $\mathcal{F}_1$ and $\mathcal{F}_2$, respectively. Then, $F = F_2 \circ F_1$ is distributed uniformly over $\mathcal{F}$. Note that for every fixed realization $F_1 = f_1$, property (4.2) for F_2 implies

$$\Pr\left(F(x) = y | F_1 = f_1\right) = \Pr\left(F_2(f_1(x)) = y | F_1 = f_1\right) = \frac{1}{|\mathcal{Y}_2|},$$

whereby we get $\Pr(F(x) = y) = 1/|\mathcal{Y}_2|$.

To verify (4.5), note that for each $x \neq x'$, we have

$$\begin{aligned}
&\Pr\left(F(x) = y, F(x') = y'\right) \\
&\quad = \Pr\left(F_1(x) = F_1(x')\right) \cdot \Pr\left(F(x) = y, F(x') = y' \mid F_1(x) = F_1(x')\right) \\
&\qquad + \Pr\left(F_1(x) \neq F_1(x')\right) \cdot \Pr\left(F(x) = y, F(x') = y' \mid F_1(x) \neq F_1(x')\right) \\
&\quad \leq \Pr\left(F_1(x) = F_1(x')\right) \cdot \Pr\left(F(x) = y, F(x') = y' \mid F_1(x) = F_1(x')\right) + \frac{\delta_2}{|\mathcal{Y}_2|} \\
&\quad \leq \delta_1 \cdot \Pr\left(F(x) = y, F(x') = y' \mid F_1(x) = F_1(x')\right) + \frac{\delta_2}{|\mathcal{Y}_2|} \\
&\quad \leq \frac{\delta_1 + \delta_2}{|\mathcal{Y}_2|},
\end{aligned}$$

where the first inequality holds by (4.5) for F_2, the second by (4.4) for F_1, and the final inequality is obtained using (4.2) for F_2 as follows:

$$\begin{aligned}
\Pr\left(F(x) = y, F(x') = y' \mid F_1(x) = F_1(x')\right)& \\
&= \mathbb{1}_{\{y=y'\}} \cdot \Pr\left(F_2(F_1(x)) = y \mid F_1(x) = F_1(x')\right) \\
&\leq \frac{1}{|\mathcal{Y}_2|}.
\end{aligned}$$

□

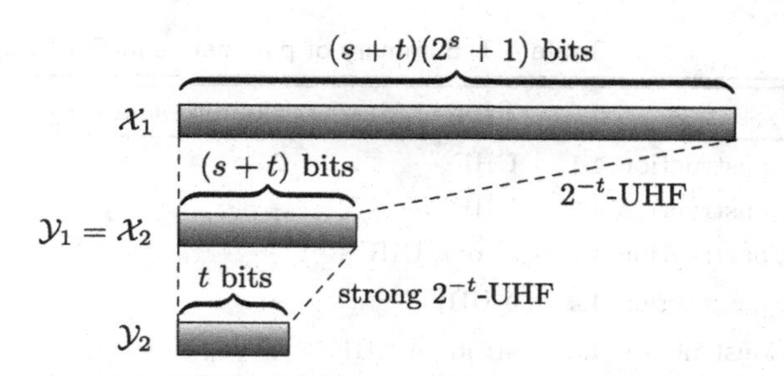

Figure 4.1 The description of Construction 4.5. The input is substantially shrunk by 2^{-t}-UHF in the first phase, and the properties of the almost strong UHF are added in the second phase.

Our ultimate construction in this chapter is obtained by applying the composition lemma to the δ-UHF of Construction 4.4 and strong UHF of Construction 4.3. We finally have a strong δ-UHF with randomness cost of order logarithmic in compression ratio.

CONSTRUCTION 4.5 (A randomness efficient strong δ-UHF) Let $\mathcal{F}_1$ be the family in Construction 4.4 with $q \leftarrow 2^{s+t}$ and $r \leftarrow 2^s + 1$ so that $\mathcal{X}_1 = \mathbb{F}_{2^{(s+t)}}^{(2^s+1)}$, $\mathcal{Y}_1 = \mathbb{F}_{2^{(s+t)}}$, and $\mathcal{F}_1 = \mathbb{F}_{2^{(s+t)}}$. Let $\mathcal{F}_2$ be the family in Construction 4.3 with $q \leftarrow 2$, $s \leftarrow s+t$, and $t \leftarrow t$ so that $\mathcal{X}_2 = \mathbb{F}_{2^{(s+t)}}$, $\mathcal{Y}_2 = \mathbb{F}_{2^t}$, and $\mathcal{F}_2 = \mathbb{F}_{2^{(s+t)}} \times \mathbb{F}_{2^t}$. Our final construction $\mathcal{F}$ comprises composition mappings $f_2 \circ f_1$ for every $f_1 \in \mathcal{F}_1$ and $f_2 \in \mathcal{F}_2$; see Figure 4.1.

Construction 4.5 yields roughly a strong δ-UHF with compression ratio r at randomness cost $\mathcal{O}(\log(r/\delta))$. Formally, we have the following characterization of its performance; the specific choice of parameters used is for convenience.

THEOREM 4.16 *Construction 4.5 constitutes a strong δ-UHF with $\delta = 2^{-(t-1)}$, compression ratio $(s+t)\big(2^s+1\big)/t$, and randomness cost $2(s/t)+3$.*

Proof By Lemma 4.14, $\mathcal{F}_1$ constitutes a δ_1-UHF with $\delta_1 = 2^{-t}$. Also, $\mathcal{F}_2$ is a strong UHF, whereby it is a strong δ_2-UHF with $\delta_2 = 2^{-t}$. Then, by Lemma 4.15, $\mathcal{F}$ constitutes a strong δ-UHF with $\delta = \delta_1 + \delta_2 = 2^{-(t-1)}$.

Furthermore, the mappings in $\mathcal{F}$ have input alphabet $\mathcal{X}$ and output alphabet $\mathcal{Y}$ with cardinalities satisfying

$$\begin{aligned}
\log|\mathcal{X}| &= \big[(s+t)\cdot(2^s+1)\big], \\
\log|\mathcal{Y}| &= t, \\
\log|\mathcal{F}| &= \log|\mathcal{F}_1| + \log|\mathcal{F}_2| \\
&= \big[(s+t) + ((s+t)+t)\big] \\
&= (2s+3t),
\end{aligned}$$

which yields the claimed compression ratio and randomness cost. □

Table 4.1 Summary of parameters in Constructions 4.1–4.5

	Type	Compression ratio	Randomness cost	δ
Construction 4.1	UHF	2	1	$\frac{1}{q}$
Construction 4.2	UHF	$\frac{s}{t}$	$\frac{s}{t}$	$\frac{1}{q^t}$
Construction 4.3	strong UHF	$\frac{s}{t}$	$\frac{s}{t}+1$	$\frac{1}{q^t}$
Construction 4.4	δ-UHF	s	1	$\frac{(r-1)}{q}$
Construction 4.5	strong δ-UHF	$\frac{(s+t)(2^s+1)}{t}$	$\frac{2s}{t}+3$	$\frac{1}{2^{(t-1)}}$

We close this section by providing a summary of the parameters of Constructions 4.1–4.5 in Table 4.1. In the table, q is the field size usually taken as 2^k for some integer k. In Constructions 4.2 and 4.3, s and t are construction parameters satisfying $s \geq t$. In these constructions, the compression ratio is (almost) the same as the randomness cost. On the other hand, in Construction 4.4, by sacrificing the collision probability δ by the factor of $(r-1)$, the compression ratio can be larger than the randomness cost by the factor of r. In Construction 4.5, s and t are construction parameters taking arbitrary integers. For fixed t, the compression ratio can be exponentially larger than the randomness cost. Alternatively, this means that the size of hash family (in bits) can be as small as the logarithmic order of the size of the input (in bits), i.e.,

$$\log |\mathcal{F}| = \mathcal{O}(\log \log |\mathcal{X}|).$$

This fact is very important for constructing the message authentication code with optimal key length in Chapter 8.

4.6 Practical Constructions

Our presentation in this chapter was driven towards our final construction of a strong δ-UHF in Construction 4.5 which roughly has randomness cost logarithmic in compression ratio. But in practice this is not the only feature that renders a scheme implementable on hardware. A specific shortcoming of this construction arises from the computational cost of implementing finite field arithmetic, which is needed by Construction 4.3. In practice, this difficulty is often overcome by replacing the outer layer of Construction 4.5 with a computationally secure construction such as a pseudorandom function; we will describe this construction in a later chapter.

Unlike Construction 4.3, the polynomial based Construction 4.4 is closer to practice. This construction is often supplemented with techniques for "randomness re-use" to reduce the randomness cost. But the real art is in identifying an appropriate finite field and replacing polynomial evaluations in Construction 4.4 with more efficient counterparts. We provide a glimpse of one such clever construction by outlining the `Poly1305` hash. Also, as an illustration of an attempt in a different direction, we outline the construction of NH (nonlinear hash) which is used in UMAC and its variants. This construction is closer to Construction 4.3 and uses a binary inner product implementation of a finite field.

After describing these two specific constructions, we describe two popular methods for reducing randomness cost.

4.6.1 The Poly1305 Hash

At a high level, Poly1305 can be viewed as an attempt towards efficient implementation of a hardware-compatible variant of Construction 4.4. In coming up with such a practical scheme, one has to keep in mind the precision of hardware. Another objective is to accommodate variable input message lengths. In Poly1305, the message is divided into ℓ blocks of 16 byte or 128 bit each and the output is 16 byte or 128 bit.[5] This corresponds to Construction 4.4 with $q = 2^{128}$ and $s = \ell$. A naive implementation would have used all possible messages and 16-byte keys (for u). Poly1305 is optimized for double-precision floating point arithmetic of 64-bit precision and allows only a subset of keys u that result in sparse forms of numbers involved, resulting in fast division.

The main idea enabling Poly1305 is to operate in a $\mathbb{F}_p$ with p that is a prime and slightly larger than 2^{128}. The operations required for Construction 4.4 are accomplished by periodically reducing results modulo p, which leads to sparse forms owing to the gap between p and 2^{128}, and the final modulo 2^{128} answer is reported. A new proof is needed to justify the security of this modification; we omit this proof and simply outline the construction.

The specific prime p used is $p = 2^{130} - 5$. Set $q = 2^{128}$, and consider the input $x_1, \ldots, x_s$ with each x_i in $\mathbb{F}_q$. For each x_i, obtain the integer $c_i \in \{1, \ldots, 2^{129}\}$ as follows:[6]

$$c_i = \sum_{j=0}^{15} x_{i,j+1} \cdot 2^{8j} + 2^{128}, \quad 1 \leq i \leq s,$$

where $x_{i,j}$, $j = 1, \ldots, 16$, denotes the jth bit of x_i. Note that all the operations are over integers. The mapping f_u corresponding to a 16-bit u is given by

$$f_u(x_1, \ldots, x_s) = \left(\sum_{i=1}^{s} c_i \cdot u^i \mod (2^{130} - 5) \right) \mod 2^{128}.$$

As mentioned earlier, only a subset $\mathcal{U}$ of $\{0,1\}^{128}$ is allowed for u. In particular, treating $u = (u_1, \ldots, u_{16})$ as a 16-byte vector corresponding to the integer

[5] The actual construction divides messages in 1-byte blocks. The simplified form here suffices for our customary treatment.

[6] For simplicity of explanation, we assumed that the length of the message is a multiple of 16 bits; in such a case, c_i does not take values less than 128. When the length of the message is not a multiple of 16 bits, the last integer c_s is computed by a different formula and may take values less than 128.

$\sum_{i=0}^{15} u_{i+1}2^{8i}$ with each byte representing an integer in the little-endian form, the first 4 bits of byte number $3, 7, 11$, and 15 and the last 2 bits of byte number $4, 8$, and 12 are set to 0. This freezes 22 out of 128 bits and leads to $\mathcal{U}$ with $|\mathcal{U}| = 2^{106}$.

A careful analysis shows that the resulting hash family constitutes a $(8s/2^{106})$-UHF. The security proof is a bit tedious, as are the details of the scheme outlined above, but we included the complete description here to point out the similarity with the basic Construction 4.4 and highlight the clever innovations needed to tailor the basic construct for practical deployment – `Poly1305` underlies the transport layer security protocols TLS and is often used in *https* protocol.

4.6.2 The Nonlinear Hash (NH)

`NH` is a building block for UMAC and its variants. In contrast to the constructions we have seen above, `NH` consists of nonlinear functions. It considers blocks of input messages with up to n blocks, where n is an even number, with each block consisting of ω bits. Note that the length of the input is variable and can be any even multiple of ω up to the factor n. This input is mapped to a fixed, 2ω-bit output. Namely, the domain $\mathcal{X}$ and the range $\mathcal{Y}$ of the mappings are set to be

$$\mathcal{X} = \{0,1\}^{2\omega} \cup \{0,1\}^{4\omega} \cup \cdots \cup \{0,1\}^{n\omega}, \quad \mathcal{Y} = \{0,1\}^{2\omega}.$$

For an input $x = (x_1, \ldots, x_\ell) \in \{0,1\}^{\ell\omega}$ with even integer $\ell \leq n$ and for random key $K = (K_1, \ldots, K_n) \in \{0,1\}^{n\omega}$, the hash value is computed by

$$\mathsf{NH}_K(x) = \sum_{i=1}^{\ell/2} (K_{2i-1} +_\omega x_{2i-1}) \cdot (K_{2i} +_\omega x_{2i}) \pmod{2^{2\omega}}, \tag{4.6}$$

where $+_\omega$ means the arithmetic addition modulo 2^ω. Note that the operations used in describing `NH` can be carried out in a 2ω-bit precision arithmetic, making it suitable for hardware implementation. In fact, it can be shown that the `NH` constitutes a $2^{-\omega}$-UHF. We remark that `NH` essentially computes the inner product $x \cdot K$, but uses a trick to replace half the multiplications with additions.

When the collision probability $2^{-\omega}$ is not acceptable and the word size ω is restricted by architectural characteristics of implementation, we can reduce the collision probability by concatenation. This requires a Toeplitz matrix based construction, which will be described in the next section.

4.6.3 Reducing Randomness Cost

As we have seen throughout this chapter, randomness is an indispensable component of the UHF; it is used to sample a function F from the UHF and to compute $F(x)$. In various situations of practical interest, we would like to construct a new UHF from a given UHF in order to reduce the collision probability or to expand the domain size. However, naive constructions will increase the size of randomness needed to sample functions from the constructed UHF. Since randomness is a precious resource, some methods for reducing randomness cost have been proposed. At a high level, randomness cost is reduced by reusing random keys in a judicious manner. In the following, we describe methods of reusing keys for two purposes: to reduce the collision probability and to expand the domain size.

Toeplitz construction

Consider a UHF $\mathcal{F} = \{f_k : k \in \mathcal{K}\}$ from $\mathcal{X}$ to $\mathcal{Y}$ with uniform randomness K on $\mathcal{K}$. By using t independent copies $K_1, \ldots, K_t$ of K, we can construct a concatenated family $(f_{K_1}(x), \ldots, f_{K_t}(x))$, and it is not difficult to see that this family constitutes a UHF from $\mathcal{X}$ to $\mathcal{Y}^t$. By concatenation, the collision probability is reduced from $1/|\mathcal{Y}|$ to $1/|\mathcal{Y}|^t$. However, this implementation requires $t \log |\mathcal{K}|$ bits of randomness. Toeplitz construction, in essence, achieves the same performance but without using independent keys and at much lower randomness cost.

Specifically, for $\mathcal{X} = \mathcal{K} = \mathbb{F}_2^n$, consider the UHF from $\mathbb{F}_2^n$ to $\mathbb{F}_2$ consisting of mappings

$$f_k(x) = \sum_{i=1}^{n} k_i x_i, \quad k \in \mathbb{F}_2^n.$$

The simple concatenation described above can be represented by the linear mappings

$$\begin{bmatrix} K_{1,1} & \cdots & K_{1,n} \\ K_{2,1} & \cdots & K_{2,n} \\ \vdots & \ddots & \vdots \\ K_{t,1} & \cdots & K_{t,n} \end{bmatrix} \begin{bmatrix} x_1 \\ \vdots \\ x_n \end{bmatrix},$$

where the entries $K_{i,j}$ of the $t \times n$ matrix are independent random bits. Thus, this construction requires tn bits of randomness. Instead, we can use a random Toeplitz matrix which will require only $(n+t-1)$ bits of randomness. Specifically, consider a UHF from $\mathbb{F}_2^n$ to $\mathbb{F}_2^t$ consisting of mappings

$$\begin{bmatrix} K_1 & \cdots & K_n \\ K_2 & \cdots & K_{n+1} \\ \vdots & \ddots & \vdots \\ K_t & \cdots & K_{n+t-1} \end{bmatrix} \begin{bmatrix} x_1 \\ \vdots \\ x_n \end{bmatrix}, \tag{4.7}$$

where the entries $K_1, \ldots, K_{n+t-1}$ of the matrix are from $\mathbb{F}_2$.

PROPOSITION 4.17 *The hash family given by (4.7) for $(K_1, \ldots, K_{n+t-1}) \in \mathbb{F}_2^{n+t-1}$ constitutes a UHF from $\mathbb{F}_2^n$ to $\mathbb{F}_2^t$.*

Proof Problem 4.9. □

While we have described this construction for a specific UHF, it can be applied to a broader class of UHFs. In particular, we can apply the same idea to `NH`, which is not even a linear UHF. In order to reduce the collision probability of basic NH from $2^{-\omega}$ to $2^{-\omega t}$, we concatenate t copies of mappings in (4.6) using Toeplitz construction to obtain the Toeplitz-NH family consisting of mappings

$$\mathsf{TNH}_K(x) = (\mathsf{NH}_{K_{1:n}}(x), \mathsf{NH}_{K_{3:n+2}}(x), \ldots, \mathsf{NH}_{K_{2(t-1):n+2(t-1)}}(x)), \tag{4.8}$$

where $K_{i:j} = (K_i, \ldots, K_j)$. Even though the NH is a nonlinear hash family and the proof argument of Proposition 4.17 does not apply directly, it can be proved that the Toeplitz-NH in (4.8) constitutes a $2^{-\omega t}$-UHF.

Merkle–Damgård construction

In the previous construction, we kept the same input x and, in effect, obtained independent copies of the hash without using independent copies. The next construction we describe works on a slightly different principle and applies roughly the same key to different messages. This allows us to obtain better compression ratio at the same randomness cost. We illustrate the main idea using Construction 4.1, but it extends to many more cases. In fact, it is most relevant in the computational security setting where it is applied to so-called *collision-resistant* hash, which is roughly a function for which any efficient algorithm can detect a collision with only a small probability. In this computational form, the Merkle–Damgård construction underlies the popular MD5 hash and also forms the basis for HMAC, a popular MAC. However, here we consider only the information-theoretic setting and work only with Construction 4.1.

Specifically, consider a mapping $f : \{0,1\}^{2k} \to \{0,1\}^k$. Its ($t$-fold) Merkle–Damgård (MD) transform is the mapping $f_{\mathsf{MD}} : \{0,1\}^{kt} \to \{0,1\}^k$ defined iteratively as follows. For $(x_1, \ldots, x_t)$ with each $x_i \in \{0,1\}^k$ and a fixed $x_0 \in \{0,1\}^k$, let

$$h_1 = f(x_1, x_0),$$
$$h_i = f(x_i, h_{i-1}), \quad 2 \le i \le t.$$

The mapping f_{MD} is given by $f_{\mathsf{MD}}(x_1, \ldots, x_t) := h_t$.

Now, consider the UHF $\mathcal{F}$ of Construction 4.1 from $\{0,1\}^{2k} \to \{0,1\}^k$ which is a UHF with compression ratio 2. Let $\mathcal{F}_{\mathsf{MD}}$ denote the family of mappings obtained by applying the MD transform to each mapping in $\mathcal{F}$.

PROPOSITION 4.18 *The family $\mathcal{F}_{\mathsf{MD}}$ described above constitutes a $(t/2^k)$-UHF from $\{0,1\}^{kt} \to \{0,1\}^k$.*

Proof Problem 4.10. □

The proof follows a case-by-case analysis and shows that for collision to occur in the iterated functions obtained by MD transform, a collision must occur for f. This idea extends readily to the computational setting as well and underlies formal security analysis of the MD5 hash. We close with the remark that even the popular SHA hash family uses a similar iterated function construction.

4.7 References and Additional Reading

A standard textbook covering the algebra reviewed in Section 4.1 is [218]. The concept of UHF was introduced by Carter and Wegman in their seminal paper [55]. The strong UHF was introduced in [347], and the almost (strong) UHF was introduced in [316] based on informal discussion in [347]. The construction of almost strong UHF given in this chapter is from [19], which is based on the composition construction shown in [316]; an idea of composition construction was first posed in [347]. A comprehensive review of various versions of UHFs can be found in [317].

`Poly1305` is a popular MAC included in several standards such as TLS. Our presentation is based on the original paper [30]. `NH` is a hash method used in UMAC [32]. The construction of a UHF from a Toeplitz matrix was introduced in [232]; see also [160] for the construction of a UHF from a modified Toeplitz matrix. For further discussion on implementation of universal hashing in the context of randomness extraction, see [163] and references therein.

Problems

4.1 Prove that the set of all functions from $\mathcal{X}$ to $\mathcal{Y}$ is a UHF.

4.2 For the UHF F in Problem 4.1 and distinct $x_1, \ldots, x_q \in \mathcal{X}$, prove

$$\Pr\left(\exists i \neq j \text{ such that} F(x_i) = F(x_j)\right) \geq 1 - e^{-\frac{q(q-1)}{2|\mathcal{Y}|}}.$$

REMARK This bound is known as the *birthday paradox*: in a class with 50 students, there exists at least one pair of students having the same birthday with probability more than 0.965. In the context of hash functions, it says that a $\mathcal{O}(\sqrt{|\mathcal{Y}|})$ query suffices to find a collision pair with high probability.

4.3 Prove that a UHF $\mathcal{F}$ from $\mathcal{X}$ to $\mathcal{Y}$ exists only if $|\mathcal{F}| \geq |\mathcal{X}|/|\mathcal{Y}|$.

HINT For a fixed $f \in \mathcal{F}$ and $y \in \mathcal{Y}$, consider the set $C = \{x \in \mathcal{X} : f(x) = y\}$ such that $|C| \geq |\mathcal{X}|/|\mathcal{Y}|$. Next, consider the random variable taking values $|\{x \in C : g(x) = y'\}|$ with equal probability for all $y' \in \mathcal{Y}$ and $g \in \mathcal{F} \setminus \{f\}$. Using the UHF property, we can derive an upper bound for the second moment of this random variable. The claim follows from the fact that the variance of this random variable is nonnegative. See also [314, Theorem 4.1].

4.4 Prove that a strong UHF $\mathcal{F}$ from $\mathcal{X}$ to $\mathcal{Y}$ exists only if $|\mathcal{F}| \geq |\mathcal{X}|(|\mathcal{Y}|-1)+1$.

HINT First show that we can find a strong UHF $\mathcal{F}'$ from $\mathcal{X}$ to $\mathcal{Y}$ of the same size as $\mathcal{F}$ such that there exist $f_0 \in \mathcal{F}'$ and $y \in \mathcal{Y}$ satisfying $f_0(x) = y$ for all $x \in \mathcal{X}$. Then, consider the random variable taking values $|\{x \in \mathcal{X} : f(x) = y\}|$ with equal probability for all $f \in \mathcal{F} \setminus \{f_0\}$. The claim follows by bounding the second moment of this random variable using the strong UHF properties and using the nonnegativity of its variance. See also [316, Theorem 4.3].

4.5 Prove that the compression scheme described at the end of Section 4.4.1 has decoding error probability less than $2^{-(k-\ell)}$.

4.6 Show that Construction 4.4 does not satisfy (4.3).

4.7 For a random variable, the quantity $H_2(Y)$ defined by

$$H_2(Y) := -\log \sum_y P_Y(y)^2$$

is called the *collision entropy* or *Rényi entropy of order* 2. Prove that $H_2(Y) \leq H(Y)$. Conclude that for $\{0,1\}^k$-valued Y, the collision probability $p_{\mathrm{col}}(Y) \geq 2^{-k}$ with equality if and only if Y is uniform on $\{0,1\}^k$.

4.8 Prove that a δ-UHF exists only if $\delta \geq (|\mathcal{X}| - |\mathcal{Y}|)/(|\mathcal{Y}|(|\mathcal{X}| - 1))$. Furthermore, provide an example of a δ-UHF attaining this bound.

REMARK The δ-UHFs attaining this bound are termed the *optimally universal hash family* and are related to balanced incomplete block designs [298, 314].

4.9 Prove Proposition 4.17.

HINT Since the function is linear, it suffices to count the number of possible values of the key such that a given nonzero vector $(x_1, \ldots, x_n)$ is mapped to the zero vector. The key $K = (K_1, \ldots, K_{n+t-1})$ can be regarded as a vector in $\mathbb{F}_2^{n+t-1}$. Then, (4.7) gives t linearly independent constraints on K.

4.10 Prove Proposition 4.18.

HINT It can be seen by making cases that a collision in f_{MD} can only occur if there is a collision in f for some index $i \in [t]$ in its iterated application. Similar analysis as the proof of Lemma 4.9 yields that collision can only happen at a given index for at most one f.

4.11 A class of functions from $\mathcal{X}$ to $\mathcal{Y}$ is called a k-wise strong universal hash family (k-wise strong UHF) if, for any distinct $x_1, \ldots, x_k \in \mathcal{X}$ and k (not necessarily distinct) elements $y_1, \ldots, y_k$,

$$\Pr\left(F(x_i) = y_i : 1 \leq i \leq k\right) = \frac{1}{|\mathcal{Y}|^k}$$

holds. When $\mathcal{X} = \mathcal{Y} = \mathbb{F}_q$, let $f_{a_1,\ldots,a_k}(x) = \sum_{i=1}^k a_i x^{i-1}$ for $a_1, \ldots, a_k \in \mathbb{F}_q$, and let $\mathcal{F} = \{f_{a_1,\ldots,a_k} : a_1, \ldots, a_k\}$. Then, prove that the class $\mathcal{F}$ of functions is a k-wise strong UHF.

REMARK The special case with $k = 2$ is an example of strong UHF described in Construction 4.3 with $s = t = 1$. For more general construction of k-wise strong UHFs, see [9, 31].

4.12 Suppose that the set $\mathcal{Y}$ is an additive group. A class of functions from $\mathcal{X}$ to $\mathcal{Y}$ is called an *XOR δ-UHF* if, for any pair of distinct elements x, x' in $\mathcal{X}$, for any element y in $\mathcal{Y}$, and for F generated uniformly from $\mathcal{F}$,

$$\Pr(F(x) - F(x') = y) \leq \delta.$$

1. Prove that $\mathcal{F}$ being a strong δ-UHF implies that $\mathcal{F}$ is an XOR δ-UHF.
2. Prove that $\mathcal{F}$ being an XOR δ-UHF implies that $\mathcal{F}$ is a δ-UHF.

See [317, Theorem 2.1].

5 Hypothesis Testing

In this chapter, we explain the hypothesis testing problem, one of the most important problems in statistics. There are two motivations for considering hypothesis testing in this book. First, when we consider the problem of detecting the presence of an adversary, for instance in Chapter 8, the problem is inherently that of hypothesis testing: the absence of the adversary corresponds to the null hypothesis and the presence of the adversary corresponds to the alternative hypothesis. Furthermore, our notions of security seen in Chapter 3 rely on the ability of an eavesdropper to distinguish between two messages, which is a hypothesis testing problem. Even the more abstract notions of security to be seen in later chapters on secure computation rely on appropriate hypothesis testing problems.

Second, we will be interested in this book in problems where one "resource" is used to enable other cryptographic primitives. For example, in Chapter 3 we saw how shared keys can enable secure communication. Similar applications will be discussed in Chapters 10, 13, and 14, where correlated random variables will be viewed as a resource. In these problems, the performance analysis of the original cryptographic scheme can be related to the hypothesis testing problem of determining whether the resource is "useful" or "useless."

In short, hypothesis testing is inherently linked to formal security analysis and is a basic building block for topics covered in this book.

5.1 Binary Hypothesis Testing: The Neyman–Pearson Formulation

We already saw binary hypothesis testing in Chapter 2 when we defined total variation distance. We review this definition first.

Let P and Q be two distributions on a finite set $\mathcal{X}$. In binary hypothesis testing, we want to test whether an observation X is generated from the *null hypothesis* P or from the *alternative hypothesis* Q. For instance, the null hypothesis may describe the behavior of a system in a regular state and the alternative hypothesis may describe the behavior of the same system when it is attacked by an adversary.

A test is described by a channel (stochastic mapping) $\mathrm{T}\colon \mathcal{X} \to \{0,1\}$, where 0 and 1 correspond to the null hypothesis and the alternative hypothesis, respectively. More specifically, $\mathrm{T}(0|x)$ and $\mathrm{T}(1|x) = 1 - \mathrm{T}(0|x)$, respectively, describe the probabilities that the null hypothesis is accepted and rejected when the observation is $X = x$.

A *deterministic test* can be described by an *acceptance region* $\mathcal{A} \subseteq \mathcal{X}$ denoting the set of observation values for which the null hypothesis is accepted. Alternatively, a deterministic set with acceptance region $\mathcal{A}$ can be expressed using the general notation above as

$$\mathrm{T}(0|x) = \mathbf{1}[x \in \mathcal{A}] .$$

In Lemma 2.10 in Chapter 2, we saw the test with acceptance region $\mathcal{A} = \{x : \mathrm{P}(x) \geq \mathrm{Q}(x)\}$. In fact, this specific test is a special case of a larger family of tests called *likelihood ratio tests*, presented as an example below.

EXAMPLE 5.1 (Neyman–Pearson test) For parameters $\gamma \geq 0$ and $0 \leq \theta \leq 1$, the Neyman–Pearson test, also known as the likelihood ratio test, is given by[1]

$$\mathrm{T}_{\mathrm{NP}}^{\gamma,\theta}(0|x) := \begin{cases} 1 & \text{if } \frac{\mathrm{P}(x)}{\mathrm{Q}(x)} > \gamma, \\ \theta & \text{if } \frac{\mathrm{P}(x)}{\mathrm{Q}(x)} = \gamma, \\ 0 & \text{if } \frac{\mathrm{P}(x)}{\mathrm{Q}(x)} < \gamma. \end{cases}$$

The test in Lemma 2.10 corresponds to $\theta = 0$ and $\gamma = 1$. In fact, when $\theta = 0$ or $\theta = 1$, then it is a deterministic test; when $\theta = 0$, we simply write $\mathrm{T}_{\mathrm{NP}}^{\gamma}$ instead of $\mathrm{T}_{\mathrm{NP}}^{\gamma,0}$. The parameters γ and θ control the tradeoff between two types of error probabilities to be discussed next.

For a given test T, we need to consider two types of errors: the first type of error is that the null hypothesis is rejected even though the observation is generated from P; the second type of error is that the null hypothesis is accepted even though the observation is generated from Q. The former is also termed a false alarm, and the latter is also termed missed detection. The probability of the first type of error, termed the *type I error probability*, is given by $1 - \mathrm{P}[\mathrm{T}]$, where

$$\mathrm{P}[\mathrm{T}] := \sum_{x \in \mathcal{X}} \mathrm{P}(x)\, \mathrm{T}(0|x);$$

and the probability of the second type of error, termed the *type II error probability*, is given by

$$\mathrm{Q}[\mathrm{T}] := \sum_{x \in \mathcal{X}} \mathrm{Q}(x)\mathrm{T}(0|x).$$

We evaluate the performance of a test T using $(1 - \mathrm{P}[\mathrm{T}], \mathrm{Q}[\mathrm{T}])$, a pair of metrics instead of a single metric. This makes the search for an "optimal" test a bit complicated to articulate – a test which fares well in one metric may perform badly in the second. For example, the test which always declares 0 as the output has type I error probability 0, but can have type II error probability as large as 1. Thus, instead of thinking of a single optimal test, we seek the class of tests which

[1] When $\mathrm{P}(x) > 0$ and $\mathrm{Q}(x) = 0$, the likelihood ratio is regarded as $+\infty$.

cannot be simultaneously improved upon with respect to both error types by any other tests. Such tests are termed *admissible tests* in statistics. In practice, we still desire a notion of optimality that justifies a specific choice of test.

There are two ways to proceed. In the first formulation, we can weigh the two errors using $\theta \in [0,1]$ and seek tests T that minimize the weighted average error $\theta(1-\mathrm{P}[\mathrm{T}]) + (1-\theta)\mathrm{Q}[\mathrm{T}]$. The resulting cost function can be viewed as the average probability of error when the unknown hypothesis itself was generated randomly, with P generated with probability θ and Q with probability $(1-\theta)$. This formulation is called the *Bayesian formulation* for hypothesis testing and the corresponding test is termed the *Bayes optimal for prior* θ. In Chapter 2, we already saw Bayesian formulation with uniform prior, namely with $\theta = 1/2$. In particular, we found that the test $T(0|x) = \mathbf{1}[(\mathrm{P}(x)/\mathrm{Q}(x)) \geq 1]$ is Bayes optimal for uniform prior.

In the second formulation, which will be our focus in this chapter, we can minimize the type II error probability under the condition that the type I error probability is smaller than a prescribed level. Specifically, for a given $0 \leq \varepsilon < 1$, define

$$\beta_\varepsilon(\mathrm{P},\mathrm{Q}) := \inf_{\substack{\mathrm{T}: \\ \mathrm{P}[\mathrm{T}] \geq 1-\varepsilon}} \mathrm{Q}[\mathrm{T}], \tag{5.1}$$

where the infimum is taken over all tests satisfying $\mathrm{P}[\mathrm{T}] \geq 1-\varepsilon$. The infimum on the right-hand side of (5.1) can be attained, and we can replace the infimum with a minimum (Problem 5.2).

In the next section, we will characterize the optimal tests that attain $\beta_\varepsilon(\mathrm{P},\mathrm{Q})$ for different values of ε. In fact, a result that is beyond the scope of this book states that the class of optimal tests for each of the two formulations coincides.

5.2 The Neyman–Pearson Lemma

Note that the optimal type II error probability $\beta_\varepsilon(\mathrm{P},\mathrm{Q})$ for given $0 \leq \varepsilon < 1$ is a linear program, which suggests a numerical recipe for evaluating it. However, a much more useful insight emerges from analysis of this expression – the minimum is attained by the Neyman–Pearson tests of Example 5.1.

LEMMA 5.2 *For any test* T *and every* $\gamma > 0$, *we have*

$$\mathrm{P}[\mathrm{T}] - \gamma\mathrm{Q}[\mathrm{T}] \leq \mathrm{P}[T_{\mathrm{NP}}^{\gamma}] - \gamma\mathrm{Q}[T_{\mathrm{NP}}^{\gamma}]. \tag{5.2}$$

In particular, if $\mathrm{P}[T_{\mathrm{NP}}^{\gamma}] \leq \mathrm{P}[\mathrm{T}]$, *then* $\mathrm{Q}[T_{\mathrm{NP}}^{\gamma}] \leq \mathrm{Q}[\mathrm{T}]$.

Proof By noting that $\mathrm{T}_{\mathrm{NP}}^{\gamma}(0|x) = \mathbf{1}[\mathrm{P}(x) - \gamma\mathrm{Q}(x) > 0]$, we have

$$\begin{aligned}
\mathrm{P}[\mathrm{T}] - \gamma\mathrm{Q}[\mathrm{T}] &= \sum_{x\in\mathcal{X}} (\mathrm{P}(x) - \gamma\mathrm{Q}(x))\mathrm{T}(0|x) \\
&\leq \sum_{x\in\mathcal{X}} (\mathrm{P}(x) - \gamma\mathrm{Q}(x))\mathrm{T}(0|x)\mathbf{1}[\mathrm{P}(x) - \gamma\mathrm{Q}(x) > 0]
\end{aligned}$$

$$\leq \sum_{x \in \mathcal{X}} (\mathrm{P}(x) - \gamma \mathrm{Q}(x)) \mathbf{1}[\mathrm{P}(x) - \gamma \mathrm{Q}(x) > 0]$$
$$= \sum_{x \in \mathcal{X}} (\mathrm{P}(x) - \gamma \mathrm{Q}(x)) \mathrm{T}_{\mathtt{NP}}^{\gamma}(0|x)$$
$$= \mathrm{P}[\mathrm{T}_{\mathtt{NP}}^{\gamma}] - \gamma \mathrm{Q}[\mathrm{T}_{\mathtt{NP}}^{\gamma}],$$

which proves (5.2). The second clam is an immediate consequence of (5.2). □

Lemma 5.2 states that the deterministic Neyman–Pearson test with type I error probability exceeding that of T must have type II error probability less than that of T. This is a variant of the *Neyman–Pearson lemma* in statistics. In fact, the result extends to randomized tests as well. When randomization is allowed, we can choose the parameters γ and θ to get $\mathrm{P}[\mathrm{T}_{\mathtt{NP}}^{\gamma,\theta}] = 1-\varepsilon$, which along with the previous lemma extended to randomized tests shows that $\beta_\varepsilon(\mathrm{P}, \mathrm{Q})$ can be attained by the randomized Neyman–Pearson test, for every $0 \leq \varepsilon < 1$. This treatment of randomized Neyman pearson tests is not needed for our purposes in this book, and we skip it.

For our application in this book, we can view $\beta_\varepsilon(\mathrm{P}, \mathrm{Q})$ as another measure of "distance" between P and Q, much like $d_{\mathtt{var}}(\mathrm{P}, \mathrm{Q})$ and $D(\mathrm{P}\|\mathrm{Q})$. The quantity $\beta_\varepsilon(\mathrm{P}, \mathrm{Q})$ captures the fundamental tradeoff between the two types of error. For a given ε, $\beta_\varepsilon(\mathrm{P}, \mathrm{Q})$ cannot be arbitrarily small. We illustrate this with an example first.

EXAMPLE 5.3 Let us consider distributions P_{XY} and Q_{XY} on $\{0,1\}^n \times \{0,1\}^n$ given by

$$\mathrm{P}_{XY}(x, y) = \frac{1}{2^n} \mathbf{1}[x = y],$$
$$\mathrm{Q}_{XY}(x, y) = \frac{1}{2^{2n}}.$$

In words, the observations (X,Y) are perfectly correlated under the null hypothesis P_{XY}, and (X,Y) are independent under the alternative hypothesis Q_{XY}. Let us consider the test given by $\mathrm{T}(0|x, y) = \mathbf{1}[x = y]$. Then, it is not difficult to see that the type I error probability is 0 and the type II error probability is $1/2^n$. In fact, this test can be described as a special case of the Neyman–Pearson test in Example 5.1 as well. Consequently, we have $\beta_0(\mathrm{P}_{XY}, \mathrm{Q}_{XY}) \leq 1/2^n$.

On the other hand, for any test T such that

$$\mathrm{P}_{XY}[\mathrm{T}] = \sum_{x \in \{0,1\}^n} \frac{1}{2^n} \mathrm{T}(0|x, x) \geq 1 - \varepsilon,$$

we have

$$\mathrm{Q}_{XY}[\mathrm{T}] = \sum_{x,y \in \{0,1\}^n} \frac{1}{2^{2n}} \mathrm{T}(0|x, y) \geq \sum_{x \in \{0,1\}} \frac{1}{2^{2n}} \mathrm{T}(0|x, x) \geq \frac{1-\varepsilon}{2^n}.$$

Thus, we have $\beta_\varepsilon(\mathrm{P}_{XY}, \mathrm{Q}_{XY}) \geq (1-\varepsilon)/2^n$ for any $0 \leq \varepsilon < 1$. This means that, even if we allow positive type I error probability, the improvement in the type II error probability is a factor of $1 - \varepsilon$.

In general, upper and lower bounds for $\beta_\varepsilon(\mathrm{P}, \mathrm{Q})$ can be established using Lemma 5.2 for any pair of distributions P and Q . In essence, the result below says that deterministic Neyman–Pearson tests are almost optimal.

THEOREM 5.4 *For any $0 \le \varepsilon < 1$ and $\gamma > 0$, we have*

$$\beta_\varepsilon(\mathrm{P}, \mathrm{Q}) \ge \frac{1}{\gamma}(1 - \mathrm{P}[T_{\mathrm{NP}}^\gamma] - \varepsilon). \tag{5.3}$$

On the other hand, for a $\gamma > 0$ satisfying $\mathrm{P}[T_{\mathrm{NP}}^\gamma] \ge 1 - \varepsilon$, we have

$$\beta_\varepsilon(\mathrm{P}, \mathrm{Q}) \le \frac{1}{\gamma}. \tag{5.4}$$

Proof For any test T satisfying $\mathrm{P}[\mathrm{T}] \ge 1 - \varepsilon$, Lemma 5.2 implies

$$\begin{aligned}
\mathrm{Q}[\mathrm{T}] &\ge \mathrm{Q}[\mathrm{T}] - \mathrm{Q}[\mathrm{T}_{\mathrm{NP}}^\gamma] \\
&\ge \frac{1}{\gamma}(\mathrm{P}[\mathrm{T}] - \mathrm{P}[\mathrm{T}_{\mathrm{NP}}^\gamma]) \\
&\ge \frac{1}{\gamma}(1 - \mathrm{P}[\mathrm{T}_{\mathrm{NP}}^\gamma] - \varepsilon),
\end{aligned}$$

which implies (5.3).

On the other hand, for any $\gamma > 0$, we have

$$\begin{aligned}
\mathrm{Q}[\mathrm{T}_{\mathrm{NP}}^\gamma] &= \sum_{x \in \mathcal{X}} \mathrm{Q}(x) \mathbf{1}\left[\frac{\mathrm{P}(x)}{\mathrm{Q}(x)} > \gamma\right] \\
&\le \sum_{x \in \mathcal{X}} \frac{1}{\gamma} \mathrm{P}(x)\, \mathbf{1}\left[\frac{\mathrm{P}(x)}{\mathrm{Q}(x)} > \gamma\right] \\
&\le \frac{1}{\gamma}.
\end{aligned}$$

Thus, if $\gamma > 0$ satisfying $\mathrm{P}[\mathrm{T}_{\mathrm{NP}}^\gamma] \ge 1 - \varepsilon$, we get (5.4). □

5.3 Divergence and Hypothesis Testing

The divergence introduced in Section 2.5 of Chapter 2 plays an important role in hypothesis testing. In the result below, we show that the optimal type II error probability $\beta_\varepsilon(\mathrm{P}, \mathrm{Q})$ can be bounded in terms of the divergence $D(\mathrm{P}\|\mathrm{Q})$.

LEMMA 5.5 *For a given $0 \le \varepsilon < 1$, we have*

$$-\log \beta_\varepsilon(\mathrm{P}, \mathrm{Q}) \le \frac{D(\mathrm{P}\|\mathrm{Q}) + h(\min\{\varepsilon, 1/2\})}{1 - \varepsilon}, \tag{5.5}$$

where $h(\cdot)$ is the binary entropy function.

Proof For an arbitrary test T satisfying $\mathrm{P}[\mathrm{T}] \ge 1 - \varepsilon$, let Z be a binary random variable denoting the output of the test T. Under the null hypothesis, the distribution of Z is given by

$$\mathrm{P} \circ \mathrm{T}(z) := \sum_{x \in \mathcal{X}} \mathrm{P}(x)\, \mathrm{T}(z|x), \quad z \in \{0, 1\},$$

and under the alternative hypothesis, it is given by

$$\mathrm{Q} \circ \mathrm{T}(z) := \sum_{x \in \mathcal{X}} \mathrm{Q}(x)\mathrm{T}(z|x), \quad z \in \{0,1\}.$$

Then, by the data processing inequality for divergence (cf. Theorem 2.17), we have

$$D(\mathrm{P} \circ \mathrm{T} \| \mathrm{Q} \circ \mathrm{T}) \leq D(\mathrm{P} \| \mathrm{Q}). \tag{5.6}$$

We can manipulate the left-hand side to obtain

$$\begin{aligned} D(\mathrm{P} \circ \mathrm{T} \| \mathrm{Q} \circ \mathrm{T}) &= \mathrm{P}[\mathrm{T}] \log \frac{\mathrm{P}[\mathrm{T}]}{\mathrm{Q}[\mathrm{T}]} + (1 - \mathrm{P}[\mathrm{T}]) \log \frac{(1 - \mathrm{P}[\mathrm{T}])}{(1 - \mathrm{Q}[\mathrm{T}])} \\ &\geq \mathrm{P}[\mathrm{T}] \log \frac{1}{\mathrm{Q}[\mathrm{T}]} - h(\mathrm{P}[\mathrm{T}]) \\ &\geq (1 - \varepsilon) \frac{1}{\mathrm{Q}[\mathrm{T}]} - h(\min\{\varepsilon, 1/2\}), \end{aligned}$$

which together with (5.6) implies

$$-\log \mathrm{Q}[\mathrm{T}] \leq \frac{D(\mathrm{P} \| \mathrm{Q}) + h(\min\{\varepsilon, 1/2\})}{1 - \varepsilon}.$$

Since this bound holds for any test T, we have (5.5). □

The key step in the proof above is the data processing inequality for divergence. Interestingly, the optimal type II error probability $\beta_\varepsilon(\mathrm{P}, \mathrm{Q})$ itself satisfies a data processing inequality of its own, which we discuss next.

Suppose that the original data X on a finite alphabet $\mathcal{X}$ is generated from either the null hypothesis P or the alternative hypothesis Q. Let W be a channel from $\mathcal{X}$ to another alphabet $\mathcal{Y}$. Then, the processed observation Y is distributed according to

$$\mathrm{P} \circ W(y) := \sum_{x \in \mathcal{X}} \mathrm{P}(x)\, W(y|x)$$

if the null hypothesis is true, and

$$\mathrm{Q} \circ W(y) := \sum_{x \in \mathcal{X}} \mathrm{Q}(x) W(y|x)$$

if the alternative hypothesis is true. But such a channel W can always be included as a preprocessing step in the design of our test T. Thus, we get the following result.

LEMMA 5.6 (Data processing inequality for $\beta_\varepsilon(\mathrm{P}, \mathrm{Q})$) *Let* P *and* Q *be two distributions on* $\mathcal{X}$. *For* $0 \leq \varepsilon < 1$ *and any channel* $W : \mathcal{X} \to \mathcal{Y}$, *we have*

$$\beta_\varepsilon(\mathrm{P}, \mathrm{Q}) \leq \beta_\varepsilon(\mathrm{P} \circ W, \mathrm{Q} \circ W). \tag{5.7}$$

Proof For an arbitrary test $\tilde{\mathrm{T}}: \mathcal{Y} \to \{0,1\}$, the composite channel $W \circ \tilde{\mathrm{T}}: \mathcal{X} \to \{0,1\}$ defined by

$$W \circ \tilde{\mathrm{T}}(z|x) := \sum_{y \in \mathcal{Y}} W(y|x)\tilde{\mathrm{T}}(z|y), \quad z \in \{0,1\}$$

can be regarded as a test for an observation on $\mathcal{X}$. Further, if the test $\tilde{\mathrm{T}}: \mathcal{Y} \to \{0,1\}$ satisfies $\mathrm{P} \circ W[\tilde{\mathrm{T}}] \geq 1-\varepsilon$, we get

$$\begin{aligned} \mathrm{P}[W \circ \tilde{\mathrm{T}}] &= \mathrm{P} \circ W[\tilde{\mathrm{T}}] \\ &\geq 1-\varepsilon \end{aligned}$$

and

$$\begin{aligned} \beta_\varepsilon(\mathrm{P},\mathrm{Q}) &\leq \mathrm{Q}[W \circ \tilde{\mathrm{T}}] \\ &= \mathrm{Q} \circ W[\tilde{\mathrm{T}}]. \end{aligned}$$

Since this bound holds for any test $\tilde{\mathrm{T}}$, we have (5.7). □

5.4 Stein's Lemma

In this section, we investigate the asymptotic behavior of the optimal type II error probability. Let P and Q be distributions on a finite alphabet $\mathcal{X}$, and let $X^n = (X_1, \ldots, X_n)$ be a sequence of i.i.d. random variables with common distribution P or Q. In other words, when the null hypothesis P is true, the observation X^n is distributed according to the product distribution P^n, and when the alternative hypothesis is true, X^n is distributed according to Q^n.

For i.i.d. observations, the optimal type II error probability converges to 0 exponentially rapidly in n, and we are interested in characterizing the optimal "exponent" K_ε given by

$$\beta_\varepsilon(\mathrm{P}^n,\mathrm{Q}^n) \simeq 2^{-nK_\varepsilon}.$$

In fact, this kind of problem appears repeatedly throughout the book, and the following argument includes some of the basic techniques used in such problems.

By using the law of large numbers introduced in Section 2.2, we can derive the following characterization of the optimal *exponent* of the type II error probability.

THEOREM 5.7 *Suppose that* $D(\mathrm{P}\|\mathrm{Q}) > 0$ *and* $D(\mathrm{P}\|\mathrm{Q})$ *takes a finite value. For every* $0 < \varepsilon < 1$, *we have*

$$\lim_{n\to\infty} -\frac{1}{n}\log \beta_\varepsilon(\mathrm{P}^n,\mathrm{Q}^n) = D(\mathrm{P}\|\mathrm{Q}). \tag{5.8}$$

Proof First, by applying (5.4) of Theorem 5.4 to P^n and Q^n, we have

$$\beta_\varepsilon(\mathrm{P}^n,\mathrm{Q}^n) \leq \frac{1}{\gamma} \tag{5.9}$$

when γ satisfies

$$\Pr\left(\log \frac{\mathrm{P}^n(X^n)}{\mathrm{Q}^n(X^n)} \le \log \gamma\right) \le \varepsilon, \tag{5.10}$$

where X^n on the left-hand side is distributed according to P^n. Note that

$$\frac{1}{n} \log \frac{\mathrm{P}^n(X^n)}{\mathrm{Q}^n(X^n)} = \frac{1}{n} \sum_{i=1}^{n} \log \frac{\mathrm{P}(X_i)}{\mathrm{Q}(X_i)} \tag{5.11}$$

and that

$$\mathbb{E}_{\mathrm{P}}\left[\log \frac{\mathrm{P}(X)}{\mathrm{Q}(X)}\right] = D(\mathrm{P}\|\mathrm{Q}). \tag{5.12}$$

Fixing $0 < \delta < D(\mathrm{P}\|\mathrm{Q})$, for $\log \gamma = n(D(\mathrm{P}\|\mathrm{Q}) - \delta)$, the law of large numbers (cf. Theorem 2.4) implies

$$\begin{aligned}\lim_{n\to\infty} \Pr\left(\log \frac{\mathrm{P}^n(X^n)}{\mathrm{Q}^n(X^n)} \le \log \gamma\right) &= \lim_{n\to\infty} \Pr\left(\frac{1}{n} \sum_{i=1}^{n} \log \frac{\mathrm{P}(X_i)}{\mathrm{Q}(X_i)} \le D(\mathrm{P}\|\mathrm{Q}) - \delta\right) \\ &= 0.\end{aligned}$$

Thus, (5.9) and (5.10) imply

$$-\frac{1}{n} \log \beta_\varepsilon(\mathrm{P}^n, \mathrm{Q}^n) \ge D(\mathrm{P}\|\mathrm{Q}) - \delta$$

for all sufficiently large n. Since $\delta > 0$ is arbitrary, we have one direction of (5.8).

Next, fix an η satisfying $0 < \eta < 1 - \varepsilon$. Then, by applying (5.3) of Theorem 5.4 to P^n and Q^n, we have

$$\beta_\varepsilon(\mathrm{P}^n, \mathrm{Q}^n) \ge \frac{1}{\gamma} \cdot \eta \tag{5.13}$$

for any γ satisfying

$$\Pr\left(\log \frac{\mathrm{P}^n(X^n)}{\mathrm{Q}^n(X^n)} \le \log \gamma\right) \ge \varepsilon + \eta,$$

or equivalently,

$$\Pr\left(\log \frac{\mathrm{P}^n(X^n)}{\mathrm{Q}^n(X^n)} > \log \gamma\right) \le 1 - \varepsilon - \eta. \tag{5.14}$$

Fix an arbitrary $\delta > 0$, and set $\log \gamma = n(D(\mathrm{P}\|\mathrm{Q}) + \delta)$. Again, by noting (5.11) and (5.12), we have

$$\begin{aligned}\lim_{n\to\infty} \Pr\left(\log \frac{\mathrm{P}^n(X^n)}{\mathrm{Q}^n(X^n)} > \log \gamma\right) &= \lim_{n\to\infty} \Pr\left(\frac{1}{n} \sum_{i=1}^{n} \log \frac{\mathrm{P}(X_i)}{\mathrm{Q}(X_i)} > D(\mathrm{P}\|\mathrm{Q}) + \delta\right) \\ &= 0,\end{aligned}$$

by the law of large numbers. This together with (5.13) and (5.14) implies that

$$-\frac{1}{n} \log \beta_\varepsilon(\mathrm{P}^n, \mathrm{Q}^n) \le D(\mathrm{P}\|\mathrm{Q}) + 2\delta$$

for all sufficiently large[2] n. Since $\delta > 0$ is arbitrary, we get the other direction of (5.8). □

Remark 5.1 In this remark, we explain the two exceptions we avoided in Theorem 5.7. First, note that $D(\mathrm{P}\|\mathrm{Q}) = 0$ if and only if $\mathrm{P} = \mathrm{Q}$. In this case, no meaningful test can be conducted. Second, when $D(\mathrm{P}\|\mathrm{Q}) = +\infty$, there exists $a \in \mathcal{X}$ such that $\mathrm{P}(a) > 0$ and $\mathrm{Q}(a) = 0$. In this case, we can consider the following test: upon observing a sequence $x^n \in \mathcal{X}^n$, $\mathrm{T}(0|x^n) = \mathbf{1}[\exists i \text{ such that } x_i = a]$. Then, the type II error probability is $\mathrm{Q}^n[\mathrm{T}] = 0$ and the type I error probability is $1 - \mathrm{P}^n[\mathrm{T}] = (1 - \mathrm{P}(a))^n$, which converges to 0. Thus, for every $0 < \varepsilon < 1$, $\beta_\varepsilon(\mathrm{P}^n, \mathrm{Q}^n) = 0$ for all sufficiently large n.

Note that Theorem 5.7 claims that the optimal exponent of the type II error probability is given by the divergence for every $0 < \varepsilon < 1$. This type of statement is termed the *strong converse* in information theory, and it means that the optimal exponent of the type II error probability cannot be improved even if we allow fairly large type I error probability. On the other hand, the inequality

$$\lim_{\varepsilon \to 0} \lim_{n \to \infty} -\frac{1}{n} \log \beta_\varepsilon(\mathrm{P}^n, \mathrm{Q}^n) \le D(\mathrm{P}\|\mathrm{Q}) \tag{5.15}$$

is termed the *weak converse*. To prove (5.15), it suffices to use Lemma 5.5 (Problem 5.5).

5.5 A Primer on Method of Types

When we consider the experiment of flipping a biased coin n times independently, the probability of outcomes only depends on the numbers of heads or tails. For instance, the probability of observing $(\mathtt{h}, \mathtt{h}, \mathtt{t})$ in three independent tosses of the same coin is the same as those of $(\mathtt{t}, \mathtt{h}, \mathtt{h})$ and $(\mathtt{h}, \mathtt{t}, \mathtt{h})$. More generally, for a sequence of i.i.d. random variables with common distribution P on $\mathcal{X}$, the probability $\mathrm{P}^n(x^n)$ of a sequence $x^n = (x_1, \ldots, x_n)$ only depends on the number of each symbol $a \in \mathcal{X}$ that occurs in x^n. Let

$$N(a|x^n) := |\{i : x_i = a\}|$$

be the number of occurrences of symbol $a \in \mathcal{X}$ in $x^n \in \mathcal{X}^n$. Then, the probability of x^n can be written as

$$\mathrm{P}^n(x^n) = \prod_{a \in \mathcal{X}} \mathrm{P}(a)^{N(a|x^n)}. \tag{5.16}$$

Since $\sum_{a \in \mathcal{X}} N(a|x^n) = n$, we can introduce the probability distribution defined by

$$\mathrm{P}_{x^n}(a) := \frac{1}{n} N(a|x^n), \quad a \in \mathcal{X}.$$

[2] We take n so that (5.14) holds and $\frac{1}{n}\log(1/\eta) \le \delta$.

The distribution P_{x^n} is termed the *type* of sequence[3] x^n. If a distribution $\bar{\mathrm{P}}$ is such that $n\bar{\mathrm{P}}(a)$ is an integer for every $a \in \mathcal{X}$, then we say that $\bar{\mathrm{P}}$ is a type on $\mathcal{X}$. The set of all types on $\mathcal{X}$ corresponding to sequences x^n of length n is denoted by $\mathcal{P}_n(\mathcal{X})$.

As we discussed above, sequences having the same type occur with the same probability under (i.i.d.) product distributions. Thus, it is convenient to partition the entire space $\mathcal{X}^n$ into subsets of sequences having the same type. For a given type $\bar{\mathrm{P}} \in \mathcal{P}_n(\mathcal{X})$, let

$$\mathcal{T}_{\bar{\mathrm{P}}}^n := \{x^n \in \mathcal{X}^n : \mathrm{P}_{x^n} = \bar{\mathrm{P}}\}$$

be the set of all sequences having the type $\bar{\mathrm{P}}$, which is referred to as the *type class* of $\bar{\mathrm{P}}$. Then,

$$\mathcal{X}^n = \bigcup_{\bar{\mathrm{P}} \in \mathcal{P}_n(\mathcal{X})} \mathcal{T}_{\bar{\mathrm{P}}}^n$$

constitutes a partition of $\mathcal{X}^n$.

EXAMPLE 5.8 For the binary alphabet $\mathcal{X} = \{0, 1\}$ with $n = 3$, there are four types; each of them can be specified as $\bar{\mathrm{P}}_i(1) = i/n$ for $i = 0, 1, 2, 3$. The type classes are given by

$$\begin{aligned}
\mathcal{T}_{\bar{\mathrm{P}}_0}^3 &= \{(000)\},\\
\mathcal{T}_{\bar{\mathrm{P}}_1}^3 &= \{(100), (010), (001)\},\\
\mathcal{T}_{\bar{\mathrm{P}}_2}^3 &= \{(110), (101), (011)\},\\
\mathcal{T}_{\bar{\mathrm{P}}_3}^3 &= \{(111)\}.
\end{aligned}$$

More generally, the types on the binary alphabet are specified as $\bar{\mathrm{P}}_i(1) = i/n$ for $i = 0, 1, \ldots, n$. Thus, the number of types on the binary alphabet is $|\mathcal{P}_n(\mathcal{X})| = n + 1$.

Now, we derive an estimate of the number of types for general alphabets.

LEMMA 5.9 *The number of types on a finite alphabet $\mathcal{X}$ is bounded as*[4]

$$|\mathcal{P}_n(\mathcal{X})| \le (n+1)^{|\mathcal{X}|}. \tag{5.17}$$

Proof For each $a \in \mathcal{X}$, $n\bar{\mathrm{P}}(a)$ can take at most $(n + 1)$ different values. Thus, we have (5.17). □

Let $\bar{\mathrm{P}} \in \mathcal{P}_n(\mathcal{X})$ be a type, and let us consider the product distribution $\bar{\mathrm{P}}^n$ on $\mathcal{X}^n$. Under this product distribution, which type class occurs with the highest probability? The following lemma claims that the peak is attained by the type class of $\bar{\mathrm{P}}$.

[3] It is sometimes called the empirical distribution too.
[4] In fact, this bound can be slightly improved; see Problem 5.7.

LEMMA 5.10 *For a type $\bar{\mathrm{P}} \in \mathcal{P}_n(\mathcal{X})$, we have*

$$\bar{\mathrm{P}}^n(\mathcal{T}^n_{\bar{\mathrm{P}}}) \geq \bar{\mathrm{P}}^n(\mathcal{T}^n_{\tilde{\mathrm{P}}})$$

for every $\tilde{\mathrm{P}} \in \mathcal{P}_n(\mathcal{X})$.

Proof For each type $\tilde{\mathrm{P}} \in \mathcal{P}_n(\mathcal{X})$, note that

$$|\mathcal{T}^n_{\tilde{\mathrm{P}}}| = \frac{n!}{\prod_{a \in \mathcal{X}} (n\tilde{\mathrm{P}}(a))!}.$$

Then, from (5.16), we can write

$$\begin{aligned}\bar{\mathrm{P}}^n(\mathcal{T}^n_{\tilde{\mathrm{P}}}) &= |\mathcal{T}^n_{\tilde{\mathrm{P}}}| \prod_{a \in \mathcal{X}} \bar{\mathrm{P}}(a)^{n\tilde{\mathrm{P}}(a)} \\ &= \frac{n!}{\prod_{a \in \mathcal{X}} (n\tilde{\mathrm{P}}(a))!} \prod_{a \in \mathcal{X}} \bar{\mathrm{P}}(a)^{n\tilde{\mathrm{P}}(a)}.\end{aligned}$$

Thus, we have

$$\frac{\bar{\mathrm{P}}^n(\mathcal{T}^n_{\tilde{\mathrm{P}}})}{\bar{\mathrm{P}}^n(\mathcal{T}^n_{\bar{\mathrm{P}}})} = \prod_{a \in \mathcal{X}} \frac{(n\bar{\mathrm{P}}(a))!}{(n\tilde{\mathrm{P}}(a))!} \cdot \bar{\mathrm{P}}(a)^{n(\tilde{\mathrm{P}}(a) - \bar{\mathrm{P}}(a))}.$$

Applying the inequality $\frac{s!}{t!} \leq s^{s-t}$ that holds for any nonnegative integers s and t, we obtain

$$\frac{\bar{\mathrm{P}}^n(\mathcal{T}^n_{\tilde{\mathrm{P}}})}{\bar{\mathrm{P}}^n(\mathcal{T}^n_{\bar{\mathrm{P}}})} \leq \prod_{a \in \mathcal{X}} n^{n(\bar{\mathrm{P}}(a) - \tilde{\mathrm{P}}(a))} = 1,$$

which implies the claim of the lemma. □

COROLLARY 5.11 *For a type $\bar{\mathrm{P}} \in \mathcal{P}_n(\mathcal{X})$, we have*

$$\bar{\mathrm{P}}^n(\mathcal{T}^n_{\bar{\mathrm{P}}}) \geq \frac{1}{(n+1)^{|\mathcal{X}|}}.$$

Proof By Lemma 5.10 and Lemma 5.9, we have

$$\begin{aligned}1 &= \sum_{\tilde{\mathrm{P}} \in \mathcal{P}_n(\mathcal{X})} \bar{\mathrm{P}}^n(\mathcal{T}^n_{\tilde{\mathrm{P}}}) \\ &\leq \sum_{\tilde{\mathrm{P}} \in \mathcal{P}_n(\mathcal{X})} \bar{\mathrm{P}}^n(\mathcal{T}^n_{\bar{\mathrm{P}}}) \\ &\leq (n+1)^{|\mathcal{X}|} \bar{\mathrm{P}}^n(\mathcal{T}^n_{\bar{\mathrm{P}}}),\end{aligned}$$

which implies the claim of the lemma. □

Now, we evaluate the size of each type class. Since the number of different types is a polynomial in n (cf. Lemma 5.9), there must be at least one type class that has an exponential size in n. In fact, most of the type classes have exponential sizes, and we can derive an estimate that is tight up to a polynomial factor in n.

LEMMA 5.12 *For a type* $\bar{\mathrm{P}} \in \mathcal{P}_n(\mathcal{X})$, *we have*

$$\frac{1}{(n+1)^{|\mathcal{X}|}} 2^{nH(\bar{\mathrm{P}})} \le |\mathcal{T}_{\bar{\mathrm{P}}}^n| \le 2^{nH(\bar{\mathrm{P}})}.$$

Proof From (5.16), we can write

$$\begin{aligned}\bar{\mathrm{P}}^n(\mathcal{T}_{\bar{\mathrm{P}}}^n) &= |\mathcal{T}_{\bar{\mathrm{P}}}^n| \prod_{a \in \mathcal{X}} \bar{\mathrm{P}}(a)^{n\bar{\mathrm{P}}(a)} \\ &= |\mathcal{T}_{\bar{\mathrm{P}}}^n| 2^{-nH(\bar{\mathrm{P}})}.\end{aligned}$$

Then, since Corollary 5.11 implies

$$\frac{1}{(n+1)^{|\mathcal{X}|}} \le \bar{\mathrm{P}}^n(\mathcal{T}_{\bar{\mathrm{P}}}^n) \le 1,$$

we have the claim of the lemma. □

Let P be a distribution (not necessarily a type) on $\mathcal{X}$. For a given $\delta > 0$, let

$$\mathcal{T}_{\mathrm{P},\delta}^n := \{x^n \in \mathcal{X}^n : |\mathrm{P}_{x^n}(a) - \mathrm{P}(a)| \le \delta, \ \forall a \in \mathcal{X}\}. \tag{5.18}$$

The set $\mathcal{T}_{\mathrm{P},\delta}^n$ is termed the P-*typical set*, or just the *typical set* if the distribution P is clear from the context. By denoting

$$\mathcal{P}_{\mathrm{P},\delta}^n := \{\bar{\mathrm{P}} \in \mathcal{P}_n(\mathcal{X}) : |\bar{\mathrm{P}}(a) - \mathrm{P}(a)| \le \delta, \ \forall a \in \mathcal{X}\},$$

we can also write the typical set as

$$\mathcal{T}_{\mathrm{P},\delta}^n = \bigcup_{\bar{\mathrm{P}} \in \mathcal{P}_{\mathrm{P},\delta}^n} \mathcal{T}_{\bar{\mathrm{P}}}^n.$$

Since $\mathrm{P}_{x^n}(a) = \frac{1}{n}\sum_{i=1}^n \mathbf{1}[x_i = a]$ can be regarded as an arithmetic mean, the law of large numbers implies that $\mathrm{P}_{x^n}(a)$ is close to $\mathrm{P}(a)$ with high probability under the product distribution P^n. Thus, the typical set is a set that occurs with high probability under P^n. In fact, by Hoeffding's inequality, we can derive the following.

LEMMA 5.13 *For i.i.d. random variables* $X^n = (X_1, \ldots, X_n)$ *with common distribution* P, *we have*

$$\Pr\left(X^n \notin \mathcal{T}_{\mathrm{P},\delta}^n\right) \le 2|\mathcal{X}|e^{-2\delta^2 n}.$$

Proof By applying Hoeffding's inequality (cf. Theorem 2.39) to random variables $\mathbf{1}[X_i = a]$ taking values in $[0,1]$, we have

$$\Pr\left(\left|\frac{1}{n}\sum_{i=1}^n \mathbf{1}[X_i = a] - \mathrm{P}(a)\right| \ge \delta\right) \le 2e^{-2\delta^2 n}$$

for every $a \in \mathcal{X}$. Thus, by the union bound, we have the claim of the lemma. □

In the rest of this section, we consider an extension of the notion of type to joint type and conditional type. Specifically, for a given pair $(x^n, y^n) \in \mathcal{X}^n \times \mathcal{Y}^n$ of sequences, let

$$N(a, b|x^n, y^n) := |\{i : (x_i, y_i) = (a, b)\}|, \quad (a, b) \in \mathcal{X} \times \mathcal{Y}$$

be the number of occurrences of a pair (a, b) in (x^n, y^n). Then, the probability distribution given by

$$\mathrm{P}_{x^n y^n}(a, b) := \frac{1}{n} N(a, b|x^n, y^n)$$

is termed the *joint type* of (x^n, y^n). It is not difficult to see that the marginal of the joint type of (x^n, y^n) is the type of x^n (or y^n), i.e.,

$$\sum_{b \in \mathcal{Y}} \mathrm{P}_{x^n y^n}(a, b) = \mathrm{P}_{x^n}(a)$$

for every $a \in \mathcal{X}$.

For a given type $\bar{\mathrm{P}} \in \mathcal{P}_n(\mathcal{X})$, a stochastic matrix $\bar{W}$ from $\mathcal{X}$ to $\mathcal{Y}$ is called a *conditional type* for $\bar{\mathrm{P}}$ if $n\bar{\mathrm{P}}(a)\bar{W}(b|a)$ is an integer for every $(a, b) \in \mathcal{X} \times \mathcal{Y}$. Also, we say that a sequence y^n has conditional type $\bar{W}$ given x^n if $\mathrm{P}_{x^n y^n}(a, b) = \bar{\mathrm{P}}(a)\bar{W}(b|a)$ for every $(a, b) \in \mathcal{X} \times \mathcal{Y}$.

For a type $\bar{\mathrm{P}} \in \mathcal{P}_n(\mathcal{X})$, let $\mathcal{W}_n(\mathcal{Y}|\bar{\mathrm{P}})$ be the set of all conditional types $\bar{W}$ for $\bar{\mathrm{P}}$. For a conditional type $\bar{W} \in \mathcal{W}_n(\mathcal{Y}|\bar{\mathrm{P}})$ and a sequence $x^n \in \mathcal{T}^n_{\bar{\mathrm{P}}}$, let

$$\mathcal{T}^n_{\bar{W}}(x^n) := \{y^n \in \mathcal{Y}^n : \mathrm{P}_{x^n y^n}(a, b) = \bar{\mathrm{P}}(a)\bar{W}(b|a), \ \forall (a, b) \in \mathcal{X} \times \mathcal{Y}\}$$

be the set of all sequences having the conditional type $\bar{W}$ given x^n, which is termed the conditional type class of $\bar{W}$ given x^n, or the $\bar{W}$-shell of x^n.

Let W be a stochastic matrix from $\mathcal{X}$ to $\mathcal{Y}$ (not necessarily a conditional type). For a given type $\bar{\mathrm{P}} \in \mathcal{P}_n(\mathcal{X})$, a conditional type $\bar{W} \in \mathcal{W}_n(\mathcal{Y}|\bar{\mathrm{P}})$, sequence $x^n \in \mathcal{T}^n_{\bar{\mathrm{P}}}$, and $y^n \in \mathcal{T}^n_{\bar{W}}(x^n)$, we can write

$$\begin{aligned} W^n(y^n|x^n) &= \prod_{a \in \mathcal{X}} \prod_{b \in \mathcal{Y}} W(b|a)^{N(a,b|x^n,y^n)} \\ &= \prod_{a \in \mathcal{X}} \left(\prod_{b \in \mathcal{Y}} W(b|a)^{N(a|x^n)\bar{W}(b|a)} \right). \end{aligned}$$

Thus, the conditional probability of y^n given x^n can be computed as the product of probabilities of sequences of length $N(a|x^n)$ having the types $\bar{W}(\cdot|a)$ under the (product) distributions $W^{N(a|x^n)}(\cdot|a)$. Similarly, the conditional type class can be regarded as the product of type classes, i.e.,

$$\mathcal{T}^n_{\bar{W}}(x^n) = \prod_{a \in \mathcal{X}} \mathcal{T}^{N(a|x^n)}_{\bar{W}(\cdot|a)}.$$

By noting these observations, the properties of types discussed above can be extended to conditional types. In particular, we will use the following fact in a later chapter.

COROLLARY 5.14 *For a sequence x^n having the type $\bar{\mathrm{P}} \in \mathcal{P}_n(\mathcal{X})$ and a conditional type $\bar{W} \in \mathcal{W}_n(\mathcal{Y}|\bar{\mathrm{P}})$, we have*

$$\bar{W}^n\left(\mathcal{T}^n_{\bar{W}}(x^n) \mid x^n\right) \geq \frac{1}{(n+1)^{|\mathcal{X}||\mathcal{Y}|}}.$$

Proof By applying Corollary 5.11 to each $a \in \mathcal{X}$, we have

$$\bar{W}^{N(a|x^n)}\left(\mathcal{T}^{N(a|x^n)}_{\bar{W}(\cdot|a)} \mid a\right) \geq \frac{1}{(N(a|x^n)+1)^{|\mathcal{Y}|}} \geq \frac{1}{(n+1)^{|\mathcal{Y}|}}$$

for every $a \in \mathcal{X}$. Thus, by taking the product of $a \in \mathcal{X}$, we have the claim of the lemma. □

5.6 Composite Hypothesis Testing

In the binary hypothesis testing formulation we have seen above, the observations are generated from exactly one distribution each under the null and alternative hypotheses. This formulation is sometimes called *simple* hypothesis testing. More generally, and often in practice, the alternative hypothesis is not captured by a single distribution but by a class of distributions. Such a problem is termed a *composite* hypothesis testing problem.

Composite hypothesis testing problems are well studied in statistics and include useful applications such as goodness-of-fit testing. While these applications are beyond the scope of this book, in Chapter 14 we do encounter a composite hypothesis testing problem. Specifically, we need to examine whether the system is under attack. In that case, the null hypothesis is exactly known since it describes regular operation of the system. However, the alternative hypothesis is not known exactly since it depends on the behavior of the attacker.

In particular, we consider the following formulation. Let P be the null hypothesis, and let $\mathcal{Q} \subset \mathcal{P}(\mathcal{X})$ be the set of alternative hypotheses, where $\mathcal{P}(\mathcal{X})$ is the set of all distributions on $\mathcal{X}$. We consider the case in which the observations are i.i.d. In other words, when the null hypothesis is true, then the observations X^n are generated from the product distribution P^n; when the alternative hypothesis is true, then the observations X^n are generated from Q^n for some $\mathrm{Q} \in \mathcal{Q}$.

As before, a test for this composite testing problem is given by a channel $\mathrm{T}\colon \mathcal{X}^n \to \{0,1\}$. For a given test T, the type I error probability is given by $1 - \mathrm{P}^n[\mathrm{T}]$, where

$$\mathrm{P}^n[\mathrm{T}] := \sum_{x^n \in \mathcal{X}^n} \mathrm{P}^n(x^n)\mathrm{T}(0|x^n).$$

On the other hand, for a fixed $\mathrm{Q} \in \mathcal{Q}$, the type II error probability is given by

$$\mathrm{Q}^n[\mathrm{T}] := \sum_{x^n \in \mathcal{X}^n} \mathrm{Q}^n(x^n)\mathrm{T}(0|x^n).$$

Since the alternative hypothesis is not known, except for the fact that it belongs to $\mathcal{Q}$, we have to consider the worst case by taking the supremum of $\mathrm{Q}^n[\mathrm{T}]$ over $\mathrm{Q} \in \mathcal{Q}$.

There is a rich history of tests for such composite hypothesis testing problems. However, we only review a specific test that leads us to the result we need for our purposes in this book. Specifically, for $\delta > 0$, we consider the following typicality test T_δ. Upon observing $x^n \in \mathcal{X}^n$,

$$\mathrm{T}_\delta(0|x^n) := \mathbf{1}\left[x^n \in \mathcal{T}^n_{\mathrm{P},\delta}\right],$$

where $\mathcal{T}^n_{\mathrm{P},\delta}$ is the P-typical set defined in (5.18).

Suppose that P is not included in the closure $\overline{\mathcal{Q}}$ of $\mathcal{Q}$ in $\mathcal{P}(\mathcal{X})$. Then, there exists $\delta > 0$ such that, for every $\mathrm{Q} \in \mathcal{Q}$,

$$|\mathrm{P}(a) - \mathrm{Q}(a)| > 2\delta \quad \text{for some } a \in \mathcal{X}. \tag{5.19}$$

LEMMA 5.15 *Suppose that* $\mathrm{P} \notin \overline{\mathcal{Q}}$ *for the closure* $\overline{\mathcal{Q}}$ *of* $\mathcal{Q}$ *in* $\mathcal{P}(\mathcal{X})$. *Then, there exists* $\delta > 0$ *such that the typicality test* T_δ *satisfies*

$$1 - \mathrm{P}^n[\mathrm{T}_\delta] \leq 2|\mathcal{X}|e^{-2\delta^2 n} \tag{5.20}$$

and

$$\mathrm{Q}^n[\mathrm{T}_\delta] \leq 2|\mathcal{X}|e^{-2\delta^2 n} \tag{5.21}$$

for every $\mathrm{Q} \in \mathcal{Q}$.

Proof Note that (5.20) follows from Lemma 5.13 for any $\delta > 0$. To prove (5.21), let $\delta > 0$ be such that (5.19) is satisfied for every $\mathrm{Q} \in \mathcal{Q}$. Then, $x^n \in \mathcal{T}^n_{\mathrm{P},\delta}$ implies that the type P_{x^n} satisfies

$$|\mathrm{P}_{x^n}(a) - \mathrm{Q}(a)| \geq |\mathrm{P}(a) - \mathrm{Q}(a)| - |\mathrm{P}_{x^n}(a) - \mathrm{P}(a)| > \delta$$

for some $a \in \mathcal{X}$. Thus, by applying Lemma 5.13 again for a Q-typical set $\mathcal{T}^n_{\mathrm{Q},\delta}$, we have

$$\begin{aligned}\mathrm{Q}^n[\mathrm{T}_\delta] &= \mathrm{Q}^n(\mathcal{T}^n_{\mathrm{P},\delta}) \\ &\leq \mathrm{Q}^n\left((\mathcal{T}^n_{\mathrm{Q},\delta})^c\right) \\ &\leq 2|\mathcal{X}|e^{-2\delta^2 n},\end{aligned}$$

where $(\mathcal{T}^n_{\mathrm{Q},\delta})^c$ is the complement of $\mathcal{T}^n_{\mathrm{Q},\delta}$. □

We remark that the test used in obtaining the result above is perhaps not practically feasible. However, our goal was to obtain simultaneous exponential decay of type I and type II error probabilities, a result we will need in later chapters.

5.7 References and Additional Reading

Binary hypothesis testing is one of the fundamental topics in statistics, and a comprehensive treatment can be found in standard textbooks on statistics such as [220]. However, the exposition in this chapter is somewhat different from the standard treatment in statistics and is motivated by applications in information-theoretic problems. The connection between hypothesis testing and information theory has been explored for a long time; for early references, see [207] and [33].

The importance of hypothesis testing in information theory was highlighted further in the information spectral method introduced by Han and Verdú in [152]; see also [151]. We will discuss the information spectral method briefly in Chapters 6 and 7. It seems that information-theoretic treatment of the Neyman–Pearson lemma, as in Section 5.2, was independently developed by two groups: Nagaoka and his collaborators in Japan [263, 254] and Verdú and his collaborators at Princeton (see [275] for references to these lecture notes, which do not seem to be publicly available).

The method of types discussed in Section 5.5 is a standard and powerful technique for handling discrete memoryless systems in information theory. Detailed expositions on this topic can be found in [87] and [88].

Problems

5.1 Describe the condition on P and Q that $\beta_0(\mathrm{P},\mathrm{Q}) = 0$.

5.2 Prove that the infimum in (5.1) can be attained.

HINT Note that the optimization problem in (5.1) can be regarded as a linear programming problem.

5.3 Prove that $\beta_\varepsilon(\mathrm{P},\mathrm{Q})$ is a concave function of Q, i.e., $\beta_\varepsilon(\mathrm{P},\mathrm{Q}) \geq \theta\beta_\varepsilon(\mathrm{P},\mathrm{Q}_1) + (1-\theta)\beta_\varepsilon(\mathrm{P},\mathrm{Q}_2)$ for a mixture $\mathrm{Q} = \theta\mathrm{Q}_1 + (1-\theta)\mathrm{Q}_2$ of two distributions Q_1 and Q_2 with weight $\theta \in [0,1]$.

5.4 Prove that $\beta_\varepsilon(\mathrm{P},\mathrm{Q})$ is a convex function of ε, i.e., $\beta_\varepsilon(\mathrm{P},\mathrm{Q}) \leq \theta\beta_{\varepsilon_1}(\mathrm{P},\mathrm{Q}) + (1-\theta)\beta_{\varepsilon_2}(\mathrm{P},\mathrm{Q})$ for $\varepsilon = \theta\varepsilon_1 + (1-\theta)\varepsilon_2$ and $\varepsilon_1, \varepsilon_2, \theta \in [0,1]$.

HINT For two tests $\mathrm{T}_1, \mathrm{T}_2$ attaining $\beta_{\varepsilon_1}(\mathrm{P},\mathrm{Q})$ and $\beta_{\varepsilon_2}(\mathrm{P},\mathrm{Q})$, respectively, consider a stochastic test that uses $\mathrm{T}_1, \mathrm{T}_2$ with probability θ and $1-\theta$.

5.5 Prove (5.15) using Lemma 5.5.

5.6 Prove that the equality of the bound in Lemma 5.5 holds for $\varepsilon = 0$ if and only if $\log \frac{\mathrm{P}(x)}{\mathrm{Q}(x)} = D(\mathrm{P}\|\mathrm{Q})$ for every $x \in \mathsf{supp}(\mathrm{P})$.

HINT To prove the "only if" part, use the strict convexity of the function 2^{-t} and Jensen's inequality.

5.7 Verify that the bound in Lemma 5.9 can improved to $|\mathcal{P}_n(\mathcal{X})| \leq (n+1)^{|\mathcal{X}|-1}$.

5.8 *Bayes optimal test*: For the Bayesian formulation with prior distribution $(\theta, 1-\theta)$, prove that the minimum of average error probability $\theta(1-\mathrm{P}[\mathrm{T}]) + (1-\theta)\mathrm{Q}[\mathrm{T}]$ is attained by the Neyman–Pearson test with $\gamma = \frac{1-\theta}{\theta}$.

HINT Use Lemma 5.2.

5.9 *Connection to the Bayes test*: Prove the following connection between the minimum average probability of error $P_{\mathrm{err}}(\mathrm{P},\mathrm{Q})$ of the Bayes test defined in Section 2.4 and the optimal type II error probability $\beta_\varepsilon(\mathrm{P},\mathrm{Q})$:

$$P_{\mathrm{err}}(\mathrm{P},\mathrm{Q}) = \min\{\varepsilon : \beta_\varepsilon(\mathrm{P},\mathrm{Q}) \le \varepsilon\}.$$

5.10 *Probability of type class*:

1. For a given sequence $x^n \in \mathcal{X}^n$ and a product distribution P^n, prove the following identity holds:
$$-\frac{1}{n}\log \mathrm{P}^n(x^n) = D(\mathrm{P}_{x^n}\|\mathrm{P}) + H(\mathrm{P}_{x^n}),$$
where P_{x^n} is the type of sequence x^n.
2. For a given type class $\mathcal{T}^n_{\bar{\mathrm{P}}}$, prove that
$$\frac{1}{(n+1)^{|\mathcal{X}|}}2^{-nD(\bar{\mathrm{P}}\|\mathrm{P})} \le \mathrm{P}^n(\mathcal{T}^n_{\bar{\mathrm{P}}}) \le 2^{-nD(\bar{\mathrm{P}}\|\mathrm{P})}.$$

REMARK This bound implies that sequences having types deviating from the underlying distribution P occur with exponentially small probability.

5.11 *Hoeffding test*: Consider the following test T between the null hypothesis P and Q. Upon observing $x^n \in \mathcal{X}^n$, the null hypothesis is accepted if and only if $D(\mathrm{P}_{x^n}\|\mathrm{P}) \le r$ for a prescribed threshold $r > 0$. Then, prove that the type I and type II error probabilities can be bounded as

$$1 - \mathrm{P}^n[\mathrm{T}] \le (n+1)^{|\mathcal{X}|}2^{-nr},$$
$$\mathrm{Q}^n[\mathrm{T}] \le (n+1)^{|\mathcal{X}|}2^{-n\min D(\bar{\mathrm{P}}\|\mathrm{Q})},$$

where the minimization is taken over the set of all types $\bar{\mathrm{P}}$ satisfying $D(\bar{\mathrm{P}}\|\mathrm{P}) \le r$.

REMARK This test is known as the Hoeffding test. It provides the optimal exponential tradeoff between the type I and type II error probabilities; for the optimality proof, see for example [87].

6 Information Reconciliation

In this chapter, we present the source coding and source coding with side-information problems. Typically, source coding is a basic topic covered in textbooks on information theory, when discussing data compression. In the context of this textbook, source coding with side-information is often used to convert correlated randomness to shared randomness, a primitive commonly referred to as *information reconciliation.*

Our exposition of source coding differs from those in standard textbooks in two aspects: first, instead of variable length coding, such as that accomplished by the Huffman code, we only consider fixed length coding. This is because it is more standard to consider fixed length coding when side-information is available, and indeed, it suffices for our information reconciliation example.

Second, instead of considering "block codes" for a sequence of i.i.d. random variables, we start with the "single-shot" regime where a single sample from a distribution needs to be encoded. Block coding results for i.i.d. random variables will be obtained as a corollary of our single-shot results. In the single-shot regime, instead of Shannon entropy, a quantity called smooth max-entropy plays a pivotal role.

6.1 Source Coding and Smooth Max-entropy

A *source* in information theory refers to a generator of data, modeled using random variables. Consider a source[1] described by a random variable X taking values in a finite set $\mathcal{X}$. The goal in data compression is to describe a source using as few bits as possible. In fact, data compression is a broader engineering problem, which includes as a component the *source coding* problem. This latter problem is our topic for this section.

Formally, a *source code* consists of encoder and decoder, where an encoder is described by a mapping $\varphi \colon \mathcal{X} \to \{0,1\}^k$ and a decoder is described by a mapping $\psi \colon \{0,1\}^k \to \mathcal{X}$; see Figure 6.1. This type of source code is referred to as a *fixed length code* since the source is encoded into a binary sequence of fixed length. When $k < \log |\mathcal{X}|$, it is not possible to assign distinct codewords to distinct source symbols, and we need to allow errors. The error probability of the source

[1] We will simply refer to X as a source.

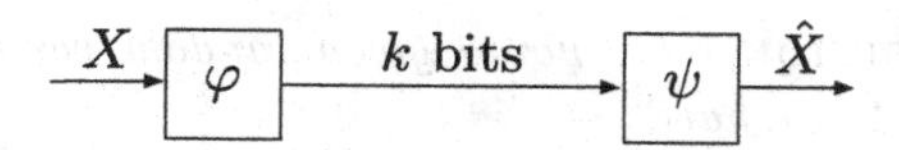

Figure 6.1 The description of source coding.

code (φ, ψ) is given by

$$P_e(\varphi, \psi) := \Pr(\psi(\varphi(X)) \neq X).$$

When $P_e(\varphi, \psi) \le \varepsilon$ for a given $0 \le \varepsilon < 1$, the source code (φ, ψ) is referred to as a (k, ε)-code for X.

EXAMPLE 6.1 For $\mathcal{X} = \{1, 2, 3, 4, 5\}$, consider the distribution $P_X(1) = 1/2$, $P_X(2) = 1/4$, $P_X(3) = 1/8$, and $P_X(4) = P_X(5) = 1/16$. For $k = 2$, consider a source code with the encoder and the decoder, respectively, given by $\varphi(1) = 00$, $\varphi(2) = 01$, $\varphi(3) = 10$, and $\varphi(4) = \varphi(5) = 11$ and $\psi(00) = 1$, $\psi(01) = 2$, $\psi(10) = 3$, and $\psi(11) = 4$. Then, $P_e(\varphi, \psi) = \Pr(X = 5)$. It can be verified that (φ, ψ) constitutes a (k, ε)-code for any $\varepsilon \ge 1/16$.

We are interested in identifying the minimum k such that a (k, ε)-code exists for a given source X. Namely, we are interested in the following quantity:

$$L_\varepsilon(X) := \min\{k : \exists (k, \varepsilon)\text{-code } (\varphi, \psi) \text{ for } X\}. \tag{6.1}$$

Our goal is to derive a characterization of the operationally defined quantity $L_\varepsilon(X)$; here, "operationally defined" means that (6.1) is defined using optimization over all possible code constructions, which makes it difficult to compute (6.1) directly. We first give an alternative expression for $L_\varepsilon(X)$, which still will be difficult to compute exactly. To that end, we bring in the notions of max-entropy and smooth max-entropy.

DEFINITION 6.2 (Max-entropy) For a given random variable X with distribution P_X on $\mathcal{X}$, the max-entropy of X is defined as

$$H_{\max}(X) := \log |\{x \in \mathcal{X} : P_X(x) > 0\}|.$$

In simpler words, max-entropy is roughly the number of bits needed to represent support of the distribution of X. A reader can verify that max-entropy coincides with Rényi entropy of order 0, although $\alpha = 0$ was not considered when Rényi entropy was defined in Chapter 2. Max-entropy is a rather "conservative" quantity, not too far from $\log |\mathcal{X}|$. Interestingly, a "smooth" version of this quantity turns out to be quite useful.

DEFINITION 6.3 (Smooth max-entropy) For a given random variable X with distribution P_X on $\mathcal{X}$ and $0 \le \varepsilon < 1$, the ε-smooth max-entropy is defined by

$$H_{\max}^{\varepsilon}(X) := \min_{\substack{\mathcal{L} \subseteq \mathcal{X}: \\ P_X(\mathcal{L}) \ge 1-\varepsilon}} \log |\mathcal{L}|.$$

When $\varepsilon = 0$, $H_{\max}^{0}(X)$ coincides with max-entropy $H_{\max}(X)$.

PROPOSITION 6.4 *For a given random variable X taking values in $\mathcal{X}$ and $0 \le \varepsilon < 1$, we have*

$$L_\varepsilon(X) = \lceil H_{\max}^\varepsilon(X) \rceil. \tag{6.2}$$

Proof Let $\mathcal{L} \subseteq \mathcal{X}$ be such that $\mathrm{P}_X(\mathcal{L}) \ge 1 - \varepsilon$ and $\log |\mathcal{L}| = H_{\max}^\varepsilon(X)$. Then, if we set $k = \lceil \log |\mathcal{L}| \rceil$, we can construct an encoder φ and a decoder ψ such that $\psi(\varphi(x)) = x$ for every $x \in \mathcal{L}$, which implies $P_{\mathrm{e}}(\varphi, \psi) \le \varepsilon$ and $L_\varepsilon(X) \le \lceil H_{\max}^\varepsilon(X) \rceil$.

To prove the inequality in the opposite direction, for a given (k, ε)-code (φ, ψ), let $\mathcal{L} = \{x \in \mathcal{X} : \psi(\varphi(x)) = x\}$. Then, this set satisfies $\mathrm{P}_X(\mathcal{L}) \ge 1 - \varepsilon$. Furthermore, since $\mathcal{L}$ is contained in the image of the mapping ψ from $\{0,1\}^k$, we have $|\mathcal{L}| \le 2^k$, which implies $k \ge H_{\max}^\varepsilon(X)$ for any (k, ε)-code. Finally, since k must be an integer, we obtain $L_\varepsilon(X) \ge \lceil H_{\max}^\varepsilon(X) \rceil$. □

As we can see from the proof of Proposition 6.4, the optimal source code can be constructed by a simple, greedy algorithm: sort the symbols in the decreasing order of probabilities, and then include symbols in the "high probability list" $\mathcal{L}$ until the total probability of the list exceeds $1 - \varepsilon$.

The computational complexity of evaluating the smooth max-entropy $H_{\max}^\varepsilon(X)$ depends on the size $|\mathcal{X}|$ of the alphabet, which is exponential in the block length when we consider the block coding. For sources with distribution having certain structures, such as the product structure for i.i.d. source, we can derive upper and lower bounds on $H_{\max}^\varepsilon(X)$ that can be evaluated more easily; see Section 6.3.

6.2 Source Coding with Side-information and Conditional Smooth Max-entropy

In this section, we consider source coding with side-information, which is an extension of the source coding problem discussed in the previous section. In source coding with side-information, in addition to the main source X to be encoded, there is another source Y that is available to the decoder as side-information; see Figure 6.2. The sources X and Y are correlated, in general, and the side-information Y helps to reduce the length of the message required to reproduce the main source X.

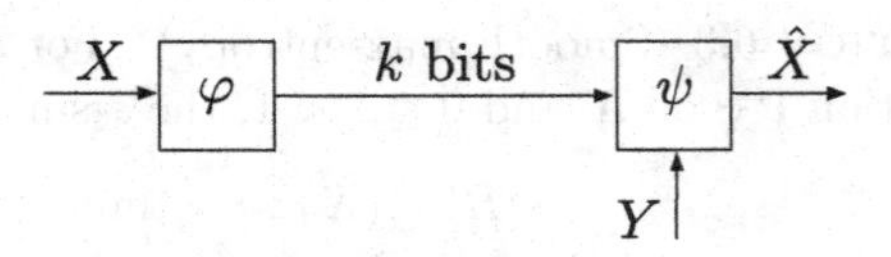

Figure 6.2 The description of source coding with side-information.

EXAMPLE 6.5 Let $X = (X_1, X_2)$ be 2 random bits, and let the side-information be $Y = X_1 \oplus X_2$. Without side-information, 2 bits must be transmitted to reproduce X at the decoder. If Y is available at the decoder, it suffices to transmit 1 bit, say $\Pi = X_1$ or $\Pi = X_2$. Then, the remaining bit can be reproduced at the decoder using Π and Y.

EXAMPLE 6.6 Let $X = (X_1, X_2, X_3)$ be 3 random bits, and let $Y = X \oplus E$, where E equals one of $\{(000), (100), (010), (001)\}$ at random. Without the side-information, 3 bits must be transmitted to reproduce X at the decoder. If Y is available at the decoder, it suffices to send 2 bits,

$$\Pi = (\Pi_1, \Pi_2) = (X_1 \oplus X_2, X_2 \oplus X_3).$$

Then, from Π and $(Y_1 \oplus Y_2, Y_2 \oplus Y_3)$, the error vector E can be determined, and $X = Y \oplus E$ can be reproduced.

Formally, a source code with side-information consists of an encoder described by a mapping $\varphi : \mathcal{X} \to \{0,1\}^k$ and a decoder described by a mapping $\psi : \{0,1\}^k \times \mathcal{Y} \to \mathcal{X}$. The error probability of this source code is given by

$$P_{\mathrm{e}}(\varphi, \psi) := \Pr(\psi(\varphi(X), Y) \neq X).$$

When $P_{\mathrm{e}}(\varphi, \psi) \leq \varepsilon$ for a given $0 \leq \varepsilon < 1$, the source code (φ, ψ) is referred to as a (k, ε)-code for X given Y. We are interested in identifying the minimum k such that a (k, ε)-code exists, that is, we are interested in the following quantity:

$$L_\varepsilon(X|Y) := \min\left\{k : \exists (k, \varepsilon)\text{-code } (\varphi, \psi) \text{ for } X \text{ given } Y\right\}.$$

In order to characterize $L_\varepsilon(X|Y)$, we introduce a conditional version of the smooth max-entropy (cf. Definition 6.3).

DEFINITION 6.7 (Smooth conditional max-entropy) For given random variables (X, Y) with joint distribution P_{XY} on $\mathcal{X} \times \mathcal{Y}$ and $0 \leq \varepsilon < 1$, the ε-smooth conditional max-entropy is defined by

$$H_{\max}^{\varepsilon}(X|Y) := \min_{\substack{\mathcal{L} \subseteq \mathcal{X} \times \mathcal{Y}: \\ \mathrm{P}_{XY}(\mathcal{L}) \geq 1 - \varepsilon}} \max_{y \in \mathcal{Y}} \log |\mathcal{L}_y|,$$

where

$$\mathcal{L}_y := \left\{x : (x, y) \in \mathcal{L}\right\}.$$

In a similar manner to Proposition 6.4, we can derive the following lower bound for $L_\varepsilon(X|Y)$.

PROPOSITION 6.8 *For a random variable (X, Y) taking values in $\mathcal{X} \times \mathcal{Y}$ and $0 \leq \varepsilon < 1$, we have*

$$L_\varepsilon(X|Y) \geq \lceil H_{\max}^{\varepsilon}(X|Y) \rceil.$$

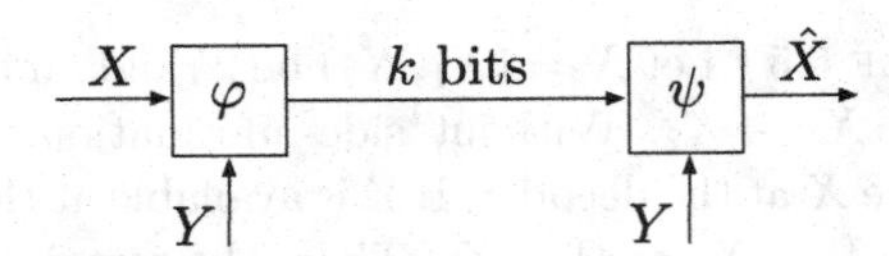

Figure 6.3 The description of source coding, where the side-information is available at both the encoder and the decoder.

Proof For a given (k, ε)-code (φ, ψ) for X given Y, let $\mathcal{L}_y = \{x : \psi(\varphi(x), y) = x\}$ and $\mathcal{L} = \cup_{y \in \mathcal{Y}} \mathcal{L}_y \times \{y\}$. Then, this set satisfies $\mathrm{P}_{XY}(\mathcal{L}) \geq 1 - \varepsilon$. Furthermore, for every $y \in \mathcal{Y}$, since $\mathcal{L}_y$ is contained in the image of the mapping $\psi(\cdot, y)$ from $\{0,1\}^k$, we have $|\mathcal{L}_y| \leq 2^k$, which implies $k \geq H_{\max}^{\varepsilon}(X|Y)$ for any (k, ε)-code. Since k must be an integer, we have the claim of the proposition. □

If the side-information is also available at the encoder, as in Figure 6.3, then it is not difficult to see that the message length $\lceil H_{\max}^{\varepsilon}(X|Y) \rceil$ is attainable; the encoder can assign distinct k-bit messages to each element of $\mathcal{L}_y$ given $Y = y$. In other words, $\mathcal{L}_y$ plays the role of a "guess list" of X given $Y = y$. Interestingly, even if the side-information is not available at the encoder, X can be sent using a message of almost the same length.

Our construction uses a universal hash family (UHF), discussed in Chapter 4. Recall that if F is distributed uniformly over a UHF from $\mathcal{X}$ to $\{0,1\}^k$, then

$$\Pr\left(F(x) = F(x')\right) \leq \frac{1}{2^k}, \quad \forall x \neq x' \in \mathcal{X}.$$

As we discussed in Section 4.4.1, the UHF can be used as a good source code as long as the message length is sufficiently larger than the size of the guess lists. For given lists $\{\mathcal{L}_y\}_{y \in \mathcal{Y}}$, our code construction is described as follows.

CONSTRUCTION 6.1 (Source code using UHF) Suppose that the sender and the receiver share F distributed uniformly over a UHF $\mathcal{F}$ from $\mathcal{X}$ to $\{0,1\}^k$.

1. Upon observing X, the encoder sends $M = F(X)$ to the decoder.
2. Upon receiving the message M and observing Y, the decoder looks for a unique $\hat{x} \in \mathcal{L}_Y$ such that $F(\hat{x}) = M$. If such $\hat{x}$ exists, then the decoder outputs $\hat{X} = \hat{x}$; otherwise, the decoder outputs a prescribed symbol $x_0 \in \mathcal{X}$.

LEMMA 6.9 *For $0 < \varepsilon < 1$ and $0 < \xi \leq \varepsilon$, let $\{\mathcal{L}_y\}_{y \in \mathcal{Y}}$ be lists such that $\mathcal{L} = \cup_{y \in \mathcal{Y}} \mathcal{L}_y \times \{y\}$ satisfies $\mathrm{P}_{XY}(\mathcal{L}) \geq 1 - (\varepsilon - \xi)$ and $\max_{y \in \mathcal{Y}} \log |\mathcal{L}_y| = H_{\max}^{\varepsilon - \xi}(X|Y)$. Further, let*

$$k = \lceil H_{\max}^{\varepsilon - \xi}(X|Y) + \log(1/\xi) \rceil. \tag{6.3}$$

Then, the error probability of Construction 6.1 is bounded as

$$\Pr(\hat{X} \neq X) \leq \varepsilon. \tag{6.4}$$

Proof Note that an error may occur for two reasons. First, if the source X is not included in the guess list $\mathcal{L}_Y$ of the decoder, the output $\hat{X}$ cannot coincide with X.[2] By the assumption for the list $\mathcal{L}$, the probability of the first error event is

$$\Pr((X,Y) \notin \mathcal{L}) \leq \varepsilon - \xi. \tag{6.5}$$

Second, if there exists $\hat{x} \neq X$ such that $\hat{x} \in \mathcal{L}_Y$ and $F(\hat{x}) = F(X)$, then the decoder may output $\hat{X}$ different from X. The probability of the second error event is

$$\begin{aligned}
&\Pr\left(\exists \hat{x} \neq X \text{ such that } \hat{x} \in \mathcal{L}_Y, F(\hat{x}) = F(X)\right) \\
&\quad = \sum_{x,y} \mathrm{P}_{XY}(x,y) \Pr\left(\exists \hat{x} \neq x \text{ such that } \hat{x} \in \mathcal{L}_y, F(\hat{x}) = F(x)\right) \\
&\quad \leq \sum_{x,y} \mathrm{P}_{XY}(x,y) \sum_{\substack{\hat{x} \in \mathcal{L}_y: \\ \hat{x} \neq x}} \Pr(F(\hat{x}) = F(x)) \\
&\quad \leq \sum_{x,y} \mathrm{P}_{XY}(x,y) \sum_{\substack{\hat{x} \in \mathcal{L}_y: \\ \hat{x} \neq x}} \frac{1}{2^k} \\
&\quad \leq \sum_{x,y} \mathrm{P}_{XY}(x,y) \frac{|\mathcal{L}_y|}{2^k} \\
&\quad \leq \sum_{x,y} \mathrm{P}_{XY}(x,y)\, 2^{H_{\max}^{\varepsilon-\xi}(X|Y)-k} \\
&\quad = 2^{H_{\max}^{\varepsilon-\xi}(X|Y)-k} \\
&\quad \leq \xi,
\end{aligned} \tag{6.6}$$

where the first inequality follows from the union bound, the second inequality follows from the property of the UHF, the fourth inequality follows from the definition of $H_{\max}^{\varepsilon-\xi}(X|Y)$, and the last inequality follows from (6.3). By combining (6.5) and (6.6), we have (6.4). □

The following corollary follows immediately from Lemma 6.9.

COROLLARY 6.10 *For $0 < \varepsilon < 1$ and $0 < \xi \leq \varepsilon$, we have*

$$L_\varepsilon(X|Y) \leq \lceil H_{\max}^{\varepsilon-\xi}(X|Y) + \log(1/\xi) \rceil.$$

6.3 Evaluation of Smooth Max-entropies

In the previous two sections, we have characterized the optimal performances of the source coding with or without side-information in terms of smooth max-entropies. Since the definition of smooth max-entropy involves optimization over sets $\mathcal{L} \subset \mathcal{X} \times \mathcal{Y}$, it is difficult to evaluate smooth max-entropy directly. In this

[2] If the source happens to be $X = x_0$, the prescribed symbol, then the error may not occur. However, for simplicity of analysis, we count this case as an error; this will only overestimate the error probability.

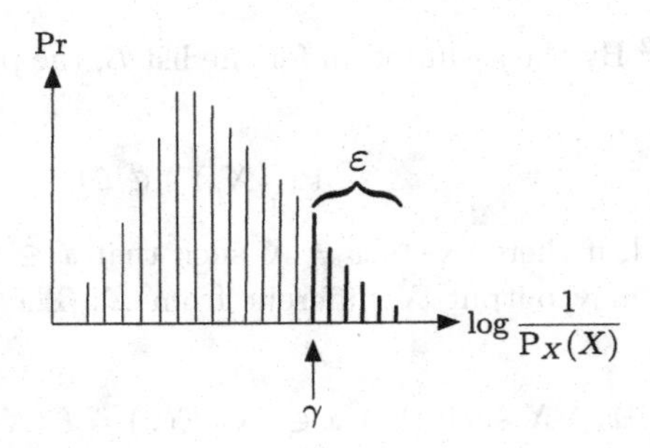

Figure 6.4 A heuristic description of Lemma 6.11. The threshold γ is the ε-upper quantile of the entropy density $\log \frac{1}{P_X(X)}$, i.e., the probability that the entropy density exceeds γ is ε.

section, we present a way to evaluate both the smooth max-entropy and smooth conditional max-entropy. First, we show that the smooth max-entropies can be bounded in terms of the ε-upper quantile of the "entropy density." Then, for a sequence of i.i.d. random variables, by using concentration bounds to evaluate the quantile, we show that smooth max-entropies can be approximated by Shannon entropies.

We start with the case without side-information. As we discussed after Proposition 6.4, in order to evaluate $H^{\varepsilon}_{\max}(X)$ exactly, we sort the symbols in decreasing order of the probabilities, and then include symbols in the list $\mathcal{L}$ until the total probability of the list exceeds $1 - \varepsilon$. The question is how many symbols do we need to include? Roughly speaking, the following lemma claims that the required size of the list is $\log|\mathcal{L}| \simeq \gamma$, where γ is the ε-upper quantile of the entropy density $\log \frac{1}{P_X(X)}$; see Figure 6.4.

LEMMA 6.11 *For a given $0 \le \varepsilon < 1$, we have*

$$H^{\varepsilon}_{\max}(X) \le \inf \left\{ \gamma \in \mathbb{R} : \Pr\left(\log \frac{1}{P_X(X)} \ge \gamma \right) \le \varepsilon \right\}, \tag{6.7}$$

and that

$$H^{\varepsilon}_{\max}(X) \ge \sup \left\{ \gamma \in \mathbb{R} : \Pr\left(\log \frac{1}{P_X(X)} \ge \gamma \right) \ge \varepsilon + \xi \right\} - \log(1/\xi) \tag{6.8}$$

for any $0 < \xi \le 1 - \varepsilon$.

Proof To prove (6.7), for any γ satisfying

$$\Pr\left(\log \frac{1}{P_X(X)} \ge \gamma \right) \le \varepsilon, \tag{6.9}$$

let

$$\mathcal{L} = \left\{ x : \log \frac{1}{P_X(x)} < \gamma \right\}.$$

Then, we have $\mathrm{P}_X(\mathcal{L}) \geq 1-\varepsilon$. Furthermore, since $\mathrm{P}_X(x) > 2^{-\gamma}$ for every $x \in \mathcal{L}$, we have

$$\begin{aligned}1 &\geq \sum_{x\in\mathcal{L}} \mathrm{P}_X(x) \\ &\geq \sum_{x\in\mathcal{L}} 2^{-\gamma} \\ &= |\mathcal{L}|2^{-\gamma},\end{aligned}$$

which implies $|\mathcal{L}| \leq 2^{\gamma}$. Consequently, we have

$$H_{\max}^{\varepsilon}(X) \leq \gamma$$

for any γ satisfying (6.9), which implies (6.7).

To prove (6.8), for any γ satisfying

$$\Pr\left(\log\frac{1}{\mathrm{P}_X(X)} \geq \gamma\right) \geq \varepsilon + \xi, \tag{6.10}$$

let

$$\mathcal{T} = \left\{x : \log\frac{1}{\mathrm{P}_X(x)} \geq \gamma\right\}.$$

Let $\mathcal{L}$ be such that $\mathrm{P}_X(\mathcal{L}) \geq 1-\varepsilon$ and $\log|\mathcal{L}| = H_{\max}^{\varepsilon}(X)$. Then, by noting that $\mathrm{P}_X(x) \leq 2^{-\gamma}$ for every $x \in \mathcal{T}$, we have

$$\begin{aligned}\varepsilon + \xi &\leq \mathrm{P}_X(\mathcal{T}) \\ &= \mathrm{P}_X(\mathcal{T}\cap\mathcal{L}^c) + \mathrm{P}_X(\mathcal{T}\cap\mathcal{L}) \\ &\leq \mathrm{P}_X(\mathcal{L}^c) + \sum_{x\in\mathcal{T}\cap\mathcal{L}} \mathrm{P}_X(x) \\ &\leq \varepsilon + \sum_{x\in\mathcal{T}\cap\mathcal{L}} 2^{-\gamma} \\ &\leq \varepsilon + |\mathcal{L}|2^{-\gamma},\end{aligned}$$

which implies

$$\log|\mathcal{L}| \geq \gamma - \log(1/\xi).$$

Since this inequality holds for any γ satisfying (6.10), we have (6.8). □

Next, we consider a sequence of i.i.d. random variables $X^n = (X_1, \ldots, X_n)$ with common distribution P_X. In this case, the smooth max-entropy $H_{\max}^{\varepsilon}(X^n)$ is known to grow linearly in n, and we are interested in characterizing the coefficient of n in the growth rate. In fact, it can be verified that the coefficient is given by a familiar quantity, the Shannon entropy. We show this below.

Note that the expectation of the entropy density equals the Shannon entropy, i.e.,

$$\mathbb{E}\left[\log\frac{1}{\mathrm{P}_X(X)}\right] = H(X).$$

Furthermore, since the alphabet $\mathcal{X}$ is finite, the variance of the entropy density is finite. Thus, by the law of large numbers (cf. Theorem 2.4), for a given $0 < \varepsilon < 1$ and an arbitrary $\delta > 0$, we have

$$\Pr\left(\frac{1}{n}\sum_{i=1}^{n} \log \frac{1}{\mathrm{P}_X(X_i)} \geq H(X) + \delta\right) \leq \varepsilon$$

for every sufficiently large n. By (6.7), this means that

$$\frac{1}{n} H_{\max}^{\varepsilon}(X^n) \leq H(X) + \delta$$

for every sufficiently large n. Since we can take $\delta > 0$ arbitrarily small by making n sufficiently large, we have

$$\limsup_{n\to\infty} \frac{1}{n} H_{\max}^{\varepsilon}(X^n) \leq H(X).$$

Similarly, by using the law of large numbers and (6.8), we can also prove the opposite inequality

$$\liminf_{n\to\infty} \frac{1}{n} H_{\max}^{\varepsilon}(X^n) \geq H(X).$$

In the following, by using the Hoeffding bound, we derive more explicit bounds on $\frac{1}{n} H_{\max}^{\varepsilon}(X^n)$.

THEOREM 6.12 *For a given $0 < \varepsilon < 1$, we have*

$$H(X) - \kappa\sqrt{\frac{\ln(2/(1-\varepsilon))}{2n}} - \frac{\log(2/(1-\varepsilon))}{n} \leq \frac{1}{n} H_{\max}^{\varepsilon}(X^n) \leq H(X) + \kappa\sqrt{\frac{\ln(1/\varepsilon)}{2n}}, \tag{6.11}$$

where $\kappa = \max_{x \in \mathtt{supp}(\mathrm{P}_X)} \log \frac{1}{\mathrm{P}_X(x)}$.

Proof Note that $\log \frac{1}{\mathrm{P}_X(X_i)}$ takes values in $[0, \kappa]$ with probability 1. To prove the inequality on the right-hand side of (6.11), by using the Hoeffding bound (see Theorem 2.39) with $\delta = \kappa\sqrt{\frac{\ln(1/\varepsilon)}{2n}}$, we have

$$\Pr\left(\frac{1}{n}\sum_{i=1}^{n} \log \frac{1}{\mathrm{P}_X(X_i)} \geq H(X) + \kappa\sqrt{\frac{\ln(1/\varepsilon)}{2n}}\right) \leq \varepsilon.$$

Thus, by (6.7), we have

$$\frac{1}{n} H_{\max}^{\varepsilon}(X^n) \leq \frac{1}{n} \inf\left\{\gamma : \Pr\left(\sum_{i=1}^{n} \log \frac{1}{\mathrm{P}_X(X_i)} \geq \gamma\right) \leq \varepsilon\right\}$$
$$\leq H(X) + \kappa\sqrt{\frac{\ln(1/\varepsilon)}{2n}}.$$

To prove the inequality on the left-hand side of (6.11), again, by using the Hoeffding bound with $\delta = \kappa\sqrt{\frac{\ln(2/(1-\varepsilon))}{2n}}$, we have

$$\Pr\left(\frac{1}{n}\sum_{i=1}^{n}\log\frac{1}{\mathrm{P}_X(X_i)} < H(X) - \kappa\sqrt{\frac{\ln(2/(1-\varepsilon))}{2n}}\right) \le \frac{1-\varepsilon}{2}.$$

Thus, by (6.8) with $\xi = \frac{1-\varepsilon}{2}$, we have

$$\begin{aligned}\frac{1}{n}H^{\varepsilon}_{\max}(X^n) &\ge \frac{1}{n}\sup\left\{\gamma : \Pr\left(\sum_{i=1}^{n}\log\frac{1}{\mathrm{P}_X(X_i)} < \gamma\right) \le \frac{1-\varepsilon}{2}\right\} - \frac{\log(2/(1-\varepsilon))}{n} \\ &\ge H(X) - \kappa\sqrt{\frac{\ln(2/(1-\varepsilon))}{2n}} - \frac{\log(2/(1-\varepsilon))}{n}.\end{aligned}$$

□

For a sequence of i.i.d. random variables $X^n = (X_1, \ldots, X_n)$, note that $\frac{1}{n}L_\varepsilon(X^n)$ represents the optimal "rate" of a source code for the source X^n, namely the minimum number of bits needed per symbol for storing n symbols in X^n. By combining Proposition 6.4 and Theorem 6.12, we obtain Shannon's source coding theorem below as an immediate corollary.

COROLLARY 6.13 *For a given $0 < \varepsilon < 1$, we have*

$$\lim_{n\to\infty}\frac{1}{n}L_\varepsilon(X^n) = H(X).$$

Next, we move to the case with side-information. In fact, the arguments are similar to the case without side-information. First, we derive bounds on the conditional smooth max-entropy in terms of the ε-upper quantile of the conditional entropy density as follows.

LEMMA 6.14 *For a given $0 \le \varepsilon < 1$, we have*

$$H^{\varepsilon}_{\max}(X|Y) \le \inf\left\{\gamma \in \mathbb{R} : \Pr\left(\log\frac{1}{\mathrm{P}_{X|Y}(X|Y)} \ge \gamma\right) \le \varepsilon\right\}, \tag{6.12}$$

and that

$$H^{\varepsilon}_{\max}(X|Y) \ge \sup\left\{\gamma \in \mathbb{R} : \Pr\left(\log\frac{1}{\mathrm{P}_{X|Y}(X|Y)} \ge \gamma\right) \ge \varepsilon + \xi\right\} - \log(1/\xi) \tag{6.13}$$

for any $0 < \xi \le 1 - \varepsilon$.

Proof The proof is similar to that of Lemma 6.11. To prove (6.12), for any γ satisfying

$$\Pr\left(\log\frac{1}{\mathrm{P}_{X|Y}(X|Y)} \ge \gamma\right) \le \varepsilon, \tag{6.14}$$

let

$$\mathcal{L}_y = \left\{x : \log\frac{1}{\mathrm{P}_{X|Y}(x|y)} < \gamma\right\}$$

for each $y \in \mathcal{Y}$, and let $\mathcal{L} = \cup_{y \in \mathcal{Y}} \mathcal{L}_y \times \{y\}$. Then, we have $\mathrm{P}_{XY}(\mathcal{L}) \geq 1 - \varepsilon$. Furthermore, in the same manner as the proof of Lemma 6.11, we can prove $|\mathcal{L}_y| \leq 2^\gamma$ for every $y \in \mathcal{Y}$. Consequently, for any γ satisfying (6.14) we have

$$H_{\max}^{\varepsilon}(X|Y) \leq \gamma,$$

which implies (6.12).

To prove (6.13), for any γ satisfying

$$\Pr\left(\log \frac{1}{\mathrm{P}_{X|Y}(X|Y)} \geq \gamma\right) \geq \varepsilon + \xi, \tag{6.15}$$

let

$$\mathcal{T}_y = \left\{x : \log \frac{1}{\mathrm{P}_{X|Y}(x|y)} \geq \gamma\right\}$$

for each $y \in \mathcal{Y}$, and let $\mathcal{T} = \cup_{y \in \mathcal{Y}} \mathcal{T}_y \times \{y\}$. Let $\mathcal{L}$ be such that $\mathrm{P}_{XY}(\mathcal{L}) \geq 1 - \varepsilon$ and $\max_{y \in \mathcal{Y}} \log |\mathcal{L}_y| = H_{\max}^{\varepsilon}(X|Y)$. Then, by noting that $\mathrm{P}_{X|Y}(x|y) \leq 2^{-\gamma}$ for every $x \in \mathcal{T}_y$, we have

$$\begin{aligned}
\varepsilon + \xi &\leq \mathrm{P}_{XY}(\mathcal{T}) \\
&= \mathrm{P}_{XY}(\mathcal{T} \cap \mathcal{L}^c) + \mathrm{P}_{XY}(\mathcal{T} \cap \mathcal{L}) \\
&\leq \mathrm{P}_{XY}(\mathcal{L}^c) + \sum_{(x,y) \in \mathcal{T} \cap \mathcal{L}} \mathrm{P}_{XY}(x, y) \\
&\leq \varepsilon + \sum_y \mathrm{P}_Y(y) \sum_{x \in \mathcal{T}_y \cap \mathcal{L}_y} \mathrm{P}_{X|Y}(x|y) \\
&\leq \varepsilon + \sum_y \mathrm{P}_Y(y) \sum_{x \in \mathcal{T}_y \cap \mathcal{L}_y} 2^{-\gamma} \\
&\leq \varepsilon + \sum_y \mathrm{P}_Y(y) |\mathcal{L}_y| 2^{-\gamma} \\
&\leq \varepsilon + \max_{y \in \mathcal{Y}} |\mathcal{L}_y| 2^{-\gamma},
\end{aligned}$$

which implies

$$\max_{y \in \mathcal{Y}} \log |\mathcal{L}_y| \geq \gamma - \log(1/\xi).$$

Since this inequality holds for any γ satisfying (6.15), we have (6.13). □

Finally, let us consider a sequence of i.i.d. random variables (X^n, Y^n) with common joint distribution P_{XY}, where $X^n = (X_1, \ldots, X_n)$ and $Y^n = (Y_1, \ldots, Y_n)$. Note that the expectation of the conditional entropy density is the conditional entropy, i.e.,

$$\mathbb{E}\left[\log \frac{1}{\mathrm{P}_{X|Y}(X|Y)}\right] = H(X|Y).$$

By using the Hoeffding bound, we derive the following bounds for conditional smooth max-entropy in terms of conditional Shannon entropy. Since the proof is exactly the same as Theorem 6.12, we leave it as an exercise.

THEOREM 6.15 *For a given $0 < \varepsilon < 1$, we have*

$$H(X|Y) - \kappa\sqrt{\frac{\ln(2/(1-\varepsilon))}{2n}} - \frac{\log(2/(1-\varepsilon))}{n} \le \frac{1}{n} H_{\max}^{\varepsilon}(X^n|Y^n)$$
$$\le H(X|Y) + \kappa\sqrt{\frac{\ln(1/\varepsilon)}{2n}}, \quad (6.16)$$

where $\kappa = \max_{(x,y)\in \mathrm{supp}(P_{XY})} \log \frac{1}{P_{X|Y}(x|y)}$.

Proof Problem 6.2. □

By using Proposition 6.8 and Corollary 6.10 together with Theorem 6.15, we obtain the following characterization of the asymptotic optimal rate $\frac{1}{n}L_\varepsilon(X^n|Y^n)$ of the source code for X^n with side-information Y^n.

COROLLARY 6.16 *For a given $0 < \varepsilon < 1$, we have*

$$\lim_{n\to\infty} \frac{1}{n} L_\varepsilon(X^n|Y^n) = H(X|Y).$$

Proof The one direction $\ge$ follows from Proposition 6.8 and the inequality on the left-hand side of (6.16). To prove the opposite inequality, we apply Corollary 6.10 with, say $\xi = \varepsilon/2$, together with the inequality on the right-hand side of (6.16). □

6.4 References and Additional Reading

An overview of fixed length source coding can be found in any standard textbook on information theory; for instance, see [75, Chapter 3]. A traditional approach to fixed length source coding is to invoke the asymptotic equipartition property from the beginning so that the asymptotic limit of the compression rate can be derived. Our treatment in this chapter is along the line of the recently popularized area of finite block length analysis [159, 274]. For an overview of finite block length analysis, see [320]. The treatment of Section 6.1 based on smooth max-entropy can be found in [337].

The source coding with side-information was initially studied by Slepian and Wolf for the class of i.i.d. sources [312]. After that, a simple proof based on so-called random binning was introduced by Cover [74]; his approach easily extends to the class of ergodic sources. In fact, random binning can be regarded as a special case of the universal hash family. The treatment of Section 6.2 is essentially from [290].

The use of quantiles of the entropy density is the basic philosophy of the information spectrum approach introduced by Han and Verdú [152]; see also [151]. The source coding with side-information was studied using the information spectrum approach in [252]. In this chapter, we followed the more modern approach: we first characterized the fundamental limit of the source coding with/without side-information in terms of the smooth max-entropies; then, we evaluated the smooth max-entropies using the information spectrum approach (cf. Lemma 6.11 and Lemma 6.14). For a more detailed connection between the smooth entropy framework and the information spectrum approach, see [98, 199, 325, 337].

Problems

6.1 Prove that the ε-smooth max-entropy can be alternatively written as [337]

$$H_{\max}^{\varepsilon}(X) = \min_{\mathrm{P}_{\tilde{X}} \in \bar{\mathcal{B}}_{\varepsilon}(\mathrm{P}_X)} H_{\max}(\tilde{X}),$$

where the minimization is taken over

$$\bar{\mathcal{B}}_{\varepsilon}(\mathrm{P}_X) = \{\mathrm{P}_{\tilde{X}} : d_{\mathrm{var}}(\mathrm{P}_{\tilde{X}}, \mathrm{P}_X) \le \varepsilon\}.$$

6.2 Prove Theorem 6.15.

6.3 Using Fano's inequality (cf. Theorem 2.29), prove the following lower bound on the minimum length of a (k, ε)-code:

$$L_{\varepsilon}(X) \ge H(X) - \varepsilon \log |\mathcal{X}| - h(\varepsilon).$$

REMARK For a sequence X^n of i.i.d. sources, this bound gives

$$\frac{1}{n} L_{\varepsilon}(X^n) \ge H(X) - \varepsilon \log |\mathcal{X}| - \frac{h(\varepsilon)}{n}.$$

However, this bound is not tight enough to prove Corollary 6.13 since the residual terms do not vanish unless we take the limit $\varepsilon \to 0$.

6.4 *Variable length code*: A *variable length code* consists of an encoder $\varphi \colon \mathcal{X} \to \{0,1\}^*$ and a decoder $\psi \colon \{0,1\}^* \to \mathcal{X}$ such that $\psi(\varphi(x)) = x$ for every $x \in \mathcal{X}$, where $\{0,1\}^*$ is the set of all finite length binary sequences.[3] For a length $\ell(x)$ of the encoded message $\varphi(x)$ and a threshold k, the probability $\Pr(\ell(X) > k)$ is termed the overflow probability of the code. Prove the following connection between the error probability of a fixed length code and the overflow probability of a *variable length code*.

1. If there exists a *variable length code* with overflow probability less than ε for threshold k, then there exists a fixed length code of length k with error probability less than ε.
2. If there exists a fixed length code of length k with error probability less than ε, then there exists a *variable length code* with overflow probability less than ε for threshold $k+1$.

REMARK Various connections between the *variable length code* and the fixed length code have been known as folklores. For the analysis of overflow probability, see [248, 336].

6.5 *Discrepancy between average and ε-upper quantile*: For a (large) integer t and $\delta = \frac{1}{t}$, let P_X be the distribution on $\mathcal{X} = \{1, \ldots, 2^t\}$ given by $\mathrm{P}_X(2^t) = 1 - \delta$ and $\mathrm{P}_X(x) = \frac{\delta}{2^t - 1}$ for $x \neq 2^t$.

[3] For instance, the Huffman code and the Lempel–Ziv code are *variable length codes*.

1. Verify that $H(X) \le 1 + h(\delta)$.
   HINT Use the Fano inequality (cf. Theorem 2.29).
2. For $\varepsilon = \frac{\delta}{2}$, verify that $H_{\max}^{\varepsilon}(X) \ge \log \frac{2^t - 1}{2}$.
   HINT Use (6.8) with $\xi = \frac{\delta}{2}$.

REMARK Even though the Shannon entropy $H(X)$ is an appropriate information measure to quantify the asymptotic limit of the source coding (cf. Corollary 6.13), for the single-shot regime, it may underestimate the coding length necessary for a given error probability.

6.6 *Compression and encryption*: Suppose that the sender and the receiver share a secret key K uniformly distributed on $\mathcal{K}$. Prove that, if $\log|\mathcal{K}| \ge \lceil H_{\max}^{\varepsilon}(X) \rceil$, then there exists an ε-reliable and perfectly secure encryption scheme for transmitting the message X (cf. Section 3.4 for the reliability of an encryption scheme).

REMARK When the message to be transmitted is biased, the length of secret key needed in an encryption scheme can be saved by a compression.

6.7 *Mixed source*: Suppose that the distributions of a sequence X^n of sources are given by

$$\mathrm{P}_{X^n}(x^n) = \theta \mathrm{P}_{X_1}^n(x^n) + (1-\theta)\mathrm{P}_{X_2}^n(x^n)$$

for some $0 < \theta < 1$ and common distributions P_{X_1} and P_{X_2} on $\mathcal{X}$ with $H(X_1) > H(X_2)$; this type of source is termed the mixed source [151]. Prove that the asymptotic limit of compression for mixed source is given as

$$\lim_{\varepsilon \to 0} \limsup_{n \to \infty} \frac{1}{n} L_\varepsilon(X^n) = \max[H(X_1), H(X_2)] = H(X_1).$$

6.8 *Second-order coding rate*: For a sequence X^n of i.i.d. sources with

$$\mathtt{V}(X) := \mathtt{Var}\left[\log \frac{1}{\mathrm{P}_X(X)}\right] > 0,$$

prove that

$$L_\varepsilon(X^n) = nH(X) + \sqrt{n\mathtt{V}(X)}\mathtt{Q}^{-1}(\varepsilon) + o(\sqrt{n}), \tag{6.17}$$

where $\mathtt{Q}^{-1}(\varepsilon)$ is the inverse of the tail probability

$$\mathtt{Q}(a) = \int_a^\infty \frac{1}{\sqrt{2\pi}} e^{-\frac{t^2}{2}} dt$$

of the Gaussian distribution.

REMARK The coefficient of $\sqrt{n}$ in (7.20) is referred to as the second-order coding rate, and it clarifies the dependence of the asymptotic limit of compression on the error probability ε; see the review article [320] for a recent development of the second-order coding rate.

7 Random Number Generation

In this chapter, we present the problem of random number generation. The purpose of random number generation is to create almost uniform bits from a given biased random number. This problem has been studied in various research fields, including information theory, cryptography, theoretical computer science, and physics. In particular, it constitutes a basic building block of cryptographic protocols called *privacy amplification*.

In a similar manner to Chapter 6, we start with the single-shot regime, and obtain the block coding version for i.i.d. sources as a simple corollary at the end. In the single-shot regime, the quantity called the smooth min-entropy plays a pivotal role. This quantity can be regarded as a kind of "dual" quantity to the smooth max-entropy discussed in Chapter 6. The main technical workhorse for this chapter is the *leftover hash lemma*, and we present many variants of it.

7.1 Random Number Generation

We consider the source model of Chapter 6. Let X be a source taking values in a finite set $\mathcal{X}$. By using X as a seed, we seek to generate random bits K distributed (almost) uniformly on $\mathcal{K} = \{0,1\}^\ell$. Namely, we want to design a function $f\colon \mathcal{X} \to \mathcal{K}$ such that the distribution P_K of the output $K = f(X)$ is "close" to the uniform distribution $\mathrm{P}_{\mathtt{unif}}$ on $\mathcal{K}$. Such a function f is referred to as an *extractor*.

In order to get some insight, consider the following simple example.

EXAMPLE 7.1 For $\mathcal{X} = \{1,2,3,4,5\}$, consider the distribution P_X given by $\mathrm{P}_X(1) = 1/3$, $\mathrm{P}_X(2) = 1/4$, $\mathrm{P}_X(3) = 1/6$, $\mathrm{P}_X(4) = 1/8$, and $\mathrm{P}_X(5) = 1/8$. For $\mathcal{K} = \{0,1\}$, let $f(1) = f(3) = 0$ and $f(2) = f(4) = f(5) = 1$. Then, the distribution P_K of the output $K = f(X)$ satisfies $\mathrm{P}_K(0) = \mathrm{P}_X(1)+\mathrm{P}_X(3) = 1/2$ and $\mathrm{P}_K(1) = \mathrm{P}_X(2)+\mathrm{P}_X(4)+\mathrm{P}_X(5) = 1/2$. Thus, f is an extractor which generates a uniform bit from the source X.

In general, it is not possible to generate exactly uniform bits from a given source; in practice as well, it suffices to generate almost uniform bits. For cryptographic applications (such as those seen in Chapter 3), it is convenient to

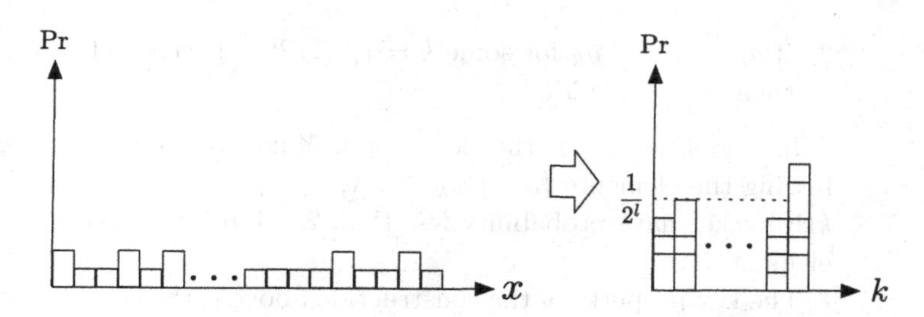

Figure 7.1 Description of Construction 7.1.

evaluate the closeness to the uniform distribution using variational distance. Thus, for a given $0 \le \varepsilon < 1$, we require the output $K = f(X)$ to satisfy

$$d_{\mathtt{var}}(\mathrm{P}_K, \mathrm{P}_{\mathtt{unif}}) \le \varepsilon, \tag{7.1}$$

where $\mathrm{P}_{\mathtt{unif}}$ is the uniform distribution on $\mathcal{K} = \{0,1\}^\ell$. An extractor satisfying (7.1) is referred to as an ε-extractor. Under the required security level, we would like to make the length ℓ of the generated bits as large as possible.

In the source coding considered in Chapter 6, max-entropy and its variants played a pivotal role. In random number generation, a similar role will be played by another notion of entropy called the min-entropy which is defined below.

DEFINITION 7.2 (Min-entropy) For a given random variable X with distribution P_X on $\mathcal{X}$, the min-entropy of X is defined as[1]

$$H_{\min}(X) = \min_{x \in \mathcal{X}} -\log \mathrm{P}_X(x)\,.$$

It can be checked that both the min-entropy and max-entropy can be obtained as limiting quantities for Rényi entropies of different order. In particular, $H_{\min}(X) = \lim_{\alpha\to\infty} H_\alpha(X)$ and $H_{\max}(X) = \lim_{\alpha\to 0} H_\alpha(X)$.

We present a construction for an extractor that allows us to extract roughly as many bits of almost uniform randomness from X as its min-entropy $H_{\min}(X)$. The construction is a greedy one and proceeds by dividing the symbols $x \in \mathcal{X}$ into bins with probabilities close to $2^{-\ell}$; see Figure 7.1 for a pictorial description.

CONSTRUCTION 7.1 (Greedy construction) Let $\mathcal{X} = \{1, \ldots, |\mathcal{X}|\}$ and $\mathcal{K} = \{1, \ldots, 2^\ell\}$.

1. Set $a_0 = 1$. For $k = 1, \ldots, 2^\ell - 1$, find the integer a_k such that[2]

$$\sum_{x=a_{k-1}}^{a_k - 1} \mathrm{P}_X(x) \le 2^{-\ell} < \sum_{x=a_{k-1}}^{a_k - 1} \mathrm{P}_X(x) + \mathrm{P}_X(a_k)\,.$$

[1] When $\mathrm{P}_X(x) = 0$, $-\log \mathrm{P}_X(x)$ is regarded as ∞; thus, the minimization is essentially taken over $x \in \mathrm{supp}(\mathrm{P}_X)$.

[2] We assume that there is no symbol with $\mathrm{P}_X(x) = 0$. That is, we restrict to the support of the distribution P_X.

2. If $a_{k-1} \le x < a_k$ for some $k = 1, \ldots, 2^\ell - 1$, then set $f(x) = k$; if $x \ge a_{2^\ell - 1}$, then set $f(x) = 2^\ell$.

In words, we divide the elements in $\mathcal{X}$ into buckets, with the kth bucket comprising the elements $\{a_{k-1}, a_{k-1}+1, \ldots, a_k - 1\}$, such that the elements in the kth bucket have probability less than $2^{-\ell}$ but will exceed $2^{-\ell}$ if we include the element a_k.

The key property of the construction above is that each bucket's probability is at most $\max_{x \in \mathcal{X}} \mathrm{P}_X(x)$ away from the uniform distribution probability $2^{-\ell}$. The next result uses this observation to evaluate the performance of Construction 7.1.

PROPOSITION 7.3 *For a random variable X, if we set*

$$\ell = \lfloor H_{\min}(X) - \log(1/\varepsilon) \rfloor,$$

then the output $K = f(X)$ of Construction 7.1 satisfies $d_{\mathtt{var}}(\mathrm{P}_K, \mathrm{P}_{\mathtt{unif}}) \le \varepsilon$.

Proof By the construction, note that

$$0 \le \frac{1}{2^\ell} - \mathrm{P}_K(k) \le \mathrm{P}_X(a_k) \le \max_x \mathrm{P}_X(x) = 2^{-H_{\min}(X)}$$

for each $k = 1, \ldots, 2^\ell - 1$. Furthermore, note that

$$\begin{aligned}
\mathrm{P}_K(2^\ell) - \frac{1}{2^\ell} &= \left(1 - \frac{1}{2^\ell}\right) - \left(1 - \mathrm{P}_K(2^\ell)\right) \\
&= \sum_{k=1}^{2^\ell - 1} \left(\frac{1}{2^\ell} - \mathrm{P}_K(k)\right).
\end{aligned}$$

From these observations, it follows that

$$\begin{aligned}
d_{\mathtt{var}}(\mathrm{P}_K, \mathrm{P}_{\mathtt{unif}}) &= \frac{1}{2}\left[\sum_{k=1}^{2^\ell - 1} \left(\frac{1}{2^\ell} - \mathrm{P}_K(k)\right) + \left(\mathrm{P}_K(2^\ell) - \frac{1}{2^\ell}\right)\right] \\
&= \sum_{k=1}^{2^\ell - 1} \left(\frac{1}{2^\ell} - \mathrm{P}_K(k)\right) \\
&\le \sum_{k=1}^{2^\ell - 1} 2^{-H_{\min}(X)} \\
&\le 2^{\ell - H_{\min}(X)} \\
&\le \varepsilon,
\end{aligned}$$

where we used $\ell = \lfloor H_{\min}(X) - \log(1/\varepsilon) \rfloor$ in the last step. □

The following simple example illustrates the performance of Construction 7.1.

EXAMPLE 7.4 Let $X^n = (X_1, \ldots, X_n)$ be a sequence of i.i.d. Bernoulli random variables with common distribution specified by $\mathrm{P}_X(1) = p \ge 1/2$. In this case, $H_{\min}(X^n) = -n \log p$. For instance, if $p = 1/\sqrt{2}$, we can generate $\lfloor n/2 - \log(1/\varepsilon) \rfloor$ almost uniform bits by Construction 7.1.

Proposition 7.3 states that min-entropy constitutes roughly a lower bound for the number of almost uniform bits that can be extracted from a given source X. However, Construction 7.1 relies on knowledge of the distribution P_X. Next, we shall consider the generation of almost uniform bits when the source distribution P_X is not known completely; we only know that the source has a sufficient amount of randomness. To formalize our setting, we introduce the following class of ν-sources.

DEFINITION 7.5 (ν-source) A source X with distribution P_X on $\mathcal{X}$ constitutes a *ν-source* if

$$H_{\min}(X) \geq \nu.$$

The class of all ν-sources on $\mathcal{X}$ is denoted by $\mathcal{P}_\nu(\mathcal{X})$.

Heuristically, a ν-source can be regarded as a source with at least ν bits of randomness. However, it is not possible to extract even one bit of uniform randomness from $\mathcal{P}_\nu(\mathcal{X})$ if we only allow deterministic extractors, as is argued in the following example.

EXAMPLE 7.6 For an arbitrary extractor $f\colon \{0,1\}^n \to \{0,1\}$ such that $|f^{-1}(0)| \geq |f^{-1}(1)|$, consider uniform distribution Q on $f^{-1}(0)$. Since $|f^{-1}(0)| \geq |f^{-1}(1)|$, Q is the distribution of an $(n-1)$-source but $\mathrm{Q}(f^{-1}(0)) = 1$, i.e., the output of the extractor is 0 with probability 1 and is extremely biased. Thus, there exists no one-bit deterministic extractor for $\mathcal{P}_{n-1}(\{0,1\}^n)$.

In view of the previous example, to find extractors that work for all ν-sources, we must take recourse to randomized extractors. The *leftover hash* lemma given in the next section shows the existence of a randomized ε-extractor for $\mathcal{P}_\nu(\mathcal{X})$.

However, the notion of randomized extractor must be carefully defined. In particular, treating a randomized extractor as a function of randomness U used to choose the random extractor and source X, we may simply declare the randomness U as output irrespective of the value of X. But this will not capture randomness extracted from X. To circumvent this conundrum, we require that the generated bits are not only almost uniform but are also almost independent of the randomness that is used to choose an extractor. In other words, the randomness for choosing an extractor is not consumed and only plays the role of a catalyst.

Formally, we can capture both the requirements of almost uniformity and almost independence for a randomized extractor succinctly as follows.

DEFINITION 7.7 (Randomized extractor) For a given source X with distribution P_X on $\mathcal{X}$, a randomized *ε-extractor* of length ℓ for P_X consists of a random mapping $F\colon \mathcal{X} \to \mathcal{K}$, selected using a distribution P_F on a set of mappings $\mathcal{F}$, such that $K = F(X)$ satisfies

$$d_{\mathtt{var}}(\mathrm{P}_{KF}, \mathrm{P}_{\mathtt{unif}} \times \mathrm{P}_F) \leq \varepsilon,$$

where $\mathrm{P_{unif}}$ is the uniform distribution on $\mathcal{K} = \{0,1\}^\ell$. If F constitutes a randomized ε-extractor for every source in $\mathcal{P}_\nu(\mathcal{X})$, we say that F is a randomized ε-extractor for ν-sources.[3]

In the next section, we present the most widely used construction for a randomized extractor.

7.2 Leftover Hashing

We saw that a randomized extractor is necessary to generate almost uniform bits from the class of ν-sources. As we discussed briefly in Section 4.4.2, a universal hash family (UHF) can be used as a good extractor. Recall that, if F is distributed uniformly over a UHF $\mathcal{F}$ from $\mathcal{X}$ to $\{0,1\}^\ell$, then

$$\Pr(F(x) = F(x')) \le \frac{1}{2^\ell}, \quad \forall x \ne x' \in \mathcal{X}.$$

In Section 4.4.2, too, we saw that the output of the UHF is close to uniform if the length of the output is sufficiently smaller than the initial randomness. But there, closeness to uniformity was evaluated using the collision probability of P_X. In this section, even when we evaluate closeness to uniformity using the variational distance, the collision probability plays an important role. Recall that the collision probability of a random variable X with distribution P_X is given by $\sum_x \mathrm{P}_X(x)^2$. In this chapter, the quantity of interest for us is the *collision entropy* given by

$$H_2(X) = -\log \sum_x \mathrm{P}_X(x)^2.$$

The reader may recognize that collision entropy is nothing but the Rényi entropy of order 2, and since $H_\alpha(X)$ is a nonincreasing function of α (see also Problem 7.1), we have the relation $H_2(X) \ge H_{\min}(X)$. Thus, for any ν-source X, the source X has $H_2(X) \ge \nu$, namely it has ν bits of randomness in the sense of collision entropy as well.

The following theorem, known as the *leftover hash lemma*, is one of the main tools used in random number generation. In this chapter, we will develop multiple variants of the leftover hash lemma. The following is the most basic form, and its proof contains the main proof ideas underlying all the variants.

THEOREM 7.8 (Leftover hash lemma) *For a mapping F chosen uniformly at random from a UHF $\mathcal{F}$, $K = F(X)$ satisfies*

$$d_{\mathtt{var}}(\mathrm{P}_{KF}, \mathrm{P_{unif}} \times \mathrm{P}_F) \le \frac{1}{2}\sqrt{2^{\ell - H_2(X)}} \tag{7.2}$$

$$\le \frac{1}{2}\sqrt{2^{\ell - H_{\min}(X)}}. \tag{7.3}$$

[3] When it is apparent that F is a randomized extractor, we just say that F is an ε-extractor.

Proof Since $H_2(X) \geq H_{\min}(X)$, it suffices to show (7.2). For each realization $F = f$, by the Cauchy–Schwarz inequality, we have

$$\begin{aligned}
d_{\mathtt{var}}(\mathrm{P}_{f(X)}, \mathrm{P}_{\mathtt{unif}}) &= \frac{1}{2}\sum_k |\mathrm{P}_{f(X)}(k) - \mathrm{P}_{\mathtt{unif}}(k)| \\
&\leq \frac{1}{2}\sqrt{2^\ell \sum_k \left(\mathrm{P}_{f(X)}(k) - \mathrm{P}_{\mathtt{unif}}(k)\right)^2}.
\end{aligned}$$

The term under $\sqrt{\cdot}$ can be evaluated as

$$\begin{aligned}
&\sum_k \left(\mathrm{P}_{f(X)}(k) - \mathrm{P}_{\mathtt{unif}}(k)\right)^2 \\
&= \sum_k \mathrm{P}_{f(X)}(k)^2 - 2\sum_k \mathrm{P}_{f(X)}(k)\,\mathrm{P}_{\mathtt{unif}}(k) + \sum_k \mathrm{P}_{\mathtt{unif}}(k)^2 \\
&= \sum_k \mathrm{P}_{f(X)}(k)^2 - \frac{1}{2^\ell} \\
&= \sum_{x,x'} \mathrm{P}_X(x)\,\mathrm{P}_X(x')\left(\mathbf{1}[f(x) = f(x')] - \frac{1}{2^\ell}\right) \\
&= \sum_x \mathrm{P}_X(x)^2\left(1 - \frac{1}{2^\ell}\right) + \sum_{x \neq x'} \mathrm{P}_X(x)\,\mathrm{P}_X(x')\left(\mathbf{1}[f(x) = f(x')] - \frac{1}{2^\ell}\right) \\
&\leq 2^{-H_2(X)} + \sum_{x \neq x'} \mathrm{P}_X(x)\,\mathrm{P}_X(x')\left(\mathbf{1}[f(x) = f(x')] - \frac{1}{2^\ell}\right).
\end{aligned}$$

Upon taking the average with respect to F, the second term is bounded as

$$\begin{aligned}
&\sum_f \mathrm{P}_F(f) \sum_{x \neq x'} \mathrm{P}_X(x)\,\mathrm{P}_X(x')\left(\mathbf{1}[f(x) = f(x')] - \frac{1}{2^\ell}\right) \\
&\quad = \sum_{x \neq x'} \mathrm{P}_X(x)\,\mathrm{P}_X(x')\left(\mathrm{P}(F(x) = F(x')) - \frac{1}{2^\ell}\right) \\
&\quad \leq 0,
\end{aligned}$$

where the inequality holds since F is distributed uniformly over a UHF. Thus, concavity of $\sqrt{\cdot}$ implies

$$\begin{aligned}
d(\mathrm{P}_{KF}, \mathrm{P}_{\mathtt{unif}} \times \mathrm{P}_F) &= \sum_f \mathrm{P}_F(f)\, d(\mathrm{P}_{f(X)}, \mathrm{P}_{\mathtt{unif}}) \\
&\leq \frac{1}{2}\sqrt{2^\ell \sum_f \mathrm{P}_F(f) \sum_k \left(\mathrm{P}_{f(X)}(k) - \mathrm{P}_{\mathtt{unif}}(k)\right)^2} \\
&\leq \frac{1}{2}\sqrt{2^{\ell - H_2(X)}}.
\end{aligned}$$

□

Note that the bound in Theorem 7.8 holds for any source. Therefore, for every $\mathrm{P}_X \in \mathcal{P}_\nu(\mathcal{X})$,

$$d_{\mathtt{var}}(\mathrm{P}_{KF}, \mathrm{P}_{\mathtt{unif}} \times \mathrm{P}_F) \leq \varepsilon$$

as long as

$$\ell \leq H_{\min}(X) - 2\log(1/2\varepsilon).$$

Thus, the random mapping F distributed uniformly over a UHF $\mathcal{F}$ constitutes an ε-extractor of length $\lfloor \nu - 2\log(1/2\varepsilon) \rfloor$ for $\mathcal{P}_\nu(\mathcal{X})$.

7.3 Leftover Hashing with Side-information

In cryptographic applications, often an adversary observes side-information Z which is correlated with the source X observed by the legitimate parties. We model this situation by assuming that the joint distribution P_{XZ} of (X, Z) is known; however, the side-information Z is only observed by the adversary. The goal of the legitimate parties is to extract almost uniform bits that are almost independent of the adversary's side-information Z. This particular variant of random number generation is often called *privacy amplification.*

EXAMPLE 7.9 For $\mathcal{X} = \{0, 1\}^2$, let $X = (X_1, X_2)$ be 2 random bits, and $Z = (X_1, \mathsf{e})$ with probability 1/2 and $Z = (\mathsf{e}, X_2)$ with probability 1/2, where e indicates that the bit is erased. In other words, exactly one of the bits X is leaked to the adversary. Even though the legitimate parties do not know which bit is leaked to the adversary, they can compute $K = X_1 \oplus X_2$, which is uniform and independent of the adversary's side-information Z.

As seen from this example, the problem of privacy amplification is very similar to the problem we saw in the previous section except that the uncertainty class of distributions is $\{\mathrm{P}_{X|Z}(\cdot|z)\}_{z\in\mathcal{Z}}$. Therefore, our UHF based scheme seen earlier can be used here as well. We note that, unlike the situation discussed in Example 7.6, it is not strictly necessary to use randomized extractors in the above mentioned model. Nonetheless, we consider randomized extractors for this setting as well and formalize the problem as follows.

DEFINITION 7.10 (Randomized extractor with side-information) For given sources (X, Z) with distribution P_{XZ} on $\mathcal{X} \times \mathcal{Z}$, a randomized *$\varepsilon$-extractor* of length ℓ for[4] P_{XZ} consists of a random mapping $F\colon \mathcal{X} \to \mathcal{K}$, selected using a distribution P_F from a set of mappings $\mathcal{F}$, such that $K = F(X)$ satisfies

$$d_{\mathtt{var}}(\mathrm{P}_{KZF}, \mathrm{P}_{\mathtt{unif}} \times \mathrm{P}_Z \times \mathrm{P}_F) \leq \varepsilon, \tag{7.4}$$

where $\mathrm{P}_{\mathtt{unif}}$ is the uniform distribution on $\mathcal{K} = \{0, 1\}^\ell$.

In cryptographic applications, the output K plays the role of a secret key. Since (7.4) guarantees that K is almost uniform and independent of Z as well as F, this key can be securely used even if the choice F of the extractor is known to the

[4] More formally, we should call it an extractor for P_{XZ} given Z, but here we are treating F as an extractor for the family of distributions $\{\mathrm{P}_{X|Z}(\cdot \mid z), z \in \mathcal{Z}\}$. However, the evaluation of security is averaged over the distribution P_Z.

adversary. Note that $d_{\mathtt{var}}(\mathrm{P}_{KZF}, \mathrm{P}_{\mathtt{unif}} \times \mathrm{P}_Z \times \mathrm{P}_F) = \mathbb{E}_F\big[d_{\mathtt{var}}(\mathrm{P}_{KZ|F}, \mathrm{P}_{\mathtt{unif}} \times \mathrm{P}_Z)\big]$, whereby the condition (7.4) implies that there exists a (deterministic mapping) f such that $K = f(X)$ satisfies

$$d_{\mathtt{var}}(\mathrm{P}_{KZ}, \mathrm{P}_{\mathtt{unif}} \times \mathrm{P}_Z) \leq \varepsilon.$$

The main goal of this section is to extend the leftover hash lemma of Theorem 7.8 to the case with side-information. In the next section, we will further extend the leftover hash lemma using a technique called "smoothing"; in that case, it is convenient to consider not only probability distributions, but also subdistributions, i.e., P_{XZ} such that $\sum_{x,z} \mathrm{P}_{XZ}(x,z) \leq 1$ but equality may not hold. Therefore, as a preparation for the next section, we consider subdistributions in this section as well. A subdistribution can be viewed as a vector with nonnegative entries with sum not exceeding 1. We can compute the total variation distance (as 1/2 of ℓ_1 distance) and further can compute the subdistribution

$$\mathrm{P}_{KZ}(k,z) = \sum_{x \in f^{-1}(k)} \mathrm{P}_{XZ}(x,z)$$

for an extractor f.

In order to develop the leftover hash lemma with side-information, we need to introduce a few conditional information measures. The first is the conditional version of the min-entropy. For the purpose of deriving the leftover hash lemma, it is convenient to define the conditional min-entropy in terms of a variational formula. Notice that, for a distribution P_{XZ}, the conditional Shannon entropy can be rewritten as

$$\begin{aligned} H(X|Z) &= H(\mathrm{P}_{XZ}) - H(\mathrm{P}_Z) \\ &= \sup_{\mathrm{Q}_Z} \big[H(\mathrm{P}_{XZ}) - H(\mathrm{P}_Z) - D(\mathrm{P}_Z \| \mathrm{Q}_Z)\big] \\ &= \sup_{\mathrm{Q}_Z} \left[-\sum_{x,z} \mathrm{P}_{XZ}(x,z) \log \frac{\mathrm{P}_{XZ}(x,z)}{\mathrm{Q}_Z(z)} \right], \end{aligned} \tag{7.5}$$

where the second equality holds since $D(\mathrm{P}_Z \| \mathrm{Q}_Z) \geq 0$ and the supremum in it is attained by $\mathrm{Q}_Z = \mathrm{P}_Z$. Inspired by the variational formula (7.5), we introduce the notion of conditional min-entropy as follows.

DEFINITION 7.11 (Conditional min-entropy) For a subdistribution P_{XZ} on $\mathcal{X} \times \mathcal{Z}$ and a distribution Q_Z on $\mathcal{Z}$,[5] the *conditional min-entropy* of P_{XZ} given Q_Z is defined as

$$H_{\min}(\mathrm{P}_{XZ}|\mathrm{Q}_Z) := \min_{(x,y) \in \mathtt{supp}(\mathrm{P}_{XZ})} -\log \frac{\mathrm{P}_{XZ}(x,z)}{\mathrm{Q}_Z(z)}.$$

Then, the conditional min-entropy of P_{XZ} given Z is defined as

$$H_{\min}(\mathrm{P}_{XZ}|Z) := \sup_{\mathrm{Q}_Z} H_{\min}(\mathrm{P}_{XZ}|\mathrm{Q}_Z). \tag{7.6}$$

[5] Although it is possible to consider subdistributions for Q_Z as well, for our purpose it suffices to consider distributions.

Unlike the variational formula of the conditional Shannon entropy, the supremum in (7.6) need not be attained by $Q_Z = P_Z$; see Problem 7.7.

In a similar spirit, we introduce the conditional collision entropy as follows.

DEFINITION 7.12 (Conditional collision entropy) For a subdistribution P_{XZ} on $\mathcal{X} \times \mathcal{Z}$ and a distribution Q_Z on $\mathcal{Z}$, the conditional collision entropy of P_{XZ} given Q_Z is defined as

$$H_2(P_{XZ}|Q_Z) := -\log \sum_{(x,z)\in \mathtt{supp}(P_{XZ})} \frac{P_{XZ}(x,z)^2}{Q_Z(z)}.$$

Then, the conditional collision entropy of P_{XZ} given Z is defined as

$$H_2(P_{XZ}|Z) := \sup_{Q_Z} H_2(P_{XZ}|Q_Z). \tag{7.7}$$

For any subdistribution P_{XZ} and distribution Q_Z, notice that

$$\sum_{(x,y)\in \mathtt{supp}(P_{XZ})} \frac{P_{XZ}(x,z)^2}{Q_Z(z)} \leq \max_{(x,y)\in \mathtt{supp}(P_{XZ})} \frac{P_{XZ}(x,z)}{Q_Z(z)}.$$

Thus, we have $H_2(P_{XZ}|Q_Z) \geq H_{\min}(P_{XZ}|Q_Z)$.

Remark 7.1 When $\mathtt{supp}(P_Z) \subseteq \mathtt{supp}(Q_Z)$ does not hold, there exists (x, z) such that $P_{XZ}(x, z) > 0$ and $Q_Z(z) = 0$. In that case, we follow a standard convention to set $H_{\min}(P_{XZ}|Q_Z)$ and $H_2(P_{XZ}|Q_Z)$ to $-\infty$. Thus, in the optimization of (7.6) and (7.7), it suffices to consider Q_Z satisfying $\mathtt{supp}(P_Z) \subseteq \mathtt{supp}(Q_Z)$.

We are ready to present the leftover hash lemma with side-information.

THEOREM 7.13 *For a given subdistribution* P_{XZ} *on* $\mathcal{X} \times \mathcal{Z}$, *and for a mapping* F *chosen uniformly at random from a UHF* $\mathcal{F}$, $K = F(X)$ *satisfies*

$$d_{\mathtt{var}}(P_{KZF}, P_{\mathtt{unif}} \times P_Z \times P_F) \leq \frac{1}{2}\sqrt{2^{\ell - H_2(P_{XZ}|Z)}} \tag{7.8}$$

$$\leq \frac{1}{2}\sqrt{2^{\ell - H_{\min}(P_{XZ}|Z)}}. \tag{7.9}$$

Proof The proof proceeds along the line of that of Theorem 7.8, with a few modifications. Since $H_2(P_{XZ}|Q_Z) \geq H_{\min}(P_{XZ}|Q_Z)$ for each Q_Z, it suffices to prove (7.8). Fix an arbitrary Q_Z satisfying $\mathtt{supp}(P_Z) \subseteq \mathtt{supp}(Q_Z)$.[6] For each $F = f$, by the Cauchy–Schwarz inequality, we have

$$d_{\mathtt{var}}(P_{f(X)Z}, P_{\mathtt{unif}} \times P_Z) = \frac{1}{2}\sum_{k,z} |P_{f(X)Z}(k,z) - P_{\mathtt{unif}}(k)\, P_Z(z)|$$

$$= \frac{1}{2}\sum_{k,z} \sqrt{Q_Z(z)} \left| \frac{P_{f(X)Z}(k,z) - P_{\mathtt{unif}}(k)\, P_Z(z)}{\sqrt{Q_Z(z)}} \right|$$

[6] For Q_Z that does not satisfy $\mathtt{supp}(P_Z) \subseteq \mathtt{supp}(Q_Z)$, the final bound (7.10) becomes trivial; see also Remark 7.1.

$$\leq \frac{1}{2}\sqrt{2^\ell \sum_{k,z} \frac{\left(\mathrm{P}_{f(X)Z}(k,z) - \mathrm{P}_{\mathtt{unif}}(k)\,\mathrm{P}_Z(z)\right)^2}{\mathrm{Q}_Z(z)}}.$$

The numerator of the term under $\sqrt{\cdot}$ can be evaluated as

$$\begin{aligned}
&\sum_k \left(\mathrm{P}_{f(X)Z}(k,z) - \mathrm{P}_{\mathtt{unif}}(k)\,\mathrm{P}_Z(z)\right)^2 \\
&\quad = \sum_k \mathrm{P}_{f(X)Z}(k,z)^2 - 2\sum_k \mathrm{P}_{f(X)Z}(k,z)\,\mathrm{P}_{\mathtt{unif}}(k)\,\mathrm{P}_Z(z) + \sum_k \mathrm{P}_{\mathtt{unif}}(k)^2\,\mathrm{P}_Z(z)^2 \\
&\quad = \sum_k \mathrm{P}_{f(X)Z}(k,z)^2 - \frac{1}{2^\ell}\mathrm{P}_Z(z)^2 \\
&\quad = \sum_{x,x'} \mathrm{P}_{XZ}(x,z)\,\mathrm{P}_{XZ}(x',z)\left(\mathbf{1}[f(x) = f(x')] - \frac{1}{2^\ell}\right) \\
&\quad = \sum_x \mathrm{P}_{XZ}(x,z)^2\left(1 - \frac{1}{2^\ell}\right) + \sum_{x \neq x'} \mathrm{P}_{XZ}(x,z)\,\mathrm{P}_{XZ}(x',z)\left(\mathbf{1}[f(x) = f(x')] - \frac{1}{2^\ell}\right) \\
&\quad \leq \sum_x \mathrm{P}_{XZ}(x,z)^2 + \sum_{x \neq x'} \mathrm{P}_{XZ}(x,z)\,\mathrm{P}_{XZ}(x',z)\left(\mathbf{1}[f(x) = f(x')] - \frac{1}{2^\ell}\right).
\end{aligned}$$

Upon taking the average with respect to F, the second term is bounded as

$$\begin{aligned}
&\sum_f \mathrm{P}_F(f) \sum_{x \neq x'} \mathrm{P}_{XZ}(x,z)\,\mathrm{P}_{XZ}(x',z)\left(\mathbf{1}[f(x) = f(x')] - \frac{1}{2^\ell}\right) \\
&\qquad = \sum_{x \neq x'} \mathrm{P}_{XZ}(x,z)\,\mathrm{P}_{XZ}(x',z)\left(\mathrm{P}(F(x) = F(x')) - \frac{1}{2^\ell}\right) \\
&\qquad \leq 0,
\end{aligned}$$

where the inequality follows from the property of the UHF. Thus, concavity of $\sqrt{\cdot}$ implies

$$\begin{aligned}
&d(\mathrm{P}_{KZF}, \mathrm{P}_{\mathtt{unif}} \times \mathrm{P}_Z \times \mathrm{P}_F) \\
&\qquad = \sum_f \mathrm{P}_F(f)\, d(\mathrm{P}_{f(X)Z}, \mathrm{P}_{\mathtt{unif}} \times \mathrm{P}_Z) \\
&\qquad \leq \frac{1}{2}\sqrt{2^\ell \sum_f \mathrm{P}_F(f) \sum_{k,z} \frac{\left(\mathrm{P}_{f(X)Z}(k,z) - \mathrm{P}_{\mathtt{unif}}(k)\,\mathrm{P}_Z(z)\right)^2}{\mathrm{Q}_Z(z)}} \\
&\qquad \leq \frac{1}{2}\sqrt{2^\ell \sum_{x,z} \frac{\mathrm{P}_{XZ}(x,z)^2}{\mathrm{Q}_Z(z)}} \\
&\qquad = \frac{1}{2}\sqrt{2^{\ell - H_2(\mathrm{P}_{XZ}|\mathrm{Q}_Z)}}.
\end{aligned} \tag{7.10}$$

□

Remark 7.2 When the set $\mathcal{Z}$ is a singleton, Theorem 7.13 implies the leftover hash lemma for a subdistribution P_X. However, it should be noted that the security requirement to be bounded is $d_{\mathtt{var}}(\mathrm{P}_{KF}, \mathrm{P}_X(\mathcal{X})\,\mathrm{P}_{\mathtt{unif}} \times \mathrm{P}_F)$ instead of $d_{\mathtt{var}}(\mathrm{P}_{KF}, \mathrm{P}_{\mathtt{unif}} \times \mathrm{P}_F)$. More specifically, we need to keep the factor $\mathrm{P}_X(\mathcal{X})$, which may be strictly smaller than 1 for subdistributions.

7.4 Smoothing

We have seen that the min-entropy is a useful measure to evaluate the amount of almost uniform bits that can be extracted from a given source. However, the min-entropy may underestimate the amount of inherent randomness. As an extreme case, consider the following example.

EXAMPLE 7.14 For $\mathcal{X} = \{0,1\}^n$, let X be the random variable on $\mathcal{X}$ with distribution $\mathrm{P}_X(\mathbf{0}) = \delta$ and $\mathrm{P}_X(x) = (1-\delta)/(2^n - 1)$, where $\mathbf{0}$ is the all-zero vector. Then, when $\delta \geq (1-\delta)/(2^n-1)$, we have $H_{\min}(X) = \log(1/\delta)$. For instance, if $\delta = 2^{-20}$ and $n = 10000$, then $H_{\min}(X) = 20$. However, since $\mathrm{P}_X(x) \leq 2^{-9999}$ for every $x \neq \mathbf{0}$ and $X = \mathbf{0}$ occurs with probability $\delta = 2^{-20}$, it is expected that almost uniform bits much longer than 20 bits can be extracted. In fact, for the security level of $\varepsilon = 2^{-10}$, it will be shown later in Example 7.17 that an ε-extractor of length 9979 bits exists.

Interestingly, underestimation as in Example 7.14 can be avoided by using the idea of "smoothing"; a similar idea was implicitly used in Chapter 6 as the smooth max-entropy. The basic idea of smoothing is that we replace a given distribution P_{XY} with a subdistribution $\tilde{\mathrm{P}}_{XZ}$ that is close to the original distribution P_{XZ}. Since the distributions are close, we can replace P_{XZ} with $\tilde{\mathrm{P}}_{XZ}$ in security analysis with small penalty terms. In particular, we can use $\tilde{\mathrm{P}}_{XZ}$ close to P_{XZ} with largest min-entropy, and address the underestimation problem given above.

In order to implement the above mentioned recipe, we introduce smoothed versions of information measures as follows.

DEFINITION 7.15 (Smooth conditional min-entropy) For distributions P_{XZ} and Q_Z, and smoothing parameter $0 \leq \varepsilon < 1$, define

$$H_{\min}^{\varepsilon}(\mathrm{P}_{XZ}|\mathrm{Q}_Z) := \sup_{\tilde{\mathrm{P}}_{XZ} \in \mathcal{B}_\varepsilon(\mathrm{P}_{XZ})} H_{\min}(\tilde{\mathrm{P}}_{XZ}|\mathrm{Q}_Z),$$

where

$$\mathcal{B}_\varepsilon(\mathrm{P}_{XZ}) := \{\tilde{\mathrm{P}}_{XZ} \in \mathcal{P}_{\mathsf{sub}}(\mathcal{X} \times \mathcal{Z}) : d_{\mathsf{var}}(\tilde{\mathrm{P}}_{XZ}, \mathrm{P}_{XZ}) \leq \varepsilon\}$$

and $\mathcal{P}_{\mathsf{sub}}(\mathcal{X} \times \mathcal{Z})$ is the set of all subdistributions on $\mathcal{X} \times \mathcal{Z}$. Then, for random variables (X, Z) with distribution P_{XZ}, the smooth conditional min-entropy of X given Z is defined as

$$H_{\min}^{\varepsilon}(X|Z) := \sup_{\mathrm{Q}_Z} H_{\min}^{\varepsilon}(\mathrm{P}_{XZ}|\mathrm{Q}_Z). \tag{7.11}$$

The smoothed version of conditional collision entropy, $H_2^{\varepsilon}(X|Z)$, is defined similarly.

The next corollary follows from Theorem 7.13 by an application of the triangle inequality.

COROLLARY 7.16 (Leftover hash lemma with smoothing) *For a given distribution* P_{XZ} *on* $\mathcal{X} \times \mathcal{Z}$, *and for a mapping* F *chosen uniformly at random from a UHF* $\mathcal{F}$, $K = F(X)$ *satisfies*

$$d_{\mathtt{var}}(P_{KZF}, P_{\mathtt{unif}} \times P_Z \times P_F) \le 2\varepsilon + \frac{1}{2}\sqrt{2^{\ell - H_2^{\varepsilon}(X|Z)}} \tag{7.12}$$

$$\le 2\varepsilon + \frac{1}{2}\sqrt{2^{\ell - H_{\min}^{\varepsilon}(X|Z)}}. \tag{7.13}$$

Proof For an arbitrary subdistribution $\tilde{P}_{XZ} \in \mathcal{B}_\varepsilon(P_{XZ})$ and the subdistributions $\tilde{P}_{KZF}$ and $\tilde{P}_Z$ induced from $\tilde{P}_{XZ}$, by the triangular inequality with respect to the variational distance, we have

$$\begin{aligned} &d_{\mathtt{var}}(P_{KZF}, P_{\mathtt{unif}} \times P_Z \times P_F) \\ &\quad \le d_{\mathtt{var}}(P_{KZF}, \tilde{P}_{KZF}) + d_{\mathtt{var}}(P_Z, \tilde{P}_Z) + d_{\mathtt{var}}(\tilde{P}_{KZF}, P_{\mathtt{unif}} \times \tilde{P}_Z \times P_F). \end{aligned}$$

By monotonicity of the variational distance,[7] the first and the second terms are bounded by ε. Then, we have the claim of the corollary by applying Theorem 7.13 to the third term and then optimizing over $\tilde{P}_{XZ} \in \mathcal{B}_\varepsilon(P_{XZ})$. □

For given random variables (X, Z) with distribution P_{XZ} and $0 \le \varepsilon < 1$, let

$$S_\varepsilon(X|Z) := \max\{\ell : \varepsilon\text{-extractor of length } \ell \text{ exists for } P_{XZ}\}.$$

While we are defining it here, this $S_\varepsilon(X|Z)$ is the fundamental quantity we have been pursuing in this chapter. Corollary 7.16 implies that

$$S_\varepsilon(X|Z) \ge \lfloor H_{\min}^{(\varepsilon-\eta)/2}(X|Z) - \log(1/4\eta^2) \rfloor \tag{7.14}$$

for $0 < \varepsilon < 1$ and any $0 < \eta \le \varepsilon$.

EXAMPLE 7.17 Returning to the source in Example 7.14, for $\delta = 2^{-20}$, $n = 10000$, and $\varepsilon = 2^{-10}$, if we take $\eta = 2^{-11}$ and $\tilde{P}_X(x) = P_X(x)\,\mathbf{1}[x \ne \mathbf{0}]$, then $\tilde{P}_X \in \mathcal{B}_{(\varepsilon-\eta)/2}(P_X)$, and thus, $H_{\min}^{(\varepsilon-\eta)/2}(X) \ge 9999$. Therefore, by (7.14), there exists an ε-extractor of length 9979 bits.

Turning our attention to an upper bound on the optimal length $S_\varepsilon(X|Z)$ of ε-extractors, we introduce a modified version of the smooth conditional min-entropy where we only consider smoothing with distributions (that is, subdistributions are not allowed).

DEFINITION 7.18 For distributions P_{XZ} and Q_Z, and smoothing parameter $0 \le \varepsilon < 1$, define

$$\bar{H}_{\min}^{\varepsilon}(P_{XZ}|Q_Z) := \sup_{\tilde{P}_{XZ} \in \bar{\mathcal{B}}_\varepsilon(P_{XZ})} H_{\min}(\tilde{P}_{XZ}|Q_Z),$$

where

$$\bar{\mathcal{B}}_\varepsilon(P_{XZ}) := \{\tilde{P}_{XZ} \in \mathcal{P}(\mathcal{X} \times \mathcal{Z}) : d_{\mathtt{var}}(\tilde{P}_{XZ}, P_{XZ}) \le \varepsilon\}$$

and $\mathcal{P}(\mathcal{X} \times \mathcal{Z})$ is the set of all distributions on $\mathcal{X} \times \mathcal{Z}$.

[7] Namely, the fact that $d_{\mathtt{var}}(P_U, Q_U) \le d_{\mathtt{var}}(P_{UV}, Q_{UV})$.

The following data processing inequality for modified smooth conditional min-entropy will be useful in our proof of the upper bound for $S_\varepsilon(X \mid Z)$.[8]

LEMMA 7.19 (Data processing inequality for $\bar{H}^\varepsilon_{\min}$) *For given random variables (X, Z), a function $f\colon \mathcal{X} \to \mathcal{K}$, $K = f(X)$, and Q_Z, we have*

$$\bar{H}^\varepsilon_{\min}(\mathrm{P}_{KZ}|\mathrm{Q}_Z) \le \bar{H}^\varepsilon_{\min}(\mathrm{P}_{XZ}|\mathrm{Q}_Z).$$

Proof For a fixed distribution $\tilde{\mathrm{P}}_{KZ} \in \bar{\mathcal{B}}_\varepsilon(\mathrm{P}_{KZ})$, define

$$\tilde{\mathrm{P}}_{XZ}(x,z) = \tilde{\mathrm{P}}_{KZ}(f(x),z)\frac{\mathrm{P}_{XZ}(x,z)}{\mathrm{P}_{KZ}(f(x),z)}.$$

In fact, it can be easily verified that $\tilde{\mathrm{P}}_{XZ}$ is a valid distribution. Then, we have

$$\begin{aligned}
d_{\mathrm{var}}(\tilde{\mathrm{P}}_{XZ},\mathrm{P}_{XZ}) &= \frac{1}{2}\sum_{x,z}|\tilde{\mathrm{P}}_{XZ}(x,z) - \mathrm{P}_{XZ}(x,z)| \\
&= \frac{1}{2}\sum_{k,z}\sum_{x\in f^{-1}(s)}\frac{\mathrm{P}_{XZ}(x,z)}{\mathrm{P}_{KZ}(k,z)}|\tilde{\mathrm{P}}_{KZ}(k,z) - \mathrm{P}_{KZ}(k,z)| \\
&= \frac{1}{2}\sum_{k,z}|\tilde{\mathrm{P}}_{KZ}(k,z) - \mathrm{P}_{KZ}(k,z)| \\
&= d_{\mathrm{var}}(\tilde{\mathrm{P}}_{KZ},\mathrm{P}_{KZ}) \\
&\le \varepsilon,
\end{aligned}$$

which implies $\tilde{\mathrm{P}}_{XZ} \in \bar{\mathcal{B}}_\varepsilon(\mathrm{P}_{XZ})$. Furthermore, by the construction of $\tilde{\mathrm{P}}_{XZ}$, we have $\tilde{\mathrm{P}}_{XZ}(x,z) \le \tilde{\mathrm{P}}_{KZ}(f(x),z)$ for every (x,z). Thus, we have

$$\begin{aligned}
H_{\min}(\tilde{\mathrm{P}}_{KZ}|\mathrm{Q}_Z) &\le H_{\min}(\tilde{\mathrm{P}}_{XZ}|\mathrm{Q}_Z) \\
&\le \bar{H}^\varepsilon_{\min}(\mathrm{P}_{XZ}|\mathrm{Q}_Z).
\end{aligned}$$

Upon taking the supremum over $\tilde{\mathrm{P}}_{KZ} \in \bar{\mathcal{B}}_\varepsilon(\mathrm{P}_{KZ})$, we have the claim of the lemma. □

By using Lemma 7.19, we can derive an upper bound on $S_\varepsilon(X|Z)$. For later reference, we summarize both lower and upper bounds in the following theorem.

THEOREM 7.20 *For given random variables (X, Z) and $0 < \varepsilon < 1$, we have*[9]

$$\lfloor H^{(\varepsilon-\eta)/2}_{\min}(X|Z) - \log(1/4\eta^2)\rfloor \le S_\varepsilon(X|Z) \le \bar{H}^\varepsilon_{\min}(\mathrm{P}_{XZ}|\mathrm{P}_Z)$$

for any $0 < \eta \le \varepsilon$.

[8] In fact, Lemma 7.19 holds for the original smooth conditional min-entropy as well. However, the proof of (7.17) in Lemma 7.24 requires that $\tilde{\mathrm{P}}_{XZ}$ is normalized, which is the reason why we introduce the modified smooth conditional min-entropy.

[9] The upper bound is valid even for $\varepsilon = 0$ as well.

Proof The lower bound is from (7.14). To prove the upper bound, for any ε-extractor f of length ℓ and $K = f(X)$, the fact that $d_{\mathtt{var}}(\mathrm{P}_{KZ}, \mathrm{P}_{\mathtt{unif}} \times \mathrm{P}_Z) \leq \varepsilon$ and Lemma 7.19 imply

$$\begin{aligned}\ell &= H_{\min}(\mathrm{P}_{\mathtt{unif}} \times \mathrm{P}_Z|\mathrm{P}_Z)\\ &\leq \bar{H}^{\varepsilon}_{\min}(\mathrm{P}_{KZ}|\mathrm{P}_Z)\\ &\leq \bar{H}^{\varepsilon}_{\min}(\mathrm{P}_{XZ}|\mathrm{P}_Z).\end{aligned}$$

□

7.5 A General Leftover Hash Lemma

In later chapters (see Chapters 10, 13, and 14) we will use leftover hashing (privacy amplification) as a building block for certain protocols. In that case, in addition to the adversary's original side-information Z, the adversary also observes an additional side-information V that is leaked by the legitimate parties during a protocol. Of course, we can regard the pair (V, Z) as the adversary's side-information, and apply the result of the previous section. However, for a given distribution P_{XZ}, the joint distribution P_{XVZ} depends on the structure of the protocol, and it is difficult to evaluate the quantity $H^{\varepsilon}_{\min}(X|V, Z)$, while $H^{\varepsilon}_{\min}(X|Z)$ can still be readily evaluated. Thus, it is convenient to decouple the effect of the side-information V revealed during the protocol from $H^{\varepsilon}_{\min}(X|V, Z)$ as follows.

LEMMA 7.21 *For a given distribution* P_{XVZ} *on* $\mathcal{X} \times \mathcal{V} \times \mathcal{Z}$ *and* $0 \leq \varepsilon < 1$, *we have*

$$H^{\varepsilon}_{\min}(X|V, Z) \geq H^{\varepsilon}_{\min}(X|Z) - \log|\mathcal{V}|.$$

Proof For an arbitrary $\tilde{\mathrm{P}}_{XZ} \in \mathcal{B}_{\varepsilon}(\mathrm{P}_{XZ})$, let

$$\tilde{\mathrm{P}}_{XVZ}(x, v, z) = \tilde{\mathrm{P}}_{XZ}(x, z)\mathrm{P}_{V|XZ}(v|x, z).$$

Then, we have

$$\begin{aligned}d_{\mathtt{var}}(\mathrm{P}_{XVZ}, \tilde{\mathrm{P}}_{XVZ}) &= \frac{1}{2}\sum_{x,v,z} |\mathrm{P}_{XZ}(x, z)\,\mathrm{P}_{V|XZ}(v|x, z) - \tilde{\mathrm{P}}_{XZ}(x, z)\mathrm{P}_{V|XZ}(v|x, z)\,|\\ &= \sum_{x,z} |\mathrm{P}_{XZ}(x, z) - \tilde{\mathrm{P}}_{XZ}(x, z)|\\ &= d_{\mathtt{var}}(\mathrm{P}_{XZ}, \tilde{\mathrm{P}}_{XZ})\\ &\leq \varepsilon,\end{aligned}$$

which implies $\tilde{\mathrm{P}}_{XVZ} \in \mathcal{B}_{\varepsilon}(\mathrm{P}_{XVZ})$. For an arbitrary Q_Z, let

$$\mathrm{Q}_{VZ}(v, z) = \frac{1}{|\mathcal{V}|}\mathrm{Q}_Z(z).$$

Then, we have

$$\max_{(x,v,z)\in\mathrm{supp}(\tilde{\mathrm{P}}_{XVZ})} \frac{\tilde{\mathrm{P}}_{XVZ}(x,v,z)}{\mathrm{Q}_{VZ}(v,z)} \le \max_{(x,z)\in\mathrm{supp}(\tilde{\mathrm{P}}_{XZ})} \frac{|\mathcal{V}|\tilde{\mathrm{P}}_{XZ}(x,z)}{\mathrm{Q}_Z(z)},$$

whereby

$$\begin{aligned} H_{\min}^{\varepsilon}(X|V,Z) &\ge H_{\min}(\tilde{\mathrm{P}}_{XVZ}|\mathrm{Q}_{VZ}) \\ &\ge H_{\min}(\tilde{\mathrm{P}}_{XZ}|\mathrm{Q}_Z) - \log|\mathcal{V}|. \end{aligned}$$

Since this inequality holds for arbitrary $\tilde{\mathrm{P}}_{XZ} \in \mathcal{B}_\varepsilon(\mathrm{P}_{XZ})$ and Q_Z, we have the claim of the lemma. □

From Lemma 7.21 and Corollary 7.16, we can immediately derive the following result; for applications to cryptographic protocols, this form of leftover hash lemma is the most useful.

COROLLARY 7.22 (General leftover hash lemma) *For a given distribution* P_{XVZ} *on* $\mathcal{X} \times \mathcal{V} \times \mathcal{Z}$, *and for a mapping* F *chosen uniformly at random from a UHF* $\mathcal{F}$, $K = F(X)$ *satisfies*

$$d_{\mathtt{var}}(\mathrm{P}_{KVZF}, \mathrm{P}_{\mathtt{unif}} \times \mathrm{P}_{VZ} \times \mathrm{P}_F) \le 2\varepsilon + \frac{1}{2}\sqrt{2^{\ell+\log|\mathcal{V}|-H_{\min}^{\varepsilon}(X|Z)}}. \tag{7.15}$$

Corollary 7.22 implies that

$$S_\varepsilon(X|V,Z) \ge \lfloor H_{\min}^{(\varepsilon-\eta)/2}(X|Z) - \log|\mathcal{V}| - \log(1/4\eta^2)\rfloor$$

for any $0 < \eta \le \varepsilon$. Thus, we can say that extractable randomness reduces by at most $\log|\mathcal{V}|$ bits when an additional $\log|\mathcal{V}|$ bit side-information V is revealed.

An advantage of this bound is that the subtracted term $\log|\mathcal{V}|$ only depends on the length of V without any reference to how V is generated. Thus, we can build the randomness extraction part and other parts of a protocol separately. Typically, subtracting $\log|\mathcal{V}|$ is not loose since the redundancy of messages exchanged during a protocol is anyway removed using data compression.

We close this section with one more technique to decouple an adversary's observation. In certain protocols, the legitimate parties sample a random variable W that is independent of (X,Z), and this W is leaked to the adversary. Since W is independent of (X,Z), it is of no use to the adversary, which is guaranteed by the following lemma.

LEMMA 7.23 *For random variables* (X,Z,W) *with joint distribution* $\mathrm{P}_{XZ}\times\mathrm{P}_W$ *and* $0 \le \varepsilon < 1$, *we have*

$$H_{\min}^{\varepsilon}(X|Z,W) \ge H_{\min}^{\varepsilon}(X|Z).$$

Proof For any Q_Z and $\tilde{\mathrm{P}}_{XZ} \in \mathcal{B}_\varepsilon(\mathrm{P}_{XZ})$, since

$$d_{\mathtt{var}}(\tilde{\mathrm{P}}_{XZ}\times\mathrm{P}_W, \mathrm{P}_{XZ}\times\mathrm{P}_W) = d_{\mathtt{var}}(\tilde{\mathrm{P}}_{XZ}, \mathrm{P}_{XZ}) \le \varepsilon,$$

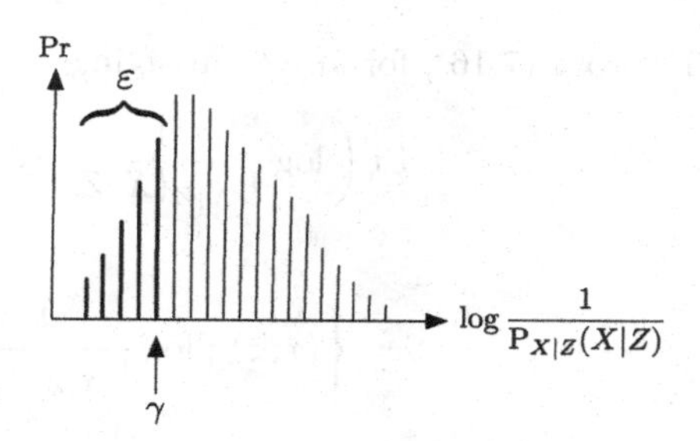

Figure 7.2 A heuristic description of Lemma 7.24. The threshold γ is the ε-lower quantile of the conditional information density $\log \frac{1}{P_{X|Z}(X|Z)}$, i.e., the probability that the conditional entropy density takes values below γ is ε.

we have

$$
\begin{aligned}
H^{\varepsilon}_{\min}(X|Z,W) &\geq H_{\min}(\tilde{P}_{XZ} \times P_W | Q_Z \times P_W) \\
&= H_{\min}(\tilde{P}_{XZ}|Q_Z).
\end{aligned}
$$

Since this inequality holds for any Q_Z and $\tilde{P}_{XZ} \in \mathcal{B}_\varepsilon(P_{XZ})$, we have the claim of the lemma. □

7.6 Evaluation of Smooth Min-entropies

All our main results involve smoothed versions of information quantities. We close this chapter by presenting methods to evaluate the smooth conditional min-entropy in Definition 7.15 and its modified version in Definition 7.18. The treatment is very similar to Section 6.3 where we evaluated smooth max-entropy.

As we have illustrated in Example 7.17, a natural method of smoothing is to truncate symbols having too large probabilities, or equivalently too small entropy densities. Following this heuristic, we first relate the smooth conditional min-entropy to the ε-lower quantile of the conditional entropy density $\log \frac{1}{P_{X|Z}(X|Z)}$; see Figure 7.2. Notice that, in contrast to Lemma 6.11 (and its conditional version, Lemma 6.14) where upper quantiles were considered, we consider the lower quantile in the following lemma.

LEMMA 7.24 *For a given $0 \leq \varepsilon < 1$, we have*

$$
H^{\varepsilon/2}_{\min}(P_{XZ}|P_Z) \geq \sup\left\{\gamma \in \mathbb{R} : \Pr\left(\log \frac{1}{P_{X|Z}(X|Z)} \leq \gamma\right) \leq \varepsilon\right\} \tag{7.16}
$$

and

$$
\bar{H}^{\varepsilon}_{\min}(P_{XZ}|P_Z) \leq \inf\left\{\gamma \in \mathbb{R} : \Pr\left(\log \frac{1}{P_{X|Z}(X|Z)} \leq \gamma\right) \geq \varepsilon + \xi\right\} + \log(1/\xi) \tag{7.17}
$$

for any $0 < \xi \leq 1 - \varepsilon$.

Proof To prove (7.16), for any γ satisfying

$$\Pr\left(\log \frac{1}{\mathrm{P}_{X|Z}(X|Z)} \le \gamma\right) \le \varepsilon, \tag{7.18}$$

let

$$\mathcal{S} = \left\{(x,z) : \log \frac{1}{\mathrm{P}_{X|Z}(x|z)} > \gamma\right\}$$

and

$$\tilde{\mathrm{P}}_{XZ}(x,z) = \mathrm{P}_{XZ}(x,z)\,\mathbf{1}[(x,z) \in \mathcal{S}]$$

be the subdistribution obtained by truncating the probabilities on $\mathcal{S}^c$. Then, we have

$$d_{\mathtt{var}}(\mathrm{P}_{XZ}, \tilde{\mathrm{P}}_{XZ}) = \frac{1}{2}\mathrm{P}_{XZ}(\mathcal{S}^c) \le \frac{\varepsilon}{2}.$$

Furthermore, since $(x,z) \in \mathcal{S}$ implies $\mathrm{P}_{X|Z}(x|z) \le 2^{-\gamma}$, we have

$$\begin{aligned} H_{\min}^{\varepsilon/2}(\mathrm{P}_{XZ}|\mathrm{P}_Z) &\ge H_{\min}(\tilde{\mathrm{P}}_{XZ}|\mathrm{P}_Z) \\ &= -\log \max_{(x,z)\in\mathcal{S}} \mathrm{P}_{X|Z}(x|z) \\ &\ge \gamma. \end{aligned}$$

Since this inequality holds for any γ satisfying (7.18), we have (7.16).

To prove (7.17), for any γ satisfying

$$\Pr\left(\log \frac{1}{\mathrm{P}_{X|Z}(X|Z)} \le \gamma\right) \ge \varepsilon + \xi, \tag{7.19}$$

let

$$\mathcal{T} = \left\{(x,z) : \log \frac{1}{\mathrm{P}_{X|Z}(x|z)} \le \gamma\right\},$$

and let

$$\mathcal{T}_z = \{x : (x,z) \in \mathcal{T}\}$$

for each $z \in \mathcal{Z}$. Then, in the same manner as in the proof of Lemma 6.11, we can prove $|\mathcal{T}_z| \le 2^\gamma$ for every $z \in \mathcal{Z}$. For arbitrarily fixed $\tilde{\mathrm{P}}_{XZ} \in \bar{\mathcal{B}}_\varepsilon(\mathrm{P}_{XZ})$, denote $a = H_{\min}(\tilde{\mathrm{P}}_{XZ}|\mathrm{P}_Z)$. Then, we have

$$\begin{aligned} \varepsilon &\ge d_{\mathtt{var}}(\mathrm{P}_{XZ}, \tilde{\mathrm{P}}_{XZ}) \\ &\ge \mathrm{P}_{XZ}(\mathcal{T}) - \tilde{\mathrm{P}}_{XZ}(\mathcal{T}) \\ &= \mathrm{P}_{XZ}(\mathcal{T}) - \sum_z \sum_{x\in\mathcal{T}_z} \mathrm{P}_Z(z)\frac{\tilde{\mathrm{P}}_{XZ}(x,z)}{\mathrm{P}_Z(z)} \\ &\ge \mathrm{P}_{XZ}(\mathcal{T}) - \sum_z \sum_{x\in\mathcal{T}_z} \mathrm{P}_Z(z)\,2^{-a} \end{aligned}$$

$$= \mathrm{P}_{XZ}(\mathcal{T}) - \sum_z \mathrm{P}_Z(z)|\mathcal{T}_z|2^{-a}$$
$$\geq \mathrm{P}_{XZ}(\mathcal{T}) - 2^{\gamma-a}$$
$$\geq \varepsilon + \xi - 2^{\gamma-a},$$

which implies

$$H_{\min}(\tilde{\mathrm{P}}_{XZ}|\mathrm{P}_Z) = a \leq \gamma + \log(1/\xi).$$

Since this inequality holds for any $\tilde{\mathrm{P}}_{XZ} \in \bar{\mathcal{B}}_\varepsilon(\mathrm{P}_{XZ})$ and γ satisfying (7.19), we have (7.17). □

Finally, consider a sequence of i.i.d. random variables (X^n, Z^n) with common distribution P_{XZ}. In exactly the same manner as Theorem 6.12, an application of the Hoeffding bound (see Theorem 2.39) to the bounds in Lemma 7.24 provides the following approximations.

THEOREM 7.25 *For a given $0 < \varepsilon < 1$, we have*

$$\frac{1}{n} H_{\min}^{\varepsilon/2}(\mathrm{P}_{X^nZ^n}|\mathrm{P}_{Z^n}) \geq H(X|Z) - \kappa\sqrt{\frac{\ln(1/\varepsilon)}{2n}}$$

and

$$\frac{1}{n} \bar{H}_{\min}^{\varepsilon}(\mathrm{P}_{X^nZ^n}|\mathrm{P}_{Z^n}) \leq H(X|Z) + \kappa\sqrt{\frac{\ln(2/(1-\varepsilon))}{2n}} + \frac{\ln(2/(1-\varepsilon))}{n},$$

where $\kappa = \max_{(x,z)\in\mathrm{supp}(\mathrm{P}_{XZ})} \log \frac{1}{\mathrm{P}_{X|Z}(x|z)}$.

For a sequence of i.i.d. random variables (X^n, Z^n), note that $\frac{1}{n} S_\varepsilon(X^n|Z^n)$ represents the optimal generation rate for block ε-extractors that use X^n as input. By combining Theorem 7.20 and Theorem 7.25, we can immediately derive the following corollary.

COROLLARY 7.26 *For a given $0 < \varepsilon < 1$, we have*

$$\lim_{n\to\infty} \frac{1}{n} S_\varepsilon(X^n|Z^n) = H(X|Z).$$

7.7 References and Additional Reading

A scientific study of random number generation was initiated by von Neumann in [341]. He proposed an algorithm to generate unbiased bits from an i.i.d. binary biased process (see Problem 7.2). Since then, his result has been extended and refined in various directions. On the other hand, Knuth and Yao studied the problem of generating a biased random process using an i.i.d. unbiased binary process [196]. These results seek to generate a target distribution without any approximation error using variable length (possibly infinite length) seed randomness. They are closely related but different from the random number generation problem considered in this chapter. For an overview of the formulation with variable length seed randomness, see [75, Section 5.11].

For cryptographic applications, it is desirable to consider random number generation from an adversarially generated seed randomness. One such model was introduced in [297]. The model of the ν-source in Definition 7.5 was introduced in [65].

The basic version of the leftover hash lemma (Theorem 7.8) was developed in the 1980s for constructing pseudorandom functions from one-way functions [168, 172, 173]. Another variant of the leftover hash lemma was developed in the context of secret key agreement [27, 28]; see also [160] for a refined evaluation of [27]. The leftover hash lemma with side-information explained in Section 7.3 was presented in [286] in the context of the quantum key distribution. The idea of smoothing in Section 7.4 was also introduced in [286]; see also [290]. Note that, in the course of proving the leftover hash lemma with min-entropy, we can derive a leftover hash lemma with collision entropy. It is known that the version with collision entropy provides a tighter bound than the version with min-entropy [162, 359]. However, for simplicity of exposition, we only considered the min-entropy version in this chapter. For another variant of the leftover hash lemma with collision entropy, see [116].

It should be noted that random number generation has been actively studied in information theory as well. The formulation in this chapter corresponds to the problem of intrinsic randomness [340]; see also [151, Chapter 2]. The greedy algorithm in Construction 7.1 is from [151, Lemma 2.1.1].

As we have mentioned in Section 6.4, the use of quantiles of the information density is the basic philosophy of the information spectrum approach introduced in [152]; see also [151]. While smooth max-entropy is related to the upper quantile of the information density, smooth min-entropy is related to the lower quantile. For a more detailed connection between the smooth entropy framework and the information spectrum approach, see [98, 199, 325, 337].

Problems

7.1 Prove that $H_2(X) \geq H_{\min}(X)$.

HINT Manipulate the definitions of $H_2(X)$ and $H_{\min}(X)$ so that we can use the fact that the minimum is less than the average.

7.2 *von Neumann's algorithm*: Let $X_1, X_2, \ldots$ be a sequence of Bernoulli random variables, i.e., $\Pr(X_i = 0) = p$ and $\Pr(X_i = 1) = 1 - p$ for each $i \in \mathbb{N}$, where $p \in (0, 1)$. Initializing the counter $j = 1$, von Neumann's algorithm outputs 0 if $(X_{2j-1}, X_{2j}) = (0, 1)$ and outputs 1 if $(X_{2j-1}, X_{2j}) = (1, 0)$; otherwise, it increments the counter j and iterates the same procedure.

1. Verify that the output is a uniformly distributed bit.
2. Prove that the expected number $\mathbb{E}[T]$ of observations $X_1, \ldots, X_T$ until the algorithm stops is given by $\frac{1}{p(1-p)}$.

7.3 For a function $f: \mathcal{X} \to \mathcal{K}$, let $K = f(X)$ be the output for a source X. Prove that, in order to satisfy $d_{\mathtt{var}}(\mathrm{P}_K, \mathrm{P}_{\mathtt{unif}}) \le \varepsilon$, $\log |\mathcal{K}| \le \frac{H(X)+h(\bar{\varepsilon})}{1-\varepsilon}$ must hold, where $\bar{\varepsilon} = \min\{\varepsilon, 1/2\}$.

HINT Use the continuity of entropy (cf. Theorem 2.34) and the fact that entropy does not increase when a function is applied.

REMARK This bound gives a simple upper bound on the length of almost uniform bits in terms of the entropy of a given source. However, it is rather loose.

7.4 For $\mathcal{X} = \{1, 2, 3\}$, let P_X be the distribution given by $\mathrm{P}_X(1) = \frac{1}{2}$ and $\mathrm{P}_X(2) = \mathrm{P}_X(3) = \frac{1}{4}$.

1. Describe a method to extract the perfectly uniform bit from a source X distributed according to P_X. Argue that it is not possible to extract the perfectly uniform 2 bits from X.
   HINT Compute $H(X)$.
2. Describe a method to extract the perfectly uniform 2 bits from two independent sources X^2 distributed according to P_X. Argue that it is not possible to extract the perfectly uniform 3 bits from X^2.
   HINT Use the upper bound in Theorem 7.20.

7.5 Let $X^n = (X_1, \ldots, X_n)$ be a sequence of Bernoulli random variables, i.e., $\Pr(X_i = 0) = p$ and $\Pr(X_i = 1) = 1 - p$ for each $i \in \mathbb{N}$, where $p \in (0, 1)$ and $p \neq \frac{1}{2}$. Prove that it is not possible to extract the perfectly uniform bit from X^n for any finite n.

HINT Use the upper bound in Theorem 7.20.

REMARK In general, it is not possible to extract even 1 bit of the uniform randomness from a very long sequence of sources, i.e., even if the efficiency is very low. This is in contrast with the source coding problem of Chapter 6, where it is always possible to realize zero-error coding if efficiency (coding rate) is sacrificed. Note that, even though von Neumann's algorithm in Problem 7.2 outputs the perfectly uniform bit, the number of observations required is unbounded.

7.6 *Discrepancy between average and ε-lower quantile:* For a (large) integer t, let P_X be the distribution on $\mathcal{X} = \{1, \ldots, 2^t\}$ given by $\mathrm{P}_X(2^t) = \frac{1}{2}$ and $\mathrm{P}_X(x) = \frac{1}{2(2^t-1)}$ for $x \neq 2^t$.

1. Verify that $H(X) = 1 + \frac{1}{2}\log(2^t - 1)$.
2. For $0 \le \varepsilon < \frac{1}{2}$, verify that $\bar{H}^{\varepsilon}_{\min}(\mathrm{P}_X) \le 1 + \log(1/(1/2 - \varepsilon))$.
   HINT Use (7.17).

REMARK For the single-shot regime, Shannon entropy may overestimate the length of uniform bits that can be extracted from a source for a given security level; cf. Problem 6.5.

7.7 For a distribution P_{XZ}, prove that the supremum in (7.6) is attained by

$$Q_Z(z) = \frac{P_Z(z) \max_x P_{X|Z}(x|z)}{\sum_z P_Z(z) \max_x P_{X|Z}(x|z)}$$

and the conditional min-entropy can be expressed as [201]

$$H_{\min}(P_{XZ}|Z) = -\log \sum_z P_Z(z) \max_x P_{X|Z}(x|z).$$

REMARK This means that the probability of correctly guessing X from Z can be written as $\sum_z P_Z(z) \max_x P_{X|Z}(x|z) = 2^{-H_{\min}(P_{XZ}|Z)}$.

7.8 *Mixed source*: Suppose that the distributions of a sequence X^n of sources are given by

$$P_{X^n}(x^n) = \theta P^n_{X_1}(x^n) + (1-\theta) P^n_{X_2}(x^n)$$

for some $0 < \theta < 1$ and common distributions P_{X_1} and P_{X_2} on $\mathcal{X}$ with $H(X_1) > H(X_2)$. Prove that the asymptotic limit of the generation rate for the mixed source is given as

$$\lim_{\varepsilon \to 0} \limsup_{n \to \infty} \frac{1}{n} S_\varepsilon(X^n) = \min[H(X_1), H(X_2)] = X_2.$$

REMARK See Problem 6.7 for a related problem.

7.9 *Second-order generation rate*: For a sequence X^n of i.i.d. sources with

$$\mathtt{V}(X) := \mathtt{Var}\left[\log \frac{1}{P_X(X)}\right] > 0,$$

prove that

$$S_\varepsilon(X^n) = nH(X) - \sqrt{n\mathtt{V}(X)}\mathtt{Q}^{-1}(\varepsilon) + o(\sqrt{n}), \tag{7.20}$$

where $\mathtt{Q}^{-1}(\varepsilon)$ is the inverse of the tail probability

$$\mathtt{Q}(a) = \int_a^\infty \frac{1}{\sqrt{2\pi}} e^{-\frac{t^2}{2}} dt$$

of the Gaussian distribution.

REMARK See Problem 6.8 for a related problem.

8 Authentication

Encryption and authentication are two basic security requirements for any secure communication protocol. We looked at encryption in Chapter 3 and we will now visit authentication. As in Chapter 3, our treatment here will be information theoretic and will assume the availability of perfect secret keys. The focus here will be on describing the basic security requirement of authentication and examining information-theoretic limits of possibilities. We shall establish lower bounds for performance and present schemes that will achieve those bounds.

In practice, encryption and authentication schemes are intertwined and are often implemented in conjunction. We will describe such practical schemes and their security guarantees in the next chapter. The authentication schemes that we will encounter in this chapter are basic constructs underlying many practical authentication schemes.

To enable message authentication, the legitimate sender appends a *tag* t to a message m. Upon receiving the pair (m, t), the receiver accepts the received message m if the received tag t is consistent with it. The communication channel used is not secure, and an adversary may change (m, t). In fact, we want only to accept a message from the legitimate sender and reject messages sent by an adversary. An authentication scheme will ensure that a forged message-tag pair is rejected with large probability. In this chapter, we assume that the legitimate sender shares a secret key K with the receiver, which is not available to the adversary. We will present schemes (i) that can authenticate messages with as large a probability of detection as possible for a given key length, or (ii) that can authenticate messages with as few bits of secret key as possible for a prescribed probability of detection.

8.1 Message Authentication Using Secret Keys

We start with the formal notion of a message authentication code; see Figure 8.1 for a depiction.

DEFINITION 8.1 (Message authentication code) Consider a sender-receiver pair where the sender seeks to transmit a message m from the set of messages $\mathcal{M}$ to the receiver. They both observe a secret key K, which is a random variable distributed uniformly over the set $\mathcal{K}$. A *message authentication code* (MAC)

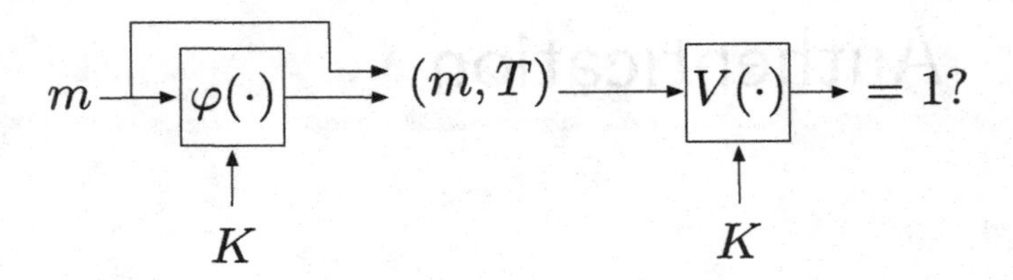

Figure 8.1 Illustration of a message authentication code (φ, V).

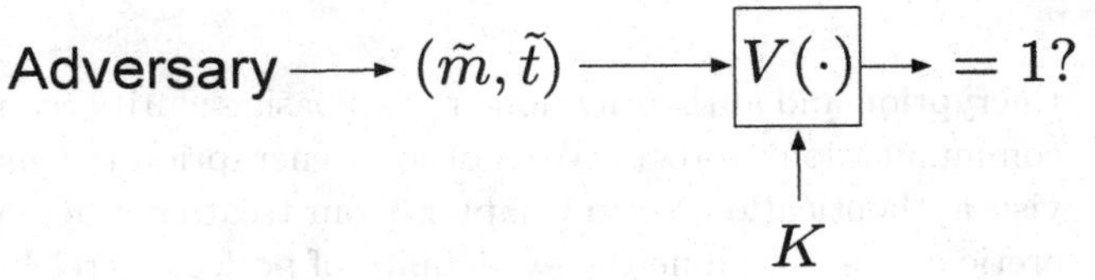

Adversary → $(\tilde{m}, \tilde{t})$ → $V(\cdot)$ → $= 1?$

K

Figure 8.2 Illustration of an impersonation attack.

consists of a pair of mappings (φ, V) where $\varphi\colon \mathcal{M} \times \mathcal{K} \to \mathcal{T}$ and $V\colon \mathcal{M} \times \mathcal{T} \times \mathcal{K} \to \{0, 1\}$. The sender appends the tag $T = \varphi(m, K)$ to message m, and upon receiving (m, T), the receiver accepts the message if $V(m, T, K) = 1$.

The purpose of authentication schemes is to detect an adversarially forged message-tag pair. In order to evaluate the performance of authentication schemes, it is important to carefully specify the power of the adversary. Note that what distinguishes an adversary from the legitimate sender in our setting is that the latter has access to the secret key K. In this chapter, we empower the adversary with two kinds of attacks.

1. *Impersonation attack*: An adversary may pretend to be the legitimate sender and create a message-tag pair (m, t). This attack will be successful if the receiver accepts the message, which will happen if the mapping V of the MAC evaluates to 1; see Figure 8.2. The success probability of an impersonation attack, denoted P_{I}, is given by

$$P_{\mathrm{I}} = \max_{(m,t) \in \mathcal{M} \times \mathcal{T}} \Pr\big(V(m, t, K) = 1\big).$$

Two important remarks are to be noted. First, the right-hand side above is the maximum of probability over all (m, t) that are fixed without the knowledge of K. Second, the attacker is allowed to insert a randomly generated message-tag pair (M, T), but the success probability of impersonation is maximized for a deterministic pair.

2. *Substitution attack*: In the previous attack, while the adversary could generate a random pair (M, T), which may even be related as $T = \varphi(M, k)$ for some k, the adversary did not have any information about the secret key K shared by the legitimate parties. In a substitution attack, we allow the adversary to intercept a legitimate message-tag pair (m, T) with $T = \varphi(m, K)$ and then

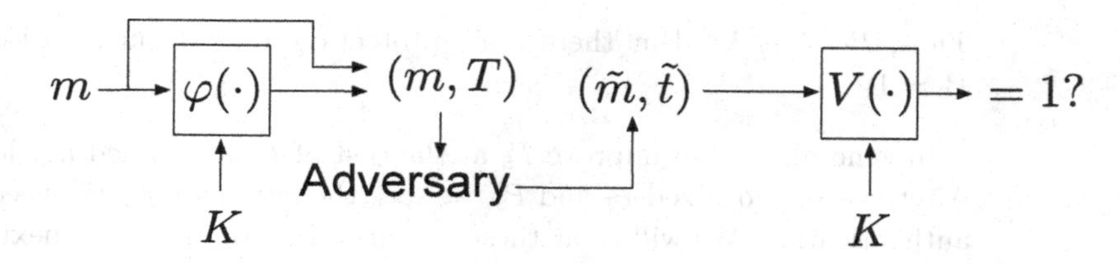

Figure 8.3 Illustration of a substitution attack.

replace it with a message $\tilde{m} \neq m$, possibly stochastically; see Figure 8.3. The success probability of this attack, denoted P_{S}, is given by

$$P_{\mathrm{S}} = \max_{m \in \mathcal{M}} \max_{P_{\tilde{M}\tilde{T}|T}} \Pr\left(V(\tilde{M}, \tilde{T}, K) = 1, \tilde{M} \neq m\right).$$

Note that $(\tilde{M}, \tilde{T})$ is generated from $T = \varphi(m, K)$ using a channel $P_{\tilde{M}\tilde{T}|T}$ which may depend on m.

If we are just concerned with minimizing the success probabilities of an adversary's attack, there is a trivial scheme that rejects any message irrespective of the tag and the key. However, such a scheme is useless since the message sent by the legitimate sender will be rejected as well. In order to exclude such useless schemes, we impose the following requirement of security on a MAC.

DEFINITION 8.2 A MAC is said to be[1] *ε-correct* if for every message m,

$$\Pr\left(V(m, \varphi(m, K), K) = 1\right) \geq 1 - \varepsilon.$$

Namely, an ε-correct MAC accepts every message from the legitimate sender with probability greater than $1 - \varepsilon$. A zero-correct MAC will be termed a *perfectly correct* MAC.

Clearly, substitution attack is more powerful than impersonation attack, whereby

$$P_{\mathrm{I}} \leq P_{\mathrm{S}}. \tag{8.1}$$

In fact, the next example provides a trivial MAC that is perfectly correct and has a very small probability of impersonation attack, but has $P_{\mathrm{S}} = 1$.

EXAMPLE 8.3 Consider the MAC which sends the secret key as the tag, i.e., $\varphi(m, k) = k$, and accepts the message-tag pair if the tag equals the shared key, i.e., $V(m, t, k) = 1$ if and only if $t = k$. This MAC is perfectly correct. Also, since K is distributed uniformly over $\mathcal{K}$, for any fixed t,

$$\Pr\left(V(m, t, K) = 1\right) = \Pr\left(K = t\right) = \frac{1}{|\mathcal{K}|}.$$

[1] The terminology of *correctness* and *soundness* is borrowed from formal verification and computational theory, and is often used in cryptography as well.

Thus, $P_{\mathrm{I}} = 1/|\mathcal{K}|$. But there is no protection against substitution attack and $P_{\mathrm{S}} = 1$.

In general, we can improve P_{S} at the cost of P_{I} for a fixed key length $\log|\mathcal{K}|$. Alternatively, for fixed P_{I} and P_{S}, we seek the minimum $\log|\mathcal{K}|$ needed to enable authentication. We will treat these scenarios separately in the next two sections and give schemes that almost attain the information-theoretic lower bounds.

Remark 8.1 In broader applications, substitution may be regarded as successful if $(\tilde{M}, \tilde{T}) \neq (m, T)$ is accepted by the receiver, i.e., the attack will be successful even if only the tag differs. While all the MAC schemes we present in this chapter satisfy this requirement as well, this extension is not needed to discuss authentication. We mention it here for later use in Section 8.5 when we discuss *nonmalleability*. Specifically, we define the success probability of a substitution attack in the wide sense as

$$P_{\mathrm{S}} = \max_{m \in \mathcal{M}} \max_{P_{\tilde{M}\tilde{T}|T}} \Pr\left(V(\tilde{M}, \tilde{T}, K) = 1, (\tilde{M}, \tilde{T}) \neq (m, T)\right).$$

See also Problem 8.3.

8.2 MAC with Optimal Attack Detection Probabilities

We examine the security requirements for a MAC again. In essence, we are seeking mappings $f_k(m) = \varphi(m, k)$, where k is the value of the secret key. To keep P_{I} small, we need to ensure that for any m, without access to K, the probability of guessing $f_K(m)$ is small. This coincides with the uniformity property of strong UHFs encountered in Chapter 4. In this section, we will show that, for a given length $\log|\mathcal{K}|$, a strong UHF can be used to find a MAC with optimal probabilities of attack detection, specifically, a MAC with the least value of the product $P_{\mathrm{I}} \cdot P_{\mathrm{S}}$.

First, we describe how to construct a MAC using a strong UHF.

CONSTRUCTION 8.1 (MAC using a strong UHF) Consider a strong UHF $\mathcal{F} = \{f_k : k \in \mathcal{K}\}$ from $\mathcal{X}$ to $\mathcal{Y}$. By setting the set of messages $\mathcal{M} = \mathcal{X}$ and the set of tags $\mathcal{T} = \mathcal{Y}$, we can define a MAC as follows:

$$\varphi(m, k) = f_k(m), \quad V(m, t, k) = \mathbf{1}[f_k(m) = t].$$

Note that this MAC requires the legitimate parties to share a random variable K distributed uniformly over $\mathcal{K}$.

The next lemma quantifies the performance of this MAC.

LEMMA 8.4 *Consider a strong UHF $\mathcal{F} = \{f_k : k \in \mathcal{K}\}$ from $\mathcal{X}$ to $\mathcal{Y}$. The MAC described in Construction 8.1, which sends $\log|\mathcal{X}|$ bits of messages using a*

$\log|\mathcal{K}|$ *bit secret key, is perfectly correct and has success probabilities of attacks satisfying*

$$P_{\mathrm{I}} = P_{\mathrm{S}} = \frac{1}{|\mathcal{T}|}.$$

Proof For the MAC of Construction 8.1, the success probability of impersonation attack is given by

$$\begin{aligned} P_{\mathrm{I}} &= \max_{(m,t)\in\mathcal{M}\in\mathcal{T}} \Pr\left(V(m,t,K)=1\right) \\ &= \max_{(m,t)\in\mathcal{M}\in\mathcal{T}} \Pr\left(f_K(m)=t\right) \\ &= \frac{1}{|\mathcal{T}|}, \end{aligned}$$

where we used the uniformity property of a strong UHF in the previous identity.

On the other hand, denoting $T = \varphi(m,K) = f_K(m)$, to bound the success probability of a substitution attack, note that for any $\mathrm{P}_{\tilde{M}\tilde{T}|T}$ (which may depend on m), we have

$$\begin{aligned} &\Pr\left(V(\tilde{M},\tilde{T},K)=1, \tilde{M}\neq m\right) \\ &\quad= \Pr\left(f_K(\tilde{M})=\tilde{T}, \tilde{M}\neq m\right) \\ &\quad= \sum_{t,\tilde{t}\in\mathcal{T}} \sum_{\substack{\tilde{m}\in\mathcal{M}:\\ \tilde{m}\neq m}} \Pr\left(f_K(m)=t\right)\mathrm{P}_{\tilde{M}\tilde{T}|T}(\tilde{m},\tilde{t}|t)\Pr\left(f_K(\tilde{m})=\tilde{t}|f_K(m)=t\right) \\ &\quad= \sum_{t,\tilde{t}\in\mathcal{T}} \sum_{\substack{\tilde{m}\in\mathcal{M}:\\ \tilde{m}\neq m}} \mathrm{P}_{\tilde{M}\tilde{T}|T}(\tilde{m},\tilde{t}|t)\Pr\left(f_K(m)=t, f_K(\tilde{m})=\tilde{t}\right) \\ &\quad\leq \sum_{t,\tilde{t}\in\mathcal{T}} \sum_{\substack{\tilde{m}\in\mathcal{M}:\\ \tilde{m}\neq m}} \mathrm{P}_{\tilde{M}\tilde{T}|T}(\tilde{m},\tilde{t}|t)\frac{1}{|\mathcal{T}|^2} \\ &\quad\leq \frac{1}{|\mathcal{T}|}, \end{aligned}$$

where the first inequality uses the strong UHF property of $\mathcal{F}$. Thus, we can bound the success probability of the substitution attack as

$$P_{\mathrm{S}} = \max_{m\in\mathcal{M}} \max_{\mathrm{P}_{\tilde{M}\tilde{T}|T}} \Pr\left(V(\tilde{M},\tilde{T},K)=1, \tilde{M}\neq m\right) \leq \frac{1}{|\mathcal{T}|},$$

which completes the proof. □

Remark 8.2 In the proof above, uniformity condition (4.2) is used only in the first part. By (8.1) and the second part of the proof, the collision condition (4.3) suffices to show that $\max\{P_{\mathrm{I}}, P_{\mathrm{S}}\} \leq 1/|\mathcal{T}|$. Even below, if we do not want to distinguish the performance of a MAC for the two security parameters P_{I} and P_{S}, we can simply use strong UHFs without requiring (4.2) to hold.

Any strong UHF can be used in Construction 8.1. In order to choose the one with appropriate parameters, we first try to ascertain a benchmark. Towards this, we will establish information-theoretic lower bounds for success probabilities of an impersonation attack and substitution attack.

Before we move to the formal derivation of these lower bounds, we remark on our strategy for deriving lower bounds.

Remark 8.3 (Lower bounds in cryptography using hypothesis testing) At a high level, our bounds below are based on realizing that the adversary cannot induce the same joint distribution between the tag and the secret key as the legitimate parties. Since the verification procedure of a MAC can distinguish between legitimate tags and those generated by an adversary, we can use it to design a hypothesis test between an adversarially generated joint distribution and the legitimate one. Our strategy for deriving lower bounds entails relating the performance of a MAC to the probabilities of errors for an appropriately constructed hypothesis test. This strategy of converting a cryptographic scheme to a hypothesis test for the correlation enabling security is used several times in this book; see lower bounds in Chapters 10, 13, and 14.

THEOREM 8.5 *Consider an ε-correct MAC (φ, V) with a secret key taking values in the set $\mathcal{K}$ and success probabilities P_{I} and P_{S} of impersonation and substitution attacks, respectively. Then, it must hold that*

$$-\log(P_{\mathrm{I}} \cdot P_{\mathrm{S}}) \leq \frac{\log|\mathcal{K}| + 2h(\varepsilon)}{1-\varepsilon}.$$

Proof We will work with a slightly more general setting and allow K to be an arbitrary random variable, not necessarily uniformly distributed. We first derive a bound for P_{I}. Fix $m \in \mathcal{M}$, and let $T = \varphi(m, K)$. Denote by $\mathcal{D}_m$ the set of tag-key pairs (t, k) for which the receiver accepts the message m, i.e.,

$$\mathcal{D}_m = \{(t,k) : V(m,t,k) = 1\}. \tag{8.2}$$

Since the MAC is ε-correct, we must have

$$\mathrm{P}_{TK}(\mathcal{D}_m) \geq 1 - \varepsilon. \tag{8.3}$$

Consider a simple impersonation attack where the adversary generates an independent copy $\bar{K}$ of K and sends $\bar{T} = \varphi(m, \bar{K})$. Since an impersonation attack can succeed with probability at most P_{I}, we must have

$$\mathrm{P}_{\bar{T}K}(\mathcal{D}_m) \leq P_{\mathrm{I}}. \tag{8.4}$$

Next, consider the hypothesis testing problem with null hypothesis P_{TK} and alternative hypothesis $P_{\bar{T}K} = P_T \times P_K$. Let $\mathcal{D}_m$ be the acceptance region of our test, namely we accept P_{TK} if we observe (t, k) in $\mathcal{D}_m$. By (8.3) and (8.4), this test has type I error probability bounded by ε and type II error probability bounded by P_{I}. Recall from Chapter 5 that $\beta_\varepsilon(P_{TK}, P_{\bar{T}K})$ denotes the minimum type II error probability when the type I error probability is less than ε. Thus, the foregoing argument yields

$$\beta_\varepsilon(P_{TK}, P_{\bar{T}K}) \leq P_{\mathrm{I}},$$

which by the divergence bound for β_ε in Lemma 5.5 gives

$$\begin{aligned} -\log P_{\mathrm{I}} &\le \frac{D(P_{TK}\|P_{\bar{T}K}) + h(\varepsilon)}{1-\varepsilon} \\ &= \frac{I(T \wedge K) + h(\varepsilon)}{1-\varepsilon}. \end{aligned} \tag{8.5}$$

We now focus on the distributions induced in a substitution attack and fix $\tilde{m} \neq m$. Note that by observing m and $T = \varphi(m, K)$, the adversary does get some information on the secret key K. In particular, since it knows the joint distribution P_{TK}, it can sample $\bar{K}$ with distribution $\mathrm{P}_{K|T}(\cdot|T)$. The random variable $(T, \bar{K})$ has the same distribution as (T, K), but $\bar{K}$ is conditionally independent of K given T. The adversary computes $\bar{T} = \varphi(\tilde{m}, \bar{K})$ and sends $(\tilde{m}, \bar{T})$ to the receiver. Since a substitution attack can succeed with probability at most P_{S}, we must have

$$\mathrm{P}_{\bar{T}K}(\mathcal{D}_{\tilde{m}}) \le P_{\mathrm{S}}, \tag{8.6}$$

where $\mathcal{D}_{\tilde{m}}$ denotes the set $\{(t, k) : V(\tilde{m}, t, k) = 1\}$.

On the other hand, consider the legitimate tag $\tilde{T} = \varphi(\tilde{m}, K)$ for $\tilde{m}$. Note that we are not claiming that the adversary can generate this tag; it is only a theoretical construct to enable our proof. Since the MAC is ε-correct, we must have

$$\mathrm{P}_{\tilde{T}K}(\mathcal{D}_{\tilde{m}}) \ge 1 - \varepsilon. \tag{8.7}$$

Consider the hypothesis testing problem with null hypothesis $P_{\tilde{T}K}$ and the alternative $P_{\bar{T}K}$, and consider the test with acceptance region $\mathcal{D}_{\tilde{m}}$ for this problem. By (8.6) and (8.7), this test has type I error probability bounded by ε and type II error probability bounded by P_{S}. Thus,

$$\beta_\varepsilon(P_{\tilde{T}K}, P_{\bar{T}K}) \le P_{\mathrm{S}},$$

whereby

$$\begin{aligned} -\log P_{\mathrm{S}} &\le \frac{D(P_{\tilde{T}K}\|P_{\bar{T}K}) + h(\varepsilon)}{1-\varepsilon} \\ &\le \frac{D(P_{\tilde{T}KT}\|P_{\bar{T}KT}) + h(\varepsilon)}{1-\varepsilon}. \end{aligned}$$

Since $\tilde{T} = \varphi(\tilde{m}, K)$ and $\bar{T} = \varphi(m, \bar{K})$, we have by the data processing inequality for divergence that

$$D(P_{\tilde{T}KT}\|P_{\bar{T}KT}) \le D(P_{KKT}\|P_{\bar{K}KT}).$$

Note that $\mathrm{P}_{\bar{K}K|T} = \mathrm{P}_{\bar{K}|T} \times \mathrm{P}_{K|T}$ and $\mathrm{P}_{\bar{K}|T} = \mathrm{P}_{K|T}$, which gives $D(P_{KKT}\|P_{\bar{K}KT}) = I(K \wedge K|T) = H(K|T)$. Combining the observations above, we obtain

$$-\log P_{\mathrm{S}} \le \frac{H(K|T) + h(\varepsilon)}{1-\varepsilon}. \tag{8.8}$$

Finally, (8.5) and (8.8) together yield

$$-\log(P_{\mathrm{I}} \cdot P_{\mathrm{S}}) \le \frac{H(K) + 2h(\varepsilon)}{1-\varepsilon}.$$

The claim follows by noting $H(K) \le \log|\mathcal{K}|$, where the equality holds for K distributed uniformly on $\mathcal{K}$. □

The following corollary is immediate from Theorem 8.5.

COROLLARY 8.6 *For any perfectly correct MAC with secret key distributed uniformly over $\mathcal{K}$, the success probabilities P_{I} and P_{S} of impersonation and substitution attacks must satisfy*

$$-\log(P_{\mathrm{I}} \cdot P_{\mathrm{S}}) \le \log|\mathcal{K}|. \tag{8.9}$$

We have our desired benchmark set. We now instantiate Construction 8.1 with a specific choice of strong UHF to obtain a MAC that matches the lower bound of Corollary 8.6. Specifically, we use the strong UHF of Construction 4.3 with $q = 2$ and $s = t$. Recall that this corresponds to strong UHF $\mathcal{F} = \{f_k : k \in \mathcal{K}\}$ from $\mathcal{X} = \mathbb{F}_{2^t}$ to $\mathcal{Y} = \mathbb{F}_{2^t}$ with $\mathcal{K} = \mathbb{F}_{2^t}^2$ and mappings f_k given by

$$f_{k_1,k_2}(x) = x \cdot k_1 + k_2. \tag{8.10}$$

Thus, by Lemma 8.4, the MAC of Construction 8.1 corresponding to this choice of strong UHF satisfies $-\log(P_{\mathrm{I}} \cdot P_{\mathrm{S}}) = 2t$ and $\log|\mathcal{K}| = 2t$. Therefore, it matches the lower bound in (8.9) and provides optimal $(P_{\mathrm{I}} \cdot P_{\mathrm{S}})$ for a $2t$-bit secret key. We have established the following corollary.

COROLLARY 8.7 *The MAC of Construction 8.1 implemented using the strong UHF of Construction 4.3 with $q = 2$ and $s = t$ is a perfectly correct MAC with the least possible value of $P_{\mathrm{I}} \cdot P_{\mathrm{S}}$ among all MACs using a $2t$-bit secret key.*

While we seem to have obtained a MAC that is information-theoretically optimal in a specific sense, it may not be satisfactory in practice. A particular shortcoming of the scheme above is that it requires a $2t$-bit secret key for sending t bits of message. In the next section, we will explore a tradeoff in another direction: the minimum key length needed for authenticating a fixed number of message bits while allowing only small success probabilities of attacks.

8.3 MAC with Optimal Length Secret Keys

At the end of the previous section, we have seen that Construction 8.1 with a specific choice of strong UHF provides a security level of $P_{\mathrm{I}} \cdot P_{\mathrm{S}} \le 1/|\mathcal{K}|$. The drawback of this construction is that a $2t$-bit key is needed to send t bits of message. In this section, by relaxing the constraint on the success probability of the substitution attack, we provide an authentication scheme that is much more efficient in the sense of consumed key length. We first describe a construction, which is almost the same as Construction 8.1 except that a strong δ-UHF is used instead of a strong UHF.

CONSTRUCTION 8.2 (MAC using a strong δ-UHF) Consider a strong δ-UHF $\mathcal{F} = \{f_k : k \in \mathcal{K}\}$ from $\mathcal{X}$ to $\mathcal{Y}$. With the set of messages $\mathcal{M} = \mathcal{X}$ and the set of tags $\mathcal{T} = \mathcal{Y}$, we can define a MAC as follows:

$$\varphi(m,k) = f_k(m), \quad V(m,t,k) = \mathbf{1}[f_k(m) = t].$$

In a similar manner to Lemma 8.4, the performance of this MAC is quantified as follows.

LEMMA 8.8 *Consider a strong δ-UHF $\mathcal{F} = \{f_k : k \in \mathcal{K}\}$ from $\mathcal{X}$ to $\mathcal{Y}$. The MAC described in Construction 8.2, which sends* $\log|\mathcal{X}|$ *bits of messages using a* $\log|\mathcal{K}|$ *bit secret key, is perfectly correct and has success probabilities of attacks satisfying*

$$P_{\mathrm{I}} = \frac{1}{|\mathcal{T}|}, \quad P_{\mathrm{S}} \le \delta.$$

Proof The proof is almost the same as that of Lemma 8.4, and is left as an exercise (Problem 8.1). □

As we have seen at the end of Section 4.5, there exists a construction of strong δ-UHFs such that the message size can be exponentially larger than the size of the hash family. In other words, there exists a MAC scheme such that

$$\log|\mathcal{K}| = \mathcal{O}(\log\log|\mathcal{M}|). \tag{8.11}$$

We can show that the order of key length in (8.11) is essentially optimal.

THEOREM 8.9 *For any ε-correct MAC (φ, V) with (possibly nonuniform) key K and such that the success probability P_{S} of substitution satisfies $P_{\mathrm{S}} + \varepsilon < 1$, it holds that*

$$\log\log|\mathcal{M}| \le 2\log|\mathcal{K}|. \tag{8.12}$$

Proof Without loss of generality, we can assume that $\mathcal{T} \subset \mathcal{K}$ since otherwise we can replace $\mathcal{T}$ with $\varphi(m,\mathcal{K})$, i.e., the range of the mapping $\varphi(m,\cdot)$ from $\mathcal{K}$ to $\mathcal{T}$, and further replace $\varphi(m,\mathcal{K})$ with a subset of $\mathcal{K}$ using a (message dependent) one-to-one map from $\varphi(m,\mathcal{K}) \subset \mathcal{T}$ to $\mathcal{K}$.

Under the assumption $P_{\mathrm{S}} + \varepsilon < 1$, we claim that $\mathcal{D}_m \neq \mathcal{D}_{\tilde{m}}$ for every pair $m \neq \tilde{m}$. To prove the claim by contradiction, suppose that there exists $m \neq \tilde{m}$ such that $\mathcal{D}_m = \mathcal{D}_{\tilde{m}}$, where $\mathcal{D}_m \subset \mathcal{K} \times \mathcal{K}$ is defined by (8.2). Then, since the MAC is ε-correct, we have

$$\Pr\left((\varphi(m,K),K) \in \mathcal{D}_m\right) \ge 1 - \varepsilon. \tag{8.13}$$

Next, noting that the substitution attack where $(m,T) = (m,\varphi(m,K))$ is replaced with $(\tilde{m},T)$ has success probability $\Pr\left((\varphi(m,K),K) \in \mathcal{D}_{\tilde{m}}\right)$, we have

$$P_{\mathrm{S}} \geq \Pr\left((\varphi(m, K), K) \in \mathcal{D}_{\tilde{m}}\right)$$
$$= \Pr\left((\varphi(m, K), K) \in \mathcal{D}_{m}\right),$$

which together with (8.13) contradicts $P_{\mathrm{S}} + \varepsilon < 1$.

Thus, the sets $\mathcal{D}_m$ are distinct for different m whereby (8.12) follows since the number of distinct subsets $\mathcal{D}_m \subseteq \mathcal{K} \times \mathcal{K}$ is at most $2^{|\mathcal{K}|^2}$. □

The bound (8.12) does not depend on the security levels P_{I} and P_{S}, other than the requirement $P_{\mathrm{S}} + \varepsilon < 1$. By a more sophisticated argument, it is possible to derive a bound on the key length that depends on both $\log \log |\mathcal{M}|$ and $(P_{\mathrm{I}}, P_{\mathrm{S}})$; see Section 8.6.

8.4 Authentication of Multiple Messages

The schemes presented in the previous two sections fall short of fulfilling the practical requirements for two reasons. The first is that the proposed schemes still rely on finite field arithmetic and may be computationally heavy for implementation; practical MAC schemes will be discussed in the next chapter. But perhaps more importantly for our treatment, the definition of security captured by impersonation and substitution attacks does not capture all attacks feasible in practice. An important class of attacks ruled out is where the adversary gets access to not just one message but multiple messages authenticated using the same secret key. We begin by noting that Construction 8.1 instantiated using (8.10) is not secure when the adversary can access multiple messages.

Indeed, if two messages m_1 and m_2 are authenticated by the same key (k_0, k_1), the adversary observes $c_1 = m_1 \cdot k_1 + k_0$ and $c_2 = m_2 \cdot k_1 + k_0$, from which it can compute $k_1 = (m_1 - m_2)^{-1}(c_1 - c_2)$. Then, for a forged message $\tilde{m} = m_2 + e$, the adversary can produce the legitimate tag $\tilde{t} = \tilde{m} \cdot k_1 + k_0$ for $\tilde{m}$ as $c_2 + k_1 e$.

In this section, we present a more stringent notion of security that renders a MAC secure even when the adversary can access multiple messages, and present schemes that remain secure under this new notion of security. We consider a MAC that authenticates up to r messages. Further, we allow that different mappings (ϕ_i, V_i) are used for each message $1 \leq i \leq r$. These mappings can be randomized using different (independent) local randomness for each message. It is important to note that this is the *synchronous* mode of communication where the sender and the receiver share the index i; a more general and practically relevant variant is the *asynchronous* mode, which is harder to analyze formally and will not be covered in this chapter.

To describe this notion of security, we need to precisely lay down the "threat model," namely the procedure the adversary is allowed to follow, and define a parameter that governs security of the scheme. This is the first occurrence of such an elaborate description of security requirements in this book, but such descriptions are commonplace in cryptography and will be seen in the coming chapters as well.

Threat model for security of a MAC with multiple messages
We consider the security of a MAC for authenticating up to r messages. In this scenario, the adversary should not be allowed to substitute the ith message by observing the message-tag pairs for the first i messages. As for the substitution attack in the single-message setting above, we consider the worst-case situation for the observed messages, which can be viewed equivalently as the adversary selecting the first i messages.

Formally, for $1 \le i \le r$ and $j \le i$, upon observing the messages $M_1, \ldots, M_{j-1}$ and tags $T_1, \ldots, T_{j-1}$, the adversary generates the next random message M_j and observes the tag $T_j = \phi_j(M_j, K)$ for it. Thus, the adversary selects the conditional distribution $\mathrm{P}_{M_j|M^{j-1}T^{j-1}}$ and the protocol determines $\mathrm{P}_{T_j|M_jM^{j-1}T^{j-1}}$. After receiving tags for i messages of its choice, the adversary substitutes the message M_i with $\tilde{M} \neq M_i$ by forging a tag $\tilde{T}$ using a conditional distribution $\mathrm{P}_{\tilde{T}\tilde{M}|M^iT^i}$ of its choice. The adversary succeeds if the receiver accepts the pair $(\tilde{M}, \tilde{T})$ for $\tilde{M} \neq M_i$, for any $i \in \{1, \ldots, r\}$. Note that the class $\mathcal{A}_i$, $1 \le i \le r$, of such adversaries is determined by the conditional distributions $\mathrm{P}_{M_i|M^{i-1}T^{i-1}}$ and $\mathrm{P}_{\tilde{T}\tilde{M}|M^iT^i}$. For a given MAC, we denote by $\mathcal{P}_{\mathcal{A}_i}$ the set of distributions $\mathrm{P}_{\tilde{M}\tilde{T}M^iT^i}$ where the distribution $\mathrm{P}_{T_j|M_jM^{j-1}T^{j-1}}$ is fixed by the MAC and the remaining conditional distribution can be arbitrarily chosen (by an adversary in $\mathcal{A}_i$).

DEFINITION 8.10 (Security of a MAC for multiple messages) For a multimessage MAC, the i-message success probability $P_{\mathrm{S},i}$ for an adversary in $\mathcal{A}_i$ is given by

$$P_{\mathrm{S},i} = \max_{\mathrm{P}_{\tilde{M}\tilde{T}M^iT^i} \in \mathcal{P}_{\mathcal{A}_i}} \Pr\left(V_i(\tilde{M}, \tilde{T}, K) = 1, \tilde{M} \neq M_i\right),$$

where V denotes the verifier function used by the MAC and K is the key. A multimessage MAC is δ-secure (under the threat model above) if

$$\max_{1 \le i \le r} P_{\mathrm{S},i} \le \delta.$$

Note that the impersonation attack is already subsumed in the class of adversaries in $\mathcal{A}_1$.

When authenticating multiple messages, the MAC is allowed to use a different mapping to encode each message M_i, which can even depend on the previous messages and tag. Interestingly, the scheme we present uses one-time pad encryption to authenticate multiple messages. We assume that the tag space $\mathcal{T}$ is an additive group. The sender and the receiver share two kinds of keys: the key K to generate the tag, and another independent set of keys $(S_1, \ldots, S_r)$ for one-time pad encryption. The sender can authenticate up to r messages using the following scheme.

CONSTRUCTION 8.3 (A MAC for multiple messages using encryption) Consider a strong δ-UHF $\mathcal{F} = \{f_k : k \in \mathcal{K}\}$ from $\mathcal{X}$ to $\mathcal{Y}$. Further, assume that the sender and the receiver share independent secret keys $K, S_1, \ldots, S_r$ where K is distributed uniformly on $\mathcal{K}$ and the S_i are distributed uniformly on $\mathcal{Y}$. For the set of messages $\mathcal{M} = \mathcal{X}$ and the set of tags $\mathcal{T} = \mathcal{Y}$, we can define a MAC as follows: for $1 \le i \le r$, the MAC (φ_i, V_i) used for the ith message m_i is

$$\varphi_i(m_i, k, s_1, \ldots, s_r) = f_k(m_i) + s_i, \quad V_i(m, t, k, s_1, \ldots, s_r) = \mathbf{1}[f_k(m) = t - s_i].$$

Note that this MAC requires the legitimate sender and receiver to share the counter i that keeps a track of the number of transmitted messages in the past.

For this scheme, the success probability of attack after i messages is given by

$$P_{\mathrm{S},i} = \max_{\mathrm{P}_{\tilde{M}\tilde{T}M^iT^i} \in \mathcal{P}_{\mathcal{A}_i}} \Pr\left(V_i(\tilde{M}, \tilde{T}, K, S_1, \ldots, S_r) = 1, \tilde{M} \neq M_i\right).$$

Clearly, $P_{\mathrm{S},i}$ is nondecreasing in i, whereby $P_{\mathrm{S},r}$ captures the worst-case success probability up to r messages. The receiver verifies each message-tag pair received, and success of any attack on an earlier message will also be bounded by the same probability.

Heuristically, Construction 8.3 is secure since the raw tags $f_K(M_1), \ldots, f_K(M_r)$ are encrypted by the secret keys $S_1, \ldots, S_r$, and thus the tags $T_1, \ldots, T_r$ do not reveal any information about the key K to the adversary. Formally, we can bound $P_{\mathrm{S},r}$ for the MAC in Construction 8.3 as follows.

THEOREM 8.11 *The MAC described in Construction 8.3, which sends up to r messages of length $\log|\mathcal{X}|$ bits using $\log|\mathcal{K}| + r\log|\mathcal{Y}|$ bits of keys, is perfectly correct and δ-secure.*

Proof Since $P_{\mathrm{S},i}$ is nondecreasing in i, it suffices to show that $P_{\mathrm{S},r} \le \delta$. For fixed strategy of the adversary, i.e., for fixed $\mathrm{P}_{\tilde{M}\tilde{T}M^rT^r} \in \mathcal{P}_{\mathcal{A}_r}$, the probability of success attack can be expanded as follows:

$$\begin{aligned}
&\Pr\left(V_r(\tilde{M}, \tilde{T}, K, S^r) = 1, \tilde{M} \neq M_r\right) \\
&= \sum_{\substack{\tilde{m}, \tilde{t}, m_r: \\ \tilde{m} \neq m_r}} \Pr\left(\tilde{M} = \tilde{m}, \tilde{T} = \tilde{t}, M_r = m_r, f_K(\tilde{m}) + S_r = \tilde{t}\right) \\
&= \sum_{\substack{\tilde{m}, \tilde{t}, m_r, t_r: \\ \tilde{m} \neq m_r}} \Pr\big(\tilde{M} = \tilde{m}, \tilde{T} = \tilde{t}, M_r = m_r, T_r = t_r, \\
&\qquad\qquad f_K(m_r) + S_r = t_r, f_K(\tilde{m}) + S_r = \tilde{t}\big) \\
&= \sum_{\substack{\tilde{m}, \tilde{t}, m_r, t_r: \\ \tilde{m} \neq m_r}} \Pr\left(\tilde{M} = \tilde{m}, \tilde{T} = \tilde{t}, M_r = m_r, T_r = t_r, f_K(\tilde{m}) - f_K(m_r) = \tilde{t} - t_r\right) \\
&\sum_{\substack{\tilde{m}, \tilde{t}, m_r, t_r: \\ \tilde{m} \neq m_r}} \Pr\left(f_K(\tilde{m}) - f_K(m_r) = \tilde{t} - t_r\right) \\
&\qquad \times \Pr\left(\tilde{M} = \tilde{m}, \tilde{T} = \tilde{t}, M_r = m_r, T = t_r \mid f_K(\tilde{m}) - f_K(m_r) = \tilde{t} - t_r\right).
\end{aligned}$$

We note that the event $\{f_K(\tilde{m}) - f_K(m_r) = \tilde{t} - t_r\}$ is computed for fixed realizations $(\tilde{m}, m_r, \tilde{t}, t_r)$ of $(\tilde{M}, M, \tilde{T}, T_r)$. We now take recourse to a basic property of a strong δ-UHF with range $\mathcal{Y}$ identified with a finite group, namely that it is also an XOR δ-UHF (see Problem 4.12) and satisfies $\Pr(F(m) - F(m') = t) \le \delta$ for every $m \neq m'$ and t. Thus,

$$\Pr\left(f_K(\tilde{m}) - f_K(m_r) = \tilde{t} - t_r\right) \le \delta,$$

whereby

$$\begin{aligned}
&\Pr\left(V(\tilde{M}, \tilde{T}, K, S_r) = 1, \tilde{M} \neq M\right) \\
&\le \delta \cdot \sum_{\substack{\tilde{m}, \tilde{t}, m_r, t_r: \\ \tilde{m} \neq m_r}} \Pr\left(\tilde{M} = \tilde{m}, \tilde{T} = \tilde{t}, M_r = m_r, T = t_r \mid f_K(\tilde{m}) - f_K(m) = \tilde{t} - t_r\right).
\end{aligned}$$

Each term in the summation above can be simplified further as follows:

$$\begin{aligned}
&\Pr\left(\tilde{M} = \tilde{m}, \tilde{T} = \tilde{t}, M_r = m_r, T = t_r \mid f_K(\tilde{m}) - f_K(m_r) = \tilde{t} - t_r\right) \\
&= \sum_{m^{r-1}, t^{r-1}} \Pr\left(\tilde{M} = \tilde{m}, \tilde{T} = \tilde{t}, M^r = m^r, T^r = t^r \mid f_K(\tilde{m}) - f_K(m_r) = \tilde{t} - t_r\right) \\
&= \sum_{m^{r-1}, t^{r-1}} \Pr\left(\tilde{M} = \tilde{m}, \tilde{T} = \tilde{t} \mid M^r = m^r, T^r = t^r, f_K(\tilde{m}) - f_K(m_r) = \tilde{t} - t_r\right) \\
&\quad \times \Pr\left(M^r = m^r, T^r = t^r \mid f_K(\tilde{m}) - f_K(m_r) = \tilde{t} - t_r\right) \\
&= \sum_{m^{r-1}, t^{r-1}} \Pr\left(\tilde{M} = \tilde{m}, \tilde{T} = \tilde{t} \mid M^r = m^r, T^r = t^r\right) \\
&\quad \times \Pr\left(M^r = m^r, T^r = t^r \mid f_K(\tilde{m}) - f_K(m_r) = \tilde{t} - t_r\right),
\end{aligned}$$

where the final identity uses the fact that an adversary from $\mathcal{A}_r$ is only allowed to generate $(\tilde{M}, \tilde{T})$ based on only (M^r, T^r); specifically, $(\tilde{M}, \tilde{T})$ and K are conditionally independent given (M^r, T^r). Thus,

$$\begin{aligned}
&\sum_{\substack{\tilde{m}, \tilde{t}, m_r, t_r: \\ \tilde{m} \neq m_r}} \Pr\left(\tilde{M} = \tilde{m}, \tilde{T} = \tilde{t}, M_r = m_r, T = t_r \mid f_K(\tilde{m}) - f_K(m_r) = \tilde{t} - t_r\right) \\
&= \sum_{m^r, t^r} \sum_{\substack{\tilde{m}, \tilde{t}: \\ \tilde{m} \neq m_r}} \Pr\left(\tilde{M} = \tilde{m}, \tilde{T} = \tilde{t} \mid M^r = m^r, T^r = t^r\right) \\
&\quad \times \Pr\left(M^r = m^r, T^r = t^r \mid f_K(\tilde{m}) - f_K(m_r) = \tilde{t} - t_r\right).
\end{aligned}$$

Note that it is not a priori clear how to take the sum on the right-hand side above since variables are intertwined on the conditioning side as well. Nonetheless, we show now that this sum is bounded above by 1. It is in this part that we use the independence and uniformity of the secret keys S_i. Towards that, we first simplify the second term in the summand by exploiting the information structure of the threat model. We have

$$\begin{aligned}
&\Pr\left(M^r = m^r, T^r = t^r \mid f_K(\tilde{m}) - f_K(m_r) = \tilde{t} - t_r\right) \\
&= \prod_{i=1}^{r} \Pr\left(M_i = m_i \mid M^{i-1} = m^{i-1}, T^{i-1} = t^{i-1}, f_K(\tilde{m}) - f_K(m_r) = \tilde{t} - t_r\right) \\
&\quad \times \Pr\left(T_i = t_i \mid M^i = m^i, T^{i-1} = t^{i-1}, f_K(\tilde{m}) - f_K(m_r) = \tilde{t} - t_r\right) \\
&= \prod_{i=1}^{r} \Pr\left(M_i = m_i \mid M^{i-1} = m^{i-1}, T^{i-1} = t^{i-1}\right) \\
&\quad \times \Pr\left(f_K(m_i) + S_i = t_i \mid M^i = m^i, T^{i-1} = t^{i-1}, f_K(\tilde{m}) - f_K(m_r) = \tilde{t} - t_r\right),
\end{aligned}$$

where we used the fact that M_i is independent of K given (M^{i-1}, T^{i-1}) in the second equality. Also, note that S_i is uniformly distributed and independent of (K, S^{i-1}), which further implies that S_i is independent of the event $\{M^i = m^i, T^{i-1} = t^{i-1}, f_K(\tilde{m}) - f_K(m_r) = \tilde{t} - t_r\}$. It follows that

$$\Pr\left(f_K(m_i) + S_i = t_i \mid M^i = m^i, T^{i-1} = t^{i-1}, f_K(\tilde{m}) - f_K(m_r) = \tilde{t} - t_r\right) = \frac{1}{|\mathcal{T}|}$$

holds for $1 \le i \le r$, which gives

$$\begin{aligned}
&\Pr\left(M^r = m^r, T^r = t^r \mid f_K(\tilde{m}) - f_K(m_r) = \tilde{t} - t_r\right) \\
&= \frac{1}{|\mathcal{T}|^r} \prod_{i=1}^{r} \Pr\left(M_i = m_i \mid M^{i-1} = m^{i-1}, T^{i-1} = t^{i-1}\right).
\end{aligned}$$

Combining all the observations above, we finally get that

$$\begin{aligned}
&\Pr\left(V(\tilde{M}, \tilde{T}, K, S_r) = 1, \tilde{M} \neq M\right) \\
&\le \delta \cdot \sum_{\substack{\tilde{m}, \tilde{t}, m^r, t^r: \\ \tilde{m} \neq m_r}} \frac{1}{|\mathcal{T}|^r} \Pr\left(\tilde{M} = \tilde{m}, \tilde{T} = \tilde{t} \mid M^r = m^r, T^r = t^r\right) \\
&\qquad \times \prod_{i=1}^{r} \Pr\left(M_i = m_i \mid M^{i-1} = m^{i-1}, T^{i-1} = t^{i-1}\right) \\
&\le \delta \cdot \sum_{m^r, t^r} \frac{1}{|\mathcal{T}|^r} \prod_{i=1}^{r} \Pr\left(M_i = m_i \mid M^{i-1} = m^{i-1}, T^{i-1} = t^{i-1}\right) \\
&= \delta,
\end{aligned}$$

where in the final identity the remaining summations are taken recursively from $i = r$ to $i = 1$. □

Note that a much simpler scheme which just uses a single message MAC with independent key can be used to authenticate r messages. But this will require $r \log |\mathcal{K}|$ bits of shared key. In contrast, Construction 8.3 requires $\log |\mathcal{K}| + r \log |\mathcal{Y}|$ bits of shared key; this is much smaller than $r \log |\mathcal{K}|$ since, in a typical construction of a strong δ-UHF, the output size $|\mathcal{Y}| = |\mathcal{T}|$ is much smaller than the key size $|\mathcal{K}|$.

In Chapter 9, we will discuss a computationally secure authentication scheme. In fact, we will see that Construction 8.3 can be readily converted into a computationally secure scheme by replacing the strong UHF with a hash family such as those described in Section 4.6 and by replacing the encryption mapping with an encryption scheme using the pseudorandom generator.

8.5 Nonmalleability

In this section, we shall explain a security notion that is closely related to authentication, termed *nonmalleability*. In Chapter 3, we discussed the secrecy of encryption schemes; in particular, we have seen that one-time pad encryption satisfies perfect secrecy.

For concreteness, consider the following situation. Suppose that two companies, say company A and company B, are bidding for a government sponsored project. The company with smaller bid will get to deliver the project. Suppose that one of the companies, say company A, sends the bidding message M to the government using a one-time pad encryption scheme where the last ℓ bits describe the bid amount, and company B intercepts the ciphertext. Even though company B cannot learn the bid of company A, company B can flip the ℓth-last bits so that company A will lose the bidding (assuming that the ℓth-last bit of the message M is 0 and the flipping will increase the amount of the bidding by company A significantly).

Roughly speaking, an encryption scheme is said to be *malleable* if an adversary can substitute the ciphertext in a manner so that the decrypted message of the substituted ciphertext satisfies a certain relation with the original message. An encryption scheme is *nonmalleable* if it is not malleable. As is illustrated above, the one-time pad encryption scheme is malleable, although it satisfies perfect secrecy.

There is no agreement on a formal definition of information-theoretic nonmalleability. However, from the example above, it is clear that it does not suffice to have just the mutual information between message and ciphertext small (or even zero). Note that if the receiver can detect a substitution attack, then the attack will fail. Thus, an authentication scheme with a small probability of successful substitution attack will be approximately nonmalleable. Interestingly, we present below a scheme that can be interpreted as a perfectly nonmalleable scheme, but it cannot detect a substitution attack at all.

CONSTRUCTION 8.4 (Perfectly nonmalleable scheme) Let $\mathcal{M} = \mathcal{C} = \mathbb{F}_q$ and $\mathcal{K} = \mathbb{F}_q \backslash \{0\} \times \mathbb{F}_q$. Further, let

$$f(m, k) = k_1 \cdot m + k_2$$

for $m \in \mathcal{M}$ and $k = (k_1, k_2) \in \mathcal{K}$. Since $k_1 \neq 0$, the decryption is given by

$$g(c, k) = k_1^{-1}(c - k_2).$$

For the uniform key $K = (K_1, K_2)$ on $\mathcal{K}$, clearly, this scheme satisfies perfect secrecy. Now, consider a substitution attack where C is replaced with $\tilde{C} = C + E$

with E taking values in $\mathbb{F}_q \backslash \{0\}$. Then, $\tilde{M} = M + K_1^{-1} \cdot E$. Thus, $\tilde{M}$ is uniformly distributed on $\mathbb{F}_q \backslash \{M\}$ irrespective of the value of $\tilde{C}$. In fact, given $M = m$, $\tilde{M}$ is distributed uniformly on $\mathbb{F}_q \backslash \{m\}$ if $E \neq 0$, showing that the adversary cannot force any relation between M and $\tilde{M}$ if it substitutes the cipher. Note that the receiver cannot detect a substitution attack in the previous scheme. Therefore, nonmalleable encryption differs from authentication.

The previous scheme circumvents the malleability problem. Any reasonable definition of nonmalleability should deem the scheme above nonmalleable, while identifying one-time pad encryption as malleable. Such definitions are available in the literature but, as mentioned earlier, there is no consensus on any one definition. In particular, the following definition fulfills both criteria.

Let (f, g) be an encryption scheme such that $g(f(m, k), k) = m$ for every message $m \in \mathcal{M}$ and key $k \in \mathcal{K}$. We allow $g(c, k)$ to output a rejection symbol $\perp \notin \mathcal{M}$ if the ciphertext c is not compatible with the key k. Suppose that a message $M \in \mathcal{M}$ is encrypted into $C = f(M, K)$ using a secret key K. Upon observing (M, C),[2] the adversary creates a substituted ciphertext $\tilde{C}$; here, the substituted ciphertext is assumed to satisfy

$$\tilde{C} -\!\!\circ\!\!- (M, C) -\!\!\circ\!\!- K, \tag{8.14}$$

and

$$\Pr(\tilde{C} \neq C) = 1, \tag{8.15}$$

i.e., the adversary changes the ciphertext to a different value with probability 1. Let $\tilde{M} = g(\tilde{C}, K)$ be the decrypted message of the substituted ciphertext.

DEFINITION 8.12 (Nonmalleable encryption) An encryption scheme (f, g) is (perfectly) nonmalleable if, for any distribution P_M, for any substituted ciphertext $\tilde{C}$ satisfying (8.14) and (8.15), and for the decrypted message $\tilde{M} = g(\tilde{C}, K)$, it holds that $I(\tilde{M} \wedge \tilde{C} | M, C) = 0$.

It can be checked that the one-time pad encryption is malleable in the sense of Definition 8.12, while Construction 8.4 is nonmalleable; see Problem 8.4. But the latter scheme consumes almost twice the length of secret key compared to the length of the message. In fact, it can be proved that this scheme has optimal key consumption (cf. Problem 8.5). In order to construct more efficient schemes, we can relax the nonmalleability requirement to an approximate one. Often, authentication schemes satisfy such a criterion and can be used as nonmalleable encryption schemes.

[2] Even though the adversary only observes the ciphertext C in practice, we provide additional power to the adversary to admit a stronger attack similar to the *chosen-plaintext attack* to be discussed in Chapter 9.

8.6 References and Additional Reading

Theoretical study of authentication schemes was initiated by Gilbert, MacWilliams, and Sloane in their paper [135]. Construction of schemes based on a universal hash family was initiated by Wegman and Carter [347], and further studied by Stinson [316]. Good expositions on authentication are available in [102, 265, 309, 318].

Information-theoretic lower bounds on authentication schemes were initially developed by Simmons [308, 309]. The approach based on hypothesis testing was introduced by Maurer [239]. Interestingly, the hypothesis testing approach suggests a decomposition of the randomness of the secret key into two parts: the exponent of the impersonation probability is upper bounded by the mutual information between the secret key and the tag (cf. (8.5)); and the exponent of the substitution attack is upper bounded by the equivocation of the key given the tag (cf. (8.8)). In fact, under certain assumptions, these upper bounds are attainable asymptotically [200].

As we have seen in Section 8.3, the length of the secret key consumed in authentication can be as small as the logarithmic order of the message length. This is in contrast with the fact that secret key encryption consumes a secret key of length as large as the message length (cf. Chapter 3). In other words, the cost of authentication can be regarded as almost free compared to the cost of secrecy. As we will see in Chapter 10, we typically assume that an authenticated public channel between the communicating parties is available in the secret key agreement problem. In practice, we create the authenticated channel by using an authentication scheme with a pre-shared short key. For instance, the authentication scheme described in Section 8.3 is used in the implementation of the quantum key distribution protocol in [342].

In Theorem 8.9, we derived a lower bound on the key length by a simple counting argument. By a more sophisticated argument, it is possible to derive a lower bound that depends on both the message size and the success probabilities of the attacks [13, 134].

In this chapter, we only considered *standalone* security of authentication schemes. Typically, authentication is used as a building block of a larger encryption system, requiring composably secure authentication schemes. Composably secure authentication has been discussed in the literature; see, for instance, [277].

Furthermore, we only considered one-way authentication schemes in which the sender transmits a message-tag pair to the receiver and there is no communication in the opposite direction. More generally, it is possible to consider interactive authentication schemes [134]; see also [132, 133, 257]. Surprisingly, by allowing multiple rounds of interaction, we can construct an authentication scheme such that the length of the secret key need not even grow logarithmically in the length of the message.

Another topic we did not cover in this chapter is authentication from correlated randomness (cf. Problem 8.6); we only considered authentication using a perfect secret key shared between the sender and the receiver. Authentication using correlation or a noisy channel has been studied in the literature; for instance, see [105, 214, 243, 289, 329, 264]. In a secret key agreement from correlated observations (see Chapter 10), if the public communication channel is not authenticated, messages exchanged between parties must be authenticated using a part of correlated observations [243].

The concept of nonmalleability was introduced by Dolev, Dwork, and Naor in [107]. Initially, nonmalleability was studied in the context of computational security, particularly for public key cryptography; for instance see [21]. Nonmalleability of secret key cryptography was studied in [188]. The information-theoretic nonmalleability in Definition 8.12 was introduced in [155], and the construction of a perfect nonmalleable encryption scheme in Construction 8.4 is from [154] (see also [153]). Approximate nonmalleability and its realization using authentication was studied in [189]; see also [246].

Problems

8.1 Prove Lemma 8.8.

8.2 Prove that, for any perfectly correct authentication scheme (φ, V) with (possibly nonuniform) key K, the success probability P_{I} of impersonation and the success probability P_{S} of substitution must satisfy (cf. [306])

$$-\log(P_{\mathrm{I}} \cdot P_{\mathrm{S}}) \le H_2(K),$$

where $H_2(K)$ is the collision entropy of K (see Problem 4.7). Note that this bound is tighter than the one in Theorem 8.5 when P_K is not uniform.

8.3 Verify that Lemma 8.4 and Lemma 8.8 are true even for substitution attack in the wide sense (cf. Remark 8.1).

HINT Consider two cases $\tilde{M} = m$ and $\tilde{M} \neq m$ separately. In the former case, $\tilde{T} \neq T$ must hold, and the event $\{f_K(m) = \tilde{T}, f_K(m) = T\}$ never occur.

8.4 *Nonmalleability*: Show that one-time pad encryption is malleable in the sense of Definition 8.12. Further, show that the encryption scheme in Construction 8.4 is nonmalleable in the sense of Definition 8.12.

8.5 *Perfect nonmalleability and two-message perfect secrecy*:

1. Consider the scheme in Construction 8.4. For any pair of messages $m_1 \neq m_2$, verify that the two ciphertexts $C_1 = f(m_1, K)$ and $C_2 = f(m_2, K)$ encrypted by the same key satisfy

$$\mathrm{P}_{C_1 C_2 | M_1 M_2}(c_1, c_2 | m_1, m_2) = \frac{1}{|\mathcal{C}|(|\mathcal{C}| - 1)} \tag{8.16}$$

for every $c_1 \neq c_2 \in \mathcal{C}$.

2. Prove that any nonmalleable scheme with $|\mathcal{C}| > 2$ satisfies (8.16) for two messages $m_1 \neq m_2$ and $c_1 \neq c_2 \in \mathcal{C}$ [189].

REMARK The criterion (8.16) is referred to as *two-message perfect secrecy*. This result implies that any nonmalleable scheme requires a secret key of length $\log |\mathcal{M}|(|\mathcal{M}| - 1)$.

8.6 *Authentication from correlation*: Suppose that the sender and the receiver observe correlated sources X and Y, respectively. In a similar manner to the model with shared secret key, an authentication scheme with correlated sources consists of a function for computing a message tag, $\varphi\colon \mathcal{M} \times \mathcal{X} \to \mathcal{T}$, and a verification function $V\colon \mathcal{M} \times \mathcal{T} \times \mathcal{Y} \to \{0,1\}$. The ε-correctness condition is given by

$$\Pr\big(V(m, \varphi(m, X), Y) = 1\big) \geq 1 - \varepsilon,$$

and the success probabilities of impersonation and substitution are defined as

$$P_{\mathrm{I}} = \max_{(m,t) \in \mathcal{M} \times \mathcal{T}} \Pr(V(m, t, Y) = 1)$$

and

$$P_{\mathrm{S}} = \max_{m \in \mathcal{M}} \max_{\mathrm{P}_{\tilde{M}\tilde{T}|T}} \Pr(V(\tilde{M}, \tilde{T}, Y) = 1, \tilde{M} \neq m),$$

respectively.

1. For any ε-correct scheme, prove that

$$-\log(P_{\mathrm{I}} \cdot P_{\mathrm{S}}) \leq \frac{I(X \wedge Y) + 2h(\varepsilon)}{1 - \varepsilon} \tag{8.17}$$

holds.

HINT To bound P_{I}, introduce an independent copy $\bar{X}$ of X; to bound P_{S}, introduce $\bar{X}$ distributed according to $\mathrm{P}_{X|T}(\cdot|T)$. Then, consider hypothesis testing in a similar manner to the proof of Theorem 8.5.

2. Let (X_0, Y_0) and (X_1, Y_1) be mutually independent correlated sources having common distribution P_{XY}. For given $0 \leq \varepsilon < 1$, let

$$\mathcal{S}(\lambda) := \left\{(x, y) : \log \frac{\mathrm{P}_{XY}(x,y)}{\mathrm{P}_X(x)\,\mathrm{P}_Y(y)} \geq \lambda\right\},$$

and let λ_ε be the supremum of λ satisfying $\mathrm{P}_{XY}(\mathcal{S}(\lambda)^c) \leq \varepsilon$. Consider the following authentication scheme for a 1-bit message (cf. [243]): for $m \in \{0,1\}$, the tag is $\varphi(m, (x_0, x_1)) = x_m$ and the verification function is $V(m, t, (y_0, y_1)) = \mathbf{1}[(t, y_m) \in \mathcal{S}(\lambda_\varepsilon)]$. Prove that this scheme is ε-correct and $\max\{P_{\mathrm{I}}, P_{\mathrm{S}}\} \leq 2^{-\lambda_\varepsilon}$.

REMARK When (X_0, Y_0) and (X_1, Y_1) are sequences of i.i.d. observations, this scheme attains the bound (8.17) asymptotically.

9 Computationally Secure Encryption and Authentication

In Chapter 3, we discussed encryption schemes when only one message is transmitted using a shared secret key. We have also shown that, in order to attain information-theoretic security, the key length must be as large as the message length itself. However, in many applications of practical usage, a more relaxed notion of security suffices. In this chapter, we digress from the main theme of the book, and discuss computationally secure encryption schemes. Under the assumption that computationally secure cryptographic primitives – such as pseudorandom generators or pseudorandom functions – are available, we construct encryption schemes that are secure (i) when the message length is larger than the key length; (ii) when multiple messages are transmitted using the same key; and (iii) when an adversary can access the encryption or decryption algorithm. Interestingly, the security of these schemes is proved by comparing them with their information-theoretic counterparts that use perfect secret keys and random functions.

To establish computational security, we show that an encryption scheme cannot be broken "efficiently." One of the most commonly used measures of efficiency is the notion of probabilistic polynomial-time (PPT) algorithm. An algorithm A is PPT if there exists a polynomial function $p(\cdot)$ on $\mathbb{N}$ such that, for every input $x \in \{0,1\}^*$ of length $|x|$, it outputs a value $A(x)$ within $p(|x|)$ steps. Of course, the number of steps depends on the computational model, such as a single-tape Turing machine or a multi-tape Turing machine. However, since the exact model of computation is irrelevant for our discussion, we do not specify the model of computation. A reader who is unfamiliar with computational complexity theory can just regard a PPT algorithm as a mathematical model of a computer program that terminates within a realistic amount of time. More concretely, one can view the number of steps of the algorithm above as the number of CPU clock cycles used.

We will remain informal in our discussion about computational complexity. The main purpose of this chapter is to highlight the similarities between information-theoretic cryptography and computational cryptography, rather than to provide a complete introduction to computational cryptography.

9.1 One-time Encryption Revisited

In Section 3.3.1, we analyzed one-time pad (OTP) encryption when the underlying key is not uniformly distributed but is only an almost uniform random variable K. In the security proof, we bounded the information leaked to an adversary who observes the encrypted message $C = m \oplus K$.[1] In Chapter 3, we saw that the notion of *semantic security* is equivalent to *indistinguishable security* where the adversary should not be able to use the encrypted output to statistically distinguish between messages m_0, m_1, for any pair of messages (m_0, m_1) of the adversary's choice. We lead in to the content of this chapter with an alternative security analysis for OTP encryption that directly gives indistinguishable security. This alternative analysis will be seen to open a completely different perspective on the security of OTP encryption, one that has fueled the success of cryptography in securing message transmission over the Internet.

Recall that in OTP encryption the legitimate sender and the receiver share an ℓ-bit random variable K, which is used to encrypt the ℓ-bit message m as $E(m, K) = m \oplus K$. The result to be presented in Theorem 9.3 shows that if, for any pair of messages (m_0, m_1), we can distinguish the random variables $E(m_0, K)$ and $E(m_1, K)$ with reasonable statistical significance, then we can also distinguish K and an ℓ-bit uniformly distributed key K_ℓ with reasonable statistical significance.

To state the formal result, we need the notion of a distinguisher. For random variables X and Y taking values in a common set $\mathcal{X}$, the total variation distance (see Definition 2.12) between P_X and P_Y can be written as

$$d_{\mathrm{var}}(\mathrm{P}_X, \mathrm{P}_Y) = \max_{\mathscr{D}:\mathcal{X}\to\{0,1\}} \left|\Pr\left(\mathscr{D}(X) = 1\right) - \Pr\left(\mathscr{D}(Y) = 1\right)\right|;$$

see Problem 9.2. Thus, the total variation distance can be captured alternatively using the notion of ε-distinguisher defined below.[2]

DEFINITION 9.1 (Distinguisher (function)) A randomized function $\mathscr{D}\colon \mathcal{X} \to \{0, 1\}$ constitutes an ε-*distinguisher* for random variables X and Y taking values in $\mathcal{X}$ if

$$|\Pr(\mathscr{D}(X) = 1) - \Pr(\mathscr{D}(Y) = 1)| > \varepsilon. \tag{9.1}$$

In other words, $d_{\mathrm{var}}(\mathrm{P}_X, \mathrm{P}_Y) \le \varepsilon$ is equivalent to the claim that there is no ε-distinguisher for X and Y.

Since we consider computational security in this chapter, we extend the notion of an ε-distinguisher to incorporate algorithms. In fact, both randomized function and randomized algorithm signify an input-to-output mapping, except that in

[1] Throughout this chapter, addition will be for vectors in $\mathbb{F}_2$ and $\oplus$ will denote coordinate wise XOR.

[2] A distinguisher is simply a randomized test for the binary hypothesis testing problem of P_X versus P_Y; see Section 2.4 in Chapter 2 and Chapter 5.

the latter case we keep track of the computational complexity of evaluating that mapping.

DEFINITION 9.2 (Distinguisher (algorithm)) A randomized algorithm $\mathscr{D}$ with inputs from the set $\mathcal{X}$ and output in $\{0,1\}$ constitutes an *ε-distinguisher* for random variables X and Y taking values in $\mathcal{X}$ if the outputs $\mathscr{D}(X)$ and $\mathscr{D}(Y)$ of the algorithm $\mathscr{D}$ for inputs X and Y, respectively, satisfy $|\Pr(\mathscr{D}(X)=1) - \Pr(\mathscr{D}(Y)=1)| > \varepsilon$.

The result below is a restatement of Theorem 3.18, using the notion of distinguishers.

THEOREM 9.3 *Let K, K_ℓ be random variables taking values in $\{0,1\}^\ell$, with K_ℓ distributed uniformly. For messages $m_0 \neq m_1$, given a 2ε-distinguisher $\mathscr{D}$ for $E(m_0,K)$ and $E(m_1,K)$, where $E(m,K) = m \oplus K$ is the encryption mapping of the OTP, we can construct an ε-distinguisher $\tilde{\mathscr{D}}$ for K and K_ℓ that invokes $\mathscr{D}$ once.*

Proof By the assumption of the theorem, there exists a distinguisher $\mathscr{D}$ such that

$$\begin{aligned} 2\varepsilon &< \big|\Pr\left(\mathscr{D}(E(m_0,K))=1\right) - \Pr\left(\mathscr{D}(E(m_1,K))=1\right)\big| \\ &\leq \big|\Pr\left(\mathscr{D}(E(m_0,K))=1\right) - \Pr\left(\mathscr{D}(K_\ell)=1\right)\big| \\ &\quad + \big|\Pr\left(\mathscr{D}(E(m_1,K))=1\right) - \Pr\left(\mathscr{D}(K_\ell)=1\right)\big|, \end{aligned}$$

whereby one of the following must hold:

$$\begin{aligned} &\big|\Pr\left(\mathscr{D}(E(m_0,K))=1\right) - \Pr\left(\mathscr{D}(K_\ell)=1\right)\big| > \varepsilon, \qquad \text{or} \\ &\big|\Pr\left(\mathscr{D}(E(m_1,K))=1\right) - \Pr\left(\mathscr{D}(K_\ell)=1\right)\big| > \varepsilon. \end{aligned} \tag{9.2}$$

Without loss of generality, assume that the former inequality holds. Since K_ℓ is the uniform random number on $\{0,1\}^l$, $m_0 \oplus K_\ell$ is distributed uniformly as well, whereby

$$\Pr\left(\mathscr{D}(K_\ell)=1\right) = \Pr\left(\mathscr{D}(m_0 \oplus K_\ell)=1\right).$$

By substituting this and $E(m_0,K) = m_0 \oplus K$ into (9.2), we obtain

$$\big|\Pr\left(\mathscr{D}(m_0 \oplus K)=1\right) - \Pr\left(\mathscr{D}(m_0 \oplus K_\ell)=1\right)\big| > \varepsilon.$$

Thus, the desired distinguisher $\tilde{\mathscr{D}}$ is given by $\tilde{\mathscr{D}}(x) = \mathscr{D}(m_0 \oplus x)$, $x \in \{0,1\}^\ell$. □

As a consequence, we obtain indistinguishable security of OTP encryption using an almost uniform key K. Indeed, if $d_{\mathrm{var}}(\mathrm{P}_K, \mathrm{P}_{K_\ell}) \leq \varepsilon$, there does not exist an ε-distinguisher for K and K_ℓ. Then, by Theorem 9.3, there does not exist a 2ε-distinguisher for $E(m_0,K)$ and $E(m_0,K_\ell)$, establishing indistinguishable security for OTP encryption.

This result was proved in Theorem 3.18 by a direct analysis of the total variation distance. The advantage of this alternative argument is that it can be applied to a computational ε-distinguisher as well.

9.2 Pseudorandom Generators and Computational Security

The proof of Theorem 9.3, though simple, leads to a profound insight. The only feature of a random key K that is critical for the security of OTP encryption is that no ε-distinguisher exists for K and a perfect key. In fact, a careful examination of the proof suggests that this reduction is computationally efficient. Namely, if we can find an efficient distinguisher for $E(m_0, K)$ and $E(m_1, K)$, then we can obtain an efficient distinguisher for K and a perfect key. This observation motivates the following definition of a pseudorandom key that can be used to enable computationally secure OTP encryption.

We start with the notion of computational indistinguishability, which is a computational counterpart of the information-theoretic notion of indistinguishability introduced in Chapter 3.

DEFINITION 9.4 (Computational indistinguishability) Consider families $\{X_n\}_{n\in\mathbb{N}}$ and $\{Y_n\}_{n\in\mathbb{N}}$ of random variables and a function $\varepsilon\colon \mathbb{N} \to [0,1]$. The families $\{X_n\}_{n\in\mathbb{N}}$ and $\{Y_n\}_{n\in\mathbb{N}}$ are *computationally ε-indistinguishable* if for any PPT algorithm $\mathscr{D}$, it holds that

$$\left| \Pr\left(\mathscr{D}(1^n, X_n) = 1\right) - \Pr\left(\mathscr{D}(1^n, Y_n) = 1\right)\right| \le \varepsilon(n), \quad \forall n \in \mathbb{N}, \tag{9.3}$$

where 1^n denotes the all 1 sequence of length n. That is, for every n, no PPT algorithm $\mathscr{D}$ is an $\varepsilon(n)$-distinguisher for random variables X_n and Y_n. For brevity, we shall simply denote the family $\{X_n\}_{n\in\mathbb{N}}$ by X_n and speak of computationally ε-indistinguishable X_n and Y_n.

Remark 9.1 Note that the running time of the algorithm is measured as a function of the input length, which is typically assumed to be available to the algorithm as an input. We indicate this dependence by using the extra input 1^n for $\mathscr{D}$ in (9.3), which is also a standard practice in complexity theory. In encryption schemes, the integer n plays the role of the security parameter and is typically taken as the key length. We include the extra input 1^n for the algorithm to indicate that we consider complexity of the algorithm as a function of n, and not its input argument; the two can differ, for instance, when the adversary only observes a small portion of the ciphertext. An uncomfortable reader may ignore this formal input 1^n and simply work with a slightly less accurate notation $\mathscr{D}(X_n)$ as in the statistical setting of the previous section.

The previous definition is valid for any function ε. In fact, most of our proofs account for the precise dependence of security parameters on the number of times we allow the adversary to access the encryption algorithm. But since our computational complexity notions are asymptotic to begin with, we simply gear our presentation towards an asymptotic theory and restrict to polynomially negligible functions ε. Note that, since the adversary may repeat the distinguishing procedure polynomially many times, it is important that the security parameter $\varepsilon(n)$ converges to 0 faster than the reciprocal of any polynomial function. Such functions ε are called negligible functions.

DEFINITION 9.5 (Negligible functions) A function $\varepsilon\colon \mathbb{N} \to [0,1]$ is said to be negligible if, for any polynomial function $p(\cdot)$ on $\mathbb{N}$, $\varepsilon(n) \le 1/p(n)$ holds for all sufficiently large n.

If a function ε is not negligible, then there exists a polynomial $p(n)$ such that $\varepsilon(n) \ge 1/p(n)$ for infinitely many[3] n.

We are now in a position to define a pseudorandom generator, an object that will replace random keys in computational security.

DEFINITION 9.6 (Pseudorandom generator) Let K_n and $K_{\ell(n)}$ be the uniform random variables on $\{0,1\}^n$ and $\{0,1\}^{\ell(n)}$, respectively. A deterministic polynomial-time algorithm G that outputs $y \in \{0,1\}^{\ell(n)}$ and takes input $x \in \{0,1\}^n$ is a *pseudorandom generator* (PRG) if the families $G(K_n)$ and $K_{\ell(n)}$ are computationally ε-indistinguishable for a negligible function ε.

Equivalently, G is a PRG if there does not exist a PPT algorithm that serves as an $\varepsilon(n)$-distinguisher for $G(K_n)$ and $K_{\ell(n)}$ for all n and a function ε that is not negligible.

Note that the notion of PRG is nontrivial only when the output length $\ell(n)$ is larger than the input length n since otherwise $G(K_n) = K_n$ will constitute a PRG. Also, since the generating algorithm is deterministic, the entropy $H(G(K_n))$ of the output cannot be larger than $H(K_n) = n$ bits. Nevertheless, we require that the output of PRG is computationally indistinguishable from the uniform random number of larger length $\ell(n)$. If we require information-theoretic indistinguishability, even $\ell(n) = n+1$ is not possible. Surprisingly, it is widely believed (supported by candidate constructions that have lasted years of attacks) that PRG with $\ell(n)$ much larger than n is possible. Note that in our asymptotic theory we can only hope to have[4] $\ell(n) = \mathcal{O}(p(n))$ for some polynomial $p(n)$ since a PRG is restricted to be a PPT algorithm.

Since the PRG can output a large key that emulates a random key, we can now strive for encryption of messages using this pseudorandom key. However, we already saw in Chapter 3 that there is no hope of information-theoretically secure encryption if the length of the random key (which is much smaller than the length of the pseudorandom key generated from it) is less than the message length. Interestingly, we can establish formal security of such encryption schemes using PRGs in terms of an alternative, computational notion of indistinguishable security.

We consider encryption schemes that take a random key K_n as an input and encrypt a message m as $E(m, K_n)$. Since there is an underlying computational model, we will use n to denote the input length, i.e., computational complexity is evaluated as a function of n (security parameter), and K_n to denote the corre-

[3] That is, for all $m \in \mathbb{N}$ we can find $n \ge m$ such that $\varepsilon(n) \ge 1/p(n)$.

[4] We have used the standard order notation from computational complexity – a function $f(n)$ is $\mathcal{O}(g(n))$ for a nonnegative function g if there exists a constant C independent of n such that $|f(n)| \le Cg(n)$ for all n sufficiently large.

sponding key. In typical applications, K_n takes values in $\{0,1\}^n$, i.e., the length of the key itself plays a role of security parameter.

DEFINITION 9.7 (Computationally indistinguishable security) An encryption scheme E is *computationally indistinguishable secure* if, for every pair of messages (m_0, m_1) with $m_0 \neq m_1$, the families $E(m_0, K_n)$ and $E(m_1, K_n)$ are computationally ε-indistinguishable for a negligible function ε.[5]

We remark without proof that the definition above not only prevents the adversary from distinguishing between pairs (m_0, m_1) but even prevents the adversary from obtaining any other information such as "is the message m_0 or not?" In Chapter 3, we saw an equivalence between indistinguishable security and semantic security. A similar equivalence can be shown between the computational variant of indistinguishable security above and a computational variant of semantic security (see Problem 9.1). Below, we will see that an OTP encryption using keys from a PRG remains computationally indistinguishable secure.

CONSTRUCTION 9.1 (OTP using PRG) Let K_n be distributed uniformly over $\{0,1\}^n$. For $\ell = \ell(n)$ parameterized by n, consider a PRG G that gives an ℓ-bit output for input key K_n. For a message $m \in \{0,1\}^\ell$, the encryption algorithm E computes

$$E(m, K_n) = m \oplus G(K_n),$$

where $\oplus$ is the bitwise modulo sum. The decryption algorithm observes the ciphertext C and outputs $C \oplus G(K_n)$.

Construction 9.1 is simply the OTP encryption of Chapter 3 applied using a PRG. To establish the computational security of this scheme, we can simply follow the steps in the proof of Theorem 9.3 to see that, if we can distinguish $E(m_0, K_n)$ and $E(m_1, K_n)$, then we can distinguish the pseudorandom key $G(K_n)$ from the uniform key K_ℓ.[6] As a corollary, we get the security of the encryption scheme in Construction 9.1.

Before we present this corollary, we need to define one more staple notion of complexity theory, that of oracle access. In our proofs below, we allow the adversary to use an encryption scheme as a black-box entity applied to an input of its choice. When we consider a computationally bounded adversary, we only take into account the cost of producing the input to the black-box; the cost of implementing the black-box is not included in the computational cost of the adversary. Oracle access formalizes such a black-box access.

DEFINITION 9.8 (Oracle access) An algorithm A with *oracle access* to an algorithm B, denoted A^B, invokes B as a subroutine, possibly multiple times,

[5] Note that the messages m_0, m_1 also vary depending on n, but we omit the dependence to keep the notation simple.

[6] In fact, this reduction is information theoretic, i.e., it does not rely on any computational assumption for the encryption scheme or for $G(K_n)$.

by passing inputs of its choice to B and using the resulting outputs for subsequent steps.

Note that if both A and B are PPT, then so is A^B, even when the computational complexity of implementing the oracle B is taken into account, since A can only invoke B polynomially many times and a sum of polynomially many polynomial terms is a polynomial term.

We now use the notion of oracle access to establish the following corollary.

COROLLARY 9.9 *If G is a pseudorandom generator, then the encryption scheme E in Construction 9.1 is computationally indistinguishable secure.*

Proof Suppose that the encryption scheme is not secure. Then, there exists a PPT algorithm $\mathscr{D}$ and a nonnegligible function $\varepsilon\colon \mathbb{N} \to \mathbb{R}$ such that $\mathscr{D}$ is a $2\varepsilon(n)$-distinguisher for $E(m_0, K_n)$ and $E(m_1, K_n)$ for infinitely many n. Then, by Theorem 9.3, there exists an algorithm $\tilde{\mathscr{D}}$ that uses $\mathscr{D}$ as oracle access once and is an $\varepsilon(n)$-distinguisher for $G(K_n)$ and K_ℓ, the uniform random variable on $\{0,1\}^\ell$. Note that the algorithm $\tilde{\mathscr{D}}$ constructed in the proof of Theorem 9.3 invokes $\mathscr{D}$ only once in an oracle access and computes a bitwise XOR with its output. Since computing a bitwise XOR for ℓ-bit vectors requires $\mathcal{O}(\ell)$ steps and $\mathscr{D}$ is a PPT algorithm, $\tilde{\mathscr{D}}$ is a PPT algorithm as well. Thus, $\tilde{\mathscr{D}}$ constitutes a PPT algorithm that is an $\varepsilon(n)$-distinguisher for $G(K_n)$ and K_ℓ for infinitely many n, where $\varepsilon(n)$ is a nonnegligible function. But this is a contradiction since G is a PRG (see Definition 9.6). □

Thus, a PRG enables us to securely transmit a message with length much larger than the length of secret key. However, the scheme in Construction 9.1 is secure only for one-time encryption. In particular, it is not secure when we transmit multiple messages using the key $G(K_n)$. Indeed, the adversary can easily distinguish the pair (m_0, m_0) from (m_0, m_1) if $m_0 \neq m_1$ simply by taking XOR of ciphertexts corresponding with the first and the second messages. This is precisely the situation in practice: we need to send multiple messages securely. In fact, in practice, the adversary can be more powerful and can even force the encryption scheme to reveal the ciphertext for a certain number of (plaintext) messages. Such an attack is called a *chosen plaintext attack* (CPA). In the next section, we will present an encryption scheme that is computationally secure even under CPA, but against a computationally bounded adversary.

9.3 CPA Secure Encryption and Pseudorandom Functions

In practice, we do not need to send just one message but a sequence of messages. Also, the adversary is often quite resourceful and can manoeuvre the sender to obtain ciphertexts for messages chosen by the adversary. We saw such a threat model in Section 8.4 where we considered authentication of multiple messages in such a setting. An astute reader may have noticed that the authentication scheme proposed there requires an independent key S_i for encrypting the tag for each message i. A similar scheme can be used for encrypting multiple messages using OTP encryption with an independent key for each message. We leave it as an exercise for the reader to verify the validity of this claim in Problem 9.3. However, this scheme requires the parties to share very large secret keys – independent keys are needed for every new message – and is not feasible in practice.

In this section, we pursue a more ambitious goal, namely that of using pseudorandom keys to enable encryption of multiple messages. But this calls for an extension of the definition of pseudorandom generators: we need to generate a new pseudorandom key for every message using the same seed. A construct termed a *pseudorandom function* will serve this purpose. Roughly, it is an algorithm F which cannot be distinguished from the random function by a PPT algorithm with oracle access to the algorithm. The notion of computational indistinguishability for algorithms is a slight generalization of Definition 9.4, and it is defined using the concept of oracle access in Definition 9.8.

DEFINITION 9.10 (Computational indistinguishability for algorithms) Families of algorithms $\{A_n\}_{n\in\mathbb{N}}$ and $\{B_n\}_{n\in\mathbb{N}}$ are *computationally ε-indistinguishable* if for any PPT algorithm $\mathscr{D}$, it holds that

$$\left|\Pr\left(\mathscr{D}^{A_n}(1^n)=1\right)-\Pr\left(\mathscr{D}^{B_n}(1^n)=1\right)\right|\le\varepsilon(n),\quad\forall n\in\mathbb{N},$$

where 1^n denotes the all 1 sequence of length n.

Note that it is implicit in the previous definition that the distinguisher $\mathscr{D}$ is only allowed to access the algorithms A_n and B_n polynomially many times. Furthermore, there is no requirement for algorithms A_n and B_n to be computationally bounded. For instance, in the definition of a pseudorandom function to be presented in Definition 9.11 below, the random function cannot be implemented by any PPT algorithm since it must store exponentially many random numbers. Nonetheless, Definition 9.10 makes sense since we do not include the computational cost of implementing the oracle in the computational cost of implementing the distinguisher.

Remark 9.2 In the previous definition we have been careful to include 1^n as input to the algorithm to indicate that algorithms do depend on input length. However, from now on, we omit this dependence for brevity. We will indicate the dependence on input length n as well as the random key by including the random key K_n as the input to algorithms. Recall that K_n is the uniform random variable taking values in $\{0,1\}^n$. Our algorithms below are randomized using the key K_n, and their complexity will be measured in terms of the parameter n.

DEFINITION 9.11 (Pseudorandom function) Let K_n be distributed uniformly over $\{0,1\}^n$. A deterministic polynomial-time algorithm F with input $(k,x)\in\{0,1\}^n\times\{0,1\}^\ell$ and output $y\in\{0,1\}^\ell$ is a *pseudorandom function* (PRF) if the algorithm $F_{K_n}(\cdot) := F(K_n, \cdot)$ is computationally ε-indistinguishable from the random function[7] $F_\ell\colon \{0,1\}^\ell \to \{0,1\}^\ell$ for a negligible function ε.

The random function can alternatively be defined as a random function that maps each input to the uniform output, independently for each input; in other words, it stores 2^ℓ uniform random numbers $U_0, \ldots, U_{2^\ell - 1}$ on $\{0,1\}^\ell$ and outputs U_x upon receiving $x \in \{0,1\}^\ell$. It follows that the output of a random function is independent of its input (see Problem 9.5). Thus, a PRF cannot be distinguished efficiently from a function that outputs a new, independent random variable in every access.

The mappings F_{K_n} can be viewed as a random mapping selected using a key that maps inputs of length ℓ bits to outputs of length ℓ bits. Such mappings are called *keyed, length-preserving mappings.* Note that the random function requires an $\ell 2^\ell$-bit secret key to be shared between the parties; the PRF seeks to mimic the random function by only using an n-bit secret key.

Next, we formally define the notion of CPA security. For simplicity, we restrict our definition to CPA secure encryption for sending one message; the extension to multiple messages requires some more work but the analysis is very similar. For this definition, we need to extend the notion of computationally ε-indistinguishable algorithms to allow the distinguisher algorithm to have oracle access to another algorithm. We omit the formal definition of such distinguishers with additional oracle access, since it is an obvious extension of Definition 9.4.

DEFINITION 9.12 (CPA security) An encryption scheme E is CPA secure if, for every pair of messages (m_0, m_1) with $m_0 \neq m_1$, the families (of random variables) $E(m_0, K_n)$ and $E(m_1, K_n)$ are computationally ε-indistinguishable by any PPT algorithm with oracle access to $E(\cdot, K_n)$, for a negligible function ε.

The notion of CPA security allows the adversary to access the encryption scheme polynomially many times before attempting to decrypt the observed encrypted messages.

We now present an encryption scheme that is CPA secure for sending one message. Note that while we are securing encryption of only one message m, the adversary can observe ciphertext for polynomially many messages before attempting to get information about m from its ciphertext.

CONSTRUCTION 9.2 (OTP using a PRF) Let K_n be distributed uniformly over $\{0,1\}^n$. For $\ell = \ell(n)$ parameterized by n, consider a PRF F that gives an ℓ-bit output for an n-bit input key K_n and ℓ-bit input. Denoting by $F_{K_n}(\cdot)$ the mapping $F(K_n, \cdot)$, for a message $m \in \{0,1\}^\ell$, the encryption scheme E computes

$$E(m, K_n) = (S, m \oplus F_{K_n}(S)),$$

[7] Namely, a function chosen uniformly at random from the set of all functions from $\{0,1\}^\ell$ to $\{0,1\}^\ell$.

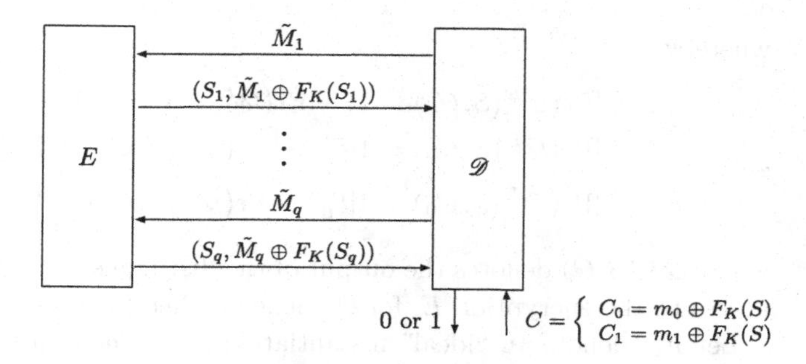

Figure 9.1 A description of the distinguisher $\mathscr{D}$ in the proof of Theorem 9.13.

where S is distributed uniformly over $\{0,1\}^\ell$. The decryption algorithm D observes the ciphertext (S, C) and outputs

$$D((S,C), K_n) = C \oplus F_{K_n}(S).$$

Note that the random variable S is generated afresh every time a new message is encrypted using E, and is shared to the receiver over the public channel. Thus, the only secret key needed for the algorithm is K_n, which can be of much smaller length than the secret key needed for independently repeating the scheme in Construction 9.1.

THEOREM 9.13 *Let K_n be distributed uniformly on $\{0,1\}^n$, F_ℓ be the random function from $\{0,1\}^\ell$ to $\{0,1\}^\ell$, and E be the encryption mapping of Construction 9.2. For messages $m_0 \neq m_1$, suppose that an algorithm $\mathscr{D}$ making q queries to the encryption mapping E is an ε-distinguisher for random variables $E(m_0, K_n)$ and $E(m_1, K_n)$. Then, there exists an algorithm $\tilde{\mathscr{D}}$ which makes $(q+1)$ queries to F_{K_n} or F_ℓ, executes $\mathscr{D}$ once, and satisfies*

$$\left|\Pr\left(\tilde{\mathscr{D}}^{F_{K_n}}(1^n) = 1\right) - \Pr\left(\tilde{\mathscr{D}}^{F_\ell}(1^n) = 1\right)\right| \geq \frac{\varepsilon - q \cdot 2^{-\ell}}{2}.$$

Proof Consider an execution of $\mathscr{D}$, depicted in Figure 9.1. For $1 \leq i \leq q$, in the ith query that $\mathscr{D}$ makes to E, it sends a message $\tilde{M}_i$ of its choice and receives $(S_i, \tilde{M}_i \oplus F_K(S_i))$. Recall that the distinguisher is presented the "challenge ciphertext" (S, C) where C is either $C_0 = m_0 \oplus F_K(S)$ or $C_1 = m_1 \oplus F_K(S)$, and it can respond after making q queries to the encryption scheme. We can view (S, C) as an input to the algorithm $\mathscr{D}$ with query access to the encryption scheme.

Denoting $\mathcal{S}_0 = \{S_1, \ldots, S_q\}$, let $\mathcal{E}_0 = \{S \in \mathcal{S}_0\}$ be the event that the key S of the challenge ciphertext was seen earlier by $\mathscr{D}$ in one of its queries to E. This event $\mathcal{E}_0$ is determined by random keys $S, S_1, \ldots, S_q$ alone and does not depend on K or any other randomness used in $\mathscr{D}$. Note that $S_1, \ldots, S_q$ are independent, with each S_i distributed uniformly over $\{0,1\}^\ell$, and are independent jointly of the key S used in $E(m_0, K)$ and $E(m_1, K)$. Therefore, by the union bound,

$$\Pr(S \in \mathcal{E}_0) \leq \frac{q}{2^\ell},$$

whereby

$$\begin{aligned}\varepsilon &< |\Pr(\mathscr{D}^E(S,C_0)=1) - \Pr(\mathscr{D}^E(S,C_1)=1)| \\ &\le |\Pr(\mathscr{D}^E(S,C_0)=1|\mathcal{E}_0^c) - \Pr(\mathscr{D}^E(S,C_1)=1|\mathcal{E}_0^c)| + \Pr(\mathcal{E}_0) \\ &\le |\Pr(\mathscr{D}^E(S,C_0)=1|\mathcal{E}_0^c) - \Pr(\mathscr{D}^E(S,C_1)=1|\mathcal{E}_0^c)| + \frac{q}{2^\ell},\end{aligned} \tag{9.4}$$

where $\mathscr{D}^E(S,C)$ denotes the output of the distinguisher algorithm $\mathscr{D}$ with oracle access to the encryption E, for challenge ciphertext (S,C).

Let $E_{\mathtt{id}}$ denote an "ideal" instantiation of E where instead of F_K it uses the random function F_ℓ, namely for an input message m, it outputs $(S, m \oplus F_\ell(S))$. Denote by $\mathscr{D}^{E_{\mathtt{id}}}(S,C)$ the output of $\mathscr{D}$ with oracle access to E_0, for challenge ciphertext (S,C).

We analyze the ability of $\mathscr{D}^{E_{\mathtt{id}}}$ to distinguish $(S, m_0 \oplus F_\ell(S))$ and $(S, m_1 \oplus F_\ell(S))$, given that the event $\mathcal{E}_0^c$ occurs. The following is easy to show and is left as an exercise (see Problem 9.6).

Denoting $G^q = (G_1, \ldots, G_q)$ with $G_i = F_\ell(S_i)$, $1 \le i \le q$, consider any distribution for the messages $\tilde{M}^q = (\tilde{M}_1, \ldots, \tilde{M}_q)$ such that the Markov relation $\tilde{M}^q \multimap\!\!- (S^q, G^q, S) \multimap\!\!- F_\ell(S)$ holds. Then, for every realization s, s^q such that $s \notin \{s_1, \ldots, s_q\}$, the conditional distribution of $F_\ell(S)$ given $\{G^q = g^q, S^q = s^q, S = s, \tilde{M}^q = \tilde{m}^q\}$ is uniform.

In particular, as a consequence of this observation, under $\mathcal{E}_0^c$ and for messages $\tilde{M}^q$ queried during the execution of $\mathscr{D}^{E_{\mathtt{id}}}$, the conditional distributions of $m_0 \oplus F_\ell(S)$ and $m_1 \oplus F_\ell(S)$ given $S^q, G^q, \tilde{M}^q, S$ are the same. Therefore,

$$\Pr(\mathscr{D}^{E_{\mathtt{id}}}(S, m_0 \oplus F_\ell(S)) = 1|\mathcal{E}_0^c) = \Pr(\mathscr{D}^{E_{\mathtt{id}}}(S, m_1 \oplus F_\ell(S)) = 1|\mathcal{E}_0^c).$$

Using this identity, we can bound the first term in (9.4) as

$$\begin{aligned}&\big|\Pr(\mathscr{D}^E(S,C_0)=1|\mathcal{E}_0^c) - \Pr(\mathscr{D}^E(S,C_1)=1|\mathcal{E}_0^c)\big| \\ &\quad \le \big|\Pr(\mathscr{D}^E(S,C_0)=1|\mathcal{E}_0^c) - \Pr(\mathscr{D}^{E_{\mathtt{id}}}(S, m_0 \oplus F_\ell(S))=1|\mathcal{E}_0^c)\big| \\ &\qquad + \big|\Pr(\mathscr{D}^{E_{\mathtt{id}}}(S, m_1 \oplus F_\ell(S))=1|\mathcal{E}_0^c) - \Pr(\mathscr{D}^E(S,C_1)=1|\mathcal{E}_0^c)\big|.\end{aligned}$$

Since we have assumed that $\mathscr{D}$ is an ε-distinguisher, the previous bound and (9.4) imply that for either $b=0$ or $b=1$, we must have

$$\begin{aligned}&\big|\Pr(\mathscr{D}^{E_{\mathtt{id}}}(S, m_b \oplus F_\ell(S))=1|\mathcal{E}_0^c) - \Pr(\mathscr{D}^E(S, m_b \oplus F_K(S))=1|\mathcal{E}_0^c)\big| \\ &\quad > \frac{1}{2}\left(\varepsilon - \frac{q}{2^\ell}\right),\end{aligned}$$

whereby we can find a fixed realization (s, s^q) of (S, S^q) such that $s \notin \{s_1, \ldots, s_q\}$ and

$$\begin{aligned}&\big|\Pr(\mathscr{D}^{E_{\mathtt{id}}}(S, m_b \oplus F_\ell(S))=1|S^q=s^q, S=s) \\ &\quad - \Pr(\mathscr{D}^E(S, m_b \oplus F_K(S))=1|S^q=s^q, S=s)\big| > \frac{1}{2}\left(\varepsilon - \frac{q}{2^\ell}\right).\end{aligned} \tag{9.5}$$

Thus, to test if $F = F_K$ or $F = F_\ell$, the required algorithm $\tilde{\mathscr{D}}$ computes the "challenge ciphertext" $(s, m_b \oplus F(s))$ using an oracle access to F, executes $\mathscr{D}$ with

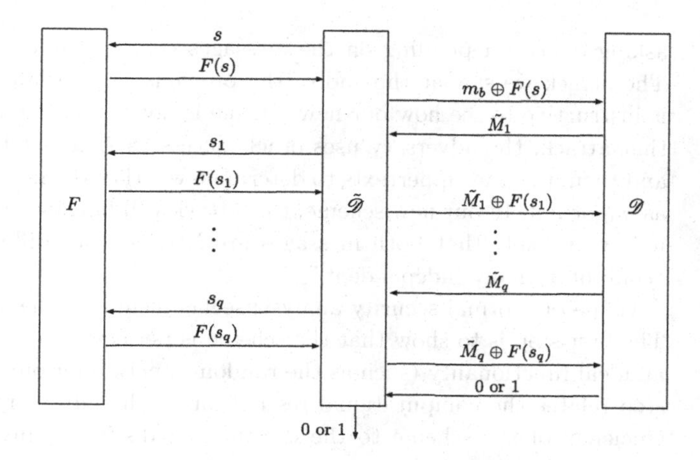

Figure 9.2 A description of the algorithm $\tilde{\mathscr{D}}$ in the proof of Theorem 9.13.

$(s, m_b \oplus F(s))$ as the input, and computes the query responses $(s_i, \tilde{M}_i \oplus F(s_i))$, $1 \le i \le q$, for $\mathscr{D}$ using query access to F; see Figure 9.2.[8] □

The phrases "ideal encryption" and "challenge ciphertext" introduced inline in the proof above rather casually are in fact staple notions for formal security proofs. However, rather than defining these abstract notions formally upfront, we prefer to slowly introduce them to the reader as the book progresses.

The reduction in Theorem 9.13 is information theoretic and does not require any computational assumptions. As a corollary, we establish the computational security of the scheme in Construction 9.2.

COROLLARY 9.14 *For $\ell(n) \ge n$, the encryption scheme in Construction 9.2 is CPA secure.*

Proof Suppose that the proposed encryption scheme is not CPA secure. Then, we can find a pair of messages (m_0, m_1) and PPT algorithm $\mathscr{D}$ with oracle access to E that serves as an $\varepsilon(n)$-distinguisher for $E(m_0, K_n, S)$ and $E(m_1, K_n, S)$, for a nonnegligible function ε. Suppose that this algorithm makes no more than $q(n)$ queries to E. Since $\ell(n) \ge n$, it follows from Theorem 9.13 that we can find an algorithm $\tilde{\mathscr{D}}$ satisfying

$$\left|\Pr\left(\tilde{\mathscr{D}}^{F_{K_n}}(1^n) = 1\right) - \Pr\left(\tilde{\mathscr{D}}^{F_{\ell(n)}}(1^n) = 1\right)\right| \ge \frac{\varepsilon - q \cdot 2^{-n}}{2}.$$

Since $\mathscr{D}$ is a PPT algorithm, $q(n)$ must be polynomial in n. Further, since ε is a nonnegligible function, so is $(\varepsilon(n) - q(n)/2^n)/2$. But this contradicts the assumption that F_{K_n} is a PRF, completing the proof. □

Note that in our definition of CPA security, we did not rule out the adversary

[8] Since (s, s^q) satisfying $s \notin \{s_1, \dots, s_q\}$ and (9.5) does not depend on $F = F_K$ or $F = F_\ell$, it can be precomputed before starting the distinguishing algorithm $\tilde{\mathscr{D}}$.

asking queries depending on the messages (m_0, m_1) it is trying to distinguish. The attack we saw at the end of the previous section falls in this class, and it is instructive to see how our new scheme is not vulnerable to such an attack. In that attack, the adversary uses oracle access to E to get the encryption of m_0 and then uses the ciphertexts to determine whether the next message is the same as m_0 or not. In our new scheme, this attack will fail since by the PRF property, it is very likely that both messages are XORed with different keys, which are "computationally independent."

While our formal security analysis seems technical, the overall idea is simple. The first step is to show that the scheme is perfectly secure if we have access to an ideal functionality (such as the random function for our scheme). The second step relates the computational restriction on the adversary and computational efficiency of the scheme to the difficulty in distinguishing a real functionality from an ideal functionality.

9.4 CCA Secure Encryption and Authentication

CPA security is a strong notion of security. But it rules out an even stronger, practically feasible attack – an attack where the adversary is allowed to decrypt ciphertexts of its choice to gather information about the encryption mapping. An attack where the adversary is given oracle access to both the encryption and the decryption algorithm is called a *chosen ciphertext attack* (CCA). Of course, we shall not allow the adversary to apply this oracle to the ciphertext being attacked. But the adversary can decrypt any other ciphertext. Note that Construction 9.2 is not secure against such an attack. Indeed, upon observing (S, C), the adversary can use the decryption oracle to decrypt $(S, C \oplus m_0)$ and determine whether the message sent was m_0 or not. Our goal in this section is to give an encryption scheme that remains secure even under such an attack. Formally, we seek encryption schemes that remain secure under the following stringent notion of security.

DEFINITION 9.15 (CCA security) An encryption scheme E is CCA secure if, for every pair of messages (m_0, m_1) with $m_0 \neq m_1$, the families $E(m_0, K_n)$ and $E(m_1, K_n)$ are computationally ε-indistinguishable by any PPT algorithm with oracle access to $E(\cdot, K_n)$ and the corresponding decryption algorithm $D(\cdot, K_n)$, for a negligible function ε.

Interestingly, the problem of designing a CCA secure encryption scheme requires a primitive we have already encountered in an earlier chapter. Note that the attack outlined above will be forbidden if the adversary is not allowed to apply the decryption oracle to a ciphertext that has not been generated by the encryption oracle. How can we forbid the adversary from forging such queries? This is exactly what a message authentication code (MAC) enables. Specifically, we can add an authentication tag for the ciphertext using a MAC, as a part of the encryption scheme, before sending the ciphertext over the public channel. Furthermore, the decryption algorithm will first validate the tag for the received authenticated ciphertext, before decrypting and returning the answer. In case authentication fails, we will allow the decryption algorithm to abort and return $\perp$. We take a pause here and note that this is the first instance of a protocol that allows abort in this book. Such protocols are commonplace in security, theoretical and practical alike.

We need a MAC that works for multiple messages since it needs to provide security against attacks entailing multiple queries by the adversary. We saw such a scheme in Example 8.3, and it looked very similar to the CPA secure encryption scheme of Construction 9.2. However, it used a strong δ-UHF in place of the PRF used in Construction 9.2. Furthermore, it required the sender and the receiver to share uniform and independent keys $S_1, \ldots, S_r$ for authenticating r messages, in addition to the key K required for the strong δ-UHF.

In this section, we switch to a computational notion of security for a MAC that will allow us to use a PRF in place of the strong δ-UHF in Construction 9.2. Note that the random function $F_\ell \colon \{0,1\}^\ell \to \{0,1\}^\ell$ constitutes a strong δ-UHF with $\delta = 2^{-\ell}$ (see Problem 9.4). Thus, it can be expected from our analysis in the previous section that we can replace the strong δ-UHF of Construction 9.2 with a PRF, if we define computational security appropriately. Also, the requirement of random shared keys $S_1, \ldots, S_r$ can be relaxed to a PRG. This is roughly the construct of our computationally secure MAC.

Formally, we define computationally secure authentication as follows.

DEFINITION 9.16 (Computationally secure MAC) A MAC (φ, V) using a secret key K_n of length n is a computationally secure MAC if for every PPT algorithm $\mathscr{A}$ with oracle access to (the algorithm implementing) φ and with output $(M, T) = \mathscr{A}(1^n)$ such that M differs from all the messages used by $\mathscr{A}$ as inputs for oracle access to φ and the message-tag pair (M, T) satisfies

$$\Pr(V(M, T) = 1) \leq \varepsilon(n), \quad \forall n \in \mathbb{N}$$

for a negligible function ε.

Below we propose a computationally secure MAC algorithm which keeps track of the number of times it has been accessed using a counter. For simplicity, we begin by assuming a bound on the "lifetime" of the MAC, namely a bound q on the total number of messages that can be authenticated using the MAC.

CONSTRUCTION 9.3 (MAC using a PRF) Let K_n and K'_n be independent and distributed uniformly over $\{0,1\}^n$. For $\ell = \ell(n)$ and $q = q(n)$ parameterized by n, consider a PRF $F_{K_n} \colon \{0,1\}^\ell \to \{0,1\}^\ell$ and a PRG $G(K'_n)$ that gives a $q\ell$-bit output for input key K'_n. Denote the output $G(K'_n)$ of the PRG by $(S_1, \ldots, S_q)$ where each S_i takes values in $\{0,1\}^\ell$.

For the ith message m_i, $1 \leq i \leq q$, the MAC algorithm computes the tag $(i, \varphi_i(m_i))$ and the verifier observes (j, t) and evaluates $V_j(m, t)$ using the mappings (φ_i, V_i) given by

$$\varphi_i(m_i) = F_{K_n}(m_i) \oplus S_i, \quad V_i(m,t) = \mathbf{1}[F_{K_n}(m) = t - S_i],$$

i.e., the message number is appended to the tag as a prefix and is used to decide which "encryption key" among $S_1, \ldots, S_q$ should be used for verification.

The next result relates the security of the proposed scheme to the pseudorandomness of F_{K_n} and $G(K'_n)$ using an information-theoretic reduction.

THEOREM 9.17 *Suppose that an algorithm $\mathscr{A}$ with oracle access to the MAC of Construction 9.3 makes q queries to the MAC to obtain tags for messages in the set $\mathcal{M}$ with $|\mathcal{M}| \le q$.*[9] *If $\mathscr{A}$ can find a message $M \notin \mathcal{M}$ and a tag T such that $\Pr(V_q(M,T) = 1) > \varepsilon$ for $\varepsilon \in (0,1)$, then one of the following must hold true.*

1. *There exists an algorithm $\tilde{\mathscr{D}}$ such that*

$$\left|\Pr\left(\tilde{\mathscr{D}}^{F_{K_n}}(1^n)\right) - \Pr\left(\tilde{\mathscr{D}}^{F_\ell}(1^n)\right)\right| > \frac{\varepsilon - 2^{-\ell}}{2},$$

where F_ℓ denotes the random function from $\{0,1\}^\ell$ to $\{0,1\}^\ell$. Furthermore, $\tilde{\mathscr{D}}$ uses the MAC algorithm to compute the tags for at most q messages, using evaluation of the PRG $G(K'_n)$ and oracle access to F, and applies the verifier V_q once.

2. *There exists an $(\varepsilon - 2^{-\ell})/2$-distinguisher $\tilde{\mathscr{D}}$ for $G(K'_n)$ and $K_{q\ell}$, where $K_{q\ell}$ denotes the uniform random variable on $\{0,1\}^{q\ell}$. Furthermore, $\tilde{\mathscr{D}}$ uses the MAC algorithm to compute the tags for at most q messages, using the evaluation of the random function F_ℓ for q inputs and $q\ell$ bits obtained from $G(K'_n)$ or $K_{q\ell}$, and applies the verifier V_q once.*

Proof Without loss of generality, we may assume that $|\mathcal{M}| = q$. Consider the algorithm $\mathscr{D}$ which executes $\mathscr{A}$ to obtain the message M and its tag T and outputs $V_q(M,T)$, using oracle access to F_{K_n} and $G(K'_n)$, indirectly when it makes oracle access to φ. We denote this overall algorithm by $\tilde{\mathscr{D}}$, whereby $\mathscr{D} = \tilde{\mathscr{D}}^{F_{K_n}, G(K'_n)}$, and by our assumption, $\Pr(\mathscr{D} = 1) > \varepsilon$.

Next, consider $\mathscr{D}' = \tilde{\mathscr{D}}^{F_\ell, G(K'_n)}$, which also executes $\mathscr{A}$ but it operates the MAC with the random function F_ℓ instead of the PRF F_{K_n}.

Finally, consider $\mathscr{D}'' = \tilde{\mathscr{D}}^{F_\ell, K_{q\ell}}$ which operates the MAC with random function F_ℓ and uniform random variable $K_{q\ell}$ instead of F_{K_n} and $G(K'_n)$, respectively. Note that each of $\mathscr{D}, \mathscr{D}'$, and $\mathscr{D}''$ has output 0 or 1. Therefore,

$$\begin{aligned}
\varepsilon &< \Pr(\mathscr{D} = 1) \\
&\le |\Pr(\mathscr{D} = 1) - \Pr(\mathscr{D}' = 1)\,| + |\Pr(\mathscr{D}' = 1) - \Pr(\mathscr{D}'' = 1)\,| + \Pr(\mathscr{D}'' = 1)\,.
\end{aligned} \tag{9.6}$$

We have already derived a bound for the last term on the right-hand side in the previous chapter. Indeed, since the random function is a strong $2^{-\ell}$-UHF (see Problem 9.4), Theorem 8.11 implies

$$\Pr(\mathscr{D}'' = 1) \le 2^{-\ell},$$

[9] The messages can be chosen adaptively based on previously received tags.

which along with the previous bound yields

$$\varepsilon - 2^{-\ell} \le |\Pr(\mathscr{D} = 1) - \Pr(\mathscr{D}' = 1)| + |\Pr(\mathscr{D}' = 1) - \Pr(\mathscr{D}'' = 1)|.$$

Therefore, either $|\Pr(\mathscr{D} = 1) - \Pr(\mathscr{D}' = 1)| > (\varepsilon - 2^{-\ell})/2$ or $|\Pr(\mathscr{D} = 1) - \Pr(\mathscr{D}'' = 1)| > (\varepsilon - 2^{-\ell})/2$. In the first case, we obtain $\tilde{\mathscr{D}}^{F,G(K_n')}$ as our desired algorithm for Claim 1 when executed with $F = F_{K_n}$ or $F = F_\ell$. Note that this algorithm needs to execute the PRG algorithm $G(K_n')$ and use its output to compute tags for the (at most) q messages.

Similarly, in the second case we obtain that $\tilde{\mathscr{D}}^{F_\ell,G}$ is an $(\varepsilon - 2^{-\ell})/2$-distinguisher for $G = G(K_n')$ and $G = K_\ell$. This distinguisher evaluates the random function F_ℓ on at most q different ℓ-bit strings obtained from G. Note that each evaluation of F_ℓ can be completed by generating an independent sample from uniform distribution on $\{0,1\}^\ell$. This completes the proof. □

Remark 9.3 (Hybrid arguments) The previous proof introduces another concept that is used heavily in theoretical cryptography. Specifically, the bound in (9.6) relates the performance of an algorithm interacting with one kind of oracle to that interacting with another kind, by modifying one feature of the oracle at a time. These elementary steps, termed "hybrid arguments," are very powerful when applied iteratively and allow us to relate the performance of algorithms with access to "ideal" primitives and those with access to "real" primitives.

The reduction in Theorem 9.17 requires no computational assumption and is information theoretic. Observe that there is no dependence on q in the result above, except through the requirement of producing a pseudorandom key of length $q\ell$ (using the random seed K_n' of length n). For proving computational security for one message, q will be set to be one more than the number of times an adversary can access the MAC oracle, and therefore, can be assumed to grow at most polynomially in n.

COROLLARY 9.18 *The MAC of Construction 9.3 is computationally secure.*

Proof See Problem 9.7. □

We now return to the problem of CCA secure encryption. We have accumulated all the basic ingredients to provide a CCA secure encryption scheme. In fact, we show that a general construction that combines a secure MAC with a CPA secure encryption scheme is CCA secure. The high-level idea is straightforward. We use the secure MAC to authenticate the ciphertext obtained from a CPA secure encryption scheme. During decryption, the tag for the received ciphertext is first verified, and only when the verification is successful is the decryption algorithm of the CPA secure scheme applied. Thus, if the adversary tries to access the decryption oracle, it can only do so for the message-tag pairs obtained from the encryption oracle and not for new ones. But for these pairs the adversary knows the result of decryption anyway since they were obtained by encrypting messages of the adversary's choice. Thus, in effect, the adversarial attack is limited to CPA for the underlying encryption scheme.

CONSTRUCTION 9.4 (CCA secure encryption) Let (E, D) be a CPA secure encryption scheme using a perfect secret key K_n of length n, where E and D, respectively, denote the encryption and the decryption mappings. Denote by $\ell = \ell(n)$ the length of ciphertexts produced by E. Let (φ, V) be a computationally secure MAC for messages of length ℓ and using a perfect secret key K'_n of length n. The new encryption scheme (E', D') uses (E, D) and (φ, V). For a message m, the encryption scheme E' outputs

$$E'(m) = (E(m), \varphi(E(m))).$$

The decryption algorithm observes the ciphertext (C, T) and outputs

$$D'(C,T) = \begin{cases} D(C), & \text{if } V(C,T) = 1, \\ \perp, & \text{if } V(C,T) = 0. \end{cases}$$

As mentioned earlier, we allow the decryption algorithm to declare an abort symbol $\perp$ to indicate suspicion of an attack. Note that we need two independent keys K_n and K'_n for randomizing the encryption scheme and the MAC. Thus, the encryption scheme (E', D') is keyed, using (K_n, K'_n) as a perfect secret key. We will parameterize its computational complexity by n.

We now establish CCA security of the encryption scheme in Construction 9.4. As we have been doing above, we highlight the information-theoretic part of the proof first in the theorem below, before utilizing it to establish CCA security as a corollary.

THEOREM 9.19 *For messages $m_0 \neq m_1$, suppose that an algorithm $\mathscr{A}$ making q queries to the encryption or the decryption oracle is an ε-distinguisher for $E'(m_0)$ and $E'(m_1)$. Then, one of the following must hold.*

1 *There exists an $\varepsilon/2$-distinguisher for $E(m_0)$ and $E(m_1)$ that makes at most q oracle queries to E and, further, makes oracle access to the MAC algorithm.*
2 *There exists an algorithm that makes at most q oracle queries to the MAC algorithm to obtain tags for messages in the set $\mathcal{M}$, makes oracle access to the encryption scheme, and finds a message $M \notin \mathcal{M}$ and a tag T such that* $\Pr(V_q(M,T) = 1) > \varepsilon/(2q)$.

Proof Denote by $\mathscr{D}$ the assumed ε-distinguisher for $E'(m_0)$ and $E'(m_1)$. Note that $\mathscr{D}$ is allowed to make up to q oracle queries to E' and D' and satisfies

$$\varepsilon < \left| \Pr\left(\mathscr{D}^{E',D'}(E'(m_0)) = 1\right) - \Pr\left(\mathscr{D}^{E',D'}(E'(m_1)) = 1\right) \right|.$$

Denote by $\mathcal{E}$ the event that there is at least one decryption query made by $\mathscr{D}$ that does not return $\perp$ from the MAC verifier V. Note that the event $\mathcal{E}$ occurs during the execution of $\mathscr{D}$ and depends on whether $\mathscr{D}$ was initiated with input $E'(m_0)$ or $E'(m_1)$. Using the previous bound, we get

$$\varepsilon \le \Pr(\mathcal{E}) + \left| \Pr\left(\mathscr{D}^{E',D'}(E'(m_0)) = 1, \mathcal{E}^c\right) - \Pr\left(\mathscr{D}^{E',D'}(E'(m_1)) = 1, \mathcal{E}^c\right) \right|.$$

Therefore, either $\Pr(\mathcal{E}) \geq \varepsilon/2$ or

$$\left|\Pr\left(\mathscr{D}^{E',D'}(E'(m_0)) = 1, \mathcal{E}^c\right) - \Pr\left(\mathscr{D}^{E',D'}(E'(m_1)) = 1, \mathcal{E}^c\right)\right| \geq \frac{\varepsilon}{2}.$$

In the former case, when $\Pr(\mathcal{E}) \geq \varepsilon/2$, note that if $\mathscr{D}$ uses its oracle access to D' to decrypt the ciphertext for a message it encrypted using an oracle query, it will not gain any additional information. In particular, we can consider without loss of generality an algorithm $\mathscr{D}$ which never uses oracle access to the decryption algorithm to decrypt a ciphertext that it has obtained using oracle access to the encryption algorithm. For such an algorithm $\mathscr{D}$, the event $\mathcal{E}$ occurs only if $\mathscr{D}$ can successfully guess the tag corresponding to φ for a new ciphertext for which it has not seen a tag before. Denote by I the number of queries to the encryption oracle before the first such successful guess is made. Then, since $1 \leq I \leq q$, by union bound we can find $1 \leq i \leq q$ such $\Pr(\mathcal{E}, I = i) \geq \varepsilon/(2q)$. Therefore, with probability greater than $\varepsilon/(2q)$, the algorithm $\mathscr{D}$ makes at most q oracle queries to the MAC (and the encryption scheme E) and outputs a new message-tag pair for the MAC that is accepted by its verifier.

In the latter case, consider a distinguisher $\tilde{\mathscr{D}}$ for $C' = E'(m_0)$ and $C' = E'(m_1)$ that executes $\mathscr{D}$ with C' as the input and declares the output as 1 only when $\mathcal{E}^c$ occurs and $\mathscr{D}$ declares the output 1. Thus,

$$\left|\Pr\left(\tilde{\mathscr{D}}^{E',D'}(E'(m_0)) = 1\right) - \Pr\left(\tilde{\mathscr{D}}^{E',D'}(E'(m_1)) = 1\right)\right| > \varepsilon/2,$$

namely $\tilde{\mathscr{D}}$ with oracle access to E' and D' is an $\varepsilon/2$-distinguisher for $E'(m_0)$ and $E'(m_1)$. But since under $\mathcal{E}^c$ the algorithm $\tilde{\mathscr{D}}$ does not successfully decrypt any message using D, it serves as an $\varepsilon/2$-distinguisher for $E(m_0)$ and $E(m_1)$ using at most q queries to E. □

The next corollary follows immediately from the previous theorem.

COROLLARY 9.20 *If both the MAC algorithm and the encryption scheme (E, D) are PPT algorithms, the encryption scheme of Construction 9.4 is CCA secure.*

Proof See Problem 9.8. □

9.5 References and Additional Reading

The paradigm of provable security for computational cryptography was introduced in a landmark paper by Goldwasser and Micali [147]. The concept of the pseudorandom number generator (PRG) was introduced in [39], and a construction based on a number-theoretic assumption was provided. A construction of the PRG from one-way permutation was presented in [361], and a construction of the PRG from the one-way function was presented in [168]; in the construction of the latter, the leftover hash lemma discussed in Chapter 7 plays a crucial role.

The concept of the pseudorandom function (PRF) was introduced in [143], and a construction of the PRF from the PRG was also provided there.

A thorough exposition on the topic can be found in the two volume textbook by Goldreich [141, 142]. Also, a more introductory but rigorous exposition can be found in the textbook by Katz and Lindell [187]. In addition to these textbooks, our exposition in this chapter is also based on the lecture notes by Trevisan [328]. We have adapted the language to make the connection between computational cryptography and information-theoretic cryptography more clear.

Problems

9.1 Define a notion of computational semantic security as a counterpart of the information-theoretic notion of semantic security given in Definition 3.9 and show that it is equivalent to computational indistinguishable security.

HINT See the proof of Proposition 3.10.

9.2 Show that for two random variables X and Y taking values in a finite set $\mathcal{X}$, the total variation distance $d_{\mathtt{var}}(\mathrm{P}_X, \mathrm{P}_Y) = \max_{\mathscr{D}:\mathcal{X}\to\{0,1\}} |\Pr(\mathscr{D}(X) = 1) - \Pr(\mathscr{D}(Y) = 1)\,|$, where the maximum is over all randomized mappings D from $\mathcal{X}$ to $\{0, 1\}$. Further, find a mapping that attains the maximum.

9.3 For a CPA secure encryption scheme with encryption mapping E, show (by providing an attack) that the multimessage scheme obtained by applying E to multiple messages is not necessarily CPA secure. Furthermore, show that if we use independently generated pseudorandom keys for different messages, then the resulting scheme is CPA secure.

9.4 Show that a random function is a PRF.

9.5 Show that the following two definitions of random function induce the same input-output joint distributions.

1. F is generated uniformly from the set of all mappings $f : \mathcal{X} \to \mathcal{Y}$, and the output for an input $x \in \mathcal{X}$ is $F(x)$.
2. $F(x)$ is distributed uniformly over $\mathcal{Y}$ for every $x \in \mathcal{X}$.

Conclude that the output and input of a random function are independent.

9.6 For the encryption scheme in Construction 9.2, let $S_1, \ldots, S_q$ be independent random variables distributed uniformly over $\{0, 1\}^\ell$. Further, let $F_\ell : \{0, 1\}^\ell \to \{0, 1\}^\ell$ be the random function, which is assumed to be independent of $(S_1, \ldots, S_q)$. Denoting $G^q = (G_1, \ldots, G_q)$ with $G_i = F_\ell(S_i)$, $1 \le i \le q$, consider any distribution for the messages $\tilde{M}^q = (\tilde{M}_1, \ldots, \tilde{M}_q)$ such that the Markov relation $\tilde{M}^q \multimap\!\!- (S^q, G^q, S) \multimap\!\!- F_\ell(S)$ holds. Then, for every realization s, s^q such that $s \notin \{s_1, \ldots, s_q\}$, the conditional distribution of $F_\ell(S)$ given $\{G^q = g^q, S^q = s^q, S = s, \tilde{M}^q = \tilde{m}^q\}$ is uniform.

9.7 Prove Corollary 9.18.

9.8 Prove Corollary 9.20.

10 Secret Key Agreement

In this chapter, we study the secret key agreement problem. As we have seen in Chapter 3, in order to realize secure communication, the legitimate parties (sender and receiver) must share a secret key that is not known to an adversary. Various ways have been proposed for the legitimate parties to share an information-theoretically secure secret key, such as the use of wireless communication, the use of biometrics information, or the use of quantum communication. However, it is typically difficult for the legitimate parties to share a perfect secret key; the copies of keys obtained by the parties may have discrepancies caused by noise, or a part of the keys may be leaked to the adversary. In other words, the parties typically obtain a correlation that cannot be used as a secret key itself, but it has a potential to be used as a seed for creating a secret key. The secret key agreement protocol is a procedure of distilling an almost perfect secret key from a given correlation.

Similarly to Chapters 6 and 7, we present a single-shot setting first, before moving to the asymptotic setting and considering secret key capacity. We comment also on the role of interactive communication, and compare interactive protocols with one-way protocols.

We remark that, in practical security, public key cryptography relies on computational assumptions for parties to share a secret key. In this chapter, we present an alternative to public key cryptography where the keys are extracted from correlated randomness. However, this alternative view has not yet led to a widely adopted practical system. Nonetheless, it is an important theoretical area of research in information-theoretic cryptography.

10.1 Information-theoretic Secret Key Agreement

We start with a model for a source (X, Y, Z) comprising correlated random variables with joint distribution P_{XYZ}. Suppose that two legitimate parties $\mathcal{P}_1$ and $\mathcal{P}_2$ observe X and Y on finite alphabets $\mathcal{X}$ and $\mathcal{Y}$, respectively, and an eavesdropper[1] observes Z. We assume that the joint distribution P_{XYZ} of (X, Y, Z)

[1] It is customary to refer to the adversary as the "eavesdropper" in the context of the secret key agreement.

is known to the parties and the adversary. The parties seek to distill a secret key from (X, Y) that is concealed from the eavesdropper.

In addition to the source (X, Y), the parties are allowed to communicate over an authenticated but public communication channel. Here, "public" means that the channel is not secure and messages sent over this channel can be observed by the eavesdropper; "authenticated" means that the adversary cannot insert or substitute any message over this channel. As we have discussed in Chapter 8, the cost of realizing an authenticated channel is much lower than the cost of realizing secure communication. In fact, a message can be authenticated by using a key of length that is logarithmic order of the message length. Thus, it is reasonable to assume the existence of a public authenticated channel. In the rest of this chapter, when we say the public channel, this also means that the channel is authenticated.

Before formally introducing the problem setting, the following examples can help us set the stage.

EXAMPLE 10.1 Let $X = (X_1, X_2, X_3)$ be 3 random bits, and let $Y = X \oplus E$, where E takes one of $\{(000), (100), (010), (001)\}$ at random (cf. Example 6.6). Let Z be a constant, i.e., the adversary does not have any observation. In order to agree on a secret key, $\mathcal{P}_1$ sends

$$\Pi = (\Pi_1, \Pi_2) = (X_1 \oplus X_2, X_2 \oplus X_3)$$

over the public channel. Then, from Π and $(Y_1 \oplus Y_2, Y_2 \oplus Y_3)$, $\mathcal{P}_2$ can identify the error vector E and recover X. Although Π is leaked to the adversary, $K = X_1 \oplus X_2 \oplus X_3$ is uniform and independent of Π. Thus, a 1-bit secret key can be generated.

EXAMPLE 10.2 Suppose that (X_1, X_2, X_3, X_4) are 4 random bits; Y takes one of $(X_1, X_2, X_3, \mathsf{e})$, $(X_1, X_2, \mathsf{e}, X_4)$, $(X_1, \mathsf{e}, X_3, X_4)$, and $(\mathsf{e}, X_2, X_3, X_4)$ at random; and Z takes one of $(X_1, \mathsf{e}, \mathsf{e}, \mathsf{e})$, $(\mathsf{e}, X_2, \mathsf{e}, \mathsf{e})$, $(\mathsf{e}, \mathsf{e}, X_3, \mathsf{e})$, and $(\mathsf{e}, \mathsf{e}, \mathsf{e}, X_4)$ at random, where e means that the bit is erased. In other words, $\mathcal{P}_2$ observes X with 1 erasure, and the adversary observes X with 3 erasures. In order to agree on a secret key, $\mathcal{P}_1$ sends $\Pi = X_1 \oplus X_2 \oplus X_3 \oplus X_4$ over the public channel. Then, from Π and Y, $\mathcal{P}_2$ can identify the missing bit and recover X. Although the adversary observes Z and Π, $K = (X_1 \oplus X_2, X_2 \oplus X_3)$ is uniform and independent of (Z, Π). Thus, a 2-bit secret key can be generated. Note that $\mathcal{P}_2$'s initial ambiguity about X is $H(X|Y) = 1$, while the adversary's initial ambiguity about X is $H(X|Z) = 3$. As we will see later, roughly speaking, the difference in the ambiguities can be turned into a secret key.

Now, we formally describe a secret key agreement protocol. As we mentioned above, in order to agree on a secret key, the parties communicate over the public channel. The transcript sent by the parties may be stochastic functions of their observations.[2] To handle such stochastic processing, in addition to the correlated

[2] In this context, the "transcript" refers to the messages exchanged during the protocol.

Figure 10.1 A description of the secret key agreement.

sources X and Y, we assume that the parties $\mathcal{P}_1$ and $\mathcal{P}_2$ observe independent local randomness U_1 and U_2, respectively. More precisely, U_1 and U_2 are independent of each other, and (U_1, U_2) are jointly independent of (X, Y, Z).

An r-round protocol proceeds as follows. First, $\mathcal{P}_1$ computes $\Pi_1 = \Pi_1(X, U_1)$ and sends Π_1 to $\mathcal{P}_2$ over the public channel. Next, $\mathcal{P}_2$ computes $\Pi_2 = \Pi_2(Y, U_2, \Pi_1)$ and sends Π_2 to $\mathcal{P}_1$. More generally, for odd i, $\mathcal{P}_1$ computes $\Pi_i = \Pi_i(X, U_1, \Pi^{i-1})$ and sends Π_i to $\mathcal{P}_2$; for even i, $\mathcal{P}_2$ computes $\Pi_i = \Pi_i(Y, U_2, \Pi^{i-1})$ and sends Π_i to $\mathcal{P}_1$, where $\Pi^{i-1} = (\Pi_1, \ldots, \Pi_i)$ is the past transcript. At the end of the protocol, $\mathcal{P}_1$ computes a key $K_1 = K_1(X, U_1, \Pi)$ and $\mathcal{P}_2$ computes a key $K_2 = K_2(Y, U_2, \Pi)$, where $\Pi = (\Pi_1, \ldots, \Pi_r)$ is the entire transcript; see Figure 10.1.

For reliability, we require that the keys K_1 and K_2 coincide with high probability. Since the transcript Π is transmitted over the public channel, it is leaked to the eavesdropper as well. Thus, for secrecy, we require that the key K_1 is almost uniform and independent of the eavesdropper's observation (Z, Π).[3] Formally, the security of a secret key agreement protocol is defined as follows.

DEFINITION 10.3 For given $0 \leq \varepsilon, \delta < 1$, a protocol π with output $(K_1, K_2) \in \mathcal{K} \times \mathcal{K}$ and transcript Π is an (ε, δ)-secure secret key agreement (SK) protocol of length ℓ if

$$\Pr(K_1 \neq K_2) \leq \varepsilon \quad \text{(reliability)} \tag{10.1}$$

and

$$d_{\mathtt{var}}(\mathrm{P}_{K_1 \Pi Z}, \mathrm{P}_{\mathtt{unif}} \times \mathrm{P}_{\Pi Z}) \leq \delta \quad \text{(security)} \tag{10.2}$$

where $\mathrm{P}_{\mathtt{unif}}$ is the uniform distribution on $\mathcal{K} = \{0,1\}^\ell$. For brevity, we refer to the output (K_1, K_2) as a secret key on $\mathcal{K}$ generated using π.

Requirement (10.1) represents the reliability of the protocol in the sense that the keys obtained by the parties should coincide with high probability; and requirement (10.2) represents the secrecy of the protocol in the sense that the key of $\mathcal{P}_1$ is close to uniform and independent of the information observed by the

[3] Since K_2 should coincide with K_1 with high probability, K_2 is almost uniform and independent of (Z, Π) as well.

adversary. Under these requirements, we shall maximize the length of the secret key as much as possible. To that end, we introduce the following quantity.

DEFINITION 10.4 For a given $0 \le \varepsilon, \delta < 1$, the optimal length of secret key that can be generated from (X, Y, Z) is defined as the supremum over ℓ such that an (ε, δ)-SK protocol of length ℓ exists, i.e.,

$$S_{\varepsilon,\delta}(X, Y|Z) := \sup\{\ell : \text{there exists an } (\varepsilon, \delta)\text{-secure SK protocol of length } \ell\}.$$

A *one-way* protocol involves only a one-way communication from $\mathcal{P}_1$ to $\mathcal{P}_2$. We can define $S^{\rightarrow}_{\varepsilon,\delta}(X, Y|Z)$ to be the maximum length of an (ε, δ)-SK generated using a one-way protocol, in the same manner as Definition 10.4. Clearly, we have $S_{\varepsilon,\delta}(X, Y|Z) \ge S^{\rightarrow}_{\varepsilon,\delta}(X, Y|Z)$.

The goal of this chapter is to derive useful bounds on the quantities $S_{\varepsilon,\delta}(X, Y|Z)$ and $S^{\rightarrow}_{\varepsilon,\delta}(X, Y|Z)$.

10.2 Information Reconciliation and Privacy Amplification

In this section, we present a basic one-way protocol for secret key agreement, which will be a stepping-stone for building other advanced protocols that we will encounter later in this chapter. In fact, the basic protocol is a simple combination of information reconciliation (source coding with side-information) discussed in Chapter 6 and privacy amplification (leftover hashing) discussed in Chapter 7.

Recall that, in the basic construction for information reconciliation (cf. Construction 6.1), the sender $\mathcal{P}_1$ transmits a message $M = G(X)$ to the receiver $\mathcal{P}_2$, for a G distributed uniformly over a UHF. Then, $\mathcal{P}_2$ recovers X from the message M and the side-information Y. For recovery, we use guess lists $\{\mathcal{L}_y\}_{y \in \mathcal{Y}}$ which represent the prior knowledge of $\mathcal{P}_2$. Specifically, $\mathcal{P}_2$ assumes $X \in \mathcal{L}_y$ when $Y = y$, and uses the hash sent by $\mathcal{P}_1$ to determine X in the guess list $\mathcal{L}_Y$.

On the other hand, in the basic construction for privacy amplification, the parties compute the value $F(X)$ of a UHF F from $\mathcal{X}$ to $\{0, 1\}^\ell$. Then, if the length ℓ of the output is chosen properly, the output will be close to uniform and independent of the observations available to the eavesdropper. Here, the eavesdropper observes the choice of UHFs (F, G), the message M, and the side-information Z.

For given lists[4] $\{\mathcal{L}_y\}_{y \in \mathcal{Y}}$, our one-way protocol is described in Protocol 10.1, and its performance is analyzed in the following theorem.

THEOREM 10.5 *For given $0 < \varepsilon, \delta < 1$ and any fixed $0 < \xi \le \varepsilon$ and $0 < \eta \le \delta$, set*

$$m = \lceil H^{\varepsilon-\xi}_{\max}(X|Y) + \log(1/\xi) \rceil$$

[4] Only $\mathcal{P}_2$ needs to know the lists; $\mathcal{P}_1$ just needs to use the hash size appropriately.

Protocol 10.1 One-way secret key agreement protocol

Input: $\mathcal{P}_1$'s input X and $\mathcal{P}_2$'s input Y

Output: $\mathcal{P}_1$'s output $K_1 \in \{0,1\}^\ell$ and $\mathcal{P}_2$'s output $K_2 \in \{0,1\}^\ell$

1. $\mathcal{P}_1$ samples F uniformly from a UHF from $\mathcal{X}$ to $\{0,1\}^\ell$ and a UHF G from $\mathcal{X}$ to $\{0,1\}^m$, and sends $\Pi = (F, G, M)$ to $\mathcal{P}_2$, where $M = G(X)$.
2. Upon receiving Π, $\mathcal{P}_2$ looks for a unique $\hat{x} \in \mathcal{L}_Y$ such that $G(\hat{x}) = M$. If such an $\hat{x}$ exists, then $\mathcal{P}_2$ sets $\hat{X} = \hat{x}$; otherwise, $\mathcal{P}_2$ sets $\hat{X} = x_0$, where $x_0 \in \mathcal{X}$ is a prescribed symbol.
3. $\mathcal{P}_1$ and $\mathcal{P}_2$ output $K_1 = F(X)$ and $K_2 = F(\hat{X})$, respectively.

and

$$\ell = \lfloor H_{\min}^{(\delta-\eta)/2}(X|Z) - m - \log(1/4\eta^2) \rfloor.$$

Let $\{\mathcal{L}_y\}_{y\in\mathcal{Y}}$ *be lists such that* $\mathcal{L} = \cup_{y\in\mathcal{Y}} \mathcal{L}_y \times \{y\}$ *satisfies*[5] $\mathrm{P}_{XY}(\mathcal{L}) \ge 1 - (\varepsilon - \xi)$ *and* $\max_{y\in\mathcal{Y}} \log|\mathcal{L}_y| = H_{\max}^{\varepsilon-\xi}(X|Y)$. *Then, Protocol 10.1 is an* (ε, δ)*-secure SK protocol of length* ℓ.

Proof Since Steps 1 and 2 of Protocol 10.1 are the same as Construction 6.1, Lemma 6.9 guarantees that

$$\Pr(\hat{X} \neq X) \le \varepsilon,$$

which implies

$$\Pr(K_1 \neq K_2) \le \varepsilon.$$

On the other hand, by applying Corollary 7.22 with $V \leftarrow M$ and $Z \leftarrow (Z, G)$ and Lemma 7.23 to account for information leaked due to G, we have[6]

$$d_{\mathtt{var}}(\mathrm{P}_{K_1 \Pi Z}, \mathrm{P}_{\mathtt{unif}} \times \mathrm{P}_{\Pi Z}) \le \delta.$$

Thus, Protocol 10.1 is (ε, δ)-secure. □

From Theorem 10.5, we immediately get the following bound on $S_{\varepsilon,\delta}^{\rightarrow}(X, Y|Z)$.

COROLLARY 10.6 *For given* $0 < \varepsilon, \delta < 1$ *and for arbitrarily fixed* $0 < \xi \le \varepsilon$ *and* $0 < \eta \le \delta$, *we have*

$$S_{\varepsilon,\delta}^{\rightarrow}(X, Y|Z) \ge H_{\min}^{(\delta-\eta)/2}(X|Z) - H_{\max}^{\varepsilon-\xi}(X|Y) - \log(1/4\eta^2) - \log(1/\xi) - 2.$$

The bound in Corollary 10.6 seems complicated at first glance. Later, in Section 10.4, for a sequence of i.i.d. random variables, we will derive a compact asymptotic formula using this bound. In that case, the leading terms are the first two terms, the conditional min-entropy and the conditional max-entropy, and these terms can be approximated by Theorem 6.15 and Theorem 7.25.

[5] That is, $X \in \mathcal{L}_Y$ with probability at least $1 - (\varepsilon - \xi)$.

[6] Note that F is included in the transcript Π.

10.3 Impossibility Bounds

In this section, we derive upper bounds (impossibility bounds) for $S_{\varepsilon,\delta}(X,Y|Z)$. Starting from a basic bound, we will gradually move to more complicated bounds. Some of these bounds and techniques are useful not only for secret key agreement, but also in other cryptographic problems; see Chapter 13 and Chapter 14.

10.3.1 A Basic Impossibility Bound

We start with a basic bound based on Fano's inequality. The following simple but fundamental property of interactive communication protocol will be used throughout the book. Heuristically, it says that correlation cannot be built by conditioning on interactive communication.

LEMMA 10.7 *For any protocol π with transcript Π,*

$$I(X \wedge Y|Z,\Pi) \le I(X \wedge Y|Z).$$

In particular, if $\mathrm{P}_{XYZ} = \mathrm{P}_{X|Z}\mathrm{P}_{Y|Z}\mathrm{P}_Z$*, then* $\mathrm{P}_{XYZ\Pi} = \mathrm{P}_{X|Z\Pi}\mathrm{P}_{Y|Z\Pi}\mathrm{P}_{Z\Pi}$.

Proof First, since Π_1 is generated from X by $\mathcal{P}_1$, we have $I(\Pi_1 \wedge Y|X,Z) = 0$. Thus, we have

$$\begin{aligned}
I(X \wedge Y|Z) &= I(X \wedge Y|Z) + I(\Pi_1 \wedge Y|X,Z) \\
&= I(X,\Pi_1 \wedge Y|Z) \\
&= I(\Pi_1 \wedge Y|Z) + I(X \wedge Y|Z,\Pi_1) \\
&\ge I(X \wedge Y|Z,\Pi_1).
\end{aligned}$$

Next, since Π_2 is generated from (Y,Π_1) by $\mathcal{P}_2$, we have $I(\Pi_2 \wedge X|Y,Z,\Pi_1) = 0$. Thus, we have

$$\begin{aligned}
I(X \wedge Y|Z,\Pi_1) &= I(X \wedge Y|Z,\Pi_1) + I(\Pi_2 \wedge X|Y,Z,\Pi_1) \\
&= I(X \wedge Y,\Pi_2|Z,\Pi_1) \\
&= I(X \wedge \Pi_2|Z,\Pi_1) + I(X \wedge Y|Z,\Pi_1,\Pi_2) \\
&\ge I(X \wedge Y|Z,\Pi_1,\Pi_2).
\end{aligned}$$

Upon repeating these steps for multiple rounds, we have the claim of the lemma. □

By combining Lemma 10.7 with Fano's inequality, we can derive the next result.

THEOREM 10.8 *For every $0 \le \varepsilon,\delta < 1$ with $0 \le \varepsilon + \delta < 1$, it holds that*

$$S_{\varepsilon,\delta}(X,Y|Z) \le \frac{I(X \wedge Y|Z) + h(\bar{\varepsilon}) + h(\bar{\delta})}{1-\varepsilon-\delta},$$

where $\bar{a} = \min\{a, 1/2\}$.

Proof Consider an (ε, δ)-secure SK protocol for generating (K_1, K_2) from (X, Y, Z) with transcript Π. By Fano's inequality (cf. Theorem 2.29), it holds that

$$H(K_1|K_2) \le h(\bar{\varepsilon}) + \varepsilon \log |\mathcal{K}|,$$

and by the continuity of the Shannon entropy (cf. Theorem 2.34), that

$$\begin{aligned}\log |\mathcal{K}| - H(K_1|Z,\Pi) &\le \mathbb{E}\big[h(d_{\mathtt{var}}(\mathrm{P}_{K|Z\Pi}, \mathrm{P}_{\mathtt{unif}}))\big] + \mathbb{E}\big[d_{\mathtt{var}}(\mathrm{P}_{K|Z\Pi}, \mathrm{P}_{\mathtt{unif}})\big] \log |\mathcal{K}| \\ &\le h(\mathbb{E}\big[d_{\mathtt{var}}(\mathrm{P}_{K|Z\Pi}, \mathrm{P}_{\mathtt{unif}})\big]) + \mathbb{E}\big[d_{\mathtt{var}}(\mathrm{P}_{K|Z\Pi}, \mathrm{P}_{\mathtt{unif}})\big] \log |\mathcal{K}| \\ &\le h(\bar{\delta}) + \delta \log |\mathcal{K}|,\end{aligned}$$

where the second inequality uses the concavity of entropy and Jensen's inequality, and the previous inequality holds since $\mathbb{E}\big[d_{\mathtt{var}}(\mathrm{P}_{K|Z\Pi}, \mathrm{P}_{\mathtt{unif}})\big] = d_{\mathtt{var}}(\mathrm{P}_{KZ\Pi}, \mathrm{P}_{\mathtt{unif}} \times \mathrm{P}_{Z\Pi})$.

Thus,

$$\begin{aligned}\log |\mathcal{K}| &\le H(K_1|Z,\Pi) + h(\bar{\delta}) + \delta \log |\mathcal{K}| \\ &= I(K_1 \wedge K_2|Z,\Pi) + H(K_1|K_2,Z,\Pi) + h(\bar{\delta}) + \delta \log |\mathcal{K}| \\ &\le I(K_1 \wedge K_2|Z,\Pi) + h(\bar{\varepsilon}) + h(\bar{\delta}) + (\varepsilon + \delta) \log |\mathcal{K}| \\ &\le I(X \wedge Y|Z,\Pi) + h(\bar{\varepsilon}) + h(\bar{\delta}) + (\varepsilon + \delta) \log |\mathcal{K}| \\ &\le I(X \wedge Y|Z) + h(\bar{\varepsilon}) + h(\bar{\delta}) + (\varepsilon + \delta) \log |\mathcal{K}|,\end{aligned}$$

where the second-last inequality uses the data processing inequality and the final inequality uses Lemma 10.7. □

We illustrate the utility of the bound in Theorem 10.8 with a simple example.

EXAMPLE 10.9 (Impossibility of growing secret key) Suppose that X comprises random n bits, $Y = X$, and Z is a constant. In other words, the parties initially share an n-bit perfect secret key. Then, by allowing communication over the public channel and small errors $\varepsilon, \delta > 0$, can the parties generate a secret key longer than n bits? We expect the answer is no, but a rigorous proof is not that trivial. Since $I(X \wedge Y) = n$ in this case, the bound in Theorem 10.8 implies that there does not exist an (ε, δ)-secure SK protocol of length exceeding $(n + 2)/(1 - \varepsilon - \delta)$. Thus, we have established the impossibility of growing secret key up to the multiplicative factor of $1/(1 - \varepsilon - \delta)$.

10.3.2 Conditional Independence Testing Bound

The basic bound in Theorem 10.8 has a drawback: it has a multiplicative dependence on $1/(1 - \varepsilon - \delta)$ (see Example 10.9). In this section, we derive a bound that has an additive loss instead of the multiplicative factor. In fact, the additive loss term is only $\mathcal{O}(\log(1/(1 - \varepsilon - \delta)))$.

Heuristically, the lower bound we are about to present can be described as follows. Consider binary hypothesis testing (cf. Chapter 5) for observation (X, Y, Z)

where the null hypothesis is the joint distribution P_{XYZ} and the alternative hypothesis is a (fixed) distribution Q_{XYZ} under which X and Y are conditionally independent given Z. That is, the alternative hypothesis distribution satisfies $Q_{XYZ} = Q_{X|Z}Q_{Y|Z}Q_Z$. The distribution Q_{XYZ} does not exist in the actual problem setting, but it is hypothetically introduced for the purpose of deriving the bound. Given an (ε, δ)-secure SK protocol π of length ℓ, we use it as a test for this hypothesis testing problem. This in turn allows us to relate the key length ℓ and the optimal exponent of type II error probability $\beta_\varepsilon(P, Q)$ given that the type I error probability is less than ε; see (5.1).

Formally, we derive the following bound.

THEOREM 10.10 *Given $0 \le \varepsilon, \delta < 1$ and $0 < \eta < 1 - \varepsilon - \delta$, it holds that*

$$S_{\varepsilon,\delta}(X, Y|Z) \le -\log \beta_{\varepsilon+\delta+\eta}(P_{XYZ}, Q_{XYZ}) + 2\log(1/\eta)$$

for every $Q_{XYZ} = Q_{X|Z}Q_{Y|Z}Q_Z$.

The first observation needed to prove Theorem 10.10 is that the separate security conditions (10.1) and (10.2) imply the following combined security criterion (cf. Proposition 3.22):

$$d(P_{K_1K_2Z\Pi}, P^{(2)}_{\mathtt{unif}} \times P_{Z\Pi}) \le \varepsilon + \delta, \tag{10.3}$$

where

$$P^{(2)}_{\mathtt{unif}}(k_1, k_2) := \frac{\mathbf{1}[k_1 = k_2]}{|\mathcal{K}|}.$$

The core of the proof of Theorem 10.10 is the following lemma.

LEMMA 10.11 *For given $0 \le \varepsilon, \delta < 1$ and $0 < \eta < 1 - \varepsilon - \delta$, for any K_1, K_2 taking values in $\mathcal{K}$ and satisfying (10.3), and for any $Q_{K_1K_2Z\Pi}$ of the form $Q_{K_1|Z\Pi}Q_{K_2|Z\Pi}Q_{Z\Pi}$, it holds that*

$$\log |\mathcal{K}| \le -\log \beta_{\varepsilon+\delta+\eta}(P_{K_1K_2Z\Pi}, Q_{K_1K_2Z\Pi}) + 2\log(1/\eta).$$

Before proving this result, we show how it leads to the proof of Theorem 10.10.

Proof of Theorem 10.10

Consider an (ε, δ)-secure SK protocol for generating (K_1, K_2) from $(X, Y, Z) \sim P_{XYZ}$ with transcript Π. Let $Q_{K_1K_2Z\Pi}$ be the joint distribution obtained by running the same protocol for $Q_{XYZ} = Q_{X|Z}Q_{Y|Z}Q_Z$. By Lemma 10.7, $Q_{K_1K_2Z\Pi}$ satisfies the assumption of Lemma 10.11. Thus, by regarding the protocol as a channel converting P_{XYZ} to $P_{K_1K_2Z\Pi}$ and Q_{XYZ} to $Q_{K_1K_2Z\Pi}$, and by applying the data processing inequality for $\beta_\varepsilon(\cdot, \cdot)$ (see Lemma 5.6), Theorem 10.10 follows from Lemma 10.11. □

Proof of Lemma 10.11

To prove Lemma 10.11, we construct a likelihood ratio test using the given SK protocol. The key idea is to consider the likelihood ratio between $\mathrm{P}^{(2)}_{\mathtt{unif}} \times \mathrm{P}_{Z\Pi}$ and $\mathrm{Q}_{K_1K_2Z\Pi}$, instead of the likelihood ratio between $\mathrm{P}_{K_1K_2Z\Pi}$ and $\mathrm{Q}_{K_1K_2Z\Pi}$. Indeed, consider the test with acceptance region defined by

$$\mathcal{A} := \left\{(k_1, k_2, z, \tau) : \log \frac{\mathrm{P}^{(2)}_{\mathtt{unif}}(k_1, k_2)}{\mathrm{Q}_{K_1K_2|Z\Pi}(k_1, k_2|z, \tau)} \geq \lambda\right\},$$

where

$$\lambda = \log |\mathcal{K}| - 2\log(1/\eta).$$

Then, a change-of-measure argument[7] yields the following bound for the type II error probability:

$$\begin{aligned}\mathrm{Q}_{K_1K_2Z\Pi}(\mathcal{A}) &= \sum_{z,\tau} \mathrm{Q}_{Z\Pi}(z, \tau) \sum_{(k_1,k_2):(k_1,k_2,z,\tau)\in\mathcal{A}} \mathrm{Q}_{K_1K_2|Z\Pi}(k_1, k_2|z, \tau) \\ &\leq 2^{-\lambda} \sum_{z,\tau} \mathrm{Q}_{Z\Pi}(z, \tau) \sum_{(k_1,k_2)} \mathrm{P}^{(2)}_{\mathtt{unif}}(k_1, k_2) \\ &= \frac{1}{|\mathcal{K}|\eta^2}. \end{aligned} \tag{10.4}$$

On the other hand, the security condition (10.3) yields the following bound for the type I error probability:

$$\begin{aligned}\mathrm{P}_{K_1K_2Z\Pi}(\mathcal{A}^c) &\leq d_{\mathtt{var}}(\mathrm{P}_{K_1K_2Z\Pi}, \mathrm{P}^{(2)}_{\mathtt{unif}} \times \mathrm{P}_{Z\Pi}) + \mathrm{P}^{(2)}_{\mathtt{unif}} \times \mathrm{P}_{Z\Pi}(\mathcal{A}^c) \\ &\leq \varepsilon + \delta + \mathrm{P}^{(2)}_{\mathtt{unif}} \times \mathrm{P}_{Z\Pi}(\mathcal{A}^c), \end{aligned} \tag{10.5}$$

where the first inequality follows from the definition of the variational distance (cf. Lemma 2.14). The last term above can be expressed as

$$\begin{aligned}\mathrm{P}^{(2)}_{\mathtt{unif}} \times \mathrm{P}_{Z\Pi}(\mathcal{A}^c) &= \sum_{z,\tau} \mathrm{P}_{Z\Pi}(z, \tau) \frac{1}{|\mathcal{K}|} \sum_{k} \mathbf{1}[(k, k, z, \tau) \in \mathcal{A}^c] \\ &= \sum_{z,\tau} \mathrm{P}_{Z\Pi}(z, \tau) \frac{1}{|\mathcal{K}|} \sum_{k} \mathbf{1}[\mathrm{Q}_{K_1K_2|Z\Pi}(k, k|z, \tau)|\mathcal{K}|^2\eta^2 > 1],\end{aligned}$$

and the inner sum can be bounded as

$$\begin{aligned}&\sum_{k} \mathbf{1}[\mathrm{Q}_{K_1K_2|Z\Pi}(k, k|z, \tau)|\mathcal{K}|^2\eta^2 > 1] \\ &\quad \leq \sum_{k} \left(\mathrm{Q}_{K_1K_2|Z\Pi}(k, k|z, \tau)|\mathcal{K}|^2\eta^2\right)^{1/2} \\ &\quad = |\mathcal{K}|\eta \sum_{k} \mathrm{Q}_{K_1K_2|Z\Pi}(k, k|z, \tau)^{1/2} \\ &\quad = |\mathcal{K}|\eta \sum_{k} \mathrm{Q}_{K_1|Z\Pi}(k|z, \tau)^{1/2}\mathrm{Q}_{K_2|Z\Pi}(k|z, \tau)^{1/2},\end{aligned}$$

[7] This is a generic name for techniques where the probability of an event is bounded by considering an alternative probability measure; we have seen such arguments earlier in the book, including in the proof of Lemma 5.2.

where the last equality holds since $Q_{K_1K_2Z\Pi}$ has the form $Q_{K_1|Z\Pi}Q_{K_2|Z\Pi}Q_{Z\Pi}$ by the assumption.

Next, an application of the Cauchy–Schwarz inequality yields

$$\begin{aligned}\sum_k Q_{K_1|Z\Pi}(k|z,\tau)^{\frac{1}{2}} Q_{K_2|Z\Pi}(k|z,\tau)^{\frac{1}{2}} \\ &\le \Big(\sum_{k_1} Q_{K_1|Z\Pi}(k_1|z,\tau)\Big)^{\frac{1}{2}} \Big(\sum_{k_2} Q_{K_2|Z\Pi}(k_2|z,\tau)\Big)^{\frac{1}{2}} \\ &= 1.\end{aligned}$$

Upon combining the bounds above, we obtain

$$P^{(2)}_{\mathtt{unif}} \times P_{Z\Pi}(\mathcal{A}^c) \le \eta,$$

which, together with (10.4) and (10.5), implies Lemma 10.11. □

EXAMPLE 10.12 (Impossibility of growing secret key revisited) Consider the setting of Example 10.9, i.e., $X = Y$ are random n bits and Z is a constant. In this case, by setting $Q_{XY}(x, y) = 1/2^{2n}$, i.e., the independent distribution, we have $\beta_{\varepsilon+\delta+\eta}(P_{XY}, Q_{XY}) \ge (1-\varepsilon-\delta-\eta)/2^n$ (see Example 5.3). Thus, Theorem 10.10 implies that there does not exist an (ε, δ)-secure SK protocol of length exceeding $n+\log(1/(1-\varepsilon-\delta-\eta))+2\log(1/\eta)$. Namely, we have shown the impossibility of growing secret key up to the additive factor of $\log(1/(1-\varepsilon-\delta-\eta))+2\log(1/\eta)$.

10.3.3 Bounds Using Monotones

The final bound that we describe in this section is based on defining a "monotone." Roughly, a monotone is a function of distribution of the observations of the parties which decreases (or increases) monotonically as the protocol proceeds. Such monotones are formed by carefully examining the steps in the standard impossibility proofs and identifying the abstract properties that enable the proof. Often, this abstraction leads to new bounds which are tighter than the original ones.

DEFINITION 10.13 (Monotone) For a given joint distribution P_{XYZ}, a nonnegative function $M_{\varepsilon,\delta}(X, Y|Z)$ of P_{XYZ} constitutes a monotone for a secret key agreement if it satisfies the following properties.

1. $M_{\varepsilon,\delta}(X, Y|Z)$ does not increase by local processing, i.e., for any X' satisfying $X' -\!\!\circ\!\!- X -\!\!\circ\!\!- (Y, Z)$,

$$M_{\varepsilon,\delta}(X, Y|Z) \ge M_{\varepsilon,\delta}(X', Y|Z); \tag{10.6}$$

similarly, for any $Y' -\!\!\circ\!\!- Y -\!\!\circ\!\!- (X, Z)$,

$$M_{\varepsilon,\delta}(X, Y|Z) \ge M_{\varepsilon,\delta}(X, Y'|Z). \tag{10.7}$$

2. $M_{\varepsilon,\delta}(X, Y|Z)$ does not increase by any interactive communication, i.e., for any protocol π with transcript Π,

$$M_{\varepsilon,\delta}(X, Y|Z) \geq M_{\varepsilon,\delta}((X, \Pi), (Y, \Pi)|(Z, \Pi)).$$

3. For any secret key (K_1, K_2) on $\mathcal{K}$ satisfying (10.1) and (10.2),

$$\log |\mathcal{K}| \leq M_{\varepsilon,\delta}((K_1, \Pi), (K_2, \Pi)|(Z, \Pi)) + \Delta(\varepsilon, \delta)$$

for a suitable $\Delta(\varepsilon, \delta) \geq 0$.

From the definition of monotones, we can immediately derive the result that a monotone bounds the function $S_{\varepsilon,\delta}(X, Y|Z)$ from above.

PROPOSITION 10.14 *For $0 \leq \varepsilon, \delta < 1$ and a monotone $M_{\varepsilon,\delta}(X, Y|Z)$ satisfying the properties in Definition 10.13, it holds that*

$$S_{\varepsilon,\delta}(X, Y|Z) \leq M_{\varepsilon,\delta}(X, Y|Z) + \Delta(\varepsilon, \delta).$$

Proof For any (ε, δ)-secure protocol generating (K_1, K_2) from (X, Y, Z) with transcript Π, properties 3, 1, and 2 imply

$$\begin{aligned}\log |\mathcal{K}| &\leq M_{\varepsilon,\delta}((K_1, \Pi), (K_2, \Pi)|(Z, \Pi)) + \Delta(\varepsilon, \delta) \\ &\leq M_{\varepsilon,\delta}((X, \Pi), (Y, \Pi)|(Z, \Pi)) + \Delta(\varepsilon, \delta) \\ &\leq M_{\varepsilon,\delta}(X, Y|Z) + \Delta(\varepsilon, \delta).\end{aligned}$$

□

We illustrate the utility of the monotone approach by a few examples. In fact, we construct appropriate monotones that will allow us to recover the bounds obtained by other approaches presented in Sections 10.3.1 and 10.3.2, showing that the monotone approach can recover all these bounds.

EXAMPLE 10.15 (Conditional mutual information) For

$$M_{\varepsilon,\delta}(X, Y|Z) = \frac{1}{1 - \varepsilon - \delta} I(X \wedge Y|Z),$$

in the manner of the proof of Theorem 10.8, we can verify that $M_{\varepsilon,\delta}(X, Y|Z)$ satisfies the properties in Definition 10.13 with $\Delta(\varepsilon, \delta) = \frac{h(\bar{\varepsilon}) + h(\bar{\delta})}{1 - \varepsilon - \delta}$ for $0 \leq \varepsilon + \delta < 1$.

EXAMPLE 10.16 (Conditional independence testing) Let

$$M_{\varepsilon,\delta}(X, Y|Z) = \inf_{\mathrm{Q}_{XYZ} \in \mathcal{Q}_{\mathrm{CI}}} \left[-\log \beta_{\varepsilon+\delta+\eta}(\mathrm{P}_{XYZ}, \mathrm{Q}_{XYZ}) \right], \tag{10.8}$$

where $\mathcal{Q}_{\mathrm{CI}}$ is the set of all conditionally independent distributions. Then, we can verify that $M_{\varepsilon,\delta}(X, Y|Z)$ satisfies[8] the properties in Definition 10.13 with $\Delta(\varepsilon, \delta) = 2 \log(1/\eta)$ for $0 < \eta < 1 - \varepsilon - \delta$.

[8] More specifically, Property 1 follows from the data processing inequality for $\beta_\varepsilon(\cdot, \cdot)$, Property 2 follows from Lemma 10.7, and Property 3 follows from Lemma 10.11.

EXAMPLE 10.17 (Intrinsic information) For

$$M_{\varepsilon,\delta}(X,Y|Z) = \frac{1}{1-\varepsilon-\delta} \inf_{Z' \multimap Z \multimap (X,Y)} I(X \wedge Y|Z'),$$

noting that

$$H(K_1|Z,\Pi) \le H(K_1|Z',\Pi), \tag{10.9}$$

we can verify that $M_{\varepsilon,\delta}(X,Y|Z)$ satisfies the properties in Definition 10.13 with $\Delta(\varepsilon,\delta) = \frac{h(\bar{\varepsilon})+h(\bar{\delta})}{1-\varepsilon-\delta}$ for $0 \le \varepsilon + \delta < 1$.

The previous example provides a monotone which improves upon the bound obtained by the conditional mutual information monotone; we will revisit this point later.

Note that the optimal key length $S_{\varepsilon,\delta}(X,Y|Z)$ itself satisfies the properties of a monotone in Definition 10.13 with $\Delta(\varepsilon,\delta) = 0$. In this trivial sense, the monotone approach gives tight bounds. In fact, any upper bound can be proved without referring to the monotone properties; however, it is conceptually helpful to recognize what properties should be verified in order to prove an upper bound.

The following bound is a further strengthening of the intrinsic information bound seen in the previous example.

PROPOSITION 10.18 *Let*

$$B(X,Y|Z) := \inf_U \left[I(X \wedge Y|U) + I(X,Y \wedge U|Z) \right],$$

where the infimum is taken over all conditional distributions $\mathrm{P}_{U|XYZ}$ *for the auxiliary random variable* U. *Then,*

$$M_{\varepsilon,\delta}(X,Y|Z) = \frac{1}{1-\varepsilon-\delta} B(X,Y|Z)$$

satisfies the properties in Definition 10.13 with $\Delta(\varepsilon,\delta) = \frac{h(\bar{\varepsilon})+h(\bar{\delta})}{1-\varepsilon-\delta}$ *for* $0 \le \varepsilon + \delta < 1$.

Proof To verify Property 1, for any X' satisfying $X' \multimap X \multimap (Y,Z)$ and any $\mathrm{P}_{U|XYZ}$, consider the joint distribution

$$\mathrm{P}_{UX'YZ}(u,x',y,z) = \sum_x \mathrm{P}_{XYZ}(x,y,z)\, \mathrm{P}_{U|XYZ}(u|x,y,z)\, \mathrm{P}_{X'|X}(x'|x).$$

Then, by the data processing inequality for conditional mutual information, we have

$$\begin{aligned} &I(X \wedge Y|U) + I(X,Y \wedge U|Z) \\ &\quad \ge I(X' \wedge Y|U) + I(X',Y \wedge U|Z) \\ &\quad \ge B(X',Y|Z). \end{aligned}$$

Since this inequality holds for any $\mathrm{P}_{U|XYZ}$, we have (10.6). We can verify (10.7) similarly.

To verify Property 2, for any protocol with inputs (X, Y) and transcript Π, and for any $P_{U|XYZ}$, consider the joint distribution $P_{UXYZ\Pi} = P_{XYZ}P_{U|XYZ}P_{\Pi|XY}$. Then, we have

$$\begin{aligned}
&I(X \wedge Y|U) + I(X, Y \wedge U|Z) \\
&\quad \geq I(X \wedge Y|U, \Pi) + I(X, Y \wedge U|Z) \\
&\quad = I(X \wedge Y|U, \Pi) + I(X, Y, \Pi \wedge U|Z) \\
&\quad \geq I(X \wedge Y|U, \Pi) + I(X, Y \wedge U|Z, \Pi) \\
&\quad = I(X \wedge Y|U, \Pi) + I(X, Y, \Pi \wedge U, \Pi|Z, \Pi) \\
&\quad = I(X \wedge Y|U') + I(X, Y, \Pi \wedge U'|Z, \Pi) \\
&\quad \geq B((X, \Pi), (Y, \Pi)|(Z, \Pi)),
\end{aligned}$$

where $U' = (U, \Pi)$, the first inequality follows from the property of the interactive communication in Lemma 10.7, and the last inequality holds since $P_{U'|XYZ\Pi}$ is a valid choice of channel in the definition of the quantity $B((X, \Pi), (Y, \Pi)|(Z, \Pi))$. Since this inequality holds for any $P_{U|XYZ}$, we have verified Property 2.

To verify Property 3, using Fano's inequality and the continuity of the Shannon entropy in the same manner as in the proof of Theorem 10.8, for any $P_{U|K_1K_2Z\Pi}$, we have

$$\begin{aligned}
&\log|\mathcal{K}| \\
&\quad \leq H(K_1|Z, \Pi) + h(\bar{\delta}) + \delta \log|\mathcal{K}| \\
&\quad = H(K_1|Z, U, \Pi) + I(K_1 \wedge U|Z, \Pi) + h(\bar{\delta}) + \delta \log|\mathcal{K}| \\
&\quad \leq H(K_1|U) + I(K_1, K_2, \Pi \wedge U|Z, \Pi) + h(\bar{\delta}) + \delta \log|\mathcal{K}| \\
&\quad = I(K_1 \wedge K_2|U) + H(K_1|K_2, U) + I(K_1, K_2, \Pi \wedge U|Z, \Pi) + h(\bar{\delta}) + \delta \log|\mathcal{K}| \\
&\quad \leq I(K_1 \wedge K_2|U) + I(K_1, K_2, \Pi \wedge U|Z, \Pi) + h(\bar{\varepsilon}) + h(\bar{\delta}) + (\varepsilon + \delta) \log|\mathcal{K}| \\
&\quad \leq I(K_1, \Pi \wedge K_2, \Pi|U) + I(K_1, K_2, \Pi \wedge U|Z, \Pi) \\
&\qquad + h(\bar{\varepsilon}) + h(\bar{\delta}) + (\varepsilon + \delta) \log|\mathcal{K}|,
\end{aligned}$$

which implies

$$\log|\mathcal{K}| \leq \frac{I(K_1, \Pi \wedge K_2, \Pi|U) + I(K_1, K_2, \Pi \wedge U|Z, \Pi) + h(\bar{\varepsilon}) + h(\bar{\delta})}{1 - \varepsilon - \delta}.$$

Since this inequality holds for any $P_{U|K_1K_2Z\Pi}$, we have verified Property 3. □

10.4 Secret Key Capacity

In this section, we consider the situation in which the legitimate parties and the eavesdropper observe a sequence of i.i.d. random variables (X^n, Y^n, Z^n) with common distribution P_{XYZ}. In this case, the optimal length $S_{\varepsilon,\delta}(X^n, Y^n|Z^n)$ of secret key that can be generated from (X^n, Y^n, Z^n) grows linearly in n, and we are interested in characterizing the coefficient of linear growth. To that end, we introduce the notion of secret key capacity.

DEFINITION 10.19 (Secret key capacity) Given $0 < \varepsilon, \delta < 1$ and a distribution P_{XYZ}, the (ε, δ)-secret key capacity for P_{XYZ} is defined as

$$C_{\varepsilon,\delta}(X, Y|Z) := \liminf_{n\to\infty} \frac{1}{n} S_{\varepsilon,\delta}(X^n, Y^n|Z^n). \tag{10.10}$$

Then, the secret key capacity for P_{XYZ} is defined as

$$C(X, Y|Z) := \lim_{\varepsilon,\delta\to 0} C_{\varepsilon,\delta}(X, Y|Z). \tag{10.11}$$

When we restrict our attention to the class of one-way protocols, the one-way secret key capacity is defined by

$$C^{\rightarrow}(X, Y|Z) := \lim_{\varepsilon,\delta\to 0} \liminf_{n\to\infty} \frac{1}{n} S^{\rightarrow}_{\varepsilon,\delta}(X^n, Y^n|Z^n). \tag{10.12}$$

Clearly, we have $C(X, Y|Z) \geq C^{\rightarrow}(X, Y|Z)$. Later, we will see examples such that this inequality is strict.

First, we derive a lower bound on the one-way secret key capacity by using the protocol given in Section 10.2.

PROPOSITION 10.20 *For any distribution* P_{XYZ}, *we have*

$$\begin{aligned} C^{\rightarrow}(X, Y|Z) &\geq H(X|Z) - H(X|Y) \qquad (10.13)\\ &= I(X \wedge Y) - I(X \wedge Z). \qquad (10.14)\end{aligned}$$

Proof For any fixed $0 < \varepsilon, \delta < 1$, by Corollary 10.6, we have

$$\begin{aligned}&\frac{1}{n} S^{\rightarrow}_{\varepsilon,\delta}(X^n, Y^n|Z^n)\\ &\geq \frac{1}{n} H_{\min}^{(\delta-\eta)/2}(X^n|Z^n) - \frac{1}{n} H_{\max}^{\varepsilon-\xi}(X^n|Y^n) - \frac{\log(1/4\eta^2) + \log(1/\xi) + 2}{n} \qquad (10.15)\end{aligned}$$

for any $0 < \xi < \varepsilon$ and $0 < \eta < \delta$. Further, by Theorem 7.25, we have

$$\frac{1}{n} H_{\min}^{(\delta-\eta)/2}(X^n|Z^n) \geq H(X|Z) - \kappa_1 \sqrt{\frac{\ln(1/(\delta-\eta))}{2n}}, \tag{10.16}$$

where $\kappa_1 = \max_{(x,z)\in\mathrm{supp}(\mathrm{P}_{XZ})} \log \frac{1}{\mathrm{P}_{X|Z}(x|z)}$. On the other hand, by Theorem 6.15, we have

$$\frac{1}{n} H_{\max}^{\varepsilon-\xi}(X^n|Y^n) \leq H(X|Y) + \kappa_2 \sqrt{\frac{\ln(1/(\varepsilon-\xi))}{2n}}, \tag{10.17}$$

where $\kappa_2 = \max_{(x,y)\in\mathrm{supp}(\mathrm{P}_{XY})} \log \frac{1}{\mathrm{P}_{X|Y}(x|y)}$. Thus, by combining (10.15)–(10.17) and taking the limit as n goes to infinity, we obtain

$$\liminf_{n\to\infty} \frac{1}{n} S_{\varepsilon,\delta}(X^n, Y^n|Z^n) \geq H(X|Z) - H(X|Y).$$

Since this inequality holds for arbitrary $0 < \varepsilon, \delta < 1$, we have (10.13). □

In a simple extension of the protocol of Section 10.2, the party $\mathcal{P}_1$ can first locally process X to obtain another random variable V, and then apply the protocol to (V, Y, Z). Furthermore, for a random variable U generated from V, $\mathcal{P}_1$ can send U to $\mathcal{P}_2$ over the public channel separately, in addition to communication used for information reconciliation. The parties can then apply Protocol 10.1 to $(V, (Y, U), (Z, U))$. Note that the standard protocol, Protocol 10.1, with this preprocessing is still a one-way protocol. Interestingly, as we will demonstrate later, this preprocessing does improve the performance of the standard protocol. In fact, we can also show that the standard protocol with this preprocessing is asymptotically optimal among the class of one-way protocols.

THEOREM 10.21 *For a distribution* P_{XYZ}, *we have*

$$C^{\rightarrow}(X, Y|Z) = \sup_{U,V} \left[H(V|Z, U) - H(V|Y, U)\right] \tag{10.18}$$

$$= \sup_{U,V} \left[I(V \wedge Y|U) - I(V \wedge Z|U)\right], \tag{10.19}$$

where the supremum is taken over auxiliary random variables (U, V) *satisfying* $U -\!\!\circ\!\!- V -\!\!\circ\!\!- X -\!\!\circ\!\!- (Y, Z)$.[9]

Proof Fix arbitrary (U, V) satisfying the Markov chain condition $U -\!\!\circ\!\!- V -\!\!\circ\!\!- X -\!\!\circ\!\!- (Y, Z)$. The party $\mathcal{P}_1$ locally generates U^n and V^n from X^n using the same channel $\mathrm{P}_{UV|X}$ applied to each sample X_i independently,[10] and sends U^n to $\mathcal{P}_2$ over the public channel; this U^n is leaked to the eavesdropper as well. Then, the parties apply Protocol 10.1 to $(V^n, (Y^n, U^n), (Z^n, U^n))$. Since this protocol is a one-way protocol, Proposition 10.20 implies

$$C^{\rightarrow}(X, Y|Z) \geq H(V|Z, U) - H(V|U, U).$$

Since this bound holds for arbitrary (U, V) satisfying the Markov chain condition, we have the inequality "$\geq$" in (10.18).

To prove the opposite inequality, for any SK protocol π generating (K_1, K_2) with one-way communication Π, by using the Fano inequality and the continuity of the conditional entropy in the same manner as in the proof of Theorem 10.8, we have

$$\begin{aligned} \log|\mathcal{K}| &\leq H(K_1|Z^n, \Pi) - H(K_1|K_2) + h(\bar{\varepsilon}) + h(\bar{\delta}) + (\varepsilon + \delta)\log|\mathcal{K}| \\ &\leq H(K_1|Z^n, \Pi) - H(K_1|Y^n, \Pi) + h(\bar{\varepsilon}) + h(\bar{\delta}) + (\varepsilon + \delta)\log|\mathcal{K}|, \end{aligned} \tag{10.20}$$

where we used the data processing inequality for mutual information in the previous step. Denoting $Z_j^n = (Z_j, \ldots, Z^n)$ and using the chain rule, we can manipulate the first two terms on the right-hand side above to obtain[11]

[9] The range of the auxiliary random variables is discrete but unbounded. By using a technique from convex analysis, we can show that the ranges $\mathcal{U}$ and $\mathcal{V}$ of the auxiliary random variables satisfy $|\mathcal{U}| \leq |\mathcal{X}|$ and $|\mathcal{V}| \leq |\mathcal{X}|$ without loss of optimality; see Section 10.6 for appropriate references.

[10] The overall channel from X^n to (U^n, V^n) constitutes a so-called *memoryless channel*.

[11] This is a very useful identity often used in multiterminal information theory.

$$
\begin{aligned}
&H(K_1|Z^n,\Pi) - H(K_1|Y^n,\Pi) \\
&\quad = H(K_1|Z^n,\Pi) - H(K_1|Y_1,Z_2^n,\Pi) \\
&\qquad + H(K_1|Y_1,Z_2^n,\Pi) - H(K_1|Y^2,Z_3^n,\Pi) \\
&\qquad\quad \vdots \\
&\qquad + H(K|Y^{n-1},Z_n,\Pi) - H(K_1|Y^n,\Pi) \\
&\quad = \sum_{j=1}^{n} \big[H(K_1|Z_j,Y^{j-1},Z_{j+1}^n,\Pi) - H(K_1|Y_j,Y^{j-1},Z_{j+1}^n,\Pi)\big] \\
&\quad = \sum_{j=1}^{n} \big[H(V_j|Z_j,U_j) - H(V_j|Y_j,U_j)\big],
\end{aligned} \tag{10.21}
$$

where the first identity follows by canceling adjacent terms, Y^0 and Z_{n+1}^n are ignored in the summation, and we set $U_j = (Y^{j-1}, Z_{j+1}^n, \Pi)$ and $V_j = (U_j, K_1)$ in the last identity. Note that we have

$$
U_j \multimap V_j \multimap X_j \multimap (Y_j, Z_j) \tag{10.22}
$$

for every $1 \le j \le n$. In fact, U_j is a function of V_j and

$$
\begin{aligned}
&I(V_j \wedge Y_j, Z_j|X_j) \\
&\quad = I(Y^{j-1}, Z_{j+1}^n, \Pi, K_1 \wedge Y_j, Z_j|X_j) \\
&\quad \le I(X^{j-1}, X_{j+1}^n, Y^{j-1}, Z_{j+1}^n, \Pi, K_1 \wedge Y_j, Z_j|X_j) \\
&\quad = I(X^{j-1}, X_{j+1}^n, Y^{j-1}, Z_{j+1}^n \wedge Y_j, Z_j|X_j) + I(\Pi, K_1 \wedge Y_j, Z_j|X^n, Y^{j-1}, Z_{j+1}^n) \\
&\quad = 0,
\end{aligned}
$$

where the first term is 0 since (X^n, Y^n, Z^n) are i.i.d., and the second term is 0 since (Π, K_1) is generated from X^n in the one-way protocol.

Next, let J be the uniform random variable on the index set $\{1, \ldots, n\}$ that is independent of other random variables. Then, we can rewrite (10.21) as

$$
\begin{aligned}
&n \sum_{j=1}^{n} \frac{1}{n} \big[H(V_j|Z_j,U_j) - H(V_j|Y_j,U_j)\big] \\
&\qquad = n\big[H(V_J|Z_J,U_J,J) - H(V_J|Y_J,U_J,J)\big].
\end{aligned} \tag{10.23}
$$

Since (X^n, Y^n, Z^n) are i.i.d., $\mathrm{P}_{X_J Y_J Z_J}$ coincides with the common distribution P_{XYZ}. Thus, by setting $U = (U_J, J)$ and $V = (U, V_J)$, we can further rewrite (10.23) as

$$
\begin{aligned}
&n\big[H(V_J|Z_J,U_J,J) - H(V_J|Y_J,U_J,J)\big] \\
&\qquad = n\big[H(V|Z,U) - H(V|Y,U)\big].
\end{aligned} \tag{10.24}
$$

Again, note that (10.22) for every $1 \le j \le n$ implies $U \multimap V \multimap X \multimap (Y, Z)$. In fact, U is a function of V and

$$\begin{aligned} I(V \wedge Y, Z|X) &= I(V_J, U_J, J \wedge Y_J, Z_J|X_J) \\ &= I(J \wedge Y_J, Z_J|X_J) + I(V_J, U_J \wedge Y_J, Z_J|X_J, J) \\ &= 0. \end{aligned}$$

By combining (10.20), (10.21), (10.23), and (10.24), we have

$$\begin{aligned} \frac{1}{n} \log |\mathcal{K}| &\leq \frac{H(V|Z,U) - H(V|Y,U) + h(\bar{\varepsilon}) + h(\bar{\delta})}{1-\varepsilon-\delta} \\ &\leq \frac{\sup_{U,V}[H(V|Z,U) - H(V|Y,U)] + h(\bar{\varepsilon}) + h(\bar{\delta})}{1-\varepsilon-\delta}. \end{aligned}$$

Since this bound holds for any (ε, δ)-secure protocol, we have

$$\frac{1}{n} S_{\varepsilon,\delta}(X^n, Y^n|Z^n) \leq \frac{\sup_{U,V}[H(V|Z,U) - H(V|Y,U)] + h(\bar{\varepsilon}) + h(\bar{\delta})}{1-\varepsilon-\delta}.$$

Consequently, by taking the limits of $n \to \infty$ and $\varepsilon, \delta \to 0$, we have the desired inequality. □

We have now seen that Protocol 10.1 with preprocessing is optimal among the class of one-way protocols. But does preprocessing help, and do we need both U and V? We illustrate the necessity of preprocessing with a few examples. The necessity of V is illustrated in the following example.

EXAMPLE 10.22 (Noisy preprocessing) For $\mathcal{X} = \mathcal{Y} = \{0,1\}$ and $\mathcal{Z} = \{0,1\}^2 \cup \{\mathsf{e}\}$, consider the distribution P_{XYZ} such that $\mathrm{P}_{XY}(x,y) = \frac{1-p}{2}$ for $x = y$, $\mathrm{P}_{XY}(x,y) = \frac{p}{2}$ for $x \neq y$, $\mathrm{P}_{Z|XY}(x,y|x,y) = 1-q$, and $\mathrm{P}_{Z|XY}(\mathsf{e}|x,y) = q$, where $0 \leq p \leq 1/2$ and $0 \leq q \leq 1$. In words, $\mathcal{P}_2$ observes X via the binary symmetric channel and the eavesdropper observes (X,Y) via the erasure channel.[12] In this case, the rate without preprocessing is $H(X|Z) - H(X|Y) = q - h(p)$.

On the other hand, consider V generated from X via the binary symmetric channel with crossover probability t; this preprocessing is referred to as *noisy preprocessing*. Then, the rate with noisy preprocessing is $H(V|Z) - H(V|Y) = q + (1-q)h(t) - h(p * t)$, where $p * t = p(1-t) + (1-p)t$ is the binary convolution operation. Interestingly, adding noise to X does improve the rate for a certain range of p; for the erasure probability $q = 0.7$, these rates are plotted in Figure 10.2 as a function of p, where $0 \leq t \leq 1/2$ is optimized for each p.

Next, we consider the role of U. Heuristically, if X contains some information that is known to the eavesdropper, or at least better known to the eavesdropper compared to $\mathcal{P}_2$, then it helps to send that information to $\mathcal{P}_2$ upfront, before starting the information reconciliation procedure, as is illustrated in the following example.

EXAMPLE 10.23 Suppose that $X = (X_1, X_2)$ comprises 2 random bits, Y takes the value (X_1, e) or (e, X_2) at random, and $Z = X_1 \oplus X_2$, where e represents

[12] As long as we consider the class of one-way protocols, the secret key capacity is unchanged even if the eavesdropper only observes X via the erasure channel.

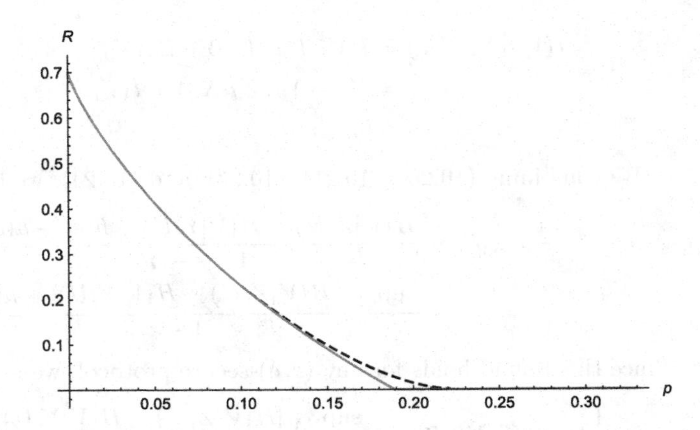

Figure 10.2 Rates obtained using Protocol 10.1 without any preprocessing (gray solid line) and with the noisy preprocessing (dashed line) in Example 10.22, for $q = 0.7$.

that bit is erased. In this case, we have $H(X|Z) - H(X|Y) = 0$. On the other hand, for $U = X_1 \oplus X_2$, which is already known to the eavesdropper, we have $H(X|Z,U) - H(X|Y,U) = 1$. Thus, the protocol that begins by sharing $X_1 \oplus X_2$ with $\mathcal{P}_2$ outperforms the one without any preprocessing.

Having characterized the secret key capacity for one-way protocols, we now move on to examine whether interactive protocols can help us generate longer keys. In the rest of this section, we present upper bounds for secret key capacity, without restricting to one-way protocols. The exact expression for secret key capacity is still unknown, but in the next section we will show how interaction can help in increasing secret key length.

We start with a simple upper bound for $C_{\varepsilon,\delta}(X,Y|Z)$.

THEOREM 10.24 *For $0 < \varepsilon, \delta < 1$ with $\varepsilon + \delta < 1$, we have*

$$C_{\varepsilon,\delta}(X,Y|Z) \leq \inf_{\bar{Z}} I(X \wedge Y|\bar{Z}), \tag{10.25}$$

where $\bar{Z}$ is an auxiliary random variable satisfying $\bar{Z} \multimap Z \multimap (X,Y)$. In particular,

$$C_{\varepsilon,\delta}(X,Y|Z) \leq I(X \wedge Y|Z). \tag{10.26}$$

Proof We first prove (10.26). By setting $\mathrm{Q}_{XYZ} = \mathrm{P}_{X|Z}\mathrm{P}_{Y|Z}\mathrm{P}_Z$, i.e., the conditionally independent distribution induced by P_{XYZ}, and by applying Theorem 10.10, we have

$$\frac{1}{n}S_{\varepsilon,\delta}(X^n,Y^n|Z^n) \leq -\frac{1}{n}\log\beta_{\varepsilon+\delta+\eta}(\mathrm{P}^n_{XYZ}, \mathrm{P}^n_{X|Z}\mathrm{P}^n_{Y|Z}\mathrm{P}^n_Z) + \frac{2\log(1/\eta)}{n} \tag{10.27}$$

for any $0 < \eta < 1 - \varepsilon - \delta$. Then, by Stein's lemma (Theorem 5.7), we get

$$\lim_{n\to\infty} -\frac{1}{n}\log\beta_{\varepsilon+\delta+\eta}(\mathrm{P}^n_{XYZ}, \mathrm{P}^n_{X|Z}\mathrm{P}^n_{Y|Z}\mathrm{P}^n_Z) = D(\mathrm{P}_{XYZ}\|\mathrm{P}_{X|Z}\mathrm{P}_{Y|Z}\mathrm{P}_Z)$$
$$= I(X \wedge Y|Z). \qquad (10.28)$$

Thus, by combining (10.27) and (10.28), we have (10.26).

To prove (10.25), fix an arbitrary $\bar{Z}$ satisfying the Markov chain condition. For any (ε, δ)-secure protocol generating (K_1, K_2) from (X^n, Y^n, Z^n) with transcript Π, since $\bar{Z}^n \multimap Z^n \multimap (K_1, \Pi)$, the data processing inequality for variational distance implies

$$d_{\mathtt{var}}\big(\mathrm{P}_{K_1\Pi\bar{Z}^n}, \mathrm{P}_{\mathtt{unif}} \times \mathrm{P}_{\Pi\bar{Z}^n}\big) \le d_{\mathtt{var}}\big(\mathrm{P}_{K_1\Pi Z^n}, \mathrm{P}_{\mathtt{unif}} \times \mathrm{P}_{\Pi Z^n}\big) \le \delta.$$

Thus, the same protocol is also (ε, δ)-secure even when the eavesdropper observes $\bar{Z}^n$ instead of Z^n, whereby

$$S_{\varepsilon,\delta}(X^n, Y^n|Z^n) \le S_{\varepsilon,\delta}(X^n, Y^n|\bar{Z}^n).$$

Consequently, (10.25) follows from (10.26) applied to $(X, Y, \bar{Z})$. □

In the proof of Theorem 10.24, we could have used the basic impossibility bound, Theorem 10.8, instead of the conditional independence testing bound, Theorem 10.10. However, in that case, we need to take the limit of $\varepsilon, \delta \to 0$. Such a statement is termed a *weak converse* in information theory. On the other hand, Theorem 10.24 holds for any $0 < \varepsilon, \delta < 1$ with $\varepsilon + \delta < 1$, and this kind of statement is termed a *strong converse.*

One might be curious whether introducing the auxiliary random variable $\bar{Z}$ provides a better upper bound. This is indeed the case, as seen in the simple example below.

EXAMPLE 10.25 Suppose that X and Y are independent uniform bits, and $Z = X \oplus Y$. In this case, we have $I(X \wedge Y|Z) = 1$. However, by taking $\bar{Z}$ to be constant, i.e., Z is discarded, we have $I(X \wedge Y|\bar{Z}) = 0$. We will encounter a nontrivial choice of $\bar{Z}$ later in Example 10.32.

As the final result for this section, we present a further strengthening of the previous upper bound.

THEOREM 10.26 *For a distribution* P_{XYZ}, *we have*

$$C(X, Y|Z) \le B(X, Y|Z) := \inf_U \big[I(X \wedge Y|U) + I(X, Y \wedge U|Z)\big],$$

where the infimum is taken over channels $\mathrm{P}_{U|XYZ}$ *with a finite*[13] *random variable* U.

Proof By Proposition 10.14 and Proposition 10.18, we have

$$S_{\varepsilon,\delta}(X^n, Y^n|Z^n) \le \frac{B(X^n, Y^n|Z^n) + h(\bar{\varepsilon}) + h(\bar{\delta})}{1 - \varepsilon - \delta}.$$

[13] While U can take finitely many values, its cardinality need not be bounded.

Thus, if we prove the following subadditivity

$$B(X^n, Y^n|Z^n) \le nB(X, Y|Z), \tag{10.29}$$

then the claim of the theorem is obtained by taking the limits $n \to \infty$ and $\varepsilon, \delta \to 0$, in that order. To prove (10.29), for any $\mathrm{P}_{U|XYZ}$, note that the product channel $\mathrm{P}^n_{U|XYZ}$ is a valid choice of channel in the definition of $B(X^n, Y^n|Z^n)$. Consequently, we have

$$\begin{aligned} &n\big[I(X \wedge Y|U) + I(X, Y \wedge U|Z)\big] \\ &\quad = I(X^n \wedge Y^n|U^n) + I(X^n, Y^n \wedge U^n|Z^n) \\ &\quad \ge B(X^n, Y^n|Z^n). \end{aligned}$$

Since this inequality holds for any $\mathrm{P}_{U|XYZ}$, we have (10.29), completing the proof. □

The upper bound in (10.25) is indeed a strict strengthening of that in Theorem 10.26, as seen in the example below.

EXAMPLE 10.27 For $\mathcal{X} = \mathcal{Y} = \{0, 1, 2, 3\}$ and $\mathcal{Z} = \{0, 1\}$, let P_{XY} be given by

$$\mathrm{P}_{XY} = \begin{bmatrix} \frac{1}{8} & \frac{1}{8} & 0 & 0 \\ \frac{1}{8} & \frac{1}{8} & 0 & 0 \\ 0 & 0 & \frac{1}{4} & 0 \\ 0 & 0 & 0 & \frac{1}{4} \end{bmatrix},$$

and

$$Z = \begin{cases} X + Y \pmod 2, & \text{if } X, Y \in \{0, 1\}, \\ X \pmod 2, & \text{if } X \in \{2, 3\}. \end{cases}$$

In this case, the upper bound (10.25) can be computed as follows. For arbitrary channel $\mathrm{P}_{\bar{Z}|Z}$ and for each $\bar{z} \in \bar{\mathcal{Z}}$, denote $p = \mathrm{P}_{\bar{Z}|Z}(\bar{z}|0)$ and $q = \mathrm{P}_{\bar{Z}|Z}(\bar{z}|1)$. Then, the conditional distribution $\mathrm{P}_{XY|\bar{Z}=\bar{z}}$ is given as

$$\mathrm{P}_{XY|\bar{Z}=\bar{z}} = \begin{bmatrix} \frac{p}{4(p+q)} & \frac{q}{4(p+q)} & 0 & 0 \\ \frac{q}{4(p+q)} & \frac{p}{4(p+q)} & 0 & 0 \\ 0 & 0 & \frac{p}{2(p+q)} & 0 \\ 0 & 0 & 0 & \frac{q}{2(p+q)} \end{bmatrix}.$$

Thus, we have

$$I(X \wedge Y|\bar{Z} = \bar{z}) = 1 + \frac{1}{2}\left(1 - h\left(\frac{p}{p+q}\right)\right) + \frac{1}{2}h\left(\frac{p}{p+q}\right) = \frac{3}{2},$$

and $I(X \wedge Y|\bar{Z}) = \sum_{\bar{z}} \mathrm{P}_{\bar{Z}}(\bar{z})\, I(X \wedge Y|\bar{Z} = \bar{z}) = \frac{3}{2}$. Since the channel $\mathrm{P}_{\bar{Z}|Z}$ is arbitrary, we have

$$\inf_{\bar{Z}} I(X \wedge Y|\bar{Z}) = \frac{3}{2}.$$

Now, let $K = \lfloor X/2 \rfloor$, which implies that $K = \lfloor Y/2 \rfloor$ as well. Further, let

$$U = \begin{cases} \mathtt{e}, & \text{if } K = 0, \\ Z, & \text{if } K = 1. \end{cases}$$

It is not difficult to verify that $I(X \wedge Y|U) = 0$. On the other hand, by noting that U is a function of K and Z, we have

$$I(X, Y \wedge U|Z) \le I(X, Y \wedge K|Z) \le H(K) = 1.$$

As a consequence, we have $B(X, Y|Z) \le 1$. In fact, K itself can be regarded as a 1-bit perfect secret key, whereby $C(X, Y|Z) = 1$.

10.5 Protocols with Interactive Communication

We obtained upper bounds for $C_{\varepsilon,\delta}(X, Y|Z)$ in the previous section, but the only protocols we have seen so far are one-way protocols. We now present interactive protocols that can outperform one-way protocols for specific distributions.

Consider a simple example first.

EXAMPLE 10.28 For $\mathcal{X} = \mathcal{Y} = \mathcal{Z} = \{0, 1, 2\}$, let $\mathrm{P}_{XYZ}(x, y, z) = 1/6$ for (x, y, z) with all distinct values and $\mathrm{P}_{XYZ}(x, y, z) = 0$ for other (x, y, z). In this case, since $\mathrm{P}_{XY} = \mathrm{P}_{XZ}$, it is not possible to generate any secret key using a one-way protocol; in fact, the eavesdropper can recover $\mathcal{P}_1$'s key with the same error probability as $\mathcal{P}_2$.

Next, consider the following protocol using interactive communication. $\mathcal{P}_1$ and $\mathcal{P}_2$ send $\Pi_1 = \mathbf{1}[X \in \{0, 1\}]$ and $\Pi_2 = \mathbf{1}[Y \in \{0, 1\}]$ to each other over the public channel. If $\Pi_1 = \Pi_2$, then $\mathcal{P}_1$ and $\mathcal{P}_2$ set $K_1 = X$ and $K_2 = Y \oplus 1$, respectively. Note that $K_1 = K_2$ under the condition $\Pi_1 = \Pi_2$, and the eavesdropper cannot learn the value of K_1 since $(X, Y) = (0, 1)$ and $(X, Y) = (1, 0)$ occur with the same probability conditioned on $Z = 2$. Since $\Pr(Z = 2) = 1/3$, a roughly $n/3$ bit secret key can be generated with $\delta = 0$ and arbitrarily small ε provided that the length n of the observations (X^n, Y^n, Z^n) is sufficiently large.

Now, let us consider a general interactive protocol. In the protocol, the parties iteratively apply the preprocessing and the information reconciliation steps used earlier in multiple rounds, and apply privacy amplification at the end.

More specifically, $\mathcal{P}_1$ first generates (U_1, V_1) from X, sends U_1 to $\mathcal{P}_2$, and sends the hashed (compressed) version of V_1 to $\mathcal{P}_2$. Upon recovering (U_1, V_1), $\mathcal{P}_2$ generates (U_2, V_2) from Y and (U_1, V_1), sends U_2 to $\mathcal{P}_1$, and sends the hashed version of V_2 to $\mathcal{P}_1$. The parties iterate these procedures r rounds, and eventually generate the secret key from $V^r = (V_1, \ldots, V_r)$.

In the ith round for odd i, $\mathcal{P}_2$ recovers V_i using the source coding with side-information for a given list $\{\mathcal{L}_{yu^iv^{i-1}}\}_{y \in \mathcal{Y}, u^i \in \mathcal{U}^i, v^{i-1} \in \mathcal{V}^{i-1}}$; and similarly for even i, where the roles of $\mathcal{P}_1$ and $\mathcal{P}_2$ are flipped. Without loss of generality, we assume

Protocol 10.2 Interactive secret key agreement protocol

Input: $\mathcal{P}_1$'s input X and $\mathcal{P}_2$'s input Y

Output: $\mathcal{P}_1$'s output $K_1 \in \{0,1\}^\ell$ and $\mathcal{P}_2$'s output $K_2 \in \{0,1\}^\ell$

1. For odd $1 \le i \le r$, the parties conduct Steps 2 and 3; for even $1 \le i \le r$, the parties conduct Steps 4 and 5.
2. $\mathcal{P}_1$ computes $(\tilde{U}_i, \tilde{V}_i) = \rho_i(X, \tilde{U}^{i-1}, \tilde{V}^{i-1})$, samples a UHF G_i from $\mathcal{V}_i$ to $\{0,1\}^{m_i}$, and sends $\Pi_i = (\tilde{U}_i, G_i, M_i)$ to $\mathcal{P}_2$, where $M_i = G_i(\tilde{V}_i)$.
3. Upon receiving Π_i, $\mathcal{P}_2$ sets $\hat{U}_i = \tilde{U}_i$, and looks for a unique $\hat{v}_i \in \mathcal{L}_{Y\hat{U}^i\hat{V}^{i-1}}$ such that $G_i(\hat{v}_i) = M_i$. If such $\hat{v}_i$ exists, then $\mathcal{P}_2$ sets $\hat{V}_i = \hat{v}_i$; otherwise, $\mathcal{P}_2$ sets $\hat{V}_i$ as a prescribed symbol in $\mathcal{V}_i$.
4. $\mathcal{P}_2$ computes $(\hat{U}_i, \hat{V}_i) = \rho_i(Y, \hat{U}^{i-1}, \hat{V}^{i-1})$, samples a UHF G_i from $\mathcal{V}_i$ to $\{0,1\}^{m_i}$, and sends $\Pi_i = (\hat{U}_i, G_i, M_i)$ to $\mathcal{P}_1$, where $M_i = G_i(\hat{V}_i)$.
5. Upon receiving Π_i, $\mathcal{P}_1$ sets $\tilde{U}_i = \hat{U}_i$, and looks for a unique $\tilde{v}_i \in \mathcal{L}_{X\tilde{U}^i\tilde{V}^{i-1}}$ such that $G_i(\tilde{v}_i) = M_i$. If such $\tilde{v}_i$ exists, then $\mathcal{P}_1$ sets $\tilde{V}_i = \tilde{v}_i$; otherwise, $\mathcal{P}_1$ sets $\tilde{V}_i$ as a prescribed symbol in $\mathcal{V}_i$.
6. $\mathcal{P}_1$ samples a UHF F from $\mathcal{V}_1 \times \cdots \times \mathcal{V}_r$ to $\{0,1\}^\ell$, and sends F to $\mathcal{P}_2$.
7. $\mathcal{P}_1$ and $\mathcal{P}_2$ output $K_1 = F(\tilde{V}^r)$ and $K_2 = F(\hat{V}^r)$.

that (U_i, V_i) are generated from X (or Y) and (U^{i-1}, V^{i-1}) by a deterministic function ρ_i; by including independent local randomness in X and Y, any stochastic preprocessing can be realized as deterministic functions (see Problem 10.4).

Since the recovery procedure in each round may cause a small probability of errors and those errors may propagate, we have to carefully keep track of generated random variables. We denote the random variables generated by $\mathcal{P}_1$ with a "tilde" and those generated by $\mathcal{P}_2$ with a "hat." The detail of the protocol is described in Protocol 10.2.

THEOREM 10.29 *For given $0 < \varepsilon, \delta < 1$, $\{\varepsilon_i\}_{i=1}^r$ with $\sum_{i=1}^r \varepsilon_i \le \varepsilon$ and for any fixed $0 < \xi_i \le \varepsilon_i$ and $0 < \eta \le \delta$, set*

$$m_i = \lceil H_{\max}^{\varepsilon_i - \xi_i}(V_i | Y, U^i, V^{i-1}) + \log(1/\xi_i) \rceil$$

for odd i,

$$m_i = \lceil H_{\max}^{\varepsilon_i - \xi_i}(V_i | X, U^i, V^{i-1}) + \log(1/\xi_i) \rceil$$

for even i, and

$$\ell = \left\lfloor H_{\min}^{(\delta-\eta)/2}(V^r | Z, U^r) - \sum_{i=1}^r m_i - \log(1/4\eta^2) \right\rfloor.$$

For odd $1 \le i \le r$, let $\{\mathcal{L}_{yu^iv^{i-1}}\}_{y\in\mathcal{Y}, u^i\in\mathcal{U}^i, v^{i-1}\in\mathcal{V}^{i-1}}$ be the list that attains $H_{\max}^{\varepsilon_i-\xi_i}(V_i|Y, U^i, V^{i-1})$; similarly for even $1 \le i \le r$. Then, Protocol 10.2 is an $(\varepsilon, \delta + 2\varepsilon)$-secure SK protocol of length ℓ.

Proof In the first round, $(\tilde{U}_1, \tilde{V}_1) = (U_1, V_1)$. Since the procedures of Steps 2 and 3 of Protocol 10.2 are the same as Construction 6.1, Lemma 6.9 guarantees that

$$\Pr(\hat{V}_1 \neq V_1) \leq \varepsilon_1,$$

which implies

$$\Pr((U_1, V_1) = (\tilde{U}_1, \tilde{V}_1) = (\hat{U}_1, \hat{V}_1)) \geq 1 - \varepsilon_1. \tag{10.30}$$

In order to analyze the error probability of the second round, introduce the event

$$E_1 = \{(U_1, V_1) = (\tilde{U}_1, \tilde{V}_1) = (\hat{U}_1, \hat{V}_1), (\hat{U}_2, \hat{V}_2) = (U_2, V_2), \tilde{U}_2 = U_2\}.$$

Since $(\hat{U}_1, \hat{V}_1) = (U_1, V_1)$ implies $(\hat{U}_2, \hat{V}_2) = (U_2, V_2)$ and $\tilde{U}_2 = U_2$, we have $\Pr(E_1) \geq 1 - \varepsilon_1$ from (10.30). Hypothetically, suppose that $\mathcal{P}_1$ receives $M_2' = G_2(V_2)$ instead of $M_2 = G_2(\hat{V}_2)$, and conduct the same recovery procedure to obtain $\tilde{V}_2'$. Then, since the procedures in this hypothetical second round are the same as Construction 6.1, Lemma 6.9 guarantees that

$$\Pr(\tilde{V}_2' \neq V_2) \leq \varepsilon_2.$$

Note that, under the event E_1, $G_2(\hat{V}_2) = G(V_2)$ and thus $\tilde{V}_2 = \tilde{V}_2'$. By noting these observations, we have

$$\begin{aligned}
\Pr((U^2, V^2) = (\tilde{U}^2, \tilde{V}^2) = (\hat{U}^2, \hat{V}^2)) &= \Pr(E_1, \tilde{V}_2' = V_2) \\
&= \Pr(E_1) - \Pr(E_1, \tilde{V}_2' \neq V_2) \\
&\geq \Pr(E_1) - \Pr(\tilde{V}_2' \neq V_2) \\
&\geq 1 - \varepsilon_1 - \varepsilon_2.
\end{aligned}$$

Repeating the analysis above for each round, we have

$$\Pr((U^r, V^r) = (\tilde{U}^r, \tilde{V}^r) = (\hat{U}^r, \hat{V}^r)) \geq 1 - \sum_{i=1}^{r} \varepsilon_i \geq 1 - \varepsilon, \tag{10.31}$$

which implies

$$\Pr(K_1 \neq K_2) \leq \varepsilon.$$

To bound the secrecy parameter, let Π' be such that $\tilde{U}_i = \hat{U}_i$ for $1 \leq i \leq r$ in $\Pi = (\Pi_1, \ldots, \Pi_r)$ are replaced by U_i, and let $K = F(V^r)$. Then, by applying Corollary 7.22 with X set to $(V_1, \ldots, V_r)$, V set to $(M_1, \ldots, M_r)$ and Z set to $(Z, U_1, \ldots, U_r, G_1, \ldots, G_r)$ and Lemma 7.23 to remove $(G_1, \ldots, G_r)$, we have

$$d_{\mathtt{var}}(\mathrm{P}_{K\Pi' ZF}, \mathrm{P}_{\mathtt{unif}} \times \mathrm{P}_{\Pi' Z} \times \mathrm{P}_F) \leq \delta.$$

Finally, since $(K_1, \Pi) = (K, \Pi')$ with probability at least $1 - \varepsilon$ by (10.31), we have

$$\begin{aligned} & d_{\mathtt{var}}(\mathrm{P}_{K_1 \Pi Z F}, \mathrm{P}_{\mathtt{unif}} \times \mathrm{P}_{\Pi Z} \times \mathrm{P}_F) \\ & \leq d_{\mathtt{var}}(\mathrm{P}_{K_1 \Pi Z F}, \mathrm{P}_{K \Pi' Z F}) + d_{\mathtt{var}}(\mathrm{P}_{K \Pi' Z F}, \mathrm{P}_{\mathtt{unif}} \times \mathrm{P}_{\Pi' Z} \times \mathrm{P}_F) + d_{\mathtt{var}}(\mathrm{P}_{\Pi Z}, \mathrm{P}_{\Pi' Z}) \\ & \leq 2\varepsilon + \delta, \end{aligned}$$

where we used the maximum coupling lemma (Lemma 2.31) in the last step – we used the fact that for two random variables W and W' on the same alphabet such that $\Pr(W \neq W') \leq \varepsilon$, their marginals must satisfy $d_{\mathtt{var}}(\mathrm{P}_W, \mathrm{P}_{W'}) \leq \varepsilon$. □

For a sequence of i.i.d. random variables, we can derive the following lower bound for the secret key capacity using Theorem 10.29.

THEOREM 10.30 *Let (U^r, V^r) be auxiliary random variables such that*

$$(U_i, V_i) -\!\!\circ\!\!- (X, U^{i-1}, V^{i-1}) -\!\!\circ\!\!- (Y, Z)$$

for odd i and

$$(U_i, V_i) -\!\!\circ\!\!- (Y, U^{i-1}, V^{i-1}) -\!\!\circ\!\!- (X, Z)$$

for even i. Then, we have

$$C(X, Y|Z) \geq H(V^r|Z, U^r) - \sum_{\substack{i=1 \\ i:\,odd}}^{r} H(V_i|Y, U^i, V^{i-1}) - \sum_{\substack{i=1 \\ i:\,even}}^{r} H(V_i|X, U^i, V^{i-1}).$$

Proof Apply Theorem 10.29 for a sequence of i.i.d. random variables

$$(X^n, Y^n, Z^n, U_1^n, \ldots, U_r^n, V_1^n, \ldots, V_r^n),$$

together with Theorem 6.15 and Theorem 7.25 in the same manner as the proof of Proposition 10.20. □

Theorem 10.30 is an abstraction of various protocols, and we can construct concrete protocols by choosing the auxiliary random variables appropriately. We illustrate specific bounds obtained using a few examples.

EXAMPLE 10.31 (Erasure correlation) Let S be the random bit on $\{0, 1\}$, and let X, Y, and Z be the output of independent erasure channels with input S, where the erasure probabilities are p_1, p_2, and q, respectively. Namely, each of X, Y, and Z is either equal to S (independently) or equal to an erasure symbol e. In this case, the simple upper bound (10.26) can be evaluated as $I(X \wedge Y|Z) = q(1 - p_1)(1 - p_2)$. In fact, this upper bound can be attained by the lower bound in Theorem 10.30. For $r = 3$, we set $U_1 = \mathbf{1}[X \neq \mathsf{e}]$, $V_1 = \bot$, $U_2 = \mathbf{1}[Y \neq \mathsf{e}]$, $V_2 = \bot$, $U_3 = \bot$, and

$$V_3 = \begin{cases} X, & \text{if } U_1 = U_2 = 1, \\ \bot, & \text{else,} \end{cases}$$

where $\bot$ means the constant random variable. In words, the parties tell each other if their observations are erasure or not, and then keep the observation only when both the observations are not erasure. Then, the lower bound becomes $H(V_3|Z, U_1, U_2) - H(V_3|Y, U_1, U_2) = q(1-p_1)(1-p_2)$. Note that, when $p_1 = p_2 = q$, the one-way secret key capacity is 0 since $\mathrm{P}_{XY} = \mathrm{P}_{XZ}$ (cf. Theorem 10.21).

Example 10.31 suggests that identifying erasure is one effective way of using interactive communication. When the discrepancy between X and Y is a bit-flip, an erasure-type event can be created by using the repetition code, as is illustrated in the following example.

EXAMPLE 10.32 (Advantage distillation) Consider the same distribution P_{XYZ} as in Example 10.22. For an integer $N \geq 2$ and $r = 3$, by regarding (X^N, Y^N, Z^N) as a single copy of correlated sources, we create the auxiliary random variables from (X^N, Y^N, Z^N) as follows: we set $U_1 = (X_i \oplus X_{i+1} : 1 \leq i \leq N-1)$, $V_1 = \perp$, $U_2 = \mathbf{1}[U_1 = (Y_i \oplus Y_{i+1} : 1 \leq i \leq N-1)]$, $V_2 = \perp$, $U_3 = \perp$, and

$$V_3 = \begin{cases} X_1, & \text{if } U_2 = 1, \\ \perp, & \text{else.} \end{cases}$$

For $N = 2$, $U_2 = 0$ means that the parities of (X_1, X_2) and (Y_1, Y_2) disagree, and the discrepancy $(X_1 \oplus Y_1, X_2 \oplus Y_2)$ is either $(0,1)$ or $(1,0)$ with probability 1/2, respectively. In that case, $H(X_1, X_2|Y_1, Y_2, U_1, U_2 = 0) = 1$ while $H(X_1, X_2|Z_1, Z_2, U_1, U_2 = 0) = q^2$, i.e., the correlation between $\mathcal{P}_1$ and $\mathcal{P}_2$ is worse than that between $\mathcal{P}_1$ and the eavesdropper. Thus, $U_2 = 0$ is regarded as an erasure-type event, and the parties create a secret key only when $U_2 = 1$. Similarly for $N \geq 3$, the choice of auxiliary random variables above seeks to "distill" the situation where the correlation between $\mathcal{P}_1$ and $\mathcal{P}_2$ is better than that between $\mathcal{P}_1$ and the eavesdropper.

For the above choice of auxiliary random variables, since $U_2 = 1$ with probability $(1-p)^N + p^N$ and V_3 is unknown to the eavesdropper only when Z^N are all erasure, the effective rate of the lower bound in Theorem 10.30 becomes

$$\begin{aligned} &\frac{1}{N}[H(V_3|Z^N, U_1, U_2) - H(V_3|Y^N, U_1, U_2)] \\ &= \frac{(1-p)^N + p^N}{N}\left[q^N - h\left(\frac{p^N}{(1-p)^N + p^N}\right)\right]. \end{aligned} \tag{10.32}$$

For $q = 0.7$, the rate (10.32) for $N = 2, 3, 4, 5$ is plotted in Figure 10.3. For large values of p, we can find that the rate (10.32) improves upon the rate of the one-way protocol in Example 10.22. We can also find that the threshold of positive rate become larger as N increases. In fact, for sufficiently large N so that $(p/(1-p))^N \leq 1/2$, note that

$$\begin{aligned} h\left(\frac{p^N}{(1-p)^N + p^N}\right) &\leq h\left(\left(\frac{p}{1-p}\right)^N\right) \\ &\leq -2\left(\frac{p}{1-p}\right)^N \cdot N \log\left(\frac{p}{1-p}\right), \end{aligned}$$

where the second inequality follows from $-a \log a \geq -(1-a)\log(1-a)$ for $0 \leq a \leq 1/2$. Thus, the rate (10.32) is positive for sufficiently large N as long as

$$q > \frac{p}{1-p}. \tag{10.33}$$

Next, by using the upper bound (10.25), we prove that the threshold (10.33) is tight in the sense that the secret key capacity is 0 whenever $q \leq \frac{p}{1-p}$. For $\bar{\mathcal{Z}} = \{0, 1, \mathsf{e}\}$, let

$$\begin{aligned}
\mathrm{P}_{\bar{Z}|Z}(\mathsf{e}|0,1) &= \mathrm{P}_{\bar{Z}|Z}(\mathsf{e}|1,0) = \mathrm{P}_{\bar{Z}|Z}(\mathsf{e}|\mathsf{e}) = 1, \\
\mathrm{P}_{\bar{Z}|Z}(0|0,0) &= \mathrm{P}_{\bar{Z}|Z}(1|1,1) = \frac{1-2p}{(1-q)(1-p)}, \\
\mathrm{P}_{\bar{Z}|Z}(\mathsf{e}|0,0) &= \mathrm{P}_{\bar{Z}|Z}(\mathsf{e}|1,1) = 1 - \frac{1-2p}{(1-q)(1-p)}.
\end{aligned}$$

Note that this is a valid channel whenever $q \leq \frac{p}{1-p}$. It is obvious that

$$I(X \wedge Y|\bar{Z} = 0) = I(X \wedge Y|\bar{Z} = 1) = 0.$$

Since

$$\mathrm{P}_{XY\bar{Z}}(0,1,\mathsf{e}) = \mathrm{P}_{XY\bar{Z}}(1,0,\mathsf{e}) = \frac{p}{2}$$

and

$$\begin{aligned}
\mathrm{P}_{XY\bar{Z}}(0,0,\mathsf{e}) &= \mathrm{P}_{XY\bar{Z}}(1,1,\mathsf{e}) \\
&= \frac{1-p}{2} \cdot q + \frac{1-p}{2} \cdot (1-q) \cdot \left(1 - \frac{1-2p}{(1-q)(1-p)}\right) \\
&= \frac{p}{2},
\end{aligned}$$

X and Y are independent given $\bar{Z} = \mathsf{e}$ and $I(X \wedge Y|\bar{Z} = \mathsf{e}) = 0$. Thus, we have $I(X \wedge Y|\bar{Z}) = 0$ and the secret key capacity is 0 whenever $q \leq \frac{p}{1-p}$.

The interactive protocol in Example 10.32 is effective when the value of p is large compared to q, i.e., the advantage of the parties over the eavesdropper is not sufficient; this is why this protocol is referred to as *advantage distillation.* However, when p is small, sending the parity without coding is not efficient. If a hashed version of the parity is sent so that a part of it is concealed from the eavesdropper, then it is possible to generate a secret key from the parity as well. Based on this observation, we can consider an alternative interactive protocol as follows.

EXAMPLE 10.33 (Hashed interactive communication) Consider the same distribution as in Examples 10.22 and 10.32. For $N = 2$ and $r = 3$, we create the auxiliary random variables from (X^2, Y^2, Z^2) as follows: we set $U_1 = \perp$, $V_1 = X_1 \oplus X_2$, $U_2 = \mathbf{1}[V_1 = Y_1 \oplus Y_2]$, $V_2 = \perp$, $U_3 = \perp$, and

$$V_3 = \begin{cases} X_1, & \text{if } U_2 = 1, \\ \perp, & \text{else.} \end{cases}$$

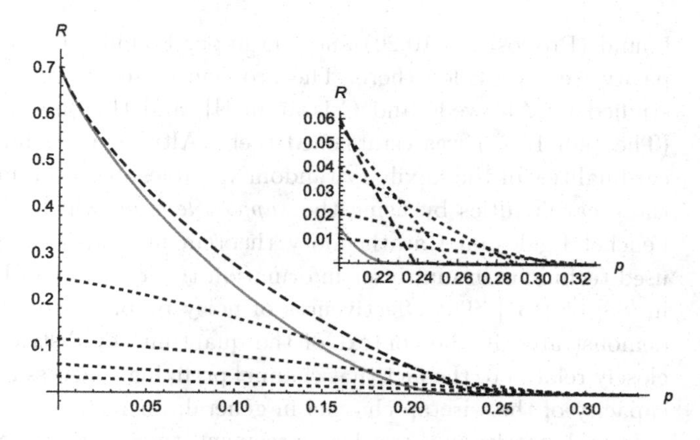

Figure 10.3 Key rates for the one-way protocol with noisy preprocessing (gray solid line) in Example 10.22, the advantage distillation (dotted lines) for $N = 2, 3, 4, 5$ in Example 10.32 and the hashed interactive communication (dashed line) in Example 10.33, for $q = 0.7$.

For these choices, of auxiliary random variables, note that $U_2 = 1$ with probability $(1-p)^2 + p^2$, V_1 is unknown to the eavesdropper when either of (Z_1, Z_2) is erased, and V_3 given V_1 is unknown to the eavesdropper when both (Z_1, Z_2) are erased. Thus, the effective rate of the lower bound in Theorem 10.30 becomes

$$\begin{aligned}
&\frac{1}{2}\left[H(V_1, V_3|Z^2, U_2) - H(V_1|Y^2) - H(V_3|Y^2, U_2)\right] \\
&\quad = \frac{1}{2}\left[H(V_1|Z^2, U_2) - H(V_1|Y^2) + H(V_3|Z^2, U_2, V_1) - H(V_3|Y^2, U_2)\right] \\
&\quad = \frac{1}{2}\Bigg[q^2 + 2(1-q)q - h((1-p)^2 + p^2) \\
&\qquad + ((1-p)^2 + p^2)\left(q^2 - h\left(\frac{p^2}{(1-p)^2 + p^2}\right)\right)\Bigg].
\end{aligned} \tag{10.34}$$

In Figure 10.3, for $q = 0.7$, the rate (10.34) is plotted and compared to the rates of the protocols in Examples 10.22 and 10.32. We can find that the rate (10.34) improves upon the rate of the one-way protocol even for small values of p.

10.6 References and Additional Reading

The idea of using public communication to generate a secret key was developed during the late 1980s to early 1990s, inspired by the quantum key distribution; see [28] and [237].[14] The information-theoretic formulation of the secret key agreement problem is due to Maurer [237]. In particular, the basic lower

[14] The utility of public feedback in the context of the wiretap channel was demonstrated in [221] earlier.

bound (Proposition 10.20) and the upper bound (10.26) on the secret key capacity were provided there. The problem of secret key agreement was further studied by Ahlswede and Csiszár in [1], and the one-way secret key capacity (Theorem 10.21) was established there. Although we did not comment on the cardinalities of the auxiliary random variables in Theorem 10.21, we can bound those cardinalities by using the *support lemma*, which is a consequence of the Fenchel–Eggleston–Carathéodory theorem in convex analysis; this is a widely used technique in multiuser information theory, and can be found, for instance, in [88] or [131]. The effectiveness of noisy preprocessing in Example 10.22 was demonstrated in the context of the quantum key distribution in [287]. This is closely related to the fact that channel prefixing is necessary to attain the secrecy capacity of the wiretap channel in general; for instance, see [268].

In early works on secret key agreement, the security of a secret key was evaluated using mutual information normalized by the block length of the independent correlated random variables, which is termed *weak secrecy*; for instance, see [1, 237]. In the past few decades, a more stringent notion of secrecy with unnormalized mutual information, termed *strong secrecy*, has become popular. In fact, it was proved in [242] that secret key capacities under both secrecy requirements are the same. In this chapter, we have evaluated the secrecy by using the variational distance since it is consistent with other problems treated in the book.

For general protocols with interactive communication, the secret key capacity is unknown in general. The upper bound in (10.25) is called the intrinsic information bound [241]. In certain interesting cases, such as Example 10.32, the intrinsic information bound is known to be tight in the sense that the secret key capacity is positive if and only if the intrinsic information bound is positive [241]. However, the intrinsic information bound is not tight in general, and an improved upper bound was introduced in [288]. A further improved upper bound was introduced in [136]; the upper bound in Theorem 10.26 is from [136]. The correlation in Example 10.27 is from [288].

The advantage distillation idea given in Example 10.32 was introduced in [237]; it uses interactive communication, but the preprocessing phase is done without any compression. The idea of using compression interactively was introduced in [136] and [345], independently; the latter work considered a protocol in the context of the quantum key distribution. Theorem 10.30 is a modification of the result in [136] so that the protocol in [345] can be derived from it. The protocol in Example 10.33 is from [345]. The tight secret key capacity result in Example 10.31 is from [253] (see also [138]). More recently, another interactive protocol was proposed in [139] to prove positivity of the secret key capacity for a broader class of correlations including the one in Example 10.32.

The treatment of the standard protocol in Section 10.2 is essentially from Renner's Ph.D. thesis [286]. In fact, a secret key agreement protocol in the context of quantum key distribution was studied in [286]; see also the references in Chapter 6 and Chapter 7.

The property of interactive communication in Lemma 10.7 first appeared in [1, 237]; the basic impossibility bound in Theorem 10.8 is also from [1, 237]. The conditional independence testing bound was shown in [332, 333]; this bound is tight enough to provide the strong converse as well as the second-order asymptotic for the degraded source [164]. The basic idea of the monotone in Section 10.3.3 was introduced in various places; for instance, see [56, 70, 136, 290]. The monotone approach is not only used in the secret key agreement, but is also relevant to other related problems, such as multiparty secure computation [278, 353] and entanglement distillation [110].

There are several important topics we could not cover in this chapter. First, we have assumed here that the public channel is authenticated. However, in practice, such a channel must be created by using an authentication scheme. Thus, it is interesting to consider the secret key agreement over a public unauthenticated channel. In such a problem, a part of the legitimate parties' observation is used to create the authenticated public channel; for instance, see [243, 244, 245].

Second, we have assumed that there is no rate constraint on the public channel. As we have mentioned in Section 10.1, this is a reasonable assumption in many applications. However, in certain applications, it is more realistic that the communication rate over the public channel is limited. For secret key agreement with rate limited communication, see [69, 89, 225, 331, 346].

Finally, we have only considered the two-party secret key agreement. However, there is a large body of work addressing multiparty secret key agreements; for instance, see [57, 58, 59, 60, 61, 62, 90, 91, 92, 136, 137, 334] and the monograph [258].

Problems

10.1 *Secret key from a card deck*: We consider a deck of n cards $\boxed{\spadesuit 1} \cdots \boxed{\spadesuit n}$. Suppose that a randomly chosen cards X are distributed to $\mathcal{P}_1$, b cards Y are distributed to $\mathcal{P}_2$, and c cards Z are distributed to an adversary without replacement, where $n = a + b + c$.

1, For $a = b = c = 1$, prove that it is not possible for $\mathcal{P}_1$ and $\mathcal{P}_2$ to agree on a 1-bit perfect secret key.

HINT Compute the mutual information between the parties' observations.

2. For $a = b = c = 2$, let us consider the following protocol [123]. $\mathcal{P}_1$ randomly picks a card $x_1 \in X$ and a card $y_1 \notin X$, and announces $\Pi_1 = \{x_1, y_1\}$; if $y_1 \in Y$, $\mathcal{P}_2$ announces acceptance of Π_1, and the parties agree on the key $K = \mathbf{1}[x_1 < y_1]$; otherwise, $\mathcal{P}_2$ randomly picks a card $y_2 \in Y$ and $x_2 \notin Y \cup \{x_1\}$, and announces $\Pi_2 = \{x_2, y_2\}$; if $x_2 \in X$, $\mathcal{P}_1$ announces acceptance of Π_2, and the parties agree on the key $K = \mathbf{1}[x_2 < y_2]$; otherwise, labeling (x_3, y_3) so that $\{x_3\} = X\backslash\{x_1\}$ and $\{y_3\} = Y\backslash\{y_2\}$, the parties agree on the key $K = \mathbf{1}[x_3 < y_3]$. Verify that the parties can agree on 1 bit of perfect secret key.

3. Generalize the protocol in 2 so that it works as long as $a + b \geq c + 2$.

10.2 *Degraded source*:
1. Prove that the secret key capacity is $C(X,Y|Z) = I(X \wedge Y|Z)$ when X, Y, and Z form a Markov chain in any order.
2. When X, Y, and Z form a Markov chain in this order,[15] prove that

$$C^{\rightarrow}(X,Y|Z) = C(X,Y|Z) = I(X \wedge Y) - I(X \wedge Z).$$

COMMENT This means that the secret key capacity can be attained by one-way protocols.

10.3 *Less noisy source*: When a pair of channels $P_{Y|X}$ and $P_{Z|X}$ satisfy $I(V \wedge Y) \geq I(V \wedge Z)$ for any $P_{VXYZ} = P_{VX}P_{Y|X}P_{Z|X}$, then $P_{Y|X}$ is defined to be *less noisy* than $P_{Z|X}$ [203].
1. Prove that $P_{Y|X}$ is less noisy than $P_{Z|X}$ if and only if the function $f(P_X) = I(X \wedge Y) - I(X \wedge Z)$ is a concave function of P_X [339].
2. Suppose that (X,Y,Z) is such that $P_{Y|X}$ is less noisy than $P_{Z|X}$. Prove that $C^{\rightarrow}(X,Y|Z) = I(X \wedge Y) - I(X \wedge Z)$.

HINT 1. Use the identities $I(X \wedge Y) = I(V \wedge Y) + I(X \wedge Y|V)$ and $I(X \wedge Z) = I(V \wedge Z) + I(X \wedge Z|V)$. 2. To eliminate U, use the identities $I(V \wedge Y) = I(U \wedge Y) + I(V \wedge Y|U)$ and $I(V \wedge Z) = I(U \wedge Z) + I(V \wedge Z|U)$, and the property of being less noisy; to eliminate V, use the property of being less noisy once more.

10.4 For correlated random variables (U,X) on $\mathcal{U} \times \mathcal{X}$, prove that there exist a random variable W on $\mathcal{W}$ and a function $f \colon \mathcal{X} \times \mathcal{W} \to \mathcal{U}$ such that W is independent of X and $(f(X,W),X)$ has the same distribution as (U,X).

HINT A naive way to construct W is to generate $W = (W_1, \ldots, W_{|\mathcal{X}|})$ on $\mathcal{U}^{|\mathcal{X}|}$ by $W_x \sim P_{U|X}(\cdot|x)$, and set $f(X,W) = W_x$. By using the conditional cumulative distribution function of U given $X = x$ in a more sophisticated manner, it is possible to construct W with $|\mathcal{W}| \leq |\mathcal{X}|(|\mathcal{U}| - 1) + 1$; this technique is known as the functional representation lemma (cf. [131, Appendix B]) or the random mapping representation (cf. [222, Section 1.2]).

[15] This case is referred to as *degraded source*.

Part II

Internal Adversary: Secure Computation

11 Secret Sharing

Secret sharing is a technique for storing data by dividing it into multiple parts, termed "shares." The shares must fulfill the following two requirements. First, even if some of the shares are lost, the original data can be restored from the remaining shares. This requirement is similar to that of resistance to "erasure errors" in coding theory. The second requirement is that of security, requiring that even if some of the shares are illegally accessed, the original data remains confidential. We will study various versions of this general problem in this chapter and provide secret sharing schemes based on algebraic constructions.

Secret sharing is an important primitive in its own right. Additionally, it is used as a building block for the secure computation considered in Chapters 12, 17, and 19. We present and analyze basic secret sharing schemes below. Also, we discuss how to prevent share holders from forging shares and connection to error correcting codes.

11.1 Threshold Schemes

The first class of secret sharing schemes we present are *threshold schemes* which ensure that the secret can be recovered only when at least more than a threshold number of shares are present, and no information about the secret can be ascertained from fewer than that threshold number of shares. Specifically, the secret data S taking values in a finite set $\mathcal{S}$ is divided into n shares so that S can be recovered if k out of n shares are collected and no information about S is leaked from any $(k-1)$ or less number of shares. Typically, the party creating the shares is termed the *dealer*, and the parties who keep the shares are termed *share holders*. Thus, a secret sharing scheme is a protocol involving $n+1$ parties.[1]

DEFINITION 11.1 (Secure threshold schemes) Let $\mathcal{N} = \{1, \ldots, n\}$ be the set of share holders. For $1 \le k \le n$, a (k, n)-threshold scheme consists of an encoding mapping $\varphi\colon \mathcal{S} \to \prod_{i \in \mathcal{N}} \mathcal{W}_i$ and reconstruction mappings $\psi_{\mathcal{I}}\colon \mathcal{W}_{\mathcal{I}} \to \mathcal{S}$, for every subset $\mathcal{I} \subseteq \mathcal{N}$ with $|\mathcal{I}| \ge k$, where $\mathcal{W}_{\mathcal{I}} = \prod_{i \in \mathcal{I}} \mathcal{W}_i$ for some finite alphabet $\mathcal{W}_i$, $i \in \mathcal{N}$. Both encoding and reconstruction mappings may be randomized.

[1] When we apply secret sharing to secure computation in the latter chapters, the dealer is one of the share holders.

A (k,n)-threshold scheme is called *secure* if the following conditions hold for every distribution P_S on $\mathcal{S}$:

$$\Pr\left(\psi_{\mathcal{I}}(W_{\mathcal{I}}) = S\right) = 1, \ \forall \mathcal{I} \subseteq \mathcal{N} \text{ such that } |\mathcal{I}| \geq k \quad \text{(reliability)} \tag{11.1}$$

where $W_{\mathcal{N}} = (W_1, \ldots, W_n) = \varphi(S)$ for secret data S and $W_{\mathcal{I}} = (W_i : i \in \mathcal{I})$ for a subset $\mathcal{I} \subseteq \mathcal{N}$; and

$$I(S \wedge W_{\mathcal{I}}) = 0, \ \forall \mathcal{I} \subseteq \mathcal{N} \text{ such that } |\mathcal{I}| < k \quad \text{(security)}. \tag{11.2}$$

EXAMPLE 11.2 For a single-bit secret data $S \in \{0,1\}$, let $W_1 \in \{0,1\}$ be a uniform bit generated independent of S, and let $W_2 = S \oplus W_1$. Then, the secret data S can be reconstructed from (W_1, W_2) as $S = W_1 \oplus W_2$. On the other hand, we have $H(S|W_1) = H(S|W_2) = H(S)$. Thus, this scheme is a secure $(2,2)$-threshold scheme. More generally, we can construct a secure (n,n)-threshold scheme by generating uniform bits $W_1, \ldots, W_{n-1}$ and setting $W_n = S \oplus W_1 \oplus \cdots \oplus W_{n-1}$. It can be verified that this scheme is secure.

We now present a general construction of (k,n)-threshold schemes. Our construction uses finite fields; we refer the reader to Section 4.1 for basic terminologies. The construction we present is based on the following basic fact from algebra.

LEMMA 11.3 (Lagrange interpolation) *Let $\mathbb{F}_q$ be the finite field of order q, and let $\alpha_1, \ldots, \alpha_k$ be distinct elements of $\mathbb{F}_q$. Then, for any given $w_1, \ldots, w_k \in \mathbb{F}_q$, there exists a unique polynomial of degree less than or equal to $k-1$ such that $f(\alpha_i) = w_i$ for every $1 \leq i \leq k$.*

Proof The existence follows from the Lagrange interpolation formula. Specifically, consider the following polynomial:

$$f(x) = \sum_{i=1}^{k} w_i \frac{(x-\alpha_1)\cdots(x-\alpha_{i-1})(x-\alpha_{i+1})\cdots(x-\alpha_k)}{(\alpha_i-\alpha_1)\cdots(\alpha_i-\alpha_{i-1})(\alpha_i-\alpha_{i+1})\cdots(\alpha_i-\alpha_k)}. \tag{11.3}$$

It can be readily verified that the polynomial above is of degree less than or equal to $k-1$ and satisfies $f(\alpha_i) = w_i$ for every $1 \leq i \leq k$.

To see uniqueness, consider two polynomials $f(x)$ and $g(x)$ of degree less than or equal to $k-1$ such that $f(\alpha_i) = g(\alpha_i) = w_i$ for every $1 \leq i \leq k$. Then, $h(x) = f(x) - g(x)$ is a polynomial of degree less than or equal to $k-1$ and $h(\alpha_i) = 0$, for every $1 \leq i \leq k$. Since a nonzero polynomial of degree less than or equal to $k-1$ can have at most $k-1$ roots, $h(x)$ must equal 0 for all x, which implies $f(x) = g(x)$. □

The scheme we present is based on polynomials on finite fields and uses the interpolation formula in the previous proof. Let $\alpha_1, \ldots, \alpha_n$ be distinct nonzero elements from $\mathbb{F}_q$, and let

$$\mathbb{P}_{k-1}(s) = \left\{ s + a_1 x + \cdots + a_{k-1} x^{k-1} : a_1, \ldots, a_{k-1} \in \mathbb{F}_q \right\}$$

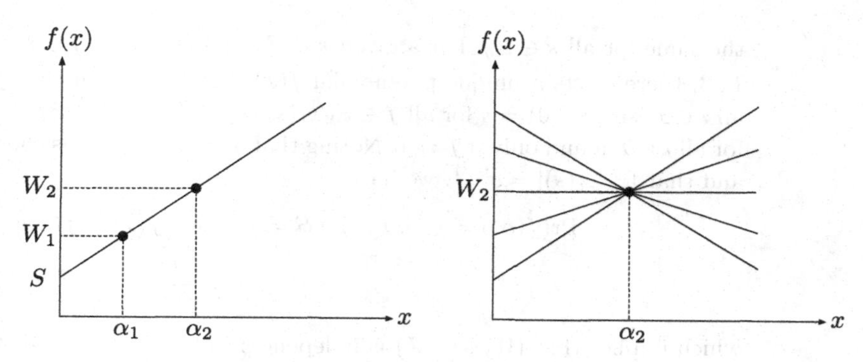

Figure 11.1 Illustration of the $(2, n)$-threshold scheme of Protocol 11.1. Given two points, the degree 1 polynomial passing through them is uniquely determined (left plot); on the other hand, there are multiple degree 1 polynomials that pass through the point (α_2, W_2) (right plot).

Protocol 11.1 A (k, n)-threshold scheme

---

1. **Distribution phase:** The dealer uniformly generates $f(x)$ from $\mathbb{P}_{k-1}(S)$ and sets the share of the ith share holder $\mathcal{P}_i$ as $W_i = f(\alpha_i)$ for $i \in \mathcal{N}$.
2. **Reconstruction phase:** A subset $(\mathcal{P}_i : i \in \mathcal{I})$ of share holders with $|\mathcal{I}| \geq k$ combine their shares and reconstruct the polynomial f using the shares $(W_i : i \in \mathcal{I})$ and applying the Lagrange interpolation formula (11.3) with $f(\alpha_i) = W_i$, $i \in \mathcal{I}$. The share holders in $\mathcal{I}$ recover $S = f(0)$.

---

be the set of all polynomials of degree less than or equal to $k - 1$ such that $f(0) = s$.

The basic idea of the scheme is to encode the secret data S into a randomly chosen polynomial $f(x)$ from $\mathbb{P}_{k-1}(S)$ and give away the $f(\alpha_i)$ as shares. That is, the secret is encoded as $f(0)$ and the shares are obtained as $W_i = f(\alpha_i)$, $i \in \mathcal{N}$. Then, when more than k shares are collected together, the polynomial can be uniquely reconstructed by the Lagrange interpolation formula. On the other hand, from $(k-1)$ shares, there are multiple polynomials that are compatible with the collected shares, which will be seen to guarantee secrecy (cf. Figure 11.1). The details of the scheme are described in Protocol 11.1.

THEOREM 11.4 *The scheme in Protocol 11.1 is a secure (k, n)-threshold scheme.*

Proof Since the polynomial $f(x)$ chosen by the dealer is of degree at most $k-1$, it can be recovered from k points $(\alpha_{i_1}, f(\alpha_{i_1})), \ldots, (\alpha_{i_k}, f(\alpha_{i_k}))$ by using the Lagrange interpolation formula (cf. Lemma 11.3), which shows that the reliability condition is satisfied.

To establish the security condition (11.2), for clarity, we first consider the special case of $|\mathcal{I}| = k - 1$. Note that, to establish the independence of $W_{\mathcal{I}}$ and S, it suffices to show that the distribution of $W_{\mathcal{I}}$ conditioned on $S = s$ remains the same for all $s \in \mathbb{F}_q$. For a given $s \in \mathbb{F}_q$ and $(y_i : i \in \mathcal{I}) \in \mathbb{F}_q^{k-1}$, by Lemma 11.3, there exists a unique polynomial $\tilde{f}(x) \in \mathbb{P}_{k-1}(s)$ such that $\tilde{f}(\alpha_i) = y_i$ for all $i \in \mathcal{I}$, since $f(0) = s$ for all $f \in \mathbb{P}_{k-1}(s)$. Namely, for $f \in \mathbb{P}_{k-1}(s)$, $f(\alpha_i) = y_i$ for all $i \in \mathcal{I}$ if and only if $f = \tilde{f}$. Noting that $f(x) \in \mathbb{P}_{k-1}(s)$ is chosen at random and that $|\mathbb{P}_{k-1}(s)| = q^{k-1}$, we have

$$\begin{aligned}\Pr(f(\alpha_i) = y_i, \forall i \in \mathcal{I} \mid S = s) &= \Pr\Big(f(x) = \tilde{f}(x)\Big) \\ &= \frac{1}{q^{k-1}},\end{aligned}$$

which implies that $(W_i : i \in \mathcal{I})$ is independent of S.

For general $|\mathcal{I}| \le k-1$, let $\mathbb{P}_{k-1}(s, y_i : i \in \mathcal{I}) \subset \mathbb{P}_{k-1}(s)$ be the set of polynomials $\tilde{f}(x)$ such that $\tilde{f}(0) = s$ and $\tilde{f}(x) = y_i$ for every $i \in \mathcal{I}$. As above, $f \in \mathbb{P}_{k-1}(s, y_i : i \in \mathcal{I})$ if and only if $f(0) = s$ and $f(\alpha_i) = y_i$ for all $i \in \mathcal{I}$. Let $\mathcal{J} \subseteq \mathcal{N}\backslash\mathcal{I}$ be an arbitrarily fixed subset with $|\mathcal{J}| = k - 1 - |\mathcal{I}|$. For given $(w_j : j \in \mathcal{J}) \in \mathbb{F}_q^{|\mathcal{J}|}$, by Lemma 11.3, there exists a unique polynomial $\tilde{f}(x) \in \mathbb{P}_{k-1}(s, y_i : i \in \mathcal{I})$ such that $\tilde{f}(\alpha_j) = w_j$ for every $j \in \mathcal{J}$. Thus, we have

$$|\mathbb{P}_{k-1}(s, y_i : i \in \mathcal{I})| = q^{|\mathcal{J}|} = q^{k-1-|\mathcal{I}|}.$$

Since $f(x) \in \mathbb{P}_{k-1}(s)$ is chosen at random, we have

$$\begin{aligned}\Pr(f(\alpha_i) = y_i, \forall i \in \mathcal{I} \mid S = s) &= \Pr(f(x) \in \mathbb{P}_{k-1}(s, y_i : i \in \mathcal{I})) \\ &= \frac{q^{k-1-|\mathcal{I}|}}{q^{k-1}} \\ &= \frac{1}{q^{|\mathcal{I}|}}\end{aligned}$$

for every $(y_i : i \in \mathcal{I}) \in \mathbb{F}_q^{|\mathcal{I}|}$, which implies that $(W_i : i \in \mathcal{I})$ is independent of S. □

As an illustration, we provide the simple example below.

EXAMPLE 11.5 Let $q = 5$, $n = 4$, $k = 3$, and $\alpha_i = i$ for $1 \le i \le 4$. For the secret data $S = 3$, if the polynomial $f(x) = 3 + 2x + 4x^2 \in \mathbb{P}_2(3)$ is picked, then the shares are given by $W_1 = 4$, $W_2 = 3$, $W_3 = 0$, $W_4 = 0$. When any three shares, say (W_1, W_2, W_4), are collected, then the polynomial can be reconstructed by the Lagrange interpolation formula as follows:

$$\begin{aligned}f(x) &= 4 \cdot \frac{(x-2)(x-4)}{(1-2)(1-4)} + 3 \cdot \frac{(x-1)(x-4)}{(2-1)(2-4)} + 0 \cdot \frac{(x-1)(x-2)}{(4-1)(4-2)} \\ &= 3 + 2x + 4x^2,\end{aligned}$$

from which the secret data $S = 3$ can be recovered. On the other hand, when any two shares are collected, the secrecy is guaranteed since there are five polynomials of degree 2 that pass through (α_{i_1}, W_{i_1}) and (α_{i_2}, W_{i_2}).

In Protocol 11.1, the size of each share is the same as the size of the original secret data S, i.e., $\log q$ bits, whereby the size of all the shares collectively is

$n \log q$. From a practical point of view, it is desirable that the size of the shares is as small as possible. In fact, it can be shown that Protocol 11.1 attains the optimal share size in the following sense.

PROPOSITION 11.6 *For any secure (k, n)-threshold scheme with shares $(W_i : i \in \mathcal{N})$ and secret S with distribution P_S, we must have for every $i \in \mathcal{N}$ that*

$$H(W_i) \geq H(S).$$

Proof For any $i \in \mathcal{N}$ and any subset $\mathcal{I} \subseteq \mathcal{N}$ such that $i \notin \mathcal{I}$ and $|\mathcal{I}| = k - 1$, the reproduction condition (11.1) implies

$$H(S|W_i, W_{\mathcal{I}}) = 0.$$

By subtracting this from the secrecy condition (11.2), we have

$$\begin{aligned} H(S) &= H(S|W_{\mathcal{I}}) - H(S|W_i, W_{\mathcal{I}}) \\ &= I(S \wedge W_i|W_{\mathcal{I}}) \\ &\leq H(W_i), \end{aligned}$$

which completes the proof. □

11.2 Ramp Schemes

As we have shown in Proposition 11.6, the size of each share in any secure threshold scheme must be as large as the size of the secret. This drawback can be circumvented by relaxing the security requirements. One such relaxation is allowing for a linear ramp-up of the leaked information as the number of colluding share holders increases after a threshold number. The schemes which satisfy such a requirement are termed *ramp secret sharing* schemes, or simply ramp schemes.

Specifically, in a ramp scheme, if k or more shares are collected, then the secret information can be recovered. When the number of collected shares t satisfies $k - \ell < t \leq k - 1$ for some parameter $1 \leq \ell \leq k$, the secret information is partially revealed depending on the number of collected shares. When the number of collected shares is less than or equal to $k - \ell$, the secret information is completely concealed. As we will see, the advantage of a ramp scheme is that the size of each share can be $H(S)/\ell$, which is smaller than the $H(S)$ requirement for threshold schemes.

DEFINITION 11.7 (Secure ramp schemes) For a given integer n, k with $1 \leq k \leq n$, and ℓ with $1 \leq \ell \leq k$, a (k, ℓ, n)-ramp scheme consists of an encoder mapping $\varphi \colon \mathcal{S}^\ell \to \prod_{i \in \mathcal{N}} \mathcal{W}_i$ and reconstruction mappings $\psi_{\mathcal{I}} \colon \mathcal{W}_{\mathcal{I}} \to \mathcal{S}^\ell$ for every subset $\mathcal{I} \subseteq \mathcal{N}$ with $|\mathcal{I}| \geq k$.

A (k, ℓ, n)-ramp scheme is called *secure* if the following requirements are satisfied for all distribution $P_{S_{\mathcal{L}}}$:

$$\Pr(\psi_{\mathcal{I}}(W_{\mathcal{I}}) = S_{\mathcal{L}}) = 1, \forall \mathcal{I} \subseteq \mathcal{N} \text{ such that } |\mathcal{I}| \geq k \quad \text{(reliability)} \qquad (11.4)$$

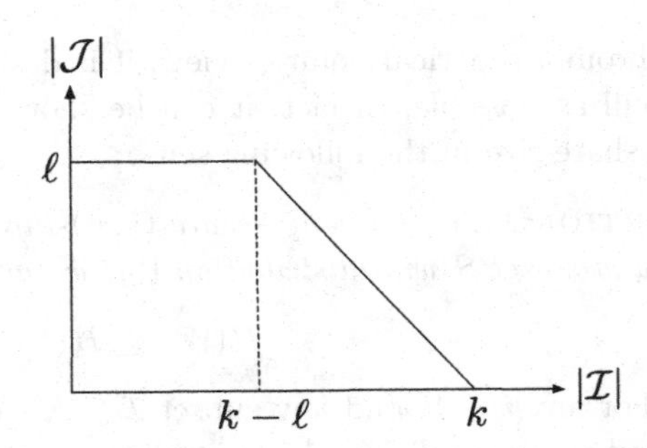

Figure 11.2 The relation between the number $|\mathcal{I}|$ of leaked shares $W_{\mathcal{I}}$ and the number $|\mathcal{J}|$ of concealed part $S_{\mathcal{J}}$ of the secret datas in a (k, ℓ, n)-ramp scheme.

where $W_{\mathcal{N}} = (W_1, \ldots, W_n) = \varphi(S_{\mathcal{L}})$ for secret data $S_{\mathcal{L}} = (S_1, \ldots, S_\ell)$, $\mathcal{L} = \{1, \ldots, \ell\}$, and $W_{\mathcal{I}} = (W_i : i \in \mathcal{I})$ for a subset $\mathcal{I} \subseteq \mathcal{N}$;

$$I(S_{\mathcal{L}} \wedge W_{\mathcal{I}}) = 0, \ \forall \mathcal{I} \subseteq \mathcal{N} \text{ such that } |\mathcal{I}| \le k - \ell; \quad \text{(security 1)}; \tag{11.5}$$

and for uniformly distributed $S_{\mathcal{L}}$,

$$I(S_{\mathcal{J}} \wedge W_{\mathcal{I}}) = 0, \ \forall \mathcal{I} \subseteq \mathcal{N}, \mathcal{J} \subseteq \mathcal{L} \text{ such that } k - \ell < |\mathcal{I}| < k, |\mathcal{J}| = k - |\mathcal{I}| \quad \text{(security 2)}. \tag{11.6}$$

The secrecy constraint (11.5) says that, when the number of leaked shares is less than $k - \ell$, then the entire secret data is concealed. On the other hand, the secrecy constraint (11.6) says that, when the number of leaked shares is between $k-\ell$ and k, a part $S_{\mathcal{J}}$ of the secret data is concealed. The size $|\mathcal{J}|$ of the concealed part decreases as the size $|\mathcal{I}|$ of leaked shares increases (cf. Fig. 11.2). This definition allows only uniform distribution for the "ramp-up" leakage condition (11.6). This is perhaps a limitation of the definition.

Remark 11.1 The secrecy requirement in (11.6) is stronger than the following alternative secrecy requirement for uniformly distributed $S_{\mathcal{L}}$:

$$H(S_{\mathcal{L}}|W_{\mathcal{I}}) = \frac{k - |\mathcal{I}|}{\ell} H(S_{\mathcal{L}}), \ \forall \mathcal{I} \subseteq \mathcal{N} \text{ such that } k - \ell < |\mathcal{I}| < k. \tag{11.7}$$

For instance, when $|\mathcal{I}| = k - \ell + 1$ shares are collected, (11.7) can be satisfied even if one of the S_j is completely revealed and other secrets $(S_i : i \in \mathcal{L}\backslash\{j\})$ are concealed. Such a situation can be a serious security compromise in some applications (see Problem 11.1). On the other hand, (11.6) requires that any subset $S_{\mathcal{J}}$ of size $|\mathcal{J}| = k - |\mathcal{I}|$ is concealed. The requirement (11.6) is sometimes referred to as the *strong security* requirement in the literature.

A (k, ℓ, n)-ramp scheme can also be constructed by using polynomials on $\mathbb{F}_q$, using a small extension of the threshold scheme we saw earlier. For $q \ge n + \ell$, let $\alpha_1, \ldots, \alpha_{n+\ell}$ be distinct elements of $\mathbb{F}_q$. Further, for given $s_1, \ldots, s_\ell \in \mathbb{F}_q$, let $\mathbb{P}_{k-1}(s_1, \ldots, s_\ell)$ be the set of all polynomials of degree at most $k - 1$ such that $f(\alpha_{n+j}) = s_j$ for $j = 1, \ldots, \ell$. The main difference from the threshold scheme is

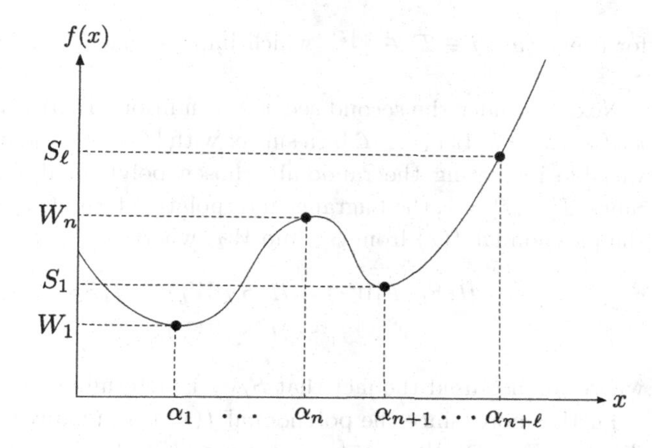

Figure 11.3 An illustration of the (k, ℓ, n)-ramp scheme of Protocol 11.2. The secret data $S_1, \ldots, S_\ell$ is embedded into ℓ points $f(\alpha_{n+1}), \ldots, f(\alpha_{n+\ell})$.

Protocol 11.2 (k, ℓ, n)-ramp scheme

1. **Distribution phase:** The dealer randomly chooses $f(x) \in \mathbb{P}_{k-1}(S_1, \ldots, S_\ell)$ and sets the share of the ith share holder $\mathcal{P}_i$ as $W_i = f(\alpha_i)$.
2. **Reconstruction phase:** A subset $(\mathcal{P}_i : i \in \mathcal{I})$ of share holders with $|\mathcal{I}| \geq k$ combine their shares and reconstruct the polynomial f using the shares $(W_i : i \in \mathcal{I})$ and applying the Lagrange interpolation formula (11.3). The share holders in $\mathcal{I}$ recover the secret as $S_\mathcal{L} = (f(\alpha_{n+1}), \ldots, f(\alpha_{n+\ell}))$.

---

that, instead of embedding the secret in a single point as $f(0)$, ℓ units of the secret are embedded into ℓ different points $f(\alpha_{n+1}), \ldots, f(\alpha_{n+\ell})$ (cf. Figure 11.3). The details of the construction are described in Protocol 11.2.

THEOREM 11.8 *The scheme in Protocol 11.2 is a secure (k, ℓ, n)-ramp scheme.*

Proof The proof of reliability follows exactly in the same manner as for the threshold scheme of the previous section. The first security condition (11.5) for $|\mathcal{I}| \leq k - \ell$ follows in a similar manner to the secrecy for Protocol 11.1. Specifically, note first that $|\mathbb{P}_{k-1}(s_\mathcal{L})| = q^{k-\ell}$. Further, let $\mathbb{P}_{k-1}(s_\mathcal{L}, y_i : i \in \mathcal{I}) \subset \mathbb{P}_{k-1}(s_\mathcal{L})$ be the set of polynomials $\tilde{f}(x)$ such that $\tilde{f}(\alpha_{n+j}) = s_j$ for $j \in \mathcal{L}$ and $\tilde{f}(\alpha_i) = y_i$ for $i \in \mathcal{I}$. Then, as in the proof of Theorem 11.4, we can verify that $|\mathbb{P}_{k-1}(s_\mathcal{L}, y_i : i \in \mathcal{I})| = q^{k-\ell-|\mathcal{I}|}$. Since $f(x) \in \mathbb{P}_{k-1}(s_\mathcal{L})$ is chosen at random, we have

$$\Pr\left((f(\alpha_i) : i \in \mathcal{I}) = (y_i : i \in \mathcal{I}) | S_\mathcal{L} = s_\mathcal{L}\right) = \Pr\left(f(x) \in \mathbb{P}_{k-1}(s_\mathcal{L}, y_i : i \in \mathcal{I})\right)$$
$$= \frac{1}{q^{|\mathcal{I}|}},$$

for every $(y_i : i \in \mathcal{I}) \in \mathbb{F}^{|\mathcal{I}|}$, which implies that $(W_i : i \in \mathcal{I})$ is independent of $S_{\mathcal{L}}$.

Next, consider the second security condition (11.5) which is for the case with $k-\ell < |\mathcal{I}| < k$. Let $\mathcal{J} \subset \mathcal{L}$ be a subset with $|\mathcal{I}|+|\mathcal{J}| = k$, and let F be the random variable indicating the randomly chosen polynomial $f(x) \in \mathbb{P}_{k-1}(S_1, \ldots, S_\ell)$. Since $|\mathcal{I}|+|\mathcal{J}| = k$, the Lagrange interpolation formula can be used to reconstruct the polynomial $f(x)$ from $S_{\mathcal{J}}$ and $W_{\mathcal{I}}$, whereby

$$\begin{aligned} H(S_{\mathcal{L}}, F|W_{\mathcal{I}}) &= H(S_{\mathcal{J}}|W_{\mathcal{I}}) + H(S_{\mathcal{L}\backslash\mathcal{J}}, F|S_{\mathcal{J}}, W_{\mathcal{I}}) \\ &= H(S_{\mathcal{J}}|W_{\mathcal{I}}), \end{aligned} \tag{11.8}$$

where we also used the fact that $S_{\mathcal{L}\backslash\mathcal{J}}$ is determined from the polynomial.

Furthermore, since the polynomial $f(x)$ is randomly chosen from a set of cardinality $|\mathbb{P}_{k-1}(S_{\mathcal{L}})| = q^{k-\ell}$, we have $H(F|S_{\mathcal{L}}) = (k-\ell)\log q$. Thus, by noting that $S_{\mathcal{L}}$ is uniform, we have

$$\begin{aligned} H(S_{\mathcal{L}}, F|W_{\mathcal{I}}) &= H(S_{\mathcal{L}}, F, W_{\mathcal{I}}) - H(W_{\mathcal{I}}) \\ &= H(S_{\mathcal{L}}, F) - H(W_{\mathcal{I}}) \\ &= H(S_{\mathcal{J}}) + H(S_{\mathcal{L}\backslash\mathcal{J}}|S_{\mathcal{J}}) + (k-\ell)\log q - H(W_{\mathcal{I}}) \\ &\geq H(S_{\mathcal{J}}) + H(S_{\mathcal{L}\backslash\mathcal{J}}|S_{\mathcal{J}}) - (\ell - |\mathcal{J}|)\log q \\ &= H(S_{\mathcal{J}}), \end{aligned}$$

where the second identity holds since $W_{\mathcal{I}}$ can be determined from F and the inequality follows from $H(W_{\mathcal{I}}) \leq |\mathcal{I}|\log q = (k - |\mathcal{J}|)\log q$. The claim follows by the previous bound combined with (11.8). □

Remark 11.2 As we can see from the proof of Theorem 11.8 for $k-\ell < |\mathcal{I}| < k$, even if $S_{\mathcal{L}}$ is not uniform, the secret $S_{\mathcal{L}}$ is partially concealed in the following sense:

$$H(S_{\mathcal{L}}, F|W_{\mathcal{I}}) = H(S_{\mathcal{L}}|W_{\mathcal{I}}) \geq H(S_{\mathcal{L}}) - (|\mathcal{I}| - (k-\ell))\log q,$$

i.e., $I(S_{\mathcal{L}} \wedge W_{\mathcal{I}}) \leq (|\mathcal{I}| - (k-\ell))\log q$.

11.3 Secret Sharing Schemes with a General Access Structure

In secure threshold schemes defined earlier, the recovery of the secret only depends on the number of shares collected – it does not matter which shares are collected. In some applications, it is desirable that the recovery can be decided more flexibly depending on the identity of the collected shares. For instance, we might deem certain share holders superior to others, and allow recovery of the secret with fewer shares when shares of superior share holders are included. Such secret sharing schemes are termed *secret sharing schemes with a general access structure.*

DEFINITION 11.9 (Secure general access structure scheme) Given a partition $2^{\mathcal{N}} = \mathscr{A}_1 \cup \mathscr{A}_0$ of the power set of $\mathcal{N} = \{1, \ldots, n\}$, an $(\mathscr{A}_1, \mathscr{A}_0)$-access structure scheme consists of an encoder mapping $\varphi \colon \mathcal{S} \to \prod_{i \in \mathcal{N}} \mathcal{W}_i$ and reconstruction mappings $\psi_{\mathcal{I}} \colon \mathcal{W}_{\mathcal{I}} \to \mathcal{S}$ for every $\mathcal{I} \in \mathscr{A}_1$.

An $(\mathscr{A}_1, \mathscr{A}_0)$-access structure scheme is called *secure* if the following requirements are satisfied for every distribution P_S of $\mathcal{S}$:

$$\Pr(\psi_{\mathcal{I}}(W_{\mathcal{I}}) = S) = 1, \ \forall \mathcal{I} \in \mathscr{A}_1 \quad \text{(reliability)} \tag{11.9}$$

where $W_{\mathcal{N}} = (W_1, \ldots, W_n) = \varphi(S)$ for the secret S and $W_{\mathcal{I}} = (W_i : i \in \mathcal{I})$ for a subset $\mathcal{I} \subseteq \mathcal{N}$; on the other hand, the secrecy constraint is given by

$$I(S \wedge W_{\mathcal{I}}) = 0, \ \forall \mathcal{I} \in \mathscr{A}_0 \quad \text{(security)}. \tag{11.10}$$

The sets $\mathscr{A}_1$ and $\mathscr{A}_0$ are called the *qualified family* and *forbidden family*, respectively.

From (11.9) and (11.10), it can be readily verified that a secure $(\mathscr{A}_1, \mathscr{A}_0)$-access structure scheme exists only if the qualified family and the forbidden family satisfy the following monotonicity conditions:[2]

$$\begin{aligned} &\mathcal{I} \subset \mathcal{I}' \text{ and } \mathcal{I} \in \mathscr{A}_1 \Longrightarrow \mathcal{I}' \in \mathscr{A}_1, \\ &\mathcal{I}' \subset \mathcal{I} \text{ and } \mathcal{I} \in \mathscr{A}_0 \Longrightarrow \mathcal{I}' \in \mathscr{A}_0. \end{aligned} \tag{11.11}$$

By monotonicity, the qualified family and the forbidden family, respectively, can be specified by the minimal sets and the maximal sets of the families:

$$\begin{aligned} \mathscr{A}_1^- &:= \{\mathcal{I} \in \mathscr{A}_1 : \forall \mathcal{I}' \text{ such that } \mathcal{I}' \subsetneq \mathcal{I}, \ \mathcal{I}' \notin \mathscr{A}_1\}, \\ \mathscr{A}_0^+ &:= \{\mathcal{I} \in \mathscr{A}_0 : \forall \mathcal{I}' \text{ such that } \mathcal{I} \subsetneq \mathcal{I}', \ \mathcal{I}' \notin \mathscr{A}_0\}. \end{aligned}$$

In fact, it suffices to construct a scheme so that the secret can be reconstructed from $W_{\mathcal{I}}$ for every $\mathcal{I} \in \mathscr{A}_1^-$ and concealed from $W_{\mathcal{I}}$ for every $\mathcal{I} \in \mathscr{A}_0^+$.

EXAMPLE 11.10 When the access structure is given by

$$\mathscr{A}_1 = \{\mathcal{I} \subseteq \mathcal{N} : |\mathcal{I}| \geq k\} \text{ and } \mathscr{A}_0 = \{\mathcal{I} \subseteq \mathcal{N} : |\mathcal{I}| < k\},$$

for some integer $1 \leq k \leq n$, then a secure $(\mathscr{A}_1, \mathscr{A}_0)$-access structure scheme is a secure (k, n)-threshold scheme.

EXAMPLE 11.11 For $n = 4$, consider the access structure $(\mathscr{A}_1, \mathscr{A}_0)$ specified by

$$\begin{aligned} \mathscr{A}_1^- &= \{\{1,2,3\}, \{1,4\}, \{2,4\}, \{3,4\}\}, \\ \mathscr{A}_0^+ &= \{\{1,2\}, \{1,3\}, \{2,3\}, \{4\}\}. \end{aligned}$$

In words, the fourth share holder is a superior share holder and can reconstruct the secret together with any one other share holder. Further, all the other three share holders need to be together in order to reconstruct the secret without the fourth share holder.

[2] In fact, the two conditions in (11.11) imply each other since $\mathscr{A}_1$ and $\mathscr{A}_0$ are a partition of $2^{\mathcal{N}}$.

In order to construct a scheme for this access structure, let $(V_1, \ldots, V_4)$ be the shares of the secret S with a $(4,4)$-threshold scheme, i.e., S can be reconstructed only if all four shares $(V_1, \ldots, V_4)$ are collected. If we distribute these shares as

$$W_1 = (V_3, V_4),\ W_2 = (V_2, V_4),\ W_3 = (V_1, V_4),\ W_4 = (V_1, V_2, V_3),$$

then we can verify that $W_{\mathcal{I}} = (V_1, \ldots, V_4)$ for every $\mathcal{I} \in \mathscr{A}_1^-$ and $W_{\mathcal{I}} \neq (V_1, \ldots, V_4)$ for every $\mathcal{I} \in \mathscr{A}_0^+$. Thus, it is a secure $(\mathscr{A}_1, \mathscr{A}_0)$-access structure scheme.

In Example 11.11, we have assigned four shares $(V_1, \ldots, V_4)$ so that V_j is included in W_i only when i is not included in the jth set of $\mathscr{A}_0^+$. This idea can be generalized to construct a $(\mathscr{A}_1, \mathscr{A}_0)$-access structure scheme for any access structure satisfying the monotonicity conditions (11.11). Indeed, for a given access structure $(\mathscr{A}_1, \mathscr{A}_0)$ specified by $(\mathscr{A}_1^-, \mathscr{A}_0^+)$, denote $\mathscr{A}_0^+ = \{\mathcal{I}_1, \ldots, \mathcal{I}_m\}$, and let $(V_1, \ldots, V_m)$ be the shares of the secret data S obtained using an (m,m)-threshold scheme. Further, let

$$W_i = (V_j : 1 \leq j \leq m, i \notin \mathcal{I}_j).$$

This particular construction is called a *multiple assignment scheme.*

THEOREM 11.12 *For an access structure $(\mathscr{A}_1, \mathscr{A}_0)$ satisfying the monotonicity conditions* (11.11) *and with $|\mathscr{A}_0^+| = m$, the multiple assignment scheme constructed using a secure (m,m)-threshold scheme is a secure $(\mathscr{A}_1, \mathscr{A}_0)$-access structure scheme.*

Proof For an $\mathcal{I} \in \mathscr{A}_1^-$ and for every $j \in [m]$, there exists at least one $i \in \mathcal{I}$ such that $i \notin \mathcal{I}_j$, since otherwise $\mathcal{I} \subseteq \mathcal{I}_j$ which contradicts $\mathcal{I} \in \mathscr{A}_1^-$. Thus, by the construction of the scheme, V_j is included in $W_{\mathcal{I}}$ for every $j \in [m]$, whereby $W_{\mathcal{I}} = (V_1, \ldots, V_m)$ and the secret data S can be reconstructed from $W_{\mathcal{I}}$.

On the other hand, for every $\mathcal{I} \in \mathscr{A}_0^+$, $W_{\mathcal{I}}$ does not include V_j if $\mathcal{I} = \mathcal{I}_j \in \mathscr{A}_0^+$. Thus, S is concealed from $W_{\mathcal{I}}$ by the security of the underlying (m,m)-threshold scheme. □

By using the multiple assignment scheme, we can realize a secure scheme for any access structure as long as the monotonicity condition (11.11) is satisfied. However, the multiple assignment scheme may not be efficient in terms of share sizes. In fact, even though the multiple assignment scheme can realize the access structure corresponding to a (k,n)-threshold scheme via an (m,m)-threshold scheme with $m = |\mathscr{A}_0^+| = \binom{n}{k-1}$ (cf. Example 11.10), each share size of the resulting scheme is $\binom{n-1}{k-1}$ times larger than the size of the secret S, much worse than the size of shares for our (k,n)-threshold scheme described earlier.

We close this section with a lower bound for sizes of shares required for a secure general access structure scheme. The share W_i (or ith share holder) is said to be *significant* if there exists $\mathcal{I} \in \mathscr{A}_0$ such that $\mathcal{I} \cup \{i\}$ belongs to $\mathscr{A}_1$. In other words, a nonsignificant share plays no role in a secret sharing scheme – we can replace it with an empty share without violating the reliability or security condition. In a similar manner to Proposition 11.6, we can derive a lower bound on the share size as follows.

PROPOSITION 11.13 *For any secure $(\mathscr{A}_1, \mathscr{A}_0)$-access structure scheme, the size of each significant share must satisfy*

$$H(W_i) \geq H(S). \tag{11.12}$$

We leave the proof as an exercise; see Problem 11.2.

When (11.12) is satisfied with equality for every significant share, the scheme is said to be *ideal*. For instance, the (k, n)-threshold scheme presented in Protocol 11.1 is ideal. In general, an ideal scheme may not exist for a given access structure, and the lower bound on the share size in Proposition 11.13 can be improved by more sophisticated arguments. On the other hand, other constructions have been proposed to improve upon the share size of the multiple assignment scheme; see Section 11.6 for references and further discussion.

11.4 Dishonest Share Holders

An alternative view of the (k, n)-threshold scheme is that $n - k$ or less shares are lost or *erased* in the recovery phase. Yet, we are required to be able to reconstruct the secret. However, the identity of the erased shares was known and the remaining shares were reported without any modification. Another share corruption model allows up to t shares to be modified adversarially to any other values, for some threshold t, and yet requires the scheme to be able to recover the secret. This is a more challenging task since the identity of corrupted shares is not known – we can view erasure corruptions of threshold schemes as knowing the identity of modified shares and ignoring them. Interestingly, we will see that polynomial based schemes introduced earlier work in this setting as well.

For concreteness, we can imagine that the share holders of up to t shares are malicious and they collaborate to report modified values of their shares. That is, a subset $\mathcal{I}$ of malicious share holders observe their original shares $W_{\mathcal{I}}$ and report $\tilde{W}_{\mathcal{I}}$ during the reconstruction phase. However, this makes the adversary a bit restrictive since the modified share values $\tilde{W}_{\mathcal{I}}$ should be determined (using a fixed stochastic mapping) from $W_{\mathcal{I}}$. In fact, we consider a more powerful adversary[3] who can modify up to t shares by observing all the shares $W_{\mathcal{N}}$. This more powerful adversary can be easily described using the notion of *Hamming distance*, defined below.

DEFINITION 11.14 (Hamming distance) For $w_{\mathcal{N}}$ and $w'_{\mathcal{N}}$ taking values in $\mathcal{S}^n$, the Hamming distance $d_H(w_{\mathcal{N}}, w'_{\mathcal{N}})$ between $w_{\mathcal{N}}$ and $w'_{\mathcal{N}}$ is given by

[3] A basic principle of theoretical security is to assume that all adversaries collaborate and implement the most challenging attack. Thus, we can think of a single adversary who can control up to t share holders.

$$d_H(w_{\mathcal{N}}, w'_{\mathcal{N}}) := \sum_{i=1}^{n} \mathbf{1}[w_i \neq w'_i].$$

With this notion, we can formally define a secure (k,n)-threshold scheme with t dishonest share holders.

DEFINITION 11.15 (Secure threshold schemes with dishonest share holders) For a given integer n, consider a secret sharing scheme with encoder mapping $\varphi\colon \mathcal{S} \to \prod_{i\in\mathcal{N}} \mathcal{W}_i$ and a reconstruction mapping $\psi\colon \mathcal{W}_{\mathcal{N}} \to \mathcal{S}$. This scheme constitutes a t-secure (k,n)-threshold scheme if the following conditions hold for every distribution P_S of S:

$$\Pr(\psi_{\mathcal{N}}(w) = S, \forall\, w \text{ such that } d_H(w, W_{\mathcal{N}}) \le t) = 1 \quad \text{(reliability)} \tag{11.13}$$

and

$$I(S \wedge W_{\mathcal{I}}) = 0, \ \forall \mathcal{I} \subseteq \mathcal{N} \text{ such that } |\mathcal{I}| < k. \quad \text{(security)}. \tag{11.14}$$

The reliability condition above says that all the vectors obtained by modifying up to t shares are mapped back to the original secret S in the reconstruction phase.

We now present a t-secure (k,n)-threshold scheme. In fact, this secret sharing scheme is almost the same at that in Protocol 11.1, with slight modification to the reconstruction phase. Namely, upon receiving all the (possibly modified) shares, each share holder outputs a polynomial $\hat{f}$ of degree no more than $k-1$ that is consistent with at least $n-t$ share values; see Protocol 11.3 for a description. The following result shows that all these polynomials must coincide with the unknown f, provided $k \le n - 2t$.

LEMMA 11.16 *If $t \le \frac{n-k}{2}$, then $\hat{f} = f$ is the unique polynomial of degree less than or equal to $k-1$ such that* (11.15) *holds for some $\mathcal{I} \subseteq \mathcal{N}$ with $|\mathcal{I}| \ge n-t$.*

Proof Suppose that there exists another polynomial $g(x)$ of degree less than or equal to $k-1$ such that

$$g(\alpha_i) = \tilde{W}_i, \ \forall\, i \in \mathcal{I}$$

Protocol 11.3 A (k,n)-threshold scheme with dishonest share holders

1. **Distribution phase:** The dealer uniformly generates $f(x)$ from $\mathbb{P}_{k-1}(S)$ and sets the share of the ith share holder $\mathcal{P}_i$ as $W_i = f(\alpha_i)$ for $i \in \mathcal{N}$.
2. **Reconstruction phase:** Upon observing $\tilde{W}_{\mathcal{N}} = (\tilde{W}_i : i \in \mathcal{N})$ (possibly modified) shares, each share holder outputs $\hat{f}$ of degree less than or equal to $k-1$ such that

$$\hat{f}(\alpha_i) = \tilde{W}_i, \ \forall\, i \in \mathcal{I} \tag{11.15}$$

for some subset $\mathcal{I} \subseteq \mathcal{N}$ with $|\mathcal{I}| \ge n-t$.

holds for some $\mathcal{I} \subseteq \mathcal{N}$ with $|\mathcal{I}| \geq n - t$. Then, denoting by $\mathcal{I}_0$ the set of honest share holders, namely $\mathcal{I}_0 = \{i : W_i = \tilde{W}_i\}$, we have

$$g(\alpha_i) = W_i = f(\alpha_i) = \quad \forall i \in \mathcal{I}_0 \cap \mathcal{I}.$$

Since $|\mathcal{I}_0| \geq n - t$ and $|\mathcal{I}| \geq n - t$, we have $|\mathcal{I}_0 \cap \mathcal{I}| \geq n - 2t$. Therefore, since $n - 2t \geq k$ by the assumption of the lemma, $f(x)$ and $g(x)$ coincide at k or more points. Furthermore, since $f(x)$ and $g(x)$ are polynomials of degree less than or equal to $k - 1$, then so is $f(x) - g(x)$. Thus, $f(x) - g(x)$ is a polynomial of degree $k - 1$ or less and has k roots, whereby it must be the zero polynomial. □

THEOREM 11.17 *For $t \leq \frac{n-k}{2}$, the secret sharing scheme in Protocol 11.3 is a t-secure (k, n)-threshold scheme.*

Proof The security condition (11.14) is guaranteed exactly in the same manner as in the proof of Theorem 11.4. The reliability condition (11.13) is guaranteed by Lemma 11.16. □

Since dishonest share holders can be treated as an adversary, it is natural to seek to protect the secret from them and use a threshold scheme with $k = t + 1$ which will forbid them, even when they collude, to get access to the secret. We state this special case of Theorem 11.17 as a corollary.

COROLLARY 11.18 *Protocol 11.3 with $k = t + 1$ is secure when $t < \frac{n}{3}$.*

Next, we would like to examine whether there is a scheme that can be secure for t exceeding $\frac{n-k}{2}$. It turns out that there is no such scheme.

THEOREM 11.19 *A t-secure (k, n)-threshold scheme exists only if $t \leq \frac{n-k}{2}$.*

Proof Consider a reconstruction function $\psi \colon \mathcal{W}_{\mathcal{N}} \to \mathcal{S}$ that correctly reconstructs the secret S when up to t forged shares are submitted.

For each $s \in \mathcal{S}$, let $\mathcal{C}_s \subset \mathcal{W}_{\mathcal{N}}$ be the set of possible vectors that can be converted from s by the (stochastic) mapping φ. That is,

$$\mathcal{C}_s := \{w_{\mathcal{N}} \in \mathcal{W}_{\mathcal{N}} \colon \Pr(\varphi(s) = w_{\mathcal{N}}) > 0)\}.$$

The reliability condition in the presence of dishonest share holders implies that for any $w_{\mathcal{N}} \in \mathcal{C}_s$ and $\hat{w}_{\mathcal{N}} \in \mathcal{W}_{\mathcal{N}}$ such that $d_H(w_{\mathcal{N}}, \hat{w}_{\mathcal{N}}) \leq t$, we have $\psi(\hat{w}_{\mathcal{N}}) = \psi(w_{\mathcal{N}}) = s$.

We first claim that, for any $s \neq s'$, there exists no pair of $w_{\mathcal{N}} \in \mathcal{C}_s$ and $\tilde{w}_{\mathcal{N}} \in \mathcal{C}_{s'}$ satisfying $d_H(w_{\mathcal{N}}, \tilde{w}_{\mathcal{N}}) \leq 2t$. To prove the claim by contradiction, suppose that there exist $w_{\mathcal{N}} \in \mathcal{C}_s$ and $\tilde{w}_{\mathcal{N}} \in \mathcal{C}_{s'}$ satisfying $d_H(w_{\mathcal{N}}, \tilde{w}_{\mathcal{N}}) \leq 2t$. Then, by the reliability condition we must have $d_H(w_{\mathcal{N}}, \tilde{w}_{\mathcal{N}}) > t$. In this case, we can find $\hat{w}_{\mathcal{N}}$ such that $d_H(\hat{w}_{\mathcal{N}}, w_{\mathcal{N}}) \leq t$ and $d_H(\hat{w}_{\mathcal{N}}, \tilde{w}_{\mathcal{N}}) \leq t$, whereby, by the reliability criterion, we must have $s = \psi_{\mathcal{N}}(w_{\mathcal{N}}) = \psi_{\mathcal{N}}(\hat{w}_{\mathcal{N}}) = \psi_{\mathcal{N}}(\tilde{w}_{\mathcal{N}}) = s'$, which is a contradiction. Indeed, we can find a subset $\mathcal{I}_1$ with $|\mathcal{I}_1| = t$ and $\mathcal{I}_2$ with $|\mathcal{I}_2| \leq t$ such that (i) for all $i \notin \mathcal{I}_1 \cup \mathcal{I}_2$, $w_i = \tilde{w}_i$ and (ii) for all $i \in \mathcal{I}_1$,

$w_i \neq \tilde{w}_i$. The required $\hat{w}_{\mathcal{N}}$ is obtained by setting $\hat{w}_i = w_i$ for $i \notin \mathcal{I}_2$ and $\hat{w}_i = w_i$ for $i \in \mathcal{I}_2$.

It follows from the previous claim that the secret S can be reconstructed from $w_{\mathcal{I}}$ for any subset $\mathcal{I} \subseteq \mathcal{N}$ with $|\mathcal{I}| \geq n - 2t$. When $t > \frac{n-k}{2}$, i.e., $k > n - 2t$, S can be reconstructed from any $k - 1 \geq n - 2t$ shares, which contradicts the security condition for the (k, n)-threshold scheme. □

Even though Theorem 11.19 guarantees the optimality of the threshold $t \leq \frac{n-k}{2}$ for perfect reliability, this threshold can be improved by allowing some nonzero probability of error. The key idea is to introduce a mechanism to detect forged shares before the start of the reconstruction process; this is in contrast to Protocol 11.3 in which forged shares are implicitly detected and corrected during the reconstruction process. We have already seen such a mechanism: it is the message authentication code (MAC) of Chapter 8. Specifically, we can enhance the basic scheme in Protocol 11.1 as follows. We assume that the jth share holder shares a secret key K_j with the dealer and the dealer appends to the ith share W_i tags T_{ij} so that only the jth share holder can verify the validity using K_j. In the reconstruction phase, each share holder finds a subset of shares for which it can validate the tag and uses this subset to reconstruct the secret using the reconstruction mapping of Protocol 11.1 applied to validated shares. The details are left as an exercise; see Problem 11.4.

11.5 Error Correction and Secret Sharing

Readers familiar with error correcting codes might have noticed the similarity between secret sharing and error correction. The basic problem of error correction entails encoding a message m as a binary vector $e(m)$ in such a manner that when some of the bits of $e(m)$ get erased or flipped due to errors, we can still recover the message m. More generally, $e(m)$ can be a vector with entries in $\mathbb{F}_q$. As noted earlier, the reliability condition for a secure (k, n)-threshold scheme, in effect, requires that the secret can be recovered as long as less than k shares are erased. In the dishonest share holder setting, a t-secure (k, n)-threshold scheme requires that the secret can be recovered as long as less than $\frac{n-k}{2}$ shares are modified. These coincide with erasure and bit-flip errors in error correction. However, an additional security requirement is placed in secret sharing.

In this section, we review Reed–Solomon codes which have a similar structure to the polynomial based schemes above. These codes will be used for multiparty secure computation in Section 19.4. This section assumes familiarity with elementary notions in error correcting codes such as linear codes over finite fields and their generator and parity check matrices; see Section 11.6 for references.

The (generalized) Reed–Solomon code $\mathcal{C} \subset \mathbb{F}_q^n$ is given by

$$\mathcal{C} = \{(f(\alpha_1), \ldots, f(\alpha_n)) : f \in \mathbb{P}_{k-1}\},$$

where $\mathbb{P}_{k-1}$ is the set of all polynomials on $\mathbb{F}_q$ of degree less than or equal to $k-1$, and $\alpha_1, \ldots, \alpha_n$ are distinct elements of $\mathbb{F}_q$. Note that the Reed–Solomon code is a linear code, i.e., a linear combination of two codewords is also a codeword (obtained by taking the same linear combination of the coefficients of the two polynomials underlying the two codewords). Therefore, we can associate a *generator matrix* with it. In fact, the generator matrix of Reed–Solomon code is given by

$$G = \begin{bmatrix} 1 & 1 & \cdots & 1 \\ \alpha_1 & \alpha_2 & \cdots & \alpha_n \\ \vdots & \vdots & \ddots & \vdots \\ \alpha_1^{k-1} & \alpha_2^{k-1} & \cdots & \alpha_n^{k-1} \end{bmatrix}.$$

We can verify that (see Problem 11.6)

$$\mathcal{C} = \{\mathbf{m}G : \mathbf{m} \in \mathbb{F}_q^k\}. \tag{11.16}$$

Let H be the parity check matrix of $\mathcal{C}$, i.e., the $(n-k) \times n$ matrix H satisfying

$$GH^T = \mathbf{O}, \tag{11.17}$$

where $\mathbf{O}$ is the zero matrix and H^T is the transpose of H.[4]

Let $w_H(\mathbf{u}) := |\{i \in \mathcal{N} : u_i \neq 0\}|$ be the Hamming weight of vector $\mathbf{u} = (u_1, \ldots, u_n)$. Since a polynomial $f(x) \in \mathbb{P}_{k-1}$ has at most $k-1$ roots, the Hamming weight of every nonzero codeword of the Reed–Solomon code is at least $n-k+1$. Further, since $\mathcal{C}$ is a linear code, it is easy to check that the Hamming distance between vectors $\mathbf{u}$ and $\mathbf{v}$ is $d_H(\mathbf{u}, \mathbf{v}) = w_H(\mathbf{u}-\mathbf{v})$, which implies that the minimum Hamming distance between two codewords in $\mathcal{C}$ is $d_{\min} = n-k+1$.[5]

For a given codeword $\mathbf{u} \in \mathcal{C}$ and arbitrary error vector $\mathbf{e} \in \mathbb{F}_q^n$, since $\mathbf{u}H^T = \mathbf{0}$, we have

$$\mathbf{s} = (\mathbf{u}+\mathbf{e})H^T = \mathbf{u}H^T + \mathbf{e}H^T = \mathbf{e}H^T.$$

The vector $\mathbf{s} \in \mathbb{F}_q^{n-k}$ is referred to as the syndrome, and it can be used to identify the codeword $\mathbf{u}$ if $w_H(\mathbf{e}) \leq (d_{\min}-1)/2$. Indeed, for a pair of error vectors $\mathbf{e} \neq \mathbf{e}'$ such that $w_H(\mathbf{e}) \leq (d_{\min}-1)/2$ and $w_H(\mathbf{e}') \leq (d_{\min}-1)/2$, we have $\mathbf{e}H^T \neq \mathbf{e}'H^T$, since otherwise $(\mathbf{e}-\mathbf{e}')H^T = \mathbf{0}$, and we have a codeword $\mathbf{e}-\mathbf{e}'$ such that $w_H(\mathbf{e}-\mathbf{e}') \leq d_{\min}-1$, which is a contradiction. Thus, if the syndrome $\mathbf{s} = \mathbf{e}H^T$ of an error vector $\mathbf{e}$ with $w_H(\mathbf{e}) \leq (d_{\min}-1)/2$ is provided, we can uniquely identify $\mathbf{e}$ from the syndrome $\mathbf{s}$.

Various algorithms have been proposed to efficiently identify the error vector from the syndrome. Since we only use the fact that the error can be identifiable

[4] If we choose $\alpha_i = \beta^i$ for an element $\beta \in \mathbb{F}_q$ with order n, then the parity check matrix can be written more explicitly.

[5] The argument above only guarantees that the minimum Hamming distance is at least $n-k+1$, but it can be shown that the minimum Hamming distance cannot be larger than $n-k+1$, which is known as the *singleton bound.*

from the syndrome in the rest of the book, we do not elaborate on these schemes; interested readers are encouraged to consult the references given in Section 11.6.

When the Reed–Solomon code above is used as an error correcting code, a k symbol message $\mathbf{m} \in \mathbb{F}_q^k$ is encoded as the codeword $\mathbf{m}G \in \mathcal{C}$. On the other hand, in the threshold secret sharing scheme of Protocol 11.1, shares of the secret $S \in \mathbb{F}_q$ are created as $W_{\mathcal{N}} = (W_1, \ldots, W_n) = (S, A_1, \ldots, A_{k-1})G$ for randomly chosen $(A_1 \ldots, A_{k-1}) \in \mathbb{F}_q^{k-1}$. The randomly chosen $(A_1 \ldots, A_{k-1})$ plays the role of a dummy message, generated to prevent information leakage.

Next, suppose that t dishonest share holders substitute their shares W_i with $\tilde{W}_i$. The tuple $\tilde{W}_{\mathcal{N}}$ of substituted shares satisfies $d_H(W_{\mathcal{N}}, \tilde{W}_{\mathcal{N}}) \leq t$, i.e., $w_H(\tilde{W}_{\mathcal{N}} - W_{\mathcal{N}}) \leq t$. Since the tuple $W_{\mathcal{N}}$ of the original shares is a codeword of the Reed–Solomon code, once the syndrome $\mathbf{s} = (\tilde{W}_{\mathcal{N}} - W_{\mathcal{N}})H^T = \tilde{W}_{\mathcal{N}}H^T$ is obtained, honest parties can identify the error vector $\mathbf{e} = \tilde{W}_{\mathcal{N}} - W_{\mathcal{N}}$ as far as the number of dishonest share holders satisfies $t \leq (n-k)/2$. In fact, this is, in essence, a rephrasing of Theorem 11.17 in the language of error correcting codes. However, it should be emphasized that there is no need for the honest parties to exchange their shares – it suffices to compute the syndrome, which only depends on the error vector and does not depend on the original share or the secret S. This observation will be used in Section 19.4 to construct a secure multiparty computation protocol against an active adversary.

11.6 References and Additional Reading

The concepts of secret sharing and threshold schemes were introduced in [35, 303]. Ramp secret sharing schemes were independently introduced in [36, 355] and are implicitly mentioned in [247]. The notion of strong secrecy in ramp secret sharing was introduced in [355]. The scheme in Protocol 11.2 is the same as [247] (see also [129, 262]). The strong secrecy (see (11.6)) of Protocol 11.2 was proved in [262].

The concept of the secret sharing scheme with general access structure was introduced in [177]; the multiple assignment scheme we present was also introduced in the same work. In Section 11.3, although we have only explained the multiple assignment using an (m, m)-threshold scheme as the basic scheme, it is possible to use threshold schemes with other parameters as the basic scheme. In general, it is a difficult problem to optimize the share sizes of a secret sharing scheme with general access structure. For a lower bound on the share size, see [54]. By using integer programming, a method was proposed in [183] to optimize the average share size in the multiple assignment construction. For other constructions of the secret sharing with general access structure, see [26, 315, 321, 322, 323, 324]. It is known that the construction of ideal secret sharing is related to matroids [45]; using the matroid theory, it can be proved that an ideal secret sharing scheme does not exist for certain access structures [233].

In Section 11.3, we restricted our attention to the case where the qualified family $\mathscr{A}_1$ and the forbidden family $\mathscr{A}_0$ are a partition of $2^{\mathcal{N}}$; this case is termed a *complete* access structure. In general, we can introduce intermediate families so that a subset in an intermediate family can partially recover the secret data; ramp secret sharing schemes can be regarded as a special case of such a structure with intermediate families. For a construction of a strongly secure ramp secret sharing scheme with general access structure, see [182]. Also, for a construction of secret sharing with nonuniform secret source, see [197].

In [247], a connection between the polynomial based threshold scheme and Reed–Solomon codes was pointed out. Furthermore, it was also pointed out that, by the associated error correction algorithm, the secret data can be reconstructed correctly even if some of the share holders submit forged shares (see also [326] for the detection of dishonest share holders). In this regard, Lemma 11.16 is tantamount to the fact that the minimum distance of the Reed–Solomon code we discussed is $n - k + 1$, a well-known fact in coding theory (e.g., see [184]). On the other hand, Theorem 11.19 can be interpreted as the singleton bound of coding theory, although the exact statement of Theorem 11.19 itself was proved in [108]. For further connections between the error correcting code and secret sharing, see [209, 234].

Problems

11.1 *Strong security*: For a trivial $(2, 2, 2)$-ramp scheme given by $(W_1, W_2) = (S_1, S_2)$, verify that it satisfies (11.7) but does not satisfy the strong security (11.6).

11.2 Prove Proposition 11.13.

11.3 Consider two independent random variables S and S' generated uniformly from $\mathbb{F}_q$, and let $W_{\mathcal{N}}$ and $W'_{\mathcal{N}}$ be the shares obtained by applying the secure (k, n)-threshold scheme of Protocol 11.1 independently to S and S', respectively. Show that $S + S'$ can be recovered from $W^+_{\mathcal{N}} = (W_i + W'_i, i \in \mathcal{N})$. Furthermore, $I(W^+_{\mathcal{I}} \wedge S + S') = 0$ for every $\mathcal{I} \subset \mathcal{N}$ with $|\mathcal{I}| \le k$.

11.4 Consider an extension of t-secure (k, n)-threshold schemes where the reliability condition (11.13) is relaxed to allow a probability of error ε. We call such a secret sharing scheme an (ε, t)-secure (k, n)-threshold scheme. Now consider an extension of Protocol 11.1 obtained using a MAC (φ, V) with probability of success of substitution attack $P_{\mathrm{S},n} \le \varepsilon$. Specifically, share holder i is given $(W_i, T_{ij}, j \in \mathcal{N} \setminus \{i\})$ where $W_i = f(\alpha_i)$ and $T_{ij} = \varphi(W_i, K_j)$ is the tag of share W_i. Then, during the reconstruction phase, each share holder first validates the tags for each W_i received from other share holders and uses only the validated W_i to reconstruct f using the Lagrange interpolation formula.

Show that for $t \le n - k$ dishonest share holders, this modified scheme is $(t(n - t)\varepsilon, t)$-secure.

REMARK The usage of a MAC in a secret sharing scheme was introduced in [281]. This usage has a drawback that the size of each share depends on the number of share holders n. A scheme such that the size of share is independent of n was studied in [211]. Typically, in the context of secret sharing with dishonest share holders, we consider substitution attack. For an efficient construction of secret sharing detecting an impersonation attack, see [178].

11.5 In the setting of Problem 11.4, show that an (ε, t)-secure (k, n)-threshold scheme exists for all $\varepsilon > 0$ only if $t \leq n - k$.

HINT Let $\mathcal{I}_0$ and $\mathcal{I}_1$ be the set of honest and dishonest share holders, respectively. Consider an attack where $\tilde{W}_{\mathcal{I}_1}$ is mapped to a constant vector. For this attack, using Fano's inequality and reliability condition, show that $H(S|W_{\mathcal{I}_0})$ can be made arbitrarily small as ε goes to 0, if $t > n - k$. This contradicts the security condition.

11.6 Show that the generator matrix of the Reed–Solomon code is given by (11.16).

12 Two-party Secure Computation for a Passive Adversary

Up to this chapter, we have mainly considered problems in which there exists an external adversary other than the legitimate parties. In this and subsequent chapters, we consider the problem of secure computation in which legitimate parties may behave as adversaries.

To motivate secure computation, consider Yao's millionaires' problem. Suppose that there are two millionaires who are interested in comparing their wealth. However, they do not want to reveal their own wealth to each other. If there is another trusted person, they can privately send the amounts of their fortune to this person, who can then tell them the outcome of the comparison. The problem of secure computation seeks to implement such a task without the help of any trusted third party; see Figure 12.1.

In this chapter, we provide an introduction to the two-party secure computation problem for the case when the adversarial party is only passively adversarial, i.e., it does not deviate from the protocol. The passive adversary, sometimes referred to as an honest-but-curious party, seeks to gain as much information as possible about the other party's input during the computation protocol, but does not deviate from the protocol description. In contrast, a more powerful adversary, an active adversary, may not follow the protocol. Considering an active adversary requires much more involved concepts and techniques; we postpone this treatment to later chapters.

12.1 Secure Computation with Interactive Communication

The secure computation problem captures a rather broad spectrum of cryptographic primitives under a single abstraction. In two-party secure computation, parties with inputs seek to compute a function of their inputs using an interactive communication protocol. To lay down the problem concretely, we need to formally define notions of *function*, *interactive communication*, and *security*. Unlike message transmission with secret key cryptography, by obtaining the output of the function, the parties necessarily must get some information about each other's input. Thus, the notion of security must be laid down carefully.

Consider two parties $\mathcal{P}_1$ and $\mathcal{P}_2$ observing inputs $x_1 \in \mathcal{X}_1$ and $x_2 \in \mathcal{X}_2$, respectively. They seek to evaluate a function $f\colon \mathcal{X}_1 \times \mathcal{X}_2 \to \mathcal{Z}$ of their inputs (x_1, x_2),

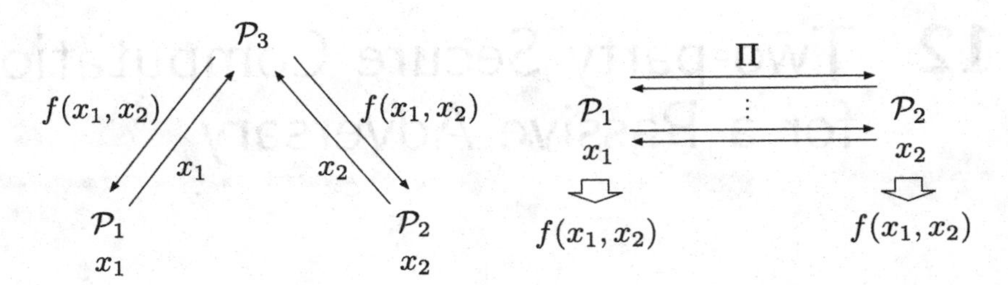

Figure 12.1 The concept of secure computation. If a trustable third party is available, the third party can compute the function on behalf of the parties (left); secure computation seeks to implement such a task by a communication between the parties (right).

namely both parties are interested in knowing the value of $f(x_1, x_2)$. Towards that goal, the parties communicate with each other using an interactive communication protocol. In each round of an interactive communication protocol, only one of the parties communicates. The communication sent depends on all the observations of the communicating party up to that round, namely it depends on the communication up to the previous round, the private (local) randomness of the party, and the public (shared) randomness available.

Formally, we allow the parties to interact using *tree protocols*, defined below. We consider *public-coin protocols* where, in addition to the inputs, both parties have access to public-coins U_0 and party $\mathcal{P}_i$, $i \in \{1, 2\}$, has access to private-coin U_i. The random variables U_0, U_1, U_2 are assumed to be mutually independent, finite valued, and uniformly distributed.

DEFINITION 12.1 (Tree protocol) A two-party tree protocol π is given by a labeled binary tree of finite depth.[1] Each vertex of the tree, except the leaf-nodes, is labeled by 1 or 2, with the root labeled as 1. This tree represents the state of the protocol; it determines which party needs to communicate and how the state changes after communication.

Specifically, the protocol starts at the root and proceeds as follows: when the protocol is at a vertex v labeled $i \in \{1, 2\}$, party $\mathcal{P}_i$ communicates one bit $\Pi_v = b_v(x_i, U_i, U_0)$. If $\Pi_v = 0$, the state of the protocol updates and it moves to the left-child of v; if $\Pi_v = 1$, the protocol moves to the right-child of v. The communication protocol terminates when it reaches a leaf-node. We denote the overall bits transmitted during the protocol by Π; this is termed the *transcript* of the protocol. Note that the transcript Π of the protocol is seen by both the parties. Since the protocol involves randomness U_0, U_1, U_2, the transcript Π is a random variable even if the inputs (x_1, x_2) are fixed.

We have illustrated a typical tree protocol in Figure 12.2. The definition of communication protocol above is the most general we consider. Certain restricted notions will be of interest as well. When U_0 is fixed to a constant, we call the

[1] In the rest of the book, when we just say protocol, that always means a tree protocol.

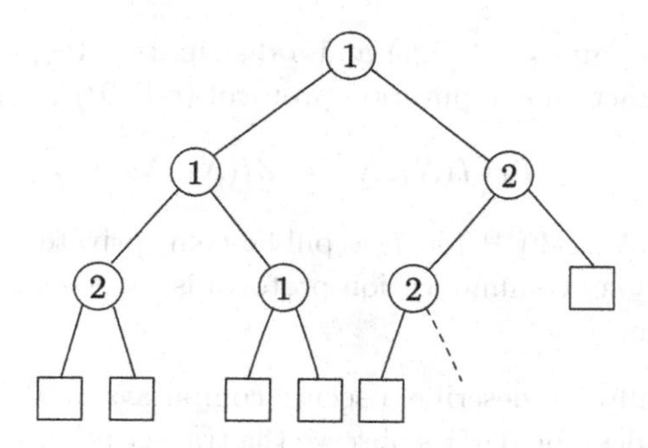

Figure 12.2 An illustration of a tree protocol.

communication protocol a *private-coin protocol* and when U_0, U_1, U_2 are all fixed to constants, we call the protocol a *deterministic protocol.*

Even though we have defined each communication Π_v to be labeled by a vertex, it is sometime more convenient to consider an alternating protocol in which the parties communicate alternately. More specifically, at an odd depth of the tree, $\mathcal{P}_1$ communicates irrespective of the history of the past communication; and at an even depth, $\mathcal{P}_2$ communicates. In such a protocol, we denote the transcript as $\Pi = (\Pi_1, \ldots, \Pi_r)$ indexed by a round (depth) i instead of by a vertex. In fact, we can convert any tree protocol into an alternating protocol by padding communication taking constant values. Furthermore, by slightly relaxing the notion of tree protocol, we may consider a protocol in which a message Π_i in each round is a bit sequence instead of a single bit. In the following, depending on the context, we use both the tree protocol with communication labeled by a vertex and the alternating nonbinary valued communication protocol.[2]

Now that we have formally described our allowed communication, we describe what it means to compute the function f using interactive communication. Simply stated, we require that each party forms a reliable estimate of $f(x_1, x_2)$ by using the transcript Π of the communication protocol and their local observations. A formal definition follows.

DEFINITION 12.2 (Function computation protocol) For a given function $f : \mathcal{X}_1 \times \mathcal{X}_2 \to \mathcal{Z}$ and a given protocol π, with transcript denoted by Π, let $\hat{Z}_1 = \hat{Z}_1(x_1, x_2) \in \mathcal{Z}$ and $\hat{Z}_2 = \hat{Z}_2(x_1, x_2) \in \mathcal{Z}$ be the parties' estimates of $f(x_1, x_2)$ given by[3]

$$\hat{Z}_1(x_1, x_2) := \hat{Z}_1(x_1, \Pi, U_1, U_0) \in \mathcal{Z},$$
$$\hat{Z}_2(x_1, x_2) := \hat{Z}_2(x_2, \Pi, U_2, U_0) \in \mathcal{Z}.$$

[2] In Section 10.1, we employed the latter notion of protocol.
[3] Note that the outputs are functions of inputs (x_1, x_2) and randomness (U_0, U_1, U_2). To avoid cumbersome notation, we write $\hat{Z}_i$ or $\hat{Z}_i(x_1, x_2)$ depending on the context.

Then, given $\varepsilon \in [0,1)$ and a distribution $\mathrm{P}_{X_1X_2}$ of the inputs, the protocol is an ε-function computation protocol (ε-FCP) for f if

$$\Pr\Big(f(X_1,X_2) = \hat{Z}_1(X_1,X_2) = \hat{Z}_2(X_1,X_2)\Big) \geq 1-\varepsilon \tag{12.1}$$

holds. An ε-FCP for f is public-coin, private-coin, or deterministic when the underlying communication protocol is so. Also, a 0-FCP for f is called a *perfect FCP* for f.

Finally, we describe a secure computation protocol. Note that, when an FCP concludes, the parties observe the transcript Π of the communication protocol in addition to their original observations. The computation of the function already requires that each party should be able to determine the function value using this transcript and other local observations. Secure computation of a function forbids the parties from determining any other information about the other party's input. Heuristically, we require that the transcript and randomness (local and public) seen by a party should be independent of the input of the other party, when conditioned on its input and the function value.

Denoting $Z = f(X_1,X_2)$ for inputs (X_1,X_2) distributed according to $\mathrm{P}_{X_1X_2}$, we can formalize a notion of *perfect security* using mutual information as follows:

$$I(X_2 \wedge \Pi, U_1, U_0|X_1, Z) = I(X_1 \wedge \Pi, U_2, U_0|X_2, Z) = 0. \tag{12.2}$$

Thus, the transcript Π does not reveal any other information about the input of the other party beyond what the function value Z already leaks. Alternatively, (12.2) can be also written as

$$H(X_2|X_1,Z) = H(X_2|X_1,Z,\Pi,U_1,U_0), \tag{12.3}$$
$$H(X_1|X_2,Z) = H(X_1|X_2,Z,\Pi,U_2,U_0). \tag{12.4}$$

The left-hand side of (12.3) is the ambiguity of $\mathcal{P}_2$'s input for $\mathcal{P}_1$ when the function is computed by the trustable third party (cf. Figure 12.1, left). The right-hand side of (12.3) is the ambiguity of $\mathcal{P}_2$'s input for $\mathcal{P}_1$ when the function is computed by an interactive protocol (cf. Figure 12.1, right). Thus, our security requires that the protocol does not leak more information than the ideal situation where the function is computed by the third party.

While the reader may readily agree with this definition of security, this notion of security that allows a nominal leakage of information mandated by the problem definition is an important advancement over less mature attempts that strive for absolute security, without keeping in mind the leakage necessary to enable function computation. To appreciate this definition, we present the following example.

EXAMPLE 12.3 (NOT, XOR, AND) Consider $\mathcal{X}_1 = \mathcal{X}_2 = \{0,1\}$ and the computation of standard Boolean gates. For the NOT gate, the parties seek to compute $f(x_1,x_2) = x_1 \oplus 1$. If the parties compute this function, then clearly $\mathcal{P}_2$ gets x_1. In other words, the input of $\mathcal{P}_1$ does not remain private. But this observation can alternatively be interpreted as saying that any protocol computing the NOT gate cannot reveal anything more than the function value itself (since the function value reveals everything). Thus, our definition of security (12.2) deems this function secure computable.[4]

Next, we consider the XOR gate given by $f(x_1, x_2) = x_1 \oplus x_2$. When the parties compute $f(x_1, x_2)$, they can simply get the other party's input by taking XOR of their input with the function value. Thus, this function is also secure computable in the sense of (12.2) just by sending X_1 and X_2 each other.

Finally, we consider the AND gate given by $f(x_1, x_2) = x_1 \wedge x_2$. When $x_2 = 1$, $f(x_1, x_2) = x_1$ and an FCP can liberally leak information about x_1 in this case. But when $x_2 = 0$, $f(x_1, x_2) = 0$ and a secure protocol must not reveal any information about x_1. However, how does $\mathcal{P}_1$ know if x_2 is 0 or 1 without breaching the security requirement for $\mathcal{P}_2$? In fact, we will see later that there is no secure computation protocol for the AND gate (without additional resources).

We now extend this definition to an approximate version, where the perfect conditional independence is replaced by an approximate version. For convenience in later chapters, we use the total variation distance to measure almost independence.

DEFINITION 12.4 (Secure computation protocol) Given $\varepsilon, \delta \in [0, 1)$ and a function $f \colon \mathcal{X}_1 \times \mathcal{X}_2 \to \mathcal{Z}$, an (ε, δ)-secure computation protocol ((ε, δ)-SCP) for f under $\mathrm{P}_{X_1X_2}$ comprises a protocol π, with transcript denoted by Π, and output $\hat{Z}_1(x_1, x_2)$ and $\hat{Z}_2(x_1, x_2)$ such that

$$\Pr\Big(f(X_1, X_2) = \hat{Z}_1(X_1, X_2) = \hat{Z}_2(X_1, X_2)\Big) \geq 1 - \varepsilon \quad \text{(reliability)} \tag{12.5}$$

and

$$\begin{aligned} d_{\mathrm{var}}\big(\mathrm{P}_{X_1X_2Z\Pi U_1U_0}, \mathrm{P}_{X_1X_2Z} \times \mathrm{P}_{\Pi U_1U_0|X_1Z}\big) \leq \delta \\ d_{\mathrm{var}}\big(\mathrm{P}_{X_1X_2Z\Pi U_2U_0}, \mathrm{P}_{X_1X_2Z} \times \mathrm{P}_{\Pi U_2U_0|X_2Z}\big) \leq \delta \end{aligned} \quad \text{(security)}, \tag{12.6}$$

where $Z = f(X_1, X_2)$ for inputs (X_1, X_2) distributed according to $\mathrm{P}_{X_1X_2}$. For $\varepsilon = \delta = 0$, we term the protocol a *perfect SCP* under $\mathrm{P}_{X_1X_2}$. When such a protocol exists for a function f, we say that f is *perfectly secure computable* under $\mathrm{P}_{X_1X_2}$. If f is perfectly secure computable under every input distribution $\mathrm{P}_{X_1X_2}$, we just say that f is secure computable.

In a similar spirit as the interpretation of the perfect security condition (12.2), the security conditions in (12.6) can be interpreted as follows. Hypothetically, suppose that the function is computed by a trusted third party, and $\mathcal{P}_1$ obtains $Z = f(X_1, X_2)$ without running the protocol. Then, suppose that $\mathcal{P}_1$ locally simulates $(\tilde{\Pi}, \tilde{U}_1, \tilde{U}_0)$ according to the conditional distribution $\mathrm{P}_{\Pi U_1U_0|X_1Z}(\cdot|X_1, Z)$. The first condition in (12.6) says that the joint distribution of the observations $(X_1, X_2, Z, \tilde{\Pi}, \tilde{U}_1, \tilde{U}_0)$ in this hypothetical situation is close to the joint

[4] We say that a function f is *secure computable* when we can find a secure computation protocol for f.

distribution of the observations $(X_1, X_2, Z, \Pi, U_1, U_0)$ in the actual protocol. In fact, it will turn out in Chapter 15 that this simulation viewpoint is a correct way to define security against an active adversary as well.

12.2 Secure Computation of Asymmetric Functions and Randomized Functions

In the previous section, we described the problem of secure computation of deterministic and symmetric functions, namely functions where both parties seek the same value $f(x_1, x_2)$ which is determined by the input pair (x_1, x_2). In this section, we present secure computation of more general functions and show how it suffices to look at deterministic symmetric functions.

Specifically, we consider the more general class of *asymmetric functions* where $f(x_1, x_2) = (f_1(x_1, x_2), f_2(x_1, x_2))$ and $\mathcal{P}_i$ seeks to know the value $f_i(x_1, x_2)$, $i = 1, 2$. The requirements for (ε, δ)-secure computation protocols in Definition 12.4 can be extended to asymmetric functions as follows:

$$\Pr\Big(f_1(X_1, X_2) = \hat{Z}_1(X_1, X_2),\ f_2(X_1, X_2) = \hat{Z}_2(X_1, X_2)\Big) \geq 1 - \varepsilon \quad \text{(reliability)}$$

and

$$\begin{aligned} d_{\mathrm{var}}\big(\mathrm{P}_{X_1X_2Z_1\Pi U_1U_0}, \mathrm{P}_{X_1X_2Z_1} \times \mathrm{P}_{\Pi U_1 U_0|X_1Z_1}\big) &\leq \delta \\ d_{\mathrm{var}}\big(\mathrm{P}_{X_1X_2Z_2\Pi U_2U_0}, \mathrm{P}_{X_1X_2Z_2} \times \mathrm{P}_{\Pi U_2 U_0|X_2Z_2}\big) &\leq \delta \end{aligned} \quad \text{(security)},$$

where $(Z_1, Z_2) = (f_1(X_1, X_2), f_2(X_1, X_2))$.

A symmetric function can be regarded as a special case of the asymmetric function such that $f_1 = f_2$. Interestingly, the result below shows that secure computation of an asymmetric function is equivalent to secure computation of a (related) symmetric function.

LEMMA 12.5 *Consider an asymmetric function with* $f_i \colon \mathcal{X}_1 \times \mathcal{X}_2 \to \mathcal{Z}_i$, $i = 1, 2$. *Let* $\tilde{f} \colon \mathcal{X}_1 \times \mathcal{Z}_1 \times \mathcal{X}_2 \times \mathcal{Z}_2 \to \mathcal{Z}_1 \times \mathcal{Z}_2$ *be the symmetric function given by*

$$\tilde{f}((x_1, k_1), (x_2, k_2)) := (f_1(x_1, x_2) + k_1, f_2(x_1, x_2) + k_2),$$

where the additions are done by regarding $\mathcal{Z}_i = \{0, 1, \ldots, |\mathcal{Z}_i| - 1\}$, $i = 1, 2$, *as an additive group. Then, the asymmetric function* (f_1, f_2) *is perfectly secure computable under* $\mathrm{P}_{X_1X_2}$ *if and only if the symmetric function* $\tilde{f}$ *is perfectly secure computable under* $\mathrm{P}_{X_1X_2} \times \mathrm{P}_{K_1K_2}$, *where* (K_1, K_2) *are independently uniformly distributed random variables.*

Proof If π is a perfectly secure computation protocol for (f_1, f_2) under $\mathrm{P}_{X_1X_2}$ with transcript Π and output (Z_1, Z_2), we have $\Pr(Z_1 = f_1(X_1, X_2), Z_2 = f_2(X_1, X_2)) = 1$. Consider a new FCP $\tilde{\pi}$ for $\tilde{f}$ which first executes π, and then $\mathcal{P}_i$ computes $Z_i + K_i$, $i = 1, 2$, and sends it to the other party. The protocol $\tilde{\pi}$ then outputs $\tilde{Z} = (Z_1 + K_1, Z_2 + K_2)$ for both parties. For this new protocol $\tilde{\pi}$, we also have $\Pr(\tilde{Z} = \tilde{f}((X_1, K_1), (X_2, K_2))) = 1$, for every joint distribution

$P_{X_1X_2K_1K_2}$. Furthermore, when (K_1, K_2) are independent from all the other random variables, we have

$$\begin{aligned}
&I(X_2, K_2 \wedge \tilde{\Pi}, U_1, U_0 \mid X_1, K_1, \tilde{Z}) \\
&\quad = I(X_2, K_2 \wedge \Pi, U_1, U_0 \mid X_1, K_1, \tilde{Z}) \\
&\quad = I(X_2, K_2 \wedge \Pi, U_1, U_0 \mid X_1, K_1, Z_1, Z_2 + K_2) \\
&\quad \le I(X_2, K_2, Z_2 + K_2 \wedge \Pi, U_1, U_0 \mid X_1, K_1, Z_1) \\
&\quad = I(X_2, K_2 \wedge \Pi, U_1, U_0 \mid X_1, K_1, Z_1) \\
&\quad = I(X_2 \wedge \Pi, U_1, U_0 \mid X_1, Z_1) \\
&\quad = 0,
\end{aligned}$$

where in the fourth step we used the fact that Z_2 is a function of (X_1, X_2) and in the fifth step we used the independence of (K_1, K_2) from all the other random variables. Similarly, we can see that $I(X_1, K_1 \wedge \tilde{\Pi}, U_2, U_0 \mid X_2, K_2, \tilde{Z}) = 0$, which shows that $\tilde{\pi}$ is a perfectly secure computation protocol for $\tilde{f}$.

Conversely, given a secure computation protocol $\tilde{\pi}$ for $\tilde{f}$ under $P_{X_1X_2} \times P_{K_1K_2}$ where K_1, K_2 are independent of each other and uniformly distributed, consider the protocol π described below.

1. $\mathcal{P}_i$ generates K_i, uniformly and independently of all the other random variables.
2. Parties execute $\tilde{\pi}$ with inputs (X_1, K_1) and (X_2, K_2) to get the transcript $\tilde{\Pi}$ and output $\tilde{Z} = (\tilde{Z}_1, \tilde{Z}_2)$.
3. $\mathcal{P}_i$ computes the output $Z_i = \tilde{Z}_i - K_i$.

We denote the protocol above by π and its transcript by[5] $\Pi = \tilde{\Pi}$; note that in addition to the local and shared randomness (U_1, U_2) and U_0, respectively, used for $\tilde{\pi}$, (K_1, K_2) will be treated as local randomness for π. Clearly, if $\tilde{\pi}$ has perfect reliability for $\tilde{f}$, then so does π for (f_1, f_2). In particular, $Z_1 = f_1(X_1, X_2)$ and $Z_2 = f_2(X_1, X_2)$ with probability 1. For security conditions, since K_1 is independent of $(X_2, X_1, Z_1, Z_2, K_2)$ and Z_2 is a function of (X_1, X_2), observe that

$$\begin{aligned}
&I(X_2 \wedge \Pi, U_1, U_0, K_1 \mid X_1, Z_1) \\
&\quad \le I(X_2 \wedge \Pi, U_1, U_0, K_1, Z_2 + K_2 \mid X_1, Z_1) \\
&\quad = I(X_2 \wedge K_1, Z_2 + K_2 \mid X_1, Z_1) + I(X_2 \wedge \Pi, U_1, U_0 \mid X_1, K_1, Z_1, Z_2 + K_2) \\
&\quad \le I(X_1, X_2 \wedge K_1, Z_2 + K_2) + I(X_2, K_2 \wedge \tilde{\Pi}, U_1, U_0 \mid X_1, K_1, \tilde{Z}) \\
&\quad = 0,
\end{aligned}$$

where the second inequality used $H(Z_1 \mid X_1, X_2) = 0$ and in the last step we used $I(X_1, X_2 \wedge K_1, Z_2 + K_2) = 0$, which holds since $(K_1, Z_2 + K_2)$ is independent of (X_1, X_2) as (K_1, K_2) are independent of (X_1, X_2) and uniform; and $I(X_2, K_2 \wedge$

[5] The transcripts of the two protocols coincide since they only differ in sampling the local randomness K_i and using it to compute the final output.

$\tilde{\Pi}, U_1, U_0 \mid X_1, K_1, \tilde{Z}) = 0$, which holds by the perfect security of $\tilde{\pi}$. Similarly, we can show that $I(X_1 \wedge \Pi, U_2, U_0, K_2 \mid X_2, Z_2) = 0$, which completes the proof. □

Thus, we can simply consider deterministic symmetric functions, instead of deterministic asymmetric functions. However, we still need to handle randomized asymmetric functions. Again, it turns out that secure computation of randomized asymmetric functions is equivalent to secure computation of deterministic asymmetric functions. To formally see this result, we note that the reliability condition (12.5) needs to be modified for randomized functions as follows.

For a given x_1, x_2, $f(x_1, x_2) = (f_1(x_1, x_2), f_2(x_1, x_2))$ is now a random variable with a fixed distribution (i.e., a channel with input (x_1, x_2)). Denoting by (Z_1, Z_2) the output of the protocol, the reliability condition now requires that

$$\mathbb{E}\big[d_{\mathtt{var}}\left(\mathrm{P}_{W_f(\cdot,\cdot|X_1,X_2)}, \mathrm{P}_{Z_1Z_2|X_1,X_2}\right)\big] \le \varepsilon,$$

where W_f denotes the channel from the input to the output of the function f and the expectation is over $(X_1, X_2) \sim \mathrm{P}_{X_1X_2}$, with $\varepsilon = 0$ for perfect reliability. Note that both local and shared randomness may be used in the protocol to produce the desired distribution $f(x_1, x_2)$; however, they must be accounted for in security analysis.

We now show how secure computation of a randomized function can be related to that of a deterministic function. Specifically, consider a randomized asymmetric function from $\mathcal{X}_1 \times \mathcal{X}_2$ to $\mathcal{Z}_1 \times \mathcal{Z}_2$. By introducing a uniform random variable S on a finite alphabet $\mathcal{S}$ as the "internal state," such a function can be described as $(f_1(x_1, x_2, S), f_2(x_1, x_2, S))$ using deterministic functions (f_1, f_2) of the expanded input (x_1, x_2, S), where S is not known by any party. The next result shows that a randomized function (f_1, f_2) is secure computable if a deterministic function is secure computable.

LEMMA 12.6 *Consider a randomized function (f_1, f_2) with $f_i \colon \mathcal{X}_i \to \mathcal{Z}_i$, $i = 1, 2$, which can be expressed as a deterministic function with uniform internal state $S \in \mathcal{S}$. Further, consider the deterministic asymmetric function $(\tilde{f}_1, \tilde{f}_2)$ given by*

$$\tilde{f}_i((x_1, s_1), (x_2, s_2)) := f_i(x_1, x_2, s_1 + s_2), \quad i = 1, 2.$$

Then, any perfectly secure computation protocol for $(\tilde{f}_1, \tilde{f}_2)$ under $\mathrm{P}_{X_1X_2} \times \mathrm{P}_{S_1S_2}$, where S_1 and S_2 are independent and both uniformly distributed over $\mathcal{S}$, also gives a perfectly secure computation protocol for (f_1, f_2).

Proof Given a perfectly secure computation protocol $\tilde{\pi}$ for $(\tilde{f}_1, \tilde{f}_2)$, consider the FPC for (f_1, f_2) where $\mathcal{P}_i$, $i = 1, 2$, first samples S_i uniformly over $\mathcal{S}$ and independently of everything else, and then the parties execute $\tilde{\pi}$ with inputs $((X_1, S_1), (X_2, S_2))$.

Since $S = S_1 + S_2$ is distributed uniformly over $\mathcal{S}$ and $\tilde{\pi}$ has perfect reliability, the distribution of the output (Z_1, Z_2) of resulting protocol has the same distribution given (X_1, X_2) as $(f_1(X_1, X_2, S), f_2(X_1, X_2, S))$ given (X_1, X_2). Thus, π also has perfect reliability.

For security, the only difference between π and $\tilde{\pi}$ is that the former has (S_1, S_2) as local randomness and the latter as inputs. Besides that, the transcripts Π and $\tilde{\Pi}$ and the outputs (Z_1, Z_2) and $(\tilde{Z}_1, \tilde{Z}_2)$ of the two protocols coincide. Thus, for security for $\mathcal{P}_1$, we have

$$\begin{aligned}
&I(X_2 \wedge \Pi, S_1, U_1, U_0 \mid X_1, Z_1) \\
&\leq I(X_1, X_2, Z_1 \wedge S_1) + I(X_2 \wedge \Pi, U_1, U_0 \mid X_1, S_1, Z_1) \\
&= I(X_1, X_2, f_1(X_1, X_2, S_1 + S_2) \wedge S_1) + I(X_2 \wedge \tilde{\Pi}, U_1, U_0 \mid X_1, S_1, \tilde{Z}_1) \\
&= 0,
\end{aligned}$$

where the final step uses

$$I(X_1, X_2, f_1(X_1, X_2, S_1 + S_2) \wedge S_1) \leq I(X_1, X_2, S_1 + S_2 \wedge S_1) = 0,$$

and the perfect security of $\tilde{\pi}$. Similarly, we can show security for $\mathcal{P}_2$, completing the proof. □

Therefore, in view of the two results above, we can restrict our attention to secure computation of deterministic and symmetric functions, which is what we focus on in the remainder of the book; a function always means a symmetric function unless otherwise stated. We remark that the two results above continue to hold in the approximate security setting as well; see Problems 12.1 and 12.2.

12.3 Perfect Secure Computation from Scratch

Earlier we saw that NOT and XOR are secure computable using trivial protocols that simply reveal the data of the parties. With a little thought, we can convince ourselves that a broader class of functions are secure computable in this manner. Roughly speaking, any function that is "equivalent" to the transcript of a tree protocol is secure computable by executing that tree protocol. But are there more interesting functions that are secure computable from scratch, namely without using any additional resources? Unfortunately, the answer is no: only trivial functions that are "essentially" equivalent to tree protocols are secure computable from scratch.

In this section, we provide a formal proof of this statement. We show that perfect SCPs exist only for "interactive functions," which are secure computable in a trivial manner. We start by showing this result for deterministic SCPs, since the proof is conceptually simpler in this case. But first, let us formalize what we mean by "interactive functions."

DEFINITION 12.7 (Equivalence of random variables) Two discrete random variables U, V with joint distribution P_{UV} are equivalent, denoted $U \equiv V$, if $H(U|V) = H(V|U) = 0$.

Table 12.1 $f(x_1, x_2)$

$x_1 \backslash x_2$	0	1	2
0	a	a	b
1	d	e	b
2	c	c	c

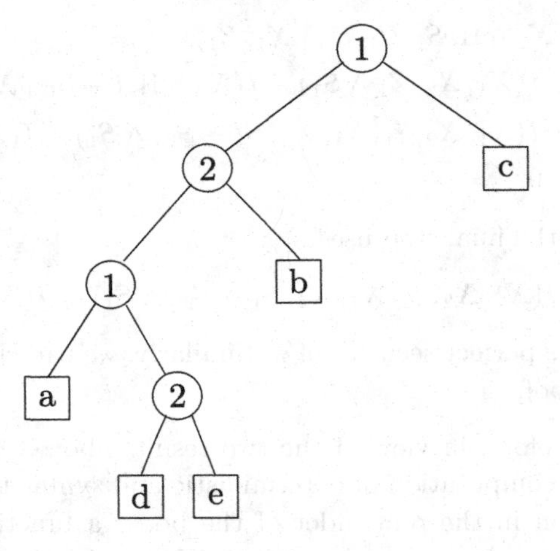

Figure 12.3 A protocol corresponding to the function in Table 12.1.

DEFINITION 12.8 (Interactive functions) A random variable $Z = f(X_1, X_2)$ is an interactive function under distribution $P_{X_1X_2}$ if there exists a deterministic tree protocol π with transcript Π satisfying $Z \equiv \Pi$. A function that is not interactive is called a noninteractive function under $P_{X_1X_2}$.

We illustrate the definition with an example.

EXAMPLE 12.9 Consider the function given in Table 12.1 and the protocol defined by output functions for each node of the protocol tree given below:

$$\Pi_{\texttt{root}} = \mathbf{1}[x_1 = 2],\ \Pi_0 = \mathbf{1}[x_2 = 2],\ \Pi_{00} = \mathbf{1}[x_1 = 1],\ \Pi_{001} = \mathbf{1}[x_2 = 1],$$

see Figure 12.3. Clearly, this protocol is a perfect FCP for f. For any distribution $P_{X_1X_2}$, since the label of a leaf uniquely determines the path to the leaf, $\Pi \equiv f(X_1, X_2)$, and thus the function $f(X_1, X_2)$, is an interactive function under $P_{X_1X_2}$.

An interactive function is one which, with probability 1, induces the same partition as an interactive communication. This property depends both on the structure of the function and on the support of the distribution $P_{X_1X_2}$. For instance, the function f defined in Table 12.2 is not an interactive function if $P_{X_1X_2}$ has full support, but it is an interactive function if $P_{X_1X_2}(2, 0) = 0$ since Table 12.2 is the same as Table 12.1 except $f(2, 0)$. When $Z = f(X_1, X_2)$ is an

Table 12.2 $f(x_1, x_2)$

$x_1 \backslash x_2$	0	1	2
0	a	a	b
1	d	e	b
2	d	c	c

interactive function under $P_{X_1X_2}$ having full support, then we just say that f is an interactive function.

Readers can take a pause and convince themselves that interactive functions are secure computable. Indeed, the parties can simply execute the interactive protocol equivalent to the function. This will reveal to the parties exactly the function value, and nothing more. Of course, this is not very impressive – interactive functions seem, in some sense, to be trivial secure computable functions.

Note that the XOR function, which was seen above to be secure computable, is not an interactive function for any distribution with full support (see Problem 12.3). In fact, there are functions beyond interactive functions that are secure computable.

When we compute XOR, since both the parties can recover the other party's input anyway, we end up computing the identity function $f(x_1, x_2) = (x_1, x_2)$, which is interactive. In general, when we compute a given function securely, we end up computing a "refinement" of the given function; to describe this formally, we need the notion of a *maximum common function.*

DEFINITION 12.10 (Maximum common function) A common function of finite valued random variables U and V with joint distribution P_{UV} is a random variable F such that there exist functions $\alpha(U)$ and $\beta(V)$ such that $P(\alpha(U) = \beta(V) = F) = 1$. The maximum common function of U and V, denoted $\mathtt{mcf}(U, V)$, is a common function of U and V for which every other common function F of U and V satisfies $H(F|\mathtt{mcf}(U, V)) = 0$.

Note that the maximum common function is uniquely defined up to the equivalence of random variables (cf. Definition 12.7); in the following, we use the most convenient labeling depending on context.

For a given joint distribution P_{UV}, an explicit form of $\mathtt{mcf}(U, V)$ can be written down as follows. To state the result, we need the notion of *connected components* of a graph, which we recall. Given an undirected graph $\mathcal{G} = (\mathcal{V}, \mathcal{E})$ with vertex set $\mathcal{V}$ and edge set $\mathcal{E}$, define $u \equiv v$ if and only if there is a path from u to v; it can be verified that the relation $u \equiv v$ is an equivalence relation on $\mathcal{V}$. Connected components of a graph are equivalence classes of this equivalence relation. Thus, two vertices have a path between them if and only if they are within the same connected component, and connected components form a partition of vertices of a graph.

LEMMA 12.11 *For given random variables (U, V) on $\mathcal{U} \times \mathcal{V}$ with joint distribution P_{UV}, let $\mathcal{G} = (\mathcal{U} \cup \mathcal{V}, \mathcal{E})$ be the bipartite graph such that $(u, v) \in \mathcal{E}$ if* *and only if* $\mathrm{P}_{UV}(u,v) > 0$. *Let* $\mathcal{C}$ *be the set of all connected components of the graph* $\mathcal{G}$. *Let* $\alpha\colon \mathcal{U} \to \mathcal{C}$ *be the map such that* $\alpha(u)$ *is the connected component containing* u. *Similarly, let* $\beta\colon \mathcal{V} \to \mathcal{C}$ *be the map such that* $\beta(v)$ *is the connected component containing* v. *Then, we have* $\mathtt{mcf}(U,V) = \alpha(U) = \beta(V)$.

Proof Since u and v with $\mathrm{P}_{UV}(u,v) > 0$ have an edge between them, they belong to the same connected component whereby $\alpha(u) = \beta(v)$. Thus, $\alpha(U) = \beta(V)$ with probability 1 and is a common function of U and V, and it suffices to show that any other common function $F = \tilde{\alpha}(U) = \tilde{\beta}(V)$ is a function of $\alpha(U) = \beta(V)$.

To that end, note that $\tilde{\alpha}(u) = \tilde{\beta}(v)$ whenever $\mathrm{P}_{UV}(u,v) > 0$ since F is a common function. Now, consider $u, u' \in \mathcal{U}$ which are in the same connected component of $\mathcal{G}$. Therefore, we can find $v_1, u_1, \ldots, v_{t-1}, u_{t-1}, v_t$, for some $t \in \mathbb{N}$, such that $\mathrm{P}_{UV}(u, v_1) > 0$, $\mathrm{P}_{UV}(u', v_t) > 0$, $\mathrm{P}_{UV}(u_i, v_i) > 0$, and $\mathrm{P}_{UV}(u_i, v_{i+1}) > 0$ for all $1 \le i \le t-1$. From the observation above, it follows that

$$\tilde{\alpha}(u) = \tilde{\beta}(v_1) = \tilde{\alpha}(u_1) = \cdots = \tilde{\beta}(v_t) = \tilde{\alpha}(u').$$

Thus, $\tilde{\alpha}$, and using a similar argument $\tilde{\beta}$ as well, must take a constant value on each connected component of $\mathcal{G}$, whereby F is a function of $\alpha(U) = \beta(V)$. □

Now, using the notion of the maximum common function, let us formally introduce the refinement of a given function.

DEFINITION 12.12 (Refinement) For a given function $f\colon \mathcal{X}_1 \times \mathcal{X}_2 \to \mathcal{Z}$ and a distribution $\mathrm{P}_{X_1X_2}$, the *refinement of* f *under* $\mathrm{P}_{X_1X_2}$ is defined by[6]

$$Z_{\mathtt{ref}} = f_{\mathtt{ref}}(X_1, X_2) = \mathtt{mcf}((X_1, Z), (X_2, Z)),$$

where $Z = f(X_1, X_2)$.

Going back to our running example of the XOR function $f(x_1, x_2) = x_1 \oplus x_2$, we can verify that its refinement is $Z_{\mathtt{ref}} = f_{\mathtt{ref}}(X_1, X_2) = (X_1, X_2)$. Clearly, the refinement $f_{\mathtt{ref}}(x_1, x_2)$ is a function from $\mathcal{X}_1 \times \mathcal{X}_2$ to a finite set $\mathcal{Z}'$. However, it should be noted that its definition depends on the support of the underlying distribution $\mathrm{P}_{X_1X_2}$ of the inputs.

Using the characterization of $\mathtt{mcf}$ in Lemma 12.11, we can derive a similar explicit characterization for the refinement of a function. We do that next, but for simplicity, restrict to *full support distributions*, namely distributions with $\{(x_1, x_2) : \mathrm{P}_{X_1X_2}(x_1, x_2) > 0\} = \mathcal{X}_1 \times \mathcal{X}_2$.

COROLLARY 12.13 *For a given function* $f\colon \mathcal{X}_1 \times \mathcal{X}_2 \to \mathcal{Z}$ *and* $z \in \mathcal{Z}$, *let* $\mathcal{G}_z = (\mathcal{X}_1 \cup \mathcal{X}_2, \mathcal{E}_z)$ *be the bipartite graph such that* $(x_1, x_2) \in \mathcal{E}_z$ *if and only if* $f(x_1, x_2) = z$. *Then, let* $\alpha_z\colon \mathcal{X}_1 \to \mathcal{C}_z$ *and* $\beta_z\colon \mathcal{X}_2 \to \mathcal{C}_z$ *be the maps such that* $\alpha_z(x_1)$ *and* $\beta_z(x_2)$ *are the connected components of* $\mathcal{G}_z$ *containing* x_1 *and* x_2,

[6] For a pair $((U_1, U_2), (V_1, V_2))$ of random vectors, the maximum common function is defined by regarding $U = (U_1, U_2)$ and $V = (V_1, V_2)$ as a pair (U, V) of random variables.

Table 12.3 $f(x_1, x_2)$ (left) in Example 12.14 and its refinement $f_{\texttt{ref}}(x_1, x_2)$ (right)

$x_1 \backslash x_2$	0	1	2	3
0	a	a	c	d
1	c	b	b	d
2	c	d	a	a
3	b	d	c	b

$x_1 \backslash x_2$	0	1	2	3
0	(a,1)	(a,1)	(c,2)	(d,1)
1	(c,1)	(b,1)	(b,1)	(d,1)
2	(c,1)	(d,2)	(a,2)	(a,2)
3	(b,2)	(d,2)	(c,2)	(b,2)

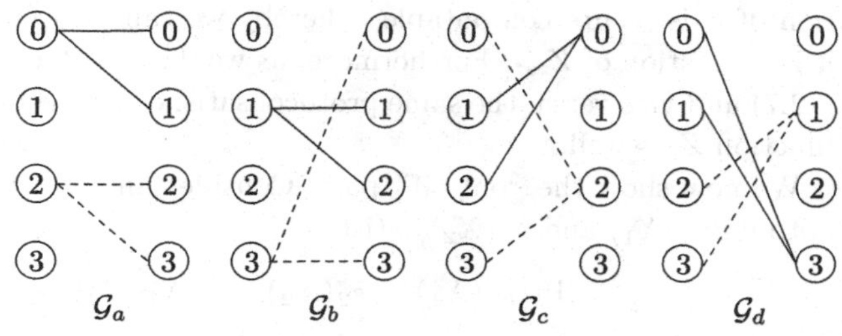

Figure 12.4 The bipartite graphs $\mathcal{G}_z$ of the function in Table 12.3. For each graph, the solid edges correspond to the first connected component and the dashed edges correspond to the second connected component.

respectively. Then, the refinement of f under a full support distribution can be written as

$$f_{\texttt{ref}}(x_1, x_2) = (f(x_1, x_2), \alpha_{f(x_1,x_2)}(x_1)) = (f(x_1, x_2), \beta_{f(x_1,x_2)}(x_2)).$$

Proof Consider the bipartite graph $\mathcal{G} = ((\mathcal{X}_1 \times \mathcal{Z}) \cup (\mathcal{X}_2 \times \mathcal{Z}), \mathcal{E})$ induced by the joint distribution of $((X_1, Z), (X_2, Z))$ with $Z = f(X_1, X_2)$ as in Lemma 12.11, i.e., the graph with edges corresponding to pairs (x_1, z_1) and (x_2, z_2) that lie in the support of $\mathrm{P}_{(X_1,Z)(X_2,Z)}$. Note that $((x_1, z_1), (x_2, z_2)) \in \mathcal{E}$ if and only if $z_1 = z_2 = z$ and $f(x_1, x_2) = z$ for some $z \in \mathcal{Z}$, whereby the graph $\mathcal{G}$ can be identified with the disjoint union of the graphs $\mathcal{G}_z$ for $z \in \mathcal{Z}$. It follows that $(z, \alpha_z(x_1))$ corresponds to the connected component containing (x_1, z) in $\mathcal{G}$, and $(z, \beta_z(x_2))$ corresponds to the connected component containing (x_2, z) in $\mathcal{G}$. Thus, the claimed corollary holds simply by applying Lemma 12.11 to random variables $U = (X_1, Z)$ and $V = (X_2, Z)$. □

EXAMPLE 12.14 Consider the function $f(x_1, x_2)$ given in Table 12.3 (left). For this function, the bipartite graphs $\mathcal{G}_z$ for $z \in \{a, b, c, d\}$ are given in Figure 12.4. Thus, Corollary 12.13 says that the refinement $f_{\texttt{ref}}(x_1, x_2)$ can be written as in Table 12.3 (right).

Once the parties compute a function $Z = f(X_1, X_2)$ securely, the parties can compute its refinement $Z_{\texttt{ref}} = f_{\texttt{ref}}(X_1, X_2)$ without leaking information beyond what was already leaked in computing f. In fact, it holds that $(X_i, Z) \equiv (X_i, Z_{\texttt{ref}})$ for $i = 1, 2$; from this fact, it can be verified that the security conditions in (12.6) imply

$$d_{\mathtt{var}}(\mathrm{P}_{X_1X_2Z_{\mathtt{ref}}\Pi U_iU_0}, \mathrm{P}_{X_1X_2Z_{\mathtt{ref}}} \times \mathrm{P}_{\Pi U_iU_0|X_iZ_{\mathtt{ref}}}) \le \delta, \ i = 1, 2, \tag{12.7}$$

and vice versa. The next result establishes this simple observation.

LEMMA 12.15 *A communication protocol* π *constitutes an* (ε, δ)*-SCP for a function* f *under* $\mathrm{P}_{X_1X_2}$ *if and only if it constitutes an* (ε, δ)*-SCP for the refinement* $f_{\mathtt{ref}}$ *of* f *under* $\mathrm{P}_{X_1X_2}$.

Proof The "if" part is straightforward: if $Z_{\mathtt{ref}}$ is secure computable, any function of it is secure computable, whereby we can compute $Z = f(X_1, X_2)$ since it is a function of $Z_{\mathtt{ref}}$. Furthermore, as we discussed above, since (12.6) implies (12.7) and vice versa, the same protocol satisfies the security requirement for the function Z as well.

We now show the "only if" part. Consider an (ε, δ)-SCP π for f. Then, the outputs $\pi_1(X_1)$ and $\pi_2(X_2)$ satisfy

$$\Pr(\pi_1(X_1) = \pi_2(X_2) = f(X_1, X_2)) \ge 1 - \varepsilon.$$

Consider $\alpha(X_1, Z)$ and $\beta(X_2, Z)$ such that $\Pr(Z_{\mathtt{ref}} = \alpha(X_1, Z) = \beta(X_2, Z)) = 1$. Define $\pi_1'(X_1) = \alpha(X_1, \pi_1(X_1))$ and $\pi_2'(X_2) = \beta(X_2, \pi_2(X_2))$. Then, we have

$$\Pr(\pi_1'(X_1) = \pi_2'(X_2) = Z_{\mathtt{ref}}) \ge 1 - \varepsilon.$$

Note that π_1' and π_2' can be treated as outputs of the same protocol π. Finally, since (12.6) implies (12.7), the same protocol π continues to satisfy the security requirement for the function $Z_{\mathtt{ref}}$. □

From here on, we only worry about secure computation of the refinement of functions. We note the following property of $f_{\mathtt{ref}}$ which reassures us that the refinement operation needs to be applied only once.

LEMMA 12.16 *Let* $Z_{\mathtt{ref}} = f_{\mathtt{ref}}(X_1, X_2)$ *denote the refinement of a function* $Z = f(X_1, X_2)$ *under* $\mathrm{P}_{X_1X_2}$. *Then, the refinement* $\mathtt{mcf}((X_1, Z_{\mathtt{ref}}), (X_2, Z_{\mathtt{ref}}))$ *of* $Z_{\mathtt{ref}}$ *under* $\mathrm{P}_{X_1X_2}$ *is equivalen to* $Z_{\mathtt{ref}}$.

Proof See Problem 12.4. □

With Lemma 12.15 at our disposal, we can expand our class of "trivial" secure computable functions. Any function for which the refinement is interactive is secure computable. It turns out that these are the only functions that are secure computable.

THEOREM 12.17 (Characterization of secure computable) *A function* $Z = f(X_1, X_2)$ *is perfectly secure computable under* $\mathrm{P}_{X_1X_2}$ *using a deterministic protocol if and only if its refinement* $Z_{\mathtt{ref}} = f_{\mathtt{ref}}(X_1, X_2)$ *under* $\mathrm{P}_{X_1X_2}$ *is an interactive function under* $\mathrm{P}_{X_1X_2}$.

Proof The "if" part is easy to see: from the definition of an interactive function, there exists a protocol π such that $\Pi \equiv Z_{\mathtt{ref}}$, and it is easy to verify that it constitutes a perfect SCP for $f_{\mathtt{ref}}$.

For the "only if" part, when f is secure computable, then so is $f_{\mathtt{ref}}$ by Lemma 12.15. Consider a perfect SCP π for $f_{\mathtt{ref}}$ under $\mathrm{P}_{X_1X_2}$ and denote by Π its random transcript.[7] The perfect security condition (12.2) gives

$$\begin{aligned} 0 &= I(X_2 \wedge \Pi | X_1, Z_{\mathtt{ref}}) = H(\Pi | X_1, Z_{\mathtt{ref}}), \\ 0 &= I(X_1 \wedge \Pi | X_2, Z_{\mathtt{ref}}) = H(\Pi | X_2, Z_{\mathtt{ref}}), \end{aligned} \tag{12.8}$$

where we used the fact that π is a deterministic protocol. Thus, Π is a common function of $(X_1, Z_{\mathtt{ref}})$ and $(X_2, Z_{\mathtt{ref}})$.

Further, by the perfect recovery condition (setting $\varepsilon = 0$ in (12.5)), we have

$$\begin{aligned} \mathtt{mcf}((X_1, \Pi), (X_2, \Pi)) &\equiv \mathtt{mcf}((X_1, \Pi, Z_{\mathtt{ref}}), (X_2, \Pi, Z_{\mathtt{ref}})) \\ &\equiv \mathtt{mcf}((X_1, Z_{\mathtt{ref}}), (X_2, Z_{\mathtt{ref}})) \\ &\equiv Z_{\mathtt{ref}}, \end{aligned}$$

where we used (12.8) and Lemma 12.16 in the last two steps. Therefore, $Z_{\mathtt{ref}}$ is equivalent to $\mathtt{mcf}((X_1, \Pi), (X_2, \Pi))$. To complete the proof, we establish that the latter is an interactive function.

Indeed, by the definition of $\mathtt{mcf}$ we get

$$\begin{aligned} \mathtt{mcf}((X_1, \Pi), (X_2, \Pi)) &\equiv (\Pi, \mathtt{mcf}((X_1, \Pi), (X_2, \Pi))) \\ &\equiv (\Pi, \alpha(X_1, \Pi)) \\ &\equiv (\Pi, \beta(X_2, \Pi)) \end{aligned}$$

for some mappings α and β. Hence, we can enhance the interactive protocol Π with one more round of communication $\alpha(X_1, \Pi)$ or $\beta(X_2, \Pi)$ to obtain a new interactive protocol that is equivalent to $Z_{\mathtt{ref}}$, which completes the proof. $\square$

A few remarks are in order. First, while the joint distribution $\mathrm{P}_{X_1X_2}$ is fixed, the characterization of secure computable functions above depends only on the support. As long as this support is fixed, the class of secure computable functions remains the same, namely interactive functions for the uniform distribution on this support.

Also, note that our proof above relies on the assumptions of perfect secrecy and deterministic protocols to get (12.8). For public-coin protocols with U_1 and U_2 constant, since U_0 is independent of the inputs (X_1, X_2), the perfect security condition in (12.2) can be written equivalently as

$$I(X_2 \wedge \Pi | X_1, Z, U_0) = I(X_1 \wedge \Pi | X_2, Z, U_0) = 0.$$

This case can also be handled by our proof by simply repeating the argument for every fixed realization of U_0. However, our proof above does not go beyond this, and in particular, cannot handle nonconstant private randomness U_1, U_2.

We now present an alternative proof that extends the result above to the case when private randomness is allowed. To enable this, we need a more elaborate

[7] Here, the random transcript means that the transcript Π computed for random inputs $(X_1, X_2) \sim \mathrm{P}_{X_1X_2}$.

understanding of the structure of an interactive function. Heuristically, an interactive function is obtained by recursively partitioning the input space $\mathcal{X}_1 \times \mathcal{X}_2$ one coordinate at a time until we reach parts where the function values are constant. In each step when we partition one coordinate, $\mathcal{X}_1$ or $\mathcal{X}_2$, it must be done without "cutting across" a set where the function value does not change – this will maintain perfect security.

For the alternative proof, we simply try to directly identify this partition recursively in a greedy fashion. At each step, we partition a coordinate in the maximal possible way, without violating security, and terminate when we cannot move forward. When we stop, either the function will be constant in each part, or not. In the former case, the function is interactive, and we can compute it securely by following the same partition that we constructed. However, in the latter case, we can show that the function is not secure computable for a distribution with full support. Those "bad" parts are termed "forbidden sets." We will show that, if the refinement of a given function f is a noninteractive function, then f must have a forbidden set; furthermore, any function with forbidden set cannot be perfectly secure computable, even when private-coin protocols are allowed. Thus, if a function is such that its refinement is not interactive, then it is not secure computable even using private-coin protocols.

We now formalize this argument. We begin by describing the notion of "forbidden sets" which in turn rely on the following equivalence relation.

DEFINITION 12.18 (f-equivalence) Given a subset S of $\mathcal{X}_1 \times \mathcal{X}_2$, consider the binary relation $\sim_S$ on $\mathcal{X}_1$ defined as $x_1 \sim_S x_1'$ if there exists $x_2 \in \mathcal{X}_2$ such that $(x_1, x_2) \in S$, $(x_1', x_2) \in S$, and $f(x_1, x_2) = f(x_1', x_2)$. This relation is clearly reflexive and symmetric. The f-equivalence $=_S$ for $\mathcal{X}_1$ (corresponding to S) is the transitive closure of $\sim_S$. Namely, $x_1 =_S x_1'$ if there exist $u_1, \ldots, u_k \in \mathcal{X}_1$ such that $u_1 = x_1$, $u_k = x_1'$, and $u_i \sim_S u_{i+1}$ for $1 \le i \le k-1$. Similarly, we can define f-equivalence for $\mathcal{X}_2$.

Note that equivalence classes of f-equivalence $=_S$ on $\mathcal{X}_1$ and $\mathcal{X}_2$ with $S = \{(x_1, x_2) : f(x_1, x_2) = z\}$ coincide with the connected components in the definition of functions α_z and β_z, respectively, used in Corollary 12.13 to characterize $f_{\mathtt{ref}}$.

Consider a distribution $\mathrm{P}_{X_1 X_2}$ with support given by S. Any perfect SCP should not distinguish between any two points $x_1 \sim_S x_1'$ since it is possible that $\mathcal{P}_2$'s input is an x_2 such that $f(x_1, x_2) = f(x_1', x_2) = z$. In this case, the function value z and the observation x_2 will not allow $\mathcal{P}_2$ to distinguish between x_1 and x_1'. Then, in a perfect SCP, too, $\mathcal{P}_2$ is also forbidden from distinguishing between the pair x_1 and x_1' using the transcript and x_2. Thus, using its observation and the transcript of a perfect SCP, $\mathcal{P}_2$ is also forbidden from distinguishing between any two x_1 and x_1' such that $x_1 =_S x_1'$. Even though this necessary condition must hold for any S, in the following, we particularly focus on sets of the form $S = A \times B$. Note that, if all $x_1 \in A$ are f-equivalent and all $x_2 \in B$ are f-equivalent, any perfect SCP should not distinguish between any two points in it.

This motivates the notion of *forbidden sets* below, sets which should not appear for a perfectly secure computable function.

DEFINITION 12.19 (Forbidden sets) For $C \subset \mathcal{X}_1$ and $D \subset \mathcal{X}_2$, the set $S = C \times D$ is a forbidden subset for f if the following hold:

1. $x_1 =_S x_1'$ for all $x_1, x_1' \in A$;
2. $x_2 =_S x_2'$ for all $x_2, x_2' \in B$;
3. there exist two points (x_1, x_2) and (x_1', x_2') in S such that $f(x_1, x_2) \neq f(x_1', x_2')$.

It can be seen that for the function given in Table 12.2, the entire domain $\mathcal{X}_1 \times \mathcal{X}_2$ is a forbidden set. Also, for the AND function, the domain $\{0, 1\} \times \{0, 1\}$ is a forbidden set.

Next, we show a connection between functions whose refinement is noninteractive (namely, functions we saw earlier to be not perfectly secure computable from scratch) and forbidden sets.

LEMMA 12.20 (Forbidden sets are forbidden) *Let $f \colon \mathcal{X}_1 \times \mathcal{X}_2 \to \mathcal{Z}$ be a function such that its refinement $f_{\mathtt{ref}}$ is noninteractive under a distribution $\mathrm{P}_{X_1X_2}$ with full support. Then, there exits a forbidden set for f.*

Proof By contraposition, we will show that when f does not have any forbidden set, then $f_{\mathtt{ref}}$ must be an interactive function under every distribution with full support. Towards that, consider the following recursive procedure.

1. Initialize $S = \mathcal{X}_1 \times \mathcal{X}_2$.
2. For $S = C \times D$, apply either of the following steps:
   i. partition C using the equivalence classes under $=_S$;
   ii. partition D using the equivalence classes under $=_S$.
3. Apply the previous step recursively to each new part obtained.

The procedure stops when further partitioning of C or D using $=_{C\times D}$ is not possible. Note that the recursive procedure above describes an interactive communication protocol where in each step the corresponding party sends the equivalence class its observation lies in. In other words, the final partition obtained corresponds to that induced by an interactive function. Furthermore, each of the final parts must have all points in C or D equivalent under $=_{C\times D}$. If forbidden sets do not exist, then these parts must all have the same value of the function. Therefore, the interactive communication protocol π described above constitutes a perfect FCP for f. Moreover, π is a deterministic protocol.

We will argue that this protocol π is secure as well. This will hold under any distribution $\mathrm{P}_{X_1X_2}$ with full support, whereby Theorem 12.17 yields that $f_{\mathtt{ref}}$ is an interactive function under any $\mathrm{P}_{X_1X_2}$ with full support. Therefore, it only remains to show that π is secure.

To see this, we note that protocol π is deterministic and so, by (12.8), security will follow upon showing that the transcript Π of protocol π is a common function of (X_1, Z) and (X_2, Z). To show that Π is a function of (X_1, Z), it suffices to show that for any (x_1, x_2) and (x_1, x_2') such that $f(x_1, x_2) = f(x_1, x_2')$, the protocol π produces the same transcript. Namely, these two points must lie in the same part of the partition obtained by our procedure. If this property does not hold, then consider the first step at which our procedure puts these two points in different parts. Note that at the beginning of this step, the corresponding set S will contain both these points. But then these two points, x_2 and x_2', are equivalent under $=_S$ since $f(x_1, x_2) = f(x_1, x_2')$, and our procedure will not assign them to different parts, which is a contradiction. Thus, this property holds. Similarly, we can show that Π is a function of (X_2, Z), which completes the proof. □

So, now we know that, if $f_{\mathtt{ref}}$ is a noninteractive function, then f must have forbidden sets. Finally, we show that if we have forbidden sets for f, then f cannot be perfectly secure computable. These observations together lead to our main result below.

THEOREM 12.21 *A function f is perfectly secure computable under $\mathrm{P}_{X_1X_2}$ with full support if and only if its refinement $f_{\mathtt{ref}}$ under $\mathrm{P}_{X_1X_2}$ is an interactive function.*

Proof We have already shown that if $f_{\mathtt{ref}}$ is an interactive function, then we can find a deterministic perfect SCP for f.

Conversely, by Lemma 12.20, we know that, if $f_{\mathtt{ref}}$ is noninteractive, then it must have a forbidden set S. To complete the proof, we show that f cannot be perfectly secure computable. Suppose this does not hold, i.e., there exists a perfect SCP π for f. In the following, we show that the transcript Π must have the same distribution for any point (x_1, x_2) in the forbidden set S; however, since the function f takes more than one value on S, this must mean that the protocol cannot guarantee perfect recovery, which is a contradiction.

To complete the proof, denoting by $\Pi^t = (\Pi_1, \ldots, \Pi_t)$ the transcript obtained in the first $t \geq 1$ rounds of communication, we show

$$\Pr(\Pi^t = \tau^t | X_1 = x_1, X_2 = x_2) = \Pr(\Pi^t = \tau^t | X_1 = x_1', X_2 = x_2')$$

for every (x_1, x_2) and (x_1', x_2') in the forbidden set $S = C \times D$. To prove the claim by induction on t, suppose that $\Pr(\Pi^{t-1} = \tau^{t-1} | X_1 = x_1, X_2 = x_2)$ remains the same for every $(x_1, x_2) \in S$.[8] If this probability is 0, we get that $\Pr(\Pi^t = \tau^t | X_1 = x_1, X_2 = x_2)$ is 0 for every $(x_1, x_2) \in S$. Thus, it suffices to consider transcript values τ such that

$$\Pr(\Pi^{t-1} = \tau^{t-1} | X_1 = x_1, X_2 = x_2) = c > 0, \quad \forall (x_1, x_2) \in S. \tag{12.9}$$

Since the forbidden set is of the form $C \times D$, the point (x_1', x_2) must lie in the forbidden set.[9] The desired claim can be proved if we show

[8] For $t = 1$, there is no inductive assumption, and we prove the claim for $t = 1$ as a special case of general $t \geq 2$ in the following.

[9] A set satisfying this property is called a *rectangle set*. Each part of the partition induced by the transcripts of a tree protocol on the input set $\mathcal{X}_1 \times \mathcal{X}_2$ is a rectangle set. See Section 12.7 for more discussion on rectangle sets.

$$\begin{aligned}\Pr\big(\Pi^t = \tau^t | X_1 = x_1, X_2 = x_2\big) &= \Pr\big(\Pi^t = \tau^t | X_1 = x_1', X_2 = x_2\big) \\ &= \Pr\big(\Pi^t = \tau^t | X_1 = x_1', X_2 = x_2'\big).\end{aligned}$$

We only show the first identity; we can show the second identity in a similar manner by flipping the roles of $\mathcal{P}_1$ and $\mathcal{P}_2$.

The interactive structure of the protocol gives

$$\mathrm{P}_{\Pi^t|X_1=x_1,X_2=x_2} = \begin{cases} \mathrm{P}_{\Pi^{t-1}|X_1=x_1,X_2=x_2}\mathrm{P}_{\Pi_t|X_1=x_1,\Pi^{t-1}}, & t \text{ odd}, \\ \mathrm{P}_{\Pi^{t-1}|X_1=x_1,X_2=x_2}\mathrm{P}_{\Pi_t|X_2=x_2,\Pi^{t-1}}, & t \text{ even}. \end{cases}$$

Thus, for even t, the inductive assumption immediately implies

$$\mathrm{P}_{\Pi^t|X_1=x_1,X_2=x_2} = \mathrm{P}_{\Pi^t|X_1=x_1',X_2=x_2}.$$

For odd t, if we show

$$\mathrm{P}_{\Pi_t|X_1=x_1,\Pi^{t-1}} = \mathrm{P}_{\Pi_t|X_1=x_1',\Pi^{t-1}}, \tag{12.10}$$

then the inductive assumption yields

$$\mathrm{P}_{\Pi^t|X_1=x_1,X_2=x_2} = \mathrm{P}_{\Pi^t|X_1=x_1',X_2=x_2}.$$

To establish (12.10), note that the definition of forbidden set implies $x_1 =_S x_1'$; in particular, we can find $u_1, \ldots, u_k \in \mathcal{X}_1$ such that $u_1 = x_1$, $u_k = x_1'$, and $u_i \sim_S u_{i+1}$ for $1 \le i \le k-1$. We use induction to show that, for every i,

$$\mathrm{P}_{\Pi_t|X_1=u_i,\Pi^{t-1}} = \mathrm{P}_{\Pi_t|X_1=u_{i+1},\Pi^{t-1}}, \tag{12.11}$$

from which (12.10) follows. Since $u_i \sim_S u_{i+1}$, there is a $y \in D$ such that (u_i, y) and (u_{i+1}, y) lie in S and $f(u_i, y) = f(u_{i+1}, y) = z$. It can be verified that the perfect security requirement enforces the following condition: for (x_1, x_2) such that $f(x_1, x_2) = z$,

$$\mathrm{P}_{\Pi|X_1=x_1,X_2=x_2} = \mathrm{P}_{\Pi|X_1=x_1,Z=z} = \mathrm{P}_{\Pi|X_2=x_2,Z=z}.$$

Thus, perfect security implies

$$\mathrm{P}_{\Pi|X_1=u_i,X_2=y} = \mathrm{P}_{\Pi|X_1=u_{i+1},X_2=y}.$$

By taking the marginal distribution for the first t rounds of communication, we get

$$\mathrm{P}_{\Pi^t|X_1=u_i,X_2=y} = \mathrm{P}_{\Pi^t|X_1=u_{i+1},X_2=y}.$$

Thus, we have

$$\begin{aligned}\mathrm{P}_{\Pi^{t-1}|X_1=u_i,X_2=y}\mathrm{P}_{\Pi_t|X_1=u_i,\Pi^{t-1}} &= \mathrm{P}_{\Pi^t|X_1=u_i,X_2=y} \\ &= \mathrm{P}_{\Pi^t|X_1=u_{i+1},X_2=y} \\ &= \mathrm{P}_{\Pi^{t-1}|X_1=u_{i+1},X_2=y}\mathrm{P}_{\Pi_t|X_1=u_{i+1},\Pi^{t-1}},\end{aligned}$$

which gives

$$\Pr\big(\Pi_t = \tau_t | X_1 = u_i, \Pi^{t-1} = \tau^{t-1}\big) = \Pr\big(\Pi_t = \tau_t | X_1 = u_{i+1}, \Pi^{t-1} = \tau^{t-1}\big)$$

for all τ satisfying (12.9). Since this holds for every $1 \le i \le k$, we get (12.10), which completes the proof. □

Remark 12.1 As we have verified in the proof of Theorem 12.21, if there exists a forbidden set S for f, then any perfectly secure protocol π must satisfy $P_{\Pi|X_1X_2}(\cdot|x_1,x_2) = P_{\Pi|X_1X_2}(\cdot|x_1',x_2')$ for any $(x_1,x_2),(x_1',x_2') \in S$, which implies

$$\Pr\left(\hat{Z}_i(X_i,\Pi) = z|(X_1,X_2) = (x_1,x_2)\right) = \Pr\left(\hat{Z}_i(X_i,\Pi) = z|(X_1,X_2) = (x_1',x_2')\right)$$

for any $z \in \mathcal{Z}$. For (x_1,x_2) and (x_1',x_2') such that $f(x_1,x_2) \neq f(x_1',x_2')$, which must exist since S is forbidden, we have

$$\Pr\left(\hat{Z}_i(X_i,\Pi) = f(X_1,X_2)|(X_1,X_2) = (x_1,x_2)\right) < 1$$

or

$$\Pr\left(\hat{Z}_i(X_i,\Pi) = f(X_1,X_2)|(X_1,X_2) = (x_1',x_2')\right) < 1.$$

By adjusting the prior distribution $P_{X_1X_2}$, this means that the protocol π is not ε-FCP for any $\varepsilon > 0$.

For both the AND function and the function in Table 12.2, since the domains constitute forbidden sets, these functions are not perfectly secure computable.

For Boolean functions, we can derive a more explicit characterization of secure computable functions.

COROLLARY 12.22 (Secure computable Boolean functions) *A Boolean function f is perfectly secure computable under $P_{X_1X_2}$ with full support if and only if there exist Boolean functions $f_1(x_1)$ and $f_2(x_2)$ such that*

$$f(x_1,x_2) = f_1(x_1) \oplus f_2(x_2). \tag{12.12}$$

Proof If f can be written as (12.12), then its refinement is an interactive function, and thus perfectly secure computable. To prove the opposite implication, fix an arbitrary element $a_2 \in \mathcal{X}_2$ such that $f(x_1,a_2) = 0$ for some x_1; if such an a_2 does not exist, then f is a constant function and the claim is trivially satisfied. For a fixed a_2 as above, let

$$C = \{x_1 \in \mathcal{X}_1 : f(x_1,a_2) = 0\},$$
$$D = \{x_2 \in \mathcal{X}_2 : \forall x_1 \in C,\ f(x_1,x_2) = 0\}.$$

Note that C is not the empty set, and $a_2 \in D$. Denoting $C^c = \mathcal{X}_1 \backslash C$ and $D^c = \mathcal{X}_2 \backslash D$, we prove the following four claims.

1. For every $(x_1,x_2) \in C \times D$, $f(x_1,x_2) = 0$. This follows from the definition of the set D.
2. For every $(x_1,x_2) \in C \times D^c$, $f(x_1,x_2) = 1$. To prove by contradiction, assume that there exists $(x_1,x_2) \in C \times D^c$ such that $f(x_1,x_2) = 0$. By the definition of the set D^c, there exists $a_1 \in C$ such that $f(a_1,x_2) = 1$. Furthermore, the definition of C implies $f(x_1,a_2) = f(a_1,a_2) = 0$. Thus, $\{x_1,a_1\} \times$

$\{x_2, a_2\}$ constitutes a forbidden subset for f, which contradicts that f is secure computable.

3. For every $(x_1, x_2) \in C^c \times D$, $f(x_1, x_2) = 1$. To prove by contradiction, assume that there exists $(x_1, x_2) \in C^c \times D$ such that $f(x_1, x_2) = 0$. Note that $x_2 \neq a_2$; otherwise, $x_1 \in C$. Fix an arbitrary $a_1 \in C$. Then, we have $f(a_1, a_2) = 0$. Furthermore, $f(a_1, x_2) = 0$ since $x_2 \in D$ and $f(x_1, a_2) = 1$ since $x_1 \in C^c$. Thus, $\{x_1, a_1\} \times \{x_2, a_2\}$ constitutes a forbidden subset for f, which contradicts that f is secure computable.
4. For every $(x_1, x_2) \in C^c \times D^c$, $f(x_1, x_2) = 0$. To prove by contradiction, assume that there exists $(x_1, x_2) \in C^c \times D^c$ such that $f(x_1, x_2) = 1$. Note that $f(x_1, a_2) = 1$ since $x_1 \in C^c$. Fix an arbitrary $a_1 \in C$. Then, we have $f(a_1, a_2) = 0$. According to the second claim, $f(a_1, x_2) = 1$. Thus, $\{x_1, a_1\} \times \{x_2, a_2\}$ constitutes a forbidden subset for f, which contradicts that f is secure computable.

Consequently, by setting $f_1(x_1) = \mathbf{1}[x_1 \in C]$ and $f_2(x_2) = \mathbf{1}[x_2 \in D]$, we can write f in the form of (12.12). □

We close this section with a negative result for the example we mentioned at the beginning of this chapter.

EXAMPLE 12.23 (Millionaires' problem) When there exists x_1, x_1', x_2, x_2' such that $f(x_1, x_2) = f(x_1, x_2') = f(x_1', x_2) = z$ and $f(x_1', x_2') \neq z$, then $\{x_1, x_1'\} \times \{x_2, x_2'\}$ constitutes a forbidden subset, and f is not secure computable. For the function in the millionaires' problem, $\mathcal{X}_1 = \mathcal{X}_2$ is a set of bounded integers and the function is comparing x_1 and x_2, i.e., $f(x_1, x_2) = 1$ if and only if $x_1 \geq x_2$. Then, for every integer a, we have $f(a-1, a) = f(a-1, a+1) = f(a, a+1) = 0$ and $f(a, a) = 1$. Thus, this function is not secure computable; in other words, the millionaires' problem cannot be solved without some additional resources.

12.4 Oblivious Transfer

The previous section would have disappointed an optimistic reader by showing that only trivial functions are secure computable from scratch. We now lift the spirits of our reader by declaring at the outset that all functions are secure computable if we assume the availability of an additional resource. This resource is access to a correlated randomness by the parties. In the remainder of this chapter, we present this result.

There are various forms of correlation that will suffice. But in this book, we focus on a specific form of correlation that is of fundamental importance. This correlation is called *oblivious transfer* (OT). In fact, OT itself can be viewed as a function, and OT can be realized from other resources, which will be discussed in Chapter 13. The previous statement is often summarized as follows:

OT is complete for secure function computation.

Specifically, OT is an asymmetric function (cf. Section 12.2) defined as follows.

DEFINITION 12.24 ($\binom{m}{1}$-Oblivious transfer) For $\mathcal{X}_1 = \{0,1\}^m$ and $\mathcal{X}_2 = \{0,1,\ldots,m-1\}$, suppose that $\mathcal{P}_1$ observes m bits denoted by $x = (x_0,\ldots,x_{m-1}) \in \mathcal{X}_1$ and $\mathcal{P}_2$ observes a number $b \in \mathcal{X}_2$. The $\binom{m}{1}$-oblivious transfer ($\binom{m}{1}$-OT) is an asymmetric function where $\mathcal{P}_2$ seeks to compute the bth bit x_b of $\mathcal{P}_1$'s input and $\mathcal{P}_1$ seeks no output. Specifically, the $\binom{m}{1}$-OT function is specified by (f_1, f_2) with domain $\mathcal{X}_1 \times \mathcal{X}_2$ and functions f_1, f_2 given by $f_1(x,b) = \emptyset$ and $f_2(x,b) = x_b$, where $\emptyset$ denotes that f_1 is constant.

Note that while $\mathcal{P}_2$ obtains x_b, it learns nothing about any other bit of $\mathcal{P}_1$. Also, $\mathcal{P}_1$ does not learn anything about b. Thus, at a high level, OT provides an advantage to each party over the other, something which simple shared randomness lacks.

We will see in the next section that, if we have unlimited access to independent copies of $\binom{4}{1}$-OT, then we can securely compute any function. For now, let us convince ourselves that this is plausible by looking at a very simple function – the AND function. In fact, for the secure computation of the AND function, we only need a $\binom{2}{1}$-OT.

The protocol is very simple. $\mathcal{P}_1$ looks at its bit x_1 and computes the list $(x_1 \wedge 0, x_1 \wedge 1)$. This is basically the list of values of the AND function for all possible values of $\mathcal{P}_2$'s input. Next, $\mathcal{P}_1$ and $\mathcal{P}_2$ can use $\binom{2}{1}$-OT for $\mathcal{P}_2$ to obtain the entry in the list corresponding to x_2, which is $x_1 \wedge x_2$. Finally, the parties can apply the same procedure in the opposite direction. Note that at the end, both parties have just learned AND of the bits and nothing else. Thus, the protocol is perfectly secure.

This protocol extends to any function easily if we have access to $\binom{\min\{|\mathcal{X}_1|,|\mathcal{X}_2|\}}{1}$-OT. Specifically, if $|\mathcal{X}_1| > |\mathcal{X}_2|$, $\mathcal{P}_1$ can compute the list $(f(x_1, y) : y \in \mathcal{X}_2)$ and $\mathcal{P}_2$ can obtain $f(x_1, x_2)$ using an OT, and then $\mathcal{P}_2$ can send back the outcome to $\mathcal{P}_1$. But this is not a satisfactory solution since it seems that the form of correlation needed can change with the problem. However, this simple illustration would hopefully convince the reader that OT is fundamental for secure computation.

Note that we do not mind requiring different "amounts" of resources for different functions, but the form should remain the same. Specifically, we would like to use independent copies of the same form of resource for solving all problems, though the number of copies used may vary across functions. Below, we show that a Boolean function can be securely computed using as many copies of $\binom{4}{1}$-OT as the number of NAND gates[10] required to realize the function.

[10] The NAND gate outputs the complement of AND when applied to two bits b_1 and b_2, i.e., it outputs $1 \oplus (b_1 \wedge b_2)$.

12.5 Oracle Access and Memoryless Channels

Before we proceed to show the completeness result mentioned above, we formalize the notion of *oracle access* in this section. Also, we relate it to the notion of *memoryless channels*, which is a basic model for noisy communication in information theory. Algorithms with oracle access were seen earlier in Chapter 9, albeit for a computational setting. An algorithm can make queries to a stochastic oracle by selecting an input, and it receives the corresponding output. The random inputs across multiple queries may be correlated arbitrarily, depending on how the algorithm selects them. However, given the sequence of inputs, the random outputs must be conditionally independent, where the conditional distribution is specified by a fixed channel W.

Recall from Chapter 2 that, for finite sets $\mathcal{A}$ and $\mathcal{B}$, a discrete channel $W : \mathcal{A} \to \mathcal{B}$ with input alphabet $\mathcal{A}$ and output alphabet $\mathcal{B}$ specifies a conditional distribution $W(\cdot|a)$ on $\mathcal{B}$ for every $a \in \mathcal{A}$. When $a \in \mathcal{A}$ is sent over the channel W, the random output takes value $b \in \mathcal{B}$ with probability $W(b|a)$. Such a channel is called a discrete memoryless channel (DMC), when multiple inputs $A_1 = a_1, \ldots, A_n = a_n$ are given to it, then the corresponding outputs $B_1, \ldots, B_n$ are independent. That is,

$$\Pr(B_1 = b_1, \ldots, B_n = b_n | A_1 = a_1, \ldots, A_n = a_n) = \prod_{t=1}^{n} W(b_t|a_t). \tag{12.13}$$

To extend the notion of oracle access to interactive protocols, we bring in another notion from information theory – that of a *two-way channel* (TWC). For finite sets $\mathcal{A}_1, \mathcal{A}_2, \mathcal{B}_1, \mathcal{B}_2$, a TWC is a discrete channel $W: \mathcal{A}_1 \times \mathcal{A}_2 \to \mathcal{B}_1 \times \mathcal{B}_2$. This channel is made available to two parties $\mathcal{P}_1$ and $\mathcal{P}_2$ where $\mathcal{P}_i$ selects the input A_i and receives the output B_i. We consider *discrete memoryless* TWCs, namely TWCs which satisfy the DMC property (12.13) when used multiple times. A *memoryless oracle* in the interactive setting will be specified by a TWC W which satisfies (12.13).

DEFINITION 12.25 (Protocols with memoryless oracle) Given a memoryless oracle $W: \mathcal{A}_1 \times \mathcal{A}_2 \to \mathcal{B}_1 \times \mathcal{B}_2$, an interactive protocol with T oracle accesses to W is a multistage protocol executed by $\mathcal{P}_1$ and $\mathcal{P}_2$. In the tth stage, the parties communicate using a tree protocol π_t and use the output of the tree protocol, along with the previous observations to select inputs (A_{1t}, A_{2t}) for the oracle access to W. The corresponding outputs of the oracle (B_{1t}, B_{2t}) are revealed to the parties; $\mathcal{P}_i$ gets B_{it}. These new observations, and all the other previous observations, constitute the input for the protocol π_{t+1} used in the next stage. Finally, the protocol terminates after T stages, when the parties compute outputs based on their local observations.

Expressed simply, protocols with oracle access can be viewed as communication protocols where, in addition to the noiseless communication channel which the parties were using in each round of an interactive protocol, they can use a

TWC for communication. Incidentally, a TWC is used to model so-called *full-duplex communication* in wireless communication. Like many notions in information theory, this was introduced by Shannon in his early works. However, understanding of reliable communication over a discrete memoryless TWC, and the design of error correcting code for such a channel, are still preliminary. It is this notion of TWC that plays a critical role in secure computing; however, we shall consider a few, very specific TWCs throughout this book. In particular, we shall use the TWC corresponding to an OT quite often. We describe it below.

The memoryless oracle corresponding to $\binom{m}{1}$-OT is given by the TWC W with input alphabets $\mathcal{A}_1 = \{0,1\}^m$ and $\mathcal{A}_2 = \{0,1,\ldots,m-1\}$ and output alphabets $\mathcal{B}_1 = \{\perp\}$ and $\mathcal{B}_2 = \{0,1\}$, and channel transition matrix W given by

$$W(\perp, y|x, b) = \mathbf{1}[y = x_b], \quad \forall y \in \mathcal{B}_2, x \in \mathcal{A}_1, b \in \mathcal{A}_2.$$

12.6 Completeness of Oblivious Transfer

In this section, we show how any function can be securely computed using interactive protocols with oracle access to OT. We assume a memoryless oracle access to $\binom{4}{1}$-OT. We state the result for Boolean functions, but it can be extended to any function. We rely on a basic result from Boolean logic: any Boolean function can be realized using a circuit consisting of only NAND gates, with some of the inputs set to appropriate constants (0 or 1).

Specifically, consider a Boolean function $f\colon \{0,1\}^m \times \{0,1\}^n \to \{0,1\}$ which can be computed using a circuit of depth h comprising only NAND gates. Some of the inputs of these NAND gates may be set to constants. The rest can correspond either to input bits from $x_1 \in \{0,1\}^m$ and $x_2 \in \{0,1\}^n$, or to outputs of other NAND gates in the circuit. When studying secure computing, we assume that constant inputs are a part of the observations of one of the parties, say a part of the input x_1 of $\mathcal{P}_1$. This assumption is without loss of generality since both parties know the function f, and thereby also its NAND circuit representation.

We now set the notation to describe the depth h of a circuit for f. For $1 \le i \le h$, the ith level of the circuit comprises N_i NAND gates f_{ij}, $1 \le j \le N_i$. Note that the final level h has exactly one gate. The inputs and the output of the gate f_{ij} are denoted by (l_{ij}, r_{ij}) and z_{ij}, respectively. An example of the circuit representation of a function is depicted in Figure 12.5.

In should be emphasized that l_{ij} and r_{ij} are not necessarily owned by $\mathcal{P}_1$ and $\mathcal{P}_2$, respectively. For level 1, the inputs of a gate may be owned by one each, or owned by one of the parties; the latter case means that the gate is just a local processing. For a level $i \ge 2$, the ownership cannot be assigned any more. In fact, in the following construction of the protocol, once a preprocessing phase is done, the ownerships of inputs are irrelevant.

Next, we extend the notion of security to protocols with oracle access. The key difference between the oracle setting and the earlier setting of tree protocols is

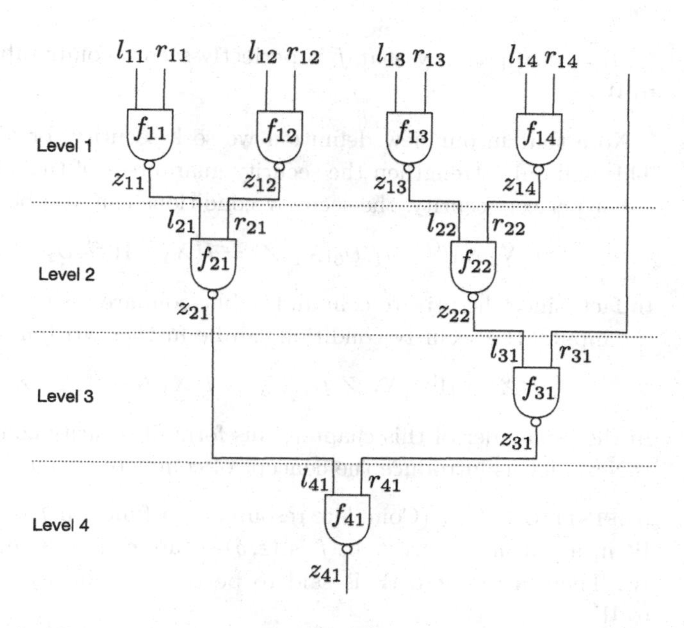

Figure 12.5 An illustration of the circuit representation of a function.

that, in the oracle setting, the two parties may get different transcripts from the protocol. For instance, when the OT is invoked during a protocol, then only $\mathcal{P}_2$ will get an output from the OT; we regard this output as a part of the transcript observed by $\mathcal{P}_2$. We denote the transcripts of parties $\mathcal{P}_1$ and $\mathcal{P}_2$ generated by an interactive protocol π with oracle access by $\Pi^{(1)}$ and $\Pi^{(2)}$, respectively. Note that these transcripts include the transcript of the interactive protocols in different stages, which are seen by both the parties, and the output received by the parties from memoryless oracle access. The parties use these transcripts to compute the output $\hat{Z}_1(x_1, x_2) = \hat{Z}_1(x_1, \Pi^{(1)}, U_1, U_0)$ and $\hat{Z}_2(x_1, x_2) = \hat{Z}_2(x_2, \Pi^{(2)}, U_2, U_0)$ as before.[11]

DEFINITION 12.26 (SCP with oracle access) Given a memoryless oracle W, an interactive protocol π with oracle access to W constitutes an (ε, δ)-SCP if, for every input distribution $\mathrm{P}_{X_1X_2}$, the outputs $\hat{Z}_1(x_1, x_2)$ and $\hat{Z}_2(x_1, x_2)$ of the protocol satisfy (12.1) and the following security conditions

$$d_{\mathrm{var}}\left(\mathrm{P}_{X_1X_2Z\Pi^{(1)}U_1U_0}, \mathrm{P}_{X_1X_2Z}\mathrm{P}_{\Pi^{(1)}U_1U_0|X_1Z}\right) \leq \delta,$$
$$d_{\mathrm{var}}\left(\mathrm{P}_{X_1X_2Z\Pi^{(2)}U_2U_0}, \mathrm{P}_{X_1X_2Z}\mathrm{P}_{\Pi^{(2)}U_2U_0|X_2Z}\right) \leq \delta,$$

where (U_1, U_2) and U_0 denote private-coin and public-coin as before and $Z = f(X_1, X_2)$. When an (ε, δ)-SCP with oracle access to W exists for f, we say that f is (ε, δ)-secure computable using oracle access to W. In particular, for

[11] Recall the notational convention in Definition 12.2.

$(\varepsilon, \delta) = (0, 0)$, we say that f is perfectly secure computable using oracle access to W.

Note that in our new definition we seek security for all input distributions. This will only strengthen the security guarantees of the scheme we present.

For perfect security, the security conditions can also be written as

$$I(X_2 \wedge \Pi^{(1)}, U_1, U_0 | X_1, Z) = I(X_1 \wedge \Pi^{(2)}, U_2, U_0 | X_2, Z) = 0.$$

In fact, since the private-coin and public-coin are assumed to be independent of the inputs, the security condition can be further written as

$$I(X_2 \wedge \Pi^{(1)} | X_1, Z, U_1, U_0) = I(X_1 \wedge \Pi^{(2)} | X_2, Z, U_2, U_0) = 0.$$

In the remainder of this chapter, this form of security condition will be used.

Next, let us introduce the concept of complete resource.

DEFINITION 12.27 (Complete resource) A function f is *reducible* to a resource W if, for arbitrary $\varepsilon, \delta > 0$, f is (ε, δ)-secure computable using oracle access to W. Then, a resource W is said to be *complete* if any function f is reducible to W.

The following theorem is one of the fundamental results on secure computing.

THEOREM 12.28 (OT is complete for a passive adversary) *For $\mathcal{X}_1 = \{0,1\}^m$ and $\mathcal{X}_2 = \{0,1\}^n$, a Boolean function $f \colon \mathcal{X}_1 \times \mathcal{X}_2 \to \{0,1\}$ is perfectly secure computable using oracle access to $\binom{4}{1}$-OT.*

Moreover, the number of oracle accesses needed is at most the number of NAND gates needed to realize f.

In Problem 12.6, we will see that $\binom{4}{1}$-OT can be obtained from three invocations of $\binom{2}{1}$-OT, which together with Theorem 12.28 imply the following corollary.

COROLLARY 12.29 *The $\binom{2}{1}$-OT is a complete resource.*

The proof of Theorem 12.28 provides a general recipe for secure computation. In fact, the recipe provided here will be a prototype for secure computation protocol against a malicious adversary to be discussed in Chapter 17. Note that the malicious adversary is very different from our current setting where the parties are *honest-but-curious*, and will not deviate from the mutually agreed protocol. Security in this honest-but-curious setting simply forbids information leakage when both parties follow the protocol.

At a high level, our procedure starts from the lowest level of the circuit and proceeds towards level h, securely computing each gate at each level. However, the problem with this naive view is that we will end up revealing the outputs of some of the intermediate gates to the parties. This will leak more information than mandated by the final function f. To fix this, the protocol we present uses a *preprocessing step* to replace original inputs with their secure "clones," in effect, and then applies the protocol to these secure clones in the manner that the final output obtained can be used to obtain the actual function output using a *postprocessing step.*

Proof of Theorem 12.28
The protocol we present uses the idea of secret sharing discussed in Chapter 11. But we only use the simplest scheme, the $(2, 2)$-threshold scheme, and the following presentation is self contained without referring to Chapter 11.

For the input (l, r) of a NAND gate, suppose that $\mathcal{P}_1$ observes $(l^{(1)}, r^{(1)})$ and $\mathcal{P}_2$ observes $(l^{(2)}, r^{(2)})$ such that $l^{(1)} \oplus l^{(2)} = l$ and $r^{(1)} \oplus r^{(2)} = r$. In other words, $\mathcal{P}_1$ and $\mathcal{P}_2$ observe the shares of the input of the NAND gate; the observations of each party are not enough to identify the inputs. In the following, we will construct a subprotocol such that $\mathcal{P}_1$ and $\mathcal{P}_2$ obtain $z^{(1)}$ and $z^{(2)}$ satisfying $z^{(1)} \oplus z^{(2)} = \mathtt{NAND}(l, r)$ as the outputs; in other words, the parties obtain the shares of the output of the NAND gate.

In the preprocessing phase, the parties first create shares for all the input bits. Then, using the above mentioned subprotocol, the parties evaluate NAND gates one by one. Finally, by exchanging the shares of the output of the final NAND gate, the parties reconstruct the output of the entire function. The details of each phase are described as follows.

1. **Preprocessing step.** $\mathcal{P}_1$ and $\mathcal{P}_2$ use their respective private-coins to sample m random coins C and n random coins D, respectively. Then, for $\mathcal{P}_1$'s input $X_1 \in \{0,1\}^m$ and $\mathcal{P}_2$'s input $X_2 \in \{0,1\}^n$, $\mathcal{P}_1$ sends $\overline{X}_1 = C \oplus X_1$ to $\mathcal{P}_2$ and $\mathcal{P}_2$ sends $\overline{X}_2 = D \oplus X_2$ to $\mathcal{P}_1$, where $\oplus$ means bitwise modulo-sum. Further, $\mathcal{P}_1$ and $\mathcal{P}_2$ retain C and D, respectively. Consequently, $\mathcal{P}_1$ obtains $(C, \overline{X}_2)$ and $\mathcal{P}_2$ obtains $(\overline{X}_1, D)$. By noting $X_1 = C \oplus \overline{X}_1$ and $X_2 = D \oplus \overline{X}_2$, $(C, \overline{X}_1)$ and $(\overline{X}_2, D)$ can be regarded as shares of X_1 and X_2 by the $(2, 2)$-threshold secret sharing scheme.
2. **Computing secretly shared gates.** In this step, the protocol follows the circuit, one NAND gate at a time, starting from level 1 and proceeding to level h. When the protocol reaches a level $1 \le i \le h$, parties $\mathcal{P}_1$ and $\mathcal{P}_2$, respectively, already have shares $(L_{ij}^{(1)}, R_{ij}^{(1)}) \in \{0,1\}^2$ and $(L_{ij}^{(2)}, R_{ij}^{(2)}) \in \{0,1\}^2$ for two input bits (l_{ij}, r_{ij}) corresponding to the gate f_{ij}, $1 \le j \le N_i$. We proceed iteratively and compute a "secret shared NAND gate" where each party gets one independent share of the output. This sets up the iterative process that can be applied to each gate.

   We now describe the computation of each secretly shared NAND gate. The NAND gate can be viewed as an AND gate with its output passed through a NOT gate. We show separately how to implement the secret shared versions of each of these gates. The secret shared NAND gate will be obtained by running each implementation in composition.

   Secret shared NOT gate. For an input v for the NOT gate, $\mathcal{P}_i$ gets the share v_i, $i = 1, 2$, such that $v = v_1 \oplus v_2$. $\mathcal{P}_1$ computes $z^{(1)} = v_1 \oplus 1$ and $\mathcal{P}_2$ sets

Protocol 12.1 Secret shared AND gate

Input: $(l^{(1)}, r^{(1)})$ owned by $\mathcal{P}_1$ and $(l^{(2)}, r^{(2)})$ owned by $\mathcal{P}_2$ such that $l = l^{(1)} \oplus l^{(2)}$ and $r = r^{(1)} \oplus r^{(2)}$

Output: $\mathcal{P}_1$ gets $z^{(1)}$ and $\mathcal{P}_2$ gets $z^{(2)}$ such that $l \wedge r = z^{(1)} \oplus z^{(2)}$

1. $\mathcal{P}_1$ uses its private-coins to sample a random coin $C_0 \in \{0, 1\}$ and sets $z^{(1)} = C_0$.
2. $\mathcal{P}_1$ and $\mathcal{P}_2$ execute a $\binom{4}{1}$-OT with $\mathcal{P}_1$'s input set to the vector

$$(C_0 \oplus l^{(1)} \wedge r^{(1)}, C_0 \oplus (l^{(1)} \oplus 1) \wedge r^{(1)},$$
$$C_0 \oplus l^{(1)} \wedge (r^{(1)} \oplus 1), C_0 \oplus (l^{(1)} \oplus 1) \wedge (r^{(1)} \oplus 1))$$

   and $\mathcal{P}_2$'s input $(l^{(2)}, r^{(2)}) = l^{(2)} + 2r^{(2)} \in \{0, 1, 2, 3\}$.
3. $\mathcal{P}_2$ sets $z^{(2)}$ to be the output of the OT.

$z^{(2)} = v_2$. Clearly, $z^{(1)}$ and $z^{(2)}$ constitute shares of the output $\neg v = v \oplus 1$ and satisfy $\neg v = z^{(1)} \oplus z^{(2)}$. Note that no communication was exchanged in this step.

Secret shared AND gate. For inputs l and r for an AND gate, $\mathcal{P}_i$ gets the shares $(l^{(i)}, r^{(i)})$, $i = 1, 2$, such that $l = l^{(1)} \oplus l^{(2)}$ and $r = r^{(1)} \oplus r^{(2)}$. The parties seek to obtain shares $z^{(1)}$ and $z^{(2)}$ of the output $z = l \wedge r$ satisfying $z^{(1)} \oplus z^{(2)} = z$. Implementation of this task is nontrivial, and we use oracle access to a $\binom{4}{1}$-OT. We describe the process in Protocol 12.1.

3. **Postprocessing step.** At the completion of the process, the parties have shares $Z_{h1}^{(1)} \in \{0, 1\}$ and $Z_{h1}^{(2)} \in \{0, 1\}$ for the output of the final gate f_{h1}. They simply reveal these shares to each other and estimate the function value as $\hat{Z} = Z_{h1}^{(1)} \oplus Z_{h1}^{(2)}$.

With the protocol described, we now analyze its reliability and security. The reliability part is easy to see since throughout we compute secret shared versions of the same gate as the circuit, with parties holding shares of inputs in the original circuit. Specifically, it can be shown using induction that each subprotocol π_{ij} for computing the gate f_{ij} reveals to the parties shares of the output of f_{ij} in the original circuit, whereby the π_{h1} reveals shares of $f(x_1, x_2)$ to the parties. Thus, the final estimate $\hat{Z}$ must coincide with Z.

Before we verify the security of this protocol, we divide the security requirement of the entire protocol into those for each phase of the protocol. Towards that, denote by Π_0 and Π_∞ the transcripts of the protocol in the preprocessing and postprocessing steps. After the preprocessing step, the parties execute subprotocols π_{ij}, $1 \leq i \leq h$, $1 \leq j \leq N_i$, one for each gate f_{ij}. Denote the transcript of each of these subprotocols by Π_{ij} and the overall transcript of this intermediate phase by $\tilde{\Pi} = (\Pi_{ij} : 1 \leq i \leq h, 1 \leq j \leq N_i)$. Thus, the transcript of the protocol is given by $\Pi = (\Pi_0, \tilde{\Pi}, \Pi_\infty)$, with components received by $\mathcal{P}_1$ and $\mathcal{P}_2$ denoted by $\Pi^{(1)}$ and $\Pi^{(2)}$.

Note that the protocols π_{ij} in the intermediate phase are applied to inputs derived after the preprocessing stage and the outputs of the subprotocols executed earlier. We denote by $(L_{ij}^{(1)}, R_{ij}^{(1)})$ and $(L_{ij}^{(2)}, R_{ij}^{(2)})$ the input pair for π_{ij}, with the former denoting $\mathcal{P}_1$'s inputs and the latter denoting $\mathcal{P}_2$'s inputs. Note that, depending on the structure of the circuit, these inputs are obtained from preprocessed inputs and/or the previously received transcripts. We denote by $\tilde{\Pi}_{ij-}$ the concatenation of the transcripts $(\Pi_{k\ell} : 1 \le k \le i-1, 1 \le \ell \le N_k)$ up to the $(i-1)$th level and the transcript $(\Pi_{i1}, \ldots, \Pi_{i(j-1)})$ up to the $(j-1)$th gate of level i.

We remark that while we do need to execute the π_{ij} in the order of levels i, there is no need to execute the π_{ij} for the same level i in any specific order. They can even be executed simultaneously.[12] However, for convenience of security analysis, we assume that, even for the same level, the π_{ij} are executed in the increasing order of $1 \le j \le N_i$.

Consider the security of the overall protocol π. We present only security against dishonest $\mathcal{P}_2$; security against dishonest $\mathcal{P}_1$ can be analyzed similarly.[13] First, the security condition of the entire protocol can be expanded as

$$\begin{aligned}
&I(X_1 \wedge \Pi^{(2)}|Z, X_2, U_2, U_0) \\
&\quad = I(X_1 \wedge \Pi_0^{(2)}|Z, X_2, U_2, U_0) + I(X_1 \wedge \tilde{\Pi}^{(2)}|Z, X_2, U_2, U_0, \Pi_0^{(2)}) \\
&\qquad\qquad + I(X_1 \wedge \Pi_\infty^{(2)}|Z, X_2, U_2, U_0, \Pi_0^{(2)}, \tilde{\Pi}^{(2)}).
\end{aligned}$$

The first and the third terms in the expression on the right-hand side will turn out to be zero by ensuring that the preprocessing and the postprocessing phases are secure. For the second term, we can further expand as

$$\begin{aligned}
&I(X_1 \wedge \tilde{\Pi}^{(2)}|Z, X_2, U_2, U_0, \Pi_0^{(2)}) \\
&\quad = \sum_{i=1}^{h}\sum_{j=1}^{N_i} I(X_1 \wedge \Pi_{ij}^{(2)}|Z, X_2, U_2, U_0, \Pi_0^{(2)}, \tilde{\Pi}_{ij-}^{(2)}).
\end{aligned}$$

Thus, to establish perfect security, it suffices to show the following:

$$\begin{aligned}
I(X_1 \wedge \Pi_0^{(2)}|Z, X_2, U_2, U_0) &= 0, \\
I(X_1 \wedge \Pi_{ij}^{(2)}|Z, X_2, U_2, U_0, \Pi_0^{(2)}, \tilde{\Pi}_{ij-}^{(2)}) &= 0, \quad 1 \le i \le h, 1 \le j \le N_i, \qquad (12.14) \\
I(X_1 \wedge \Pi_\infty^{(2)}|Z, X_2, U_2, U_0, \Pi_0^{(2)}, \tilde{\Pi}^{(2)}) &= 0.
\end{aligned}$$

For the preprocessing step, $\Pi_0^{(2)} = C \oplus X_1$. Since C is independent of (X_1, X_2, U_2, U_0), we obtain

$$I(X_1 \wedge \Pi_0^{(2)}|Z, X_2, U_2, U_0) = 0.$$

[12] Since we consider a passive adversary in this chapter, simultaneous execution of subprotocols does not cause any problem. However, for an active adversary, it may cause a problem; see Section 15.8.

[13] In our protocol, $\mathcal{P}_2$ nominally receives more information, i.e., the outputs of OTs, than $\mathcal{P}_1$.

For the postprocessing step, from $\Pi_0^{(\xi)}$ and $\tilde{\Pi}^{(\xi)}$, $\mathcal{P}_\xi$ has access to $Z^{(\xi)}$, $\xi = 1, 2$, such that $Z^{(1)} \oplus Z^{(2)} = Z$. Furthermore, $\Pi_\infty^{(1)} = Z^{(2)}$ and $\Pi_\infty^{(2)} = Z^{(1)}$, whereby

$$\begin{aligned} & I(X_1 \wedge \Pi_\infty^{(2)} | Z, X_2, U_2, U_0, \Pi_0^{(2)}, \tilde{\Pi}^{(2)}) \\ & = I(X_1 \wedge Z^{(1)} | Z, X_2, U_2, U_0, \Pi_0^{(2)}, \tilde{\Pi}^{(2)}, Z^{(2)}) \\ & = 0. \end{aligned}$$

Finally, while executing the step corresponding to f_{ij}, the transcript Π_{ij} is exchanged. Note that $\Pi_{ij}^{(1)} = \emptyset$ and $\Pi_{ij}^{(2)} = C_{ij} \oplus M_{ij}$, where C_{ij} is independent of all the random variables encountered up to the jth gate of stage i and $M_{ij} = (L_{ij}^{(1)} \oplus L_{ij}^{(2)}) \wedge (R_{ij}^{(1)} \oplus R_{ij}^{(2)})$ depends on the shares of the inputs of the AND. That is, C_{ij} is sampled from private randomness of $\mathcal{P}_1$ and is independent jointly of

$$\mathcal{I}_{ij-}^{(2)} := (Z, X_2, U_2, U_0, \Pi_0^{(2)}, \tilde{\Pi}_{ij-}^{(2)}).$$

Thus,

$$I(X_1 \wedge \Pi_{ij}^{(2)} | \mathcal{I}_{ij-}^{(2)}) = I(X_1 \wedge C_{ij} \oplus M_{ij} | \mathcal{I}_{ij-}^{(2)}) = 0.$$

Therefore, all the terms in (12.14) are 0, whereby the protocol is secure. □

A shortcoming of the protocol is that we have assumed perfectly secure OT instances. Replacing this with an (ε, δ)-secure OT will require slight changes to the proof, though a similar template works. We will see this variant of the proof later, when we encounter the malicious adversary setting.

12.7 Classification of Deterministic Symmetric Functions

In Section 12.6, we have shown that $\binom{2}{1}$-OT is a complete function, i.e., any other function can be securely computed by a protocol with oracle access to $\binom{2}{1}$-OT. The purpose of this section is to provide a characterization of symmetric functions that are complete. Technically speaking, $\binom{2}{1}$-OT is not included in the class of complete functions provided in this section since it is not symmetric. However, a characterization of complete asymmetric functions is known in the literature; see references in Section 12.8.

In order to provide a characterization of complete functions, we need to introduce a few new concepts. The first one, a rectangle set, has already appeared in earlier sections implicitly and is defined formally below.

DEFINITION 12.30 (Rectangle) A subset $S \subseteq \mathcal{X}_1 \times \mathcal{X}_2$ is a *rectangle* if, whenever there exist $(x_1, x_2), (x_1', x_2') \in S$ with $x_1 \neq x_1'$ and $x_2 \neq x_2'$, it holds that $(x_1, x_2'), (x_1', x_2) \in S$.

A rectangle set has the following alternative characterization; the proof is left as an exercise (Problem 12.8).

LEMMA 12.31 *A subset $S \subseteq \mathcal{X}_1 \times \mathcal{X}_2$ is a* rectangle *if and only if it has a form $S = A \times B$ for some $A \subseteq \mathcal{X}_1$ and $B \subseteq \mathcal{X}_2$.*

We call a function rectangular if inverse images $f^{-1}(z)$ induced by the function are rectangles.

DEFINITION 12.32 (Rectangular function) A function $f\colon \mathcal{X}_1 \times \mathcal{X}_2 \to \mathcal{Z}$ is *rectangular* if the inverse image $f^{-1}(z) = \{(x_1, x_2) : f(x_1, x_2) = z\}$ is a rectangle for each $z \in \mathcal{Z}$.

In fact, we can verify that the class of interactive functions is a subclass of rectangular functions; we leave the proof as an exercise (Problem 12.9).

LEMMA 12.33 *An interactive function under a full support distribution is a rectangular function.*

We have already seen that the AND function is a representative function that is not secure computable without any additional resource. In fact, it will turn out that it also plays an important role in secure computing other functions as well.

DEFINITION 12.34 (AND-minor) A function $f\colon \mathcal{X}_1 \times \mathcal{X}_2 \to \mathcal{Z}$ has an *AND-minor* if there exist $x_1 \neq x_1'$ and $x_2 \neq x_2'$ such that $f(x_1', x_2) = f(x_1, x_2') = f(x_1, x_2)$ and $f(x_1', x_2') \neq f(x_1, x_2)$.

When we considered the secure computability of a function f, its refinement $f_{\mathtt{ref}}$ played an important role since we effectively compute the refinement $f_{\mathtt{ref}}$ when we securely compute the original function f. For the same reason, when we can use a function f as a resource, we can effectively use its refinement $f_{\mathtt{ref}}$ as well. The following lemma characterizes the existence of an AND-minor in a function in terms of its refinement.

LEMMA 12.35 *A function $f\colon \mathcal{X}_1 \times \mathcal{X}_2 \to \mathcal{Z}$ does not have an AND-minor if and only if its refinement $f_{\mathtt{ref}}$ under a full support distribution is rectangular.*

Proof Recall that we can write

$$f_{\mathtt{ref}}(x_1, x_2) = (f(x_1, x_2), \alpha_{f(x_1,x_2)}(x_1)) = (f(x_1, x_2), \beta_{f(x_1,x_2)}(x_2))$$

by using the functions α_z and β_z defined in Corollary 12.13.

First, we prove that, if f does not have an AND-minor, then $f_{\mathtt{ref}}$ is rectangular. For an arbitrarily fixed (x_1, x_2) with $f(x_1, x_2) = z$, let

$$\mathcal{X}_1(x_2, z) := \{x_1 : f(x_1, x_2) = z\},$$
$$\mathcal{X}_2(x_1, z) := \{x_2 : f(x_1, x_2) = z\}.$$

We will show below that

$$\mathcal{X}_1(x_2, z) = \mathcal{X}_1(x_2', z), \quad \forall x_2' \in \mathcal{X}_2(x_1, z), \tag{12.15}$$
$$\mathcal{X}_2(x_1, z) = \mathcal{X}_2(x_1', z), \quad \forall x_1' \in \mathcal{X}_1(x_2, z). \tag{12.16}$$

But before we establish these claims, we observe that they imply that for a given (x_1, x_2) with $f(x_1, x_2) = z$ and $\alpha_z(x_1) = \beta_z(x_2) = c$,

$$f_{\mathtt{ref}}^{-1}(z, c) = \mathcal{X}_1(x_2, z) \times \mathcal{X}_2(x_1, z), \tag{12.17}$$

and thus $f_{\mathtt{ref}}$ is rectangular. Indeed, for any $(x_1', x_2') \in \mathcal{X}_1(x_2, z) \times \mathcal{X}_2(x_1, z)$, since $x_1' \in \mathcal{X}_1(x_2, z) = \mathcal{X}_1(x_2', z)$ implies

$$f(x_1', x_2') = f(x_1', x_2) = f(x_1, x_2) = z,$$

(x_1', x_2') is in the same connected component as (x_1, x_2) in the graph $\mathcal{G}_z$, which implies $f_{\mathtt{ref}}(x_1', x_2') = (z, c)$. Thus, we have $\mathcal{X}_1(x_2, z) \times \mathcal{X}_2(x_1, z) \subseteq f_{\mathtt{ref}}^{-1}(z, c)$. On the other hand, for any $(x_1', x_2') \in f_{\mathtt{ref}}^{-1}(z, c)$, $\alpha_z(x_1) = \alpha_z(x_1') = c$ implies that there exists a path $x_1 = x_1^{(1)}, x_2^{(1)}, x_1^{(2)}, \ldots, x_1^{(m)}$ such that $f(x_1^{(i)}, x_2^{(i)}) = f(x_1^{(i+1)}, x_2^{(i)}) = z$ for $1 \le i \le m - 1$. Then, since $x_2^{(i-1)}, x_2^{(i)} \in \mathcal{X}_2(x_1^{(i)}, z)$ for $2 \le i \le m - 1$, (12.15) implies

$$\mathcal{X}_1(x_2^{(1)}, z) = \mathcal{X}_1(x_2^{(2)}, z) = \cdots = \mathcal{X}_1(x_2^{(m-1)}, z).$$

Since $\mathcal{X}_1(x_2, z) = \mathcal{X}_1(x_2^{(1)}, z)$ and $x_1' = x_1^{(m)} \in \mathcal{X}_1(x_2^{(m-1)}, z)$, we have $x_1' \in \mathcal{X}_1(x_2, z)$. Similarly, we have $x_2' \in \mathcal{X}_2(x_1, z)$ using (12.16), and thus $f_{\mathtt{ref}}^{-1}(z, c) \subseteq \mathcal{X}_1(x_2, z) \times \mathcal{X}_2(x_1, z)$.

To prove (12.15) by contraposition, suppose that it does not hold. Then, there exists $x_1' \neq x_1$ such that either $x_1' \in \mathcal{X}_1(x_2, z) \backslash \mathcal{X}_1(x_2', z)$ or $x_1' \in \mathcal{X}_1(x_2', z) \backslash \mathcal{X}_1(x_2, z)$. If the former is true (the latter case is proved similarly), we have $f(x_1', x_2) = f(x_1, x_2') = f(x_1, x_2) = z$ and $f(x_1', x_2') \neq z$, which contradict that f does not have an AND-minor. Thus, (12.15) must hold; similarly, (12.16) must hold.

Next, we shall show that, if $f_{\mathtt{ref}}$ is rectangular, then f does not have an AND-minor. To that end, suppose that there exist $x_1 \neq x_1'$ and $x_2 \neq x_2'$ such that $f(x_1, x_2') = f(x_1', x_2) = f(x_1, x_2) = z$. Then, since x_1, x_1', x_2, x_2' are in the same connected component in $\mathcal{G}_z$, we have $f_{\mathtt{ref}}(x_1, x_2') = f_{\mathtt{ref}}(x_1', x_2) = f_{\mathtt{ref}}(x_1, x_2)$ holds. Since $f_{\mathtt{ref}}$ is rectangular, we have $f_{\mathtt{ref}}(x_1', x_2') = f_{\mathtt{ref}}(x_1, x_2)$, which implies $f(x_1', x_2') = f(x_1, x_2)$. Thus, f does not have an AND-minor. □

As a consequence of Lemma 12.35, we can obtain the following information-theoretic characterization of functions with AND-minor.

COROLLARY 12.36 *A function $f \colon \mathcal{X}_1 \times \mathcal{X}_2 \to \mathcal{Z}$ does not have an AND-minor if and only if*

$$I(X_1 \wedge X_2 | \mathtt{mcf}((X_1, Z), (X_2, Z))) = 0$$

for all independent inputs (X_1, X_2) and $Z = f(X_1, X_2)$.

We are now ready to state a characterization of complete symmetric functions.

THEOREM 12.37 (Characterization of complete symmetric functions for a passive adversary) *A symmetric function f is a complete resource for secure computation under a passive adversary if and only if its refinement $f_{\mathtt{ref}}$ under a full support distribution is not rectangular.*

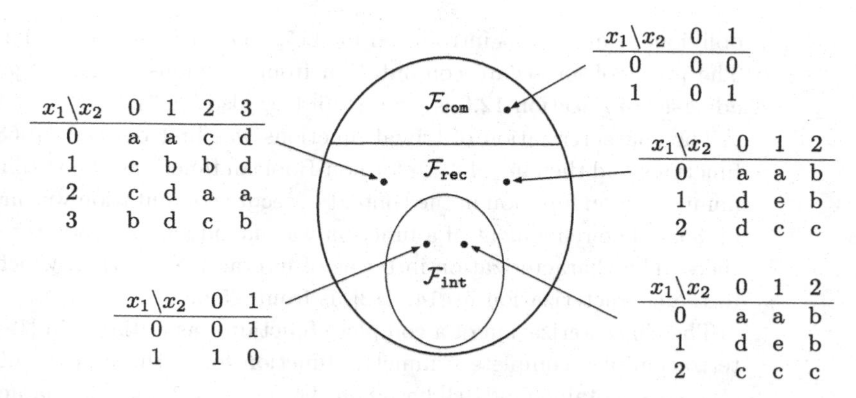

Figure 12.6 The inclusive relation among the three classes of functions and examples of representative functions of each class.

Since we have proved that the OT is a complete resource against a passive adversary in the previous section, in order to prove Theorem 12.37, it suffices to prove that the OT can be securely computed from f if and only if its refinement is not rectangular. We postpone the proof of this claim to the next chapter, in which the construction of an OT from other functions is studied.

By Lemma 12.35, Theorem 12.37 can be equivalently stated as follows: a function f is complete if and only if it contains an AND-minor. In fact, containing an AND-minor plays an important role in the proof of the "if" part, and non-rectangular characterization plays an important role in the proof of the "only if" part.

We conclude this section by classifying the set of all functions into three classes: the first class is the set $\mathcal{F}_{\texttt{int}}$ of all functions such that their refinements are interactive functions; the second class is the set $\mathcal{F}_{\texttt{rec}}$ of all functions such that their refinements are rectangular functions; and the third class is the set $\mathcal{F}_{\texttt{com}}$ of all functions that are a complete resource for a passive adversary. Theorem 12.21 says that the class $\mathcal{F}_{\texttt{int}}$ coincides with the set of all functions that are perfectly secure computable without any resource. On the other hand, Theorem 12.37 says that functions in the class $\mathcal{F}_{\texttt{rec}}$ are not a complete resource, and $\mathcal{F}_{\texttt{com}}$ coincides with the complement of $\mathcal{F}_{\texttt{rec}}$. The inclusion relation among the classes is described in Figure 12.6. It should be emphasized that the class $\mathcal{F}_{\texttt{int}}$ is a strict subset of $\mathcal{F}_{\texttt{rec}}$.

12.8 References and Additional Reading

The problem of secure computation was introduced in Yao's landmark paper [360]. In the 1980s, some basic techniques for proving the feasibility of secure computation, such as the use of secret sharing and zero-knowledge proofs (only used for an active adversary; see Chapter 17), were developed. In particular, the oblivious transfer was introduced in [115, 280]; see Chapter 13 for more details. The protocol for secure computation from oblivious transfer against a passive adversary in Section 12.6 is from [146] (see also [142]).

The characterization of trivial functions was first derived in [68] for Boolean functions and then in [14, 212] for non-Boolean functions. The utility of the maximum common function in the context of secure computation was first pointed out in [353]. The refinement of a function was introduced in [230], termed the *kernel* there. The characterization in terms of interactive function, which is equivalent to the characterization in [14, 212], is from [259].

The characterization of a complete function was initiated in [194]; the characterization of a complete symmetric function for both passive and active adversaries was obtained in [194] based on the result in [192]. The characterization of a complete one-output asymmetric function for both passive and active adversaries was obtained in [195]. The characterization of a complete randomized symmetric or one-output asymmetric function for a passive adversary was also obtained in [195]. The completeness result for channels, which correspond to one-input/one-output randomized asymmetric functions, was obtained in [83, 260]. The characterization of complete functions has gained attention more recently, and has been actively studied in [229, 230, 231]. In particular, the characterization of complete deterministic asymmetric functions [204] and complete randomized asymmetric functions [205] has been obtained.

Problems

12.1 Show that Lemma 12.5 on the equivalence between secure computation of deterministic asymmetric and deterministic symmetric functions continues to hold when perfect reliability and security are replaced with approximate reliability with parameter ε and approximate security with parameter δ.

12.2 Show that Lemma 12.6 continues to hold when perfect reliability and security are replaced with approximate reliability with parameter ε and approximate security with parameter δ.

12.3 Show that the XOR function is not an interactive function for any distribution $\mathrm{P}_{X_1X_2}$ on $\{0,1\}^2$ with full support.

12.4 Prove Lemma 12.16.

12.5 *OT from one-sided AND*: Consider the following protocol that implements the $\binom{2}{1}$-OT using two instances of the one-sided AND function (only $\mathcal{P}_2$ gets the output): $\mathcal{P}_1$ inputs K_0 and $\mathcal{P}_2$ inputs $\overline{B}$ to the first invocation of the AND function, and $\mathcal{P}_1$ inputs K_1 and $\mathcal{P}_2$ inputs B to the second invocation of the AND function. Verify that this protocol securely computes the $\binom{2}{1}$-OT.

12.6 *Conversion of OT*: Consider the following protocol that implements the $\binom{4}{1}$-OT from three invocations of the $\binom{2}{1}$-OT [41]. For inputs $(a_0, \ldots, a_3) \in \{0,1\}^4$ and $b \in \{0,1,2,3\}$ of the $\binom{4}{1}$-OT, $\mathcal{P}_1$ randomly generates $S_0, S_1 \in \{0,1\}$ and sets (a_0, S_0), $(a_1 \oplus S_0, S_0 \oplus S_1)$, $(a_2 \oplus S_1, a_3 \oplus S_1)$ as the inputs to the three invocations of the $\binom{2}{1}$-OT. $\mathcal{P}_2$ sets (c_1, c_2, c_3) as the inputs to the invocations of the $\binom{2}{1}$-OT, where $c_i = 1$ for $1 \le i \le b$ and $c_i = 0$ otherwise. Let $\hat{K}_1, \hat{K}_2, \hat{K}_3$ be the outputs of the invocations of the $\binom{2}{1}$-OT. Then, $\mathcal{P}_2$ outputs $\hat{K}_1$ when $b = 0$, $\hat{K}_2$ when $b = 1$, and $\hat{K}_1 \oplus \hat{K}_2 \oplus \hat{K}_3$ when $b = 2$ or $b = 3$. Verify that this protocol securely computes the $\binom{4}{1}$-OT. Also, provide an extension of the protocol above to the $\binom{m}{1}$-OT using $(m-1)$ invocations of the $\binom{2}{1}$-OT.

12.7 *Secure computation of the inner product*: Consider the following protocol that computes the inner product $\oplus_{i=1}^{n} x_{1i} \wedge x_{2i}$ of $x_1 = (x_{11}, \ldots, x_{1n})$ and $x_2 = (x_{21}, \ldots, x_{2n})$ using n invocations of the $\binom{2}{1}$-OT [18]. $\mathcal{P}_1$ randomly generates $S_1, \ldots, S_{n-1} \in \{0, 1\}$, $\mathcal{P}_1$ sets $S_n = \oplus_{i=1}^{n-1} S_i$ and inputs $(S_i, x_{1i} \oplus S_i)$ to the ith invocation of the OT. $\mathcal{P}_2$ inputs x_{2i} to the ith invocation of the OT, and outputs $\oplus_{i=1}^{n} \hat{K}_i$, where $\hat{K}_i$ is the output of the ith invocation of the OT. Verify that this protocol securely computes the inner product function.

HINT Note the relation $\hat{K}_i = (x_{1i} \wedge x_{2i}) \oplus S_i$.

12.8 Prove Lemma 12.31.

12.9 Prove Lemma 12.33.

12.10 Prove Corollary 12.36.

HINT See Lemma 13.26 for a more general version.

13 Oblivious Transfer from Correlated Randomness

In Chapter 12, we introduced the oblivious transfer (OT) function and established that it is a complete primitive that enables the parties to securely compute any function. It is known that a computationally secure OT can be constructed from several one-way functions; see Section 12.8. However, since the secure computation protocol using OT developed in Chapter 12 is information-theoretically secure provided that the OT used is IT secure, it is desirable to have an IT secure implementation of OT. In fact, it turns out that IT secure OT can be constructed from correlated randomness, just as the secret key agreement can be accomplished from correlated randomness as seen in Chapter 10.

Compared to the secret key agreement problem, the theory of OT generation from correlated randomness is rather immature. In this chapter, we provide various constructions for string oblivious transfer, which we introduce in the next section. It turns out that the construction of OT is closely related to the secret key agreement; in fact, OT enables the parties to share two independent secret keys in such a manner that the receiver $\mathcal{P}_2$ can learn only one of the secret keys of its choice, and the choice of $\mathcal{P}_2$ is concealed from the sender $\mathcal{P}_1$. By utilizing such a connection, we provide impossibility results for OT construction as well. We also discuss OT capacity, which is defined as the asymptotic limit of OT constructions, in a similar manner to the secret key capacity introduced in Chapter 10.

Finally, building on tools developed in this chapter, we prove the characterization of complete functions for a passive adversary (Theorem 12.37).

13.1 Information-theoretic Construction of Oblivious Transfer

Recall that in the $\binom{m}{1}$-oblivious transfer (OT) function, $\mathcal{P}_1$ has input $x = (x_0, \ldots, x_{m-1}) \in \{0,1\}^m$, $\mathcal{P}_2$ has input $b \in \{0, 1, \ldots, m-1\}$, and $\mathcal{P}_2$ obtains x_b as the output. We have shown in Chapter 12 that any function is perfectly securely computable using oracle access to $\binom{2}{1}$-OT. Heuristically, we can regard $\binom{2}{1}$-OT as a counterpart of the perfectly secure bit channel in secure communication. In Chapter 3, we have seen that a perfectly secure channel can be implemented from a perfectly secure secret key via the one-time pad encryption scheme. A natural question arises: is there a counterpart of secret keys in secure

Protocol 13.1 $\binom{2}{1}$-OT from ROT

Input: $\mathcal{P}_1$'s input (K_0, K_1) and $\mathcal{P}_2$'s input B
Output: $\mathcal{P}_2$'s output $\hat{K}$

1. The parties invoke ROT to obtain (S_0, S_1) and (C, S_C), respectively.
2. $\mathcal{P}_2$ sends $\Pi_1 = B \oplus C$ to $\mathcal{P}_1$.
3. $\mathcal{P}_1$ sends $\Pi_{2,0} = K_0 \oplus S_{\Pi_1}$ and $\Pi_{2,1} = K_1 \oplus S_{\bar{\Pi}_1}$ to $\mathcal{P}_2$, where $\bar{\Pi}_1 = \Pi_1 \oplus 1$.
4. $\mathcal{P}_2$ computes $\hat{K} = \Pi_{2,B} \oplus S_C$ as the output.

computation? The answer is in the affirmative. As we will see below, randomized oblivious transfer (ROT) plays the role of secret keys in secure computation.[1]

DEFINITION 13.1 (Randomized oblivious transfer) Randomized oblivious transfer (ROT) is a source comprising correlated random variables $((S_0, S_1), (C, S_C))$ such that $\mathcal{P}_1$ observes (S_0, S_1), $\mathcal{P}_2$ observes (C, S_C), and S_0, S_1, C are independently uniformly distributed on $\{0, 1\}$.

It is easy to see that, using $\binom{2}{1}$-OT, the parties can create ROT. Indeed, $\mathcal{P}_1$ samples (S_0, S_1), $\mathcal{P}_2$ samples C, and then the parties invoke $\binom{2}{1}$-OT so that $\mathcal{P}_2$ can obtain S_C. Conversely, it is also possible to implement $\binom{2}{1}$-OT from the ROT and interactive communication between the parties; the implementation is described in Protocol 13.1.

PROPOSITION 13.2 *Protocol 13.1 perfectly securely implements*[2] $\binom{2}{1}$*-OT.*

Proof It is not difficult to verify that the protocol is "correct," i.e., $\hat{K} = K_B$ with probability 1. Since $\binom{2}{1}$-OT is an asymmetric function, we need to discuss security for $\mathcal{P}_1$ and $\mathcal{P}_2$ separately. To establish the security for $\mathcal{P}_2$, we need to verify that $\mathcal{P}_2$'s input B is not leaked to $\mathcal{P}_1$. This is guaranteed by the security of one-time pad encryption since C is the uniform bit that is independent of the random variables available to $\mathcal{P}_1$. To establish the security for $\mathcal{P}_1$, we need to verify that $K_{\bar{B}}$ is not leaked to $\mathcal{P}_2$. Note that the protocol is designed so that $\Pi_{2,B} = K_B \oplus S_C$ and $\Pi_{2,\bar{B}} = K_{\bar{B}} \oplus S_{\bar{C}}$. Since $S_{\bar{C}}$ plays the role of secret key that is unknown to $\mathcal{P}_2$, the information of $K_{\bar{B}}$ is not leaked from $\Pi_{2,\bar{B}}$ by the security of one-time pad encryption. □

In secure communication, a secret key is a more flexible resource and is easier to realize compared to secure communication itself. For instance, when two parties are in the same location in advance, they can agree on a secret key for a future secure communication. In a similar spirit, ROT is a more flexible resource and is easier to realize compared to OT itself. For instance, if a trusted third party (authoritative association) is available at some point in time, it can distribute

[1] Although we can consider $\binom{m}{1}$-randomized oblivious transfer, we restrict our attention to $\binom{2}{1}$-randomized oblivious transfer, which we call randomized oblivious transfer.

[2] Since we have not yet provided a formal definition of a secure implementation of OT, we will explain the meaning of "secure implementation" in the proof.

ROT to $\mathcal{P}_1$ and $\mathcal{P}_2$ so that the parties can later conduct secure computation by using that ROT and Protocol 13.1. Distribution of ROT by a trusted third party is sometimes referred to as the commodity model; see Section 13.7.

Even if a trusted third party is not available, there is a hope of realizing information-theoretically secure computation. If the parties can create a certain noisy correlation, they can implement OT and realize secure computation. We do not pursue how to create such a correlation physically; interested readers can see the references provided in Section 13.7. The main focus of this chapter is to exhibit how we can implement OT using a given correlation. In fact, instead of implementing OT for 1 bit at a time, it is more efficient to implement OT for bit strings (rather than individual bits), which we will introduce in the next section.

13.2 String Oblivious Transfer from Correlation

Suppose that the sender $\mathcal{P}_1$ and the receiver $\mathcal{P}_2$ observe correlated randomness X and Y, respectively, which are distributed according to P_{XY}. The sender generates two uniform random strings K_0 and K_1 on $\{0,1\}^l$, and the receiver generates a uniform bit B, distributed uniformly on $\{0,1\}$, as inputs to an OT protocol. The random variables (X,Y), K_0, K_1, and B are assumed to be mutually independent. The goal of an OT protocol is for $\mathcal{P}_2$ to obtain K_B in such a manner that B is concealed from $\mathcal{P}_1$ ($\mathcal{P}_2$'s security) and $K_{\bar{B}}$ is concealed from $\mathcal{P}_2$ ($\mathcal{P}_1$'s security), where $\bar{B} = B \oplus 1$. For that purpose, the parties use (X,Y) as a resource, and communicate over the noiseless channel. More specifically, $\mathcal{P}_2$ (possibly stochastically) generates message Π_1 from (Y,B), $\mathcal{P}_1$ generates message Π_2 from (X,K_0,K_1,Π_1), and so forth.[3] The entire communication of r rounds protocol is denoted by $\Pi = (\Pi_1,\ldots,\Pi_r)$. At the end of the protocol, $\mathcal{P}_2$ computes an estimate $\hat{K} = \hat{K}(Y,\Pi,B)$ of K_B.

A protocol realizes $(\varepsilon,\delta_1,\delta_2)$-OT of length l (for a passive adversary) if

$$\Pr(\hat{K} \neq K_B) \leq \varepsilon, \tag{13.1}$$

$$d_{\mathtt{var}}(\mathrm{P}_{K_{\bar{B}}Y\Pi B}, \mathrm{P}_{K_{\bar{B}}} \times \mathrm{P}_{Y\Pi B}) \leq \delta_1, \tag{13.2}$$

$$d_{\mathtt{var}}(\mathrm{P}_{BK_0K_1X\Pi}, \mathrm{P}_B \times \mathrm{P}_{K_0K_1X\Pi}) \leq \delta_2. \tag{13.3}$$

The first requirement above is referred to as ε-correctness of the protocol; the second and the third requirements are referred to as δ_1-security for $\mathcal{P}_1$ and δ_2-security for $\mathcal{P}_2$, respectively. When δ_1 and δ_2 are 0, respectively, i.e., the perfect security case, the security conditions can be equivalently described by $I(K_{\bar{B}} \wedge Y,\Pi,B) = 0$ and $I(K_0,K_1,X,\Pi \wedge B) = 0$, respectively. In fact, the schemes described in the next section attain perfect security for $\mathcal{P}_2$.

For given security parameters $(\varepsilon,\delta_1,\delta_2)$, we shall construct an OT protocol of as large a length as possible; denote by $L_{\varepsilon,\delta_1,\delta_2}(X,Y)$ the largest l such that a protocol realizing an $(\varepsilon,\delta_1,\delta_2)$-OT of length l exists.

[3] We initiate the communication at $\mathcal{P}_2$ for the convenience of presentation.

In this chapter, we mainly consider a source model in which the parties observe correlated random variables without any inputs. When a noisy channel from $\mathcal{P}_1$ to $\mathcal{P}_2$ is given, the parties can first emulate a source by $\mathcal{P}_1$ locally generating the input X with a prescribed distribution. Thus, the protocols in this chapter can be used in the channel model as well if the parties are semi-honest (passive adversary). However, when $\mathcal{P}_1$ is malicious (active adversary), $\mathcal{P}_1$ may not generate the input with the prescribed distribution. Such attacks will be discussed in Chapter 15.

13.3 Construction of Oblivious Transfer from Correlation

We present four constructions and build tools to address a broader class of correlation. Different classes of correlation and the construction of OT using them is covered in different sections below.

13.3.1 Erasure Correlation

We start with the basic erasure correlation. Erasure correlation is a very basic correlation where the first party observes random bits and the second observes an independently "erased" version of these bits. Understanding how OT can be constructed from this basic correlation sheds light on the kind of correlation needed to construct OT. Even though the construction is quite simple, it contains some important ideas on how noise is useful for OT construction. The more sophisticated constructions in later sections use the observations of this section.

Suppose that $\mathcal{P}_1$ and $\mathcal{P}_2$ observe an i.i.d. sequence of erasure correlation $X^n = (X_1, \ldots, X_n)$ and $Y^n = (Y_1, \ldots, Y_n)$, i.e., the joint distribution P_{XY} of each coordinate (X_i, Y_i) is given by

$$\mathrm{P}_{XY}(x,y) = \begin{cases} \frac{1-p}{2}, & \text{if } x = y, \\ \frac{p}{2}, & \text{if } y = \mathsf{e}, \\ 0, & \text{else} \end{cases}$$

for $(x, y) \in \{0,1\} \times \{0,1,\mathsf{e}\}$, where $0 \le p \le 1$ is the erasure probability.

There are two important features of erasure correlation.

1. When $p = \frac{1}{2}$, the erasure event $\{Y_i = \mathsf{e}\}$ of each coordinate is completely concealed from $\mathcal{P}_1$, i.e., for $x \in \{0,1\}$,

$$\Pr\left(Y_i \neq \mathsf{e} \middle| X_i = x\right) = \Pr\left(Y_i = \mathsf{e} \middle| X_i = x\right) = \frac{1}{2}.$$

2. When $\{Y_i \neq \mathsf{e}\}$ occurs, X_i is known to $\mathcal{P}_2$; on the other hand, when $\{Y_i = \mathsf{e}\}$ occurs, X_i is completely unknown to $\mathcal{P}_2$.

Because of Feature 2 above, the X_i with coordinates among the good indices $\{i : Y_i \neq \mathtt{e}\}$ can be used as a secret key that is known to $\mathcal{P}_2$ while the X_i with coordinates among the bad indices $\{i : Y_i = \mathtt{e}\}$ can be used as a secret key that is unknown to $\mathcal{P}_2$.

When the erasure probability is $p = \frac{1}{2}$, $\mathcal{P}_2$ can disclose to $\mathcal{P}_1$ the sets of good and bad indices $\{i : Y_i \neq \mathtt{e}\}$ and $\{i : Y_i = \mathtt{e}\}$, respectively, without revealing which sets contain good or bad indices to $\mathcal{P}_1$. However, when $p \neq \frac{1}{2}$, $\mathcal{P}_1$ can predict which set contains good indices and which one contains bad indices from the sizes of the sets. We can avoid this by discarding some indices of the larger set so that the sizes of both the sets are equal. For the sake of formal security analysis described later, let us consider a channel $\mathrm{P}_{V|Y}$ from $\mathcal{Y} = \{0, 1, \mathtt{e}\}$ to $\mathcal{V} = \{0, 1, 2\}$ given by

$$\begin{aligned} \mathrm{P}_{V|Y}(0|y) &= \tfrac{p}{1-p}, && \text{for } y \neq \mathtt{e}, \\ \mathrm{P}_{V|Y}(1|y) &= 1, && \text{for } y = \mathtt{e}, \\ \mathrm{P}_{V|Y}(2|y) &= \tfrac{1-2p}{1-p}, && \text{for } y \neq \mathtt{e} \end{aligned} \tag{13.4}$$

when $0 \leq p \leq \frac{1}{2}$ and

$$\begin{aligned} \mathrm{P}_{V|Y}(0|y) &= 1, && \text{for } y \neq \mathtt{e}, \\ \mathrm{P}_{V|Y}(1|y) &= \tfrac{1-p}{p}, && \text{for } y = \mathtt{e}, \\ \mathrm{P}_{V|Y}(2|y) &= \tfrac{2p-1}{p}, && \text{for } y = \mathtt{e} \end{aligned} \tag{13.5}$$

when $\frac{1}{2} \leq p \leq 1$. We can easily verify that the output $V^n = (V_1, \ldots, V_n)$ of the channel $\mathrm{P}^n_{V|Y}$ with input Y^n satisfies the following condition:

$$\Pr\left(V_i = 0 \middle| X_i = x\right) = \Pr\left(V_i = 1 \middle| X_i = x\right) = \min[p, 1-p]. \tag{13.6}$$

Note that $V_i = 0$ and $V_i = 1$ imply $Y_i \neq \mathtt{e}$ and $Y_i = \mathtt{e}$, respectively. The event $\{V_i = 2\}$ corresponds to discarding the index i. Now, the sets of indices $\{i : V_i = 0\}$ and $\{i : V_i = 1\}$ can be disclosed without revealing to $\mathcal{P}_1$ which one contains good and which one contains bad indices.

Using the notation above, OT protocol from erasure correlation can be constructed as in Protocol 13.2.

The performance of Protocol 13.2 is characterized below.

THEOREM 13.3 *Protocol 13.2 realizes $(\varepsilon, 0, 0)$-OT of length l for*

$$\varepsilon = \sum_{\substack{0 \leq l_0, l_1 \leq n: \\ l_0 < l \vee l_1 < l}} P_{\mathtt{mlt}}(l_0, l_1; n, \bar{p}, \bar{p}), \tag{13.7}$$

where $\bar{p} = \min[p, 1-p]$ and

$$P_{\mathtt{mlt}}(l_0, l_1; n, \bar{p}, \bar{p}) := \frac{n!}{l_0! l_1! (n - l_0 - l_1)!} \bar{p}^{l_0} \bar{p}^{l_1} (1 - 2\bar{p})^{n - l_0 - l_1} \tag{13.8}$$

is the pmf of the multinomial distribution.

Protocol 13.2 OT construction from erasure correlation (X, Y)

Input: $\mathcal{P}_1$'s input $(K_0, K_1) \in \{0,1\}^l \times \{0,1\}^l$ and $\mathcal{P}_2$'s input $B \in \{0,1\}$

Output: $\mathcal{P}_2$'s output $\hat{K} \in \{0,1\}^l$

1. For $\mathrm{P}_{V|Y}$ in (13.4) or (13.5), $\mathcal{P}_2$ generates $V_i \sim \mathrm{P}_{V|Y}(\cdot|Y_i)$ for $1 \le i \le n$, and sets $\tilde{\mathcal{I}}_b = \{i : V_i = b\}$, $b = 0, 1$. If $|\tilde{\mathcal{I}}_0| < l$ or $|\tilde{\mathcal{I}}_1| < l$, abort the protocol; otherwise, $\mathcal{P}_2$ sets $\mathcal{I}_B$ as the first l indices from $\tilde{\mathcal{I}}_0$ and $\mathcal{I}_{\bar{B}}$ as the first l indices from $\tilde{\mathcal{I}}_1$, and sends $\Pi_1 = (\mathcal{I}_0, \mathcal{I}_1)$ to $\mathcal{P}_1$.
2. $\mathcal{P}_1$ sends $\Pi_2 = (\Pi_{2,0}, \Pi_{2,1})$ to $\mathcal{P}_2$, where
$$\Pi_{2,b} = K_b \oplus X_{\mathcal{I}_b}, \quad b = 0, 1,$$
where $X_{\mathcal{I}_b} = (X_i : i \in \mathcal{I}_b)$.
3. $\mathcal{P}_2$ recovers $\hat{K} = \Pi_{2,B} \oplus Y_{\mathcal{I}_B}$.

Proof When the protocol is not aborted, clearly $\hat{K} = K_B$ and $K_{\bar{B}}$ is completely concealed from $\mathcal{P}_2$, i.e., (13.2) is satisfied for $\delta_1 = 0$, since $Y_i \ne \mathsf{e}$ for $i \in \mathcal{I}_B$ and $Y_i = \mathsf{e}$ for $i \in \mathcal{I}_{\bar{B}}$. Thus, complete security for $\mathcal{P}_1$ is attained and $\Pr(\hat{K} \ne K_B)$ coincides with the probability of abortion, $\Pr(|\tilde{\mathcal{I}}_0| < l \vee |\tilde{\mathcal{I}}_1| < l)$, which is given by the right-hand side of (13.7).

To verify security for $\mathcal{P}_2$, note that $\mathcal{I}_0$ and $\mathcal{I}_1$ are the first l indices of $\{i : \tilde{V}_i = 0\}$ and $\{i : \tilde{V}_i = 1\}$, where

$$\tilde{V}_i = \begin{cases} V_i \oplus B, & \text{if } V_i \in \{0,1\}, \\ V_i, & \text{if } V_i = 2. \end{cases}$$

Thus, $\Pi_1 = (\mathcal{I}_0, \mathcal{I}_1)$ is a function of $\tilde{V}^n = (\tilde{V}_1, \ldots, \tilde{V}_n)$, and it suffices to show $I(X^n, \tilde{V}^n \wedge B) = 0$. Note that $I(X^n \wedge B) = 0$. Using (13.6), we have[4]

$$\mathrm{P}_{\tilde{V}_i|X_i B}(v|x, 0) = \mathrm{P}_{V_i|X_i}(v|x) = \mathrm{P}_{V_i|X_i}(v \oplus 1|x) = \mathrm{P}_{\tilde{V}_i|X_i B}(v|x, 1)$$

for $v \in \{0,1\}$. Also, we have $\mathrm{P}_{\tilde{V}_i|X_i B}(2|x, b) = \mathrm{P}_{V_i|X_i}(2|x)$, which together with the observation above implies $I(\tilde{V}^n \wedge B|X^n) = 0$. □

When l/n is strictly smaller than $\bar{p}$, the right-hand side of (13.7) converges to 0 as n becomes large. Thus, erasure correlation with erasure probability p enables us to construct string OT of length $l \approx n \min[p, 1-p]$ asymptotically. In fact, we will see that this performance is optimal asymptotically in n in Section 13.5.

Remark 13.1 In Step 1 of Protocol 13.2, if we define $\tilde{\mathcal{I}}_0 = \{i : Y_i \ne \mathsf{e}\}$, $\tilde{\mathcal{I}}_1 = \{i : Y_i = \mathsf{e}\}$, $\mathcal{I}_B$ as the first l indices from $\tilde{\mathcal{I}}_0$, and $\mathcal{I}_{\bar{B}}$ as the first l indices from $\tilde{\mathcal{I}}_1$, then such a protocol is not secure unless $p = 1/2$. As an example, consider the following combinatorial erasure correlation. Let X be the uniform sequence on $\{0,1\}^3$, and let Y be either of $(X_1, \mathsf{e}, \mathsf{e})$, $(\mathsf{e}, X_2, \mathsf{e})$, or $(\mathsf{e}, \mathsf{e}, X_3)$ with

[4] This identity is true even if B is not uniform as long as B is independent of (X, V).

probability 1/3 each. In this case, if we decide $\mathcal{I}_B$ and $\mathcal{I}_{\bar{B}}$ of length 1 based on the rule mentioned above, then $\mathcal{I}_{\bar{B}}$ will never be $\{3\}$ while $\mathcal{I}_B$ can be $\{3\}$ with probability 1/3. Thus, B can be leaked to $\mathcal{P}_1$ from $(\mathcal{I}_0, \mathcal{I}_1)$.

If we choose $\mathcal{I}_B$ as a random subset of $\tilde{\mathcal{I}}_0$ and $\mathcal{I}_{\bar{B}}$ as a random subset of $\tilde{\mathcal{I}}_1$, then the issue mentioned above is resolved. We have followed an alternative approach and introduced the auxiliary channel $\mathrm{P}_{V|Y}$ in Protocol 13.2 for the ease of formal analysis.

13.3.2 Generalized Erasure Correlation

The second construction we present is OT protocol from a generalized erasure correlation (GEC). A distribution P_{XY} is a GEC if the forward channel $W = \mathrm{P}_{Y|X}$ satisfies the following: there exist two disjoint subsets $\mathcal{Y}_0$ and $\mathcal{Y}_1$ with $\mathcal{Y} = \mathcal{Y}_0 \cup \mathcal{Y}_1$ and two channels $W_0\colon \mathcal{X} \to \mathcal{Y}_0$ and $W_1\colon \mathcal{X} \to \mathcal{Y}_1$ such that, for every $x \in \mathcal{X}$,

$$W(y|x) = \begin{cases} (1-p)W_0(y|x), & \text{if } y \in \mathcal{Y}_0, \\ pW_1(y|x), & \text{if } y \in \mathcal{Y}_1. \end{cases} \tag{13.9}$$

The parameter $0 \le p \le 1$ is termed the erasure probability of the GEC. As the name indicates, erasure correlation seen earlier is a special case of GEC, i.e., $\mathcal{X} = \{0,1\}$, P_X is the uniform distribution, $\mathcal{Y}_0 = \{0,1\}$, $\mathcal{Y}_1 = \{\mathsf{e}\}$, $W_0(x|x) = 1$, and $W_1(\mathsf{e}|x) = 1$. Another important example of GEC is the binary symmetric erasure correlation (GSEC), given below:

EXAMPLE 13.4 The binary symmetric erasure correlation (BSEC) is given by

$$\mathrm{P}_{XY}(x,y) = \begin{cases} \frac{(1-p)(1-q)}{2}, & \text{if } y \ne \mathsf{e}, y = x, \\ \frac{p}{2}, & \text{if } y = \mathsf{e}, \\ \frac{(1-p)q}{2}, & \text{if } y \ne \mathsf{e}, y \ne x \end{cases}$$

for crossover probability $0 \le q \le 1$. For this correlation, by letting $\mathcal{Y}_0 = \{0,1\}$ and $\mathcal{Y}_1 = \{\mathsf{e}\}$, the forward channel can be decomposed as in (13.9), where W_0 is the binary symmetric channel with crossover probability q and W_1 with $W_1(\mathsf{e}|x) = 1$ is a trivial, useless channel.

Suppose that $\mathcal{P}_1$ and $\mathcal{P}_2$ observe an i.i.d. sequence (X^n, Y^n) distributed according to a GEC P_{XY}. The basic idea of the construction of OT from a GEC is very similar to that from erasure correlation. When $p = \frac{1}{2}$, note that the generalized erasure event $\{Y_i \in \mathcal{Y}_1\}$ of each coordinate is completely concealed from $\mathcal{P}_1$. To handle the case with $p \ne \frac{1}{2}$ in a similar manner to erasure correlation, let us consider a channel $\mathrm{P}_{V|Y}$ from $\mathcal{Y}$ to $\mathcal{V} = \{0,1,2\}$ given by

$$\begin{aligned} \mathrm{P}_{V|Y}(0|y) &= \tfrac{p}{1-p}, && \text{for } y \in \mathcal{Y}_0, \\ \mathrm{P}_{V|Y}(1|y) &= 1, && \text{for } y \in \mathcal{Y}_1, \\ \mathrm{P}_{V|Y}(2|y) &= \tfrac{1-2p}{1-p}, && \text{for } y \in \mathcal{Y}_0 \end{aligned} \tag{13.10}$$

Protocol 13.3 OT construction from generalized erasure correlation (X, Y)

Input: $\mathcal{P}_1$'s input $(K_0, K_1) \in \{0,1\}^l \times \{0,1\}^l$ and $\mathcal{P}_2$'s input $B \in \{0,1\}$
Output: $\mathcal{P}_2$'s output $\hat{K} \in \{0,1\}^l$

1. For $\mathrm{P}_{V|Y}$ in (13.10) or (13.11), $\mathcal{P}_2$ generates $V_i \sim \mathrm{P}_{V|Y}(\cdot|Y_i)$ for $1 \le i \le n$, and sets $\tilde{\mathcal{I}}_0 = \{i : V_i = 0\}$ and $\tilde{\mathcal{I}}_1 = \{i : V_i = 1\}$; if $|\tilde{\mathcal{I}}_0| < m$ or $|\tilde{\mathcal{I}}_1| < m$, abort the protocol; otherwise, $\mathcal{P}_2$ sets $\mathcal{I}_B$ as the first m indices from $\tilde{\mathcal{I}}_0$ and $\mathcal{I}_{\bar{B}}$ as the first m indices from $\tilde{\mathcal{I}}_1$, and sends $\Pi_1 = (\mathcal{I}_0, \mathcal{I}_1)$ to $\mathcal{P}_1$.
2. $\mathcal{P}_1$ randomly picks UHFs $F : \mathcal{X}^m \to \{0,1\}^l$ and $G : \mathcal{X}^m \to \{0,1\}^\kappa$, computes $S_b = F(X_{\mathcal{I}_b})$ and $C_b = G(X_{\mathcal{I}_b})$ for $b = 0, 1$, and sends $\Pi_2 = (\Pi_{2,0}, \Pi_{2,1}, \Pi_{2,2})$ and (F, G), where
$$\Pi_{2,b} = K_b \oplus S_b \text{ for } b = 0, 1 \text{ and } \Pi_{2,2} = (C_0, C_1).$$
3. Using the guess list $\mathcal{L} = \mathcal{L}^*(\mathrm{P}_{X_{\mathcal{I}_B} Y_{\mathcal{I}_B}})$, $\mathcal{P}_2$ reproduces $\hat{X}_{\mathcal{I}_B} \in \mathcal{L}_{Y_{\mathcal{I}_B}}$ such that $G(\hat{X}_{\mathcal{I}_B}) = C_B$ then computes $\hat{K} = \Pi_{2,B} \oplus F(\hat{X}_{\mathcal{I}_B})$.

when $0 \le p \le \frac{1}{2}$ and

$$\begin{array}{ll} \mathrm{P}_{V|Y}(0|y) = 1, & \text{for } y \in \mathcal{Y}_0, \\ \mathrm{P}_{V|Y}(1|y) = \frac{1-p}{p}, & \text{for } y \in \mathcal{Y}_1, \\ \mathrm{P}_{V|Y}(2|y) = \frac{2p-1}{p}, & \text{for } y \in \mathcal{Y}_1 \end{array} \tag{13.11}$$

when $\frac{1}{2} \le p \le 1$. Then, we can verify that the output V^n of the channel $\mathrm{P}^n_{V|Y}$ with input Y^n satisfies (13.6).

For a GEC, $V_i = 0$ does not necessarily imply that X_i is completely known to $\mathcal{P}_2$; similarly, $V_i = 1$ does not necessarily imply that X_i is completely concealed from $\mathcal{P}_2$. Thus, in addition to the ideas used in the OT protocol from erasure correlation, we need further to use the information reconciliation protocol of Construction 6.1 in Section 6.2 and the privacy amplification protocol given in Section 7.3 to guarantee the correctness and security for $\mathcal{P}_1$. An OT protocol using a GEC is described in Protocol 13.3.

The performance of Protocol 13.3 is given below.

THEOREM 13.5 *Given GEC* P_{XY}, *let* (X^m, Y_0^m, Y_1^m) *be i.i.d. random variables distributed according to* $\mathrm{P}_{XY_0Y_1}$ *given by* $\mathrm{P}_{XY_0Y_1}(x, y_0, y_1) = \mathrm{P}_X(x)\, W_0(y_0|x)\, W_1(y_1|x)$. *Then, Protocol 13.3 realizes* $(\varepsilon, \delta, 0)$*-OT of length* l *for* $\varepsilon = \eta + \xi + P_{\mathtt{abt}}$, $\delta = 4\eta + 2\xi$, *and*

$$l = \left\lfloor H^\eta_{\min}(X^m|Y_1^m) - H^\eta_{\max}(X^m|Y_0^m) - \nu(\xi) \right\rfloor,$$

where $\nu(\xi) = \log(1/4\xi^3) + 1$,

$$P_{\mathtt{abt}} = \sum_{\substack{0 \le m_0, m_1 \le n: \\ m_0 < m \vee m_1 < m}} P_{\mathtt{mlt}}(m_0, m_1; n, \bar{p}, \bar{p}), \tag{13.12}$$

$\bar{p} = \min[p, 1-p]$, *and* $\mathrm{P}_{\mathtt{mlt}}$ *denotes the pmf of the multinomial distribution given in* (13.8).

Proof Since the channel $\mathrm{P}_{V|Y}$ satisfies (13.6), perfect security for $\mathcal{P}_2$ can be proved using exactly the same argument as Theorem 13.3. For a fixed $V^n = v^n$, note that $\mathrm{P}_{X^nY^n|V^n}$ can be decomposed as

$$\mathrm{P}_{X^nY^n|V^n}(x^n, y^n|v^n) = \prod_{i \in \tilde{\mathcal{I}}_0} \mathrm{P}_{XY_0}(x_i, y_i) \prod_{i \in \tilde{\mathcal{I}}_1} \mathrm{P}_{XY_1}(x_i, y_i) \prod_{i \in \tilde{\mathcal{I}}_2} \mathrm{P}_{XY|V}(x_i, y_i|v_i), \tag{13.13}$$

where $\tilde{\mathcal{I}}_2 = \{1, \ldots, n\} \backslash (\tilde{\mathcal{I}}_0 \cup \tilde{\mathcal{I}}_1)$. Let $\mathcal{A} \subset \mathcal{V}^n$ be the set of all v^n such that the protocol is not aborted. Note that the probability of abortion $\Pr(V^n \notin \mathcal{A})$ is given by (13.12). When the protocol is aborted, there is no need to consider security for $\mathcal{P}_1$. Thus, it suffices to show that, when $v^n \in \mathcal{A}$,

$$\Pr\left(K_b \neq \hat{K} | V^n = v^n, B = b\right) \leq \eta + \xi \tag{13.14}$$

and

$$d_{\mathtt{var}}\left(\mathrm{P}_{K_{\bar{b}}Y^n\Pi_2FG|V^nB}(\cdot,\cdot,\cdot|v^n,b), \mathrm{P}_{K_{\bar{b}}} \times \mathrm{P}_{Y^n\Pi_2FG|V^nB}(\cdot,\cdot|v^n,b)\right) \leq 4\eta + 2\xi \tag{13.15}$$

hold for each v^n and b.

Note that (13.13) implies that $(X_{\mathcal{I}_b}, Y_{\mathcal{I}_b})$ has the same distribution as the i.i.d. random variables (X^m, Y_0^m). Thus, by the information reconciliation result in Lemma 6.9, if we set

$$\kappa = \left\lceil H_{\max}^{\eta}(X^m|Y_0^m) + \log(1/\xi) \right\rceil, \tag{13.16}$$

then we have

$$\Pr\left(X_{\mathcal{I}_b} \neq \hat{X}_{\mathcal{I}_b} | V^n = v^n, B = b\right) \leq \eta + \xi,$$

which implies (13.14).

Again, note that (13.13) implies that $(X_{\mathcal{I}_{\bar{b}}}, Y_{\mathcal{I}_{\bar{b}}})$ has the same distribution as the i.i.d. random variables (X^m, Y_1^m). Denote $\tilde{\Pi}_2 = (\Pi_{2,b}, C_b, C_{\bar{b}})$. Since $((X_i, Y_i) : i \notin \mathcal{I}_{\bar{b}})$ is independent of $(X_{\mathcal{I}_{\bar{b}}}, Y_{\mathcal{I}_{\bar{b}}})$, $(\Pi_{2,b}, C_b, G)$ is independent of $(X_{\mathcal{I}_{\bar{b}}}, Y_{\mathcal{I}_{\bar{b}}})$. Thus, by the results on leftover hashing in Corollary 7.22 and Lemma 7.23 (the former is used to remove $C_{\bar{b}}$ by subtracting κ and the latter is used to remove $(\Pi_{2,b}, C_b, G)$), upon setting

$$l \leq \left\lfloor H_{\min}^{\eta}(X^m|Y_1^m) - \kappa - \log(1/4\xi^2) \right\rfloor, \tag{13.17}$$

we have

$$d_{\mathtt{var}}\left(\mathrm{P}_{S_{\bar{b}}Y^n\tilde{\Pi}_2FG|V^nB}(\cdot,\cdot,\cdot|v^n,b), \mathrm{P}_{\mathtt{unif}} \times \mathrm{P}_{Y^n\tilde{\Pi}_2FG|V^nB}(\cdot,\cdot|v^n,b)\right) \leq 2\eta + \xi. \tag{13.18}$$

By applying the relation between the security of one-time pad encryption and a secret key in Theorem 3.18, the left-hand side of (13.15) is upper bounded by two times the left-hand side of (13.18). Thus, by combining (13.16) and (13.18), we have the desired result. □

13.3.3 Correlation with Degenerate Reverse Channel

Upon taking a close look at the construction of the OT protocol from a GEC, we notice that the existence of a channel $\mathrm{P}_{V|Y}$ satisfying

$$\Pr\left(V=0|X=x\right)=\Pr\left(V=1|X=x\right),\ \forall x\in\mathcal{X} \tag{13.19}$$

is crucial for guaranteeing the security for $\mathcal{P}_2$. In fact, it is not necessary for the correlation to be a GEC as long as (13.19) is satisfied. By this observation, Protocol 13.3 can be used for a correlation other than a GEC as follows.

THEOREM 13.6 *Given a correlation* P_{XY} *and channel* $\mathrm{P}_{V|Y}$ *satisfying* (13.19), *let* (X^m, Y_0^m, Y_1^m) *be i.i.d. random variables such that the marginals are given by* $\mathrm{P}_{XY_0}=\mathrm{P}_{XY|V=0}$ *and* $\mathrm{P}_{XY_1}=\mathrm{P}_{XY|V=1}$.[5] *Then, Protocol 13.3 realizes* $(\varepsilon,\delta,0)$*-OT of length* l *for* $\varepsilon=\eta+\xi+P_{\mathtt{abt}}$, $\delta=4\eta+2\xi$, *and*

$$l=\left\lfloor H_{\min}^{\eta}(X^m|Y_1^m)-H_{\max}^{\eta}(X^m|Y_0^m)-\nu(\xi)\right\rfloor,$$

where $\nu(\xi)=\log(1/4\xi^3)+1$,

$$P_{\mathtt{abt}}=\sum_{\substack{0\le m_0,m_1\le n:\\ m_0<m\vee m_1<m}} P_{\mathtt{mlt}}(m_0,m_1;n,\bar{p},\bar{p}),$$

$\bar{p}:=\mathrm{P}_V(0)=\mathrm{P}_V(1)$, *and* $P_{\mathtt{mlt}}$ *is the pmf of the multinomial distribution given in* (13.8).

Proof This can be proved in exactly the same manner as Theorem 13.5. □

Now, the problem is to determine under what circumstances we can find a channel $\mathrm{P}_{V|Y}$ satisfying (13.19).

DEFINITION 13.7 A reverse channel $\mathrm{P}_{X|Y}$ is *degenerate* if the rows of the stochastic matrix $[\mathrm{P}_{X|Y}(x|y)]_{x\in\mathcal{X},y\in\mathcal{Y}}$ are linearly dependent.

When the reverse channel of a given correlation P_{XY} is degenerate, there exist two distinct distributions Q_0 and Q_1 on $\mathcal{Y}$ such that

$$\sum_y \mathrm{Q}_0(y)\mathrm{P}_{X|Y}(x|y)=\sum_y \mathrm{Q}_1(y)\mathrm{P}_{X|Y}(x|y)\,. \tag{13.20}$$

Denote $\mathtt{s}(\mathrm{Q}_b)=\mathtt{supp}(\mathrm{Q}_b)$ for $b\in\{0,1\}$, and let

$$\bar{p}:=\min_{y\in\mathtt{s}(\mathrm{Q}_0)\cup\mathtt{s}(\mathrm{Q}_1)}\frac{\mathrm{P}_Y(y)}{\mathrm{Q}_0(y)+\mathrm{Q}_1(y)}. \tag{13.21}$$

[5] It is well defined since (13.19) implies $\mathrm{P}_{X|V=0}=\mathrm{P}_{X|V=1}$.

Note that $0 \le \bar{p} \le \frac{1}{2}$. Indeed, otherwise we have $P_Y(y) > \frac{Q_0(y)+Q_1(y)}{2}$ for every $y \in \mathcal{Y}$, which leads to a contradiction by taking the summation over $y \in \mathcal{Y}$. Define $P_{V|Y}$ by

$$P_{V|Y}(0|y) = \frac{Q_0(y)}{P_Y(y)}\bar{p},$$
$$P_{V|Y}(1|y) = \frac{Q_1(y)}{P_Y(y)}\bar{p},$$
$$P_{V|Y}(2|y) = 1 - \frac{Q_0(y) + Q_1(y)}{P_Y(y)}\bar{p}$$

for $y \in \mathtt{supp}(P_Y)$. Then, we can verify that $P_{V|Y}$ is a valid conditional distribution, and it satisfies $P_V(0) = P_V(1) = \bar{p}$ and $P_{Y|V}(\cdot|v) = Q_v$ for $v = 0, 1$, which together with (13.20) imply (13.19). For a GEC, the above construction coincides with the construction in the previous section (Problem 13.1).

Let us verify the above construction with a concrete example.

EXAMPLE 13.8 Let us consider asymmetric erasure correlation given by

$$P_{XY}(0,0) = \frac{1-p}{2},$$
$$P_{XY}(0,\mathsf{e}) = \frac{p}{2},$$
$$P_{XY}(1,\mathsf{e}) = \frac{q}{2},$$
$$P_{XY}(1,1) = \frac{1-q}{2}$$

for some erasure probabilities $0 \le p, q \le 1$. When $p \ne q$, the erasure probabilities depend on the input x, and this is not a GEC. By noting that the reverse channel is given by

$$P_{X|Y}(0|0) = 1,$$
$$P_{X|Y}(0|\mathsf{e}) = \frac{p}{p+q},$$
$$P_{X|Y}(1|\mathsf{e}) = \frac{q}{p+q},$$
$$P_{X|Y}(1|1) = 1,$$

we find that $Q_0(0) = \frac{p}{p+q}$ and $Q_0(1) = \frac{q}{p+q}$, $Q_1(\mathsf{e}) = 1$, whereby Q_0 and Q_1 satisfy (13.20). By substituting the above choice of Q_b into (13.21), we have

$$\bar{p} = \min\left[\frac{p+q}{2}, \frac{(1-p)(p+q)}{2p}, \frac{(1-q)(p+q)}{2q}\right].$$

When the reverse channel of a given correlation is not degenerate, is it not possible to construct an OT protocol? The following proposition partially answers this question; it claims that, at least, preceding strategies of construction are not enough. For simplicity of notation, we state the proposition in the single-shot regime.

PROPOSITION 13.9 *Let π be an OT protocol that involves communication Π_1 from $\mathcal{P}_2$ to $\mathcal{P}_1$ and communication Π_2 from $\mathcal{P}_1$ to $\mathcal{P}_2$. Suppose that the protocol π is perfectly secure for $\mathcal{P}_2$, i.e.,*

$$I(X, \Pi_1 \wedge B) = 0, \tag{13.22}$$

and is ε-correct, i.e.,

$$\Pr\left(\hat{K}(Y, \Pi, B) \neq K_B\right) \leq \varepsilon. \tag{13.23}$$

If the reverse channel $\mathrm{P}_{X|Y}$ is not degenerate, then we have[6]

$$\Pr\left(\hat{K}(Y, \Pi, \bar{B}) \neq K_{\bar{B}}\right) \leq \varepsilon. \tag{13.24}$$

Proof First, (13.22) implies $I(\Pi_1 \wedge B) = 0$ and $I(X \wedge B|\Pi_1) = 0$. Furthermore, the latter implies that

$$\mathrm{P}_{X|\Pi_1 B}(x|\tau_1, 0) = \mathrm{P}_{X|\Pi_1 B}(x|\tau_1, 1) \tag{13.25}$$

for every x and τ_1. Since $X -\!\!\circ\!\!- Y -\!\!\circ\!\!- (\Pi_1, B)$, we can write

$$\mathrm{P}_{X|\Pi_1 B}(x|\tau_1, b) = \sum_y \mathrm{P}_{X|Y}(x|y)\, \mathrm{P}_{Y|\Pi_1 B}(y|\tau_1, b)\,. \tag{13.26}$$

Since the stochastic matrix $\mathrm{P}_{X|Y}$ is full rank by assumption, (13.25) and (13.26) imply

$$\mathrm{P}_{Y|\Pi_1 B}(y|\tau_1, 0) = \mathrm{P}_{Y|\Pi_1 B}(y|\tau_1, 1)$$

for every y and τ_1, which implies $I(Y \wedge B|\Pi_1) = 0$. This together with $I(\Pi_1 \wedge B) = 0$ implies $I(Y, \Pi_1 \wedge B) = 0$, i.e., (Y, Π_1) and B are independent. Thus, (K_0, K_1, X, Y, Π, B) and $(K_0, K_1, X, Y, \Pi, \bar{B})$ have the same distribution, which together with (13.23) implies (13.24). □

Proposition 13.9 claims that, for any protocols satisfying the assumptions of the proposition, a dishonest $\mathcal{P}_2$ can recover $K_{\bar{B}}$ with probability $1 - \varepsilon$ as well, when the channel is not degenerate.

13.3.4 The Alphabet Extension Technique

In this section, we present a method to construct OT when the reverse channel is not degenerate. Note that even the binary symmetric correlation constitutes such an example (Problem 13.2).

We first explain the method with a channel model, i.e., $\mathcal{P}_1$ can choose the input distribution. The main idea is to create a GEC by using two instances of

[6] The transcript Π in (13.24) is the one computed from $\mathcal{P}_2$'s observation B in protocol π; (13.24) says that if $\mathcal{P}_2$ virtually computes $\hat{K}(Y, \Pi, \bar{B})$ using flipped input $\bar{B} = B \oplus 1$ after protocol π is conducted, then it coincides with $K_{\bar{B}}$ with probability at least $1 - \varepsilon$.

channel $W = \mathrm{P}_{Y|X}$. For distinct symbols x_0 and x_1 in $\mathcal{X}$, let $\tilde{W} = \mathrm{P}_{\tilde{Y}|\tilde{X}}$ be a virtual channel from $\tilde{\mathcal{X}} = \{(x_0, x_1), (x_1, x_0)\}$ to $\tilde{\mathcal{Y}} = \mathcal{Y}^2$ defined by

$$\tilde{W}(y, y'|x, x') = W(y|x)W(y'|x'). \tag{13.27}$$

For any input distribution $\mathrm{P}_{\tilde{X}}$ on $\tilde{\mathcal{X}}$, we can verify that $\mathrm{P}_{\tilde{X}\tilde{Y}} = \mathrm{P}_{\tilde{X}}\tilde{W}$ is a GEC. In fact, by denoting the erasure set as $\tilde{\mathcal{Y}}_1 = \{(y, y) : y \in \mathcal{Y}\}$, the erasure probability

$$p = \tilde{W}(\tilde{\mathcal{Y}}_1|\tilde{x}) = \sum_{y \in \tilde{\mathcal{Y}}_1} \tilde{W}(y|\tilde{x}) \tag{13.28}$$

does not depend on $\tilde{x}$ for all $\tilde{x} \in \tilde{\mathcal{X}}$. Thus, defining $\tilde{\mathcal{Y}}_0 = \tilde{\mathcal{Y}} \backslash \tilde{\mathcal{Y}}_1$, we can decompose $\tilde{W}$ into two channels $\tilde{W}_0 \colon \tilde{\mathcal{X}} \to \tilde{\mathcal{Y}}_0 = \tilde{\mathcal{Y}} \backslash \tilde{\mathcal{Y}}_1$ and $\tilde{W}_1 \colon \tilde{\mathcal{X}} \to \tilde{\mathcal{Y}}_1$ by setting

$$\tilde{W}_b(\tilde{y}|\tilde{x}) = \frac{1}{\tilde{W}(\tilde{\mathcal{Y}}_b|\tilde{x})} \tilde{W}(\tilde{y}|\tilde{x}) \mathbf{1}[\tilde{y} \in \tilde{\mathcal{Y}}_b], \quad b = 0, 1.$$

EXAMPLE 13.10 For binary symmetric channel W with crossover probability q, we can take $x_0 = 0$ and $x_1 = 1$. Then, the virtual channel $\tilde{W}$ constructed above is essentially the binary symmetric erasure channel with erasure probability $2q(1 - q)$.

Once the parties emulate a GEC as above, we can apply Protocol 13.3; the performance of the protocol can be evaluated using Theorem 13.5.

For the source model, $\mathcal{P}_1$ can send a preprocessing message $\Pi_0 = (\Pi_{0,1}, \ldots, \Pi_{0,n/2})$ with

$$\Pi_{0,i} = \mathbf{1}[(X_{2i-1}, X_{2i}) \in \tilde{\mathcal{X}}]$$

to create a virtual GEC and use Protocol 13.3 again.

13.4 Impossibility Results

In the construction of OT protocols in Section 13.3, we have seen similar techniques to those applied in the secret key agreement. In this section, we present a more direct connection between OT and the secret key agreement. More specifically, we present two reduction methods to construct a secret key agreement protocol from a given OT protocol. These reductions in turn lead to impossibility bounds for OT construction using the impossibility bounds for the secret key agreement derived in Section 10.3. Recall that $L_{\varepsilon,\delta_1,\delta_2}(X, Y)$ denotes the largest length l such that a protocol realizing $(\varepsilon, \delta_1, \delta_2)$-OT of length l exists. We will derive an upper bound for $L_{\varepsilon,\delta_1,\delta_2}(X, Y)$ in this section.

13.4.1 Bound Based on Correlation

The first reduction is based on the following idea. When we construct OT protocols from the correlation (X, Y), X and Y should have a sufficient amount

Protocol 13.4 Reduction 1 of a secret key to OT

Input: $\mathcal{P}_1$'s observation X, $\mathcal{P}_2$'s observation Y, and eavesdropper's observation V_0

Output: $\mathcal{P}_1$'s output S_1 and $\mathcal{P}_2$'s output S_2

1. $\mathcal{P}_1$ generates two random strings $K_0, K_1 \in \{0,1\}^l$, and $\mathcal{P}_2$ generates a random bit B.
2. The parties run the OT protocol π with communication Π, and $\mathcal{P}_2$ obtains an estimate $\hat{K}$ of K_B.
3. $\mathcal{P}_2$ sends B to $\mathcal{P}_1$.
4. $\mathcal{P}_1$ and $\mathcal{P}_2$ output $S_1 = K_B$ and $S_2 = \hat{K}$, respectively.

of correlation so that K_B is conveyed to $\mathcal{P}_2$ reliably while $K_{\bar{B}}$ is uncorrelated with the communication Π. In order to formalize this heuristic, consider a virtual situation in which $\mathcal{P}_1$ observes X, $\mathcal{P}_2$ observes Y, and an eavesdropper observes $V_0 = \mathtt{mcf}(X,Y)$.[7] For a given OT protocol π, we can construct a secret key agreement protocol as in Protocol 13.4, which leads to the following proposition.

PROPOSITION 13.11 *If the protocol π realizes $(\varepsilon, \delta_1, \delta_2)$-OT of length l, then Protocol 13.4 realizes an $(\varepsilon, \delta_1 + 4\delta_2)$-SK of length l. In particular,*

$$L_{\varepsilon,\delta_1,\delta_2}(X,Y) \le S_{\varepsilon,\delta_1+4\delta_2}(X,Y|V_0).$$

Proof The reliability of Protocol 13.4 follows from the correctness $\Pr(K_B \neq \hat{K}) \le \varepsilon$ of the protocol π. On the other hand, by the security for $\mathcal{P}_2$ in π, the overall observation (K_0, K_1, X, Π) of $\mathcal{P}_1$ in π has roughly the same distribution even when B is replaced by $\bar{B}$. Thus, the eavesdropper cannot determine K_B from (B, Π); otherwise, $\mathcal{P}_2$ would have learned $K_{\bar{B}}$, which violates the security for $\mathcal{P}_1$ in π.

Formally, observe that along with the triangular inequality the condition in (13.3) implies

$$\begin{aligned} d_{\mathtt{var}}(\mathrm{P}_{K_0K_1X\Pi|B=0}, \mathrm{P}_{K_0K_1X|\Pi B=1}) &\le \sum_{b\in\{0,1\}} d_{\mathtt{var}}(\mathrm{P}_{K_0K_1X\Pi|B=b}, \mathrm{P}_{K_0K_1X\Pi}) \\ &\le 2\delta_2. \end{aligned} \tag{13.29}$$

Then, we have

$$\begin{aligned} &d_{\mathtt{var}}(\mathrm{P}_{K_BV_0\Pi B}, \mathrm{P}_{\mathtt{unif}} \times \mathrm{P}_{V_0\Pi B}) \\ &\quad = \frac{1}{2}\sum_b d_{\mathtt{var}}(\mathrm{P}_{K_bV_0\Pi|B=b}, \mathrm{P}_{\mathtt{unif}} \times \mathrm{P}_{V_0\Pi|B=b}) \\ &\quad \le \frac{1}{2}\sum_b \Big[d_{\mathtt{var}}(\mathrm{P}_{K_bV_0\Pi|B=\bar{b}}, \mathrm{P}_{\mathtt{unif}} \times \mathrm{P}_{V_0\Pi|B=\bar{b}}) \end{aligned}$$

[7] See Definition 12.10 for $\mathtt{mcf}(X,Y)$.

$$+ d_{\mathtt{var}}(\mathrm{P}_{K_b V_0 \Pi | B=b}, \mathrm{P}_{K_b V_0 \Pi | B=\bar{b}}) + d_{\mathtt{var}}(\mathrm{P}_{V_0 \Pi | B=b}, \mathrm{P}_{V_0 \Pi | B=\bar{b}})\Big]$$

$$= d_{\mathtt{var}}(\mathrm{P}_{K_{\bar{B}} V_0 \Pi B}, \mathrm{P}_{\mathtt{unif}} \times \mathrm{P}_{V_0 \Pi B})$$

$$+ \frac{1}{2} \sum_b \Big[d_{\mathtt{var}}(\mathrm{P}_{K_b V_0 \Pi | B=b}, \mathrm{P}_{K_b V_0 \Pi | B=\bar{b}}) + d_{\mathtt{var}}(\mathrm{P}_{V_0 \Pi | B=b}, \mathrm{P}_{V_0 \Pi | B=\bar{b}})\Big]$$

$$\leq \delta_1 + 4\delta_2,$$

where the first inequality uses the triangular inequality twice and the second inequality uses (13.2) and (13.29) together with the fact that V_0 is a function of X. □

By combining Proposition 13.11 with the bound on secret key length in Theorem 10.10, we have the following bound for OT length.

COROLLARY 13.12 *For a given correlation* (X, Y),

$$L_{\varepsilon,\delta_1,\delta_2}(X, Y) \leq -\log \beta_\eta(\mathrm{P}_{XYV_0}, \mathrm{P}_{X|V_0} \mathrm{P}_{Y|V_0} \mathrm{P}_{V_0}) + 2\log(1/\xi)$$

for any $\xi > 0$ *with* $\eta = \varepsilon + \delta_1 + 4\delta_2 + \xi < 1$, *where* $V_0 = \mathtt{mcf}(X, Y)$.

13.4.2 Bound Based on Uncertainty

Before providing the second bound, we introduce the notion of a sufficient statistic.

DEFINITION 13.13 (Minimum sufficient statistic) For a pair (X_1, X_2) of random variables, a sufficient statistic for X_2 given X_1 is a random variable U such that there exists a function $U = g(X_1)$ such that the Markov chain $X_1 -\!\!\circ\!\!- U -\!\!\circ\!\!- X_2$ holds. The minimum sufficient statistic for X_2 given X_1, denoted by $\mathtt{mss}(X_2|X_1)$, is a sufficient statistic for X_2 given X_1 such that it is a function of every sufficient statistic U for X_2 given X_1, i.e., $H(\mathtt{mss}(X_2|X_1)|U) = 0$.

Even though we have defined the minimum sufficient statistic operationally, an explicit characterization is available (see Problem 13.4).

We now derive another bound using a different reduction. Heuristically, our second reduction is based on the following idea. To construct OT protocols from a correlation (X, Y), X should have a sufficient amount of uncertainty given Y so that $K_{\bar{B}}$ is concealed from $\mathcal{P}_2$. In order to formalize this heuristic, consider a virtual situation in which $\mathcal{P}_1$ observes X, $\mathcal{P}_2$ observes (V_1, Y), and an eavesdropper observes Y, where $V_1 = \mathtt{mss}(Y|X)$. For a given OT protocol π, we can construct a secret key protocol as in Protocol 13.5, which leads to the following proposition.

PROPOSITION 13.14 *If the protocol* π *realizes* $(\varepsilon, \delta_1, \delta_2)$*-OT of length* l, *then Protocol 13.5 realizes an* $(\varepsilon + 2\delta_2, \delta_1)$*-SK of length* l. *In particular,*

$$L_{\varepsilon,\delta_1,\delta_2}(X, Y) \leq S_{\varepsilon+2\delta_2,\delta_1}(X, (V_1, Y)|Y),$$

where $V_1 = \mathtt{mss}(Y|X)$.

Protocol 13.5 Reduction 2 of a secret key to OT

Input: $\mathcal{P}_1$'s observation X, $\mathcal{P}_2$'s observation (V_1, Y), and the eavesdropper's observation Y

Output: $\mathcal{P}_1$'s output S_1 and $\mathcal{P}_2$'s output S_2

1. $\mathcal{P}_1$ generates two random strings $K_0, K_1 \in \{0,1\}^l$, and $\mathcal{P}_2$ generates a random bit B.
2. The parties run the OT protocol π with communication Π.
3. $\mathcal{P}_2$ sends B to $\mathcal{P}_1$.
4. $\mathcal{P}_2$ samples $\tilde{Y} \sim \mathrm{P}_{Y|V_1 B \Pi}(\cdot|V_1, \bar{B}, \Pi)$.
5. $\mathcal{P}_1$ and $\mathcal{P}_2$ output $S_1 = K_{\bar{B}}$ and $S_2 = \hat{K}(\tilde{Y}, \bar{B}, \Pi)$, respectively.

Proof The security of Protocol 13.5 follows from the security for $\mathcal{P}_1$ in the protocol π. On the other hand, for reliability, we shall show that $\tilde{K} = \hat{K}(\tilde{Y}, \bar{B}, \Pi)$ satisfies

$$\Pr(K_{\bar{B}} \neq \tilde{K}) \leq \varepsilon + 2\delta_2. \tag{13.30}$$

Heuristically, Protocol 13.5 entails $\mathcal{P}_2$ emulating $\tilde{Y}$ by pretending that the protocol π was executed for $\bar{B}$ instead of B. Since the communication of $\mathcal{P}_1$ is oblivious of the value of B, plugging $\tilde{Y}$ into $\hat{K}(\cdot, \bar{B}, \Pi)$ will lead to an estimate of $K_{\bar{B}}$, provided that $\tilde{Y}$ preserves the joint distribution.

Formally, we have

$$\begin{aligned}
&\Pr(K_{\bar{B}} \neq \tilde{K}) \\
&= \frac{1}{2} \sum_{k,b,v,\tau} \mathrm{P}_{K_{\bar{b}} V_1 \Pi | B}(k, v, \tau | b) \cdot \Pr\left(\hat{K}(Y, \bar{b}, \tau) \neq k | V_1 = v, B = \bar{b}, \Pi = \tau\right) \\
&\leq \frac{1}{2} \sum_{k,b,v,\tau} \mathrm{P}_{K_{\bar{b}} V_1 \Pi | B}(k, v, \tau | \bar{b}) \cdot \Pr\left(\hat{K}(Y, \bar{b}, \tau) \neq k | V_1 = v, B = \bar{b}, \Pi = \tau\right) + 2\delta_2 \\
&= \frac{1}{2} \sum_{k,b,v,\tau} \mathrm{P}_{K_b V_1 \Pi | B}(k, v, \tau | b) \cdot \Pr\left(\hat{K}(Y, b, \tau) \neq k | V_1 = v, B = b, \Pi = \tau\right) + 2\delta_2 \\
&= \Pr(K_B \neq \hat{K}) + 2\delta_2,
\end{aligned}$$

where the inequality uses (13.29), and the last equality uses the Markov relation $Y \multimap\!\!- (V_1, B, \Pi) \multimap\!\!- (K_0, K_1)$, which holds in view of the interactive communication property; then (13.30) follows from (13.1). □

By combining Proposition 13.14 with the bound on secret key length in Theorem 10.10, we have the following bound on OT length.

COROLLARY 13.15 *For a given correlation (X, Y),*

$$L_{\varepsilon,\delta_1,\delta_2}(X, Y) \leq -\log \beta_\eta(\mathrm{P}_{V_1 V_1 Y}, \mathrm{P}_{V_1|Y}\mathrm{P}_{V_1|Y}\mathrm{P}_Y) + 2\log(1/\xi)$$

for any $\xi > 0$ with $\eta = \varepsilon + \delta_1 + 2\delta_2 + \xi < 1$, where $V_1 = \mathtt{mss}(Y|X)$.

Proof Problem 13.6. □

13.5 Oblivious Transfer Capacity

In this section, we investigate the asymptotic behavior of $L_{\varepsilon,\delta_1,\delta_2}(X^n, Y^n)$, the maximum length of OT that can be constructed from a given i.i.d. correlation (X^n, Y^n). In fact, it is demonstrated that the constructions presented in Section 13.3 are optimal for many cases, and yet other cases remain open for this basic problem.

DEFINITION 13.16 For a given i.i.d. correlation (X^n, Y^n) distributed according to P_{XY}, the oblivious transfer capacity is defined as

$$C_{\mathtt{OT}}(X,Y) := \lim_{\varepsilon,\delta_1,\delta_2 \to 0} \liminf_{n\to\infty} \frac{1}{n} L_{\varepsilon,\delta_1,\delta_2}(X^n, Y^n).$$

For a given channel, the oblivious transfer capacity $C_{\mathtt{OT}}(W)$ is defined similarly.

First, we shall derive an upper bound on the OT capacity by using the impossibility bounds in Section 13.4.

THEOREM 13.17 *For a given* P_{XY}*, the OT capacity is upper bounded as*

$$C_{\mathtt{OT}}(X,Y) \le \min\left[I(X \wedge Y|V_0), H(V_1|Y)\right],$$

where $V_0 = \mathtt{mcf}(X,Y)$ *and* $V_1 = \mathtt{mss}(Y|X)$.

Proof This follows from Corollary 13.12 and Corollary 13.15 along with Stein's lemma (cf. Theorem 5.7). □

Remark 13.2 By noting that V_0 and V_1 are functions of X, we can verify that $I(X \wedge Y|V_0) \le I(X \wedge Y)$ and $H(V_1|Y) \le H(X|Y)$. Thus, we can also derive a relaxed upper bound from Theorem 13.17:

$$C_{\mathtt{OT}}(X,Y) \le \min\left[I(X \wedge Y), H(X|Y)\right].$$

Next, we examine the OT capacity of the erasure correlation.

THEOREM 13.18 *For the erasure correlation* P_{XY} *with erasure probability* $0 \le p \le 1$*, the OT capacity is given by*

$$C_{\mathtt{OT}}(X,Y) = \min[p, 1-p].$$

Proof The converse part follows from Theorem 13.17 by plugging the erasure correlation into the upper bound. On the other hand, the OT capacity for the erasure correlation can be attained by Protocol 13.2; for arbitrary $\gamma > 0$, by applying Theorem 13.3 with $l = n(\min[p, 1-p] - \gamma)$, we can verify that the correctness error ε is vanishing as $n \to \infty$ by the law of large numbers. □

Next, we consider the GEC.

THEOREM 13.19 *For a GEC with erasure probability* $0 \le p \le 1$*, let* $\mathrm{P}_{XY_0Y_1}$ *be such that* $\mathrm{P}_{XY_0} = \mathrm{P}_X \times W_0$ *and* $\mathrm{P}_{XY_1} = \mathrm{P}_X \times W_1$*. Then, the OT capacity is lower bounded as*

$$C_{\mathtt{OT}}(X,Y) \ge \min[p, 1-p] \cdot \left[H(X|Y_1) - H(X|Y_0)\right]. \tag{13.31}$$

In particular, when W_0 is a noiseless channel, i.e., $H(X|Y_0) = 0$, and $0 \le p \le \frac{1}{2}$,

$$C_{\mathrm{OT}}(X,Y) = pH(X|Y_1); \tag{13.32}$$

when W_1 is a useless channel, i.e., $I(X \wedge Y_1) = 0$, and $\frac{1}{2} \le p \le 1$,

$$C_{\mathrm{OT}}(X,Y) = (1-p)I(X \wedge Y_0). \tag{13.33}$$

Proof The lower bound follows from Theorem 13.5 by noting that the smooth min-entropy and max-entropy converge to the conditional Shannon entropies for i.i.d. distributions (cf. Theorem 6.15 and Theorem 7.25). To verify (13.32), since the lower bound in (13.31) specializes to (13.32) under the given assumption, it suffices to derive the matching upper bound. Let $E = 0$ and $E = 1$ represent a nonerasure event $\{Y \in \mathcal{Y}_0\}$ and erasure event $\{Y \in \mathcal{Y}_1\}$, respectively. Then, the upper bound in Theorem 13.17 together with Remark 13.2 imply

$$\begin{aligned} C_{\mathrm{OT}}(X,Y) &\le H(X|Y) \\ &= H(X|Y,E) \\ &= pH(X|Y_1), \end{aligned}$$

where the last equality follows from the fact that W_0 is noiseless. Similarly, (13.33) follows from Theorem 13.17 as

$$\begin{aligned} C_{\mathrm{OT}}(X,Y) &\le I(X \wedge Y) \\ &= I(X \wedge Y, E) \\ &= (1-p)I(X \wedge Y_0) \end{aligned}$$

under the assumption that W_1 is a useless channel. □

EXAMPLE 13.20 For BSEC given in Example 13.4, note that W_1 is a useless channel. Thus, when the erasure probability satisfies $\frac{1}{2} \le p \le 1$, we have

$$C_{\mathrm{OT}}(X,Y) = (1-p)(1-H(q)).$$

For $0 < p < \frac{1}{2}$, the OT capacity of BSEC is not known.

Next, we study the OT capacity of a more general correlation using Theorem 13.6.

THEOREM 13.21 *For a given correlation P_{XY} and channel $\mathrm{P}_{V|Y}$ satisfying (13.19), the OT capacity is lower bounded as*

$$C_{\mathrm{OT}}(X,Y) \ge \bar{p}\big[H(X|Y,V=1) - H(X|Y,V=0)\big],$$

where $\bar{p} = \mathrm{P}_V(0) = \mathrm{P}_V(1)$. In particular, when $\mathrm{P}_{XY|V=0}$ and $\mathrm{P}_{XY|V=2}$ are noiseless, i.e., $H(X|Y,V=0) = H(X|Y,V=2) = 0$, then

$$C_{\mathrm{OT}}(X,Y) = \mathrm{P}_V(1)\, H(X|Y,V=1); \tag{13.34}$$

when $\mathrm{P}_{XY|V=1}$ and $\mathrm{P}_{XY|V=2}$ are useless, i.e., $I(X \wedge Y|V=1) = I(X \wedge Y|V=2) = 0$, then

$$C_{\mathrm{OT}}(X,Y) = \mathrm{P}_V(0)\, I(X \wedge Y|V=0). \tag{13.35}$$

Proof The lower bound follows from Theorem 13.6. On the other hand, (13.34) and (13.35) can be verified in a similar manner to (13.32) and (13.33) by replacing E with V. □

Remark 13.3 In Theorem 13.19 and Theorem 13.21, there might exist multiple choices of erasure set $\mathcal{Y}_1$ or channel $\mathrm{P}_{V|Y}$. The lower bounds hold for arbitrary choices. On the other hand, the tightness results hold if there exist choices satisfying the noiseless and the useless channel assumptions.

EXAMPLE 13.22 For the asymmetric erasure correlation given in Example 13.8, we can verify that $\mathrm{P}_{XY|V=0}$ and $\mathrm{P}_{XY|V=2}$ are noiseless when $p \le \frac{1}{2}$ and $q \le \frac{1}{2}$. In this case, $\bar{p} = \mathrm{P}_V(1) = \frac{p+q}{2}$ and

$$C_{\mathrm{OT}}(X,Y) = \frac{p+q}{2} H\left(\frac{p}{p+q}\right).$$

By using the alphabet extension technique of Section 13.3.4 to create the GEC along with Theorem 13.19, we can derive the following lower bound for the OT capacity of an arbitrary channel W.

THEOREM 13.23 *For a given channel W, the OT capacity is lower bounded as*

$$C_{\mathrm{OT}}(W) \ge \frac{p}{2}\big[H(\tilde{X}|\tilde{Y}_1) - H(\tilde{X}|\tilde{Y}_0)\big],$$

where $\mathrm{P}_{\tilde{X}\tilde{Y}_0\tilde{Y}_1}$ is the distribution induced by the GEC created by (13.27), *and p is the erasure probability given by* (13.28).

EXAMPLE 13.24 For the binary symmetric channel with crossover probability $0 \le q \le 1$, using the alphabet extension given in Example 13.10, we can show the following lower bound for OT capacity:

$$C_{\mathrm{OT}}(W) \ge q(1-q)\left(1 - H\left(\frac{q^2}{(1-q)^2 + q^2}\right)\right).$$

Even though there is no known example such that the lower bound in Theorem 13.23 is tight, it is useful to determine whether the OT capacity of a given correlation is positive or not (see Problem 13.8).

13.6 Proof of Characterization of Complete Functions

In Section 12.7 we saw the notion of complete functions, namely those functions which when available as an oracle can enable secure computing of any other function. In Theorem 12.37, we provided a characterization of complete symmetric functions. We now prove this theorem.

13.6.1 The Construction of a Secure Computing Protocol

For a given function f having an AND-minor, we show that the bit OT is reduced to secure computing of f, i.e., there exists a protocol to implement bit OT by using f as an oracle access. Recall that a function f has an AND-minor if there exist $x_1 \neq x_1'$ and $x_2 \neq x_2'$ such that $f(x_1, x_2') = f(x_1', x_2) = f(x_1, x_2)$ and $f(x_1', x_2') \neq f(x_1, x_2)$. Since we consider a passive adversary and the parties follow the protocol, it suffices to consider protocols that use only the input pairs (x_1, x_1') and (x_2, x_2'), which is the same as the AND function. Thus, without loss of generality, we can assume that a given function is the AND function, $f(x_1, x_2) = x_1 \wedge x_2$.

When the AND function is available through an oracle access, the parties can create a virtual erasure channel with erasure probability 1/2 as follows. First, $\mathcal{P}_1$ generates a random bit X_1 and $\mathcal{P}_2$ generates a random bit X_2. Then the parties invoke the AND function f with inputs (X_1, X_2). Next, $\mathcal{P}_1$ picks an input s to the virtual channel, and sends $s \oplus X_1$ to $\mathcal{P}_2$. When $X_2 = 1$, $\mathcal{P}_2$ can obtain the input s by subtracting the output $f(X_1, X_2)$ of the AND function; when $X_2 = 0$, since X_1 is a random bit unknown to $\mathcal{P}_2$, $\mathcal{P}_2$ has no information about the input s. Thus, this virtual channel can be identified with an erasure channel with erasure probability 1/2.

By repeating the procedure above a sufficient number of times and using Protocol 13.2, we can implement $(\varepsilon, 0, 0)$-secure bit OT for arbitrary $\varepsilon > 0$. Since the OT is a complete resource, f is complete as well.

13.6.2 Impossibility Results on OT Generation

As a preparation for our final result which establishes the impossibility of OT generation from functions without AND-minor, we prove three technical results. The first is an alternative expression of the upper bound on the OT capacity (cf. Theorem 13.17).

LEMMA 13.25 *For given random variables (Y_1, Y_2), it holds that*

$$I(Y_1 \wedge Y_2 | V_0) = \min_{\substack{P_{V|Y_1 Y_2}: \\ H(V|Y_1) = H(V|Y_2) = 0}} I(Y_1 \wedge Y_2 | V), \tag{13.36}$$

where $V_0 = \mathtt{mcf}(Y_1, Y_2)$.

Proof Note that V satisfying $H(V|Y_1) = H(V|Y_2) = 0$ is a common function of (Y_1, Y_2). Thus, we can write

$$I(Y_1 \wedge Y_2 | V) = I(Y_1 \wedge Y_2) - H(V).$$

Since V_0 is the maximal common function, we have $H(V_0) \geq H(V)$ for any common function V, which proves the claim of the lemma. □

The second result is a generalization of the property of interactive communication (cf. Lemma 10.7) to rectangular functions; note that an interactive communication is a special case of rectangular function (cf. Lemma 12.33).

LEMMA 13.26 *For any rectangular function f, $((X_1, Y_1), (X_2, Y_2))$, U, and $Z = f(X_1, X_2)$, we have*

$$I(X_1, Y_1 \wedge X_2, Y_2|U) \geq I(X_1, Y_1 \wedge X_2, Y_2|U, Z).$$

Proof Let $Q_{X_1Y_1X_2Y_2U}$ be the U-conditionally independent distribution induced by $P_{X_1Y_1X_2Y_2U}$, i.e.,

$$Q_{X_1Y_1X_2Y_2U} = P_{X_1Y_1|U}P_{X_2Y_2|U}P_U.$$

Then, we can write

$$I(X_1, Y_1 \wedge X_2, Y_2|U) = D(P_{X_1Y_1X_2Y_2U} \| Q_{X_1Y_1X_2Y_2U}).$$

Since f is rectangular, the inverse image can be written as $f^{-1}(z) = \mathcal{A}_z \times \mathcal{B}_z$. Thus, for z, x_1, x_2 such that $z = f(x_1, x_2)$, we can write

$$\begin{aligned}
&Q_{X_1Y_1X_2Y_2|UZ}(x_1, y_1, x_2, y_2|u, z) \\
&\quad= \frac{Q_{X_1Y_1X_2Y_2|U}(x_1, y_1, x_2, y_2|u)\mathbf{1}[(x_1, x_2) \in \mathcal{A}_z \times \mathcal{B}_z]}{Q_{X_1Y_1X_2Y_2|U}(\mathcal{A}_z \times \mathcal{Y}_1 \times \mathcal{B}_z \times \mathcal{Y}_2|u)} \\
&\quad= \frac{Q_{X_1Y_1|U}(x_1, y_1|u)\mathbf{1}[x_1 \in \mathcal{A}_z]Q_{X_2Y_2|U}(x_2, y_2|u)\mathbf{1}[x_2 \in \mathcal{B}_z]}{Q_{X_1Y_1|U}(\mathcal{A}_z, \mathcal{Y}_1|u)Q_{X_2Y_2|U}(\mathcal{B}_z, \mathcal{Y}_2|u)} \\
&\quad= Q_{X_1Y_1|UZ}(x_1, y_1|u, z)Q_{X_1Y_2|UZ}(x_2, y_2|u, z),
\end{aligned}$$

i.e., $Q_{X_1Y_1X_2Y_2UZ}$ can be factorized as $Q_{X_1Y_2|UZ}Q_{X_2Y_2|UZ}Q_{UZ}$. Thus, we have

$$\begin{aligned}
&D(P_{X_1Y_1X_2Y_2U} \| Q_{X_1Y_1X_2Y_2U}) \\
&\quad= D(P_{X_1Y_1X_2Y_2UZ} \| Q_{X_1Y_1X_2Y_2UZ}) \\
&\quad\geq D(P_{X_1Y_1X_2Y_2|UZ} \| Q_{X_1Y_1X_2Y_2|UZ}|P_{UZ}) \\
&\quad= D(P_{X_1Y_1X_2Y_2|UZ} \| Q_{X_1Y_1|UZ} \times Q_{X_2Y_2|UZ}|P_{UZ}) \\
&\quad= D(P_{X_1Y_1X_2Y_2|UZ} \| P_{X_1Y_1|UZ} \times P_{X_2Y_2|UZ}|P_{UZ}) \\
&\qquad+ D(P_{X_1Y_1|UZ} \| Q_{X_1Y_1|UZ}|P_{UZ}) + D(P_{X_2Y_2|UZ} \| Q_{X_2Y_2|UZ}|P_{UZ}) \\
&\quad\geq I(X_1, Y_1 \wedge X_2, Y_2|U, Z).
\end{aligned}$$

□

Suppose that the parties observe (Y_1, Y_2) as inputs, and run an interactive protocol with oracle access to a function f. More specifically, in the tth stage, the parties communicate Π_t interactively and invoke f with inputs (X_{1t}, X_{2t}) to obtain $Z_t = f(X_{1t}, X_{2t})$. At the end of the protocol, $\mathcal{P}_1$'s observation Y_1' comprises Y_1, $\Pi = (\Pi_1, \ldots, \Pi_r)$, $X_1^r = (X_{11}, \ldots, X_{1r})$, and $Z^r = (Z_1, \ldots, Z_r)$; $\mathcal{P}_2$'s observation Y_2' comprises Y_2, $\Pi = (\Pi_1, \ldots, \Pi_r)$, $X_1^r = (X_{21}, \ldots, X_{2r})$, and $Z^r = (Z_1, \ldots, Z_r)$. The result below says that the quantity in (13.36) is a monotone under the above described protocol if the refinement of f is rectangular.

LEMMA 13.27 *Suppose that a function f is such that its refinement $f_{\texttt{ref}}$ is rectangular. Then, the random variable (Y_1', Y_2') above, which is obtained from (Y_1, Y_2) using a protocol with oracle access to f, satisfies*

$$I(Y_1 \wedge Y_2|\texttt{mcf}(Y_1, Y_2)) \geq I(Y_1' \wedge Y_2'|\texttt{mcf}(Y_1', Y_2')).$$

Proof For $i = 1, 2$ and $t = 1, \ldots, r$, let $Y_i[t]$ be Y_i, $(\Pi_1, \ldots, \Pi_i)$, $(X_{i1}, \ldots, X_{it})$, and $(Z_1, \ldots, Z_t)$; for $t = 0$, let $Y_i[t] = Y_i$. For $t = 1, \ldots, r$, we shall inductively prove

$$I(Y_1[t-1] \wedge Y_2[t-1] \mid V_0[t-1]) \geq I(Y_1[t] \wedge Y_2[t] \mid V_0[t]),$$

where $V_0[t] = \mathtt{mcf}(Y_1[t], Y_2[t])$ for $t = 0, 1, \ldots, r$. Since Π_t is an interactive communication and X_{it} is obtained from $Y_i[t-1]$ and Π_t, by Lemma 10.7 we get

$$\begin{aligned} I(Y_1[t-1] \wedge Y_2[t-1] \mid V_0[t-1]) &\geq I(Y_1[t-1] \wedge Y_2[t-1] \mid V_0[t-1], \Pi_t) \\ &= I(X_{1t}, Y_1[t-1] \wedge X_{2t}, Y_2[t-1] \mid V_0[t-1], \Pi_t). \end{aligned}$$

Denote $\tilde{Z}_t = f_{\mathtt{ref}}(X_{1t}, X_{2t})$. Then, since $f_{\mathtt{ref}}$ is rectangular, Lemma 13.26 implies

$$\begin{aligned} &I(X_{1t}, Y_1[t-1] \wedge X_{2t}, Y_2[t-1] \mid V_0[t-1], \Pi_t) \\ &\quad \geq I(X_{1t}, Y_1[t-1] \wedge X_{2t}, Y_2[t-1] \mid V_0[t-1], \Pi_t, \tilde{Z}_t). \end{aligned}$$

Since Z_t is a function of $\tilde{Z}_t$ and $(V_0[t-1], \Pi_t, \tilde{Z}_t)$ is a common function of $(Y_1[t], Y_2[t])$,[8] by Lemma 13.25, we have

$$\begin{aligned} &I(X_{1t}, Y_1[t-1] \wedge X_{2t}, Y_2[t-1] \mid V_0[t-1], \Pi_t, \tilde{Z}_t) \\ &\quad = I(Y_1[t] \wedge Y_2[t] \mid V_0[t-1], \Pi_t, \tilde{Z}_t) \\ &\quad \geq I(Y_1[t] \wedge Y_2[t] \mid V_0[t]), \end{aligned}$$

which completes the proof. □

Proof of Theorem 12.37

We now go back to the proof of the impossibility part of Theorem 12.37. For a given function f such that its refinement $f_{\mathtt{ref}}$ is rectangular and arbitrary $\varepsilon, \delta_1, \delta_2 > 0$, suppose that there exists a protocol π that implements an $(\varepsilon, \delta_1, \delta_2)$-secure bit OT using f as oracle access. For independent private-coins (U_1, U_2) and inputs (K_0, K_1) and B to the bit OT, let $Y_1 = ((K_0, K_1), U_1)$ and $Y_2 = (U_2, B)$, which are independent. After running the protocol π, let Y_1' and Y_2' be $\mathcal{P}_1$ and $\mathcal{P}_2$'s observations. Viewing (Y_1', Y_2') as a correlated resource, the parties can implement an $(\varepsilon, \delta_1, \delta_2)$-secure bit ROT trivially without any communication. Thus, by applying Proposition 13.11,[9] the parties can generate an $(\varepsilon, \delta_1 + 4\delta_2)$-SK of one bit in the situation where $\mathcal{P}_1$ observes Y_1', $\mathcal{P}_2$ observes Y_2', and the eavesdropper observes $V_0' = \mathtt{mcf}(Y_1', Y_2')$. By applying Theorem 10.8, we have[10]

$$1 \leq S_{\varepsilon, \delta_1 + 4\delta_2}(Y_1', Y_2' \mid V_0') \leq \frac{I(Y_1' \wedge Y_2' \mid V_0') + h(\varepsilon) + h(\delta_1 + 4\delta_2)}{1 - \varepsilon - \delta_1 - 4\delta_2}. \tag{13.37}$$

[8] Here, note that the refinement $f_{\mathtt{ref}}$ is defined with respect to a full support distribution on $\mathcal{X}_1 \times \mathcal{X}_2$; $\tilde{Z}_t$ is a common function of $((X_{1t}, Z_t), (X_{2t}, Z_t))$ even if (X_{1t}, X_{2t}) does not have full support.

[9] Even though Proposition 13.11 is for OT, it can be applied to ROT as well.

[10] Since the length of the key is just one bit, it is better to use Theorem 10.8 rather than Theorem 10.10.

But, by Lemma 13.27, we have

$$I(Y_1' \wedge Y_2' \mid V_0') \leq I(Y_1 \wedge Y_2 \mid \mathtt{mcf}(Y_1, Y_2)) = 0,$$

which contradicts (13.37) for sufficiently small $\varepsilon, \delta_1, \delta_2$. □

In the above proof, a crucial observation is that "useful" correlation cannot be created from scratch by protocols with oracle access to f, which was proved in Lemma 13.27. In fact, Lemma 13.27 says more than that, namely, useful correlation cannot be increased by protocols with oracle access to f. By using this observation together with some variation of Lemma 13.27, we can strengthen Theorem 13.17; the same upper bounds on the OT capacity hold even if we allow protocols with oracle access to f.

13.7 References and Additional Reading

The concept of "precomputing" OT, i.e., ROT, was formally introduced by Beaver in [16]; Protocol 13.1 is also available there. The concept of the commodity model was also introduced by Beaver [17]; see also the trusted initializer model introduced by Rivest [292]. These are candidate resources for realizing information-theoretic two-party secure computation in practice; for instance, see [327] and references therein.

There are various ways to realize OT from certain physical assumptions. One notable approach is the use of quantum communication. However, unlike quantum key distribution, it has been shown that a secure implementation of oblivious transfer (against an active adversary) cannot be guaranteed only from the laws of quantum physics [226]. Under additional assumptions, such as those satisfied by the bounded storage model [94] or the noisy storage model [202], secure implementation of OT has been proposed. An alternative approach is the use of a physically unclonable function (PUF); for instance, see [46, 266, 294] and references therein.

The study of constructing OT from correlation was initiated in [82]. The connection between the erasure channel (Rabin's OT) and one-out-of-two OT was noticed in [78]. The concept of GEC was introduced in [2]; the construction in Section 13.3.2 is also from [2], but the idea of using information reconciliation and privacy amplification was used earlier in the literature. The construction in Section 13.3.3 is a new observation in this book, though the role of degenerate channels in the context of the privacy problem is known [7, 50]; see also [29] for the role of degenerate channels in multiuser information theory. The idea of alphabet extension first appeared in [79].

Impossibility bounds for OT construction were first obtained in [103]. More unified treatment for deriving impossibility bounds on two-party secure computation including OT was provided in [353]. Impossibility bounds with nonperfect security was studied in [349]. The bounds in Section 13.4 are from [333].

The concept of OT capacity was introduced in [260], and further studied in [2]. The upper bound of OT capacity in Theorem 13.17 is derived based on the fact that both correlation and uncertainty are required to construct OT. By adjusting the tension between these two factors, we can derive a tighter upper bound on the OT capacity [278, 284] (see also Problem 13.12).

In this chapter, we only considered a passive adversary, i.e., the cheating party tries to gain information but follows the protocol. The concept of OT capacity can also be defined for protocols with an active adversary (see Chapter 15). In fact, it can be proved that the lower bound on OT capacity in Theorem 13.19 holds even for an active adversary [112, 273].

Further, we only considered the construction of string OT, and OT capacity was defined as the maximum rate of a string OT that can be constructed from a given correlation. An alternative definition of OT capacity could be the maximum rate of bit OTs that can be constructed from a given correlation. For such a problem, the standard techniques of information reconciliation and privacy amplification cannot be used and it is more challenging to construct an efficient protocol. In [158, 174], the construction of a protocol having positive rate was proposed using techniques from multiparty secure computation [175].

Problems

13.1 Verify that a GEC has degenerate reverse channel, and verify that the construction in Section 13.3.3 coincides with that in Section 13.3.2 for a GEC.

13.2 For a binary symmetric channel with crossover probability $q \neq 1/2$ and input distribution P_X with $0 < \mathrm{P}_X(0) < 1$, verify that the reverse channel is not degenerate.

13.3 *Reversibility of bit OT*: Let P_{XY} be the typewriter correlation on $\mathcal{X} = \mathcal{Y} = \{0, 1, 2, 3\}$ given by

$$\mathrm{P}_{XY}(x, y) = \begin{cases} \frac{1}{8}, & \text{if } x = y \text{ or } y - x = 1 \pmod 4, \\ 0, & \text{else.} \end{cases}$$

1. Verify that the reverse channel of the typewriter correlation is degenerate.
2. Provide a single-shot perfect construction of bit OT from a single use of the typewriter correlation [2, 352].
3. Construct a bit OT from $\mathcal{P}_1$ to $\mathcal{P}_2$ using a single use of the bit OT from $\mathcal{P}_2$ to $\mathcal{P}_1$ [352].

   COMMENT This means that the direction of the bit OT can be reversed; string OT does not have the same property. See also [190] for reversibility of more general functions.

13.4 *An explicit characterization of the minimum sufficient statistic*: For a pair (X_1, X_2) of random variables with joint distribution $\mathrm{P}_{X_1X_2}$,[11] let $\sim$ be an equivalence relation on $\mathcal{X}_1$ defined by

$$x_1 \sim x_1' \iff \mathrm{P}_{X_2|X_1}(x_2|x_1) = \mathrm{P}_{X_2|X_1}(x_2|x_1'), \ \forall x_2 \in \mathcal{X}_2.$$

[11] Without loss of generality, assume that $\mathrm{supp}(\mathrm{P}_{X_1}) = \mathcal{X}_1$.

Then, prove that $\mathtt{mss}(X_2|X_1) \equiv [X_1]$, where $[x_1]$ is the equivalence class of x_1 with respect to $\sim$.

COMMENT From this characterization, it can be readily verified that $\mathtt{mss}(X_2|X_1, V) \equiv \mathtt{mss}(X_2|X_1)$ when $V \multimap\!\!- X_1 \multimap\!\!- X_2$.

13.5 *Double Markov*: For a triplet (X_1, X_2, V) of random variables, suppose that $X_1 \multimap\!\!- X_2 \multimap\!\!- V$ and $X_2 \multimap\!\!- X_1 \multimap\!\!- V$. Then, prove that $\mathtt{mss}(V|X_1, X_2)$ is a function of $\mathtt{mcf}(X_1, X_2)$. In other words, there exists a common function U of X_1 and X_2 such that $(X_1, X_2) \multimap\!\!- U \multimap\!\!- V$. In fact, we find that $\mathtt{mcf}(X_1, X_2)$ itself is a sufficient statistic of V given (X_1, X_2).

13.6 Prove Corollary 13.15.

HINT by noting $X \multimap\!\!- V_1 \multimap\!\!- Y$, use the data processing inequality with respect to $\beta_\varepsilon(\cdot, \cdot)$.

13.7 *An upper bound on the OT capacity of channels*: For a given channel W, prove that the OT capacity is upper bounded as

$$C_{\mathrm{OT}}(W) \le \max_{\mathrm{P}_X} \min \big[I(X \wedge Y), H(X|Y)\big].$$

13.8 *Characterization of complete channels*: For a given channel W, prove that the OT capacity $C_{\mathrm{OT}}(W)$ is positive if and only if there exist $x, x' \in \mathcal{X}$ such that corresponding rows of the matrix W are not identical and $W(y|x)W(y|x') > 0$ for some $y \in \mathcal{Y}$. Further show that, for a given correlation P_{XY}, the OT capacity $C_{\mathrm{OT}}(X, Y)$ is positive if and only if there exist $x, x' \in \mathtt{supp}(\mathrm{P}_X)$ satisfying the above conditions [2].

COMMENT This result gives the condition such that a given channel is a complete resource for secure computation against a passive adversary.

13.9 *Monotones*:

1. For the parties' initial observations (Y_1, Y_2) and any interactive protocol π with transcript Π, prove that

$$I(Y_1 \wedge Y_2|\mathtt{mcf}(Y_1, Y_2)) \ge I(Y_1' \wedge Y_2'|\mathtt{mcf}(Y_1', Y_2')), \tag{13.38}$$

$$H(\mathtt{mss}(Y_2|Y_1)|Y_2) \ge H(\mathtt{mss}(Y_2'|Y_1')|Y_2'), \tag{13.39}$$

$$H(\mathtt{mss}(Y_1|Y_2)|Y_1) \ge H(\mathtt{mss}(Y_1'|Y_2')|Y_1'), \tag{13.40}$$

where $Y_i' = (Y_i, \Pi)$.

2. For any random variables Y_1', Y_2', Z_1, Z_2 satisfying $Y_1' \multimap\!\!- Z_1 \multimap\!\!- Z_2$ and $Z_1 \multimap\!\!- Z_2 \multimap\!\!- Y_2'$, prove

$$I(Y_1', Z_1 \wedge Y_2', Z_2|\mathtt{mcf}((Y_1', Z_1), (Y_2', Z_2))) \ge I(Z_1 \wedge Z_2|\mathtt{mcf}(Z_1, Z_2)),$$
$$H(\mathtt{mss}(Y_2', Z_2|Y_1', Z_1)|Y_2', Z_2) \ge H(\mathtt{mss}(Z_2|Z_1)|Z_2),$$
$$H(\mathtt{mss}(Y_1', Z_1|Y_2', Z_2)|Y_1', Z_1) \ge H(\mathtt{mss}(Z_1|Z_2)|Z_1).$$

COMMENT The inequalities above imply that (Y_1, Y_2) can be securely converted into (Z_1, Z_2) only if

$$I(Y_1 \wedge Y_2 | \mathtt{mcf}(Y_1, Y_2)) \geq I(Z_1 \wedge Z_2 | \mathtt{mcf}(Z_1, Z_2)), \qquad (13.41)$$
$$H(\mathtt{mss}(Y_2|Y_1)|Y_2) \geq H(\mathtt{mss}(Z_2|Z_1)|Z_2), \qquad (13.42)$$
$$H(\mathtt{mss}(Y_1|Y_2)|Y_1) \geq H(\mathtt{mss}(Z_1|Z_2)|Z_1). \qquad (13.43)$$

More generally, a quantity $M(\cdot \wedge \cdot)$ satisfying

$$M(Y_1 \wedge Y_2) \geq M(Y_1', \wedge Y_2'),$$
$$M(Y_1', Z_1 \wedge Y_2', Z_2) \geq M(Z_1 \wedge Z_2)$$

is termed *monotone*, and provides a necessary condition for secure computation [353].

3. Suppose that a function f is such that its refinement $f_{\mathtt{ref}}$ is rectangular. Then, for the random variable (Y_1', Y_2') that is obtained from (Y_1, Y_2) by a protocol with oracle access to f (cf. Lemma 13.27), prove (13.38)–(13.40).
COMMENT This result has two consequences. The first is that the monotones are useful to derive impossibility bounds for protocols with oracle access to functions that are not complete. The second is that the monotones are not useful to prove nontriviality of incomplete functions; we can prove that a given function is nontrivial if one of (13.41)–(13.43) is violated, but incomplete functions satisfy (13.41)–(13.43) even if they are not interactive functions.

13.10 *External/internal information identity*: For any triplet (U, X_1, X_2) of random variables, verify that the following identity holds:

$$I(U \wedge X_1, X_2) - I(X_1 \wedge X_2) = I(U \wedge X_1 | X_2) + I(U \wedge X_2 | X_1) - I(X_1 \wedge X_2 | U).$$

COMMENT When U is a transcript of some protocol, the quantities $I(U \wedge X_1, X_2)$ and $I(U \wedge X_1|X_2) + I(U \wedge X_2|X_1)$ are termed *external information* and *internal information*, respectively, in the context of theoretical computer science; for example see [44].

13.11 *Residuals of common information*: For a pair (X_1, X_2) of random variables, quantities given by

$$C_{\mathtt{W}}(X_1 \wedge X_2) = \min_{\substack{P_{U|X_1X_2}: \\ X_1 -\!\!\circ\!\!- U -\!\!\circ\!\!- X_2}} I(U \wedge X_1, X_2),$$
$$C_{\mathtt{GK}}(X_1 \wedge X_2) = H(\mathtt{mcf}(X_1, X_2))$$

are termed the *Wyner common information* [354] and the *Gács–Körner common information* [130], respectively.

1. Prove that

$$C_{\mathtt{GK}}(X_1 \wedge X_2) \leq I(X_1 \wedge X_2) \leq C_{\mathtt{W}}(X_1 \wedge X_2) \leq \min[H(X_1), H(X_2)].$$

2. Prove that the following claims are equivalent.
   i. $C_{\mathtt{W}}(X_1 \wedge X_2) - I(X_1 \wedge X_2) = 0$.
   ii. $I(X_1 \wedge X_2) - C_{\mathtt{GK}}(X_1 \wedge X_2) = I(X_1 \wedge X_2 | \mathtt{mcf}(X_1, X_2)) = 0$.

   HINT Use the identity in Problem 13.10.

13.12 *Monotone region*: For a pair (Y_1, Y_2) of random variables, let $\mathfrak{T}(Y_1 \wedge Y_2) \subseteq \mathbb{R}_+^3$ be the set of all triplets (r_0, r_1, r_2) such that

$$\begin{aligned} r_0 &\geq I(Y_1 \wedge Y_2 | U), \\ r_1 &\geq I(U \wedge Y_1 | Y_2), \\ r_2 &\geq I(U \wedge Y_2 | Y_1) \end{aligned}$$

hold for some U taking values in a finite set $\mathcal{U}$.[12]

1. Prove the following identities:

$$\begin{aligned} \min\{r_0 : (r_0, 0, 0) \in \mathfrak{T}(Y_1 \wedge Y_2)\} &= I(Y_1 \wedge Y_2 | \mathtt{mcf}(Y_1, Y_2)), \\ \min\{r_1 : (0, r_1, 0) \in \mathfrak{T}(Y_1 \wedge Y_2)\} &= H(\mathtt{mss}(Y_2|Y_1)|Y_2), \\ \min\{r_2 : (0, 0, r_2) \in \mathfrak{T}(Y_1 \wedge Y_2)\} &= H(\mathtt{mss}(Y_1|Y_2)|Y_1), \\ \min\{r_1 + r_2 : (0, r_1, r_2) \in \mathfrak{T}(Y_1 \wedge Y_2)\} &= C_{\mathtt{W}}(Y_1 \wedge Y_2) - I(Y_1 \wedge Y_2). \end{aligned}$$

2. For the parties' initial observation (Y_1, Y_2) and any interactive protocol π with transcript Π, prove that

$$\mathfrak{T}(Y_1 \wedge Y_2) \subseteq \mathfrak{T}(Y_1' \wedge Y_2'), \tag{13.44}$$

   where $Y_i' = (Y_i, \Pi)$.

3. For any random variables Y_1', Y_2', Z_1, Z_2 satisfying $Y_1' -\!\!\circ\!\!- Z_1 -\!\!\circ\!\!- Z_2$ and $Z_1 -\!\!\circ\!\!- Z_2 -\!\!\circ\!\!- Y_2'$, prove

$$\mathfrak{T}(Y_1', Z_1 \wedge Y_2', Z_2) \subseteq \mathfrak{T}(Z_1 \wedge Z_2).$$

   COMMENT A monotone region is a generalization of the monotones seen in Problem 13.9, and it provides a necessary condition for secure computation [278].

4. Suppose that a function f is such that its refinement $f_{\mathtt{ref}}$ is rectangular. Then, for the random variable (Y_1', Y_2') that is obtained from (Y_1, Y_2) by a protocol with oracle access to f (cf. Lemma 13.27), prove that (13.44) holds.

13.13 *Characterizations of complete functions by information measures*: For a given function $f : \mathcal{X}_1 \times \mathcal{X}_2 \to \mathcal{Z}$, prove that the following claims are equivalent.

1. The function f is not complete.
2. The refinement $f_{\mathtt{ref}}$ of f is rectangular.
3. For any inputs (X_1, X_2) and $Z = f(X_1, X_2)$, it holds that

$$C_{\mathtt{W}}(X_1 \wedge X_2) - I(X_1 \wedge X_2) \geq C_{\mathtt{W}}(X_1, Z \wedge X_2, Z) - I(X_1, Z \wedge X_2, Z).$$

[12] By using the support lemma, we can restrict the cardinality of $\mathcal{U}$ to be $|\mathcal{U}| \leq |\mathcal{Y}_1||\mathcal{Y}_2| + 2$.

In particular, for any independent inputs, it holds that[13]

$$C_{\mathtt{W}}(X_1, Z \wedge X_2, Z) - I(X_1, Z \wedge X_2, Z) = 0.$$

4. For any inputs (X_1, X_2) and $Z = f(X_1, X_2)$, it holds that

$$I(X_1 \wedge X_2 | \mathtt{mcf}(X_1, X_2)) \geq I(X_1, Z \wedge X_2, Z | \mathtt{mcf}((X_1, Z), (X_2, Z))).$$

In particular, for any independent inputs, it holds that

$$I(X_1, Z \wedge X_2, Z | \mathtt{mcf}((X_1, Z), (X_2, Z))) = 0.$$

5. For any inputs (X_1, X_2) and $Z = f(X_1, X_2)$, it holds that

$$\mathfrak{T}(X_1 \wedge X_2) \subseteq \mathfrak{T}(X_1, Z \wedge X_2, Z).$$

In particular, for any independent inputs, it holds that

$$\mathfrak{T}(X_1, Z \wedge X_2, Z) = \mathbb{R}^3_+.$$

[13] This characterization is from [344].

14 Bit Commitment from Correlated Randomness

In some board games, such as Chess or Go, a professional match may be suspended until the next day. In such a case, in order to ensure fairness, the last player to make a move on the first day, say $\mathcal{P}_1$, notes and places their next move in a sealed envelope. When the match is resumed, the sealed envelope is opened and the move noted there is played. This procedure is termed the sealed move, and it provides fairness in the following sense.

1. The opponent, $\mathcal{P}_2$, cannot learn $\mathcal{P}_1$'s next move until the envelope is opened the next day, which is termed the *hiding property*.
2. Even if $\mathcal{P}_1$ comes up with a better move during the night, $\mathcal{P}_1$ must resume the match with the move sealed in the envelope, which is termed the *binding property*.

The hiding and the binding properties are guaranteed since a trusted third party keeps the envelope safely.

Bit commitment is a cryptographic primitive that has a similar functionality to the sealed move explained above, without requiring a trusted third party. Bit commitment is used as a building block for other cryptographic protocols. For instance, it will be used to prevent a party from deviating from a protocol in secure computation considered in later chapters. In this chapter, we present how to construct bit commitment from a correlation shared by the parties. We remark that this chapter is the first time in the book where we encounter an active adversary, a regular feature of the book from here on.

14.1 Bit Commitment

Formally, we introduce the ideal function $f_{\mathtt{BC}}$ computed in bit commitment in Protocol 14.1. Unlike the ideal functions described in previous chapters, such as the AND function or the OT function, $f_{\mathtt{BC}}$ has two phases of operations: the commit phase and the reveal phase. For a multiphase function, a new input is taken in at every phase and an output may be declared. The key point is that the inputs for the previous phases are available in the next phase; in other words, an internal state is maintained across the phases. For $f_{\mathtt{BC}}$, when the commit phase of $f_{\mathtt{BC}}$ is invoked, the sender $\mathcal{P}_1$ sends a command $(\mathtt{commit}, \mathrm{id}, C^l)$, where

Protocol 14.1 Ideal function f_{BC} of bit commitment

1. **Commit phase**
2. If $\mathcal{P}_1$ sends $(\texttt{commit}, \text{id}, C^l)$ to f_{BC}, then $\mathcal{P}_2$ receives $(\texttt{committed}, \text{id})$ from f_{BC}.
3. **Reveal phase**
4. If $\mathcal{P}_1$ sends $(\texttt{reveal}, \text{id})$ to f_{BC}, then $\mathcal{P}_2$ receives $(\texttt{revealed}, \text{id}, C^l)$ for the value C^l committed with identity id from f_{BC}.
5. If either party sends the abortion command $\perp$ to f_{BC}, then the parties receive the abort notification $(\perp, \perp)$ from f_{BC}.

$C^l \in \{0,1\}^l$ is the bit sequence to be committed, and id is the identity of the memory that stores the committed bit sequence; the receiver $\mathcal{P}_2$ receives the output $(\texttt{committed}, \text{id})$. When the reveal phase of f_{BC} is invoked, the sender $\mathcal{P}_1$ sends the command $(\texttt{reveal}, \text{id})$; if a bit sequence has been committed to id before, $\mathcal{P}_2$ receives the value $(\texttt{revealed}, \text{id}, C^l)$. The identity id of the committed bit sequence becomes relevant when multiple instances of bit commitment are invoked in a protocol. When bit commitment is invoked only once in a protocol, we omit the input identity.

In later sections, we consider implementations of the bit commitment. When we consider the bit commitment, we assume that a dishonest party is active,[1] i.e., the dishonest party may intentionally abort the protocol. Thus, we add the abortion option to the reveal phase of the ideal function f_{BC}, i.e., either party can send the abortion command $\perp$ in the reveal phase. When this command is invoked by one of the parties, the ideal function f_{BC} returns the abortion notification $(\perp, \perp)$ to both parties.

The concept of "abortion" is difficult to define in a real world. For instance, a protocol may be aborted by timeout of a communication. For simplicity, in this book, we assume that an abortion is only caused by one of the parties and it can be recognized by both the parties; once a protocol is aborted, the rest of the steps are not executed.

14.2 Example Applications

We now outline some applications of bit commitment. We only provide a high-level description of these applications, relegating the more detailed discussion to later chapters.

14.2.1 Coin Flipping

One direct application of bit commitment is the coin flipping function f_{CF} described as follows: when both the parties are honest, without any input, both the

[1] A more detailed account of the active adversary will be given in the next chapter.

Protocol 14.2 An implementation π_{CF} of coin flipping f_{CF} using f_{BC}

Output: $\mathcal{P}_1$'s output C_1 and $\mathcal{P}_2$'s output C_2

1. $\mathcal{P}_1$ randomly generates $B_1 \in \{0, 1\}$.
2. The parties run the commit phase of f_{BC}, and $\mathcal{P}_1$ commits to B_1.
3. $\mathcal{P}_2$ randomly generates $B_2 \in \{0, 1\}$, and sends it to $\mathcal{P}_1$.
4. If $\mathcal{P}_2$ did not abort in Step 3, then $\mathcal{P}_1$ outputs $C_1 = B_1 \oplus B_2$; otherwise, $\mathcal{P}_1$ outputs $C_1 = B_1$.
5. The parties run the reveal phase of f_{BC}, and $\mathcal{P}_2$ receives the committed value B_1.
6. If $\mathcal{P}_1$ did not abort in Step 5, then $\mathcal{P}_2$ outputs $C_2 = B_1 \oplus B_2$; otherwise, $\mathcal{P}_2$ outputs $C_2 = B_2$.

parties receive output (C, C) with $C \sim \mathrm{Ber}(1/2)$; when either party is dishonest, the honest party receives output $C \sim \mathrm{Ber}(1/2)$.

Suppose that the parties are in different locations, and they need to implement the function f_{CF} using interactive communication. For instance, we can consider a naive protocol in which $\mathcal{P}_1$ randomly generates $C \in \{0, 1\}$, and sends it to $\mathcal{P}_2$. Clearly, such a protocol is not secure since a dishonest $\mathcal{P}_1$ can bias the outcome toward one of the values.

A more sophisticated protocol for coin flipping using bit commitment is described in Protocol 14.2. This protocol is designed so that the final outcome depends on the randomness B_1 and B_2 generated by both the parties. Since $\mathcal{P}_1$'s choice B_1 is concealed until $\mathcal{P}_2$ sends B_2 to $\mathcal{P}_1$, dishonest $\mathcal{P}_2$ cannot adjust the choice of B_2 so that $B_1 \oplus B_2$ is biased toward one of the values. On the other hand, since $\mathcal{P}_1$ has committed to the value of B_1 before $\mathcal{P}_2$ sends B_2 to $\mathcal{P}_1$, dishonest $\mathcal{P}_1$ cannot adjust the choice of B_1 so that $B_1 \oplus B_2$ is biased toward one of the values. Heuristically, it is convincing that Protocol 14.2 provides better security than the naive protocol mentioned earlier.

In Chapter 15, we will introduce a formal definition of security against an active adversary. While the protocol above seems secure, formally, it is not secure. We describe below an attack to illustrate a subtlety of security against an active adversary. Specifically, Protocol 14.2 does not emulate the perfect function f_{CF} since dishonest $\mathcal{P}_1$ can conduct the following attack. If $B_1 \oplus B_2 = 1$, $\mathcal{P}_1$ aborts the protocol in Step 5. Then, $\mathcal{P}_2$ outputs $C_2 = B_1 \oplus B_2$ if the protocol is not aborted and $C_2 = B_2$ if the protocol is aborted. In such a case, the distribution of the output by $\mathcal{P}_2$ is

$$\begin{aligned} \Pr(C_2 = 0) &= \Pr(B_1 \oplus B_2 = 0) + \Pr(B_1 \oplus B_2 = 1, B_2 = 0) \\ &= \frac{1}{2} + \frac{1}{2} \cdot \frac{1}{2} \\ &= \frac{3}{4}. \end{aligned} \tag{14.1}$$

Protocol 14.3 A protocol for sampling a different pair of bits

Output: $\mathcal{P}_1$'s output (B_1, B_2) and $\mathcal{P}_2$'s output `accept/reject`

1. $\mathcal{P}_1$ samples $(B_{1,i}, B_{2,i}) \in \{(0,1), (1,0)\}$ for $1 \le i \le n$.
2. The parties run $2n$ instances of the commit phase of f_{BC}, and $\mathcal{P}_1$ commits to $(B_{1,i}, B_{2,i})$ for $1 \le i \le n$.
3. $\mathcal{P}_2$ randomly selects a subset $\mathcal{I} \subset \{1, \ldots, n\}$ of size $n-1$, and sends $\mathcal{I}$ to $\mathcal{P}_1$.
4. The parties run the reveal phase of f_{BC}, and $\mathcal{P}_1$ reveals $(B_{1,i}, B_{2,i})$ for $i \in \mathcal{I}$.
5. If the revealed values satisfy $B_{1,i} \neq B_{2,i}$ for all $i \in \mathcal{I}$, then $\mathcal{P}_2$ outputs `accept`; otherwise, $\mathcal{P}_2$ outputs `reject`.
6. $\mathcal{P}_1$ outputs $(B_1, B_2) = (B_{1,i}, B_{2,i})$ for the remaining $i \notin \mathcal{I}$.

Thus, $\mathcal{P}_1$ can force the output of $\mathcal{P}_2$ to be biased. Of course, since this "unfair" abortion by $\mathcal{P}_1$ can be detected by $\mathcal{P}_2$, there is no need for $\mathcal{P}_2$ to continue the protocol from a practical perspective. Nevertheless, from a cryptographic perspective, Protocol 14.2 does not emulate perfect coin flipping. Such delicate issues of security will be formulated and discussed at length in the next chapter. In fact, this kind of unfairness is unavoidable in many secure computation problems; for more details, see Problem 14.1 and Section 15.7.

14.2.2 Zero-knowledge Proof

In some cryptographic protocols that we will see in later chapters, one of the parties needs to sample a codeword from a linear code, namely, a binary sequence satisfying some linear (parity check) equations, and then it needs to convince the other party that the former party is following the protocol description. As a simple example, consider a situation in which $\mathcal{P}_1$ is supposed to sample a pair of bits (B_1, B_2) from $\{(0,1), (1,0)\}$. $\mathcal{P}_1$ would like to convince $\mathcal{P}_2$ that it indeed sampled from the prescribed set, i.e., $B_1 \neq B_2$, without revealing the actual value of (B_1, B_2).

A protocol for meeting the requirements above approximately and using f_{BC} is described in Protocol 14.3. Since $\mathcal{P}_1$ does not know the sample set $\mathcal{I}$ in advance, when $\mathcal{P}_2$ accepts, the probability of $B_1 = B_2$ is vanishing as n becomes large. On the other hand, since $(B_{1,i}, B_{2,i})$ for the remaining $i \notin \mathcal{I}$ is not revealed, the output (B_1, B_2) is concealed from $\mathcal{P}_2$. This is an example of a zero-knowledge proof, a topic to be discussed in detail in Chapter 16.

14.3 Standalone Security of Implemented Bit Commitment

Like other cryptographic primitives, such as secure communication or oblivious transfer, information-theoretically secure bit commitment cannot be implemented

Protocol 14.4 One bit commitment from randomized oblivious transfer

Input: $\mathcal{P}_1$'s observation $C \in \{0,1\}$ and (K_0, K_1); and $\mathcal{P}_2$'s observation (B, K), where $K = K_B$

Output: `Accept/Reject`

1. **Commit phase**
2. $\mathcal{P}_1$ sends $\Pi = C \oplus K_0 \oplus K_1$ to $\mathcal{P}_2$.
3. **Reveal phase**
4. $\mathcal{P}_1$ sends (K_0, K_1, C) to $\mathcal{P}_2$.
5. $\mathcal{P}_2$ checks if $K_B = K$ is satisfied; if not, $\mathcal{P}_2$ outputs `reject`.
6. $\mathcal{P}_2$ checks if $C \oplus K_0 \oplus K_1 = \Pi$ is satisfied; if not, $\mathcal{P}_2$ outputs `reject`.
7. If both the tests are passed, $\mathcal{P}_2$ accept C, and outputs `accept`.

from scratch. In the next section, we consider the problem of constructing a protocol realizing bit commitment from randomized oblivious transfer (ROT) and other correlations; for ROT, see Section 13.1. Before delving into the general theory, consider a simple protocol to implement bit commitment using one instance of ROT; see Protocol 14.4. Even though the security of this protocol is not satisfactory (cf. Proposition 14.2), it illustrates how correlation can be used to implement the bit commitment. In fact, it satisfies the hiding property described at the beginning of this chapter: when $\mathcal{P}_1$ is honest and $\mathcal{P}_2$ is dishonest, the information about C is not leaked to $\mathcal{P}_2$ in the commit phase; this property is guaranteed since $K_{\bar{B}}$ is unknown to $\mathcal{P}_2$, and thus, Π can be regarded as the one-time pad encryption of C. How about the binding property? This is more subtle. When $\mathcal{P}_1$ is malicious, there is no guarantee that Π coincides with $C \oplus K_0 \oplus K_1$ since $\mathcal{P}_1$ may prepare for an attack to be conducted later in the reveal phase of the protocol. Furthermore, in the reveal phase, if $\mathcal{P}_1$ wants to reveal 0, then $\mathcal{P}_1$ may send $(\tilde{K}_0, \tilde{K}_1, 0)$; if $\mathcal{P}_1$ wants to reveal 1, then $\mathcal{P}_1$ may send $(\hat{K}_0, \hat{K}_1, 1)$. The attack is regarded as successful if both $(\tilde{K}_0, \tilde{K}_1, 0)$ and $(\hat{K}_0, \hat{K}_1, 1)$ pass the tests in Steps 5 and 6. We need to introduce a security definition that takes into account this kind of attack by a dishonest party.

More generally, we consider the implementation of the bit commitment $f_{\mathtt{BC}}$ when the parties observe, a priori, correlated random variables (X, Y). A basic protocol $\pi_{\mathtt{BC}}$ implementing $f_{\mathtt{BC}}$ can be described as follows. At the beginning of the protocol, $\mathcal{P}_1$ observes (C^l, X) and $\mathcal{P}_2$ observes Y. To implement the commit phase of $f_{\mathtt{BC}}$, the parties communicate interactively; denote the transcript by $\Pi = (\Pi_1, \ldots, \Pi_r)$. To implement the reveal phase of $f_{\mathtt{BC}}$, $\mathcal{P}_1$ reveals (C^l, X) to $\mathcal{P}_2$. Then, $\mathcal{P}_2$ conducts a test $\mathrm{T}(C^l, X, Y, \Pi)$, and accepts C^l as the output of the implementation if the outcome of the test is 0; otherwise, $\mathcal{P}_2$ rejects C^l. When C^l is rejected, the output is the abort message $(\perp, \perp)$.

As we discussed at the beginning of this chapter, the bit commitment protocol should satisfy the hiding property and the binding property. In addition to these requirements, the protocol should not abort if both the parties are honest. Thus, we consider the following three requirements for the security of implemented bit commitment.

DEFINITION 14.1 (Standalone security of implemented bit commitment) A protocol π_{BC} realizes an $(\varepsilon_{\mathtt{s}}, \varepsilon_{\mathtt{h}}, \varepsilon_{\mathtt{b}})$-BC of length l if, for any distribution P_{C^l} of committed sequence, the following conditions are satisfied.

- *Correctness*: When both the parties are honest, C^l is accepted by $\mathcal{P}_2$ with high probability, i.e.,
$$\Pr\left(\mathrm{T}(C^l, X, Y, \Pi) = 1\right) \leq \varepsilon_{\mathtt{s}}.$$
- *Hiding*: After the commit phase, for any misbehavior by $\mathcal{P}_2$, almost no information about C^l is leaked to $\mathcal{P}_2$, i.e.,
$$d_{\mathtt{var}}\left(\mathrm{P}_{C^l Y \Pi}, \mathrm{P}_{C^l} \times \mathrm{P}_{Y\Pi}\right) \leq \varepsilon_{\mathtt{h}}.$$
- *Binding*: For any misbehavior by $\mathcal{P}_1$, $\mathcal{P}_1$ cannot fool $\mathcal{P}_2$ to accept more than two values, i.e., for any $(\tilde{C}^l, \tilde{X})$ and $(\hat{C}^l, \hat{X})$,
$$\Pr\left(\mathrm{T}(\tilde{C}^l, \tilde{X}, Y, \Pi) = 0,\ \mathrm{T}(\hat{C}^l, \hat{X}, Y, \Pi) = 0,\ \tilde{C}^l \neq \hat{C}^l\right) \leq \varepsilon_{\mathtt{b}}.$$

For given security parameters $(\varepsilon_{\mathtt{s}}, \varepsilon_{\mathtt{h}}, \varepsilon_{\mathtt{b}})$, we shall construct a bit commitment protocol of as large a length as possible; denote by $L_{\varepsilon_{\mathtt{s}},\varepsilon_{\mathtt{h}},\varepsilon_{\mathtt{b}}}(X, Y)$ the largest l such that a protocol realizing an $(\varepsilon_{\mathtt{s}}, \varepsilon_{\mathtt{h}}, \varepsilon_{\mathtt{b}})$-BC of length l exists.

Remark 14.1 For the hiding condition and the binding condition, even though we used the same notation Π as the case where both the parties are honest, the dishonest party may optimize the communication strategy so as to increase the probability of successful cheating as much as possible. The hiding condition and the binding condition need to be satisfied even for such behaviors.

Remark 14.2 For the binding condition, neither of $\tilde{C}^l$ and $\hat{C}^l$ needs to coincide with the original bit sequence C^l. In fact, when $\mathcal{P}_1$ is malicious, the input C^l does not have any meaning since $\mathcal{P}_1$ may not use C^l from the beginning. For more detail on an active adversary, see Chapter 15.

Using the security of Protocol 14.4, as a warm-up exercise for verifying the security definition in Definition 14.1, let us conduct a security analysis of Protocol 14.4.

PROPOSITION 14.2 *Protocol 14.4 is a perfectly correct, perfectly hiding, and $\frac{1}{2}$-binding implementation of the bit commitment.*

Proof The perfectly correct property requires that, when both $\mathcal{P}_1$ and $\mathcal{P}_2$ are honest and follow the protocol, $\mathcal{P}_1$'s committed value C is accepted by $\mathcal{P}_2$ with probability 1. It is not difficult to see that Protocol 14.4 satisfies this property since both the tests in Step 5 and Step 6 will be passed if $\mathcal{P}_1$ follows the protocol.

Perfect hiding is guaranteed since $K_{\bar{B}}$ is unknown to $\mathcal{P}_2$, and thus, Π can be regarded as the one-time pad encryption of C.

When $\mathcal{P}_2$ is honest and $\mathcal{P}_1$ is malicious (actively dishonest), $\mathcal{P}_1$ may deviate from the protocol description so that both 0 and 1 can be accepted by $\mathcal{P}_2$ in the reveal phase. The property of ε-binding requires that the success probability, i.e., missed detection probability under such an attack by $\mathcal{P}_1$, is smaller than ε.

Heuristically, if $\mathcal{P}_1$ attempts to reveal $C \oplus 1$ in the reveal phase, then $\mathcal{P}_1$ needs to send $(K_0 \oplus 1, K_1, C \oplus 1)$ or $(K_0, K_1 \oplus 1, C \oplus 1)$ to pass the test in Step 6. However, if say $(K_0 \oplus 1, K_1, C \oplus 1)$ is sent, then the attack will be detected by the test in Step 5; the probability of missed detection is $\Pr(B = 1) = \frac{1}{2}$.

More generally, when $\mathcal{P}_1$ is malicious, there is no guarantee that Π coincides with $C \oplus K_0 \oplus K_1$ since $\mathcal{P}_1$ may prepare for an attack to be conducted later in the reveal phase of the protocol. Furthermore, in the reveal phase, if $\mathcal{P}_1$ wants to reveal 0, then $\mathcal{P}_1$ may send $(\tilde{K}_0, \tilde{K}_1, 0)$; if $\mathcal{P}_1$ wants to reveal 1, then $\mathcal{P}_1$ may send $(\hat{K}_0, \hat{K}_1, 1)$. The success probability of $\mathcal{P}_1$'s attack is given by the probability that both $(\tilde{K}_0, \tilde{K}_1, 0)$ and $(\hat{K}_0, \hat{K}_1, 1)$ pass the tests in Steps 5 and 6, which is

$$\Pr\big(\tilde{K}_B = K,\ \hat{K}_B = K,\ \tilde{K}_0 \oplus \tilde{K}_1 = \Pi,\ 1 \oplus \hat{K}_0 \oplus \hat{K}_1 = \Pi\big).$$

Denote by $\mathcal{E}_1$ the event such that $\tilde{K}_B = K$ and $\hat{K}_B = K$, and by $\mathcal{E}_2$ the event such that $\tilde{K}_0 \oplus \tilde{K}_1 = \Pi$ and $1 \oplus \hat{K}_0 \oplus \hat{K}_1 = \Pi$. Note that $\mathcal{E}_2$ occurs only when $\tilde{K}_0 \oplus \tilde{K}_1 \neq \hat{K}_0 \oplus \hat{K}_1$, which further implies $\tilde{K}_0 \neq \hat{K}_0$ or $\tilde{K}_1 \neq \hat{K}_1$. Thus, the success probability is bounded as

$$\begin{aligned}\Pr(\mathcal{E}_1 \wedge \mathcal{E}_2) &= \Pr\big((\{\tilde{K}_0 \neq \hat{K}_0\} \vee \{\tilde{K}_1 \neq \hat{K}_1\}) \wedge \mathcal{E}_1 \wedge \mathcal{E}_2\big) \\ &\leq \Pr\big((\{\tilde{K}_0 \neq \hat{K}_0\} \vee \{\tilde{K}_1 \neq \hat{K}_1\}) \wedge \mathcal{E}_1\big).\end{aligned}$$

Note that, if $\{\tilde{K}_0 \neq \hat{K}_0\}$ occurs, then $\mathcal{E}_1$ occurs only when $B = 1$. Similarly, if $\{\tilde{K}_1 \neq \hat{K}_1\}$ occurs, then $\mathcal{E}_1$ occurs only when $B = 0$. If both $\{\tilde{K}_0 \neq \hat{K}_0\}$ and $\{\tilde{K}_1 \neq \hat{K}_1\}$ occur, then $\mathcal{E}_1$ cannot occur. By noting these facts, the success probability $\Pr(\mathcal{E}_1 \wedge \mathcal{E}_2)$ is bounded above by $\frac{1}{2}$. □

14.4 Schemes for Implementing Bit Commitment Protocols

14.4.1 Bit Commitment from Oblivious Transfer

We first present a construction of bit commitment when the parties can access n instances of $\binom{2}{1}$-randomized oblivious transfer (ROT), i.e., $\mathcal{P}_1$ observes two bit strings (K_0^n, K_1^n) and $\mathcal{P}_2$ observes two bit strings (B^n, K^n) such that $K_i = K_{B_i,i}$. This construction involves some important ideas that will be used to construct bit commitment from more general correlation. More specifically, a standard construction of bit commitment involves two main ideas: (i) we generate a secret key to conceal the committed bit sequence from $\mathcal{P}_2$ in the commit phase; and (ii) we use a version of authentication code to prevent $\mathcal{P}_1$ from substituting the committed bit sequence in the reveal phase. A protocol realizing bit commitment from ROT is described in Protocol 14.5.

The performance of Protocol 14.5 can be evaluated as follows.

Protocol 14.5 Bit commitment from randomized oblivious transfer

Input: $\mathcal{P}_1$'s observation C^l and (K_0^n, K_1^n); and $\mathcal{P}_2$'s observation (B^n, K^n)

Output: `Accept/Reject`

1. **Commit phase**
2. $\mathcal{P}_2$ randomly generates $G_0, G_1 : \{0,1\}^n \to \{0,1\}^\nu$ from a UHF, and sends them to $\mathcal{P}_1$.
3. $\mathcal{P}_1$ sends $\Pi_0 = G_0(K_0^n)$ and $\Pi_1 = G_1(K_1^n)$ to $\mathcal{P}_2$.
4. $\mathcal{P}_1$ randomly generates $F : \{0,1\}^n \to \{0,1\}^l$ from a UHF, and sends F and $\Pi_2 = C^l + F(K_0^n \oplus K_1^n)$ to $\mathcal{P}_2$.
5. **Reveal phase**
6. $\mathcal{P}_1$ sends (K_0^n, K_1^n, C^l) to $\mathcal{P}_2$.
7. $\mathcal{P}_2$ checks if $G_0(K_0^n) = \Pi_0$ and $G_1(K_1^n) = \Pi_1$ are satisfied; if not, $\mathcal{P}_2$ outputs `reject`.
8. $\mathcal{P}_2$ checks if
$$K_{B_i,i} = K_i, \quad \forall\, 1 \le i \le n$$
is satisfied; if not, $\mathcal{P}_2$ outputs `reject`.
9. $\mathcal{P}_2$ checks if $C^l + F(K_0^n + K_1^n) = \Pi_2$ is satisfied; if not, $\mathcal{P}_2$ outputs `reject`.
10. If all the tests are passed, $\mathcal{P}_2$ accepts C^l, and outputs `accept`.

THEOREM 14.3 *For a given $\delta > 0$, Protocol 14.5 realizes an $(\varepsilon_{\mathtt{s}}, \varepsilon_{\mathtt{h}}, \varepsilon_{\mathtt{b}})$-BC of length l with $\varepsilon_{\mathtt{s}} = 0$, $\varepsilon_{\mathtt{h}} = 2\xi$, $\varepsilon_{\mathtt{b}} = 2^{-\delta n+2} + 2^{-\nu+2nh(\delta)+1}$, and*

$$l = \lfloor n - 2\nu - \log(1/4\xi^2) \rfloor.$$

Before proving Theorem 14.3, we recall the following result: for a given sequence $k^n \in \{0,1\}^n$ and the Hamming distance $d_H(\cdot,\cdot)$, let

$$\mathcal{B}_{\delta n}(k^n) = \{\tilde{k}^n \in \{0,1\}^n : d_H(k^n, \tilde{k}^n) \le \delta n\}$$

be the Hamming ball of radius δn around k^n. Then, it is known that, for $0 \le \delta \le 1/2$ (see Problem 14.2),

$$|\mathcal{B}_{\delta n}(k^n)| \le 2^{nh(\delta)}. \tag{14.2}$$

Proof of Theorem 14.3

When both the parties are honest, it is clear that C^l is accepted. Thus, $\varepsilon_{\mathtt{s}} = 0$ is attainable. For the hiding condition, $F(K_0^n + K_1^n)$ can be regarded as a secret key of length l under the situation that $\mathcal{P}_2$ observes $(K^n, B^n, \Pi_0, \Pi_1, G_0, G_1)$. By noting that $(K_{i,\bar{B}_i} : 1 \le i \le n)$ is independent of (K^n, B^n, G_0, G_1) and uniformly distributed on $\{0,1\}^n$, we can verify that $H_{\min}(K_0^n + K_1^n | K^n, B^n, G_0, G_1) = n$. Note also that the range of (Π_0, Π_1) is 2ν bits. Thus, by the leftover hash lemma (cf. Corollary 7.22), the key is ξ-secure. Since C^l is encrypted by the generated key, by noting the security of one-time pad encryption with a generated key (cf. Theorem 3.18), $\varepsilon_{\mathtt{h}} = 2\xi$ is attainable.

In the rest of the proof, we establish the validity of the binding condition. Note that misbehavior by $\mathcal{P}_1$ is successful when there exists $(\tilde{C}^l, \tilde{K}_0^n, \tilde{K}_1^n)$ and $(\hat{C}^l, \hat{K}_0^n, \hat{K}_1^n)$ with $\tilde{C}^l \neq \hat{C}^l$ such that $\mathcal{P}_2$ accept both the tuples. Roughly, in order to fool $\mathcal{P}_2$ to accept both the tuples in Step 9 of the reveal phase, $\mathcal{P}_1$ must reveal $(\tilde{K}_0^n, \tilde{K}_1^n)$ and $(\hat{K}_0^n, \hat{K}_1^n)$ satisfying either $\tilde{K}_0^n \neq \hat{K}_0^n$ or $\tilde{K}_1^n \neq \hat{K}_1^n$. Then, in order to pass the test in Step 8, $(\tilde{K}_0^n, \tilde{K}_1^n)$ and $(\hat{K}_0^n, \hat{K}_1^n)$ must be close (in the Hamming distance) to K_0^n and K_1^n, respectively. However, since the number of sequences close to K_0^n and K_1^n is limited, such substitution can be detected by the UHF test in Step 7. A formal analysis is conducted as follows.

Let $\mathcal{E}_2$ be the event such that $\tilde{C}^l \neq \hat{C}^l$ and

$$\tilde{C}^l + F(\tilde{K}_0^n + \tilde{K}_1^n) = \Pi_2,$$
$$\hat{C}^l + F(\hat{K}_0^n + \hat{K}_1^n) = \Pi_2$$

occur simultaneously. For $b = 0, 1$, let $\mathcal{E}_b$ be the event such that

$$\begin{aligned}
G_b(\tilde{K}_b^n) &= \Pi_b, \\
G_b(\hat{K}_b^n) &= \Pi_b, \\
\tilde{K}_{b,i} &= K_i, \quad \forall 1 \leq i \leq n \text{ with } B_i = b, \\
\hat{K}_{b,i} &= K_i, \quad \forall 1 \leq i \leq n \text{ with } B_i = b
\end{aligned}$$

hold. Then, the success probability of $\mathcal{P}_1$'s attack can be written as

$$p_{\text{succ}} = \Pr\left(\mathcal{E}_0 \wedge \mathcal{E}_1 \wedge \mathcal{E}_2\right).$$

Note that $\mathcal{E}_2$ occurs only when either $\tilde{K}_0^n \neq \hat{K}_0^n$ or $\tilde{K}_1^n \neq \hat{K}_1^n$. Thus, we have

$$\begin{aligned}
p_{\text{succ}} &= \Pr\left(\{\tilde{K}_0^n = \hat{K}_0^n\} \wedge \{\tilde{K}_1 = \hat{K}_1\} \wedge \mathcal{E}_0 \wedge \mathcal{E}_1 \wedge \mathcal{E}_2\right) \\
&\quad + \Pr\left((\{\tilde{K}_0^n \neq \hat{K}_0^n\} \vee \{\tilde{K}_1 \neq \hat{K}_1\}) \wedge \mathcal{E}_0 \wedge \mathcal{E}_1 \wedge \mathcal{E}_2\right) \\
&= \Pr\left((\{\tilde{K}_0^n \neq \hat{K}_0^n\} \vee \{\tilde{K}_1 \neq \hat{K}_1\}) \wedge \mathcal{E}_0 \wedge \mathcal{E}_1 \wedge \mathcal{E}_2\right) \\
&\leq \Pr\left((\{\tilde{K}_0^n \neq \hat{K}_0^n\} \vee \{\tilde{K}_1 \neq \hat{K}_1\}) \wedge \mathcal{E}_0 \wedge \mathcal{E}_1\right) \\
&\leq \Pr\left(\{\tilde{K}_0^n \neq \hat{K}_0^n\} \wedge \mathcal{E}_0\right) + \Pr\left(\{\tilde{K}_1^n \neq \hat{K}_1^n\} \wedge \mathcal{E}_1\right).
\end{aligned} \tag{14.3}$$

Next, we evaluate each term of the above bound as follows. For $b = 0, 1$, let $\tilde{\mathcal{E}}_b$ be the events such that $\tilde{K}_{b,i} = K_i$ for every $1 \leq i \leq n$ satisfying $B_i = b$; define the event $\hat{\mathcal{E}}_b$ similarly by replacing $\tilde{K}_{b,i}$ with $\hat{K}_{i,b}$. Then, we have

$$\begin{aligned}
&\Pr\left(\{\tilde{K}_0^n \neq \hat{K}_0^n\} \wedge \mathcal{E}_0\right) \\
&\quad = \Pr\left(\{\tilde{K}_0^n, \hat{K}_0^n \in \mathcal{B}_{\delta n}(K_0^n)\} \wedge \{\tilde{K}_0^n \neq \hat{K}_0^n\} \wedge \mathcal{E}_0\right)
\end{aligned}$$

$$+\Pr\left((\{\tilde{K}_0^n \notin \mathcal{B}_{\delta n}(K_0^n)\} \vee \{\hat{K}_0^n \notin \mathcal{B}_{\delta n}(K_0^n)\}) \wedge \{\tilde{K}_0^n \neq \hat{K}_0^n\} \wedge \mathcal{E}_0\right)$$
$$\leq \Pr\left(\{\tilde{K}_0^n, \hat{K}_0^n \in \mathcal{B}_{\delta n}(K_0^n)\} \wedge \{\tilde{K}_0^n \neq \hat{K}_0^n\} \wedge \{G_0(\tilde{K}_0^n) = G_0(\hat{K}_0^n)\}\right)$$
$$+\Pr\left(\{\tilde{K}_0^n \notin \mathcal{B}_{\delta n}(K_0^n)\} \wedge \tilde{\mathcal{E}}_0\right) + \Pr\left(\{\hat{K}_0^n \notin \mathcal{B}_{\delta n}(K_0^n)\} \wedge \hat{\mathcal{E}}_0\right). \quad (14.4)$$

By using the property of UHF, we can bound the first term above as[2]

$$\Pr\left(\{\tilde{K}_0^n, \hat{K}_0^n \in \mathcal{B}_{\delta n}(K_0^n)\} \wedge \{\tilde{K}_0^n \neq \hat{K}_0^n\} \wedge \{G_0(\tilde{K}_0^n) = G_0(\hat{K}_0^n)\}\right)$$
$$\leq \Pr\left(\exists \tilde{k}_0^n, \hat{k}_0^n \in \mathcal{B}_{\delta n}(K_0^n) \text{ such that } \tilde{k}_0^n \neq \hat{k}_0^n, G_0(\tilde{k}_0^n) = G_0(\hat{k}_0^n)\right)$$
$$\leq \sum_{k_0^n} P_{K_0^n}(k_0^n) \sum_{\substack{\tilde{k}_0^n, \hat{k}_0^n \in \mathcal{B}_{\delta n}(k_0^n) \\ \tilde{k}_0 \neq \hat{k}_0}} \Pr\left(G_0(\tilde{k}_0^n) = G_0(\hat{k}_0^n)\right)$$
$$\leq 2^{-\nu + 2nh(\delta)}, \quad (14.5)$$

where we used (14.2) in the last inequality. On the other hand, by denoting $\mathcal{I}(\tilde{K}_0^n, K_0^n) = \{i : \tilde{K}_{0,i} \neq K_{0,i}\}$, we have

$$\Pr\left(\{\tilde{K}_0^n \notin \mathcal{B}_{\delta n}(K_0^n)\} \wedge \tilde{\mathcal{E}}_0\right)$$
$$= \Pr\left(\{B_i = 1 \ \forall i \in \mathcal{I}(\tilde{K}_0^n, K_0^n)\} \wedge \{\tilde{K}_0^n \notin \mathcal{B}_{\delta n}(K_0^n)\}\right)$$
$$\leq 2^{-\delta n}. \quad (14.6)$$

Similarly, we have

$$\Pr\left(\{\hat{K}_0^n \notin \mathcal{B}_{\delta n}(K_0^n)\} \wedge \hat{\mathcal{E}}_0\right) \leq 2^{-\delta n}. \quad (14.7)$$

By combining (14.4)–(14.7), we can bound the first term of (14.3) by $2^{-\delta n+1} + 2^{-\nu+2nh(\delta)}$; we can bound the second term of (14.3) similarly. Consequently, we have the desired bound on p_{succ}. □

14.4.2 Adversarial Correlation Test

As we have seen in the security analysis of the binding property of Protocol 14.5, it is crucial that the number of coordinates $\mathcal{P}_1$ can substitute is limited because of the test in Step 8 of the reveal phase. This test heavily relies on the fact that the correlation shared by the parties is ROT. In this section, for a given correlation, we consider a test that detects the event when $\mathcal{P}_1$ replaces a sufficient portion of the observation. In this section, we assume certain familiarity with the method of types; see Section 5.5 for terminology and basic results.

[2] Since $\tilde{K}_0^n$ and $\hat{K}_0^n$ are chosen by $\mathcal{P}_1$ after G_0 is transmitted, $\tilde{K}_0^n$ and $\hat{K}_0^n$ may be correlated to G_0. Thus, we take the union bound before applying the property of UHF.

Consider an i.i.d. observation (X^n, Y^n) sampled from P_{XY}, and suppose that X^n is substituted to $\hat{X}^n$ which is sufficiently far apart from X^n. At a first glance, one might expect that the substitution can be detected by checking if the type $\mathtt{tp}(\hat{X}^n, Y^n)$ is close to P_{XY}. However, this is not the case in general. When there exist two symbols $x \neq x'$ such that $\mathrm{P}_{Y|X}(\cdot|x) = \mathrm{P}_{Y|X}(\cdot|x')$, then $\mathcal{P}_1$ can swap $X_i = x$ and $X_j = x'$ without being detected. To prevent such a substitution, we use the minimum sufficient statistic $V = f(X) := \mathtt{mss}(Y|X)$ defined in Definition 13.13 of Section 13.4.2. The sequence V^n obtained by $V_i = f(X_i)$ does not have the above mentioned problem, and substitution can be detected by an appropriate statistical test. In fact, for the construction of bit commitment from ROT, there was no need to consider the minimum sufficient statistic since $\mathtt{mss}(Y|X) = X$ (Problem 14.3). In the following, we will discuss how to detect $\mathcal{P}_1$'s substitution in detail.

For a formal analysis of detectability, the following lemma is useful.

LEMMA 14.4 *For $V = f(X) = \mathtt{mss}(Y|X)$ and any random variable $\hat{V}$ satisfying $\hat{V} -\!\!\circ\!\!- X -\!\!\circ\!\!- Y$ and $\mathrm{P}_{VY} = \mathrm{P}_{\hat{V}Y}$, we have $\hat{V} \equiv V$.*

Proof Let $v_0 \in \mathcal{V}$ be a symbol such that $\mathrm{P}_{Y|V}(\cdot|v_0)$ cannot be written as a convex combination of $\mathrm{P}_{Y|X}(\cdot|x)$ with $x \notin f^{-1}(v_0)$. Such a v_0 must exist since $\{\mathrm{P}_{Y|V}(\cdot|v) : v \in \mathcal{V}\}$ are all distinct probability vectors; we can pick v_0 such that $\mathrm{P}_{Y|V}(\cdot|v_0)$ is an extreme point of the convex hull $\mathtt{conv}(\{\mathrm{P}_{Y|V}(\cdot|v) : v \in \mathcal{V}\})$.

Since $\mathrm{P}_{Y|V}(\cdot|v_0) = \mathrm{P}_{Y|\hat{V}}(\cdot|v_0)$ by the assumption $\mathrm{P}_{VY} = \mathrm{P}_{\hat{V}Y}$, $f(x) \neq v_0$ implies that $\mathrm{P}_{\hat{V}|X}(v_0 \mid x) = 0$. Indeed since otherwise we can write

$$\begin{aligned}
\mathrm{P}_{Y|V}(y|v_0) &= \mathrm{P}_{Y|\hat{V}}(y|v_0) \\
&= \sum_x \mathrm{P}_{Y|X}(y|x)\, \mathrm{P}_{X|\hat{V}}(x|v_0) \\
&= \mathrm{P}_{Y|V}(y|v_0) \sum_{x \in f^{-1}(v_0)} \mathrm{P}_{X|\hat{V}}(x|v_0) \\
&\quad + \sum_{x \notin f^{-1}(v_0)} \mathrm{P}_{Y|X}(y|x)\, \mathrm{P}_{X|\hat{V}}(x|v_0),
\end{aligned}$$

whereby $\mathrm{P}_{Y|V}(\cdot|v_0)$ can be written as a convex combination of $\mathrm{P}_{Y|X}(\cdot|x)$ with $x \notin f^{-1}(v_0)$, which contradicts the choice of v_0 mentioned above. Furthermore, since $\mathrm{P}_V(v_0) = \mathrm{P}_{\hat{V}}(v_0)$, all symbols $x \in f^{-1}(v_0)$ must be (deterministically) mapped to v_0 by the channel $\mathrm{P}_{\hat{V}|X}$.

By removing $f^{-1}(v_0)$ from $\mathcal{X}$, we can repeat the same argument as above to eventually conclude that $\mathrm{P}_{\hat{V}|X}$ maps all x to $f(x)$, which means $\hat{V} \equiv V$. □

For the time being, let us assume that the substituted sequence $\hat{V}^n$ is generated from X^n via a memoryless channel that is chosen from the following class:

$$\mathcal{W}(\delta) := \left\{ W \in \mathcal{P}(\mathcal{V}|\mathcal{X}) : \sum_{\hat{v},x} \mathrm{P}_X(x)\, W(\hat{v}|x) \mathbf{1}[\hat{v} \neq f(x)] \geq \delta \right\},$$

where $\mathcal{P}(\mathcal{V}|\mathcal{X})$ is the set of all channels from $\mathcal{X}$ to $\mathcal{V}$. Note that the constraint in $\mathcal{W}(\delta)$ covers the case when $\mathcal{P}_1$ tries to substitute a sufficient portion of the observation V^n. The memoryless assumption will be removed later.

EXAMPLE 14.5 Consider the binary symmetric correlation given by $\mathrm{P}_{XY}(x, y) = \frac{1-q}{2}$ for $x = y$ and $\mathrm{P}_{XY}(x, y) = \frac{q}{2}$ for $x \neq y$. In this case, $\mathtt{mss}(Y|X)$ is X itself. The class $\mathcal{W}(\delta)$ consists of channels satisfying $W(1|0) + W(0|1) \geq 2\delta$.

Since the detection procedure must work for any substitution channel $W \in \mathcal{W}(\delta)$, we consider the composite hypothesis testing between the null hypothesis P_{VY}, i.e., there is no substitution, and the set $\mathcal{P}(\delta)$ of alternative hypotheses that consists of $\mathrm{P}_{\hat{V}Y}$ induced from P_{XY} for each $W \in \mathcal{W}(\delta)$.

Recall that, for a given test T, the type I error probability is defined as

$$\alpha[\mathrm{T}] := \sum_{v^n, y^n} \mathrm{P}^n_{VY}(v^n, y^n) \mathbf{1}\big[\mathrm{T}(v^n, y^n) = 1\big],$$

and the type II error probability is defined as

$$\beta[\mathrm{T}|\mathcal{W}(\delta)] := \sup_{W \in \mathcal{W}(\delta)} \sum_{\hat{v}^n, y^n} \mathrm{P}^n_{\hat{V}Y}(\hat{v}^n, y^n) \mathbf{1}\big[\mathrm{T}(\hat{v}^n, y^n) = 0\big]. \tag{14.8}$$

As we have discussed in Section 5.6, an effective test for composite hypothesis testing is the typicality test: for a given threshold $\zeta > 0$, the typicality test T_ζ accepts the null hypothesis if the type $\mathrm{P}_{\hat{v}^n y^n}$ of the observation $(\hat{v}^n, y^n)$ satisfies

$$|\mathrm{P}_{\hat{v}^n y^n}(a, b) - \mathrm{P}_{VY}(a, b)| \leq \zeta, \quad \forall (a, b) \in \mathcal{V} \times \mathcal{Y}.$$

If a channel W is such that the induced joint distribution $\mathrm{P}_{\hat{V}Y}$ coincides with P_{VY}, then Lemma 14.4 implies that $\hat{V} = V$ must hold, i.e.,

$$\sum_{\hat{v}, x} \mathrm{P}_X(x) W(\hat{v}|x) \mathbf{1}[\hat{v} \neq f(x)] = 0.$$

Thus, for any $\delta > 0$, P_{VY} is not included in the set $\mathcal{P}(\delta)$. Since the constraint of $\mathcal{P}(\delta)$ is linear, it is a closed set. Thus, as we have shown in Lemma 5.15, there exists $\zeta > 0$ such that the typicality test T_ζ satisfies

$$\alpha[\mathrm{T}_\zeta] \leq 2|\mathcal{V}||\mathcal{Y}|e^{-2\zeta^2 n},$$

and

$$\beta[\mathrm{T}_\zeta|\mathcal{W}(\delta)] \leq 2|\mathcal{V}||\mathcal{Y}|e^{-2\zeta^2 n}. \tag{14.9}$$

In the same manner, when $\mathcal{P}_1$ substitutes a significant fraction of the observation, namely when $W \in \mathcal{W}(\delta)$, such a substitution can be detected by the typicality test with error probabilities converging to 0 exponentially. When the channel W is not included in the class $\mathcal{W}(\delta)$, then the substituted sequence $\hat{V}^n$ satisfies $d_H(\hat{V}^n, V^n) < n\delta$ with high probability. In such a case, the substitution can be detected by the hash test as in Step 7 of the reveal phase in Protocol 14.5.

Now, let us consider an arbitrary channel W_n that generates $\hat{V}^n$ from X^n, and may not be memoryless. The following lemma enables us to reduce the general case to the above discussed memoryless case.

LEMMA 14.6 *For given $\delta, \zeta > 0$, the typicality test T_ζ, and any channel W_n, we have*[3]

$$\Pr\left(\mathrm{T}_\zeta(\hat{V}^n, Y^n) = 0, d_H(\hat{V}^n, V^n) \ge n\delta'\right) \le \mathtt{poly}(n) \cdot \beta[\mathrm{T}_\zeta|\mathcal{W}(\delta)],$$

where $\beta[\mathrm{T}_\zeta|\mathcal{W}(\delta)]$ is defined by (14.8), $\delta' = \delta/(\min_x \mathrm{P}_X(x))$ and $\mathtt{poly}(n)$ can be explicitly given as $(n+1)^{2|\mathcal{X}||\mathcal{V}|}$.

Proof For a permutation $\sigma \in \mathcal{S}_n$ on $\{1, \ldots, n\}$, let $\sigma(x^n) = (x_{\sigma(1)}, \ldots, x_{\sigma(n)})$. By noting that $\mathrm{P}_{X^nY^n}$, T_ζ, and d_H are invariant under the permutation, we have

$$\begin{aligned}
&\Pr\left(\mathrm{T}_\zeta(\hat{V}^n, Y^n) = 0, d_H(\hat{V}^n, V^n) \ge n\delta'\right)\\
&= \sum_{x^n, y^n} \mathrm{P}_{X^nY^n}(x^n, y^n) \sum_{\hat{v}^n} W_n(\hat{v}^n|x^n) \mathbf{1}\left[\mathrm{T}_\zeta(\hat{v}^n, y^n) = 0,\ d_H(\hat{v}^n, v^n) \ge n\delta'\right]\\
&= \sum_{\sigma \in \mathcal{S}_n} \frac{1}{|\mathcal{S}_n|} \sum_{x^n, y^n} \mathrm{P}_{X^nY^n}(\sigma(x^n), \sigma(y^n)) \sum_{\hat{v}^n} W_n(\sigma(\hat{v}^n)|\sigma(x^n))\\
&\qquad \times \mathbf{1}\left[\mathrm{T}_\zeta(\sigma(\hat{v}^n), \sigma(y^n)) = 0,\ d_H(\sigma(\hat{v}^n), \sigma(v^n)) \ge n\delta'\right]\\
&= \sum_{\sigma \in \mathcal{S}_n} \frac{1}{|\mathcal{S}_n|} \sum_{x^n, y^n} \mathrm{P}_{X^nY^n}(x^n, y^n) \sum_{\hat{v}^n} W_n(\sigma(\hat{v}^n)|\sigma(x^n))\\
&\qquad \times \mathbf{1}\left[\mathrm{T}_\zeta(\hat{v}^n, y^n) = 0,\ d_H(\hat{v}^n, v^n) \ge n\delta'\right]\\
&= \sum_{x^n, y^n} \mathrm{P}_{X^nY^n}(x^n, y^n) \sum_{\hat{v}^n} W_n^{\mathtt{sym}}(\hat{v}^n|x^n) \mathbf{1}\left[\mathrm{T}_\zeta(\hat{v}^n, y^n) = 0,\ d_H(\hat{v}^n, v^n) \ge n\delta'\right],
\end{aligned} \tag{14.10}$$

where

$$W_n^{\mathtt{sym}}(\hat{v}^n|x^n) := \sum_{\sigma \in \mathcal{S}_n} \frac{1}{|\mathcal{S}_n|} W_n(\sigma(\hat{v}^n)|\sigma(x^n)).$$

Note that the symmetrized channel $W_n^{\mathtt{sym}}(\hat{v}^n|x^n)$ only depends on the joint type of $(\hat{v}^n, x^n)$. Furthermore, when x^n has type $\mathrm{P}_{\bar{X}}$ and $\hat{v}^n \in \mathcal{T}^n_{\bar{W}}(x^n)$ for conditional type $\bar{W}$, the condition $d_H(\hat{v}^n, v^n) \ge \delta'$ can be written as

$$\sum_{\hat{v}, x} \mathrm{P}_{\bar{X}}(x)\, \bar{W}(\hat{v}|x) \mathbf{1}[\hat{v} \ne f(x)] \ge \delta'.$$

This inequality implies that at least one symbol x satisfies $\sum_{\hat{v}} \bar{W}(\hat{v}|x)\mathbf{1}[\hat{v} \ne f(x)] \ge \delta'$, which further implies $\bar{W} \in \mathcal{W}(\delta)$. By noting these facts and by classifying $\mathcal{X}^n \times \mathcal{Y}^n$ into type class $\mathcal{T}^n_{\bar{X}\bar{Y}}$ of joint type $\mathrm{P}_{\bar{X}\bar{Y}}$ on $\mathcal{X} \times \mathcal{Y}$, we can further upper bound (14.10) by

[3] As we can see from the proof, the bound holds for any test T as long as it is invariant under permutation.

$$\sum_{\mathrm{P}_{\bar{X}\bar{Y}} \in \mathcal{P}_n(\mathcal{X}\times\mathcal{Y})} \sum_{(x^n,y^n)\in \mathcal{T}^n_{\bar{X}\bar{Y}}} \mathrm{P}_{X^nY^n}(x^n,y^n)$$
$$\times \sum_{\bar{W}\in\mathcal{W}_n(\mathcal{V}|\mathrm{P}_{\bar{X}})\cap\mathcal{W}(\delta)} \sum_{\hat{v}^n\in\mathcal{T}^n_{\bar{W}}(x^n)} W_n^{\mathtt{sym}}(\mathcal{T}^n_{\bar{W}}(x^n)|x^n)\frac{1}{|\mathcal{T}^n_{\bar{W}}(x^n)|}\mathbf{1}\big[\mathrm{T}_\varsigma(\hat{v}^n,y^n)=0\big]. \tag{14.11}$$

Since (cf. Corollary 5.14)

$$\frac{1}{|\mathcal{T}^n_{\bar{W}}(x^n)|} \le (n+1)^{|\mathcal{V}||\mathcal{X}|}\bar{W}^n(\hat{v}^n|x^n)$$

holds for $\hat{v}^n \in \mathcal{T}^n_{\bar{W}}(x^n)$, we can upper bound (14.11) by

$$\begin{aligned}
&(n+1)^{|\mathcal{V}||\mathcal{X}|} \sum_{\mathrm{P}_{\bar{X}\bar{Y}} \in \mathcal{P}_n(\mathcal{X}\times\mathcal{Y})} \sum_{(x^n,y^n)\in \mathcal{T}^n_{\bar{X}\bar{Y}}} \mathrm{P}_{X^nY^n}(x^n,y^n)\\
&\qquad\times \sum_{\bar{W}\in\mathcal{W}_n(\mathcal{V}|\mathrm{P}_{\bar{X}})\cap\mathcal{W}(\delta)} \sum_{\hat{v}^n\in\mathcal{T}^n_{\bar{W}}(x^n)} \bar{W}^n(\hat{v}^n|x^n)\mathbf{1}\big[\mathrm{T}_\varsigma(\hat{v}^n,y^n)=0\big]\\
&\le (n+1)^{|\mathcal{V}||\mathcal{X}|} \sum_{\bar{W}\in\mathcal{P}_n(\mathcal{V}|\mathcal{X})\cap\mathcal{W}(\delta)} \sum_{\mathrm{P}_{\bar{X}\bar{Y}} \in \mathcal{P}_n(\mathcal{X}\times\mathcal{Y})} \sum_{(x^n,y^n)\in \mathcal{T}^n_{\bar{X}\bar{Y}}} \mathrm{P}_{X^nY^n}(x^n,y^n)\\
&\qquad\times \sum_{\hat{v}^n} \bar{W}^n(\hat{v}^n|x^n)\mathbf{1}\big[\mathrm{T}_\varsigma(\hat{v}^n,y^n)=0\big]\\
&\le (n+1)^{2|\mathcal{V}||\mathcal{X}|} \max_{\bar{W}\in\mathcal{P}_n(\mathcal{V}|\mathcal{X})\cap\mathcal{W}(\delta)} \sum_{\mathrm{P}_{\bar{X}\bar{Y}} \in \mathcal{P}_n(\mathcal{X}\times\mathcal{Y})} \sum_{(x^n,y^n)\in \mathcal{T}^n_{\bar{X}\bar{Y}}} \mathrm{P}_{X^nY^n}(x^n,y^n)\\
&\qquad\times \sum_{\hat{v}^n} \bar{W}^n(\hat{v}^n|x^n)\mathbf{1}\big[\mathrm{T}_\varsigma(\hat{v}^n,y^n)=0\big]\\
&\le (n+1)^{2|\mathcal{V}||\mathcal{X}|} \sup_{\bar{W}\in\mathcal{W}(\delta)} \sum_{x^n,y^n} \mathrm{P}_{X^nY^n}(x^n,y^n) \sum_{\hat{v}^n} \bar{W}^n(\hat{v}^n|x^n)\mathbf{1}\big[\mathrm{T}_\varsigma(\hat{v}^n,y^n)=0\big]\\
&= (n+1)^{2|\mathcal{V}||\mathcal{X}|}\beta[\mathrm{T}_\varsigma|\mathcal{W}(\delta)],
\end{aligned}$$

completing the proof, where $\mathcal{P}_n(\mathcal{V}|\mathcal{X}) = \cup_{\mathrm{P}_{\bar{X}}\in\mathcal{P}_n(\mathcal{X})}\mathcal{W}_n(\mathcal{V}|\mathrm{P}_{\bar{X}})$. □

By Lemma 14.6, if the missed detection probability $\beta[\mathrm{T}_\varsigma|\mathcal{W}(\delta)]$ for memoryless substitution is exponentially small, which is guaranteed by (14.9), then the probability of missed detection for an attack that substitutes a sufficient portion of the observation is small. Thus, by combining with the hash test which is effective for a small portion of substitution, we can guarantee the binding condition; the complete analysis of the scheme will be discussed in the next section.

14.4.3 Bit Commitment from General Correlation

By using the typicality test T_ς described in the previous section, we can construct the bit commitment from a correlation P_{XY} as in Protocol 14.6.

The performance of Protocol 14.6 can be evaluated as follows.

Protocol 14.6 Bit commitment from a correlation P_{XY}

Input: $\mathcal{P}_1$'s observation C^l and X^n; and $\mathcal{P}_2$'s observation Y^n
Output: `Accept/Reject`

1. **Commit phase**
2. $\mathcal{P}_2$ randomly generates $G : \mathcal{V}^n \to \{0,1\}^\nu$ from UHF, and sends G to $\mathcal{P}_1$.
3. $\mathcal{P}_1$ sends $\Pi_1 = G(V^n)$ to $\mathcal{P}_2$, where $V_i = \mathtt{mss}(Y_i|X_i)$.
4. $\mathcal{P}_1$ randomly generates $F : \mathcal{V}^n \to \{0,1\}^l$ from UHF, and sends F and $\Pi_2 = C^l + F(V^n)$ to $\mathcal{P}_2$.
5. **Reveal phase**
6. $\mathcal{P}_1$ reveals (V^n, C^l) to $\mathcal{P}_2$.
7. $\mathcal{P}_2$ checks if $G(V^n) = \Pi_1$ is satisfied; if not, $\mathcal{P}_2$ outputs `reject`.
8. $\mathcal{P}_2$ checks if $\mathrm{T}_\varsigma(V^n, Y^n) = 0$ is satisfied; if not, $\mathcal{P}_2$ outputs `reject`.
9. $\mathcal{P}_2$ checks if $C^l + F(V^n) = \Pi_2$ is satisfied; if not, $\mathcal{P}_2$ outputs `reject`.
10. If all the tests are passed, $\mathcal{P}_2$ accepts C^l, and outputs `accept`.

THEOREM 14.7 *We retain the same notation as in Lemma 14.6. Protocol 14.6 realizes an* $(\varepsilon_{\mathtt{s}}, \varepsilon_{\mathtt{h}}, \varepsilon_{\mathtt{b}})$*-BC of length* l *with* $\varepsilon_{\mathtt{s}} = \alpha[\mathrm{T}_\varsigma]$, $\varepsilon_{\mathtt{h}} = 4\eta + 2\xi$, $\varepsilon_{\mathtt{b}} = b_n(\delta')^2 \cdot 2^{-\nu} + 2\mathtt{poly}(n) \cdot \beta[\mathrm{T}_\varsigma|\mathcal{W}(\delta)]$, *and*

$$l = \left\lfloor H_{\min}^{\eta}(V^n|Y^n) - \nu - \log(1/4\xi^2) \right\rfloor,$$

where

$$b_n(\delta') := \sum_{i=0}^{\lceil n\delta' \rceil - 1} (|\mathcal{V}| - 1)^i \binom{n}{i} \le |\mathcal{V}|^{n\delta'} 2^{nh(\delta')}$$

is the cardinality of the Hamming ball of radius $\lceil n\delta' \rceil - 1$ *in* $\mathcal{V}^n$.

Proof The proof is very similar to that of Theorem 14.3 except that we use the typicality test T_ς. When both the parties are honest, C^l is rejected only when $\mathrm{T}_\varsigma(V^n, Y^n) = 1$, which occurs with probability $\varepsilon_{\mathtt{s}} = \alpha[\mathrm{T}_\varsigma]$. For the hiding condition, $F(V^n)$ can be regarded as a secret key of length l under the situation that $\mathcal{P}_2$ observes (Y^n, Π_1). Thus, by the leftover hash lemma (cf. Corollary 7.22), the key is $(2\eta + \xi)$-secure. Then, since C^l is encrypted by the generated key, $\varepsilon_{\mathtt{h}} = 4\eta + 2\xi$ is attainable (cf. Theorem 3.18).

Regarding the binding condition, an attack by $\mathcal{P}_1$ is successful when there exist $(\tilde{C}^l, \tilde{V}^n)$ and $(\hat{C}^l, \hat{V}^n)$ with $\tilde{C}^l \neq \hat{C}^l$ such that $\mathcal{P}_2$ accept both the pairs. Let $\mathcal{E}_2$ be the event such that $\tilde{C}^l \neq \hat{C}^l$, $\tilde{C}^l + F(\tilde{V}^n) = \Pi_2$, and $\hat{C}^l + F(\hat{V}^n) = \Pi_2$ occur simultaneously. Let $\mathcal{E}_1$ be the event such that $G(\tilde{V}^n) = \Pi_1$, $G(\hat{V}^n) = \Pi_1$, $\mathrm{T}_\varsigma(\tilde{V}^n, Y^n) = 0$, and $\mathrm{T}_\varsigma(\hat{V}^n, Y^n) = 0$ occur simultaneously. Then, the success probability of $\mathcal{P}_1$'s cheating is given by

$$\begin{aligned} p_{\mathtt{succ}} &= \Pr\left(\mathcal{E}_1 \wedge \mathcal{E}_2\right) \\ &= \Pr\left(\{\tilde{V}^n \neq \hat{V}^n\} \wedge \mathcal{E}_1 \wedge \mathcal{E}_2\right) \\ &\le \Pr\left(\{\tilde{V}^n \neq \hat{V}^n\} \wedge \mathcal{E}_1\right), \end{aligned}$$

where the second equality holds since the event $\mathcal{E}_2$ occurs only when $\tilde{V}^n \neq \hat{V}^n$. Then, we can further evaluate

$$\begin{aligned}
&\Pr\left(\{\tilde{V}^n \neq \hat{V}^n\} \wedge \mathcal{E}_1\right) \\
&\quad = \Pr\left(\{d_H(\tilde{V}^n, V^n) < n\delta', d_H(\hat{V}^n, V^n) < n\delta'\} \wedge \{\tilde{V}^n \neq \hat{V}^n\} \wedge \mathcal{E}_1\right) \\
&\qquad + \Pr\left((\{d_H(\tilde{V}^n, V^n) \geq n\delta'\} \vee \{d_H(\hat{V}^n, V^n) \geq n\delta'\}) \wedge \{\tilde{V}^n \neq \hat{V}^n\} \wedge \mathcal{E}_1\right) \\
&\quad \leq \Pr\left(\{d_H(\tilde{V}^n, V^n) < n\delta', d_H(\hat{V}^n, V^n) < n\delta'\} \wedge \{\tilde{V}^n \neq \hat{V}^n, G(\tilde{V}^n) = G(\hat{V}^n)\}\right) \\
&\qquad + \Pr\left(T_\delta(\tilde{V}^n, Y^n) = 0, d_H(\tilde{V}^n, V^n) \geq n\delta'\right) \\
&\qquad + \Pr\left(T_\delta(\hat{V}^n, Y^n) = 0, d_H(\hat{V}^n, V^n) \geq n\delta'\right).
\end{aligned} \tag{14.12}$$

Note that, by Lemma 14.6, the second and the third terms of (14.12) are bounded by $\texttt{poly}(n) \cdot \beta[\mathrm{T}_\varsigma | \mathcal{W}(\delta)]$. On the other hand, in a similar manner to (14.5), the first term of (14.12) is bounded by $b_n(\delta')^2 \cdot 2^{-\nu}$. Thus, we have the desired bound on $\varepsilon_{\mathtt{b}}$. □

14.5 Impossibility Result

In the construction of bit commitment protocols in Section 14.4, we have seen that the generation of a secret key plays an important role in guaranteeing the hiding property of bit commitment. In this section, in a similar spirit to Section 13.4, we present a reduction that constructs a secret key from a given bit commitment protocol. Then, this reduction in turn yields an impossibility bound for the bit commitment construction via the impossibility bound for a secret key agreement in Section 10.3. Recall that $L_{\varepsilon_{\mathtt{s}},\varepsilon_{\mathtt{h}},\varepsilon_{\mathtt{b}}}(X, Y)$ is the largest length l such that a protocol realizing an $(\varepsilon_{\mathtt{s}}, \varepsilon_{\mathtt{h}}, \varepsilon_{\mathtt{b}})$-BC exists.

Heuristically, the reduction is based on the following idea. Consider a virtual situation in which $\mathcal{P}_1$ observes X, $\mathcal{P}_2$ observes (V, Y) for $V = \texttt{mss}(Y|X)$ (cf. Definition 13.13), and an eavesdropper observes Y. For a given bit commitment protocol π, $\mathcal{P}_1$ and $\mathcal{P}_2$ run the commit phase of the protocol π with communication Π. Then, the hiding property guarantees that the committed bit sequence C^l is concealed from the eavesdropper observing Y and communication Π. On the other hand, we can show that, roughly, the committed bit sequence is the unique sequence that is compatible with $\mathcal{P}_2$'s observation (V, Y) and Π, which guarantees the reliability of the secret key. More specifically, let $(\hat{c}^l, \hat{x}) = (\hat{c}^l(v, \tau), \hat{x}(v, \tau))$ be the function of (v, τ) given by

$$(\hat{c}^l, \hat{x}) = \underset{(c^l, x)}{\operatorname{argmax}} \Pr\left(\mathrm{T}(c^l, x, Y, \Pi) = 0 | V = v, \Pi = \tau\right). \tag{14.13}$$

Then, $(\hat{C}^l, \hat{X}) = (\hat{c}^l(V, \Pi), \hat{x}(V, \Pi))$ must satisfy $\hat{C}^l = C^l$ with high probability; otherwise, $\mathcal{P}_1$ can cheat in protocol π. Formally, we can construct a secret key protocol as in Protocol 14.7.

Protocol 14.7 Reduction 1 of secret key to bit commitment

Input: $\mathcal{P}_1$'s observation X, $\mathcal{P}_2$'s observation (V, Y), and the eavesdropper's observation Y

Output: $\mathcal{P}_1$'s output S_1 and $\mathcal{P}_2$'s output S_2

1. $\mathcal{P}_1$ generates the random string $C^l \in \{0,1\}^l$.
2. The parties run the BC protocol π with communication Π.
3. $\mathcal{P}_2$ computes $(\hat{C}^l, \hat{X}) = (\hat{c}^l(V,\Pi), \hat{x}(V,\Pi))$ given in (14.13).
4. $\mathcal{P}_1$ and $\mathcal{P}_2$ output $S_1 = C^l$ and $S_2 = \hat{C}^l$, respectively.

PROPOSITION 14.8 *If the protocol π realizes an $(\varepsilon_{\mathtt{s}}, \varepsilon_{\mathtt{h}}, \varepsilon_{\mathtt{b}})$-BC of length l, then Protocol 14.7 realizes an $(\varepsilon_{\mathtt{s}} + \varepsilon_{\mathtt{b}}, \varepsilon_{\mathtt{h}})$-secure secret key protocol of length l. In particular,*

$$L_{\varepsilon_{\mathtt{s}},\varepsilon_{\mathtt{h}},\varepsilon_{\mathtt{b}}}(X,Y) \le S_{\varepsilon_{\mathtt{s}}+\varepsilon_{\mathtt{b}},\varepsilon_{\mathtt{h}}}(X,(V,Y)|Y).$$

Proof The security of the generated key C^l follows from the hiding property of protocol π. To prove reliability, we shall show that

$$\Pr\left(\hat{C}^l \neq C^l\right) \le \varepsilon_{\mathtt{s}} + \varepsilon_{\mathtt{b}}. \tag{14.14}$$

To that end, we note that the definition of $(\hat{c}^l, \hat{x})$ can be rewritten as

$$(\hat{c}^l, \hat{x}) = \operatorname*{argmax}_{c^l,x} \sum_y \mathrm{P}_{Y|V\Pi}(y|v,\tau) \Pr\left(\mathrm{T}(c^l, x, y, \tau) = 0\right).$$

Then, we can evaluate the probability that $(\hat{C}^l, \hat{X})$ passes the test T as follows:

$$\begin{aligned}
&\Pr\left(\mathrm{T}(\hat{C}^l, \hat{X}, Y, \Pi) = 0\right) \\
&= \sum_{v,\tau} \mathrm{P}_{V\Pi}(v,\tau) \sum_y \mathrm{P}_{Y|V\Pi}(y|v,\tau) \Pr\left(\mathrm{T}(\hat{c}^l(v,\tau), \hat{x}(v,\tau), y, \tau) = 0\right) \\
&\ge \sum_{v,\tau} \mathrm{P}_{V\Pi}(v,\tau) \sum_{c^l,x} \mathrm{P}_{C^lX|V\Pi}(c^l,x|v,\tau) \sum_y \mathrm{P}_{Y|V\Pi}(y|v,\tau) \Pr\left(\mathrm{T}(c^l,x,y,\tau) = 0\right) \\
&= \Pr\left(\mathrm{T}(C^l, X, Y, \Pi) = 0\right) \\
&\ge 1 - \varepsilon_{\mathtt{s}},
\end{aligned}$$

where the first inequality follows from the definition of $(\hat{c}^l(v,\tau), \hat{x}(v,\tau))$, and the second equality follows from the Markov relation $(C^l, X) -\!\!\circ\!\!- (V,\Pi) -\!\!\circ\!\!- Y$; this Markov relation follows from the interactive communication property in Lemma 10.7, together with the fact that $(C^l, X) -\!\!\circ\!\!- V -\!\!\circ\!\!- Y$ holds by the definition of sufficient statistics. The inequality above along with the binding condition implies

$$\begin{aligned}
1 - \varepsilon_{\mathtt{s}} &\le \Pr\left(\hat{C}^l = C^l\right) + \Pr\left(\mathrm{T}(\hat{C}^l, \hat{X}, Y, \Pi) = 0, \hat{C}^l \neq C^l\right) \\
&\le \Pr\left(\hat{C}^l = C^l\right) + \varepsilon_{\mathtt{b}},
\end{aligned}$$

which completes the proof of (14.14). □

Upon a close inspection of the proof of Proposition 14.8, we notice that we only used the binding condition for dishonest behavior by $\mathcal{P}_1$ in the reveal phase; see Problem 14.4.

By combining Proposition 14.8 together with the bound on secret key length in Theorem 10.10, we have the following bound on bit commitment length.

COROLLARY 14.9 *For a given correlation (X, Y) and $(\varepsilon_{\mathtt{s}}, \varepsilon_{\mathtt{h}}, \varepsilon_{\mathtt{b}})$ satisfying $\varepsilon_{\mathtt{s}} + \varepsilon_{\mathtt{h}} + \varepsilon_{\mathtt{b}} < 1$,*

$$L_{\varepsilon_{\mathtt{s}},\varepsilon_{\mathtt{h}},\varepsilon_{\mathtt{b}}}(X, Y) \le -\log \beta_\eta(\mathrm{P}_{VVY}, \mathrm{P}_{V|Y}\mathrm{P}_{V|Y}\mathrm{P}_Y) + 2\log(1/\xi)$$

for any $\xi > 0$ with $\eta = \varepsilon_{\mathtt{s}} + \varepsilon_{\mathtt{h}} + \varepsilon_{\mathtt{b}} + \xi$, where $V = \mathtt{mss}(Y|X)$ and $\beta(\cdot,\cdot)$ is the optimal type II error probability (cf. Section 5.1).

14.6 Bit Commitment Capacity

In this section, we investigate asymptotic behavior of $L_{\varepsilon_{\mathtt{s}},\varepsilon_{\mathtt{h}},\varepsilon_{\mathtt{b}}}(X^n, Y^n)$, the maximum length of bit commitment that can be constructed from a given i.i.d. correlation (X^n, Y^n). In fact, we prove that the construction presented in Protocol 14.6 is asymptotically optimal.

DEFINITION 14.10 For a given i.i.d. correlation (X^n, Y^n) distributed according to P_{XY}, the bit commitment capacity is defined as

$$C_{\mathtt{BC}}(X, Y) := \lim_{\varepsilon_{\mathtt{s}},\varepsilon_{\mathtt{h}},\varepsilon_{\mathtt{b}} \to 0} \liminf_{n\to\infty} \frac{1}{n} L_{\varepsilon_{\mathtt{s}},\varepsilon_{\mathtt{h}},\varepsilon_{\mathtt{b}}}(X^n, Y^n).$$

The bit commitment capacity can be characterized as follows.

THEOREM 14.11 *For a given P_{XY}, the bit commitment capacity is given by*

$$C_{\mathtt{BC}}(X, Y) = H(V|Y),$$

where $V = \mathtt{mss}(Y|X)$.

Proof The converse part follows from Corollary 14.9 along with the Stein lemma (cf. Theorem 5.7). On the other hand, this bound can be attained by Protocol 14.6. Fix arbitrarily small $\delta > 0$ in Theorem 14.7. As we have discussed in Section 14.4.2, the error probabilities $\alpha[T_\zeta]$ and $\beta[T_\zeta|\mathcal{W}(\delta)]$ of the statistical test converge to 0; in particular, the latter converges to 0 exponentially. Furthermore, if we take the rate ν of the hash test to be larger than $2(\delta' \log|\mathcal{V}| + h(\delta'))$, then the first term of $\varepsilon_{\mathtt{b}}$ also converges to 0. Note that we can take ν arbitrarily small by taking sufficiently small δ. Finally, by noting that the smooth min-entropy converges to the conditional Shannon entropy for i.i.d. distributions (cf. Theorem 7.25), we get the desired lower bound on the bit commitment capacity. ☐

The bit commitment capacity is completely determined by Theorem 14.11. This is in contrast with the OT capacity, in which only partial results are known.

Roughly speaking, the bit commitment capacity of correlation (X, Y) is given by $\mathcal{P}_2$'s ambiguity about $\mathcal{P}_1$'s observation. However, as we discussed in Section 14.4.2, in order to prevent $\mathcal{P}_1$'s attack in the reveal phase, we must remove the "redundant part" from X and only use the minimum sufficient statistic V. As an extreme example, let $X = (U, K_0, K_1)$ and $Y = (B, K_B)$, where $((K_0, K_1), (B, K_B))$ is the ROT and $U \in \{0,1\}^m$ is the uniform random variable that is independent of the ROT. Then, $\mathtt{mss}(Y|X) = (K_0, K_1)$; even though $\mathcal{P}_2$'s ambiguity about X is $m + 1$ bits, the redundant part U is useless since there is no way for $\mathcal{P}_2$ to detect substitution of U.

14.7 References and Additional Reading

Bit commitment was introduced by Blum in [38]. Since then, it has been extensively used as a building block for secure computation; we will see more on this in the next few chapters.

From the early stage of research on secure computation, it has been known that bit commitment can be constructed from oblivious transfer; in fact, the latter is a complete primitive. Construction of bit commitment from noisy correlation first appeared in [79]. The concept of the bit commitment capacity was introduced in [169, 350]. Protocol 14.6 is based on the one in [330, Chapter 8] (see also [171]).

The adversarial correlation test considered in Section 14.4.2 is an important primitive to detect substitution of an observation, and it is also used in other cryptographic protocols such as authentication using correlation [243] or digital signature using correlation [127]. In particular, Lemma 14.4 is from [127]. Since it is an important primitive, we have treated it as a separate problem in Section 14.4.2.

The construction of bit commitment from various models has been studied in the literature, such as the bounded storage model [111, 307], the unfair noisy channel [80, 95], and the elastic noisy channel [191]; see also Problem 14.11.

In this chapter, we considered string bit commitment from correlation, and the bit commitment capacity was defined as the maximum rate of string bit commitment that can be constructed from a given correlation. An alternative definition of bit commitment capacity could be the maximum rate of bit commitment that can be constructed from a given correlation. However, unlike bit OT, it has been shown in [282] that it is not possible to construct bit commitments with positive rate.

Under computational assumptions, various constructions of bit commitment have been developed, for example see [141, Section 4.8.2]. In contrast to the encryption scheme under computational assumption (cf. Chapter 9), it is possible to construct bit commitment schemes that are partially information-theoretically secure, i.e., computationally hiding and information-theoretically binding, or information-theoretically hiding and computationally binding; for example, see [96].

The constructions of such bit commitments can be based on various assumptions, such as existence of the pseudorandom generator [255] (see Problem 14.12), intractability of the discrete logarithm problem [270] (see Problem 14.13), or existence of one-way permutation [256].

Problems

14.1 Modify Step 6 of Protocol 14.2 as follows: if $\mathcal{P}_1$ aborted in Step 5, then $\mathcal{P}_2$ outputs $C_2 = B_2'$, where B_2' is any random variable satisfying $B_1 \multimap B_2 \multimap B_2'$. For this modified protocol, verify that $\mathcal{P}_1$ can force the output of $\mathcal{P}_2$ to be biased and ensure that $\max[\Pr(C_2 = 0), \Pr(C_2 = 1)] \geq \frac{3}{4}$.

HINT By noting that B_2' is independent of $B_1 \oplus B_2$, conduct a similar calculation as in (14.1).

14.2 Prove the bound in (14.2).

14.3 For the ROT correlation (X, Y), verify that $\mathtt{mss}(Y|X) = X$.

14.4 Verify that the reduction in Proposition 14.8 is valid even if we consider a weaker adversary in the sense that $\mathcal{P}_1$ follows a protocol in the commit phase and misbehaves only in the reveal phase.

14.5 *Strong converse*: Verify that the bit commitment capacity in Theorem 14.11 is unchanged even if we do not take the limit $\varepsilon_{\mathtt{s}}, \varepsilon_{\mathtt{h}}, \varepsilon_{\mathtt{b}} \to 0$ in Definition 14.10 as long as $\varepsilon_{\mathtt{s}} + \varepsilon_{\mathtt{h}} + \varepsilon_{\mathtt{b}} < 1$.

COMMENT Such a statement is termed the strong converse of bit commitment capacity [333].

14.6 Suppose that $\mathcal{P}_1$ and $\mathcal{P}_2$ share n instances (X^n, Y^n) of bit ROTs (X, Y). Prove that

$$L_{\varepsilon_{\mathtt{s}},\varepsilon_{\mathtt{h}},\varepsilon_{\mathtt{b}}}(X^n, Y^n) \leq n + \log(1/(1 - \varepsilon_{\mathtt{s}} - \varepsilon_{\mathtt{h}} - \varepsilon_{\mathtt{b}} - \eta)) + 2\log(1/\eta)$$

using Corollary 14.9, where $0 < \eta < 1 - \varepsilon_{\mathtt{s}} - \varepsilon_{\mathtt{h}} - \varepsilon_{\mathtt{b}}$.

14.7 In Theorem 14.3, if the security parameters $(\varepsilon_{\mathtt{s}}, \varepsilon_{\mathtt{h}}, \varepsilon_{\mathtt{b}})$ are constant (do not change with n), then verify that it suffices to set the hash length ν constant.

COMMENT This means that, from n instances of bit ROTs, the string bit commitment of length $n - \mathcal{O}(1)$ can be constructed. Compare this lower bound with the upper bound in Problem 14.6; they match up to the $\mathcal{O}(1)$ term.

14.8 For a general correlation (X, Y) satisfying

$$\mathtt{V}(V|Y) := \mathtt{Var}\left[\log \frac{1}{\mathrm{P}_{V|Y}(V|Y)}\right] > 0,$$

prove that

$$L_{\varepsilon_{\mathtt{s}},\varepsilon_{\mathtt{h}},\varepsilon_{\mathtt{b}}}(X^n, Y^n) \leq nH(V|Y) - \sqrt{n\mathtt{V}(V|Y)}\mathtt{Q}^{-1}(\varepsilon_{\mathtt{s}} + \varepsilon_{\mathtt{h}} + \varepsilon_{\mathtt{b}}) + o(\sqrt{n})$$

using Corollary 14.9, where $\mathtt{Q}^{-1}(\varepsilon)$ is the inverse of the tail probability

$$\mathtt{Q}(a) = \int_a^\infty \frac{1}{\sqrt{2\pi}} e^{-\frac{t^2}{2}} dt$$

of the Gaussian distribution.

COMMENT In contrast to the ROT (cf. Problem 14.7), there is a backoff factor of the order of $\sqrt{n}$ in general; it is not known whether the $\sqrt{n}$ order backoff factor is tight or not.

14.9 *Extreme inputs*: For a given channel W from $\mathcal{X}$ to $\mathcal{Y}$, an input $x \in \mathcal{X}$ is an *extreme input* if, for any distribution P on $\mathcal{X}\backslash\{x\}$,

$$W(\cdot|x) \neq \sum_{x'} P(x')W(\cdot|x').$$

Let $\mathcal{X}_{\mathsf{e}}(W)$ be the set of all extreme inputs for the channel W. Verify that $\mathcal{X}_{\mathsf{e}}(W) = \mathcal{X}$ implies $\mathtt{mss}(Y|X) = X$ for any input distribution P_X, but the opposite implication is not necessarily true.

HINT Consider the reverse erasure channel as a counterexample of the opposite implication.

14.10 *Commitment capacity of a noisy channel*: For a memoryless channel W^n, we can formulate the problem of constructing the bit commitment and the bit commitment capacity in a similar manner to the case with the source; the only difference is that $\mathcal{P}_1$ has freedom to choose the input of the channel. Prove that the bit commitment capacity $C_{\mathsf{BC}}(W)$ of channel W is given by

$$C_{\mathsf{BC}}(W) = \max\{H(X|Y) : \mathrm{P}_X \in \mathcal{P}(\mathcal{X}_{\mathsf{e}}(W))\},$$

where the maximum is taken over all input distributions such that the support is on the extreme input set $\mathcal{X}_{\mathsf{e}}(W)$ of the channel W [350].

14.11 *Unfair noisy channel*: Suppose that the crossover probability p of the binary symmetric channel (BSC) W_p is unknown among $p_1 \leq p \leq p_2$, and is controlled by a dishonest party; such a channel is termed the unfair noisy channel [95]. For $0 \leq p_1 \leq 1/2$ and $p_2 \leq p_1 * p_1$, prove that the bit commitment capacity of such an unfair noisy channel can be lower bounded (achievability bound) by $h(p_1) - h(q)$, where $q \in [0, 1/2]$ is the value satisfying $p_1 * q := p_1(1-q) + (1-p_1)q = p_2$ [80].[4]

COMMENT Heuristically, when $\mathcal{P}_2$ is dishonest, $\mathcal{P}_2$ may decrease the error level so that $\mathcal{P}_2$ can learn as much information about the committed string as possible in the commit phase; thus we must consider the least level of noise p_1, and the randomness that can be used to generate a secret key in the commit phase is $h(p_1)$. On the other hand, when $\mathcal{P}_1$ is dishonest, $\mathcal{P}_1$ may pick the noise level p_1 in the commit phase, and then flip qn positions of the input string of the noisy channel in the reveal phase so that $\mathcal{P}_1$ can change the committed string; thus we need to take the size of hash function as large as $2^{nh(q)}$, and the rate $h(q)$ must be subtracted.

14.12 *Bit commitment using a pseudorandom generator*: Let $G\colon \{0,1\}^n \to \{0,1\}^{3n}$ be a pseudorandom generator (cf. Definition 9.6). Consider the following

[4] In [80], the matching converse is also proved for a restricted class of protocols.

bit commitment scheme. $\mathcal{P}_2$ randomly generates $Y \in \{0,1\}^{3n}$, and sends it to $\mathcal{P}_1$. In order to commit to $C \in \{0,1\}$, $\mathcal{P}_1$ randomly generates $X \in \{0,1\}^n$, and sends $\Pi = G(X)$ if $C = 0$ and $\Pi = G(X) \oplus Y$ if $C = 1$. In order to reveal C, $\mathcal{P}_1$ sends (C, X) to $\mathcal{P}_2$, and $\mathcal{P}_2$ verifies if $G(X) \oplus Y \cdot C = \Pi$. Verify that this scheme is 2^{-n}-binding and computationally hiding, i.e., (Π, Y) given $C = 0$ and $C = 1$ are computationally indistinguishable [255].

HINT To verify computational hiding, note that it is equivalent to the OTP using a pseudorandom generator, where the two messages to be distinguished are $(0, \ldots, 0)$ and Y. To verify 2^{-n}-binding, note that, in order for $\mathcal{P}_1$ to cheat, $\mathcal{P}_1$ need to find two seeds $s_0, s_1 \in \{0,1\}^n$ such that $Y = G(s_0) \oplus G(s_1)$; there are 2^{2n} possible choices of (s_1, s_2), and Y is randomly generated from $\{0,1\}^{3n}$.

14.13 *Pedersen's commitment scheme*: Consider the following bit commitment scheme. Let $\mathbb{G}_q$ be a group of prime order q such that the discrete logarithm on $\mathbb{G}_q$ is difficult, and let g be a publicly known generator of $\mathbb{G}_q$ (since it is prime order, any element other than 1 will do). $\mathcal{P}_2$ randomly generates $h \neq 1$ from $\mathbb{G}_q$, and sends it to $\mathcal{P}_1$. In order to commit to $c \in \{0, 1, \ldots, p-1\}$, for randomly generated $r \in \{0, 1, \ldots, p-1\}$, $\mathcal{P}_1$ sends $\Pi = g^c h^r$ to $\mathcal{P}_2$. In order to reveal c, $\mathcal{P}_1$ sends (c, r) to $\mathcal{P}_2$, and $\mathcal{P}_2$ verifies if $g^c h^r = \Pi$. Verify that this scheme is perfectly hiding and computationally binding (under the difficulty of discrete logarithm problem) [270].

HINT Since g is a generator, $h = g^k$ for some k, which is not known to $\mathcal{P}_1$. To verify perfect hiding, note that $\Pi = g^{c+kr}$ for randomly generated r. To verify computational binding, assume that $\mathcal{P}_1$ finds $s_1 \neq s_2$ such that $g^{s_1+kr_1} = g^{s_1}h^{r_1} = g^{s_2}h^{r_2} = g^{s_2+kr_2}$ for some r_1, r_2. Then, this implies $k = (s_1 - s_2)(r_2 - r_1)^{-1}$, i.e., $\mathcal{P}_1$ can find the discrete logarithm k.

15 Active Adversary and Composable Security

In Chapter 12 we discussed two-party secure computation for a passive adversary, namely a dishonest party tries to glean as much information as possible about the other party's input but does not deviate from the protocol description. We now move to the stronger notion of *active adversary*, which need not follow the protocol. In fact, we have already encountered an active adversary in Chapter 14 – a substitution attack where the committing party changes the bit in the reveal phase is an example of an active attack. Defining security for an active adversary is quite involved since we need to keep in mind various ways in which an adversary can deviate from the protocol. Our first goal in this chapter is to build towards this notion of security, illustrating via examples the shortcomings of some first attempts at definitions.

For secure computation with a passive adversary, we established in Chapter 12 the completeness of the oblivious transfer (OT), i.e., any function can be securely computed provided the parties have oracle access to the ideal OT (secure for passive adversaries). The protocol we presented combined, that is *composed*, several basic modules: secure computation protocols for secret shared NOT, XOR, and AND gates. In the proof of security of the composed protocol, we used the chain rule for mutual information to relate the security of the composed protocols to that of the individual components. However, security of composition of protocols for an active adversary is much more involved. Our second goal in this chapter is to develop a notion of *composable security* such that the composition of secure protocols under this notion is secure too.

This chapter serves as a preparation for Chapter 17, in which the completeness of OT for secure computation under an active adversary is presented.

15.1 Functions and Protocols

When analyzing security for complicated protocols comprising several subprotocols, it is important to clearly lay down the allowed rules for the interaction of protocols. We need to carefully specify the "information flow" between the protocols to identify possible dishonest behavior. We begin this chapter by reviewing the notions of functions, ideal function oracle and composition of protocols, emphasizing the allowed information flow between the protocols and possible dishonest behavior. In system security, such specifications are sometimes called *threat models.*

Functions and ideal function oracle

We are mainly interested in secure computation of the following classes of functions.

1. *Deterministic function with inputs*; A deterministic function $f \colon \mathcal{X}_1 \times \mathcal{X}_2 \to \mathcal{Z}_1 \times \mathcal{Z}_2$ consists of the inputs (x_1, x_2) of the parties and the outputs (z_1, z_2). For instance, the oblivious transfer (OT) is a deterministic function described as $f_{\mathrm{OT}}((k_0, k_1), b) = (\emptyset, k_b)$.
2. *Stochastic function without inputs*; A stochastic function f without any inputs generates outputs (Z_1, Z_2) according to a prescribed distribution $\mathrm{P}_{Z_1 Z_2}$. For instance, the randomized oblivious transfer (ROT) is a stochastic function that generates $((S_0, S_1), (C, S_C))$, where S_0, S_1 and C are uniformly distributed on $\{0,1\}^l$ and $\{0,1\}$, respectively. Other examples include noisy correlations such as erasure correlation, namely $(Z_1, Z_2) \in \{0,1\} \times \{0,1,\mathsf{e}\}$ distributed according to $\mathrm{P}_{Z_1 Z_2}(z_1, z_2) = \frac{1-p}{2}$ if $z_1 = z_2 \neq \mathsf{e}$ and $\mathrm{P}_{Z_1 Z_2}(z_1, z_2) = \frac{p}{2}$ if $z_2 = \mathsf{e}$.
3. *Stochastic function with inputs*; Upon receiving inputs (x_1, x_2), a stochastic function f with inputs generates outputs (Z_1, Z_2) according to a prescribed distribution $\mathrm{P}_{Z_1 Z_2 | X_1 X_2}(\cdot, \cdot | x_1, x_2)$. For instance, a noisy channel $W(z_2|x_1)$ can be regarded as a stochastic function with one input by $\mathcal{P}_1$ and one output for $\mathcal{P}_2$.

In general, this notion of functions coincides with that of *two-way channels* in information theory where both the parties provide inputs to the channel and both observe the outputs. Thus, the general goal in secure computation is to generate a specific correlation while a dishonest party can try to generate a different correlation to glean additional information.

The first step in studying the secure computation of a function is to get a clear idea of what an *ideal function oracle* will do. In its simplest form, an ideal function oracle behaves exactly the same as the computed function, as an oracle to which both parties provide an input and get the corresponding output. However, often we need to add other nuances to capture additional features or even to facilitate a formal security analysis. We will see some interesting examples as we proceed in this chapter; for instance, in order to allow protocol implementations with abort, we will introduce a dummy input called "abort" in the ideal function oracle. This ideal function oracle will be our gold standard for security, and our notion of security will merely claim that a protocol is roughly as secure as the ideal function oracle. Thus, under the definition of security we are developing, a secure protocol can at best be as secure as the ideal function oracle we defined; for brevity, we will simply call an ideal function oracle an *ideal function.*

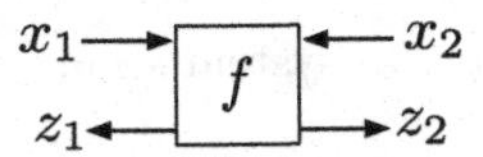

Figure 15.1 Ideal function f with input (x_1, x_2) and output (z_1, z_2).

For clear depiction, we represent ideal functions using block diagrams such as that shown in Figure 15.1, where the left and right sides denote $\mathcal{P}_1$ and $\mathcal{P}_2$, respectively. With an abuse of notation, we denote the ideal function corresponding to a function f by f itself, and sometimes even use the term function and ideal function interchangeably. For example, when we say oracle access to a function f, we mean oracle access to the ideal function f.

Protocols with oracle access

Earlier in Chapter 12, we saw that, when the ideal function for OT is available, we can implement a secure computation protocol for any function (under a passive adversary) by using oracle access to the ideal function for OT. In this chapter, too, we will implement protocols using oracle access to ideal functions.

An *interactive communication protocol* for two parties can be viewed as a sequence of computational steps with each step involving an input and an output. Each step is executed by one of the parties which completes the computation required in the step. Note that the input for the computation depends on the past observations of the executing party during the protocol, including its input for the overall protocol. A part of the output of each step is released to both the parties as a part of the *transcript*. When a party is dishonest, it can deviate from the prescribed computation for the step it executes and report a wrong answer as the output. There is no limitation on this deviation – the party is computationally unbounded and can report any random variable that depends on its observed information up to that step.

In *protocols with oracle access* to a function g, some of the steps are executed by no single party. Instead, the parties provide their part of the input to an oracle and receive their respective parts of the output. This computation is completed by a third party (the oracle) which is trusted to do the computation involved in g. In this oracle access step, even a dishonest party cannot completely deviate from the computation involved. But it can modify its part of the input and wrongly report its part of the output for the future steps. We depict this flow of information in Figure 15.2.

As mentioned in Chapter 12, we can view a protocol π with oracle access to g as the protocol π accessing a two-way channel. The combined protocol is denoted by $\pi \diamond g$. Our notation does not depict the number of times the oracle g is accessed; π may internally access g any finite number of times. Note that, from the outside, the protocol appears just like a function, with inputs from both parties and outputs to report to each party. However, there are many internal steps which reveal additional information to the parties.

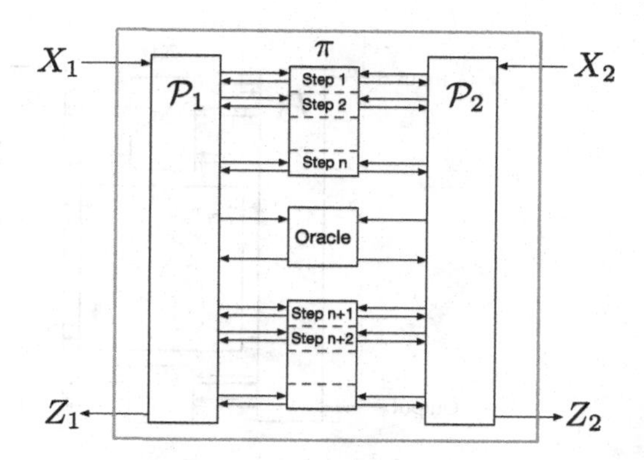

Figure 15.2 The flow of information during the execution of protocols. From the outside, the protocol has input (X_1, X_2) and output (Z_1, Z_2), but internally there are many interactive steps.

In most cases of practical interest, it is too optimistic to assume the existence of an arbitrary ideal function. The goal in secure computation is to implement protocols that behave very similarly to the ideal functions of interest; we will call such protocols *real protocols* to distinguish them from ideal functions. However, we have already seen an impossibility result claiming that only trivial functions (for passive security) can be securely implemented from scratch. Thus, we consider real protocols that may internally use oracle access to other, simpler ideal functions which serve as a resource to enable approximately more complex ideal functions.

Remark 15.1 Later in the chapter, we need to consider oracle access to real protocols implemented to mimic ideal functions. In this case, too, the oracle is given input and provides output. However, since this oracle is a protocol, it is internally implemented by parties. In particular, a dishonest party can use information it gets from the transcript generated internally during oracle access to a protocol.

Remark 15.2 Throughout this chapter, we use (X_1, X_2) and (Z_1, Z_2) to denote the input and the output of the function f. When we implement a protocol $\pi^f \diamond g$ for computing f which uses oracle access to g, we denote the input and the output of g by (Y_1, Y_2) and (V_1, V_2), respectively.

Composition of protocols

When a protocol π_1 executes another protocol π_2 as a subroutine, the *composed protocol* is denoted by $\pi_1 \circ \pi_2$. In $\pi_1 \circ \pi_2$, the outer protocol π_1 executes the inner protocol π_2 exactly once, unlike our previous notation $\pi \diamond g$ where π is given multiple oracle accesses to g. Noting that a protocol also has input and output like a function, we follow the convention of treating a protocol in a similar manner to an oracle. When π_1 accesses π_2 in the composed protocol $\pi_1 \circ \pi_2$,

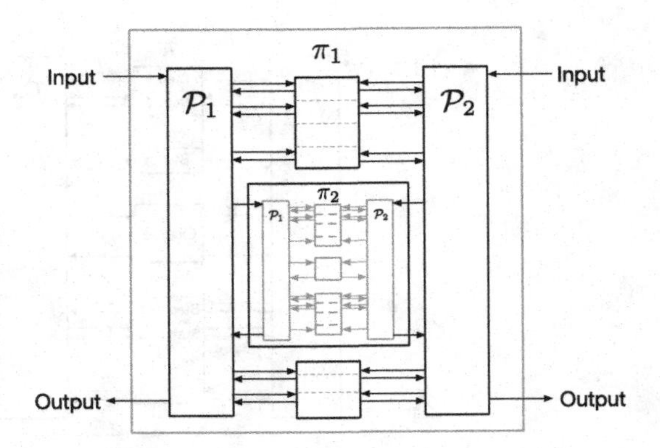

Figure 15.3 The composition of protocols π_1 and π_2. In our convention, the outer protocol π_1 treats π_2 like an oracle.

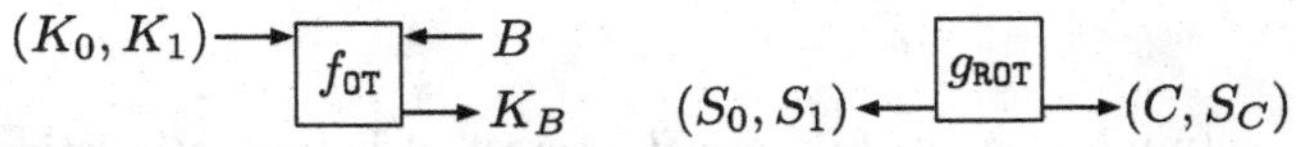

$(K_0, K_1) \to f_{\text{OT}} \leftarrow B$, $\to K_B$ $\qquad (S_0, S_1) \leftarrow g_{\text{ROT}} \to (C, S_C)$

Figure 15.4 Ideal functions for OT and ROT.

it provides input to π_2 and proceeds with the received output. Note that the protocol π_2 can use parties for computation internally, but in our convention, the *outer protocol* π_1 will treat the *inner protocol* π_2 as an oracle; we illustrate this in Figure 15.3. It is important to note that while the same parties are executing the two protocols involved in a composition, they are not supposed to exchange the internal information involved in π_1 with π_2. However, when a party is dishonest, it can attack by combining all the information it receives in the different protocols it is executing. For instance, the dishonest party may utilize the information obtained during the execution of π_2 in order to attack during an execution of π_1.

15.2 Admissible and Nonadmissible Attacks

Before presenting the definition of security for active adversaries, we outline the challenge in coming up with such a definition. For passive adversaries, perfect security of a real protocol is tantamount to the requirement that it "behaves identically" to the ideal function. However, this bar is too high for active adversaries, and we must allow certain attacks. We will illustrate via examples.

Consider Protocol 15.1 which implements OT using oracle access to random OT (ROT). Recall that ROT is a stochastic function with no inputs (namely, correlated random variables) and outputs given by $V_1 = (S_0, S_1)$ and $V_2 = (C, S_C)$, where C, S_0, S_1 are independent random bits. We denote the OT function by f_{OT} and the ROT function by g_{ROT}; see Figure 15.4 for a depiction of ideal functions for both OT and ROT.

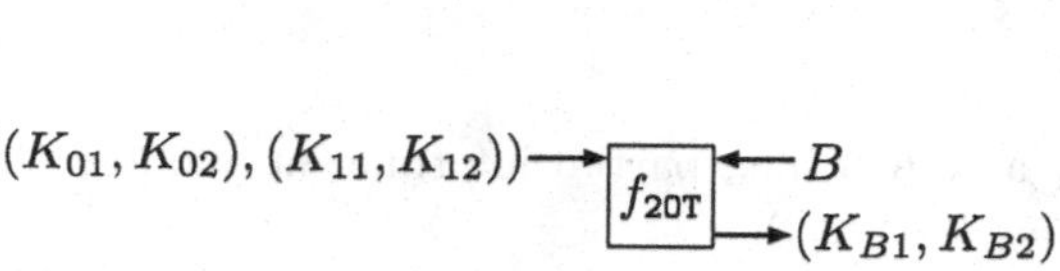

$$((K_{01}, K_{02}), (K_{11}, K_{12})) \rightarrow \boxed{f_{2OT}} \leftarrow B, \quad \rightarrow (K_{B1}, K_{B2})$$

Figure 15.5 Ideal functions for 2OT.

Protocol 15.1 Bit OT f_{OT} from bit randomized OT g_{ROT}

Input: $\mathcal{P}_1$'s input (K_0, K_1) and $\mathcal{P}_2$'s input B

Output: $\mathcal{P}_2$'s output $\hat{K}$

1. The parties invoke g_{ROT} to obtain (S_0, S_1) and (C, S_C), respectively.
2. $\mathcal{P}_2$ sends $\Pi_1 = B \oplus C$ to $\mathcal{P}_1$.
3. $\mathcal{P}_1$ sends $\Pi_{2,0} = K_0 \oplus S_{\Pi_1}$ and $\Pi_{2,1} = K_1 \oplus S_{\overline{\Pi}_1}$ to $\mathcal{P}_2$, where $\overline{\Pi}_1 = \Pi_1 \oplus 1$.
4. $\mathcal{P}_2$ computes $\hat{K} = \Pi_{2,B} \oplus S_C$ as the output.

Protocol 15.2 string OT $f_{2\text{OT}}$ from two instances of bit OT g_{OT}

Input: $\mathcal{P}_1$'s input $(K_0, K_1) = ((K_{0,1} K_{0,2}), (K_{1,1}, K_{1,2}))$ and $\mathcal{P}_2$'s input B

Output: $\mathcal{P}_2$'s output $\hat{K} = (\hat{K}_1, \hat{K}_2)$

1. The parties invoke two instances of g_{OT}: for the first invocation, inputs are $(K_{0,1}, K_{1,1})$ and B; for the second invocation, inputs are $(K_{0,2}, K_{1,2})$ and B.
2. $\mathcal{P}_2$ outputs $\hat{K}_1 = g_{\text{OT}}((K_{0,1}, K_{1,1}), B)$ and $\hat{K}_2 = g_{\text{OT}}((K_{0,2}, K_{1,2}), B)$.

When both the parties follow the protocol, we can verify that $\hat{K} = K_B$. Also, clearly B is not leaked to $\mathcal{P}_1$ since C plays the role of the secret key for one-time pad encryption. Similarly, $\mathcal{P}_2$ cannot learn $K_{\bar{B}}$ since $S_{\bar{C}}$ plays the role of the secret key for one-time pad encryption, where $\bar{B} = B \oplus 1$ and $\bar{C} = C \oplus 1$, respectively.

Next, we consider a slightly different problem where we want to implement a 2-bit OT (2OT) using oracle access to OT. Recall that 2OT is similar to OT, except that inputs K_0 and K_1 are now two bit strings each; see Figure 15.5.

Consider Protocol 15.2 which implements 2OT using oracle access to OT by simply invoking two instances of OT with the same input B from $\mathcal{P}_2$. When both the parties follow the protocol, by similar reasoning to above, we can verify that Protocol 15.2 does not leak unnecessary information. Consequently, both Protocol 15.1 and Protocol 15.2 are secure against a passive adversary.

However, the situation becomes different when we consider active adversaries where a dishonest party is allowed to deviate from the protocol. For instance, in Protocol 15.1, instead of sending $\Pi_1 = B \oplus C$, $\mathcal{P}_2$ may send $\Pi_1 = C$ and obtain $\hat{K} = K_0$ irrespective of the value of B. Alternatively, in Protocol 15.2, instead of using the same input B for two invocations of g_{OT}, $\mathcal{P}_2$ may input 0 for the first invocation and 1 for the second invocation. This allows $\mathcal{P}_2$ to get

K_{01} and K_{12}, i.e., partial information about each string $K_0 = (K_{01}, K_{02})$ and $K_1 = (K_{11}, K_{12})$.

These two attacks are very different in nature. The first attack for Protocol 15.1 is an *admissible attack* since, even when using the ideal function f_{OT}, an active adversary $\mathcal{P}_2$ can surmount this attack. On the other hand, the second attack for Protocol 15.2 is a *nonadmissible attack* since this cannot be executed by a dishonest $\mathcal{P}_2$ when we use the ideal function $f_{2\mathrm{OT}}$.

Our definition of security under an active adversary should distinguish between these two types of attack. Specifically, it should allow admissible attacks, namely attacks that cannot be avoided even when using the ideal function, but forbid nonadmissible attacks – the attacks that cannot be executed when using the ideal function.

15.3 Perfect Standalone Security

We now introduce the notions of the *view of the protocol* and a *simulator*. The former enables us to formally capture the ability of the adversary to leak information and affect the honest party's output; and the latter enables us to formally define what is an admissible attack. These concepts are profound theoretical constructs, and yet, are very close to system security.

The view of a protocol

The *view of a protocol* denotes roughly the information seen by the parties as the protocol proceeds. More explicitly, the view $\mathtt{V}[\pi]$ of a protocol π is the information revealed by the parties executing the protocol. It is convenient to think of an external *environment* not involved in the protocol to which the information is revealed. It is this environment that provides the inputs to the parties which in turn use it to execute the protocol and obtain the outputs from the protocol; subsequently, the parties reveal these outputs to the environment.

In our convention, the protocol has inputs and outputs, which we expect the parties to reveal to the environment. But during the execution of the protocol, the parties get information about other additional inputs and outputs that are internal to the protocol. When the parties are honest, they conceal this internal information from the environment and reveal only the outputs. But when a party is dishonest, it can even reveal a part of the internal information it gets during the execution. In doing so, a dishonest party can use private randomness and compute any function (using unbounded computational power). However, it is limited by the protocol in correlating the result of its computed output with the observations of the other party. We depict this scenario in Figure 15.6.

For example, when π is the ideal function f, since $\mathcal{P}_1$ and $\mathcal{P}_2$ only observe (X_1, Z_1) and (X_2, Z_2) during the course of the protocol, the view is given by $\mathtt{V}[f] = ((X_1, Z_1), (X_2, Z_2))$, where $(Z_1, Z_2) = f(X_1, X_2)$. However, for a real protocol $\pi^f \diamond g$ implementing f using g, the view $\mathtt{V}[\pi^f \diamond g]$ can differ based on the

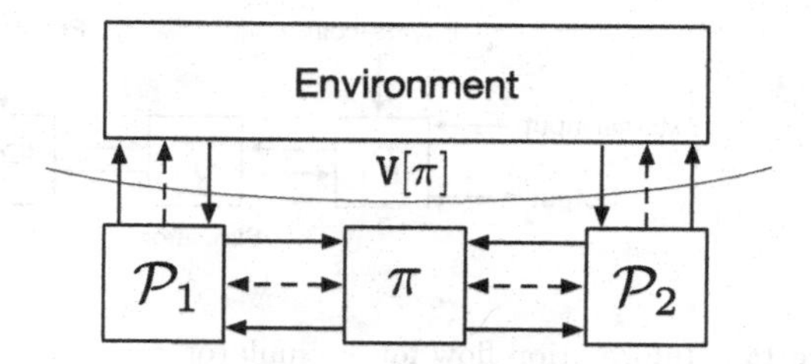

Figure 15.6 The depiction of the view of a protocol. The dashed line indicates internal input/output which the parties are expected to keep concealed from the environment.

behavior of the parties. Consider $\pi^f \diamond g$ which takes (X_1, X_2) as input, the same as f, and outputs (Z_1, Z_2). In addition, the protocol $\pi^f \diamond g$ reveals the transcript $\Pi = (\Pi^{(1)}, \Pi^{(2)})$ where $\Pi^{(i)}$ denotes the transcript seen by $\mathcal{P}_i$. Note that $\Pi^{(1)}$ and $\Pi^{(2)}$ may differ since the oracle g need not give the same output to both the parties. Furthermore, when one of the parties is dishonest, (Z_1, Z_2) need not coincide with the function output $f(X_1, X_2)$.

If both the parties are honest, they should ignore the extra information Π revealed by the protocol since this is internal to the protocol. In this case $\mathtt{V}[\pi^f \diamond g] = ((X_1, Z_1), (X_2, Z_2))$. However, when a party, say $\mathcal{P}_1$, is dishonest, it takes the transcript into consideration and $\mathtt{V}[\pi^f \diamond g] = ((X_1, Z_1, \Pi^{(1)}), (X_2, Z_2))$. Thus, there are three possible views of the protocol, one each for the case when both parties are honest, $\mathcal{P}_1$ is dishonest, and $\mathcal{P}_2$ is dishonest.

In summary, the overall flow of information is as follows: the environment engages parties to run a protocol and passes inputs to the parties. The parties get information about each other's observations and internal randomness of the protocol by engaging in the protocol. Finally, the parties reveal some information to the environment. Ideally, this information should be the output of the outer protocol. But a dishonest party can manipulate the information it gets between the steps (for instance, the transcript of an internal protocol which is used as an oracle) and release additional information to the environment beyond the protocol output.

Simulator

Towards formalizing the notion of admissible attacks, the notion of a simulator is essential. Roughly speaking, an admissible attack is one where the dishonestly modified view can be obtained even from the ideal function, with some additional *local processing* at the dishonest party. Such a protocol which is used for local processing after interacting with an ideal function is called a *simulator*. Note that we associate a simulator with a combination of an ideal function and a specific attack strategy of the dishonest party – it simulates the "correlation" used in the attack using local processing and oracle access to the ideal function.

Formally, for a function f, a simulator is a private-coin protocol with oracle access to the ideal function f and with no interaction between the parties beyond

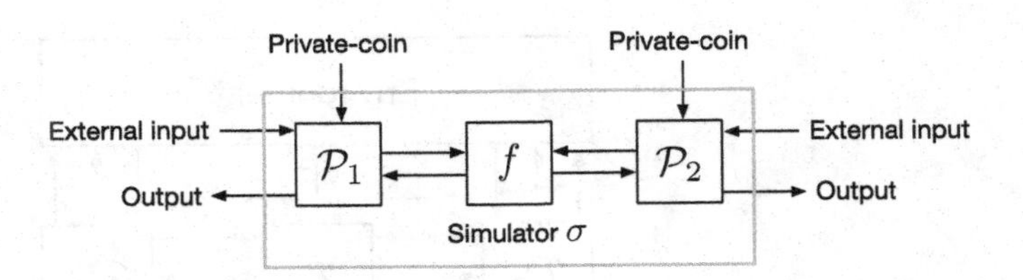

Figure 15.7 Information flow for a simulator σ.

that. In particular, the parties use their respective private-coins U_1 and U_2 to obtain the inputs (X_1, X_2) for oracle access to the ideal function, obtain the oracle output (Z_1, Z_2), and declare the final output which depends only on (U_1, Z_1) and (U_2, Z_2) for parties $\mathcal{P}_1$ and $\mathcal{P}_2$, respectively.

Later, we will encounter the notion of a simulator that interacts with another outer protocol. In this case, the inputs to the oracle and the final output may depend additionally on external inputs received from the outer protocol. Throughout, we stick to our foregoing convention – the simulator and all the other protocols exchange information only via the parties. We depict the flow of information for a simulator in Figure 15.7.

All the components needed for defining the security of a protocol are ready. The view of the protocol, in essence, represents the "correlation" that can be generated between the parties using the protocol. When using the ideal function, the parties can generate some correlation that is allowed when computing the stochastic function. An attack is tantamount to generating a correlation that is different from the one obtained by the ideal function. However, even when using the ideal function, a dishonest party can use local processing to generate a different correlation. Such attacks must be admissible; anything else is nonadmissible.

We treat the function f as the ideal function, and let $\pi^f \diamond g$ be a real protocol implementing f. Our notion of security says that a real protocol is secure, if for every attack by a dishonest party, we can find a simulator that interacts with the ideal f and produces the same view.

DEFINITION 15.1 (Perfect standalone security) A protocol $\pi^f \diamond g$ perfectly standalone securely implements f, if for every attack by a dishonest party, there exists a simulator σ such that it retains the input and the output for the honest party and, for every input distribution $\mathrm{P}_{X_1X_2}$, it holds that

$$\mathtt{V}[\pi^f \diamond g] \equiv \mathtt{V}[\sigma \circ f],$$

where $U \equiv V$ means that the distributions are the same, i.e., $\mathrm{P}_U = \mathrm{P}_V$.

If such a protocol $\pi^f \diamond g$ exists, we say that f is perfectly standalone securely computable using (oracle access to) g.

Thus, to establish security, we will need to construct a simulator for every possible attack by a dishonest party. Some remarks are in order.

Remark 15.3 (Honest parties and simulators) It is important to note that we insist that the simulator *retains the input and the output for the honest party.* Namely, in the simulator protocol, the honest party must pass the original input to the ideal function f and report the observed output from f without modification. Thus, while we defined the simulator as a two-party protocol, since at most one party can be dishonest, we only need to construct a local processing protocol for one (the dishonest one) of the parties. In particular, when both parties are honest, the simulator must be the *trivial simulator* that applies the supplied inputs to f and declares the observed output. In this case, security is tantamount to the *correctness* condition requiring the accuracy of the computed function.

Remark 15.4 (Universal simulators) In Definition 15.1, we require the existence of a simulator σ that is universal, i.e., it works for every input distribution $P_{X_1X_2}$. In fact, for standalone security, we can make do with the more modest requirement of a distribution dependent simulator. However, later we will consider composable security where the protocol must remain secure even when combined with other protocols. In that more general setting, the input distribution will be determined by an earlier interaction between the parties, whereby a dishonest party may not know the input distribution of the other party. Keeping this general notion in mind, even for standalone security, we insist on universal simulators that work for all distributions.

The definition of standalone security above is rather profound, with simple operational interpretation. However, it is quite abstract, and it may not be clear at a first glance how to use it for establishing the security of a given protocol. A particular sticky point is the phrase "for every attack" which suggests that one needs to enumerate all possible attacks and design a simulator for all of them. However, it is possible to account for all possible attacks by carefully considering the information flow.

The next result provides an alternative definition of security which is more directly applicable but perhaps not so readily interpretable as our original notion of perfect standalone security. Heuristically, this alternative form is obtained upon making the following observation: in an admissible attack, the dishonest party can replace the input to the ideal function f and manipulate the corresponding output using its local (private) randomness. A protocol is secure if we can express its input-output distribution using these operations.

Before providing the alternative definition, we extend the notion of the view to the case where a stochastic function $f(\cdot,\cdot|X_1,X_2)$ with output (Z_1,Z_2) is implemented using another stochastic function $g(\cdot,\cdot|Y_1,Y_2)$ with output (V_1,V_2). In a protocol $\pi^f \diamond g$ implementing f, the parties communicate interactively. Within the protocol, the parties are allowed to invoke function g with inputs (Y_{1t},Y_{2t}), which may depend on the past history of communication and each party's input, to obtain outputs (V_{1t},V_{2t}), several times for $1 \le t \le n$. The overall transcript is denoted by $\Pi = (\Pi^{(1)},\Pi^{(2)})$, with $\Pi^{(i)}$ denoting the part of the transcript seen by $\mathcal{P}_i$. As our convention, we include all the inputs and the outputs for the oracle g as well as the interactive communication during the protocol in the transcript Π. Note that, since the parties make oracle access to g, $\Pi^{(1)}$ and $\Pi^{(2)}$ need not coincide. At the end of the protocol, the parties $\mathcal{P}_1$ and $\mathcal{P}_2$ output Z_1 and Z_2, respectively.

When both the parties are honest, the view $\mathtt{V}[\pi^f \diamond g]$ of the parties in the real protocol is $((X_1, Z_1), (X_2, Z_2))$. When $\mathcal{P}_1$ is dishonest, the view $\mathtt{V}[\pi^f \diamond g]$ of the real protocol is $((X_1, Z_1, \Pi^{(1)}), (X_2, Z_2))$. Similarly, when $\mathcal{P}_2$ is dishonest, the view $\mathtt{V}[\pi^f \diamond g]$ in the real protocol is $((X_1, Z_1), (X_2, Z_2, \Pi^{(2)}))$. The next result provides an explicit condition for the existence of a simulator required for perfect standalone security in Definition 15.1.

LEMMA 15.2 (Perfect standalone security – alternative form) *A protocol $\pi^f \diamond g$ perfectly standalone securely implements f if and only if the following hold.*

1. *Correctness. When both the parties are honest, the output (Z_1, Z_2) of the protocol $\pi^f \diamond g$ has the distribution $f(\cdot, \cdot | X_1, X_2)$ for every input distribution $\mathrm{P}_{X_1 X_2}$.*
2. *Security for $\mathcal{P}_1$. When $\mathcal{P}_1$ is honest, for every attack by $\mathcal{P}_2$ and every input distribution $\mathrm{P}_{X_1 X_2}$, there exists a joint distribution $\mathrm{P}_{X_1 X_2 X_2' Z_1 Z_2 Z_2' \Pi^{(2)}}$, with dummy input and output (X_2', Z_2') for $\mathcal{P}_2$ and $\mathrm{P}_{X_1 X_2 Z_1 Z_2 \Pi^{(2)}}$ denoting the distribution of input, output, and transcript under the attack, such that*[1]

$$(i)\ X_2' -\!\!\circ\!\!- X_2 -\!\!\circ\!\!- X_1,$$
$$(ii)\ (Z_1, Z_2') \sim f(\cdot, \cdot | X_1, X_2'),\ \text{and}$$
$$(iii)\ (Z_2, \Pi^{(2)}) -\!\!\circ\!\!- (X_2, X_2', Z_2') -\!\!\circ\!\!- (X_1, Z_1).$$

3. *Security for $\mathcal{P}_2$. When $\mathcal{P}_2$ is honest, for every attack by $\mathcal{P}_1$ and every input distribution $\mathrm{P}_{X_1 X_2}$, there exists a joint distribution $\mathrm{P}_{X_1 X_2 X_1' Z_1 Z_2 Z_1' \Pi^{(1)}}$, with dummy input and output (X_1', Z_1') for $\mathcal{P}_1$ and $\mathrm{P}_{X_1 X_2 Z_1 Z_2 \Pi^{(1)}}$ denoting the distribution of input, output, and transcript under the attack, such that*

$$(i)\ X_1' -\!\!\circ\!\!- X_1 -\!\!\circ\!\!- X_2,$$
$$(ii)\ (Z_1', Z_2) \sim f(\cdot, \cdot | X_1', X_2),\ \text{and}$$
$$(iii)\ (Z_1, \Pi^{(1)}) -\!\!\circ\!\!- (X_1, X_1', Z_1') -\!\!\circ\!\!- (X_2, Z_2).$$

Proof We prove the "if" part first. To that end, suppose that the conditions above are satisfied. When both the parties are honest, the correctness condition guarantees that the view of the real protocol and the view of the ideal function coincide for every input distribution.

[1] For the universality of the simulator mentioned in Remark 15.4, we need to ensure that the conditional distributions used to generate dummy inputs and outputs do not depend on the input distribution $\mathrm{P}_{X_1 X_2}$.

Protocol 15.3 Simulator σ with dishonest $\mathcal{P}_2$ in Lemma 15.2

Input: (X_1, X_2)

Output: $((X_1, \tilde{Z}_1), (X_2, \tilde{Z}_2, \tilde{\Pi}^{(2)}))$

1. $\mathcal{P}_2$ samples $\tilde{X}_2' \sim \mathrm{P}_{X_2'|X_2}(\cdot|X_2)$.
2. The parties invoke f to obtain $(\tilde{Z}_1, \tilde{Z}_2') \sim f(\cdot, \cdot|X_1, \tilde{X}_2')$.
3. $\mathcal{P}_1$ outputs $(X_1, \tilde{Z}_1)$.
4. $\mathcal{P}_2$ samples $(\tilde{Z}_2, \tilde{\Pi}^{(2)}) \sim \mathrm{P}_{Z_2\Pi^{(2)}|X_2X_2'Z_2'}\left(\cdot, \cdot|X_2, \tilde{X}_2', \tilde{Z}_2'\right)$.
5. $\mathcal{P}_2$ outputs $(X_2, \tilde{Z}_2, \tilde{\Pi}^{(2)})$.

When $\mathcal{P}_1$ is honest, for an attack by $\mathcal{P}_2$, consider a simulator σ described in Protocol 15.3,[2] which is constructed using the joint distribution $\mathrm{P}_{X_1X_2X_2'Z_1Z_2'\Pi^{(2)}}$ appearing in Condition 2. Note that this distribution depends on the attack, whereby the resulting simulator also depends on the attack. However, to ensure universality, we assume that none of the conditional distributions used for local sampling at $\mathcal{P}_2$ depends on the input distribution. Using the Markov relations in Condition 2, the random variables $\mathtt{V}[\pi^f \diamond g] = ((X_1, Z_1), (X_2, Z_2, \Pi^{(2)}))$ and $\mathtt{V}[\sigma \circ f] = ((X_1, \tilde{Z}_1), (X_2, \tilde{Z}_2, \tilde{\Pi}^{(2)}))$ have the same distributions for every input distribution. Recall that $(Z_1, Z_2, \Pi^{(2)})$ denotes the random variables produced under attack and $(\tilde{Z}_1, \tilde{Z}_2, \tilde{\Pi}^{(2)})$ denotes the simulated random variables. When $\mathcal{P}_2$ is honest and $\mathcal{P}_1$ is dishonest, we can construct a simulator similarly.

For the "only if" part, suppose that $\pi^f \diamond g$ is secure in the sense of Definition 15.1. The correctness conditions coincide by definition (see Remark 15.3). Consider the case when $\mathcal{P}_2$ is the dishonest party; in particular, consider a specific attack for which σ is the simulator. Without loss of generality, we can consider the "worst-case" attack where $\mathcal{P}_2$ reveals all its observations as a part of the view.

Let (X_1, X_2') be the input used by the simulator for accessing f and (Z_1, Z_2') be the corresponding outputs. Here, too, we have followed the convention highlighted in Remark 15.3 – the honest party must pass the original input to f in the simulator protocol. Since there are no additional resources besides local randomness available to $\mathcal{P}_2$, the Markov relation $X_2' \multimap X_2 \multimap X_1$ must hold.

Furthermore, after obtaining Z_2', once again $\mathcal{P}_2$ must take recourse to local processing for generating any additional information. In particular, the part revealed by $\mathcal{P}_2$ in the view of the simulator must be conditionally independent of $\mathcal{P}_1$'s observations (X_1, Z_1) conditioned on (X_2, X_2', Z_2'). But since the view of the simulator coincides with the view of the protocol $\pi^f \diamond g$, the Markov relation $(Z_2, \Pi^{(2)}) \multimap (X_2', X_2, Z_2') \multimap (X_1, Z_1)$ must hold. The case when $\mathcal{P}_1$ is dishonest can be handled similarly. □

The proof above is, in fact, simply a further elaboration of the information flow of the simulator and relates the information flow of the real protocol to that of the

[2] We put a "tilde" on simulated random variables to differentiate them from those in the real protocol.

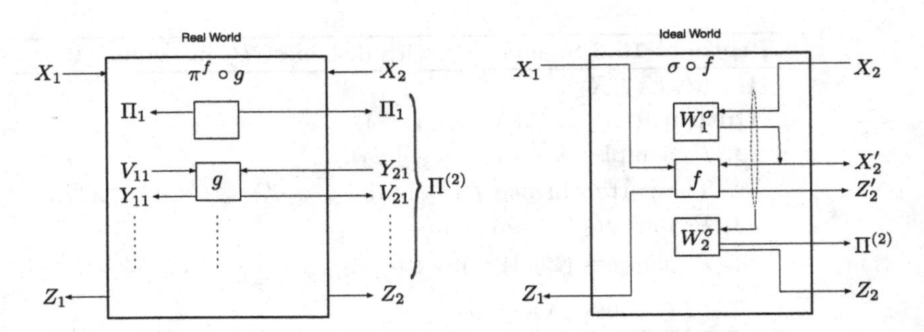

Figure 15.8 Information flow in the real world and the ideal world when $\mathcal{P}_2$ is dishonest.

ideal function with a simulator. We depict this information flow in Figure 15.8; we use the phrases *real world* and *ideal world* to describe the information flow of the real protocol and the ideal function with a simulator, respectively. Different parts of the simulator protocol are represented using channels W_i^σ.

Remark 15.5 (Protocols with abort) For now we assume that the dishonest party follows the "format" of the protocol, namely it respects the number of rounds, the order of communication, etc. In the case when a party deviates even from the format, we can simply abort the protocol. Security of protocols with abort is slightly more involved and will be treated in the next section.

We motivated the alternative form of security definition by promising that we will not need to enumerate all possible attacks when verifying security. However, it may seem that Lemma 15.2 still requires us to consider all possible attacks separately. In fact, in most applications we can identify the specific inputs and outputs a dishonest player can exploit and show Markov conditions required by Lemma 15.2 for a generic attack.

We illustrate this methodology by applying Lemma 15.2 to establish the security of specific protocols. We begin with the simple case of deterministic protocols and prove the security of Protocol 15.1 for implementing OT using oracle access to ROT. Note that the oracle ROT is a stochastic function without any inputs.

PROPOSITION 15.3 *Protocol 15.1 perfectly standalone securely implements the OT function f_{OT}.*

Proof The correctness condition is easy to verify. When $\mathcal{P}_1$ is honest, let Π denote the transcript of the protocol during an attack by $\mathcal{P}_2$; for brevity we have omitted superscripts to distinguish $\mathcal{P}_1$'s and $\mathcal{P}_2$'s observed parts of the transcript. The transcript Π comprises three parts $(\Pi_1, \Pi_{2,0}, \Pi_{2,1})$, where Π_1 is generated by $\mathcal{P}_2$ using a stochastic function of (B, C, S_C), and $\Pi_{2,0} = K_0 \oplus S_{\Pi_1}$ and $\Pi_{2,1} = K_1 \oplus S_{\bar{\Pi}_1}$ are computed by the honest party $\mathcal{P}_1$. To establish security using Lemma 15.2, we identify the dummy input for $\mathcal{P}_2$ to simulate the output and the transcript.

Specifically, set $B' = \Pi_1 \oplus C$, and let $K_{B'} = f_{\text{OT}}((K_0, K_1), B')$ denote the output of the ideal function to $\mathcal{P}_2$ corresponding to this dummy input by $\mathcal{P}_2$ and input (K_0, K_1) by $\mathcal{P}_1$. We show that the condition of Lemma 15.2 for honest $\mathcal{P}_1$ is satisfied with the dummy input B' and the corresponding dummy output $K_{B'}$; namely,

$$(C, S_C, \Pi, \hat{K}) \multimap\!\!- (B', B, K_{B'}) \multimap\!\!- (K_0, K_1). \tag{15.1}$$

Indeed, by the independence of (C, S_C) and $((K_0, K_1), B)$, we have

$$I(C, S_C, \Pi_1 \wedge K_0, K_1 \mid B) \le I(C, S_C \wedge K_0, K_1, B) = 0,$$

where the inequality uses the fact that Π_1 is a stochastic function of (B, C, S_C). Furthermore, since B' is a function of Π_1 and C, we get

$$\begin{aligned} I(C, S_C, \Pi_1 \wedge K_0, K_1 \mid B, B') &\le I(B', C, S_C, \Pi_1 \wedge K_0, K_1 \mid B) \\ &= I(C, S_C, \Pi_1 \wedge K_0, K_1 \mid B) \\ &= 0. \end{aligned}$$

The previous inequality further gives

$$\begin{aligned} I(C, S_C, \Pi_1 \wedge K_0, K_1 \mid B, B', K_{B'}) &\le I(C, S_C, \Pi_1 \wedge K_0, K_1, K_{B'} \mid B, B') \\ &= I(C, S_C, \Pi_1 \wedge K_0, K_1 \mid B, B') \\ &= 0, \end{aligned}$$

where the first identity holds since $K_{B'}$ is a function of (K_0, K_1) and B'. Next, noting that $\Pi_{2,B'} = K_{B'} \oplus S_C$ can be computed from $(K_{B'}, S_C)$, we obtain from the previous inequality that

$$I(C, S_C, \Pi_1, \Pi_{2,B'} \wedge K_0, K_1 \mid B, B', K_{B'}) = 0.$$

Also, since $S_{\bar{C}}$ is a uniformly random bit independent of $(B, K_0, K_1, C, S_C, B', K_{B'}, \Pi_1)$, we get by a "one-time pad argument" that $\Pi_{2,\bar{B}'} = K_{\bar{B}'} \oplus S_{\bar{C}}$ is a uniformly random bit that is independent of $(B, C, S_C, K_0, K_1, K_{B'}, \Pi_1)$. Therefore, by the previous identity we further obtain

$$I(C, S_C, \Pi \wedge K_0, K_1 \mid B, B', K_{B'}) = 0,$$

where $\Pi = (\Pi_1, \Pi_{2,B'}, \Pi_{2,\bar{B}'})$. Finally, since $\hat{K}$ generated by $\mathcal{P}_2$ (even when it is dishonest) must be a stochastic function of (B, C, S_C, Π), the previous identity yields

$$I(C, S_C, \Pi, \hat{K} \wedge K_0, K_1 \mid B, B', K_{B'}) = 0,$$

which is the same as (15.1).[3]

When $\mathcal{P}_2$ is honest, let $\Pi = (\Pi_1, \Pi_{2,0}, \Pi_{2,1})$ be the transcript resulting from an attack by $\mathcal{P}_1$. Now, since $\mathcal{P}_2$ is honest, $\Pi_1 = B \oplus C$, but $(\Pi_{2,0}, \Pi_{2,1})$ is arbitrarily generated from $(\Pi_1, K_0, K_1, S_0, S_1)$. We set $K_0' = \Pi_{2,0} \oplus S_{\Pi_1}$ and $K_1' = \Pi_{2,1} \oplus S_{\overline{\Pi}_1}$

[3] Note that the specific attack of choosing $\Pi_1 = C$ described in Section 15.2 is covered in the construction of the simulator above.

Protocol 15.4 2-bit string OT $f_{2\text{OT}}$ from three instances of bit OT g_{OT}

Input: $\mathcal{P}_1$'s input $(K_0, K_1) = ((K_{0,1}K_{0,2}), (K_{1,1}, K_{1,2}))$ and $\mathcal{P}_2$'s input B
Output: $\mathcal{P}_2$'s output $\hat{K} = (\hat{K}_1, \hat{K}_2)$

1. $\mathcal{P}_1$ selects random bits C_0, C_1 and sets $(Y_0, Y_1) = ((Y_{0,1}, Y_{0,2}, Y_{0,3}), (Y_{1,1}, Y_{1,2}, Y_{1,3}))$ as
$$Y_0 = (K_{0,1} \oplus C_0, C_0, K_{0,2} \oplus C_0),$$
$$Y_1 = (K_{1,1} \oplus C_1, C_1, K_{1,2} \oplus C_1).$$
2. The parties invoke three instances of g_{OT}: for the ith invocation, inputs are $(Y_{0,i}, Y_{1,i})$ and B.
3. $\mathcal{P}_2$ outputs
$$\hat{K} = \big(g_{\text{OT}}((Y_{0,1}, Y_{1,1}), B) \oplus g_{\text{OT}}((Y_{0,2}, Y_{1,2}), B), \\ g_{\text{OT}}((Y_{0,3}, Y_{1,3}), B) \oplus g_{\text{OT}}((Y_{0,2}, Y_{1,2}), B)\big).$$

as the dummy input for $\mathcal{P}_1$. To show that this is a valid dummy input, we must establish the Markov relation

$$(K_0', K_1') \multimap\!\!- (K_0, K_1) \multimap\!\!- B. \tag{15.2}$$

Towards that, note first that Π_1 is independent of (K_0, K_1, S_0, S_1, B) and (S_0, S_1) is independent of (K_0, K_1, B), which gives $(S_0, S_1, \Pi_1) \multimap\!\!- (K_0, K_1) \multimap\!\!- B$, whereby

$$(S_0, S_1, \Pi) \multimap\!\!- (K_0, K_1) \multimap\!\!- B.$$

Also, since (K_0', K_1') is a function of (S_0, S_1, Π), we have $(S_0, S_1, \Pi, K_0', K_1') \multimap\!\!- (K_0, K_1) \multimap\!\!- B$, which implies (15.2). In fact, the previous relation also gives

$$(S_0, S_1, \Pi) \multimap\!\!- (K_0, K_1, K_0', K_1') \multimap\!\!- B.$$

Note that since there is no output for $\mathcal{P}_1$ in f_{OT}, this validates Condition 3 for security in Lemma 15.2. The only remaining condition is Condition 2 which requires that the output of $\mathcal{P}_2$ is $\hat{K} = f_{\text{OT}}((K_0', K_1'), B)$. This holds since the output $\hat{K} = \Pi_{2,B} \oplus S_C$ computed by $\mathcal{P}_2$ coincides with K_B'. □

Next, we illustrate the use of Lemma 15.2 for a stochastic protocol which uses multiple oracle accesses. Recall that Protocol 15.2 for implementing 2OT from OT is not secure since there is a nonadmissible attack possible by $\mathcal{P}_2$ (where $\mathcal{P}_2$ can get partial information about both 2-bit strings). We present an alternative protocol that implements 2OT by invoking three instances of OT. The protocol is described in Protocol 15.4, and we can prove its security by using Lemma 15.2 below.

PROPOSITION 15.4 *Protocol 15.4 perfectly standalone securely implements the 2-bit OT function* $f_{2\text{OT}}$.

Proof Once again, we leave it as an exercise for the reader to verify the correctness condition.

When $\mathcal{P}_1$ is honest but $\mathcal{P}_2$ is dishonest, let B_i, $1 \le i \le 3$, denote $\mathcal{P}_2$'s input for the ith invocation of g_{OT}. Note that these inputs must be generated using stochastic functions of B, whereby $(B_1, B_2, B_3) -\!\!\circ\!\!- B -\!\!\circ\!\!- (K_0, K_1)$. We set $\mathcal{P}_2$'s dummy input B' as the majority of (B_1, B_2, B_3), which is a valid dummy input since $(B_1, B_2, B_3) -\!\!\circ\!\!- B -\!\!\circ\!\!- (K_0, K_1)$ implies $B' -\!\!\circ\!\!- B -\!\!\circ\!\!- (K_0, K_1)$. The dummy output received by $\mathcal{P}_2$ when an ideal function is invoked with inputs $((K_0, K_1), B')$ is given by $K_{B'} = f_{2\mathrm{OT}}((K_0, K_1), B')$. In the simulator (which operates in the ideal world), $\mathcal{P}_2$ needs to sample the "real world outputs" $V_i = g_{\mathrm{OT}}((Y_{0,i}, Y_{1,i}), B_i)$, $1 \le i \le 3$, from $(B, B', K_{B'})$. Towards that end, $\mathcal{P}_2$ can sample an independent copy of (C_0, C_1). Note that this is feasible since (C_0, C_1) is independent of other random variables and can be sampled locally at $\mathcal{P}_2$. Set

$$V_1 = \begin{cases} K_{B',1} \oplus C_{B'}, & \text{if } B' = B_1, \\ C_{\overline{B}'}, & \text{if } \overline{B}' = B_1, \end{cases}$$

$$V_2 = \begin{cases} C_{B'}, & \text{if } B' = B_2, \\ C_{\overline{B}'}, & \text{if } \overline{B}' = B_2, \end{cases}$$

$$V_3 = \begin{cases} K_{B',2} \oplus C_{B'}, & \text{if } B' = B_3, \\ C_{\overline{B}'}, & \text{if } \overline{B}' = B_3. \end{cases}$$

It can be verified that (V_1, V_2, V_3) is distributed identically to the output of the three instances of OT invoked in our protocol. Specifically, when $B' = B_i$, V_i coincides with $g_{\mathrm{OT}}((Y_{0,i}, Y_{1,i}), B_i)$. On the other hand, when $\overline{B}' = B_i$, which can happen for only one of $i \in \{1, 2, 3\}$ since B' is the majority of (B_1, B_2, B_3), $g_{\mathrm{OT}}((Y_{0,i}, Y_{1,i}), B_i)$ is a uniform bit independent of the other two outputs V_j, $j \ne i$. Thus, (V_1, V_2, V_3) has the desired distribution. Furthermore, by construction,

$$(V_1, V_2, V_3) -\!\!\circ\!\!- (B, B_1, B_2, B_3, B', K_{B'}) -\!\!\circ\!\!- (K_0, K_1).$$

Therefore, since (B_1, B_2, B_3) is obtained as a stochastic function of (B, B'), we must have

$$(B_1, B_2, B_3, V_1, V_2, V_3) -\!\!\circ\!\!- (B', B, K_{B'}) -\!\!\circ\!\!- (K_0, K_1).$$

Finally, by noting that the output $\hat{K}$ is generated from $(B, B_1, B_2, B_3, V_1, V_2, V_3)$ in the real protocol, we have

$$(\hat{K}, B_1, B_2, B_3, V_1, V_2, V_3) -\!\!\circ\!\!- (B', B, K_{B'}) -\!\!\circ\!\!- (K_0, K_1).$$

Consequently, all the conditions in Lemma 15.2 are satisfied.

When $\mathcal{P}_2$ is honest but $\mathcal{P}_1$ is dishonest, $\mathcal{P}_1$ may generate (Y_0, Y_1) in an arbitrary manner from (K_0, K_1). We set $\mathcal{P}_1$'s dummy input as

$$K_0' = (g_{\mathrm{OT}}((Y_{0,1}, Y_{1,1}), 0) \oplus g_{\mathrm{OT}}((Y_{0,2}, Y_{1,2}), 0),$$
$$g_{\mathrm{OT}}((Y_{0,3}, Y_{1,3}), 0) \oplus g_{\mathrm{OT}}((Y_{0,2}, Y_{1,2}), 0)),$$

$$K_1' = \big(g_{\mathrm{OT}}((Y_{0,1}, Y_{1,1}), 1) \oplus g_{\mathrm{OT}}((Y_{0,2}, Y_{1,2}), 1),$$
$$g_{\mathrm{OT}}((Y_{0,3}, Y_{1,3}), 1) \oplus g_{\mathrm{OT}}((Y_{0,2}, Y_{1,2}), 1)\big).$$

Clearly, $(K_0', K_1') \multimap\!\!\!\!- (K_0, K_1) \multimap\!\!\!\!- B$ and (K_0', K_1') is a valid dummy input. Further, the output $\hat{K}$ of the protocol satisfies $\hat{K} = f_{2\mathrm{OT}}((K_0', K_1'), B)$. Since $\mathcal{P}_1$ has no output, this completes verification of the conditions in Lemma 15.2. □

We close this subsection by discussing the connection between the two notions of security: passive and active security. Recall that a protocol $\pi^f \diamond g$ passively securely implements f if the following conditions are satisfied.

1. *Correctness:* $(Z_1, Z_2) \sim f(\cdot, \cdot | X_1, X_2)$.
2. *Security for* $\mathcal{P}_1$*:* $\Pi^{(2)} \multimap\!\!\!\!- (X_2, Z_2) \multimap\!\!\!\!- (X_1, Z_1)$.
3. *Security for* $\mathcal{P}_2$*:* $\Pi^{(1)} \multimap\!\!\!\!- (X_1, Z_1) \multimap\!\!\!\!- (X_2, Z_2)$.

Even though active security sounds like a stronger requirement than passive security, their relation is subtle. In general, if a protocol is actively secure this does not imply that it is passively secure, as is illustrated in the following example.

EXAMPLE 15.5 Let f be a function such that only $\mathcal{P}_2$ receives $X_1 \wedge X_2$ for inputs $X_1, X_2 \in \{0, 1\}$. Consider a trivial protocol in which $\mathcal{P}_1$ sends $\Pi = X_1$ to $\mathcal{P}_2$. This protocol is actively secure since $\mathcal{P}_2$ can simulate Π by using $X_2' = 1$ as the dummy input to obtain $X_1 = f(X_1, 1)$. However, this protocol is not passively secure since $\Pi \multimap\!\!\!\!- (X_2, Z_2) \multimap\!\!\!\!- X_1$ does not hold for $Z_2 = f(X_1, X_2)$, unless $X_2 = 1$ with probability 1.

Interestingly, active security implies passive security for symmetric deterministic functions where $\mathcal{P}_1$ and $\mathcal{P}_2$ receive the same output $Z_1 = Z_2$.

PROPOSITION 15.6 *Consider a deterministic function* f *with outputs* $Z_1 = Z_2 = f(X_1, X_2)$*. If a protocol* $\pi^f \diamond g$ *actively securely implements* f*, then it passively securely implements* f *too.*

Proof The correctness condition is the same for both passive and active security. Suppose that $\mathcal{P}_1$ is honest and $\mathcal{P}_2$ is passively dishonest, i.e., $\mathcal{P}_2$ follows the protocol description. Then, by the "only if" part of Lemma 15.2 and the assumption of symmetry, there exist dummy input-output (X_2', Z_2') such that the following hold: (i) $X_2' \multimap\!\!\!\!- X_2 \multimap\!\!\!\!- X_1$; (ii) $Z_1 = Z_2' = f(X_1, X_2')$; and (iii) $\Pi^{(2)} \multimap\!\!\!\!- (X_2', X_2, Z_2') \multimap\!\!\!\!- (X_1, Z_1)$. Further, from the perfect correctness condition under an active adversary, when both parties follow the protocol, we have $Z_1 = Z_2 = f(X_1, X_2)$. It follows that $Z_1 = Z_2 = Z_2' = f(X_1, X_2)$, whereby

$$\begin{aligned}
I(\Pi^{(2)} \wedge X_1, Z_1 | X_2, Z_2) &= I(\Pi^{(2)} \wedge X_1, Z_1 | X_2, Z_2') \\
&\le I(\Pi^{(2)}, X_2' \wedge X_1, Z_1 | X_2, Z_2') \\
&= I(X_2' \wedge X_1, Z_1 | X_2, Z_2') + I(\Pi^{(2)} \wedge X_1, Z_1 | X_2, Z_2', X_2') \\
&= I(X_2' \wedge X_1, Z_1 | X_2, Z_2'),
\end{aligned}$$

where in the previous identity we used condition (iii) above. For the term on the right-hand side, we have

$$\begin{aligned} I(X_2' \wedge X_1, Z_1|X_2, Z_2') &= I(X_2' \wedge X_1|X_2, Z_2') \\ &\leq I(X_2' \wedge X_1, Z_2'|X_2) \\ &= I(X_2' \wedge X_1|X_2) + I(X_2' \wedge Z_2'|X_2, X_1) \\ &= 0, \end{aligned}$$

where the final identity follows from condition (i) above and the fact that $Z_2' = f(X_1, X_2)$.

Upon combining the bounds above, we obtain $I(\Pi^{(2)} \wedge X_1, Z_1|X_2, Z_2) = 0$, which coincides with the passive security requirement $\Pi^{(2)} \multimap\!\!\!- (X_2, Z_2) \multimap\!\!\!- (X_1, Z_1)$. The case with passively dishonest $\mathcal{P}_1$ can be verified in a similar manner. □

15.4 Approximate Standalone Security and Protocols with Abort

In the previous section, we were aspiring to a rather ideal form of security, namely perfect security. Indeed, such an ideal form of security was possible because of the assumed availability of a perfect oracle g as a resource. However, it is of practical importance to consider oracle resources that are not perfect, and yet may be used to enable secure computing against an active adversary. In such cases, one must take recourse to a milder (yet quite powerful) form of security.

For concreteness, we consider the specific application of *protocols with abort.* A (real) protocol may be aborted when the honest party detects an attack from the other party, or when nonideal behavior of an oracle resource is detected. However, if a protocol is allowed to abort, even a dishonest party can intentionally engage abort if the situation is not favorable for the dishonest party. Thus, a careful security analysis is required for protocols with abort.

As an example of a protocol with abort, we implement the bit OT function f_{OT} using n instances of an erasure channel g_{EC}^n with erasure probability $1/2$ as a resource. We describe the implementation in Protocol 15.5. As we will see later, this protocol is secure against an active adversary.[4] Note that we allow $\mathcal{P}_2$ to declare abort if the random resource (erasure channel) does not exhibit "typical" behavior, i.e., the number of erasures is too small. Also, $\mathcal{P}_1$ can declare abort if the communication received from $\mathcal{P}_2$ is not as expected, and it detects an attack by $\mathcal{P}_2$.

A careful reader may object to the aforementioned protocol with abort. While it promises to implement the ideal function f_{OT}, it may declare the abort symbol $\perp$ as an output which was not an output for f_{OT}. This is where the earlier remark on careful description of ideal functions applies. Specifically, in order to consider

[4] Even though Protocol 15.5 is much less efficient than Protocol 13.2, the latter is secure only against a passive adversary. In fact, we can construct a more efficient protocol that is secure against an active adversary (cf. Section 15.9).

Protocol 15.5 Implementation of f_{OT} from n instances of erasure channel g^n_{EC}

Input: $\mathcal{P}_1$'s input (K_0, K_1), $\mathcal{P}_2$'s input B, and security parameter $\nu > 0$
Output: $\mathcal{P}_2$'s output $\hat{K}$

1. $\mathcal{P}_1$ randomly selects a sequence $Y^n = (Y_1, \ldots, Y_n) \in \{0,1\}^n$, and sends the sequence over g^n_{EC}; $\mathcal{P}_2$ receives V^n.
2. $\mathcal{P}_2$ sets the good indices $\tilde{\mathcal{I}}_0 = \{i : V_i \neq \mathsf{e}\}$ and the bad indices $\tilde{\mathcal{I}}_1 = \{i : V_i = \mathsf{e}\}$. If either $|\tilde{\mathcal{I}}_0| < (1/2 - \nu)n$ or $|\tilde{\mathcal{I}}_1| < (1/2 - \nu)n$ occurs, $\mathcal{P}_2$ declares an abort symbol $\perp$.
3. $\mathcal{P}_2$ selects subsets $\mathcal{I}_B \subseteq \tilde{\mathcal{I}}_0$ and $\mathcal{I}_{\overline{B}} \subseteq \tilde{\mathcal{I}}_1$ such that $|\mathcal{I}_0| = |\mathcal{I}_1| = (1/2 - \nu)n$, and sends the subsets $\Pi_1 = (\mathcal{I}_0, \mathcal{I}_1)$ to $\mathcal{P}_1$.
4. If $\mathcal{I}_0 \cap \mathcal{I}_1 \neq \emptyset$, $\mathcal{P}_1$ declares an abort symbol $\perp$.
5. $\mathcal{P}_1$ sends the following to $\mathcal{P}_2$:

$$\Pi_{2,0} = K_0 \oplus \bigoplus_{i \in \mathcal{I}_0} Y_i, \qquad \Pi_{2,1} = K_1 \oplus \bigoplus_{i \in \mathcal{I}_1} Y_i.$$

6. $\mathcal{P}_2$ outputs

$$\hat{K} = \Pi_{2,B} \oplus \bigoplus_{i \in \mathcal{I}_B} V_i.$$

the security of protocols with abort, we need to modify the ideal function f so that abort is a valid output. Towards that, we set the following convention for functions with abort.

We expand both the sets of inputs and outputs of the parties *for the ideal function* f to allow $\perp$ as an input and an output. Either party is allowed to input the abort command $\perp$. When this command is invoked by one of the parties, the ideal function f returns the abort notification $(\perp, \perp)$ to both the parties irrespective of the input of the other party.

With this modification to the description of ideal functions (to accommodate real implementations with abort), we can proceed to formalize the security of protocols with abort. As before, our focus will be on the information flow and how a dishonest party can reveal extra information as its view. The view of the parties in the real protocol is defined in a similar manner to the case of protocol without abort, with a minor change. When any one party declares the abort symbol $\perp$ during the execution of the protocol, the outputs (Z_1, Z_2) equal the abort notification output $(\perp, \perp)$. Furthermore, even if the protocol is aborted, the dishonest party can observe the transcript Π that is communicated before the declaration of abort and can include it in the view of the dishonest party.

The most important change in the definition of security is that we only require the views in the real and ideal worlds to agree approximately. More specifically, the view $\mathtt{V}[\pi^f \circ g]$ in the real protocol need not have exactly the same distribution as the view $\mathtt{V}[\sigma \circ f]$ of the simulated protocol. Instead, the protocol is considered secure even if the former is statistically close to the latter. This relaxation enables us to construct a protocol with acceptable security guarantees even if the implementation of a perfectly secure protocol is not possible. In fact, when the ideal function g used as a resource is a stochastic function, such as the erasure channel above, it is difficult to implement a perfectly secure protocol since g has statistical deviation. For instance, the receiver may observe all the symbols without any erasure, even though such an event occurs with a very small probability. The notion of standalone security below accommodates such rare events in the allowed statistical approximation of ε.

DEFINITION 15.7 (Standalone security) For $\varepsilon \in [0, 1)$, a protocol $\pi^f \diamond g$ standalone ε-securely implements f if, for every attack by a dishonest party, there exists a simulator σ such that it retains the input and the output for the honest party, and for every input distribution $\mathrm{P}_{X_1X_2}$, it holds that

$$d_{\mathtt{var}}\big(\mathtt{V}[\pi^f \diamond g], \mathtt{V}[\sigma \circ f]\big) \le \varepsilon.$$

If such a protocol $\pi^f \diamond g$ exists, we say that f is standalone ε-securely computable using (oracle access to) g.

Remark 15.6 (Abort input is available only in the ideal world) Before proceeding, we make an important remark about our convention for the abort input symbol $\perp$: it is available only in the ideal world where it can be engaged by a dishonest party. In particular, when a party, honest or dishonest, outputs $\perp$ in the real world, it must be simulated in the ideal world by the dishonest party providing the input $\perp$ to the ideal function. When both parties are honest, the abort output cannot be engaged by either party and must be accounted for in the probability of error in the "correctness" condition.

In a similar manner to Lemma 15.2, we can provide a more useful alternative form of the security definition above. The only difference is that the Markov relations $X \multimap\!\!- Y \multimap\!\!- Z$ will be replaced with their approximate counterparts $d_{\mathtt{var}}(\mathrm{P}_{XYZ}, \mathrm{P}_{XY}\mathrm{P}_{Z|Y}) \le \varepsilon$.

LEMMA 15.8 (Standalone security – alternative form) *Consider a stochastic function f, and let W_f denote the channel corresponding to the stochastic function $f(\cdot, \cdot|X_1, X_2)$. The protocol $\pi^f \diamond g$ standalone 3ε-securely implements f if the following conditions are satisfied.*

1. *Correctness. When both parties are honest, for every input distribution $\mathrm{P}_{X_1X_2}$, the output (Z_1, Z_2) of $\pi^f \diamond g$ satisfies $d_{\mathtt{var}}(\mathrm{P}_{Z_1Z_2X_1X_2}, W_f\mathrm{P}_{X_1X_2}) \le \varepsilon$.*[5]
2. *Security for $\mathcal{P}_1$. When $\mathcal{P}_1$ is honest, for every attack by $\mathcal{P}_2$ and every input distribution $\mathrm{P}_{X_1X_2}$, there exists a joint distribution $\mathrm{P}_{X_1X_2X_2'Z_1Z_2Z_2'\Pi^{(2)}}$, with*

[5] The notation $W_f\mathrm{P}_{X_1X_2}$ means that the observations are distributed according to $\mathrm{P}_{X_1X_2}(x_1, x_2)\, W_f(z_1, z_2|x_1, x_2)$.

Protocol 15.6 Simulator σ with dishonest $\mathcal{P}_2$ in Lemma 15.8

Input: (X_1, X_2)
Output: $((X_1, \tilde{Z}_1), (X_2, \tilde{Z}_2, \tilde{\Pi}^{(2)}))$
1. $\mathcal{P}_2$ samples $\tilde{X}_2' \sim \mathrm{P}_{X_2'|X_2}(\cdot|X_2)$.
2. The parties invoke the ideal function f to obtain $(\tilde{Z}_1, \tilde{Z}_2') \sim f(\cdot, \cdot|X_1, \tilde{X}_2')$.
3. $\mathcal{P}_1$ outputs $(X_1, \tilde{Z}_1)$.
4. $\mathcal{P}_2$ samples $(\tilde{Z}_2, \tilde{\Pi}^{(2)}) \sim \mathrm{P}_{Z_2\Pi^{(2)}|X_2X_2'Z_2'}(\cdot, \cdot|X_2, \tilde{X}_2', \tilde{Z}_2')$.
5. $\mathcal{P}_2$ outputs $(X_2, \tilde{Z}_2, \tilde{\Pi}^{(2)})$.

---

dummy input and output (X_2', Z_2') *for* $\mathcal{P}_2$ *and* $\mathrm{P}_{X_1X_2Z_1Z_2\Pi^{(2)}}$ *denoting the distribution of input, output, and transcript under the attack, such that*[6]

(i) $d_{\mathtt{var}}(\mathrm{P}_{X_2'X_1X_2}, \mathrm{P}_{X_2'|X_2}\mathrm{P}_{X_1X_2}) \le \varepsilon$,
(ii) $d_{\mathtt{var}}(\mathrm{P}_{Z_1Z_2'X_1X_2'}, W_f\mathrm{P}_{X_1X_2'}) \le \varepsilon$, *and*
(iii) $d_{\mathtt{var}}(\mathrm{P}_{Z_2\Pi^{(2)}X_2X_2'Z_2'X_1Z_1}, \mathrm{P}_{Z_2\Pi^{(2)}|X_2X_2'Z_2'}\mathrm{P}_{X_2X_2'Z_2'X_1Z_1}) \le \varepsilon$.

3. *Security for* $\mathcal{P}_2$. *When* $\mathcal{P}_2$ *is honest, for every attack by* $\mathcal{P}_1$ *and every input distribution* $\mathrm{P}_{X_1X_2}$, *there exists a joint distribution* $\mathrm{P}_{X_1X_2X_1'Z_1Z_2Z_1'\Pi^{(1)}}$, *with dummy input and output* (X_1', Z_1') *for* $\mathcal{P}_1$ *and* $\mathrm{P}_{X_1X_2Z_1Z_2\Pi^{(1)}}$ *denoting the distribution of input, output, and transcript under the attack, such that*

   (i) $d_{\mathtt{var}}(\mathrm{P}_{X_1'X_1X_2}, \mathrm{P}_{X_1'|X_1}\mathrm{P}_{X_1X_2}) \le \varepsilon$,
   (ii) $d_{\mathtt{var}}(\mathrm{P}_{Z_1'Z_2X_1'X_2}, W_f\mathrm{P}_{X_1'X_2}) \le \varepsilon$, *and*
   (iii) $d_{\mathtt{var}}(\mathrm{P}_{Z_1\Pi^{(1)}X_1X_1'Z_1'X_2Z_2}, \mathrm{P}_{Z_1\Pi^{(1)}|X_1X_1'Z_1'}\mathrm{P}_{X_1X_1'Z_1'X_2Z_2}) \le \varepsilon$.

Proof When both the parties are honest, the correctness condition implies that security in the sense of Definition 15.7 holds for the trivial simulator that simply declares the input and the output.

When $\mathcal{P}_1$ is honest, for an attack by $\mathcal{P}_2$, consider the simulator described in Protocol 15.6, which is constructed using the joint distribution $\mathrm{P}_{X_1X_2X_2'Z_1Z_2'\Pi^{(2)}}$ appearing in Condition 2. In the real world, the view of the protocol under attack by $\mathcal{P}_2$ is given by $((X_1, Z_1), (X_2, Z_2, \Pi^{(2)}))$. In the ideal world, the view of the simulated protocol is given by $((X_1, \tilde{Z}_1), (X_2, \tilde{Z}_2, \tilde{\Pi}^{(2)}))$. Therefore, to prove that $\pi^f \diamond g$ is standalone 3ε-secure, it suffices to show that

$$d_{\mathtt{var}}(\mathrm{P}_{X_1\tilde{Z}_1X_2\tilde{Z}_2\tilde{\Pi}^{(2)}}, \mathrm{P}_{X_1Z_1X_2Z_2\Pi^{(2)}}) \le 3\varepsilon.$$

Indeed, this condition can be seen to hold as follows:

$$\begin{aligned} & d_{\mathtt{var}}(\mathrm{P}_{X_1\tilde{Z}_1\tilde{X}_2'X_2\tilde{Z}_2'\tilde{Z}_2\tilde{\Pi}^{(2)}}, \mathrm{P}_{X_1Z_1X_2'X_2Z_2'Z_2\Pi^{(2)}}) \\ & \le d_{\mathtt{var}}(\mathrm{P}_{X_1\tilde{Z}_1\tilde{X}_2'X_2\tilde{Z}_2'\tilde{Z}_2\tilde{\Pi}^{(2)}}, \mathrm{P}_{X_1Z_1X_2'X_2Z_2'}\mathrm{P}_{Z_2\Pi^{(2)}|X_2'X_2Z_2'}) \end{aligned}$$

[6] In applications of this lemma below, we exhibit universal simulators by ensuring that the conditional distributions used to generate dummy inputs and outputs do not depend on the input distribution $\mathrm{P}_{X_1X_2}$.

$$
\begin{aligned}
&+ d_{\text{var}}\big(\mathrm{P}_{X_1Z_1X_2'X_2Z_2'}\mathrm{P}_{Z_2\Pi^{(2)}|X_2'X_2Z_2'}, \mathrm{P}_{X_1Z_1X_2'X_2Z_2'Z_2\Pi^{(2)}}\big)\\
&= d_{\text{var}}\big(\mathrm{P}_{X_1\tilde{Z}_1\tilde{X}_2'X_2\tilde{Z}_2'}\mathrm{P}_{Z_2\Pi^{(2)}|X_2'X_2Z_2'}, \mathrm{P}_{X_1Z_1X_2'X_2Z_2'}\mathrm{P}_{Z_2\Pi^{(2)}|X_2'X_2Z_2'}\big)\\
&\quad + d_{\text{var}}\big(\mathrm{P}_{X_1Z_1X_2'X_2Z_2'}\mathrm{P}_{Z_2\Pi^{(2)}|X_2'X_2Z_2'}, \mathrm{P}_{X_1Z_1X_2'X_2Z_2'Z_2\Pi^{(2)}}\big)\\
&= d_{\text{var}}\big(\mathrm{P}_{X_1\tilde{Z}_1\tilde{X}_2'X_2\tilde{Z}_2'}, \mathrm{P}_{X_1Z_1X_2'X_2Z_2}\big)\\
&\quad + d_{\text{var}}\big(\mathrm{P}_{X_1Z_1X_2'X_2Z_2}\mathrm{P}_{Z_2\Pi^{(2)}|X_2'X_2Z_2'}, \mathrm{P}_{X_1Z_1X_2'X_2Z_2'Z_2\Pi^{(2)}}\big)\\
&\le d_{\text{var}}\big(\mathrm{P}_{X_1\tilde{Z}_1\tilde{X}_2'X_2\tilde{Z}_2'}, \mathrm{P}_{X_1Z_1X_2'X_2Z_2}\big) + \varepsilon,
\end{aligned}
$$

where we used (iii) of Condition 2. It remains to bound the first term on the right-hand side. Note that by (i) of Condition 2 and the construction of the simulator, we have

$$d_{\text{var}}\big(\mathrm{P}_{\tilde{X}_2'X_1X_2}, \mathrm{P}_{X_2'X_1X_2}\big) \le \varepsilon. \tag{15.3}$$

It follows that

$$
\begin{aligned}
&d_{\text{var}}\big(\mathrm{P}_{X_1\tilde{Z}_1\tilde{X}_2'X_2\tilde{Z}_2}, \mathrm{P}_{X_1Z_1X_2'X_2Z_2}\big)\\
&\quad \le d_{\text{var}}\big(W^f\mathrm{P}_{X_1\tilde{X}_2'X_2}, W^f\mathrm{P}_{X_1X_2'X_2}\big) + d_{\text{var}}\big(W^f\mathrm{P}_{X_1X_2'X_2}, \mathrm{P}_{X_1Z_1X_2'X_2Z_2}\big)\\
&\quad \le 2\varepsilon,
\end{aligned}
$$

where the final inequality is by (15.3) and (ii) of Condition 2. The proof of security for $\mathcal{P}_1$ is completed upon combining the bounds above; security for $\mathcal{P}_2$ can be completed similarly. ☐

Remark 15.7 In a similar manner to Lemma 15.2, we can prove the "only if" counterpart for Lemma 15.8 at the cost of a constant factor loss in the security parameter ε (cf. Problem 15.1).

We demonstrate the usefulness of Lemma 15.8 by establishing the security of Protocol 15.5. The proof proceeds along similar lines to that of Proposition 15.4; however, the main new component is that the simulation may not be perfect. Generally speaking, a secure protocol is constructed so that a dishonest party, say $\mathcal{P}_2$, can only conduct admissible attacks unless some miraculous event occurs. When such an event does not occur, we can simulate the dishonest party's view; when that event does occur, we give up simulation, and generate the dishonest party's view arbitrarily. In the analysis of such a construction of the simulator, the following lemma is useful (the proof is very easy, and kept as an exercise; see Problem 15.2).

LEMMA 15.9 *For given random variables (X, Y), let $E = \kappa(X) \in \{0,1\}$ be an event computed from X such that $\Pr(E = 0) \le \varepsilon$. Let $\mathrm{P}_{X\tilde{Y}}$ be a distribution such that $\mathrm{P}_{\tilde{Y}|X}(\cdot|x) = \mathrm{P}_{Y|X}(\cdot|x)$ if $\kappa(x) = 1$ and $\mathrm{P}_{\tilde{Y}|X}(\cdot|x)$ is arbitrary if $\kappa(x) = 0$. Then, we have*

$$d_{\text{var}}(\mathrm{P}_{XY}, \mathrm{P}_{X\tilde{Y}}) \le \varepsilon.$$

Now, we establish the security of Protocol 15.5.

PROPOSITION 15.10 *Protocol 15.5 standalone ε-securely implements the bit OT function f_{OT} with $\varepsilon = \mathcal{O}(2^{-\nu^2 n})$.*

Proof As mentioned earlier, since we allow protocols with abort, we consider an extended version of f_{OT} which allows the inputs and the output to take the value $\perp$. When both parties are honest, Protocol 15.5 can yield an incorrect output only if the abort $\perp$ is declared in Step 2. This, in turn, can happen only if the events $\{|\tilde{\mathcal{I}}_0| < (1/2-\nu)n\}$ or $\{|\tilde{\mathcal{I}}_1| < (1/2-\nu)n\}$ occur. By the Hoeffding bound (cf. Theorem 2.39), we can see that one of these events occurs with probability ε less than $\mathcal{O}(2^{-\nu^2 n})$, which completes verification of the soundness condition of Lemma 15.8.

When $\mathcal{P}_1$ is honest and $\mathcal{P}_2$ is dishonest, $\mathcal{P}_2$ may generate Π_1 maliciously and reveal extra information in the view using its observations

$$(B, V^n, \Pi_1, \Pi_{2,0}, \Pi_{2,1}, \hat{K}).$$

Note that the outputs (Z_1, Z_2) are not explicitly defined in the protocol, but we set $Z_1 = Z_2 = \perp$ if $\perp$ is declared in Steps 2 or 4; otherwise, $Z_1 = \emptyset$ and $Z_2 = \hat{K}$. Also, when $\mathcal{I}_0 \cap \mathcal{I}_1 \neq \emptyset$ and $\mathcal{P}_1$ declares the abortion of the protocol in Step 4, we set $(\Pi_{2,0}, \Pi_{2,1}) = (\perp, \perp)$. Furthermore, if the sets $\mathcal{I}_0$ and $\mathcal{I}_1$ generated by $\mathcal{P}_2$ are such that we can find indices $i_0 \in \mathcal{I}_0$ and $i_1 \in \mathcal{I}_1$ satisfying $V_{i_0} = V_{i_1} = \mathsf{e}$, the output $\hat{K}$ in Step 6 cannot be generated. Since this situation cannot arise when both parties are honest, we did not include it in the description of the protocol. But for formal security analysis, we need to carefully specify the behavior of the protocol in all these corner cases as well; we set the output of the protocol as $Z_1 = \emptyset$ and Z_2 arbitrarily, say 0, in this case.[7] The inputs X_1 and X_2 equal (K_0, K_1) and B, respectively, in the real protocol. Moreover, the transcripts $\Pi^{(1)}$ and $\Pi^{(2)}$ equal $(\Pi_1, \Pi_{2,0}, \Pi_{2,1}, Y^n)$ and $(\Pi_1, \Pi_{2,0}, \Pi_{2,1}, V^n)$, respectively. In our current case of honest $\mathcal{P}_1$ and dishonest $\mathcal{P}_2$, $\Pi^{(2)}$ and Z_2 may be adversarially generated (while respecting the information constraints of the protocol).

We now identify the distribution required in Condition 2. Specifically, consider the distribution $\mathrm{P}_{X_1 X_2 Y^n V^n \Pi_1 \Pi_2 Z_1 Z_2}$ corresponding to the attack by $\mathcal{P}_2$, where $\Pi_2 = (\Pi_{2,0}, \Pi_{2,1})$. Set the dummy input X_2' as

$$X_2' = \begin{cases} \perp, & \text{if } \Pi_1 = \perp \text{ or } \mathcal{I}_0 \cap \mathcal{I}_1 \neq \emptyset, \\ 0, & \text{else if } V_i \neq \mathsf{e}\ \forall i \in \mathcal{I}_0,\ \exists j \in \mathcal{I}_1 \text{ such that } V_j = \mathsf{e}, \\ 1, & \text{else if } V_i \neq \mathsf{e}\ \forall i \in \mathcal{I}_1,\ \exists j \in \mathcal{I}_0 \text{ such that } V_j = \mathsf{e}, \\ \perp, & \text{else.} \end{cases}$$

For convenience, we denote the binary value taken by X_2' when it does not equal $\perp$ as B'.

Since V^n is independent of (X_1, X_2) and $(\Pi_1, \mathcal{I}_0, \mathcal{I}_1)$ must be generated as a stochastic function of (X_2, V^n), we have $(\Pi_1, \mathcal{I}_0, \mathcal{I}_1) -\!\!\circ\!\!- (X_2, V^n) -\!\!\circ\!\!- X_1$ whereby

$$(\Pi_1, V^n, \mathcal{I}_0, \mathcal{I}_1) -\!\!\circ\!\!- X_2 -\!\!\circ\!\!- X_1.$$

[7] Since $\mathcal{P}_1$ cannot recognize $V_{i_0} = V_{i_1} = \mathsf{e}$, the protocol cannot be aborted in this case; in fact, this is an admissible attack, and can be simulated.

Further, since X_2' is a function of $(X_2, \Pi_1, V^n, \mathcal{I}_0, \mathcal{I}_1)$, the previous Markov relation implies that $X_2' -\!\!\circ\!\!- X_2 -\!\!\circ\!\!- X_1$. Thus, X_2' satisfies (i) of Condition 2 with $\varepsilon = 0$.

Let the corresponding dummy output be given by $(Z_1, Z_2') = f_{\mathrm{OT}}(X_1, X_2')$ with $Z_1 = \emptyset$ and $Z_2' = K_{B'}$ when there is no abort. When X_2' equals $\perp$, we have $Z_1 = Z_2' = \perp$. Furthermore, since $\mathcal{P}_1$ does not have an output when there is no abort, it can be verified that the distribution $\mathrm{P}_{X_1 Z_1}$ is identical to the distribution of the input and the output of $\mathcal{P}_1$ when the real protocol is executed under the attack; thus (ii) of Condition 2 is satisfied with $\varepsilon = 0$. Note that here we used the fact that all possible ways for inducing an abort by $\mathcal{P}_2$ in the real world are captured as an abort input in our definition of X_2' in the ideal world.

It remains to verify (iii) of Condition 2. Note that our earlier observations give

$$(X_2', \Pi_1, V^n) -\!\!\circ\!\!- X_2 -\!\!\circ\!\!- X_1,$$

whereby

$$\begin{aligned} &I(\Pi_1 V^n \wedge X_1 Z_1 \mid X_2 X_2' Z_2') \\ &\quad \leq I(\Pi_1 V^n \wedge X_1 \mid X_2 X_2') + I(\Pi_1 V^n \wedge X_1 Z_1 Z_2' \mid X_1 X_2 X_2') \\ &\quad = 0, \end{aligned} \tag{15.4}$$

where the final identity uses the Markov relation above and the fact that (Z_1, Z_2') is a stochastic function of (X_1, X_2'); namely the Markov relation $(\Pi_1, V^n) -\!\!\circ\!\!- (X_2, X_2', Z_2') -\!\!\circ\!\!- X_1$ holds. Thus, to verify (iii), it remains to bound

$$d_{\mathrm{var}}(\mathrm{P}_{Z_2 \Pi_2 \Pi_1 V^n X_1 Z_1 X_2 X_2' Z_2'}, \mathrm{P}_{Z_2 \Pi_2 | X_2 X_2' Z_2' \Pi_1 V^n} \mathrm{P}_{X_1 Z_1 X_2 X_2' Z_2' \Pi_1 V^n}). \tag{15.5}$$

Denote the event $\{|\tilde{\mathcal{I}}_0| \geq (1-2\nu)n\}$ by $\{E = 0\}$. By the Hoeffding bound, note that this event can happen with probability at most $\mathcal{O}(2^{-(1/2-2\nu)^2 n})$. Since this is the "miraculous event" described just before Lemma 15.9,[8] the distribution of the remaining view (Z_2, Π_2) of $\mathcal{P}_2$ is irrelevant. When $\mathcal{I}_0 \cap \mathcal{I}_1 \neq \emptyset$ and $\mathcal{P}_1$ declares abortion of the protocol in Step 4, $(Z_2, \Pi_2) = (\perp, \perp)$ is uniquely determined from $(X_2, X_2', Z_2', \Pi_1, V^n)$. When $X_2' = B' \in \{0,1\}$, note that $\Pi_{2,B'}$ is uniquely determined from $Z_2' = K_{B'}$ and $(V_i : i \in \mathcal{I}_{B'})$. On the other hand, since there exists $i \in \mathcal{I}_{\overline{B'}}$ such that $V_i = \mathtt{e}$, $\Pi_{2,\overline{B'}}$ is a random bit independent of other random variables. When there exists $i_0 \in \mathcal{I}_0$ and $i_1 \in \mathcal{I}_1$ satisfying $V_{i_0} = V_{i_1} = \mathtt{e}$, then both $\Pi_{2,0}$ and $\Pi_{2,1}$ are random bits independent of other random variables. Also, note that Z_2 is a stochastic function of Π_2 and (X_2, Π_1, V^n). Noting these facts and invoking Lemma 15.9, (15.5) can be upper bounded by $\mathcal{O}(2^{-(1/2-2\nu)^2 n})$.

When $\mathcal{P}_2$ is honest and $\mathcal{P}_1$ is dishonest, $\mathcal{P}_1$ can use the transcript $\Pi^{(1)} = (Y^n, \Pi_1, \Pi_2)$ produced under attack to reveal additional information as a part of the view. Consider an attack by $\mathcal{P}_1$ where it observes Π_1 and produces Π_2 as

[8] When $|\tilde{\mathcal{I}}_0| \geq (1-2\nu)n$ occurs, $\mathcal{P}_2$ can set $\mathcal{I}_0$ and $\mathcal{I}_1$ so that $V_i \neq \mathtt{e}$ for every $i \in \mathcal{I}_0 \cup \mathcal{I}_1$ and $\mathcal{I}_0 \cap \mathcal{I}_1 = \emptyset$.

a stochastic function of (X_1, Y^n, Π_1), with joint distribution $\mathrm{P}_{X_1 Y^n \Pi_1 \Pi_2}$. Define the dummy input X_1' as follows:

$$X_1' = \begin{cases} \perp, & \text{if } \Pi_1 = \perp, \\ (K_0', K_1'), & \text{otherwise,} \end{cases}$$

where

$$K_0' = \Pi_{2,0} \oplus \bigoplus_{i \in \mathcal{I}_0} Y_i, \qquad K_1' = \Pi_{2,1} \oplus \bigoplus_{i \in \mathcal{I}_1} Y_i.$$

Note that we have not included the possible abort in Step 4, which will not happen when $\mathcal{P}_2$ is honest. Further, set $(Z_1', Z_2) = f_{\mathrm{OT}}(X_1', X_2)$. Since Π_1 is independent of (Y^n, X_1, X_2) when $\mathcal{P}_2$ is honest, (i) and (ii) of Condition 2 can be verified easily. In fact, following similar analysis as above, we get that the Markov relation $(X_1', \Pi_1, Y^n) -\!\!\circ\!\!- X_1 -\!\!\circ\!\!- X_2$ holds, which using a similar bound to (15.4) yields the Markov relation $(\Pi_1, Y^n) -\!\!\circ\!\!- (X_1, X_1', Z_1') -\!\!\circ\!\!- (X_2, Z_2)$. Part (iii) can be verified by using the facts that Π_2 can be obtained as a function of X_1', Π_1, and Y^n when abort does not occur and that Z_1 is either $\perp$ or $\emptyset$. ☐

Remark 15.8 For protocols with abort, one might want to argue security of a given protocol conditioned on the event that the protocol is not aborted. However, such a requirement is too strict. In fact, when we construct the bit OT from the erasure channel (not necessarily Protocol 15.5), $\mathcal{P}_2$ may declare abort whenever there is at least one erasure among $V^n = (V_1, \ldots, V_n)$. In other words, $\mathcal{P}_2$ continues the protocol only when $V_i \neq \mathsf{e}$ for every $1 \leq i \leq n$. Of course, such a miracle occurs with negligible probability; however, no protocol can be secure conditioned on the miraculous event since $\mathcal{P}_1$ has no "information advantage" over $\mathcal{P}_2$ under this event.

15.5 Composable Security and the Sequential Composition Theorem

In practice, an implementation $\pi^f \diamond g$ of f does not exist in isolation. It interacts with other protocols, or multiple copies of this protocol are executed within a larger protocol. The notion of standalone security described in the previous section is not equipped to handle security in such scenarios where there is composition of protocols. In this section, we will introduce a new notion of security that will allow us to handle these issues, at least partially. In particular, we will be able to handle a simple type of composition, termed *sequential composition.* Interestingly, this more general definition will be seen to follow from the previous notion of standalone security.

Our new notion of security considers the use of f inside an *outer protocol* π, that makes oracle access to f; see Figure 15.9. To evaluate the security of an implementation $\pi^f \diamond g$ of f, we compare how it will change the information revealed by π when, instead of f, it is interacting with $\pi^f \diamond g$. We allow the outer

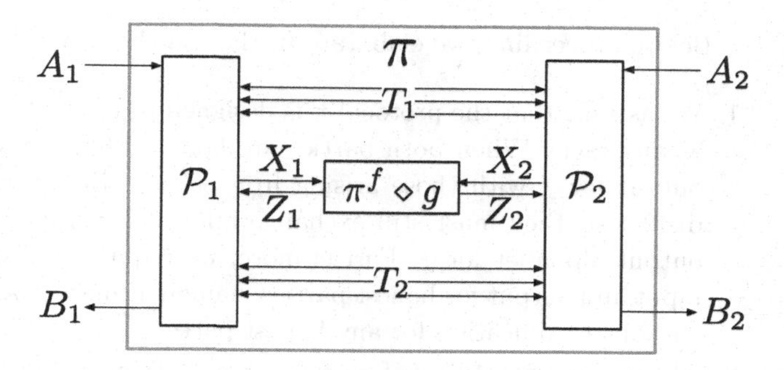

Figure 15.9 A description of composition of protocols: the outer protocol π invokes $\pi^f \diamond g$ as a subroutine.

protocol to be chosen adversarially by the dishonest party. To be more specific, consider a protocol π with inputs (A_1, A_2) and outputs (B_1, B_2). In protocol π, the parties first communicate with each other and exchange the transcript T_1. Note that this need not be just an interactive communication, and the protocol may make oracle access to some other functions (two-way channels). Specifically, both parties may see different parts of the transcripts $T_1^{(1)}$ and $T_1^{(2)}$. Using this transcript T_1, the parties generate input (X_1, X_2) for $\pi^f \diamond g$ and obtain (Z_1, Z_2). Next, the parties use these inputs and outputs, along with the transcript T_1 seen earlier, to produce the transcript T_2 of the next phase. Finally, the protocol reveals the output (B_1, B_2). During the course of the protocol, the parties obtain the inputs and outputs for oracle access to g used in $\pi^f \diamond g$. We treat these inputs and outputs, too, as a part of the transcript of the protocol π. While an honest party will not reveal them as a part of the view, a dishonest party can reveal these parts of the transcript in the view.

Our new notion of security requires that the view of the protocol π does not change much if it is interacting with $\pi^f \diamond g$ or f, together with a simulator.

DEFINITION 15.11 (Sequentially composable security) For $\varepsilon \in [0, 1)$, a protocol $\pi^f \diamond g$ sequentially composably ε-securely implements f if for every attack by a dishonest party, there exists a simulator σ such that it retains the input and the output for the honest party, and for every protocol π invoking $\pi^f \diamond g$ as a subroutine, it holds that

$$d_{\mathtt{var}}(\mathtt{V}[\pi \circ \pi^f \diamond g], \mathtt{V}[\pi \circ \sigma \circ f]) \leq \varepsilon.$$

If such a protocol $\pi^f \diamond g$ exists, we say that f is sequentially composably ε-securely computable using g.

In other words, the outer protocol π is used to produce a view to distinguish the *real protocol* π^f and the *ideal protocol* f. Note that π and π^f are actual entities while the ideal protocol f and the simulator σ are theoretical constructs required for our security analysis.

Before proceeding, we elaborate further on the flow of information.

1. We assume that the protocol π is designed to interact with the ideal protocol as an oracle. When both parties are honest, they can replace the input and output for f with those arising from $\pi^f \diamond g$. However, even when they are dishonest, they must still exchange information with π using the input and output alphabet for f. Furthermore, we require the simulator to retain the input and output for honest parties, namely it must pass the input and output without modification for any honest party.
2. The notion of the view of a protocol, as before, changes based on which party is dishonest. When the parties are honest, they only reveal the legitimate input and output of π to the environment, which constitutes $\mathtt{V}[\pi \circ \pi^f \diamond g]$ (or $\mathtt{V}[\pi \circ \sigma \circ f]$). But when they are dishonest, they can use the internal information of various protocols they engage with and reveal additional information to the environment.
3. The protocol π invoking $\pi^f \diamond g$ need not be an interactive communication protocol. In particular, it is allowed oracle access to two-way channels (for example, it may access OT in the middle). Thus, the transcript of the protocol may differ for the two parties. Indeed, this is the power of the definition of security above: the outer protocol may be interacting with many other functionalities and still we can establish the equivalence of the two views. Note that we need not consider honest behavior and dishonest behavior for the outer protocol π; since the outer protocol can be arbitrary, we can regard dishonest behavior (of some prescribed protocol) as legitimate behavior of π.
4. An important rule to follow for interaction with the outer protocol is that it can only happen through the input and the output of f. Even when using the implementation $\pi^f \diamond g$ of f, an honest party will respect this rule and will not use the transcript of $\pi^f \diamond g$ (which may include the inputs and outputs for g) to interact with the outer protocol π directly. However, a dishonest party may use this transcript to generate the random variables involved in its interaction with π.

We now provide a justification of our definition of sequentially composable security by establishing how it ensures security of a protocol under sequential composition with other protocols. This result, called the *sequential composition theorem*, is the main result of this section. This result will allow us to establish security of protocols that make oracle access to other functions, which in turn make oracle access to some other functions internally, and so on concatenated finitely many times. The utility of the definition of sequentially composable security given above is that the security of a nested protocol is automatically guaranteed, with a tractable degradation of the security parameter, by the security of subroutines constituting the entire protocol. The specific application of interest to us in this book is the case where we implement an ideal function h using oracle access to f and implement f using oracle access to g. We would like to analyze the security of an implementation of h using g that simply replaces f with its implementation using g. The following theorem provides such security guarantees, and has a much wider applicability.

THEOREM 15.12 (Sequential composition theorem) *Suppose that $\pi^f \diamond g$ sequentially composably ε_f-securely implements f using g and $\pi^h \circ f$ sequentially composably ε_h-securely implements h using f. Then, $\pi^h \circ \pi^f \diamond g$ sequentially composably $(\varepsilon_f + \varepsilon_h)$-securely implements h using g.*

Proof We shall prove that, for arbitrary protocol π invoking $\pi^h \circ \pi^f \diamond g$ as a subroutine, there exists a simulator σ such that

$$d_{\mathtt{var}}(\mathtt{V}[\pi \circ \pi^h \circ \pi^f \diamond g], \mathtt{V}[\pi \circ \sigma \circ h]) \le \varepsilon_f + \varepsilon_h. \tag{15.6}$$

To prove this, we note that the protocol $\pi \circ \pi^h$ can be regarded as an outer protocol for an ideal function f. Since, $\pi^f \diamond g$ sequentially composably ε_f-securely implements f, we have

$$d_{\mathtt{var}}(\mathtt{V}[\pi \circ \pi^h \circ \pi^f \diamond g], \mathtt{V}[\pi \circ \pi^h \circ \sigma_1 \circ f]) \le \varepsilon_f \tag{15.7}$$

for some simulator σ_1. When both the parties are honest, σ_1 is the trivial simulator that retains the inputs and the outputs for both parties. Thus, $\pi^h \circ \sigma_1 \circ f$ coincides with $\pi^h \circ f$. When one of the parties is dishonest, σ_1 in $\pi^h \circ \sigma_1 \circ f$ is, in essence, a local operation by the dishonest party. Therefore, for any attack by a dishonest party during the execution of $\pi \circ \pi^h \circ \sigma_1 \circ f$, the simulator σ_1 can be incorporated as part of the attack to form a new "composed attack".[9] Thus, for this particular attack and for the case when both parties are honest,

$$\mathtt{V}[\pi \circ \pi^h \circ f] \equiv \mathtt{V}[\pi \circ \pi^h \circ \sigma_1 \circ f],$$

whereby our assumption that $\pi^h \circ f$ sequentially composably ε_h-securely implements h yields

$$\begin{aligned} d_{\mathtt{var}}(\mathtt{V}[\pi \circ \pi^h \circ \sigma_1 \circ f], \mathtt{V}[\pi \circ \sigma \circ h]) &= d_{\mathtt{var}}(\mathtt{V}[\pi \circ \pi^h \circ f], \mathtt{V}[\pi \circ \sigma \circ h]) \\ &\le \varepsilon_h \end{aligned} \tag{15.8}$$

for some simulator σ.

By combining (15.7) and (15.8) with the triangular inequality, we obtain

$$\begin{aligned} &d_{\mathtt{var}}(\mathtt{V}[\pi \circ \pi^h \circ \pi^f \diamond g], \mathtt{V}[\pi \circ \sigma \circ h]) \\ &\le d_{\mathtt{var}}(\mathtt{V}[\pi \circ \pi^h \circ \pi^f \diamond g], \mathtt{V}[\pi \circ \pi^h \circ \sigma_1 \circ f]) + d_{\mathtt{var}}(\mathtt{V}[\pi \circ \pi^h \circ \sigma_1 \circ f], \mathtt{V}[\pi \circ \sigma \circ h]) \\ &\le \varepsilon_f + \varepsilon_h, \end{aligned}$$

which completes the proof. □

The basic idea in the proof above is very simple: the simulators arising from the security guarantees of the outer function are either trivial when both parties

[9] By a slight abuse of notation, we use π^h to describe the composed attack $\pi^h \circ \sigma_1$, which is a valid attack for π^h.

are honest or can be assumed to be a part of the attack for the inner function. This proof can be easily extended to multiple (nested) sequential compositions. In general, however, we may compose protocols in a more complicated way, invoking several functions in parallel. The proof above does not extend to such concurrent compositions. We will discuss this point in Section 15.8.

Hopefully, the reader is now convinced that our new enhanced notion of security is of conspicuous practical use. Interestingly, this seemingly more general definition of security is implied by our earlier notion of standalone security.

LEMMA 15.13 (Standalone security implies sequential composable security) *If a protocol $\pi^f \diamond g$ standalone ε-securely implements f, then it sequentially composably ε-securely implements f.*

Proof Consider an outer protocol π as depicted in Figure 15.9. The protocol π produces a transcript T_1 in the first phase, and the party $\mathcal{P}_i$ uses $(A_i, T_1^{(i)})$, $i = \{1, 2\}$, to generate the input X_i for $\pi^f \diamond g$. Denote these past observations of party i before engaging the oracle access to $\pi^f \diamond g$ by C_i, i.e., $C_i := (A_i, T_1^{(i)})$.

Since $\pi^f \diamond g$ standalone ε-securely implements f, when both parties are honest, the trivial simulator σ (which retains the inputs and outputs) produces the view $(X_1, X_2, \tilde{Z}_1, \tilde{Z}_2)$ with $\mathrm{P}_{\tilde{Z}_1\tilde{Z}_2|X_1X_2} \equiv W_f$ that satisfies

$$d_{\mathtt{var}}(\mathrm{P}_{Z_1Z_2X_1X_2|C_1=c_1,C_2=c_2}, \mathrm{P}_{\tilde{Z}_1\tilde{Z}_2X_1X_2|C_1=c_1,C_2=c_2}) \le \varepsilon$$

for every realization (c_1, c_2) of (C_1, C_2), which further yields

$$d_{\mathtt{var}}(\mathrm{P}_{Z_1Z_2X_1X_2C_1C_2}, \mathrm{P}_{\tilde{Z}_1\tilde{Z}_2X_1X_2C_1C_2}) \le \varepsilon. \tag{15.9}$$

The protocol π proceeds with the second phase to produce the transcript T_2, with each party using its previous observations (C_i, X_i, Z_i) as the input. Denote by $(\tilde{B}_1, \tilde{B}_2, \tilde{T}_2)$ the outputs and transcript obtained by using $\mathrm{P}_{\tilde{Z}_1\tilde{Z}_2X_1X_2C_1C_2}$ as the input in the second phase of the protocol, but with the original description of the protocol. Since the variational distance does not increase by applying the same channel (protocol), (15.9) implies

$$d_{\mathtt{var}}(\mathrm{P}_{B_1B_2T_2Z_1Z_2X_1X_2C_1C_2}, \mathrm{P}_{\tilde{B}_1\tilde{B}_2\tilde{T}_2\tilde{Z}_1\tilde{Z}_2X_1X_2C_1C_2}) \le \varepsilon,$$

whereby, when both parties are honest,

$$d_{\mathtt{var}}(\mathtt{V}[\pi \circ \pi^f \diamond g], \mathtt{V}[\pi \circ \sigma \circ f]) \le \varepsilon.$$

Similar arguments apply when one of the parties is dishonest. Specifically, consider the case when $\mathcal{P}_1$ is honest, but $\mathcal{P}_2$ is dishonest. Then, $\mathcal{P}_2$ uses an attack to produce the input X_2 as a stochastic function of C_2. Furthermore, denote by Π_f the transcript produced by $\pi^f \diamond g$. Note that since $\mathcal{P}_1$ is honest and will only engage with $\pi^f \diamond g$ using X_1 as the input, the Markov relation $(\Pi_f^{(2)}, Z_1, Z_2) \multimap\!\!- (X_1, X_2, C_2) \multimap\!\!- C_1$ holds. Thus, the distribution $\mathrm{P}_{\Pi_f^{(2)}Z_1Z_2|X_1X_2C_1C_2}$ used under attack equals $\mathrm{P}_{\Pi_f^{(2)}Z_1Z_2|X_1X_2C_2}$. In particular, the attack is determined by C_2 alone, which is available to $\mathcal{P}_2$. Therefore, since $\pi^f \diamond g$ standalone securely implements f, we can find a simulator σ_{C_2} that produces $(\tilde{Z}_1, \tilde{Z}_2, \tilde{\Pi}_f^{(2)})$ for the input distribution $\mathrm{P}_{X_1X_2|C_1=c_1,C_2=c_2}$ satisfying[10]

$$d_{\mathrm{var}}(\mathrm{P}_{\Pi_f^{(2)} Z_1Z_2X_1X_2|C_1=c_1,C_2=c_2}, \mathrm{P}_{\tilde{\Pi}_f^{(2)} \tilde{Z}_1\tilde{Z}_2X_1X_2|C_1=c_1,C_2=c_2}) \le \varepsilon$$

for every realization (c_1, c_2) of (C_1, C_2), which gives

$$d_{\mathrm{var}}(\mathrm{P}_{\Pi_f^{(2)} Z_1Z_2X_1X_2C_1C_2}, \mathrm{P}_{\tilde{\Pi}_f^{(2)} \tilde{Z}_1\tilde{Z}_2X_1X_2C_1C_2}) \le \varepsilon. \tag{15.10}$$

Next, we note that in the real world, a dishonest $\mathcal{P}_2$ may use $\Pi_f^{(2)}$ and (C_2, X_2, Z_2) to modify the distribution of the transcript T_2 and the output B_2. Let

$$\mathrm{P}_{\tilde{B}_1\tilde{B}_2\tilde{T}_2|\tilde{\Pi}_f\tilde{Z}_1\tilde{Z}_2X_1X_2C_1C_2} = \mathrm{P}_{B_1B_2T_2|\Pi_fZ_1Z_2X_1X_2C_1C_2},$$

namely, we assume that the same channel is used to produce the transcript T_2 and the output (B_1, B_2) in the real and the ideal world. Then, since the variational distance is not increased by the same channel, (15.10) implies

$$d_{\mathrm{var}}(\mathrm{P}_{B_1B_2\Pi_f^{(2)} Z_1Z_2X_1X_2C_1C_2}, \mathrm{P}_{\tilde{B}_1\tilde{B}_2\tilde{\Pi}_f^{(2)} \tilde{Z}_1\tilde{Z}_2X_1X_2C_1C_2}) \le \varepsilon.$$

Thus, we have shown

$$d_{\mathrm{var}}(\mathtt{V}[\pi \circ \pi^f \diamond g], \mathtt{V}[\pi \circ \sigma_{C_2} \circ f]) \le \varepsilon,$$

which completes the proof of sequentially composable security (the case when $\mathcal{P}_1$ is dishonest can be handled similarly). □

Thus, if we establish standalone security for a set of protocols, they can safely be used as replacements for the ideal functions they respectively implement, even in a nested implementation where these functions are called within each other. Of course, the security parameter will degrade linearly in the number of nested calls. Nonetheless, one can take this into account when designing the real protocols and provision for a fixed number of nested calls. However, as we will see in a later section, standalone security does not necessarily imply similar security under concurrent compositions discussed earlier.

15.6 Security for Multiphase Functions

Until now we have only considered functions that take the input and immediately reveal the output. However, functions such as bit commitment (BC) do not fall into this category. They have multiple phases, and after each phase the parties can provide a new input, interact, and obtain another output. Further, the inputs and outputs of any of the previous phases remain available in the subsequent phases. When considering sequentially composable security for multiphase functions, the outer protocol π is now allowed to manipulate the real

[10] There is an important technical point here. The simulator σ_{C_2} does not depend on the input distribution since we assumed in the definition of standalone security that the simulator is universal, i.e., it does not depend on the input distribution.

protocol in any manner it seeks. For instance, the outer protocol may terminate after any phase and declare the output. Sequentially composable security requires that even when interacting using any such protocol, the view generated by the adversary cannot differ much in the ideal world or the real world. In other words, the adversary will not be able to determine with reasonable probability if it is interacting with the real protocol or with the output of a simulator accessing the ideal function as an oracle.

It is easy to check that Theorem 15.12 extends to multiphase functions as well, when we impose the notion of security in Definition 15.11. However, to readily establish security of multiphase functions under Definition 15.11, we would like to develop the counterparts of Lemma 15.8 and Lemma 15.2. Recall that these results allow us to establish standalone security, which by Lemma 15.13 imply sequentially composable security as well. Unfortunately, it is unclear whether such simple conditions can be derived for multiphase functions too. Nonetheless, we provide some clarity below by discussing the information flow for multiphase functions, and an alternative sufficient condition of security; we leave the formal proof of sufficiency as an exercise for the reader. For simplicity, we only consider implementing protocols using oracle access to single-phase functions, although in Chapter 17 we will encounter protocols that use oracle access to multiphase functions.

Specifically, consider a k-phase function f with f_i denoting the effective function for phase i, $1 \le i \le k$. Each phase i can have its own input (X_{1i}, X_{2i}) and output (Z_{1i}, Z_{2i}). The outer protocol π allows the parties to interact between different phases, and use the interaction to choose the input for the next phase.[11] Figure 15.10 illustrates the interaction of π with f; here, T_i represents the part of the transcript of π exchanged between the $(i-1)$th and the ith phase. Once again, it is important to note that the outer protocol interacts with f or π^f in each phase only through the inputs and the outputs for all the phases up to that point. In particular, when interacting with π^f, it does not directly use the transcript of interaction with π^f. An honest party respects this information constraint, but a dishonest party may manipulate it.

Alternatively, we can view π as a generator of correlated random variables, which are then used to generate inputs for f_i, $1 \le i \le k$. Since the output of the protocol depends on the input and the part of the transcript seen by each party, it suffices to consider outer protocols π where the output comprises the entire transcript. Further, for each phase j, we need to consider protocols π that declare the output after phase j, and do not interact with f after the jth phase.

Specifically, the outer protocol π takes input (A_1, A_2) and declares an output (B_1, B_2). The protocol interacts with the function f in phases, with the transcript T_i exchanged before executing the phase i function f_i. Note that we allow oracle access to other functions in forming T_i, namely $\mathcal{P}_1$ and $\mathcal{P}_2$ may observe different parts $T_i^{(1)}$ and $T_i^{(2)}$ of the transcript respectively. For generality, we assume that

[11] This situation differs from π making oracle access to different functions since different phases of the same function are allowed to share the inputs as internal variables.

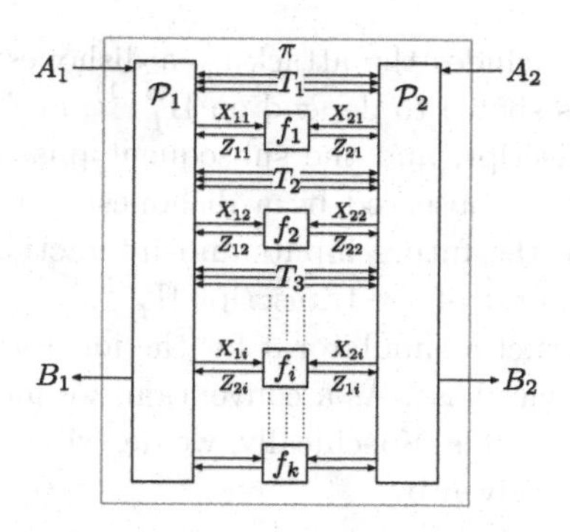

Figure 15.10 An outer protocol π interacts with the ideal protocol for a multiphase function f.

the output (B_1, B_2) is declared after phase j upon making access to f_j; a final round of interaction with transcript T_{j+1} is exchanged by π before declaring the output.

When executing this protocol using an implementation $\pi^f \diamond g$ of f, an honest party, say $\mathcal{P}_1$, will reveal only its input-output pair (A_1, B_1) as a part of its view. Further, an honest party $\mathcal{P}_1$ will not use the transcript $\Pi_{f,i}^{(1)}$ of the inner protocol $\pi_f \diamond g$ to interact with the outer protocol π. The outer protocol must interact with the honest party only through the inputs and outputs of $\pi_f \diamond g$ in various phases.

Note that a dishonest party, say $\mathcal{P}_2$, may additionally reveal its parts of the transcripts $T_i^{(2)}$, $1 \le i \le j+1$, as well as the transcript $\Pi_{f,i}^{(2)}$ it sees when it accesses the protocol $\pi^f \diamond g$ during the ith phase of f. The notion of sequentially composable security requires that an adversary can almost simulate this extra information using the same outer protocol π by interacting with the ideal function using simulators.

For brevity, we denote the input and output of the outer protocol π as $A = (A_1, A_2)$ and $B = (B_1, B_2)$, respectively, and those for phase i as $X_i = (X_{1i}, X_{2i})$ and $Z_i = (Z_{1i}, Z_{2i})$. Then, the channel (stochastic mapping) that describes the distribution of the transcript T_{i+1} and the input X_{i+1} for phase $i+1$ conditioned on the observations up to phase i can be explicitly written as

$$\begin{aligned} W_1 &= \mathrm{P}_{X_1 T_1 | A}, \\ W_i &:= \mathrm{P}_{X_i T_i | X^{i-1} Z^{i-1} \Pi_f^{i-1} T^{i-1} A}, \quad 2 \le i \le j, \\ W_{j+1} &:= \mathrm{P}_{B T_{j+1} | X^j Z^j \Pi_f{}^j T^j A}. \end{aligned}$$

Note that the description of these channels includes the attack by a dishonest party. With a slight abuse of terminology, we call $W = (W_1, \ldots, W_j, W_{j+1})$ the *attack channels*.

When interacting with the real protocol $\pi^f \diamond g$, for each phase i, the protocol $\pi^{f_i} \diamond g$ implementing f_i uses the input X_i generated by W_i and outputs $(\Pi_{f,i}, Z_i)$.

This, too, includes the attack by a dishonest party. We clarify once again that while W_i is shown to depend on $\Pi_f^{(i-1)}$, an honest party will not reveal its part of the transcript, and the subsequent phases can only depend on Z^{i-1} and a part of $\Pi_f^{(i-1)}$ observed by a dishonest party. However, a dishonest party can manipulate the future inputs and interaction using π by making them depend even on its part of the transcript Π_f^{i-1}.

To construct a simulator σ for the ideal world, it suffices to construct a simulator for each phase. As a convention, we indicate random variables in the ideal world with a tilde. Specifically, we describe the simulator used in phase i by the channel W_i^σ given by

$$W_i^\sigma = \mathrm{P}_{\tilde{\Pi}_{f,i}\tilde{Z}_i|\tilde{X}^i\tilde{Z}^{i-1}\tilde{T}^i\tilde{\Pi}_f^{i-1}A}.$$

Note that the simulator must retain the input and the output for the honest party, and does not reveal the transcript of the inner protocol. Thus, the channel above only produces the parts of $\tilde{\Pi}_{f,i}$ and $\tilde{Z}_i$ for the dishonest party as the output and depending only on the observations at the dishonest party. In doing so, the dishonest party is allowed to modify the incoming input (create a dummy input as in Lemma 15.8) before making access to the ideal function. We call W^σ *simulator channels* for the ideal world. The overall behavior of the parties in the ideal world is determined by the channels W and W^σ.

Finally, after interacting with π^f using the outer protocol π, a dishonest party can reveal any function of its observed random variables as a part of the view of π. In particular, the worst case corresponds to the revelation of all the observed random variables; it suffices to consider this worst case attack. Figure 15.11 illustrates the real world versus ideal world paradigm.

Using these observations, we can establish the following alternative definition of sequentially composable security. In fact, we only present it as a sufficient condition, though an equivalence can be established. We denote by $(\tilde{X}, \tilde{Z}, \tilde{T}, \tilde{\Pi}_f, \tilde{B})$ the ideal world counterparts for the random variables (X, Z, T, Π_f, B) from the real world. For brevity, given attack channels W and simulator channels W^σ, we denote by $W \circ \pi^f \diamond g$ and $W \circ W^\sigma \circ f$, respectively, the joint distributions of all the random variables in the real world and the ideal world. Throughout, we follow the convention that if a party is honest we will simply set its component of T_i and $\Pi_{f,i}$ as the constant random variable $\emptyset$.

LEMMA 15.14 (Sequentially composable security – alternative form) *For $\varepsilon \in [0, 1)$, a protocol $\pi^f \diamond g$ sequentially composably ε-securely implements a k-phase function f if, with at most one dishonest party, for every j and every choice of attack channels W and input distribution $\mathrm{P}_{A_1A_2}$, there exist simulator channels W^σ such that*

$$d_{\mathtt{var}}(\mathrm{P}_{W\circ\pi^f\diamond g}, \mathrm{P}_{W\circ W^\sigma\circ f}) \leq \varepsilon.$$

In a nutshell, the form of security established above is defined as the difference between the correlated random variables that can be generated in the

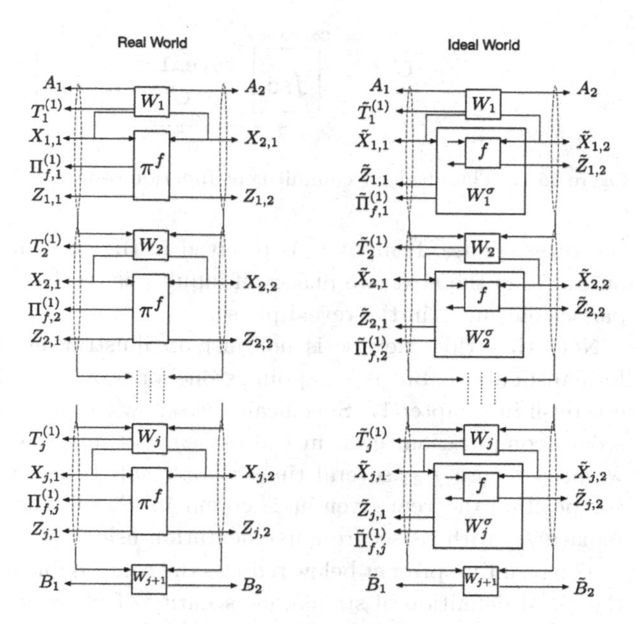

Figure 15.11 Random variables generated in the real world and the ideal world when $\mathcal{P}_1$ is dishonest. The inputs and transcript for π^f given by $\mathcal{P}_2$ in the real world are kept internal since $\mathcal{P}_2$ is honest.

real world and the ideal world. In both worlds, except when accessing f or its implementation $\pi^f \diamond g$, the parties are allowed to generate arbitrary correlations. This is the power of the notion of sequentially composable security.

We remark that the previous lemma is not necessarily easier to apply than the original definition. However, we present it merely as a clarification of the information flow, which can help in completing formal security analysis. Besides these observations, we do not have any general result to offer for simplifying the security definition for multiphase functions (such as Lemma 15.8 for single-phase functions). Instead, we illustrate how to apply this definition by considering sequentially composably secure implementation of bit commitment (BC) using ideal OT.

In Chapter 14, we exhibited a protocol that implements bit commitment using OT, but established a seemingly different form of security. Specifically, we established specific *correctness*, *hiding*, and *binding* conditions, which correspond to a notion of standalone security. Our main observation in this section is that these three conditions correspond to verifying the condition of Lemma 15.14 for the cases when both parties are honest, $\mathcal{P}_2$ is dishonest, and $\mathcal{P}_1$ is dishonest. However, we will see that the binding condition needs to be strengthened in comparison to the form we had in Chapter 14.

We have depicted our ideal bit commitment in Figure 15.12. As has been our convention, we allow an additional $\perp$ as input and output in the ideal world. We follow the convention that either party can input $\perp$ in the commit phase or

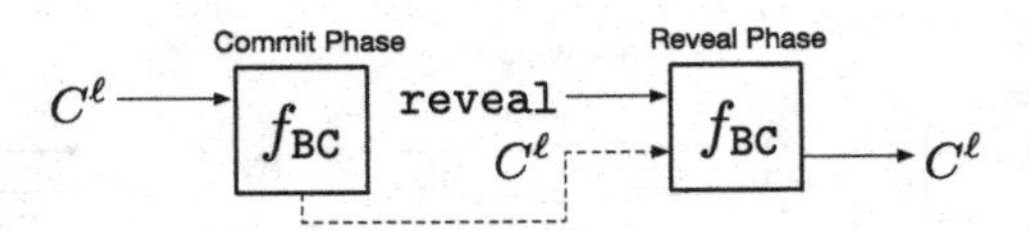

Figure 15.12 The ideal bit commitment function of length ℓ.

the reveal phase. If input $\perp$ is received in the commit phase, both parties will output $\perp$ in the next two phases; if input $\perp$ is received in the reveal phase, both parties output $\perp$ in the reveal phase.

Note that this exercise is not just an illustration of the application of the formalism above, but is a stepping stone for proving an important result that we establish in Chapter 17. Specifically, we show that g_{OT} is complete for two-party secure computation, even in the presence of an active adversary. In doing so, we need to use f_{BC} several times. Our result below, along with the sequential composition theorem given in Theorem 15.12, will allow us to show that we can replace f_{BC} with its secure implementation using g_{OT}.

The result we present below reduces the composable notion of security above to the usual definition of standalone security of bit commitment, which we define first. Recall that in the standalone mode the implementation $\pi_{\text{BC}} \diamond g$ takes an input and produces an output. In addition, internally, the parties generate a transcript Π_{BC}. The correctness and hiding conditions below are straightforward; the binding condition requires some thought. In particular, we need to keep in mind an attack where $\mathcal{P}_1$ does not commit to anything specific in the first step.

DEFINITION 15.15 (Standalone secure bit commitment) Consider a protocol $\pi_{\text{BC}} \diamond g$ that implements f_{BC}, and denote by Z the output of $\mathcal{P}_2$ in the reveal phase. The protocol $\pi_{\text{BC}} \diamond g$ standalone $(\varepsilon_s, \varepsilon_h, \varepsilon_b)$-securely implements the bit commitment using g if the following conditions are satisfied for every input distribution P_{C^ℓ}.

- *Correctness*: When both the parties are honest, we have
$$\Pr\big(Z \neq C^\ell\big) \leq \varepsilon_{\mathtt{s}}.$$
- *Hiding*: When $\mathcal{P}_2$ is dishonest and $\mathcal{P}_1$ is honest, for every attack by $\mathcal{P}_1$ there exists a distribution $\mathrm{Q}_{\Pi^{(2)}_{\text{BC},1}}$ that does not depend on the input distribution P_{C^ℓ} such that
$$d_{\mathtt{var}}\big(\mathrm{P}_{C^\ell \Pi^{(2)}_{\text{BC},1}}, \mathrm{P}_{C^\ell} \times \mathrm{Q}_{\Pi^{(2)}_{\text{BC},1}}\big) \leq \varepsilon_{\mathtt{h}}.$$
- *Binding*: When $\mathcal{P}_1$ is dishonest and $\mathcal{P}_2$ is honest, for every attack by $\mathcal{P}_1$, there exists a random variable $\hat{C}^\ell = \hat{C}^\ell(\Pi_{\text{BC},1})$ such that the output Z of $\mathcal{P}_2$ in the second phase satisfies[12]
$$\Pr\Big(Z = \hat{C}^\ell \text{ or } Z = \perp\Big) \geq 1 - \varepsilon_b.$$

[12] Namely, by looking at the transcript of both the parties in the commit phase, we can identify a $\hat{C}^\ell$ to which $\mathcal{P}_1$ essentially commits.

The following result is the counterpart of Lemma 15.13 for bit commitment; thanks to this result, it suffices to verify the set of concrete conditions in Definition 15.15 to guarantee sequential composable security.

LEMMA 15.16 *If a protocol $\pi_{\mathtt{BC}} \diamond g$ standalone $(\varepsilon_s, \varepsilon_h, \varepsilon_b)$-securely implements a bit commitment using g, then $\pi_{BC} \diamond g$ sequentially composably* $\max\{\varepsilon_s, \varepsilon_h, \varepsilon_b\}$-*securely implements the bit commitment.*

Proof The key idea of the proof is that the outer protocol can be subsumed as a part of the attack by the dishonest party, reducing the sequentially security requirement to a standalone form. In completing the proof of reduction, we explicitly use the structure of $f_{\mathtt{BC}}$, i.e., the proof argument cannot be directly applied for an other function f, but the result is valid for oracle access to an arbitrary g that is not necessarily the ideal OT. For simplicity, we will consider outer protocols that interact with all phases of functions, i.e., we set $j = k = 2$ in the general formalism described at the beginning of this section.

When both $\mathcal{P}_1$ and $\mathcal{P}_2$ are honest, since there is no output in the commit phase, the simulator channel W_1^σ need not output anything and is trivial. Thus, we need only focus on the reveal phase. In the reveal phase, the only possible input is `reveal`. Since the parties are honest, they will not reveal the internal transcript to the outer protocol. The outer protocol will simply provide input C^ℓ to the inner protocol in the commit phase and give the input `reveal` in the reveal phase. The only information received by the outer protocol is the output Z from $\mathcal{P}_2$ at the end of the reveal phase. Consider the trivial simulator channel W_2^σ whose output is Z. For this simulator, it can be verified that

$$\begin{aligned} d_{\mathtt{var}}(\mathrm{P}_{W \circ \pi^f \diamond g}, \mathrm{P}_{W \circ W^\sigma \circ f}) &\le \mathbb{E}\big[d_{\mathtt{var}}\left(\mathrm{P}_{Z|X_{1,1}=C^\ell, X_{2,1}=\mathtt{reveal}}, \mathbf{1}\big[Z = C^\ell\big]\right)\big] \\ &= \mathbb{E}\big[\Pr\big(Z \ne C^\ell \mid X_{1,1} = C^\ell, X_{2,1} = \mathtt{reveal}\big)\big], \end{aligned}$$

where the distribution $\mathrm{P}_{Z|X_{1,1}=C^\ell, X_{2,1}=\mathtt{reveal}}$ corresponds to the real protocol $\pi^f \circ g$ and $\mathbf{1}\big[Z = C^\ell\big]$ is the corresponding distribution for $f_{\mathtt{BC}}$, and the expectation is over the input distribution $\mathrm{P}_{C^\ell|AT_1}$ produced by the outer protocol. Thus, the condition of Lemma 15.14 can be seen to hold with $\varepsilon = \varepsilon_s$ by applying the soundness condition of standalone security to the input distribution $\mathrm{P}_{C^\ell|AT_1}$, for every realization of (A, T_1).

When $\mathcal{P}_1$ is honest but $\mathcal{P}_2$ is dishonest, since $\mathcal{P}_1$ is honest, it will only interact with the outer protocol using the inputs and the outputs of $\pi_{\mathtt{BC}}$. Further, $\mathcal{P}_1$ has no output in the commit phase, and it executes the reveal phase using the input `reveal`. Note that dishonest $\mathcal{P}_2$ can influence this input through the transcript T_2 of the outer protocol.[13] At the end of the reveal phase, $\mathcal{P}_2$ gets the input C^ℓ of $\mathcal{P}_1$ for the commit phase.

Driven by these observations, we use the following simulator channels. For any fixed realization (a, τ_1) of (A, T_1), consider the distribution $\mathrm{P}_{C^\ell|A=a, T_1=\tau_1}$

[13] Depending on the dishonest behavior of $\mathcal{P}_2$ in the outer protocol, $\mathcal{P}_1$ may not invoke the reveal phase of the inner protocol, i.e., the inner protocol is terminated at phase $j = 1$.

for the input. By the hiding condition, we can find a fixed distribution $Q_{\tilde{\Pi}^{(2)}_{\text{BC},1}}$ such that

$$d_{\text{var}}\big(P_{C^\ell \Pi^{(2)}_{\text{BC},1}|A=a,T_1=\tau_1}, P_{C^\ell|A=a,T_1=\tau_1} \times Q_{\tilde{\Pi}^{(2)}_{\text{BC},1}}\big) \le \varepsilon_{\text{h}} \tag{15.11}$$

for every (a, τ_1). Consider the simulator channel W_1^σ for the commit phase that interacts with an ideal implementation of f_{BC} in the commit phase and simply samples the transcript $\tilde{\Pi}^{(2)}_{\text{BC},1}$ from $Q_{\tilde{\Pi}^{(2)}_{\text{BC},1}}$ independently of every other random variable. For this simulator, denoting by W^2 the adversarial channels up to the input of the reveal phase, it is easy to see that

$$\begin{aligned} d_{\text{var}}\big(P_{W^2\circ\pi^f\diamond g}, P_{W^2\circ W_1^\sigma\circ f}\big) &= \mathbb{E}\Big[d_{\text{var}}\big(P_{\Pi^{(2)}_{\text{BC},1}C^\ell|AT_1}, P_{\tilde{\Pi}^{(2)}_{\text{BC},1}C^\ell|AT_1}\big)\Big] \\ &\le \varepsilon_{\text{h}}, \end{aligned}$$

where we used (15.11) for the final bound. For constructing the simulator W_2^σ for the reveal phase, we note that, in the ideal world, $\mathcal{P}_2$ will obtain C^ℓ from the ideal bit commitment. Consider a simple simulator that samples $\tilde{\Pi}^{(2)}_{\text{BC},2}$ from the conditional distribution $P_{\Pi^{(2)}_{\text{BC},2}|\Pi^{(2)}_{\text{BC},1}AT_1C^\ell T_2}$ using the observations seen by $\mathcal{P}_2$ earlier. For this simulator, it is easy to see by the chain rule for total variation distance that

$$\begin{aligned} &d_{\text{var}}\big(P_{W\circ\pi^f\diamond g}, P_{W\circ W^\sigma\circ f}\big) \\ &\quad\le d_{\text{var}}\big(P_{W^2\circ\pi^f\diamond g}, P_{W^2\circ W_1^\sigma\circ f}\big) \\ &\qquad + \mathbb{E}\Big[d_{\text{var}}\big(P_{\Pi^{(2)}_{\text{BC},2}|\Pi^{(2)}_{\text{BC},1}C^\ell AT_1T_2}, P_{\tilde{\Pi}^{(2)}_{\text{BC},2}|\tilde{\Pi}^{(2)}_{\text{BC},1}\tilde{C}^\ell\tilde{A}\tilde{T}_1\tilde{T}_2}\big)\Big] \\ &\quad\le \varepsilon_{\text{h}}, \end{aligned}$$

where the expectation in the first inequality is over $P_{\Pi^{(2)}_{\text{BC},1}C^\ell AT_1T_2}$, the distribution from the real world,[14] and the second inequality holds since $P_{\Pi^{(2)}_{\text{BC},2}|\Pi^{(2)}_{\text{BC},1}C^\ell AT_1T_2}$ and $P_{\tilde{\Pi}^{(2)}_{\text{BC},2}|\tilde{\Pi}^{(2)}_{\text{BC},1}\tilde{C}^\ell\tilde{A}\tilde{T}_1\tilde{T}_2}$ coincide with probability 1 by our construction. We have established the condition of Lemma 15.14 with $\varepsilon = \varepsilon_h$.

Finally, we consider the case when $\mathcal{P}_2$ is honest but $\mathcal{P}_1$ is dishonest. The simulator W_1^σ for the commit phase observes $\Pi_{\text{BC},1}$ and outputs $\hat{C}^\ell$ (guaranteed by the assumption), which in turn is committed as input to f_{BC}. Note that this requires $\mathcal{P}_1$ to observe both $\Pi^{(1)}_{\text{BC},1}$ and $\Pi^{(2)}_{\text{BC},1}$. Similarly, to construct the simulator for the reveal phase, we need to capture the case where $\mathcal{P}_1$ produces $\Pi_{\text{BC},2}$ to make $\mathcal{P}_2$ generate an output different from $\hat{C}^\ell$ in the reveal phase, but $\mathcal{P}_2$ detects the attack and declares $\perp$. We need to simulate this attack by enabling $\mathcal{P}_1$ to input $\perp$ in the reveal phase, which also requires $\mathcal{P}_1$ to have access to some information seen by $\mathcal{P}_2$ in the real world.

We circumvent this difficulty by running a local copy of the protocol at $\mathcal{P}_1$. Note that since there are no inputs for $\mathcal{P}_2$ and no output in the commit phase,

[14] The overall set of random variables has observations from the real world as well as the ideal world; this view is said to be of the *hybrid world*.

an honest $\mathcal{P}_2$ does not interact with the outer protocol before producing the final output. Thus, the internal transcript $\Pi^{(2)}_{\text{BC}}$ does not depend at all on (A, T_1, T_2), when conditioned on $\Pi^{(1)}_{\text{BC}}$. Therefore, in our simulator, $\mathcal{P}_1$ can indeed run a local copy of the inner protocol $\pi_{\text{BC}} \circ g$ without requiring the knowledge of A_2 and $T_2^{(2)}$. $\mathcal{P}_1$ can then use the local copy to generate $\Pi_{\text{BC},1}$ first to obtain $\hat{C}^\ell$ for W_1^σ, and then $\mathcal{P}_1$ continues to generate the output Z for $\mathcal{P}_2$ in the real world. If this output is $\hat{C}^\ell$, the simulator W_2^σ generates the input **reveal** for the reveal phase of f_{BC}; otherwise it generates the input $\perp$.

For this simulator, since there is no output for $\mathcal{P}_2$ in the commit phase, it is easy to check that

$$d_{\text{var}}(\mathrm{P}_{W \circ \pi^f \circ g}, \mathrm{P}_{W \circ W^\sigma \circ f}) \le \Pr\Big(Z \ne \hat{C}^\ell \text{ and } Z \ne \perp\Big),$$

where Z denotes the output of $\mathcal{P}_2$ in the real protocol. The proof is completed by noticing that the right-hand side above is less than ε_b by our condition. □

In Chapter 14, we presented Protocol 14.5 which implements bit commitment using randomized OT. Recall that the notion of security for bit commitment, specifically the binding condition, used there was slightly different. The main reason for presenting that different definition earlier is that we find it heuristically more appealing. However, that definition may not yield sequentially composable security. Interestingly, we show that Protocol 14.5 satisfies the new binding condition in Definition 15.15 as well.

THEOREM 15.17 (Composably secure bit commitment from OT) *Protocol 14.5 sequentially composably* $\max\{\varepsilon_s, \varepsilon_h, \varepsilon_b\}$*-securely implements a bit commitment with* $\varepsilon_{\text{s}} = 0$, $\varepsilon_{\text{h}} = \xi$, $\varepsilon_{\text{b}} = 2^{-\delta n + 2} + 2^{-\nu + 2nh(\delta) + 1}$, *and*

$$m = \lfloor n - m - 2\nu - \log(1/4\xi^2) \rfloor.$$

Proof By Lemma 15.16, it suffices to establish that Protocol 14.5 standalone $(\varepsilon_s, \varepsilon_h, \varepsilon_b)$-securely implements the bit commitment. The correctness and hiding conditions in Definition 15.15 coincide with those in Definition 14.1 and follow from Theorem 14.3. The binding condition does not directly follow from Theorem 14.3, but can be obtained readily from its proof.

Indeed, an attack by $\mathcal{P}_1$ will entail generating the transcript (Π_0, Π_1, Π_2) adversarially in the commit phase, and then revealing $(\tilde{K}_0^n, \tilde{K}_1^n, \tilde{C}^\ell)$ in the reveal phase. The transcript $\Pi_{\text{BC},1}$ is then given by $(K_0^n, K_1^n, B^n, \Pi_0, \Pi_1, \Pi_2, G_0, G_1, F)$. Recall that hash functions G_0, G_1 are generated by $\mathcal{P}_2$ and F is generated by $\mathcal{P}_1$; in particular F need not be sampled from the prescribed distribution when $\mathcal{P}_1$ is dishonest. We consider realizations τ of $\Pi_{\text{BC},1}$ such that there exist $\hat{K}_0^n$ and $\hat{K}_1^n$ such that

$$G_0(\hat{K}_0^n) = \Pi_0, \tag{15.12}$$

$$G_1(\hat{K}_1^n) = \Pi_1, \tag{15.13}$$

$$\hat{K}_{B_i} = K_{B_i}, \; 1 \le i \le n. \tag{15.14}$$

Note that for any transcript τ for which we cannot find such $\hat{K}_0^n$ and $\hat{K}_1^n$, we must have $\Pr(Z = \perp \mid \Pi_{\text{BC},1} = \tau) = 1$. Set $\hat{C}^\ell := \Pi_2 + F(\hat{K}_0^n + \hat{K}_1^n)$. Then, when $Z \neq \perp$, we must have

$$G_0(\hat{K}_0^n) = \Pi_0, \tag{15.15}$$

$$G_1(\hat{K}_1^n) = \Pi_1, \tag{15.16}$$

$$\hat{K}_{B_i} = K_{B_i},\ 1 \leq i \leq n, \tag{15.17}$$

$$\hat{C}^\ell = \Pi_2 + F(\hat{K}_0^n + \hat{K}_1^n). \tag{15.18}$$

Therefore, $\{Z \neq \perp\} \cap \{Z \neq \hat{C}^\ell\}$ can only happen when $\tilde{C}^\ell$ is accepted, namely when (15.12)–(15.14) and (15.15)–(15.18) hold simultaneously. By the definition of $\hat{C}^\ell$, this can happen only if $\tilde{K}_0^n \neq \hat{K}_0^n$ or $\tilde{K}_1^n \neq \hat{K}_1^n$ and the events $\mathcal{E}_b$, $b = 0, 1$, hold, where $\mathcal{E}_b$ is the event such that

$$\begin{aligned}
G_b(\tilde{K}_b^n) &= \Pi_b, \\
G_b(\hat{K}_b^n) &= \Pi_b, \\
\tilde{K}_{b,i} &= K_i, \quad \forall 1 \leq i \leq n \text{ with } B_i = b, \\
\hat{K}_{b,i} &= K_i, \quad \forall 1 \leq i \leq n \text{ with } B_i = b.
\end{aligned}$$

Upon combining the observations above, we have

$$\Pr\Big(Z \neq \hat{C}^\ell \text{ and } Z \neq \perp\Big) \leq \Pr\Big(\{\tilde{K}_0^n \neq \hat{K}_0^n\} \wedge \mathcal{E}_0\Big) + \Pr\Big(\{\tilde{K}_1^n \neq \hat{K}_1^n\} \wedge \mathcal{E}_1\Big),$$

where the right-hand side appeared in (14.3) and was shown to be bounded by ε_b in the proof of Theorem 14.3. This completes the proof. □

In essence, the proof above says that given the transcript of the commit phase, $\mathcal{P}_1$ is inherently committing to $\hat{K}_0^n$ and $\hat{K}_1^n$ such that $\hat{K}_{B_i} = K_{B_i}$ for every $1 \leq i \leq n$. This observation implies the binding condition given in this chapter as well as Chapter 14.

With this illustration, a reader is hopefully convinced that there is flexibility in defining standalone security. But one should prefer a definition that can be shown to imply composable security rather than a more heuristically appealing one.

15.7 Complete Fairness and Relaxation

In previous sections, we have defined an implemented protocol to be secure if the dishonest party can only conduct misbehaviors that are admissible in the ideal model. In other words, an implemented protocol is secure if it prevents the dishonest party from conducting any nonadmissible misbehavior. However, there is one misbehavior that is difficult to prevent in any implemented protocol.

In the ideal function, the parties receive their outputs simultaneously from a trusted party. Since the function is implemented by interactive communication in the real model, the dishonest party may abort the protocol immediately after obtaining the output. Such an unfair abort is difficult to prevent. In fact, there exist functions such that complete fairness cannot be attained. For this reason, we introduce a compromised security. In the ideal model, when the ideal function is invoked, it first provides the output to $\mathcal{P}_1$; then $\mathcal{P}_1$ has an option to decide whether the ideal function should provide the output to $\mathcal{P}_2$.[15] This modification in the ideal model enables us to simulate unfair abort of an implemented protocol. In other words, we relax security so that the unfair abort is an admissible misbehavior. Even though this modification is irrelevant for functions such that only one of the parties receives output, for general functions, it seems to be a big compromise. However, since the unfair abort is recognized by the honest party, in typical applications in practice, there are some other incentives (such as a trade will be repeated in future) for the dishonest party not to conduct the unfair abort. In fact, there are interesting research threads exploring game-theoretic formulations to avoid unfair aborts.

15.8 Concurrent Composition

In this section, we discuss the differences between sequential composition and concurrent composition. Different compositions of protocols can be categorized as follows.[16]

1. *Sequential composition*: When two protocols $\pi^{f_1} \diamond g_1$ and $\pi^{f_2} \diamond g_2$ are used in another protocol in such a manner that $\pi^{f_2} \diamond g_2$ is invoked after the invocation of $\pi^{f_1} \diamond g_1$ has terminated; see Figure 15.13 (left).
2. *Parallel self-composition*: When a protocol $\pi^f \diamond g$ is invoked twice in another protocol and the same steps in the two invocations of $\pi^f \diamond g$ are conducted one after the other; see Figure 15.13 (right).
3. *Concurrent self-composition*: Two instances of protocol $\pi^f \diamond g$ are invoked in a similar manner to parallel self-composition, but the same step for the two invocations need not alternate.
4. *Concurrent general composition*: Two protocols $\pi^{f_1} \diamond g_1$ and $\pi^{f_2} \diamond g_2$ are invoked by another protocol, and the order of each step of the two invocations is arbitrary. Moreover, steps of the inner protocols may be arbitrarily nested with steps of the outer protocols.

As we have seen in the previous section, standalone security implies security under sequential composition. As is illustrated in the following example, the same implication does not hold even for parallel composition.

[15] We could have provided this option to $\mathcal{P}_2$ instead of $\mathcal{P}_1$. This choice is just for convenience of presentation.

[16] For simplicity of presentation, we only consider composition of two protocols, but composition of multiple protocols can be treated similarly.

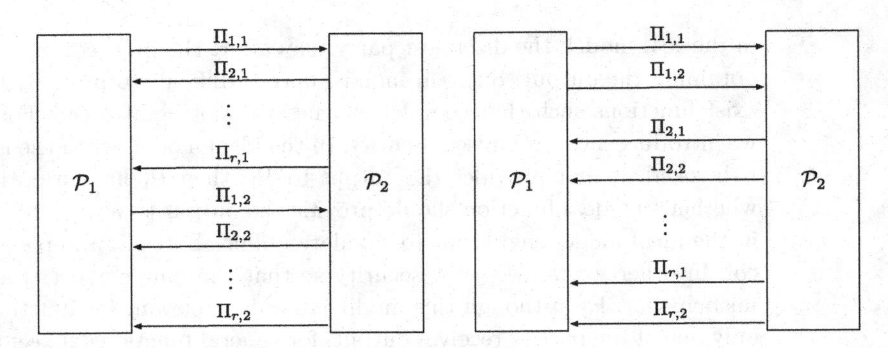

Figure 15.13 Descriptions of sequential composition (left) and parallel composition (right) of two copies π_1 and π_2 of the same protocol; the transcript of π_i is denoted by $(\Pi_{1,i}, \Pi_{2,i}, \ldots, \Pi_{r,i})$.

Protocol 15.7 Implementation π^f of function f in (15.19)

Input: (X_1, X_2)

Output: (Z_1, Z_2)

1. $\mathcal{P}_1$ sends $\Pi_1 = 1$ if $X_1 = 1$, and the parties terminate the protocol with output 1; otherwise, $\mathcal{P}_1$ sends $\Pi_1 = 0$.
2. $\mathcal{P}_2$ sends $\Pi_2 = 1$ if $X_2 = 2$, and the parties terminate the protocol with output 2; otherwise, $\mathcal{P}_2$ sends $\Pi_2 = 0$, and the parties terminate the protocol with output 3.

Consider a function $f\colon \{1,3\} \times \{2,4\} \to \{1,2,3\}$ given by

$x_1 \backslash x_2$	2	4
1	1	1
3	2	3

(15.19)

In other words, this function computes $f(x_1, x_2) = \min\{x_1, x_2\}$. Since this function is decomposable, it can be securely computed without any resource for a passive adversary. In fact, Protocol 15.7 is secure even when an active adversary is present.

LEMMA 15.18 (Standalone security of Protocol 15.7) *Protocol 15.7 is a standalone secure implementation of function f in* (15.19).

Proof The protocol is apparently correct. We shall prove security for $\mathcal{P}_1$; security for $\mathcal{P}_2$ is left as an exercise. For a dishonest $\mathcal{P}_2$, an execution of the real protocol π^f induces distribution $\mathrm{P}_{Z_1 Z_2 \Pi | X_1 X_2}$ of the parties' outputs and transcript given the parties' inputs. Note that $\mathrm{P}_{Z_1 Z_2 | X_1 X_2}$ can differ from the ideal function since $\mathcal{P}_2$ is malicious. Using this distribution, we can construct the simulator described in Protocol 15.8. When $\mathcal{P}_1$'s input is $X_1 = 1$, $\mathcal{P}_1$'s output in both the real protocol and the simulated protocol are $Z_1 = 1$. When $\mathcal{P}_1$'s input is $X_1 = 3$, from the construction of the dummy input X_2', we have $Z_1 = \tilde{Z}_1$.

Protocol 15.8 Simulator σ with dishonest $\mathcal{P}_2$ in Lemma 15.18

Input: (X_1, X_2)

Output: $((X_1, Z_1), (X_2, Z_2, \Pi)$

1. $\mathcal{P}_2$ samples $\tilde{Z}_1 \sim \mathrm{P}_{Z_1|X_1X_2}(\cdot|3, X_2)$, and sets a dummy input

$$X_2' = \begin{cases} 2, & \text{if } \tilde{Z}_1 = 2, \\ 4, & \text{if } \tilde{Z}_1 = 3. \end{cases}$$

2. The parties invoke f with inputs (X_1, X_2') to obtain (Z_1, Z_2').
3. $\mathcal{P}_1$ outputs (X_1, Z_1).
4. $\mathcal{P}_2$ sets

$$\tilde{X}_1 = \begin{cases} Z_2', & \text{if } Z_2' = 1, \\ 3, & \text{otherwise.} \end{cases}$$

5. $\mathcal{P}_2$ samples $(Z_2, \Pi) \sim \mathrm{P}_{Z_2\Pi|X_1X_2}(\cdot, \cdot|\tilde{X}_1, X_2)$.
6. $\mathcal{P}_2$ outputs (X_2, Z_2, Π).

Thus, the distribution of Z_1 in the simulated protocol is the same as that in the real protocol. Furthermore, we can also verify that $\tilde{X}_1 = X_1$. Thus, (Z_2, Π) sampled by $\mathcal{P}_2$ in the simulated protocol has the same distribution as that in the real protocol. Thus, the distribution $\mathrm{P}_{Z_1Z_2\Pi|X_1X_2}$ can be simulated by interacting with the ideal function, whereby the protocol is perfectly standalone secure. □

While the protocol above is secure when we want to compute f once, interestingly, two parallel invocations of Protocol 15.7 to compute two copies of f with input $(X_{1,1}, X_{2,1})$ and $(X_{1,2}, X_{2,2})$ is not secure.

LEMMA 15.19 (Vulnerability of Protocol 15.7 for parallel composition) *Two parallel invocation of Protocol 15.7 is not secure.*

Proof Note that each step of the two invocations is executed alternately as in Figure 15.13 (right). Consider the following attack by a dishonest $\mathcal{P}_2$. In the first step, honest $\mathcal{P}_1$ sends $\Pi_{1,1}$ and $\Pi_{1,2}$ decided from inputs $X_{1,1}$ and $X_{1,2}$, respectively. Upon receiving $(\Pi_{1,1}, \Pi_{1,2})$, $\mathcal{P}_2$ sends

$$(\Pi_{2,1}, \Pi_{2,2}) = \begin{cases} (\emptyset, \emptyset) & \text{if } (\Pi_{1,1}, \Pi_{1,2}) = (1, 1), \\ (\emptyset, 1) & \text{if } (\Pi_{1,1}, \Pi_{1,2}) = (1, 0), \\ (1, \emptyset) & \text{if } (\Pi_{1,1}, \Pi_{1,2}) = (0, 1), \\ (0, 0) & \text{if } (\Pi_{1,1}, \Pi_{1,2}) = (0, 0), \end{cases}$$

where $\emptyset$ indicates that $\mathcal{P}_2$ sends nothing. Then, the correspondence between $\mathcal{P}_1$'s input and output in the real protocol is given by

$$
\begin{array}{cc}
(X_{1,1}, X_{1,2}) & (Z_{1,1}, Z_{1,2}) \\
\hline
(1,1) & (1,1) \\
(1,3) & (1,2) \\
(3,1) & (2,1) \\
(3,3) & (3,3)
\end{array}
\tag{15.20}
$$

The reader can verify that there is no value of input $(X_{2,1}, X_{2,2})$ for which the output above is consistent for all values of the other part of the input. Thus, heuristically it seems that the protocol is not secure. To see this formally, we show below that we cannot find a simulator that can satisfy security conditions.

Consider a simulator in the ideal model. In the process of simulation, the parties invoke two instances of f (not necessarily simultaneously) with $\mathcal{P}_1$'s inputs $(X_{1,1}, X_{1,2})$ and $\mathcal{P}_2$'s dummy inputs $(X'_{2,1}, X'_{2,2})$. The correspondence between $\mathcal{P}_1$'s outputs $(Z_{1,1}, Z_{1,2})$ against the parties' inputs $((X_{1,1}, X_{1,2}), (X'_{2,1}, X'_{2,2}))$ is given as follows:

$$
\begin{array}{c|cccc}
(X_{1,1}, X_{1,2})\backslash(X'_{2,1}, X'_{2,2}) & (2,2) & (2,4) & (4,2) & (4,4) \\
\hline
(1,1) & (1,1) & (1,1) & (1,1) & (1,1) \\
(1,3) & (1,2) & (1,3)_* & (1,2) & (1,3)_* \\
(3,1) & (2,1) & (2,1) & (3,1)_* & (3,1)_* \\
(3,3) & (2,2)_* & (2,3)_* & (3,2)_* & (3,3)
\end{array}
\tag{15.21}
$$

Note that the outputs with $*$ in (15.21) disagree with the outputs in the real protocol given by (15.20). Let $J \in \{1, 2\}$ be the random variable describing the index such that $f(X_{1,J}, X'_{2,J})$ is invoked first. Note that J and $X'_{2,J}$ must be chosen independent of $(X_{1,1}, X_{1,2})$.[17] Suppose that $(X_{1,1}, X_{1,2})$ is uniformly distributed on $\{1, 3\}^2$. Then, for any choice of J, $X'_{2,J}$, and $X_{2,3-J}$, we can verify from the table in (15.21) that the outputs with $*$ occur with probability at least 1/4, which means that there exists no simulator that (perfectly) emulates the view in the real protocol. □

The crux of the vulnerability identified in the proof of Lemma 15.19 is that $\mathcal{P}_2$'s messages $\Pi_{2,1}$ and $\Pi_{2,2}$ can depend on both $(\Pi_{1,1}, \Pi_{1,2})$; in the sequential composition, $\Pi_{2,2}$ can depend on $(\Pi_{1,1}, \Pi_{1,2})$, but $\Pi_{2,1}$ can only depend on $\Pi_{1,1}$ (see also Problem 15.4).

15.9 Oblivious Transfer from Correlated Randomness: Active Adversary

In this section, we revisit the problem of constructing string oblivious transfer from correlated randomness; in contrast to the treatment in Chapter 13, here, we consider security against an active adversary. In this section, for consistency with Chapter 13, a resource shared between the parties is denoted by (X, Y).

[17] In the ideal world, there is no communication beyond the execution of the two copies of the ideal function; $X_{2,3-J}$ may depend on $f(X_{1,J}, X'_{2,J})$.

For $\mathcal{P}_1$'s input $(K_0, K_1) \in \{0,1\}^l \times \{0,1\}^l$ and $\mathcal{P}_2$'s input $B \in \{0,1\}$, an OT protocol using resource (X, Y) is formulated in the same manner as in Section 13.2. More specifically, $\mathcal{P}_1$ and $\mathcal{P}_2$ run an interactive protocol with transcript Π, and $\mathcal{P}_2$ outputs $\hat{K}$ as an estimate of K_B. Here, the protocol may be aborted, and the output $\hat{K}$ may equal the abort symbol $\perp$ in addition to a value in $\{0,1\}^l$. The security requirements in (13.1)–(13.3) need to be modified as well. In the case of string OT, the requirements in Lemma 15.8 can be translated as follows.

DEFINITION 15.20 (Security of OT for an active adversary) A protocol $\pi_{\mathtt{OT}}$ is an ε-secure implementation $f_{\mathtt{OT}}$ of length l if the following conditions are satisfied.

- *Correctness*: When both the parties are honest, then $\Pr(\hat{K} \neq K_B) \leq \varepsilon$.
- *Security for* $\mathcal{P}_1$: When $\mathcal{P}_1$ is honest, for every attack by $\mathcal{P}_2$ and every input distribution $\mathrm{P}_{K_0K_1B}$, there exists a joint distribution $\mathrm{P}_{K_0K_1BB'K_{B'}\hat{K}Y\Pi}$, with dummy input and output $(B', K_{B'})$ for $\mathcal{P}_2$ and $\mathrm{P}_{K_0K_1B\hat{K}Y\Pi}$ denoting the distribution of the input, the output, $\mathcal{P}_2$'s part of the correlated resource, and the transcript under the attack, such that[18]
  1. $d_{\mathtt{var}}(\mathrm{P}_{K_{B'}K_0K_1B}, \mathrm{P}_{B'|B}\mathrm{P}_{K_0K_1B}) \leq \varepsilon$,
  2. $d_{\mathtt{var}}(\mathrm{P}_{\hat{K}Y\Pi K_0K_1BB'K_{B'}}, \mathrm{P}_{\hat{K}Y\Pi|BB'K_{B'}}\mathrm{P}_{K_0K_1BB'K_{B'}}) \leq \varepsilon$,

  where $K_{B'} = \perp$ if $B' = \perp$.
- *Security for* $\mathcal{P}_2$: When $\mathcal{P}_2$ is honest, for every attack by $\mathcal{P}_1$ and every input distribution $\mathrm{P}_{K_0K_1B}$, there exists a joint distribution $\mathrm{P}_{K_0K_1K_0'K_1'B\hat{K}X\Pi}$, with dummy input (K_0', K_1') for $\mathcal{P}_1$ and $\mathrm{P}_{K_0K_1B\hat{K}X\Pi}$ denoting the distribution of the input, the output, $\mathcal{P}_1$'s part of the correlated resource, and the transcript under the attack, such that
  1. $d_{\mathtt{var}}(\mathrm{P}_{K_0'K_1'K_0K_1B}, \mathrm{P}_{K_0'K_1'|K_0K_1}\mathrm{P}_{K_0K_1B}) \leq \varepsilon$,
  2. $\Pr(\hat{K} \neq K_B') \leq \varepsilon$,
  3. $d_{\mathtt{var}}(\mathrm{P}_{X\Pi K_0K_1K_0'K_1'B\hat{K}}, \mathrm{P}_{X\Pi|K_0K_1K_0'K_1'}\mathrm{P}_{K_0K_1K_0'K_1'B\hat{K}}) \leq \varepsilon$,

  where $K_B' = \perp$ if $(K_0', K_1') = \perp$.

For simplicity of exposition, we focus on the construction of string OT from erasure correlation with erasure probability $p \geq 1/2$.[19] In Protocol 13.2, since $\mathcal{P}_2$'s transcript Π_1 is such that (X, Π_1) is independent of B, $\mathcal{P}_2$'s input does not leak to $\mathcal{P}_1$ even if $\mathcal{P}_1$ deviates from the protocol procedures. Thus, if we are only working against an actively dishonest $\mathcal{P}_1$, there is no need to modify Protocol 13.2; the only change we need is security analysis, i.e., in the construction of the simulator.

On the other hand, $\mathcal{P}_1$'s information may leak to $\mathcal{P}_2$ if $\mathcal{P}_2$ deviates from the procedures in Protocol 13.2. Specifically, instead of constructing $(\mathcal{I}_0, \mathcal{I}_1)$ as indicated by the protocol, $\mathcal{P}_2$ may include a half of good indices, i.e., indices with $Y_i \neq \mathtt{e}$, into $\mathcal{I}_0$ and the other half of good indices into $\mathcal{I}_1$ (see Figure 15.14); then, $\mathcal{P}_2$ can partially learn both the strings K_0 and K_1. As we discussed in Section 15.2, this is a nonadmissible attack, and must be prevented.

[18] Since $\mathcal{P}_1$ has no output, we omit the second condition in Lemma 15.8.
[19] The case with $p < 1/2$ requires additional modification; see Problem 15.10.

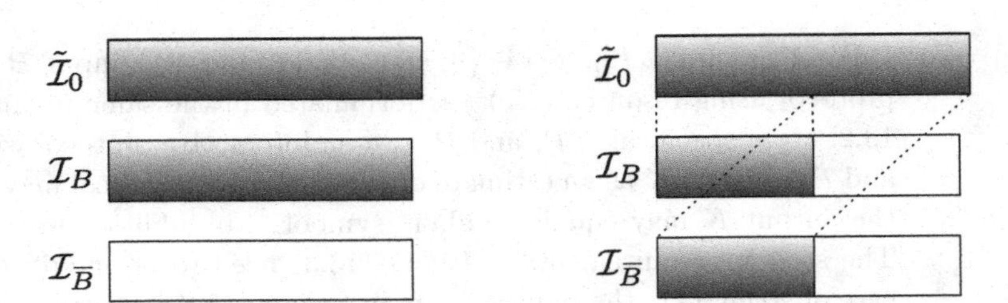

Figure 15.14 The descriptions of a possible attack by $\mathcal{P}_2$ in Protocol 13.2. $\tilde{\mathcal{I}}_0$ is the set of good indices, i.e., indices with $Y_i \neq \mathtt{e}$. $\mathcal{P}_2$ is supposed to set $(\mathcal{I}_0, \mathcal{I}_1)$ so that $\mathcal{I}_B$ consists of only good indices and $\mathcal{I}_{\bar{B}}$ consists of only bad indices (left); when $\mathcal{P}_2$ is actively dishonest, $\mathcal{P}_2$ may distribute good indices to both $\mathcal{I}_B$ and $\mathcal{I}_{\bar{B}}$ (right).

The problem here is that $\mathcal{P}_2$ may distribute the "budget" of good indices into $\mathcal{I}_0$ and $\mathcal{I}_1$ so that $\mathcal{P}_2$ can partially learn both the strings. There are two approaches to tackle this problem. The first approach is to let $\mathcal{P}_2$ distribute the budget as it wants, and prevent the leakage by privacy amplification. Since the budget of good indices is limited by the erasure probability, at least one of $\mathcal{I}_0$ or $\mathcal{I}_1$ must contain a sufficient number (roughly half the length of $|\mathcal{I}_0| = |\mathcal{I}_1|$) of bad indices, i.e., indices with $Y_i = \mathtt{e}$. Thus, a secret key that is unknown to $\mathcal{P}_2$ can be extracted by the privacy amplification. In this section, we will explain this approach in detail.

The second approach is to test whether one of $\mathcal{I}_0$ or $\mathcal{I}_1$ consists of (almost) only good indices. Again, since the budget of good indices is limited by the erasure probability, if the test is passed, then one of $\mathcal{I}_0$ or $\mathcal{I}_1$ must consist of (almost) only bad indices, which prevents the leakage of one of the strings to $\mathcal{P}_2$. In order to eliminate a small amount of leakage, we need to use privacy amplification in this approach as well. However, the length we sacrifice by the privacy amplification is negligible compared to the first approach. This test requires a sophisticated cryptographic primitive, termed the *interactive hashing*, which is beyond the scope of the book; interested readers are encouraged to see the references in Section 15.10.

As we discussed above, we need to modify Protocol 13.2 so that the protocol is secure even if $\mathcal{P}_2$ distributes good indices into $\mathcal{I}_0$ and $\mathcal{I}_1$. One approach is to add the privacy amplification procedure. Specifically, $\mathcal{P}_2$ picks $\mathcal{I}_0$ and $\mathcal{I}_1$ of length m, and $\mathcal{P}_1$ extracts l-bit sequences from $X_{\mathcal{I}_0}$ and $X_{\mathcal{I}_1}$. The modified protocol is described in Protocol 15.9, where the channel $P_{V|Y}$ is defined in the same manner as for Protocol 13.2.

THEOREM 15.21 *For $1/2 \le p < 1$ and fixed $0 < \nu < (1-p)/3$, let $m = (1-p-\nu)n$ and*

$$l = \left\lfloor \frac{(1-p-3\nu)n}{2} - 2\log(1/\xi) \right\rfloor.$$

Then, Protocol 15.9 is an ε-secure implementation f_{OT} of length l with $\varepsilon = 2\xi + \mathcal{O}(2^{-\nu^2 n})$.

Protocol 15.9 OT construction from erasure correlation (X, Y) against an active adversary

Input: $\mathcal{P}_1$'s input $(K_0, K_1) \in \{0,1\}^l \times \{0,1\}^l$ and $\mathcal{P}_2$'s input $B \in \{0,1\}$
Output: $\mathcal{P}_2$'s output $\hat{K} \in \{0,1\}^l$

1. $\mathcal{P}_2$ generates $V_i \sim \mathrm{P}_{V|Y}(\cdot|Y_i)$ for $1 \le i \le n$, and sets $\tilde{\mathcal{I}}_0 = \{i : V_i = 0\}$ and $\tilde{\mathcal{I}}_1 = \{i : V_i = 1\}$; if $|\tilde{\mathcal{I}}_0| < m$ or $|\tilde{\mathcal{I}}_1| < m$, abort the protocol; otherwise, $\mathcal{P}_2$ sets $\mathcal{I}_B$ as the first m indices from $\tilde{\mathcal{I}}_0$ and $\mathcal{I}_{\bar{B}}$ as the first m indices from $\tilde{\mathcal{I}}_1$, and sends $\Pi_1 = (\mathcal{I}_0, \mathcal{I}_1)$ to $\mathcal{P}_1$.
2. If $\mathcal{I}_0 \cap \mathcal{I}_1 \neq \emptyset$, then $\mathcal{P}_1$ aborts the protocol; otherwise, $\mathcal{P}_1$ randomly picks $F_0, F_1 \colon \{0,1\}^m \to \{0,1\}^l$ from a UHF, and sends $\Pi_2 = (\Pi_{2,0}, \Pi_{2,1}, F_0, F_1)$ to $\mathcal{P}_2$, where
$$\Pi_{2,b} = K_b \oplus F_b(X_{\mathcal{I}_b}), \quad \text{for } b = 0, 1.$$
3. $\mathcal{P}_2$ recovers $\hat{K} = \Pi_{2,B} \oplus F_B(Y_{\mathcal{I}_B})$.

Proof The proof proceeds along the line of Proposition 15.10. When both $\mathcal{P}_1$ and $\mathcal{P}_2$ are honest, the protocol is aborted only when $|\tilde{\mathcal{I}}_0| < m$ or $|\tilde{\mathcal{I}}_1| < m$ in Step 1, which occur with probability $\mathcal{O}(2^{-\nu^2 n})$ by the Hoeffding bound. If the protocol is not aborted, it is easy to verify that $\hat{K} = K_B$. Thus, the correctness is satisfied with $\varepsilon = \mathcal{O}(2^{-\nu^2 n})$.

When $\mathcal{P}_1$ is honest and $\mathcal{P}_2$ is dishonest, $\mathcal{P}_2$ may generate $\Pi_1 = (\mathcal{I}_0, \mathcal{I}_1)$ maliciously. If the protocol is aborted, we set the dummy input $B' = \perp$; otherwise, we set
$$B' = \begin{cases} 0, & \text{if } |\{i \in \mathcal{I}_0 : Y_i \neq \mathtt{e}\}| \ge |\{i \in \mathcal{I}_1 : Y_i \neq \mathtt{e}\}|, \\ 1, & \text{if } |\{i \in \mathcal{I}_0 : Y_i \neq \mathtt{e}\}| < |\{i \in \mathcal{I}_1 : Y_i \neq \mathtt{e}\}|. \end{cases}$$
Since Y is independent of $((K_0, K_1), B)$ and Π_1 must be generated as a stochastic function of (B, Y), we have $\Pi_1 -\!\!\circ\!\!- (B, Y) -\!\!\circ\!\!- (K_0, K_1)$, whereby $(\Pi_1, Y) -\!\!\circ\!\!- B -\!\!\circ\!\!- (K_0, K_1)$. Furthermore, since B' is a function of (B, Π_1, Y), the previous Markov chain implies that $B' -\!\!\circ\!\!- B -\!\!\circ\!\!- (K_0, K_1)$. Thus, B' satisfies Condition 15.20 with $\varepsilon = 0$.

To verify Condition 15.20, in a similar manner to the proof of Proposition 15.10, we have $(\Pi_1, Y) -\!\!\circ\!\!- (B, B', K_{B'}) -\!\!\circ\!\!- (K_0, K_1)$. Thus, it remains to bound
$$d_{\mathtt{var}}\big(\mathrm{P}_{\hat{K}Y\Pi K_0 K_1 BB'K_{B'}}, \mathrm{P}_{\hat{K}\Pi_2|\Pi_1 YBB'K_{B'}} \mathrm{P}_{K_0 K_1 BB'K_{B'}}\big). \tag{15.22}$$
When the protocol is aborted, denoted by the event $\mathcal{A}$, Π_2 is not sent and $\hat{K} = \perp$. Thus, it suffices to bound (15.22) under the event $\mathcal{A}^c$. Let $\mathcal{E}$ be the event that $|\tilde{\mathcal{I}}_0| > (1-p+\nu)n$, which occurs with probability $\mathcal{O}(2^{-n\nu^2})$. Then, we can expand the variational distance in (15.22) as
$$\begin{aligned} & d_{\mathtt{var}}\big(\mathrm{P}_{\hat{K}Y\Pi K_0 K_1 BB'K_{B'}}, \mathrm{P}_{\hat{K}\Pi_2|\Pi_1 YBB'K_{B'}} \mathrm{P}_{K_0 K_1 BB'K_{B'}}\big) \\ & \quad \le \Pr(\mathcal{E}) \\ & \quad + \mathbb{E}[d_{\mathtt{var}}\big(\mathrm{P}_{\hat{K}Y\Pi K_0 K_1 BB'K_{B'}}, \mathrm{P}_{\hat{K}\Pi_2|\Pi_1 YBB'K_{B'}} \mathrm{P}_{K_0 K_1 BB'K_{B'}}\big)|\mathcal{A}^c \cap \mathcal{E}^c]. \end{aligned} \tag{15.23}$$

In order to bound the third term of (15.23), under the event $\mathcal{E}^c$, we note that $\mathcal{I}_{\bar{B}'}$ must contain at least

$$m - \frac{(1-p+\nu)n}{2} = \frac{(1-p-3\nu)n}{2}$$

erasure indices. This means that $H_{\min}(X_{\mathcal{I}_{\bar{B}'}}|Y_{\mathcal{I}_{B'}}) \geq \frac{(1-p-3\nu)n}{2}$. Thus, by the leftover hash lemma (cf. Theorem 7.13), we can regard $\Pi_{2,\bar{B}'} = K_{\bar{B}'} \oplus F_{\bar{B}'}(X_{\mathcal{I}_{\bar{B}'}})$ as encrypted by a ξ-secure key; also, (F_0, F_1) are independent of (K_0, K_1). Thus, we have

$$\mathrm{E}[d_{\mathtt{var}}(\mathrm{P}_{Y\Pi_{2,\bar{B}'}F_0F_1K_0K_1BB'K_{B'}}, \mathrm{P}_{\Pi_{2,\bar{B}'}F_0F_1|\Pi_1YBB'K_{B'}}\mathrm{P}_{K_0K_1BB'K_{B'}})|\mathcal{A}^c \cap \mathcal{E}^c]$$
$$\leq 2\xi.$$

On the other hand, $\Pi_{2,B'} = K_{B'} \oplus F_{B'}(X_{\mathcal{I}_{B'}})$ is a function of the dummy output $K_{B'}$ and $F_{B'}(X_{\mathcal{I}_{B'}})$. Thus, $\Pi_{2,B'}$ can be generated from $(\Pi_1, Y, B, B', K_{B'}, F_0, F_1)$.[20] Furthermore, $\hat{K}$ is a stochastic function of (Π, Y, B). By noting these observations, the third term of (15.23) can be bounded by 2ξ.

When $\mathcal{P}_2$ is honest and $\mathcal{P}_1$ is dishonest, in exactly the same manner as in the proof of Theorem 13.3, we can show that (Π_1, X) is independent of B. Define the dummy input as $(K_0', K_1') = \perp$ if $\Pi_1 = \perp$ or $\Pi_2 = \perp$,[21] and

$$K_b' = \Pi_{2,b} \oplus F_b(X_{\mathcal{I}_b}) \text{ for } b = 0, 1$$

otherwise. It is not difficult to see that $\hat{K} = K_B'$. Also, the remaining Markov chain constraints are satisfied with $\varepsilon = 0$, which we leave as an exercise. □

The oblivious transfer capacity $\tilde{C}_{\mathtt{OT}}(X, Y)$ of correlation (X, Y) against the active adversary is defined in a similar manner as in Section 13.5. Theorem 15.21 implies that $\tilde{C}_{\mathtt{OT}}(X, Y) \geq \frac{1-p}{2}$ for the erasure correlation with $\frac{1}{2} \leq p \leq 1$. This is half the capacity $C_{\mathtt{OT}}(X, Y) = 1-p$ against the passive adversary. In fact, by using the above mentioned second approach with the interactive hashing, we can show that $\tilde{C}_{\mathtt{OT}}(X, Y) = 1 - p$ for the erasure correlation with $\frac{1}{2} \leq p \leq 1$.

15.10 References and Additional Reading

In early studies of secure computation, there was no consensus on a formal definition of security. The basics of a simulation based security definition was developed in [15, 250]. Compared to the alternative approach of enumerating every possible requirement that must be satisfied by secure protocols, an important aspect of simulation based security is that it can guarantee security against attacks that have not been recognized yet. However, constructions of simulators are inspired

[20] Even though $X_{\mathcal{I}_{B'}}$ is not available to $\mathcal{P}_2$ in the real protocol, $X_{\mathcal{I}_{B'}}$ can be generated from $Y_{\mathcal{I}_{B'}}$ independently of (K_0, K_1) when we construct a simulator.

[21] $\Pi_2 = \perp$ indicates that $\mathcal{P}_1$ aborts in Step 2. Since $\mathcal{P}_2$ is honest, $\mathcal{P}_1$ never aborts because of $\mathcal{I}_0 \cap \mathcal{I}_1 \neq \emptyset$; however, $\mathcal{P}_1$ may maliciously abort for no reason.

by possible attacks. Another important aspect of the simulation based security is its suitability for composability of protocols; formal security notions of composability matured around 2000. The exposition of sequential composition in this chapter is based on [52] (see also [142]). Information-theoretic conditions of security (Lemma 15.2 and Lemma 15.8) are from [85, 86]. Construction of oblivious transfer in Protocol 15.4 and its generalization can be found in [42].

The complete fairness of protocols is a difficult issue in cryptography. The relaxation in Section 15.7 is one of the widely used solutions (e.g. see [142]). Nonetheless, there are active research efforts towards identifying alternative solutions; for instance, see [149].

As we have seen in Section 15.8, standalone security does not guarantee security of concurrently composed protocols. The counterexample there is from [11] (see also [213] for other counterexamples). A widely used notion of security is *universal composable security* [53], which guarantees that a protocol is secure under concurrent composition. In fact, it is known that universal composable security is almost the minimal requirement for guaranteeing security under concurrent composition [224], though there are some other frameworks to guarantee security against concurrent composition [240, 271].

As illustrated in the counterexample in Section 15.8, decomposable functions are not securely computable from scratch when we consider concurrent composition. However, decomposable functions are securely computable under concurrent composition if the bit commitment functionality is available [229].

For an active adversary, the use of the privacy amplification in the construction of the string OT was introduced in [43]; see also [170] for an adaptation to the erasure channel. Interactive hashing was introduced in [267], and it has been used in various applications, including the construction of oblivious transfer in the bounded storage model [48]. From the viewpoint of the string OT construction, one of the most notable breakthroughs was made in [299] (see also [49, 84]); the use of interactive hashing in [299] enables us to double the lower bound (achieavable rate) on the OT capacity derived by the privacy amplification approach. This breakthrough led to the determination of the OT capacity of the generalized erasure channel for the high erasure probability regime [273]; for the low erasure probability regime, see [112]. See also [251] for another application of interactive hashing to the construction of string OT in an erasure wiretap model. For a construction of a constant round interactive hashing protocol, see [100].

The result in [299] implies that the string OT capacity of bit OT is 1; this is attained by allowing a vanishing leakage of $\mathcal{P}_1$'s inputs. If we do not allow vanishing leakage or the abort option, then a tighter impossibility bound can be derived [210], which implies that the string OT capacity of bit OT under such strict constraints is upper bounded by 1/2. Note that for a passive adversary the string OT capacity of bit OT can be attained with zero leakage and no abortion.

Problems

15.1 *Simulator implies an approximate Markov chain:* For a joint distribution P_{ABC}, suppose that there exists a channel (simulator) $\mathrm{Q}_{C|B}$ such that

$$d_{\mathtt{var}}(\mathrm{P}_{ABC}, \mathrm{P}_{AB}\mathrm{Q}_{C|B}) \le \varepsilon.$$

Prove that $d_{\mathtt{var}}(\mathrm{P}_{ABC}, \mathrm{P}_{AB}\mathrm{P}_{C|B}) \le \varepsilon$.

15.2 Verify Lemma 15.9.

15.3 Verify the security of Protocol 15.8 for a dishonest $\mathcal{P}_1$.

15.4 For the two sequential composition of Protocol 15.7, $\mathcal{P}_2$ can choose the following message:

$$\Pi_{2,1} = \begin{cases} \bot, & \text{if } \Pi_{1,1} = 1, \\ 1, & \text{if } \Pi_{1,1} = 0 \end{cases}$$

and

$$\Pi_{2,2} = \begin{cases} \bot, & \text{if } \Pi_{1,2} = 1, \\ 1, & \text{if } \Pi_{1,1} = 1, \Pi_{1,2} = 0, \\ 0, & \text{if } \Pi_{1,1} = 0, \Pi_{1,2} = 0. \end{cases}$$

Prove that there exists a simulator for this misbehavior by $\mathcal{P}_2$.

HINT Write down the function table as in (15.21); then you will find that there will be one less $*$ in the table compared to (15.21). Pick the first dummy input $X'_{2,1}$ appropriately, and then pick the second dummy input $X'_{2,2}$ using knowledge of the output $f(X_{1,1}, X'_{2,1})$.

15.5 For the protocol of computing the one-sided AND function (only $\mathcal{P}_2$ gets the output) in Example 15.5, verify security against a dishonest $\mathcal{P}_1$.

HINT A dishonest $\mathcal{P}_1$ need not send $\Pi_1 = X_1$; what will be the correct choice of dummy input X'_1 for the simulator?

15.6 Consider the two-sided AND function $f(x_1, x_2) = x_1 \wedge x_2$ (both $\mathcal{P}_1$ and $\mathcal{P}_2$ get the output).[22]

1. Consider the following protocol from scratch: $\mathcal{P}_1$ sends $\Pi_1 = X_1$ to $\mathcal{P}_2$, $\mathcal{P}_2$ sets $Z_2 = \Pi_1 \wedge X_2$, $\mathcal{P}_2$ sends $\Pi_2 = Z_2$ to $\mathcal{P}_1$, and $\mathcal{P}_1$ sets $Z_1 = \Pi_2$. Describe a reason why this protocol is not secure.

   HINT Consider an attack in which dishonest $\mathcal{P}_2$ sends $\Pi_2 = 0$ (so that $\mathcal{P}_1$ is forced to output 0); can we simulate this attack with the ideal function?
2. Consider the following protocol using the bit OT: $\mathcal{P}_1$ inputs $(0, X_1)$ and $\mathcal{P}_2$ inputs X_2 to the bit OT, $\mathcal{P}_2$ sets the outcome of the bit OT as Z_2, $\mathcal{P}_2$ sends $\Pi_2 = Z_2$ to $\mathcal{P}_1$, and $\mathcal{P}_1$ sets $Z_1 = \Pi_2$. Describe a reason why this protocol is not secure.

[22] We assume that $\mathcal{P}_2$ receives the output first, and can decide whether $\mathcal{P}_1$ will receive the output or not; see Section 15.7 for the modification of the ideal function to make the unfair abort admissible.

HINT Consider an attack in which dishonest $\mathcal{P}_2$ inputs 1 to the bit OT (so that $\mathcal{P}_2$ can always learn X_1) and sends $\Pi_2 = 0$ (so that $\mathcal{P}_1$ is forced to output 0); can we simulate this attack with the ideal function?

15.7 Let us consider the two-sided XOR function $f(x_1, x_2) = x_1 \oplus x_2$ (both $\mathcal{P}_1$ and $\mathcal{P}_2$ get the output).

1. Consider the following protocol from scratch: $\mathcal{P}_1$ sends $\Pi_1 = X_1$ to $\mathcal{P}_2$, $\mathcal{P}_2$ sets $Z_2 = \Pi_1 \oplus X_2$, $\mathcal{P}_2$ sends $\Pi_2 = X_2$ to $\mathcal{P}_1$, and $\mathcal{P}_1$ sets $Z_1 = X_1 \oplus \Pi_2$. Describe a reason why this protocol is not secure.

   HINT Consider an attack in which dishonest $\mathcal{P}_2$ sends $\Pi_2 = \Pi_1 = X_1$; can we simulate this attack with the ideal function?
2. Consider the following protocol with the bit commitment: $\mathcal{P}_2$ commits to $\Pi_2 = X_2$, $\mathcal{P}_1$ sends $\Pi_1 = X_1$ to $\mathcal{P}_2$, $\mathcal{P}_2$ sets $Z_2 = \Pi_1 \oplus X_2$, $\mathcal{P}_2$ reveals Π_2 to $\mathcal{P}_1$, and $\mathcal{P}_1$ sets $Z_1 = X_1 \oplus \Pi_2$. Verify that this protocol is a secure implementation of the two-sided XOR.

15.8 Identify a vulnerability in the protocol for implementing $\binom{2}{1}$-OT in Problem 12.5, namely show that it is not secure for an active adversary.

15.9 *Conversion of OT*: Verify that the protocol in Problem 12.6 is a secure implementation of the $\binom{4}{1}$-OT for an active adversary.

HINT Let $((K_{1,0}, K_{1,1}), C_1)$, $((K_{2,0}, K_{2,1}), C_2)$, and $((K_{3,0}, K_{3,1}), C_3)$ be the inputs and $\hat{K}_1$, $\hat{K}_2$, and $\hat{K}_3$ be the outputs of the three invocations of the $\binom{2}{1}$-OT. When $\mathcal{P}_1$ is dishonest, specify the dummy input (A'_0, A'_1, A'_2, A'_3) of the $\binom{4}{1}$-OT as a function of $(K_{1,0}, K_{1,1})$, $(K_{2,0}, K_{2,1})$, and $(K_{3,0}, K_{3,1})$; when $\mathcal{P}_2$ is dishonest, specify the dummy input B' as a function of (C_1, C_2, C_3), and consider how to simulate $(\hat{K}_1, \hat{K}_2, \hat{K}_3)$ using the dummy output of the $\binom{4}{1}$-OT.

15.10 For an active adversary and the erasure correlation with $p < \frac{1}{2}$, consider a string OT protocol that attains the asymptotic rate of $\frac{p}{2}$.

HINT Protocol 15.9 with $m \simeq pn$ and $l \simeq \frac{pn}{2}$ does not work; for instance, when $p = \frac{1}{4}$, there are $|\tilde{\mathcal{I}}_0| \simeq \frac{3n}{4}$ good indices, and the sets $(\mathcal{I}_0, \mathcal{I}_1)$ of size $|\mathcal{I}_0| = |\mathcal{I}_1| \simeq \frac{n}{4}$ can be filled only with good indices. To avoid such an attack, consider $(\mathcal{I}_0, \mathcal{I}_1)$ with $|\mathcal{I}_0| = |\mathcal{I}_1| = \frac{n}{2}$ irrespective of the value of p; if $\mathcal{P}_2$ is honest, $\mathcal{P}_2$ is supposed to fill $\mathcal{I}_B$ only with good indices and $\mathcal{I}_{\overline{B}}$ with the remaining indices. Then, even if $\mathcal{P}_2$ is dishonest, at least one of $\mathcal{I}_0$ or $\mathcal{I}_1$ contains $\frac{pn}{2}$ bad indices; the privacy amplification enables $\mathcal{P}_1$ to extract a pair of secret keys $(F_0(X_{\mathcal{I}_0}), F_1(X_{\mathcal{I}_1}))$ of length $\frac{pn}{2}$ such that one of them is unknown to $\mathcal{P}_2$.

By testing if one of $\mathcal{I}_0$ or $\mathcal{I}_1$ contains (almost) all bad indices with the interactive hashing, it is also possible to construct a string OT protocol that attains the asymptotic rate of p [112].

16 Zero-knowledge Proof

In this chapter, we will study zero-knowledge proofs (ZKPs), an important concept in computational complexity theory which is a very useful tool for secure computing. We can introduce ZKPs using an anecdote called "lady tasting tea," which narrates an incident that is often credited as the motivation for Ronald Fisher to build the theory of statistics. In a social gathering, Ronald Fisher offered his colleague Muriel Bristol a cup of tea with milk and tea already mixed. Bristol declined claiming that she preferred the flavor when the milk was poured into the cup before the tea, but Fisher insisted that the order of pouring could not affect the flavor. The two decided to settle the claim by conducting an experiment. Fisher prepared 8 cups of tea with 4 cups prepared by pouring the milk before the tea and the remaining 4 prepared by pouring the tea before the milk. Then, he randomly shuffled the cups and asked Bristol to taste the tea and divide the cups into two groups corresponding to tea of each kind. The outcome of the experiment is not important, but the reader must be convinced that the experiment was well designed. If Bristol was able to divide the cups correctly, then she must really have been able to distinguish tea based on the order in which milk and tea are poured – the probability that a random guess would correctly divide the cups is just $1/70$.

This incident is supposed to have led Fisher to start thinking about the theory of experimental design and statistics. But this has all the elements of a ZKP. If Bristol is able to divide the cups into two groups correctly, Fisher will be convinced that she really knows how to distinguish the two flavors. However, Fisher himself does not learn how to distinguish the flavors. Further, if Bristol's claim is false, Bristol cannot deceive Fisher. Also, we note that the experiment required interaction between the two parties and randomization. These are some common features of the ZKPs we will see in this chapter.

16.1 Interactive Proofs and Zero-knowledge Proofs

We briefly review some background material on computational complexity; for a detailed account, see the textbooks cited in Section 16.5. A fundamental type of problem in theoretical computer science is the classifying problem – such as "Can we find an Eulerian trail for a given graph?" (Eulerian trail problem) or "Can we find an input satisfying a given Boolean formula?" (satisfiability problem) – based on the computational resources required to answer those problems. Formally, a problem with different possible input instances, termed a *language*, is described by a set $L \subseteq \{0,1\}^*$ of binary strings. For instance, in the case of the Eulerian trail problem, an instance $x \in L$ is regarded as the encoded string representing a graph that contains an Eulerian trail. Such an instance is termed a "yes" instance, and L is the set of all yes instances of a given problem. For a given string $x \in \{0,1\}^*$, we are interested in deciding whether $x \in L$ or not.[1]

A language $L \subseteq \{0,1\}^*$ is said to be in class $\mathtt{P}$ if there exists a polynomial-time algorithm A such that

$$x \in L \Longleftrightarrow A(x) = 1, \tag{16.1}$$

namely the algorithm returns the output 1 if and only if the input x is a yes instance. On the other hand, a language L is said to be in class $\mathtt{NP}$ if there exists a polynomial $p\colon \mathbb{N} \to \mathbb{N}$ and a polynomial-time algorithm A such that

$$x \in L \Longleftrightarrow \exists w \in \{0,1\}^{p(|x|)} \text{ such that } A(x,w) = 1, \tag{16.2}$$

namely we can find a *witness* w of polynomial length which allows A to verify that x is a yes instance in polynomial time.

If $L \in \mathtt{P}$ holds, then an algorithm A satisfying (16.1) also satisfies (16.2) with "null witness" – a string fixed independent of x. Thus, we have $\mathtt{P} \subseteq \mathtt{NP}$. Proving or disproving strict inclusion is the famous $\mathtt{P}$ versus $\mathtt{NP}$ problem; the reader can earn one million US dollars by resolving this problem. Roughly speaking, the class $\mathtt{P}$ is the set of all problems that are efficiently solvable; on the other hand, the class $\mathtt{NP}$ is the set of all problems that are efficiently verifiable given a solution.

To compare the computational difficulty of problems, we use the following criterion. For two languages L and L', we say that L is polynomial-time reducible to L', denoted by $L \leq_p L'$, if there is a polynomial-time computable function $f\colon \{0,1\}^* \to \{0,1\}^*$ such that for every $x \in \{0,1\}^*$, $x \in L$ if and only if $f(x) \in L'$. We say that L' is $\mathtt{NP}$-hard if $L \leq_p L'$ for every $L \in \mathtt{NP}$. We say that L' is $\mathtt{NP}$-complete if L' is $\mathtt{NP}$-hard and $L' \in \mathtt{NP}$. Roughly speaking, the class $\mathtt{NP}$-complete consists of problems that are the most difficult in $\mathtt{NP}$.

EXAMPLE 16.1 (G3C) A graph $G = (V,E)$ is said to be 3-colorable if there exists a function $\phi\colon V \to \{R,G,B\}$ such that $\phi(u) \neq \phi(v)$ for every $(u,v) \in E$, i.e., all adjacent vertices are assigned with different colors (see Figure 16.1). For a given graph, deciding whether the graph is 3-colorable or not is known as the graph 3-colorable (G3C) problem. We can easily verify that this problem is in $\mathtt{NP}$; if a color assignment is given, the validity of the coloring can be efficiently verified. In fact, it is known that G3C is $\mathtt{NP}$-complete.

[1] This formulation is very similar to the binary hypothesis testing formulation we saw earlier, but the representation by strings allows us to focus on computational aspects. In fact, since our test procedures in this chapter will often be random, this connection will appear even more prominently.

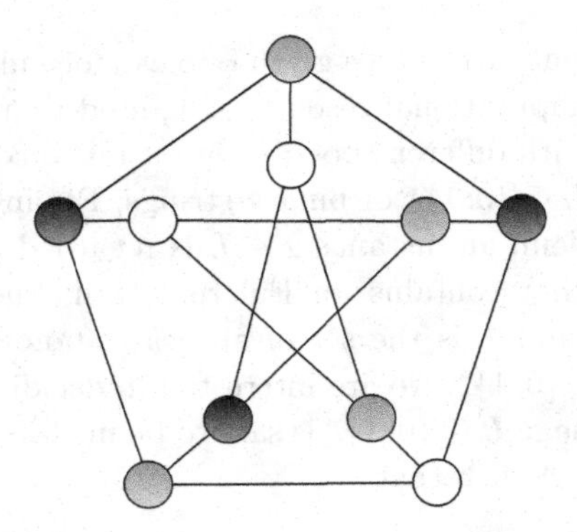

Figure 16.1 An example of a 3-colorable graph.

The class NP is the set of all problems with "succinct" proofs for the fact that the input is a yes instance, given by the witness. This witness can be viewed as something presented by the "prover" who knows that $x \in L$ to a verifier. For some problems, it will not be clear how to construct such proofs. One such problem is the graph non-homomorphism problem. Given a graph $G = (V, E)$ and a graph $H = (V, \tilde{E})$, the two graphs G and H are said to be homomorphic, denoted $G \simeq H$, if there exists a permutation σ on V such that $(x, y) \in E$ if and only if $(\sigma(x), \sigma(y)) \in \tilde{E}$. If there exists no such permutation, then G and H are said to be non-homomorphic. It is known that the graph homomorphism problem is NP because the permutation σ can be used as a witness. However, it is not known whether the graph non-homomorphism problem is NP or not. More generally, the class consisting of negation of NP problems is called coNP; for problems in coNP, it is not known whether succinct proofs exist, and the inclusive relation of NP and coNP is also not known. Nevertheless, there is a succinct proof even for the non-homomorphic problem if we allow interaction between the two parties, the prover $\mathcal{P}_{\mathrm{p}}$ and the verifier $\mathcal{P}_{\mathrm{v}}$.

Similar to the experiment in the "lady tasting tea" story at the beginning of this chapter, an interactive proof system consists of communication between the prover and the verifier. The prover is assumed to have unbounded computational power, i.e., the prover can decide whether a given input belongs to a language irrespective of the difficulty of the language. The verifier is assumed to have bounded computational power, i.e., the verifier must be realized by a polynomial-time algorithm. A protocol for the graph non-homomorphism problem is described in Protocol 16.1. Suppose that $G_0 \not\simeq G_1$. Then, $\mathcal{P}_{\mathrm{p}}$ can check that $G_0 \simeq H$ or $G_1 \simeq H$ and send c satisfying $c = b$. On the other hand, when $G_0 \simeq G_1$, both $G_0 \simeq H$ and $G_1 \simeq H$ hold. Thus, $\mathcal{P}_{\mathrm{p}}$ cannot fool $\mathcal{P}_{\mathrm{v}}$ in to accepting by sending $c = b$ with success probability higher than $\frac{1}{2}$. In fact, we can make the success probability of false acceptance arbitrarily small by repeating this protocol.

Protocol 16.1 Interactive proof of graph non-homomorphism

Input: Two graphs $G_0 = (V_0, E_0)$ and $G_1 = (V_1, E_1)$ with $V_0 = V_1 = V$
Output: Accept/Reject

1. $\mathcal{P}_v$ randomly generates $b \in \{0, 1\}$.
2. $\mathcal{P}_v$ randomly generates permutation σ on V, and sends $H = \sigma(G_b)$ to $\mathcal{P}_p$.
3. If $G_0 \simeq H$, then $\mathcal{P}_p$ sends $c = 0$ to $\mathcal{P}_v$; if $G_1 \simeq H$, then $\mathcal{P}_p$ sends $c = 1$ to $\mathcal{P}_v$.
4. If $c = b$, then $\mathcal{P}_v$ accepts. Otherwise, reject.

In general, an interactive proof protocol π for a language L can be described formally as follows. For a given instance $x \in \{0, 1\}^*$, the prover and the verifier communicate interactively: $\mathcal{P}_v$ sends a (possibly stochastically generated) message $\Pi_1 = \Pi_1(x)$, $\mathcal{P}_p$ sends $\Pi_2 = \Pi_2(x, \Pi_1)$, and so on. For r rounds of protocol, let $\Pi = (\Pi_1, \ldots, \Pi_r)$ be the entire transcript. At the end of the protocol, $\mathcal{P}_v$ outputs $Z = Z(x, \Pi) \in \{0, 1\}$.

DEFINITION 16.2 (Interactive proof) For a given language L, a protocol π is (ε, δ)-IP (interactive proof) if the following hold.

1. *Completeness*: Given $x \in L$, the output is $Z(x, \Pi) = 1$ with probability at least $1 - \varepsilon$.
2. *Soundness*: Given $x \notin L$, the output is $Z(x, \Pi) = 0$ with probability $1 - \delta$ even if $\mathcal{P}_p$ arbitrarily deviates from the prescribed protocol.[2]

For instance, if we repeat Protocol 16.1 n times, then it is a $(0, 2^{-n})$-IP for the graph non-homomorphism problem.

If you want to convince someone of a statement by a standard proof, i.e., by explaining reasons why the statement is true, you may reveal more than the truth of the statement itself. However, as we have seen in the story at the beginning of this chapter, it is sometimes possible to convince someone of a statement without revealing anything more. Such a proof is termed a *zero-knowledge proof* (ZKP). In fact, the interactive proof of graph homomorphism described in Protocol 16.1 is also a ZKP;[3] when the statement is true, i.e., $G_0 \simeq G_1$, then the verifier $\mathcal{P}_v$ already knows the value of the message c before receiving it from $\mathcal{P}_p$ since $b = c$ must hold.

Formally, a ZKP is defined as follows.

[2] At a high level, the correctness condition seen in Chapters 8, 13, 14, and 15 requires the protocol to behave appropriately when the input is as expected. In contrast, the soundness condition requires the protocol to identify inputs that deviate from expected ones.

[3] More precisely, Protocol 16.1 is zero-knowledge only for an honest verifier. When the verifier might not follow the protocol, additional procedure to "bind" the verifier's dishonest behavior is needed.

DEFINITION 16.3 (Zero-knowledge proof) For a given language L, a protocol π is an $(\varepsilon, \delta, \gamma)$-ZKP if it is (ε, δ)-IP and there exists a simulator $\Pi_{\mathtt{sim}}$ such that the following holds for every $x \in L$:

$$d_{\mathtt{var}}\big(\mathrm{P}_{\Pi(x)}, \mathrm{P}_{\Pi_{\mathtt{sim}}(x)}\big) \le \gamma. \tag{16.3}$$

In particular, when $\gamma = 0$, the protocol is said to be a *perfect* ZKP.[4] When we emphasize that a sufficiently small positive γ is allowed, the protocol is said to be a *statistical* ZKP. In fact, the ZKP for the graph non-homomorphism problem in Protocol 16.1 is a perfect ZKP.

We can also define a *computational* ZKP by replacing the requirement in (16.3) with computational indistinguishability (cf. Definition 9.4); more specifically, for a negligible function $\gamma \colon \mathbb{N} \to [0,1]$, we require that

$$\big|\Pr\left(D(\Pi(x)) = 1\right) - \Pr\left(D(\Pi_{\mathtt{sim}}(x)) = 1\right)\big| \le \gamma(|x|)$$

holds for any PPT distinguisher D.

16.2 Zero-knowledge Proof for an NP Language

One of the most fundamental results on ZKPs is the fact that every NP language has a computational zero-knowledge proof. The proof of this claim in the literature essentially uses the computational assumption only to realize bit commitment. Thus, we state a variant of this claim as follows.

THEOREM 16.4 *Every language in* NP *has a perfect ZKP if bit commitment is available.*

To prove Theorem 16.4, it suffices to show a problem in the NP-complete class has a ZKP; the ZKP for other languages in the NP class can be constructed in conjunction with polynomial reduction to an NP-complete problem. Below, we provide a zero-knowledge proof for the G3C problem described in Example 16.1. A ZKP for G3C is described in Protocol 16.2.

The protocol satisfies completeness with $\varepsilon = 0$. It is also sound with

$$\delta = \left(\frac{|E|-1}{|E|}\right)^n,$$

as can be seen from the arguments below. If the graph $G = (V, E)$ is not 3-colorable, then there exists at least one edge $(u, v) \in E$ such that $\phi(u) = \phi(v)$ for any color assignment $\phi \colon V \to \{R, G, B\}$. Then, the probability of misdetection by $\mathcal{P}_{\mathrm{v}}$ is $\frac{|E|-1}{|E|}$ in each repetition. Finally, because of the hiding property of bit commitment, $\mathcal{P}_{\mathrm{v}}$ learns only $\sigma(\phi(u))$ and $\sigma(\phi(v))$ in each round. However, these values are just two different colors randomly assigned to u and v. Thus, $\mathcal{P}_{\mathrm{v}}$ can simulate the message from $\mathcal{P}_{\mathrm{p}}$ by randomly sampling two different colors for (u, v).

[4] Of course, ε and δ must be sufficiently small to make the definition nontrivial.

Protocol 16.2 Zero-knowledge proof of the graph 3-colorable problem

Input: A graph $G = (V, E)$ and a 3-coloring $\phi \colon V \to \{R, G, B\}$
Output: Accept/Reject

1. $\mathcal{P}_p$ and $\mathcal{P}_v$ repeat Steps 2–6 n times.
2. $\mathcal{P}_p$ randomly generates a permutation σ on $\{R, G, B\}$.
3. By invoking the commit phase of bit commitment, $\mathcal{P}_p$ commits to $\sigma(\phi(v))$ for every $v \in V$.
4. $\mathcal{P}_v$ randomly chooses $(u, v) \in E$, and sends it to $\mathcal{P}_p$.
5. By invoking the reveal phase of the bit commitment, $\mathcal{P}_p$ reveals the values $\sigma(\phi(u))$ and $\sigma(\phi(v))$ to $\mathcal{P}_v$.
6. $\mathcal{P}_v$ checks if $\sigma(\phi(u)) \neq \sigma(\phi(v))$.
7. If all the tests are passed, $\mathcal{P}_v$ accepts $\mathcal{P}_p$'s proof.

Note that bit commitment is needed to prevent the dishonest prover from deceiving the verifier in to erroneously accepting the claim; without bit commitment, the prover can decide the color assignment of u and v after receiving the query from the verifier.

16.3 Multi-prover Interactive Proofs

In Section 16.2, we provided a ZKP for an NP language under the assumption that bit commitment is available. In order to realize bit commitment, we need to assume the existence of a noisy correlation between the parties; alternatively, we can provide a ZKP for a computationally bounded verifier by using computationally secure bit commitment.[5]

In this section, we explain another approach for constructing ZKPs in which the availability of bit commitment is replaced by two provers. Recall that bit commitment is used to prevent the dishonest prover from deceiving the verifier in to accepting the claim falsely. How does increasing the number of provers help in preventing such a cheating? Heuristically, this follows the same principle that is used to interrogate two suspects for a crime separately – without allowing them to communicate, it is easier to detect inconsistency in their alibi. In general, an interactive two-prover protocol π for a language L is described as follows. For a given instance x, the two provers and the verifier communicate interactively: $\mathcal{P}_v$ sends messages $\Pi_{1,1} = \Pi_{1,1}(x)$ to $\mathcal{P}_{p,1}$ and $\Pi_{1,2} = \Pi_{1,2}(x)$ to $\mathcal{P}_{p,2}$, respectively; $\mathcal{P}_{p,1}$ sends reply $\Pi_{2,1} = \Pi_{2,1}(x, \Pi_{1,1}, U)$ and $\mathcal{P}_{p,2}$ sends reply $\Pi_{2,2} = \Pi_{2,2}(x, \Pi_{1,2}, U)$, respectively, where U is the shared randomness; and so on. Based on the entire transcript Π exchanged during the protocol, $\mathcal{P}_v$ makes a decision $Z = Z(x, \Pi) \in \{0, 1\}$. The completeness, the soundness, and

[5] Note that the verifier is computationally bounded; otherwise, the verifier can find a proof by itself.

the zero-knowledge property are defined in the same manner as the single-prover case. Here, it should be emphasized that the two provers cannot communicate during the protocol; the only resource they have for coordination is the shared randomness U, which is generated and shared before the protocol starts.

Going back to our running example of the NP-complete problem – the G3C problem – in order to circumvent the use of the bit commitment, the verifier's query and the test are constructed as follows. The verifier samples a pair of vertices $(x_1, x_2) \in V \times V$ according to the distribution

$$P_{X_1X_2}(x_1, x_2) = \frac{1}{2} P_{V,\mathtt{unif}}(x_1, x_2) + \frac{1}{2} P_{E,\mathtt{unif}}(x_1, x_2), \tag{16.4}$$

where $P_{V,\mathtt{unif}}$ is the uniform distribution on $\{(x, x) : x \in V\}$ and $P_{E,\mathtt{unif}}$ is the uniform distribution on E. Then, the verifier sends queries x_1 and x_2 to the provers, and the provers send replies $y_1 = \sigma(\phi(x_1))$ and $y_2 = \sigma(\phi(x_2))$, respectively, where σ is a preshared random permutation on $\{R, G, B\}$. The verifier accepts the replies if $\omega(x_1, x_2, y_1, y_2) = 1$ for the predicate defined by

$$\omega(x_1, x_2, y_1, y_2) = \begin{cases} 1, & \text{if } \{x_1 = x_2, y_1 = y_2\} \text{ or } \{x_1 \neq x_2, y_1 \neq y_2\}, \\ 0, & \text{else.} \end{cases}$$

In order to pass this test, if the verifier picks the same vertex $x_1 = x_2$, then the provers must send the same replies $y_1 = y_2$; on the other hand, if the verifier picks a pair of adjacent vertices $(x_1, x_2) \in E$, then the provers must send different replies $y_1 \neq y_2$. Since the verifier chooses the two types of queries at random using the distribution in (16.4), the soundness error, i.e., the probability of successful cheating by the provers, is strictly smaller than 1 unless the graph G is 3-colorable. In fact, we repeat this procedure n times in order to decrease the soundness error to as small as desired. The detail is described in Protocol 16.3, where we assume that the two provers share random permutations $\sigma_1, \ldots, \sigma_n$ on $\{R, G, B\}$ as shared randomness. Note that, when the queries are the same $x_1 = x_2$, then the reply is just a random color assigned to that vertex; on the other hand, when the queries are different $(x_1, x_2) \in E$, then the replies are just random different colors assigned to those vertices. In any case, the message from the provers can be simulated by the verifier, and thus it is zero-knowledge.

16.4 Parallel Repetition

In this section, we consider an abstract game with one verifier and two provers, a generalization of the setup of the previous section, and present a result known as the "parallel repetition theorem." This result is not directly related to the topic of this book, but is very close to the topic of this chapter. Before describing the formulation, we start with a motivating example. In Section 16.3, we presented the two-prover ZKP problem. Typical protocols, such as Protocol 16.3, repeat the communication between the verifier and the provers multiple times in order to make the soundness error sufficiently small. If we conduct such a

Protocol 16.3 Two-prover zero-knowledge proof of the graph 3-colorable problem

Input: A graph $G = (V, E)$, a 3-coloring $\phi: V \to \{R, G, B\}$, and random permutations $\sigma_1, \ldots, \sigma_n$

Output: Accept/Reject

1. $\mathcal{P}_{\text{p},1}$, $\mathcal{P}_{\text{p},2}$ and $\mathcal{P}_{\text{v}}$ repeat Steps 2–6 n times.
2. $\mathcal{P}_{\text{v}}$ samples $(x_{1,i}, x_{2,i})$ according to distribution $\mathrm{P}_{X_1X_2}$, and sends $x_{1,i}$ and $x_{2,i}$ to $\mathcal{P}_{\text{p},1}$ and $\mathcal{P}_{\text{p},2}$, respectively.
3. $\mathcal{P}_{\text{p},1}$ and $\mathcal{P}_{\text{p},2}$ send replies $y_{1,i} = \sigma_i(\phi(x_{1,i}))$ and $y_{2,i} = \sigma_i(\phi(x_{2,i}))$ to $\mathcal{P}_{\text{v}}$, respectively.
4. $\mathcal{P}_{\text{v}}$ checks if $\omega(x_{1,i}, x_{2,i}, y_{1,i}, y_{2,i}) = 1$.
5. If all the tests are passed, $\mathcal{P}_{\text{v}}$ accepts the proof.

repetition sequentially as in Protocol 16.3, then we need to increase the number of communication rounds in order to decrease the soundness error.

Alternatively, if we conduct the repetition in parallel (parallel repetition), then it suffices to have just one round of communication between the verifier and the provers. For instance, in a parallel repetition version of Protocol 16.3, the verifier can send n queries $(x_{1,1}, \ldots, x_{1,n})$ to $\mathcal{P}_{\text{p},1}$ and $(x_{2,1}, \ldots, x_{2,n})$ to $\mathcal{P}_{\text{p},2}$ at once, and then the provers can send replies $(y_{1,1}, \ldots, y_{1,n})$ and $(y_{2,1}, \ldots, y_{2,n})$ at once. However, can such a protocol with parallel repetition guarantee the soundness of the protocol? As we will see below, this question is not so trivial.

Consider the following game $\mathbb{G}$ between one verifier $\mathcal{P}_{\text{v}}$ and two provers $\mathcal{P}_{\text{p},1}$ and $\mathcal{P}_{\text{p},2}$. The verifier $\mathcal{P}_{\text{v}}$ samples a pair of queries (X_1, X_2) according to a prescribed distribution $\mathrm{P}_{X_1X_2}$ on finite alphabet $\mathcal{X}_1 \times \mathcal{X}_2$, and sends the queries to each prover. The provers $\mathcal{P}_{\text{p},1}$ and $\mathcal{P}_{\text{p},2}$ send replies Y_1 and Y_2 based on (X_1, U) and (X_2, U), respectively, where U is the shared randomness. For a given predicate $\omega: \mathcal{X}_1 \times \mathcal{X}_2 \times \mathcal{Y}_1 \times \mathcal{Y}_2 \to \{0, 1\}$, the provers win the game if their replies satisfy $\omega(X_1, X_2, Y_1, Y_2) = 1$. The *value of the game* is given by the maximum winning probability

$$\rho(\mathbb{G}) := \max\left\{\mathbb{E}[\omega(X_1, X_2, Y_1, Y_2)] : Y_1 -\!\!\circ\!\!- (X_1, U) -\!\!\circ\!\!- (X_2, U) -\!\!\circ\!\!- Y_2\right\}. \quad (16.5)$$

Note that the Markov chain constraint in the maximum comes from the fact that the reply of each prover must be decided based on each query and the shared randomness. In fact, the shared randomness as well as the local randomness do not improve the winning probability, and we can write (left as an exercise; Problem 16.3)

$$\rho(\mathbb{G}) := \max\left\{\mathbb{E}[\omega(X_1, X_2, f_1(X_1), f_2(X_2))] : f_1 \in \mathcal{F}(\mathcal{Y}_1|\mathcal{X}_1), f_2 \in \mathcal{F}(\mathcal{Y}_2|\mathcal{X}_2)\right\}, \quad (16.6)$$

where $\mathcal{F}(\mathcal{Y}_i|\mathcal{X}_i)$ is the set of all functions from $\mathcal{X}_i$ to $\mathcal{Y}_i$.

In n parallel repetitions of the game, the verifier samples the sequence of queries X_1^n and X_2^n according to the product distribution $\mathrm{P}^n_{X_1X_2}$. The provers send replies Y_1^n and Y_2^n based on X_1^n and X_2^n. The provers win the game if predicates for each coordinate are satisfied, namely the predicate $\omega^{\wedge n}$ for the parallel repetition game $\mathbb{G}^{\wedge n}$ is given by

$$\omega^{\wedge n}(x_1^n, x_2^n, y_1^n, y_2^n) := \bigwedge_{i=1}^n \omega(x_{1,i}, x_{2,i}, y_{1,i}, y_{2,i}).$$

Then, the value of the game $G^{\wedge n}$ is defined similarly as

$$\rho(\mathbb{G}^{\wedge n}) := \max\left\{\mathbb{E}[\omega^{\wedge n}(X_1^n, X_2^n, f_1(X_1^n), f_2(X_2^n)] : f_1 \in \mathcal{F}(\mathcal{Y}_1^n|\mathcal{X}_2^n), f_2 \in \mathcal{F}(\mathcal{Y}_2^n|\mathcal{X}_2^n)\right\},$$

where $\mathcal{F}(\mathcal{Y}_i^n|\mathcal{X}_i^n)$ is the set of all functions from $\mathcal{X}_i^n$ to $\mathcal{Y}_i^n$. The question of parallel repetition is "how does $\rho(\mathbb{G}^{\wedge n})$ relate to $\rho(\mathbb{G})$?" A naive strategy for the provers in the game $\mathbb{G}^{\wedge n}$ is to use the same optimal strategy of $\mathbb{G}$ for each coordinate; then, we have a trivial lower bound

$$\rho(\mathbb{G}^{\wedge n}) \geq \rho(\mathbb{G})^n.$$

At a first glance, it may seem that the opposite inequality also holds, i.e., there is no better strategy than to use the optimal strategy for the single game for each coordinate. However, this turns out to be not true, as is demonstrated in the following example.

EXAMPLE 16.5 Consider the game with $\mathcal{X}_1 = \mathcal{X}_2 = \mathcal{Y}_1 = \mathcal{Y}_2 = \{0,1\}$, $\mathrm{P}_{X_1X_2}(0,0) = \mathrm{P}_{X_1X_2}(0,1) = \mathrm{P}_{X_1X_2}(1,0) = 1/3$, and

$$\omega(x_1, x_2, y_1, y_2) = \mathbf{1}\big[(x_1 \vee y_1) \neq (x_2 \vee y_2)\big].$$

In this game, at least one of the provers receives 0. In order to win the game, when both the provers receive 0, exactly one prover must respond 1, and when one of the provers receives 1, the other must respond with 0.

The value of the game is clearly $\rho(\mathbb{G}) \geq 2/3$; for instance, both the provers can always respond 0. In order to verify the optimality of this protocol, namely in order to show $\rho(\mathbb{G}) \leq 2/3$, we check all deterministic strategies. If both the provers reply with 0 irrespective of the queries, then they lose the game when $(x_1, x_2) = (0, 0)$. If one of the provers, say $\mathcal{P}_{\mathrm{p},1}$, responds $y_1 = 1$ for query $x_1 = 0$, then the provers lose the game when $(x_1, x_2) = (0, 1)$. Thus, $\rho(\mathbb{G}) = 2/3$.

If the game is repeated twice, then the provers can choose their reply based on two queries $(x_{1,1}, x_{1,2})$ and $(x_{2,1}, x_{2,2})$, respectively. If they set $(y_{1,1}, y_{1,2}) = (x_{1,2}, x_{1,1})$ and $(y_{2,1}, y_{2,2}) = (x_{2,2}, x_{2,1})$, then such a strategy attains

$$\rho(\mathbb{G}^{\wedge 2}) = \rho(\mathbb{G}) = 2/3.$$

In fact, this strategy wins each coordinate with probability 2/3, and when it wins one of the coordinates, then it wins the other one too since, by symmetry,

$$\omega(x_{1,1}, x_{2,1}, x_{1,2}, x_{2,2}) = \omega(x_{1,2}, x_{2,2}, x_{1,1}, x_{2,1}).$$

Even though the naive strategy is not optimal in general, we can show that too much improvement is not possible in the sense that $\rho(\mathbb{G}^{\wedge n})$ converges to 0 exponentially in n if $\rho(\mathbb{G}) < 1$.

THEOREM 16.6 (Parallel repetition) *There exists a function $C\colon [0,1] \to [0,1]$ satisfying $C(t) < 1$ if $t < 1$ such that, for any game $\mathbb{G}$,*

$$\rho(\mathbb{G}^{\wedge n}) \le C(\rho(\mathbb{G}))^{\frac{n}{\log |\mathcal{Y}_1||\mathcal{Y}_2|}}.$$

The statement of Theorem 16.6 holds for any game $\mathbb{G}$ with the same universal function $C(\cdot)$ and universal exponent that depends only on the cardinality of the response set $\mathcal{Y}_1 \times \mathcal{Y}_2$.

The parallel repetition theorem implies that the parallel repetition version of Protocol 16.3 still guarantees soundness. Apart from this, the parallel repetition theorem has many interesting consequences in theoretical computer science as well as quantum physics. The proof of the parallel repetition theorem is not straightforward, and is beyond the scope of this book. For further resources on this topic, see Section 16.5.

16.5 References and Additional Reading

For the background on computational complexity theory, see [311] or [5]. The concepts of the interactive proof and zero-knowledge proof were introduced in [148]. Most *randomized experiments* in statistics can be interpreted as interactive proofs, and they usually satisfy the zero-knowledge property as well. The story at the beginning of the chapter can be found in [296].

Technically speaking, the zero-knowledge requirement considered in Definition 16.3 is referred to as a *statistically zero-knowledge proof.* For more detail about statistical zero-knowledge proofs, see [338]. If we replace the variational distance criterion with computational indistinguishability (cf. Definition 9.4) in Definition 16.3, such a zero-knowledge proof is referred to as a *computationally zero-knowledge proof.* It was proved in [145] that all languages in NP have a computationally zero-knowledge proof. On the other hand, instead of the zero-knowledge requirement, we can also relax the soundness requirement in the sense that any computationally bounded prover cannot fool the verifier into accepting an instance that is not included in the language. Such a proof system is referred to as a *zero-knowledge argument.* In fact, the computationally zero-knowledge proof and the zero-knowledge argument are closely related to the computational security of the bit commitment (cf. Section 14.7): if a computationally hiding and information-theoretically binding bit commitment is available, then we can construct a computationally zero-knowledge proof; if an information-theoretically hiding and computationally binding bit commitment is available, then we can construct a statistically zero-knowledge argument. For more detail of various types of zero-knowledge proof, see [140, 141].

Multi-prover interactive proofs were introduced in [24]. The parallel repetition theorem was proved first in [285]; a simplified proof and an extension to the nonsignaling case was provided in [167]. The counterexample in Example 16.5 is from [167], which is a slight modification of the one in [128].

In certain applications, it is not enough to prove that a given instance is included in the language. For instance, in password identification, it is not enough to prove that there exists a password corresponding to the login ID; it is required to prove that the user indeed "knows" the password. Such a proof system is referred to as the *proof of knowledge*, and it has applications in user identification and the signature scheme [117, 302]; see also Problem 16.2. An abstract construction of proof of knowledge, which consists of commitment, challenge, and response, is referred to as the Σ protocol [93]. For a formal definition of the proof of knowledge, see [141, Section 4.7].

Even though interaction is a key feature of the zero-knowledge proof, in practice, it is desirable to conduct it noninteractively, i.e., one-way communication from the prover to the verifier. Under the random oracle model [20], it is possible to replace the verifier's challenge message with an output from a cryptographic hash function, and convert a Σ protocol to a noninteractive proof of knowledge, which is known as the Fiat–Shamir heuristic [119].

Problems

16.1 *Zero-knowledge proof for quadratic nonresidue*: For a nonnegative integer m, let $\mathbb{Z}_m^*$ be the set of all $1 \le x \le m-1$ such that x and m do not have a common factor. An element $a \in \mathbb{Z}_m^*$ is a quadratic residue modulo m if there exists $x \in \mathbb{Z}_m^*$ such that $a = x^2 \pmod{m}$; if a is not a quadratic residue, it is a quadratic nonresidue. Consider the following protocol to prove that, for a given $a \in \mathbb{Z}_m^*$, a is a quadratic nonresidue. (i) $\mathcal{P}_\mathrm{v}$ randomly samples $r \in \mathbb{Z}_m^*$ and $b \in \{0,1\}$, and sends $t = a^b \cdot r^2 \pmod{m}$ to $\mathcal{P}_\mathrm{p}$. (ii) If t is a quadratic residue, then $\mathcal{P}_\mathrm{p}$ sends $c = 0$ to $\mathcal{P}_\mathrm{v}$; otherwise, $\mathcal{P}_\mathrm{p}$ sends $c = 1$ to $\mathcal{P}_\mathrm{v}$. (iii) If $c = b$, $\mathcal{P}_\mathrm{v}$ accepts, otherwise, rejects. Convince yourself that this protocol is a zero-knowledge protocol (for honest verifier) [148].

16.2 *Proof of knowledge for discrete logarithm*: Let $\mathbb{G}$ be a group of prime order q such that the discrete logarithm on $\mathbb{G}$ is difficult, and let g be a publicly known generator of $\mathbb{G}$. Consider the following protocol to prove that, for a given $y \in \mathbb{G}$, $\mathcal{P}_\mathrm{p}$ knows $x \in \mathbb{Z}_q$ satisfying $y = g^x$. (i) The prover $\mathcal{P}_\mathrm{p}$ randomly samples $r \in \mathbb{Z}_q$, and sends $h = g^r$ to the verifier $\mathcal{P}_\mathrm{v}$. (ii) $\mathcal{P}_\mathrm{v}$ randomly samples $c \in \mathbb{Z}_q$, and sends it to $\mathcal{P}_\mathrm{v}$. (iii) $\mathcal{P}_\mathrm{p}$ sends $z = r + cx$ to $\mathcal{P}_\mathrm{v}$. (iv) $\mathcal{P}_\mathrm{v}$ accepts if $g^z = hy^c$ holds. Convince yourself that this protocol is a zero-knowledge protocol (for honest verifier) [302].

HINT $\mathcal{P}_\mathrm{v}$ can simulate the communication by sampling z and c at random and by computing $h = g^z (y^c)^{-1}$.

16.3 Verify that (16.5) and (16.6) attain the same value.

HINT Since the strategies in (16.5) subsume that in (16.6), it suffices to prove that (16.6) is as large as (16.5). To that end, for a given randomized strategy, explicitly write down the expectation in order to identify the optimal realization of local/shared randomness.

16.4 *CHSH game*: For $\mathcal{X}_1 = \mathcal{X}_2 = \mathcal{Y}_1 = \mathcal{Y}_2 = \{0, 1\}$, let $\mathrm{P}_{X_1 X_2}$ be the uniform distribution on $\{0,1\}^2$, and let predicate ω be given by $\omega(x_1, x_2, y_1, y_2) = \mathbf{1}[y_1 \oplus y_2 = x_1 \wedge x_2]$. Prove that the value of this game is $\rho(\mathbb{G}) \leq 3/4$.

COMMENT This game corresponds to a Bell type experiment that is used to demonstrate that quantum physics cannot be explained by the hidden variable theory [71].

16.5 *Expected value of parallel repetition game*: In n parallel repetitions of the game, let $N_\omega(x_1^n, x_2^n, y_1^n, y_2^n) := \sum_{i=1}^n \omega(x_{1,i}, x_{2,i}, y_{1,i}, y_{2,i})$ be the total number of winning coordinates. Then, let

$$\rho(\mathbb{G}^n) := \max\big\{\mathbb{E}[N_\omega(X_1^n, X_2^n, f_1(X_1^n), f_2(X_2^n))] : f_1 \in \mathcal{F}(\mathcal{Y}_1^n|\mathcal{X}_2^n), \\ f_2 \in \mathcal{F}(\mathcal{Y}_2^n|\mathcal{X}_2^n)\big\}.$$

Prove that $\rho(\mathbb{G}^n) = n\rho(\mathbb{G})$.

HINT Introduce shared randomness $U_i = (X_1^{i-1}, X_{1,i+1}^n, X_2^{i-1}, X_{2,i+1}^n)$ for each $i = 1, \ldots, n$, and use the expression (16.5).

COMMENT Instead of winning all the games, we can quantify the probability that the provers win significantly more numbers of coordinates than the expectation; for such concentration type results, see [4, 47, 217, 283, 335].

17 Two-party Secure Computation for an Active Adversary

We return to the two-party secure computation problem of Chapter 12. But this time we face a stronger adversary – an active adversary that may deviate from the protocol. Our main result in this chapter shows that oblivious transfer (OT) is complete for secure computation even with an active adversary. We follow the notions of security discussed in Chapter 15 and provide a protocol using ideal OT that sequentially composably securely implements a function. By Lemma 15.13, it suffices to establish standalone security for the protocol. In fact, we restrict the discussion to deterministic functions f and exhibit a protocol using OT that standalone securely implements f.

The overall protocol is very similar to the one we saw in Chapter 12 for a passive adversary. However, now we cannot trust the other party to follow the protocol. Therefore, it must "prove" each step using a zero-knowledge proof. In fact, these steps entail computation of (some other) Boolean functions with binary inputs held by both the parties, or even a single party (such as the NOT function). A little thought will convince the reader that, if these inputs and the function output are not committed, say using bit commitment (BC), it will be impossible to prove that the claimed function has been computed. Thus, we often commit the intermediate bits computed by one of the parties. The overall protocol is quite involved; we build it slowly by describing various components separately. Of particular interest is secure *local processing*, namely secure computation of functions that depend only on the input of one of the parties. We highlight secure computation protocols for such functions first, before utilizing them in devising secure computation protocols for general functions.

In the construction of our protocol, we use ideal BC often. However, in this chapter we only assume the availability of one *resource*: unlimited instances of OT. In Chapter 14, we presented Protocol 14.5 which implements BC using oracle access to randomized OT (ROT); ROT in turn can be implemented easily from OT. Note that we need to use both BC (and thereby ROT inside BC) several times, which requires this protocol to be sequentially composable secure. In Chapter 15, we presented a protocol that sequentially composably securely implements ROT from OT and also established in Section 15.6 that Protocol 14.5 sequentially composably securely implements BC from OT. Thus, we assume throughout this chapter that an ideal BC is available as an additional resource.

Providing a complete proof of security for a general function, using oracle access to ideal OT (and ideal BC), is a mammoth task. The proof is technically involved and tedious, and the authors feel that it is not appropriate to cover it in an introductory textbook such as this one. Instead, we focus on secure computing of a single function in this chapter, namely the AND function given by $f_{\mathtt{AND}}(x_1, x_2) = x_1 \wedge x_2$. A secure computation protocol for the AND function using OT captures all the key components needed for secure computation of an arbitrary (single-phase) function. Recall that, in Section 12.4, we presented a simple protocol to compute the AND function once.[1] At a high level, all we need to do is replace each step in that protocol with one that is secure even for an active adversary. But this, as we shall see, is quite complicated.

Throughout this chapter, we will establish security under Definition 15.11. We will build ideas in steps, starting with simple functions and building all the way to a secure computing protocol for the AND function. However, we will not be able to combine these components directly using a "composition theorem." This is because the composition framework in Chapter 15 cannot handle ideal functions with "committed inputs." Nonetheless, the analysis of the final protocol will draw from the analysis of these components, which we highlight separately. We remark that Chapter 15 is a companion for this chapter, and we recommend that the reader quickly reviews that chapter before proceeding with the current chapter.

17.1 The AND Function

We begin by describing the AND function, for which we seek a secure computation protocol. Guided by the "best practices" established in Chapter 15, we need to clearly describe the ideal AND function. As a first attempt, we can consider a simple form of ideal function where the parties observe inputs $(x_1, x_2) \in \{0, 1\}^2$ and both simultaneously get $x_1 \wedge x_2$ as outputs. Since our resource (ideal OT) is "asymmetric" – only one party gets the output – it will be very difficult to ensure that both parties get the output at the same time, securely. We strive for a lesser goal where the parties get the output in order, with, say, $\mathcal{P}_2$ getting the output before $\mathcal{P}_1$.

But this opens the door for a new attack: $\mathcal{P}_2$ can abort the protocol after it gets the output. Such attacks compromise the *fairness* of protocols (see Section 15.7 for a discussion) and are difficult to circumvent. We make such attacks admissible in this book by allowing a dishonest party $\mathcal{P}_2$ to abort even in the ideal world.

With this understanding, we are in a position to describe our ideal AND function, which we depict in Figure 17.1. This is a two-phase function, with $\mathcal{P}_2$ getting

[1] When we want to compute the AND function as a component of a large circuit, we need to use the more involved protocol described in Protocol 12.1.

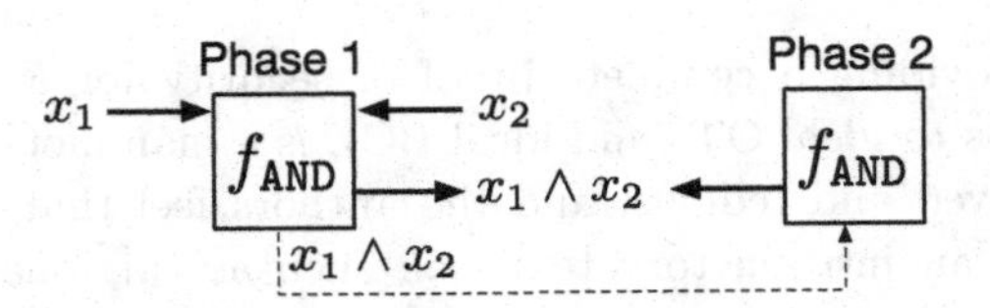

Figure 17.1 The ideal AND function.

$x_1 \wedge x_2$ in Phase 1, and $\mathcal{P}_1$ getting it in Phase 2. Note that there is no new input for Phase 2, and it must use the same inputs as the previous phase.

Following the convention from Chapter 15, we allow a dishonest party to use an abort input $\perp$ in the ideal world in any phase. In particular, a dishonest $\mathcal{P}_2$ can send an abort input $\perp$ after getting the value of AND, depriving $\mathcal{P}_1$ of the output. But the approach above has a possible vulnerability: *a dishonest party may not use the same input bit for the two phases.* Note that while it is admissible for a dishonest party to run the protocol with any input of its choice, it is not allowed for a dishonest party to change the input in the second phase of AND. Thus, in our construction, this has to be ruled out.

With these considerations in mind, we proceed to implement AND as follows.

1. In the first step, both parties commit their input bits using (a variant of) BC.
2. In the second step, the parties run a protocol to compute AND of the commited inputs. Note that there is no new input for this step.

The second step is technically involved – it requires each party $\mathcal{P}_i$ to prove to the other that the party $\mathcal{P}_i$ is indeed using the committed inputs for computation.

Next, we observe that the AND function can be easily computed using OT. Indeed, note that the output of OT using input $K_0 = 0$ and $K_1 = x_1$ for $\mathcal{P}_1$ and input $B = x_2$ for $\mathcal{P}_2$ gives the output $K_B = x_1 \wedge x_2$, i.e.,

$$f_{\text{OT}}((0, x_1), x_2) = x_1 \wedge x_2.$$

Therefore, if we can ensure that $\mathcal{P}_1$ uses 0 as the value of K_0, computing AND is the same as computing OT. But such an OT is not the same as the OT we are given as a resource – we now need to work with committed inputs. We will call such a function the *committed input oblivious transfer* (COT). In the COT, a party can commit to use any value that may not be known to the other party; in particular, a party can commit to use a fixed value, say 0. Thus, once the COT is realized, then we can force $\mathcal{P}_1$ to use $K_0 = 0$.

Remark 17.1 (Convention and notation for committed inputs) In this chapter, as seen above, we often need to work with *committed inputs.* In fact, such an input is not an input in the conventional sense and is viewed as an internal state or an input from a previous phase. We will denote a committed input x by $\boxed{x}$. Further, when we say $\mathcal{P}_1$ commits an input x, we inherently enable a reveal phase where it will be revealed to $\mathcal{P}_2$. However, the reveal phase need not be engaged during the protocol. For concreteness, the reader can imagine a reveal phase at the end, once all other computations are done. Thus, we will refer to the committed inputs without concretely describing when the corresponding reveal phase is engaged.

The proposed implementation of AND:
In summary, our strategy for implementing the AND function is as follows.

1. $\mathcal{P}_1$ commits x_1 and $\mathcal{P}_2$ commits x_2.
2. Both parties commit (separately) a 0 too.
3. Parties use COT applied with input $(K_0, K_1) = (\boxed{0}, \boxed{x_1})$ for $\mathcal{P}_1$ and input $B = \boxed{x_2}$ for $\mathcal{P}_2$; $\mathcal{P}_2$ gets the committed output $K_B = x_1 \wedge x_2$.
4. Parties use COT applied with input $B = \boxed{x_1}$ for $\mathcal{P}_1$ and input $(K_0, K_1) = (\boxed{0}, \boxed{x_2})$ for $\mathcal{P}_2$; $\mathcal{P}_1$ gets the committed output $K_B = x_1 \wedge x_2$.

Before we start providing implementation for each component, there are further elaborations needed for this high-level plan, which we provide below.

Remark 17.2 Working with committed inputs requires the parties to provide a *proof* that indeed they are working with the committed input. When the parties provide such a proof, they may reveal the committed input (using the reveal phase). But then they will not be able to use this committed input again. Thus, at the outset the parties commit multiple copies of the input.

Remark 17.3 While it is desirable to make the security proof "modular", where we just prove the security of each component and use a composition theorem to establish security of the combined protocol, we must treat the entire protocol as a multiphase protocol that shares committed inputs. This is because the composition framework in Chapter 15 cannot handle ideal functions with committed inputs.

With these clarifications, our plan is ready. We depict the overall protocol in Figure 17.2. We present the various components needed in separate sections below, before tying everything together in Section 17.5.

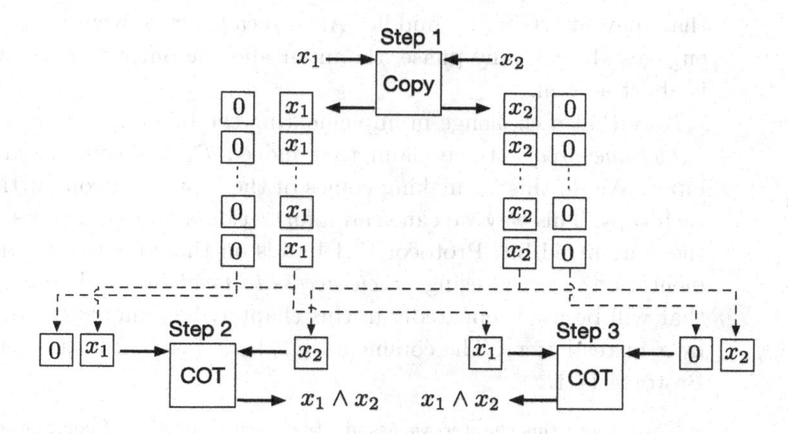

Figure 17.2 The proposed implementation of the AND function.

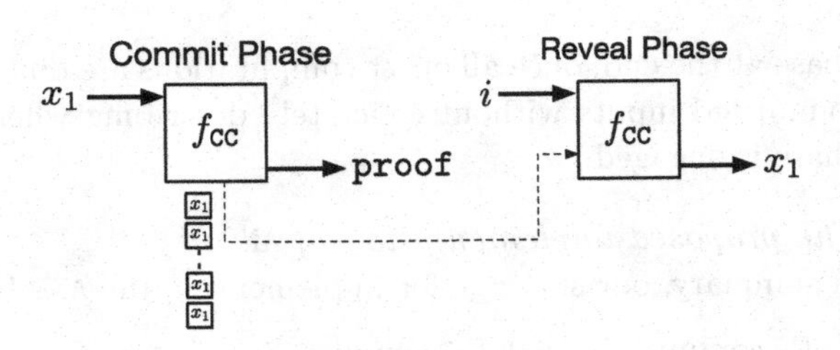

Figure 17.3 The ideal n-CC function.

17.2 The Committed Copy Function

The first component we need to implement AND is a committed copy function. This function takes as an input a single bit $x_1 \in \{0, 1\}$ for, say, $\mathcal{P}_1$ and commits its n copies. The party $\mathcal{P}_2$ receives a proof that the committed values are indeed a copy of each other. These values are kept hidden from $\mathcal{P}_2$ in the commit phase. In the reveal phase, $\mathcal{P}_2$ can ask to reveal any of the n committed values by providing an index $i \in [n]$ as the input.

Formally, the protocol runs in two phases. In the *commit phase*, $\mathcal{P}_1$ makes n copies of the input and commits them, and $\mathcal{P}_2$ receives `proof` to indicate that indeed $\mathcal{P}_1$ has committed to copies without cheating. In the *reveal phase*, $\mathcal{P}_2$ inputs $i \in [n]$ and $\mathcal{P}_1$ reveals the ith copy.[2] Note that, when $\mathcal{P}_1$ is honest, the output will coincide with x_1 for every input i.

We depict the ideal *n-committed copy* (n-CC) function in Figure 17.3. As was our convention in Chapter 15, we allow abort input and output for both phases. An honest $\mathcal{P}_1$ will not use the abort input $\perp$ and will commit copies of the input x_1, giving `proof` as output to $\mathcal{P}_2$. The output $\perp$ in the proof phase will never occur when $\mathcal{P}_1$ is honest. However, a dishonest $\mathcal{P}_1$ can use a $\perp$ input, in which case the output for $\mathcal{P}_2$ will be $\perp$ in the commit phase. Recall that this notion of $\perp$ input and output is useful when establishing the security of real protocols that may abort in the middle. Also, recall our convention that, when a party engages abort in any phase, the input and the output for all subsequent phases is abort as well.

Note that a challenge in implementing the function above is to enable a *zero-knowledge proof* of the claim that indeed $\mathcal{P}_1$ has committed to copies of the input. We do this by making copies of the input and committing to "shares" of each copy. This way we can send information about the shares without revealing the committed bit. Protocol 17.1 builds on this idea to provide a secure implementation of n-CC using *oracle access to ideal BC*, and constitutes a basic tool that will be used repeatedly in this chapter. For simplicity, we denote the input by x instead of x_1. The comments $\{\dots\}$ are inserted for readers' convenience in Protocol 17.1.

[2] Throughout this chapter, we assume that each invocation of commitment has a unique ID and the committing party cannot cheat to reveal the value corresponding to a different ID.

Protocol 17.1 Committed copy using BC

Input: $\mathcal{P}_1$'s input x and parameters n, m, ℓ

Output: A proof for $\mathcal{P}_2$ that $\mathcal{P}_1$ has committed n copies of an input

1. $\mathcal{P}_1$ samples independent random (uniform) bits $U_1, \ldots, U_m$ and creates "shares" $x_{i0} = U_i$ and $x_{i1} = U_i \oplus x$, $1 \le i \le m$.

   Commit phase
2. $\mathcal{P}_1$ makes $2m$ accesses to an ideal BC oracle to commit $(x_{i0}, x_{i1})_{i=1}^m$ to $\mathcal{P}_2$. {We assume that $\mathcal{P}_2$ can ask $\mathcal{P}_1$ to reveal any of these $2m$ BCs by providing an index.}

   Proof
3. $\mathcal{P}_2$ selects a pair of locations (s, t) randomly without replacement from the set $[m]$ and sends it to $\mathcal{P}_1$.
4. If s or t is already selected, $\mathcal{P}_1$ declares the abort symbol $\perp$.
5. $\mathcal{P}_1$ sends $c_0 = \boxed{x_{s0}} \oplus \boxed{x_{t0}}$ and $c_1 = \boxed{x_{s1}} \oplus \boxed{x_{t1}}$ to $\mathcal{P}_2$.
   {In this step, $\mathcal{P}_1$ can cheat and may not use the committed values $\boxed{x_{s0}}, \boxed{x_{t0}}, \boxed{x_{s1}}, \boxed{x_{t1}}$. }
6. $\mathcal{P}_2$ checks $c_0 = c_1$; if it is not satisfied, $\mathcal{P}_2$ declares the abort symbol $\perp$.
7. $\mathcal{P}_2$ sends a bit $B \sim \texttt{Ber}(1/2)$ to $\mathcal{P}_1$, and $\mathcal{P}_1$ reveals $(\boxed{x_{sB}}, \boxed{x_{tB}})$ to $\mathcal{P}_2$ (using the ideal BCs).
8. $\mathcal{P}_2$ checks $c_B = x_{sB} \oplus x_{tB}$; if it is not satisfied, $\mathcal{P}_2$ declares the abort symbol $\perp$.
9. Repeat Steps 3–8 above ℓ times and declare `proof` if $\perp$ is not declared.
10. Let $\mathcal{L}$ denote the locations in $[m]$ that are not chosen in any of the ℓ pairs selected in Steps 3–8.
    {The values at locations in $\mathcal{L}$ constitute the committed copies. Note that $|\mathcal{L}| = m - 2\ell$.}

    Reveal phase
11. $\mathcal{P}_2$ inputs $i \in \mathcal{L}$ and $\mathcal{P}_1$ reveals $(\boxed{x_{i0}}, \boxed{x_{i1}})$ (using the ideal BC).

We now establish the security of this protocol under Definition 15.11. We remark that our proofs in this chapter can perhaps be viewed as a (detailed) proof-sketch. Complete arguments such as those provided for implementation of BC from OT in Section 15.6 will be very tedious, and perhaps distract the reader from the key ideas. We omit them and urge the reader to try to write a complete proof, at least for a few functions.

LEMMA 17.1 (Secure n-CC) *Protocol 17.1 is a sequentially composable ε-secure implementation of n-CC using g_{BC} with $n = m - 2\ell$ and $\varepsilon = (1 - \frac{1}{2m})$.*

Remark 17.4 The security parameter $\varepsilon = (1 - \frac{1}{2m})$ is not satisfactory – we would have liked to see this parameter improve with ℓ. However, this is not the case for the current protocol since we are trying to implement n-CC. In fact, we do not need to implement n-CC for our protocol for AND; we only need to implement an approximate version of it. Lemma 17.1 may be viewed as an intermediate statement towards a more useful statement that we will present later.

Proof *When both $\mathcal{P}_1$ and $\mathcal{P}_2$ are honest*: It can be verified that the protocol is perfectly secure when the parties are honest.

When $\mathcal{P}_1$ is honest and $\mathcal{P}_2$ is dishonest: We note that the bits revealed to $\mathcal{P}_2$ in Steps 3–9 are independent of the input X (for any input distribution). Therefore, the transcript $\Pi_1^{(2)}$ of $\mathcal{P}_2$ up to the commit phase can be simulated simply by outputting random bits. Thus, the protocol is perfectly secure against dishonest $\mathcal{P}_2$.

When $\mathcal{P}_1$ is dishonest and $\mathcal{P}_2$ is honest: This is the interesting part of the proof. Let $\tilde{x} = \mathtt{maj}(x_{10} \oplus x_{11}, x_{20} \oplus x_{21}, \ldots, x_{m0} \oplus x_{m1})$, where $\mathtt{maj}$ denotes the *majority* function. Even when $\mathcal{P}_1$ is dishonest, $\mathcal{P}_1$ is committing to at least $m/2$ copies of $\tilde{x}$. However, $\mathcal{P}_2$ does not know where these "genuine" committed copies are stored. Specifically, the set $\mathcal{I} = \{i \in [m] : x_{i0} \oplus x_{i1} \neq \tilde{x}\}$ corresponds to the set of indices containing wrong values.

If $\mathcal{P}_2$ detects this inconsistency and declares abort $\perp$ in the real world, this can be simulated in the ideal world simply by $\mathcal{P}_1$ using the input $\perp$ in the commit phase. The ideal world and real world differ only when $\mathcal{P}_1$ has committed to inconsistent values (i.e., $|\mathcal{I}| > 0$) and $\mathcal{P}_2$ accepts the proof by declaring $\mathtt{proof}$ at the end of the commit phase. In this case, the real world and the ideal world will deviate.

Formally, denote by $\pi^{f_{\mathrm{CC}}}$ the protocol in Protocol 17.1. Consider the simulator σ for the ideal world where $\mathcal{P}_1$ executes Protocol 17.1 locally (executing the steps for both the parties) and generates the pairs (s_i, t_i) and bit B_i, $1 \le i \le \ell$, used in ℓ repetitions of Steps 3–8. If $\mathcal{P}_1$ gets an abort output $\perp$ from this local execution, it uses the input $\perp$ for the ideal n-CC f_{CC}; otherwise it uses the majority input $\tilde{x}$ for f_{CC}.

Denote by X $\mathcal{P}_1$'s input and by Z and $\tilde{Z}$, respectively, the output of $\mathcal{P}_2$ after the commit phase in the real and the ideal worlds (using simulator σ). As we discussed above, since deviation of the real protocol and the ideal protocol occurs only if $Z \neq \tilde{Z}$, we have

$$d_{\mathtt{var}}(\mathtt{V}[\pi^{f_{\mathrm{CC}}} \diamond g_{\mathrm{BC}}], \mathtt{V}[\sigma \circ f_{\mathrm{CC}}]) \le \Pr(Z \neq \tilde{Z}) = \Pr(\mathcal{E}),$$

where $\mathcal{E}$ is the event that in the local execution of Protocol 17.1 as part of σ, $\mathcal{P}_1$ has committed to copies with $|\mathcal{I}| > 0$ and the output is not $\perp$ after the commit phase.

We now derive a bound for $\Pr(\mathcal{E})$. In the first iteration of Steps 3–8, $\mathcal{P}_2$ successfully declares $\perp$ if exactly one of s or t is in $\mathcal{I}$ and $\mathcal{P}_1$ fails to guess B correctly. Note that even if $s \in \mathcal{I}$ and $t \notin \mathcal{I}$, $\mathcal{P}_1$ can send $c_0 = x_{s0} \oplus x_{t0}$ and $c_1 = x_{s1} \oplus x_{t1} \oplus 1$ so that the test in Step 6 will be passed. However, such an attack will be detected with probability $1/2$ in Step 8. Consequently, the probability that $\mathcal{P}_2$ misdeclares $\perp$ in the first iteration is at most $1 - \frac{1}{2}\frac{|\mathcal{I}|(m-|\mathcal{I}|)}{m(m-1)}$. In the worst case $|\mathcal{I}| = 1$, and this probability is $1 - \frac{1}{2m}$. Even though the probability of misdeclaration can be made smaller by taking into account the ℓ iterations, we do not conduct such an analysis here. In fact, in the worst case of $|\mathcal{I}| = 1$, further iterations do not reduce the error by much.

The proof is completed upon noting that the reveal phases in the real and ideal worlds coincide; the only deviation can happen in the commit phase and it can go undetected only when `proof` is declared erroneously (namely, $\mathcal{P}_2$ accepted the proof, while the copies are not all the same). □

We remind the reader that the previous result assumes the availability of an ideal BC. When this is implemented using BC with an ideal OT, there will be further degradation of security. However, by Theorem 14.3, Lemma 15.16, and the sequential composition theorem for multiphase functions discussed in Chapter 15, we only need to replace ε with $\varepsilon + N\varepsilon'$, where ε' can be made as small as we want and N is the number of instances of BC used. We ignore these additional terms throughout this chapter.

The approximate committed copy function

As remarked earlier, the previous result is not satisfactory – the security parameter $\varepsilon = 1 - \frac{1}{2m}$ is unacceptable. The reason for this poor security is that it is very difficult to detect $\mathcal{P}_1$'s misbehavior if $\mathcal{P}_1$ creates a small number of inconsistent pairs, say just 1. Luckily, we only need an approximate version of the copy function where we are guaranteed that at most a fraction $\delta \in (0, 1/2)$ of the committed copies are inconsistent (have values differing from the input). Furthermore, the positions of inconsistent copies can be chosen by the committing party; this is for the convenience of simulating misbehavior of a real protocol. In other words, we are defining a partial inconsistency as an admissible attack. We call this function the *(n, δ)-committed copy* ((n, δ)-CC) function and depict the ideal (n, δ)-CC function in Figure 17.4. Interestingly, Protocol 17.1 provides high security for this function for appropriate choices of parameters m and ℓ, when n is sufficiently large.

LEMMA 17.2 (Secure (n, δ)-CC) *Protocol 17.1 is a sequentially composable $\varepsilon_{\mathsf{cc}}$-secure implementation of (n, δ)-CC using g_{BC} with $m = 2n$, $\ell = n/2$, where $\varepsilon_{\mathsf{cc}} := (2n+1)e^{-\frac{n\delta}{4}}$.*

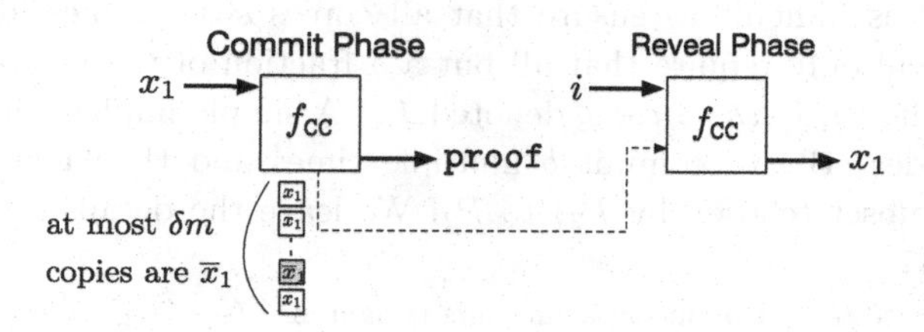

Figure 17.4 The ideal (n, δ)-CC function; at most δn copies are $\bar{x}_1 = x_1 \oplus 1$.

Proof The proof is similar to that of Lemma 17.1. Previously, we bounded the simulation error by the probability of the event $\mathcal{E}$ that $|\mathcal{I}| > 0$ and the output is not $\perp$. Since our goal now is to implement (n, δ)-CC, the simulation fails only if $|\mathcal{I}| > \delta n$ and it is undetected.[3] Thus, using the same reasoning as the proof of Lemma 17.1, we can bound the simulation error by the probability of the event $\mathcal{E}$ that $|\mathcal{I}| > \delta n$ and the output is not $\perp$ after the commit phase.

Since direct evaluation of $\Pr(\mathcal{E})$ is cumbersome, we first evaluate the probability of event $\mathcal{E}$ under the assumption that ℓ iterations of pair (s, t) in Step 3 are sampled with replacement. Note that, in each iteration, $\mathcal{P}_2$ successfully declares $\perp$ if exactly one of s or t is in $\mathcal{I}$ and $\mathcal{P}_1$ fails to guess B correctly. Thus, under the assumption of sampling with replacement, we can upper bound the probability of event $\mathcal{E}$ above as

$$\begin{aligned}\left\{(1-\tilde{\delta})^2 + \tilde{\delta}^2 + 2(1-\tilde{\delta})\tilde{\delta}\frac{1}{2}\right\}^\ell &\leq \exp\left(-\ell\{1-(1-\tilde{\delta})^2 - \tilde{\delta}^2 - (1-\tilde{\delta})\tilde{\delta}\}\right) \\ &= \exp(-\ell\tilde{\delta}(1+\tilde{\delta})) \\ &\leq e^{-\ell\tilde{\delta}},\end{aligned}$$

where $\tilde{\delta} := \frac{|\mathcal{I}|}{m} > \frac{\delta}{2}$ and we used $t \leq e^{-(1-t)}$ in the first inequality. Then, by using the connection between the sampling with and without replacement given in Problem 17.3, we have

$$\Pr(\mathcal{E}) \leq (m+1)e^{-\ell\tilde{\delta}}.$$

Thus, by noting that $\tilde{\delta} > \frac{\delta}{2}$ and substituting $m = 2n$ and $\ell = \frac{n}{2}$, we have the desired bound $\varepsilon_{\mathsf{cc}}$ on simulation error.

The proof is completed in the manner of Lemma 17.1. We remark that $\mathcal{P}_1$ is committed to n copies of the majority $\tilde{x}$ of $\{x_{i0} \oplus x_{i1} : i \in \mathcal{L}\}$ with at most $n\delta$ values that equal $\tilde{x} \oplus 1$. In the reveal phase, it is an admissible attack to get an inconsistent answer for $n\delta$ copies. □

The committed copy of a value

We close this section by pointing to another useful function: $\mathcal{P}_1$ wants to commit n copies of a fixed, public value $b \in \{0, 1\}$. This will help $\mathcal{P}_1$ prove to $\mathcal{P}_2$ that $\mathcal{P}_1$ has indeed used the public value of b in a local step. We need multiple copies so that the proof can be correct with reasonable probability and the same value can be retained for future use. As in the case of the committed copy function, it is difficult to ensure that all copies store values identical to b. Instead, we will only require that all but a δ fraction of values are b. We call this function the (n, δ)*-value copy*, denoted f_{VC}. A simple implementation will entail $\mathcal{P}_1$ using ideal BC to commit b multiple times and then revealing a randomly chosen subset (chosen by $\mathcal{P}_2$) to $\mathcal{P}_2$. We leave the details as an exercise; see Problem 17.1.

[3] Note that $n\delta$ inconsistent pairs remain in $\mathcal{L}$ (see Step 10 in Protocol 17.1) only if $|\mathcal{I}| > \delta n$.

17.3 Proving Linear Relations

Many applications in cryptography require the parties to carry out linear operations on several bits they have. When done locally, the party doing the computation needs to prove to the other that the party has indeed completed the desired linear operation and is using the output for the subsequent protocol. In fact, we can view the copy function as committing to $x_1, \ldots, x_n$ that satisfy $x_i \oplus x_j = 0$ for (almost) every pair i and j. In this section, we present a primitive that can be used to complete such proofs.

We can formalize the above mentioned function as a "standalone utility" by requiring the parties to first commit to the values of the bits in the first phase and then proving that the committed values satisfy linear relations in the second phase. Note that we will need to allow verifications of multiple linear relations. When implementing such a protocol using ideal BC, we will need to copy a previously committed value to prove relations and retain the original value for future use. However, it is difficult to copy securely from a single committed value – we do not know how to guarantee that the copy is identical to the committed value. Instead, we can implement the committed copy function of the previous section at the outset with a sufficiently large n and a sufficiently small δ, inherently committing to the majority of the values stored. Then, we can use these committed copies to prove linear relations.

For brevity,[4] we will not formalize the function described above as an ideal function. Instead, we will simply describe how to prove a linear relation among bits $x_1, \ldots, x_r$ committed using (n, δ)-CC with n copies of each.

Remark 17.5 (A notation for committed copies) We will denote a bit b committed using (n, δ)-CC by $\boxed{b}$ (with a bold box), omitting the dependence on n and δ from the notation; both these parameters will remain fixed.

The first step in our procedure is to make additional copies of b using $\boxed{b}$. This additional copy function can be formally introduced as the second phase (before the reveal phase) for the committed copy function we saw earlier. But, once again, we skip the description of an ideal function of this function and provide a protocol in Protocol 17.2 for doing this. Note that we will make $2n$ new copies: n out of the resulting $3n$ copies will be used for proving the linear relation, n will be retained for future use, and n will be "destroyed" in proving that indeed these copies are genuine.

Protocol 17.2 provides (almost) the same security guarantees as Protocol 17.1. Note that an inconsistent copy created during the execution Protocol 17.1 may be caught during the execution of Protocol 17.2, resulting in an abort. This behavior is apt – while some inconsistency is tolerable, we do not expect an

[4] As we mentioned at the beginning of this chapter, the composition framework in Chapter 15 cannot handle ideal functions with inputs committed by (n, δ)-CC. Thus, for the purpose of this chapter, it is easier to directly present a protocol for proving linear relation.

Protocol 17.2 A protocol for creating additional copies from a committed copy $\boxed{x}$

Input: $\mathcal{P}_1$'s input $\boxed{x}$ obtained using the commit phase of Protocol 17.1
Output: Two new copies $\boxed{x}$

1. $\mathcal{P}_1$ computes $x = \texttt{maj}(\boxed{x_{10}} \oplus \boxed{x_{11}}, \ldots, \boxed{x_{n0}} \oplus \boxed{x_{n1}})$, where $(\boxed{x_{i0}}, \boxed{x_{i1}})$, $1 \le i \le n$, are the shares of the ith committed copy created by Protocol 17.1.
2. $\mathcal{P}_1$ samples independent random (uniform) bits $U_{n+1}, \ldots, U_{4n}$ and creates additional shares $x_{i0} = U_i$ and $x_{i1} = U_i \oplus x$, $n + 1 \le i \le 4n$.
3. $\mathcal{P}_1$ makes $6n$ accesses to an ideal BC oracle to commit $(x_{i0}, x_{i1})_{i=n+1}^{4n}$ to $\mathcal{P}_2$.
4. Repeat Steps 3–8 of Protocol 17.1 $\ell = n$ times applied to the $m = 4n$ shares $(\boxed{x_{i0}}, \boxed{x_{i1}})_{i=1}^{4n}$ to $\mathcal{P}_2$. Declare `proof` if $\bot$ is not declared.
5. Retain $2n$ out of $4n$ copies which were not selected in any of the $2n$ copies of pairs selected in the previous step.

honest $\mathcal{P}_1$ to commit inconsistent values, and it is desirable to catch any such inconsistency at any stage of the protocol.

Similar to Lemma 17.2, we have the following guarantees for Protocol 17.2.

LEMMA 17.3 (Security of additional copies) *Bits revealed to $\mathcal{P}_2$ by Protocol 17.2 are independent of the input $\boxed{x}$ of Protocol 17.2. Furthermore, the probability that there are more than δn inconsistent bits in the $2n$ output copies and $\mathcal{P}_2$ will not declare $\bot$ is at most $(4n+1)e^{-\frac{n\delta}{4}}$.*

Note that we only allow δn inconsistent values in the final $2n$ copies, resulting in at most a fraction δ of inconsistent values in either of the new copies of $\boxed{x}$. Thus, we retain the "accuracy" of δ throughout, with an additional probability $(4n+1)e^{-\frac{n\delta}{4}}$ of failure added every time further copies are made.

With this basic primitive, we are in a position to provide our protocol for proving linear relations between committed values. Specifically, given committed copies $\boxed{x_i}$, $1 \le i \le r$, the goal is to show that the committed values satisfy $\oplus_{i=1}^{r} a_i x_i = b$ for fixed bits $a_1, \ldots, a_r$ and b known to both parties. However, we need a *zero-knowledge proof*; namely, we do not want to reveal the values $(x_1, \ldots, x_r)$ to $\mathcal{P}_2$. Protocol 17.3 accomplishes this, and is very similar to Protocol 17.1.

The following result establishes the security properties of Protocol 17.3.

LEMMA 17.4 (Security of proving linear relations) *Protocol 17.3 satisfies the following security properties.*

1. *The bits revealed to $\mathcal{P}_2$ are independent of the input $(\boxed{x_1}, \ldots, \boxed{x_r})$.*
2. *The probability that any one of $\boxed{x_1}, \ldots, \boxed{x_r}$ stored for future use has more than δ fraction of inconsistent values and $\bot$ is not declared is at most $r(4n+1)e^{-\frac{n\delta}{4}}$.*

Protocol 17.3 A protocol for proving linear relations between committed bits

---

Input: $\mathcal{P}_1$'s input $\boxed{x_i}$, $1 \le i \le r$, obtained using Protocol 17.1 or Protocol 17.2, bits a_i, $1 \le i \le r$, and b

Output: A proof for $\mathcal{P}_2$ that $\oplus_{i=1}^r a_i x_i = b$

1. $\mathcal{P}_1$ uses Protocol 17.2 to obtain two new copies for each $\boxed{x_i}$, $1 \le i \le r$ and saves one copy for future use. The following steps are applied to the other copy.
2. $\mathcal{P}_1$ sends $2n$ bits (c_{s0}, c_{s1}) to $\mathcal{P}_1$, given by $c_{s0} = \oplus_{i=1}^r a_i \boxed{x_{is0}}$ and $c_{s1} = \oplus_{i=1}^r a_i \boxed{x_{is1}}$, where $(\boxed{x_{is0}}, \boxed{x_{is0}})$ denotes the shares of the sth copy stored in $\boxed{x_i}$, $1 \le i \le r$, $1 \le s \le n$.
3. $\mathcal{P}_2$ checks $c_{s0} \oplus c_{s1} = b$ for all $1 \le s \le n$; if it is not satisfied, $\mathcal{P}_2$ declares the abort symbol $\perp$.
4. $\mathcal{P}_2$ sends random uniform bits $B_1, \ldots, B_n$ to $\mathcal{P}_1$, and $\mathcal{P}_1$ reveals bits $\boxed{x_{isB_s}}$, $1 \le i \le r$ and $1 \le s \le n$, to $\mathcal{P}_2$ (using the ideal BCs).
5. $\mathcal{P}_2$ checks $c_{sB_s} = \oplus_{i=1}^r a_i x_{isB_s}$ for all $1 \le s \le n$; if it is not satisfied, $\mathcal{P}_2$ declares the abort symbol $\perp$.

---

3. *Denote by* $\tilde{x}$ *the value* $\mathtt{maj}(\boxed{x_{10}} \oplus \boxed{x_{11}}, \ldots, \boxed{x_{n0}} \oplus \boxed{x_{n1}})$ *computed from copies stored in* $\boxed{x}$. *The probability that* $\oplus_{i=1}^r a_i \tilde{x}_i \ne b$ *and* $\mathcal{P}_2$ *does not declare* $\perp$ *is at most* $2^{-(1-r\delta)n} + r(4n+1)e^{-\frac{n\delta}{4}}$.

Proof The first property holds since only one share of each copy of the x_i is revealed to $\mathcal{P}_2$.

For the second, by Lemma 17.3, if one of the copies created for future use in Step 1 has more than δ fraction of values differing from the majority, $\perp$ is not declared with probability at most $r(4n+1)e^{-\frac{n\delta}{4}}$.

Finally, $\mathcal{P}_1$ can provide a "fake" proof of linear relation to $\mathcal{P}_2$ either if for some $i \in [r]$ the copy of x_i made for the proof is not "genuine," namely it has more than a δ fraction of inconsistent values or does not coincide with the original majority value, or if even while using a genuine copy $\mathcal{P}_1$ has provided a fake proof. As argued above, the probability of $\mathcal{P}_2$ accepting a copy that is not genuine is at most $r(4n+1)e^{-\frac{n\delta}{4}}$. Further, to bound the probability of a successful acceptance of a fake proof despite having genuine copies, we note first that there are at least $(1 - r\delta)n$ locations for which copies of x_i are consistent with the majority value for every $1 \le i \le r$. Denote this set of locations by $\mathcal{S}$. If the x_i do not satisfy the desired linear relation, $c_{s0} \oplus c_{s1} = b$ will hold for $s \in \mathcal{S}$ only if $c_{s0} \ne \oplus_{i=1}^r x_{is0}$ or $c_{s1} \ne \oplus_{i=1}^r x_{is1}$. This in turn will not be detected only if B_s did not coincide with the "faulty" share, which can happen with probability 1/2. Thus, the probability that the proof is wrong is at most $2^{-(1-r\delta)n}$. □

17.4 The Committed Oblivious Transfer Function

The final component needed for implementing AND is an implementation of COT, a "function" we depicted in Figure 17.2 earlier. COT is the same as OT, except that no input is given and the parties must use previously committed values. For our purpose, we will use values $(\boxed{k_0}, \boxed{k_1})$ and $\boxed{b}$ obtained using Protocol 17.1 or Protocol 17.2 as inputs for $\mathcal{P}_1$ and $\mathcal{P}_2$, respectively. As in the previous section, we do not formalize COT as an ideal function, but simply present it as the next phase of our protocol.

Our implementation of COT uses a *linear code*. A reader can refer to Section 11.5 and references in Section 11.6 to review the basic concepts needed. We recall the key definitions and properties below.

DEFINITION 17.5 (Linear code) A subset $\mathcal{C} \subset \{0,1\}^m$ constitutes an (m,k,d)-code if the following properties hold: (i) for every $c_1, c_2 \in \mathcal{C}$, $c_1 \oplus c_2 \in \mathcal{C}$; (ii) $|\mathcal{C}| = 2^k$; and (iii) $\min\{d_H(c_1,c_2) : c_1, c_2 \in \mathcal{C}\} = d$, where $d_H(u,v) = \sum_{i=1}^m \mathbf{1}[u_i \neq v_i]$ denotes the Hamming distance between binary vectors u and v.

LEMMA 17.6 (Properties of linear codes) *For an (m,k,d)-code $\mathcal{C}$, the following properties hold.*

1. *Parity check matrix. There exists an $m \times (m-k)$ matrix H such that $\mathcal{C} = \{x \in \{0,1\}^m : xH = 0\}$.*
2. *Nearest neighbor decoding. For any binary vector $u \in \{0,1\}^m$, let $\phi(u) = \mathrm{argmin}\{d_H(u,c) : c \in \mathcal{C}\}$ be the nearest neighbor decoding rule, where the tie is broken in an arbitrary manner. If u is such that $d_H(c,u) < d/2$ for some codeword $c \in \mathcal{C}$, then $\phi(u) = c$. In other words, it is guaranteed that the (n,k,d)-code can correct up to $\lfloor d/2 \rfloor$ errors by the nearest neighbor decoding.*
3. *Positive rate code. For every $\theta, \varepsilon \in (0,1/2)$, there exists an $(m, m(1-h(\theta)-\varepsilon), \theta m)$-code for every m sufficiently large, where $h(x) = -x\log x - (1-x)\log(1-x)$ is the* binary entropy function.[5]

Besides these standard properties, we need the following simple observation, whose proof is left as an exercise (see Problems 17.2 and 17.3).

LEMMA 17.7 *For $\theta \in (0,1/2)$, consider an $(m,k,\theta m)$-code $\mathcal{C}$, and let $I_1, \ldots, I_\ell$ be independent and distributed uniformly on $[m]$. For two distinct codewords $u, v \in \mathcal{C}$, we have*

$$\Pr(u_{I_j} = v_{I_j}, \forall 1 \le j \le \ell) \le e^{-\theta \ell}.$$

Moreover, let $\mathcal{I} \subset [m]$ be a random subset of size $|\mathcal{I}| = \ell$, i.e., sampling without replacement, then, for any distinct codewords $u, v \in \mathcal{C}$, we have

$$\Pr(u_j = v_j,\ \forall j \in \mathcal{I}) \le (m+1)e^{-\theta \ell}.$$

Lemma 17.7 guarantees that, if randomly sampled bits coincide, the two codewords must be the same with high probability, and it will be used later

[5] This bound is known as the Gilbert–Varshamov bound in coding theory.

to prove that a party is using a committed codeword. Besides linear codes, another tool we shall use in our implementation of COT is the universal hash family (UHF) of Chapter 4. In particular, we shall use a special instance of random linear mapping $F\colon \{0,1\}^k \to \{0,1\}$ that is described as $F(x) = (\oplus_{i=1}^k B_i x_i)$ using k random bits $B_1, \ldots, B_k$. We first note that this family of mappings constitutes a *universal hash family* (see Definition 4.8). Indeed, for distinct $x, x' \in \{0,1\}^k$, the hash collision event $F(x) = F(x')$ can be seen to occur with probability at most 1/2. By the *leftover hash lemma* (Corollary 7.22), we obtain the following property of random linear mappings, which will be used in our analysis.

LEMMA 17.8 *Let X be distributed uniformly over $\mathcal{X} \subset \{0,1\}^k$ and $F\colon \{0,1\}^k \to \{0,1\}$ be the random linear mapping specified above. Then, for any random variable Z (that may be correlated to X) taking values in $\mathcal{Z}$, we have*

$$d_{\mathtt{var}}\left(\mathrm{P}_{F(X)FZ}, \mathrm{P}_{\mathtt{unif}} \times \mathrm{P}_{FZ}\right) \le 2^{(1+\log|\mathcal{Z}|-\log|\mathcal{X}|)/2-1},$$

where $\mathrm{P}_{\mathtt{unif}}$ denotes the uniform distribution $\mathtt{Ber}(1/2)$.

All our tools are ready. We now describe our protocol, first heuristically, before providing the complete description in Protocol 17.4.

A heuristic description of COT protocol

$\mathcal{P}_1$ has committed bits (K_0, K_1) as input and $\mathcal{P}_2$ has committed bit B. We could have used OT to send K_B from $\mathcal{P}_1$ to $\mathcal{P}_2$, but there is no guarantee that the parties will use their committed inputs for engaging OT. This is the main difference between OT and COT: we must now force the parties to use their committed inputs.

Our overall strategy is as follows: we implement a randomized OT first, which gives random bits (U_0, U_1) to $\mathcal{P}_1$ and (B, U_B) to $\mathcal{P}_2$. $\mathcal{P}_1$ can then simply encrypt its input (K_0, K_1) using one-time pad encryption and send $C_0 = K_0 \oplus U_0$ and $C_1 = K_1 \oplus U_1$) to $\mathcal{P}_2$, of which $\mathcal{P}_2$ can decrypt only $K_B = C_B \oplus U_B$.

But this does not seem to have addressed any of our original misgivings. How do we know that $\mathcal{P}_1$ has used the committed inputs for computing C_0, C_1? And how do we know that $\mathcal{P}_2$ has used $\mathcal{P}_2$'s committed input B? The protocol below has been designed keeping these issues in mind.

Specifically, $\mathcal{P}_1$ can prove to $\mathcal{P}_2$ that $\mathcal{P}_1$ has used the committed inputs using Protocol 17.3 (proof of linear relation), once $\mathcal{P}_1$ has committed (U_0, U_1) using Protocol 17.1 (committed copy). On the other hand, in order for $\mathcal{P}_2$ to prove to $\mathcal{P}_1$ that $\mathcal{P}_2$ has used the committed input, $\mathcal{P}_2$ will need partial information about $U_{\bar{B}}$, where $\bar{B} = B \oplus 1$. To enable this, instead of using single bits U_0 and U_1, we use vectors of larger length and "leak" some information about $U_{\bar{B}}$. We need to do this without revealing $U_{\bar{B}}$ completely to $\mathcal{P}_2$ and without $\mathcal{P}_1$ getting to know B.

All this is enabled using linear codes: U_0 and U_1 are selected as two random codewords and OT is used for $\mathcal{P}_2$ to get a large fraction of bits of U_B and a small fraction of bits of $U_{\bar{B}}$. Note that this gives $\mathcal{P}_2$ at least a small fraction

Protocol 17.4 The COT protocol

Input: $\mathcal{P}_1$'s input $(\boxed{K_0}, \boxed{K_1})$ and $\mathcal{P}_2$'s input $(\boxed{B})$ obtained using Protocol 17.1, parameters m, θ, σ where $\sigma < \theta/2$ and $\theta < 1/4$

Output: $\mathcal{P}_2$ gets the output $K_{\boxed{B}}$

1. $\mathcal{P}_1$ generates U_0 and U_1 independently and distributed uniformly over $\mathcal{C}$, an $(m, (\frac{1}{2} + 2\sigma)m, \theta m)$-code known to both parties.
   {We set θ so that $(1 - 2h(\theta)) > (1/2 + 2\sigma)$ for sufficiently large m, whereby such a linear code exists by Lemma 17.6.}
2. $\mathcal{P}_1$ creates committed copies of each bit of U_0 and U_1 using Protocol 17.1, i.e., $\mathcal{P}_1$ commits to $(\boxed{U_{i1}}, \ldots, \boxed{U_{im}})$, $i \in \{0, 1\}$.
3. $\mathcal{P}_1$ proves to $\mathcal{P}_2$ using Protocol 17.3 that $\boxed{U_0}H = 0$ and $\boxed{U_1}H = 0$, namely that U_0 and U_1 belong to $\mathcal{C}$.
   {Note that new committed copies of U_{0j} and U_{1j}, $1 \le j \le m$, are created for future use as a part of Protocol 17.3.}

   Execute OT with codewords as input
4. $\mathcal{P}_2$ generates random subsets $\mathcal{I}_0, \mathcal{I}_1 \subset [m]$ of cardinality $|\mathcal{I}_0| = |\mathcal{I}_1| = \frac{\sigma}{2}m$ such that $\mathcal{I}_0 \cap \mathcal{I}_1$.
5. $\mathcal{P}_1$ and $\mathcal{P}_2$ use m instances of ideal OT; for the jth instance, $\mathcal{P}_1$ uses inputs $(\boxed{U_{0j}}, \boxed{U_{1j}})$ and $\mathcal{P}_2$ uses $\boxed{B}$ or $\boxed{\overline{B}}$ depending on $j \notin \mathcal{I}_0$ or $j \in \mathcal{I}_0$, respectively. Denote by $\widetilde{W}_j$ the output (for $\mathcal{P}_2$) of the jth instance.

   Proof that $\mathcal{P}_1$ used committed codewords for OT
6. $\mathcal{P}_2$ sends $\mathcal{I} = \mathcal{I}_0 \cup \mathcal{I}_1$ to $\mathcal{P}_1$, who reveals committed values $(\boxed{U_{0j}}, \boxed{U_{1j}})$ for $j \in \mathcal{I}$.
7. $\mathcal{P}_2$ checks if $U_{\overline{B}j} = \widetilde{W}_j$ for $j \in \mathcal{I}_0$ and $U_{Bj} = \widetilde{W}_j$ for $j \in \mathcal{I}_1$.
8. $\mathcal{P}_2$ finds the codeword W such that $d_H(W, \widetilde{W}) < \theta m/2$; if such a codeword does not exist, $\mathcal{P}_2$ declares the abortion $\perp$.
   {Since $|\mathcal{I}_0| = \sigma m < \theta m/2$, W must coincide with the codeword U_B if both the parties are honest.}

   Proof that $\mathcal{P}_2$ used the committed input B
9. $\mathcal{P}_2$ commits each bit W_j, $1 \le j \le m$, using Protocol 17.1; then $\mathcal{P}_2$ proves to $\mathcal{P}_1$ using Protocol 17.3 that the committed bits $\boxed{W} = (\boxed{W_1}, \ldots, \boxed{W_m})$ satisfy $\boxed{W}H = 0$.
10. $\mathcal{P}_1$ generates a random subset $\mathcal{I}_2 \subset [m] \setminus \mathcal{I}$ of cardinality σm.
11. $\mathcal{P}_1$ sends $\mathcal{I}_2$ to $\mathcal{P}_2$ and reveals $(\boxed{U_{0j}}, \boxed{U_{1j}})$, $j \in \mathcal{I}_2$.
12. $\mathcal{P}_2$ proves to $\mathcal{P}_1$ that $\boxed{W_j} = U_{\boxed{B}_j}$, $j \in \mathcal{I}_2$, without revealing $\boxed{B}$ as follows.
    For each $j \in \mathcal{I}_2$,

    i if $U_{0j} = U_{1j}$, $\mathcal{P}_2$ reveals W_j to $\mathcal{P}_1$;
    ii if $U_{0j} = 0$ and $U_{1j} = 1$, $\mathcal{P}_2$ uses Protocol 17.3 to prove that $\boxed{W_j} \oplus \boxed{B} = 0$;

iii if $U_{0j} = 1$ and $U_{1j} = 0$, $\mathcal{P}_2$ uses Protocol 17.3 to prove that $\boxed{W_j} \oplus \boxed{B} = 1$.

Hash and send the true input

13. $\mathcal{P}_1$ generates a random linear function $F\colon \{0,1\}^m \to \{0,1\}$ and sends it to $\mathcal{P}_2$.
14. $\mathcal{P}_1$ sends $C_0 = \boxed{K_0} \oplus F(\boxed{U_0})$ and $C_1 = \boxed{K_1} \oplus F(\boxed{U_1})$ to $\mathcal{P}_2$, and uses Protocol 17.3 to prove that $C_i = \boxed{K_i} \oplus F(\boxed{U_i})$, $i \in \{0,1\}$.
15. $\mathcal{P}_2$ obtains the output $K_{\boxed{B}} = C_{\boxed{B}} \oplus F(\boxed{W})$.

---

of bits about both committed inputs, which allows $\mathcal{P}_2$ to verify that all the subsequent steps in the algorithm use the same committed bits. Further, a linear code facilitates the reconstruction of U_B from its large fraction of bits.

Now, since a part of information about $U_{\overline{B}}$ is revealed to $\mathcal{P}_2$, we use a random linear hash to extract a random bit that is unknown to $\mathcal{P}_2$. An important observation is that this step, as well as the fact that committed inputs are indeed codewords, can be verified by proving linear relations using Protocol 17.3.

The final missing piece is a process for $\mathcal{P}_2$ to prove to $\mathcal{P}_1$ that $\mathcal{P}_2$ has used the committed input B. We shall see that even this step can be cast as a verification of linear relations and be done using Protocol 17.3.

We now analyze the possible security vulnerabilities of Protocol 17.4. This analysis will be used to provide formal security guarantees of our protocol for computing AND. For convenience, we make a list of possible vulnerabilities, categorized by which party is dishonest. Specifically, the following vulnerabilities arise.

When $\mathcal{P}_1$ is honest, but $\mathcal{P}_2$ is dishonest

When $\mathcal{P}_2$ is dishonest, $\mathcal{P}_2$ may use a value of B different from the committed one provided as the input. In our protocol, $\mathcal{P}_2$ is forced to use the committed input $\boxed{B}$ by the following logic. By the proof in Step 9 (see Protocol 17.4), it is guaranteed that $\boxed{W}$ committed by $\mathcal{P}_2$ is a codeword in $\mathcal{C}$. Then, by the proof in Step 12, it is guaranteed that $\boxed{W_j} = U_{\boxed{B}j}$ for every $j \in \mathcal{I}_2$. Thus, by Lemma 17.7, except with probability[6]

$$(m+1)e^{-(\theta-2\sigma)\frac{\sigma}{2}m}, \tag{17.1}$$

$\boxed{W}$ coincides with $U_{\boxed{B}}$.

However, still, there is a possibility that $\mathcal{P}_2$ may cheat in Step 5 so that $\mathcal{P}_2$ can get more than necessary information about $U_{\overline{B}}$ and learn about $K_{\overline{B}}$, which

[6] Note that $\mathcal{I}_2$ is randomly sampled from $[m]\backslash\mathcal{I}$. The Hamming distance between two distinct codewords $u, v \in \mathcal{C}$ restricted to coordinates in $[m]\backslash\mathcal{I}$ is $(\theta - 2\sigma)m$ in the worst case. We use Lemma 17.7 by taking this into account.

is prevented as follows. Note that U_B is randomly chosen from the code $\mathcal{C}$ of size $2^{(\frac{1}{2}+2\sigma)m}$. If $\mathcal{P}_2$ invokes less than $(\frac{1}{2}+\frac{\sigma}{2})m$ OTs with input B in Step 5, then $\mathcal{P}_2$ learns about U_B at most $(\frac{1}{2}+\frac{3\sigma}{2})m$ bits in total.[7] Then, the probability that $\mathcal{P}_2$ can correctly guess U_B and commit to $W = U_B$ in Step 9 is bounded by $2^{-\frac{\sigma}{2}m}$; cf. Problem 17.4.[8] Thus, except with probability

$$2^{-\frac{\sigma}{2}m}, \tag{17.2}$$

$\mathcal{P}_2$ must have invoked more than $(\frac{1}{2}+\frac{\sigma}{2})m$ OTs with input B in Step 5. This implies that $\mathcal{P}_2$ learns less than $(\frac{1}{2}+\frac{\sigma}{2})m$ bits about $U_{\overline{B}}$ in Step 5 and Step 6. Since $U_{\overline{B}}$ is randomly chosen from the code $\mathcal{C}$ of size $2^{(\frac{1}{2}+2\sigma)m}$, even if the extra σm bits revealed to $\mathcal{P}_2$ in Step 11 are omitted, $\frac{\sigma}{2}m$ bits of min-entropy remain. This remaining min-entropy can be used to create a secure key and prevent the leakage of $K_{\overline{B}}$ in Step 14. The leakage analysis is postponed to the end of this section. In the following, we first analyze the probability of accepting faulty proofs.

Note that the proofs in Step 9 and Step 12 are conducted by Protocol 17.3. We need to take into account that the probability of accepting a faulty proof where the copies saved for future use contain more than δ fraction of inconsistent values and the probability of accepting a faulty proof where the claimed linear relation is not satisfied.[9] Since the proof in Step 9 involves one linear relation with m variables, by Lemma 17.4, the probability of accepting a faulty proof is bounded by

$$2^{-(1-m\delta)n} + 2m(4n+1)e^{-\frac{n\delta}{4}}. \tag{17.3}$$

Similarly, since the proof in Step 12 involves σm linear relations with two variables each, by Lemma 17.4, the probability of accepting a faulty proof is bounded by

$$\sigma m\left(2^{-(1-2\delta)n} + 4(4n+1)e^{-\frac{n\delta}{4}}\right). \tag{17.4}$$

When $\mathcal{P}_1$ is dishonest, but $\mathcal{P}_2$ is honest
The analysis of this part is similar to the previous one.

First, a faulty proof might be accepted in Step 2 even though one of the copies $\boxed{U_{ij}}$, $1 \le j \le m$ and $i \in \{0,1\}$ created contains more than δ fraction of inconsistent values. Since Protocol 17.1 is invoked $2m$ times, by Lemma 17.2, the probability of accepting a faulty proof is bounded by

$$2m(2n+1)e^{-\frac{n\delta}{4}}. \tag{17.5}$$

Next, a faulty proof might be accepted in Step 3 even though either $\boxed{U_0}$ or $\boxed{U_1}$ is not a codeword in $\mathcal{C}$. Since the proof using Protocol 17.3 involves two linear relations with m variables each, by Lemma 17.4, the probability of accepting a faulty proof is bounded by

$$2\left(2^{-(1-m\delta)n} + 2m(4n+1)e^{-\frac{n\delta}{4}}\right). \tag{17.6}$$

[7] Note that $\mathcal{P}_2$ learns additional σm bits in Step 6.
[8] Note that extra σm bits of U_B is revealed to $\mathcal{P}_2$ in Step 11 after $\mathcal{P}_2$ commits to W in Step 9.
[9] In the following analysis, when Protocol 17.3 is invoked, we always take into account these two cases of faulty proofs without explicitly mentioning so.

Next, $\mathcal{P}_1$ may not use $(\boxed{U_0}, \boxed{U_1})$ as inputs for OT in Step 5. Such a deviation is prevented by the sampling test in Steps 6 and 7. Let $\widetilde{U}_0, \widetilde{U}_1 \in \{0,1\}^m$ be the actual inputs $\mathcal{P}_1$ used to invoke OTs in Step 5. If $d_H(\widetilde{U}_i, U_i) \geq (\frac{\theta}{2} - \frac{\sigma}{2})m$ for either $i = 0$ or $i = 1$, then the probability of passing the test in Steps 6 and 7 is bounded by[10]

$$2(m+1)e^{-(\frac{\theta}{2}-\frac{\sigma}{2})\frac{\sigma}{2}m}. \tag{17.7}$$

Thus, except with probability (17.7), $d_H(\widetilde{U}_i, U_i) < (\frac{\theta}{2} - \frac{\sigma}{2})m$ for both $i = 0$ and $i = 1$. Then, since this and $d_H(\widetilde{W}, \widetilde{U}_B) \leq \frac{\sigma}{2}m$ imply $d_H(\widetilde{W}, U_B) < \frac{\theta}{2}m$ by the triangular inequality, W recovered in Step 8 coincides with U_B; cf. Lemma 17.6.

Finally, $\mathcal{P}_1$ may provide a faulty proof in Steps 13–14. Note that this proof entails verifying two linear relations of length $(m+1)$ each. Proceeding using Lemma 17.4 as before, a faulty proof can be accepted in these steps with probability at most

$$2\left(2^{-(1-(m+1)\delta)n)} + (m+1)(4n+1)e^{-\frac{n\delta}{4}}\right). \tag{17.8}$$

The information leakage bound

As we discussed above, dishonest $\mathcal{P}_2$ may deviate from the protocol to get more information than necessary about $U_{\overline{B}}$ so that $\mathcal{P}_2$ can learn about $K_{\overline{B}}$. For our formal analysis, we need to show that the information revealed in the real world can be simulated in the ideal world as well. To that end, it suffices to bound the total variation distance between the conditional probability distribution of $F(U_{\overline{B}})$ given the information revealed to $\mathcal{P}_2$ and a random bit. Then, the simulator can just sample a random bit locally.

As we discussed above, we need to consider the leakage under the event $\mathcal{E}$ that $\mathcal{P}_2$ learns less than $(\frac{1}{2} + \frac{\sigma}{2})m$ bits about $U_{\overline{B}}$ in Step 5 and Step 6.[11] Denote by Z the transcript of the protocol seen by $\mathcal{P}_2$ until Step 13. Since $U_{\overline{B}}$ is distributed uniformly over the set $\mathcal{C}$, a set of cardinality $2^{(\frac{1}{2}+2\sigma)m}$, independently of K_0, K_1, and B, even if the extra σm bits revealed to $\mathcal{P}_2$ in Step 11 are omitted, there remains $\frac{\sigma}{2}m$ bits of min-entropy. Then, Lemma 17.8 yields

$$d_{\mathtt{var}}\left(\mathrm{P}_{F(U_{\overline{B}})FZK_0K_1B|\mathcal{E}}, \mathrm{P}_{\mathtt{unif}} \times \mathrm{P}_{FZK_0K_1B|\mathcal{E}}\right) \leq 2^{-\frac{\sigma}{2}m-1}, \tag{17.9}$$

where $\mathrm{P}_{\mathtt{unif}}$ denotes the distribution $\mathtt{Ber}(1/2)$. Thus, $F(U_{\overline{B}})$ acts like a secret key for encrypting $K_{\overline{B}}$.

In fact, we can show that almost no information about $K_{\overline{B}}$ is revealed. Let $\Pi^{(2)}$ denote the transcript observed by $\mathcal{P}_2$, which includes $C_{\overline{B}} = F(U_{\overline{B}}) \oplus K_{\overline{B}}$, F, and $C_B = F(U_B) \oplus K_B$ in addition to Z above.

It is important to note that we need to consider inputs (K_0, K_1, B) such that (K_0, K_1) may be correlated with B. We need this to prove sequential composable

[10] The bound (17.7) can be derived in the same manner as Lemma 17.7.

[11] In other words, we need to consider the leakage under the event that $\mathcal{P}_2$ invokes less than $(\frac{1}{2} - \frac{\sigma}{2})m$ OTs with input $\overline{B}$ in Step 5.

security for our protocol for AND, since the outer protocol may give correlated inputs to our protocol.

Specifically, we can apply the following result on the security of one-time pad encryption (see Chapter 3). Note that Lemma 17.9 is a slight extension of Theorem 3.18 in the sense that the message M may be correlated to an adversary's observation V.

LEMMA 17.9 *Consider random variables M, U, and V with M, V taking values in $\{0,1\}$ and M conditionally independent of U given V. Then, for $C = M \oplus U$,*

$$d_{\mathtt{var}}(\mathrm{P}_{MCV}, \mathrm{P}_{M|V} \times \mathrm{P}_{CV}) \le d_{\mathtt{var}}(\mathrm{P}_{UV}, \mathrm{P}_{\mathtt{unif}} \times \mathrm{P}_V).$$

Proof Since M and U are conditionally independent given V, we have

$$\mathrm{P}_{MC|V}(m,c|v) = \mathrm{P}_{M|V}(m|v)\,\mathrm{P}_{U|V}(m\oplus c|v),$$
$$\mathrm{P}_{C|V}(c|v) = \mathrm{P}_{M|V}(0|v)\,\mathrm{P}_{U|V}(c|v) + \mathrm{P}_{M|V}(1|v)\,\mathrm{P}_{U|V}(c\oplus 1|v).$$

Thus,

$$\begin{aligned}
&d_{\mathtt{var}}(\mathrm{P}_{MCV}, \mathrm{P}_{M|V} \times \mathrm{P}_{CV}) \\
&\quad= \frac{1}{2}\sum_{v}\mathrm{P}_V(v)\sum_{m,c}\left|\mathrm{P}_{MC|V}(m,c|v) - \mathrm{P}_{M|V}(m|v)\,\mathrm{P}_{C|V}(c|v)\right| \\
&\quad= \frac{1}{2}\sum_{m,v}\mathrm{P}_{MV}(m,v)\sum_{c}\left|\mathrm{P}_{U|V}(m\oplus c|v) - \mathrm{P}_{C|V}(c|v)\right| \\
&\quad= \frac{1}{2}\sum_{m,v}\mathrm{P}_{MV}(m,v)\sum_{c}\big|\mathrm{P}_{U|V}(m\oplus c|v) - \mathrm{P}_{M|V}(0|v)\,\mathrm{P}_{U|V}(0\oplus c|v) \\
&\qquad\qquad\qquad\qquad\qquad -\mathrm{P}_{M|V}(1|v)\,\mathrm{P}_{U|V}(1\oplus c|v)\big| \\
&\quad= \sum_{m,v}\mathrm{P}_{MV}(m,v)\,(1-\mathrm{P}_{M|V}(m|v))\left|\mathrm{P}_{U|V}(0|v) - \mathrm{P}_{U|V}(1|v)\right| \\
&\quad\le \frac{1}{2}\sum_{v}\mathrm{P}_V(v)\left|\mathrm{P}_{U|V}(0|v) - \mathrm{P}_{U|V}(1|v)\right| \\
&\quad\le d_{\mathtt{var}}(\mathrm{P}_{UV}, \mathrm{P}_{\mathtt{unif}} \times \mathrm{P}_V),
\end{aligned}$$

where in the second-last inequality we used $x(1-x) \le 1/4$ for every $x \in [0,1]$. □

Applying this lemma to our application with $K_{\overline{B}}$ in the role of the message M, $F(U_{\overline{B}})$ in the role of U, and $(\Pi^{(2)}, B)$ in the role of V, we obtain

$$\begin{aligned}
d_{\mathtt{var}}\left(\mathrm{P}_{K_{\overline{B}}\Pi^{(2)}B|\mathcal{E}}, \mathrm{P}_{K_{\overline{B}}|B\mathcal{E}} \times \mathrm{P}_{\Pi^{(2)}B|\mathcal{E}}\right) &\le d_{\mathtt{var}}\left(\mathrm{P}_{F(U_{\overline{B}})FZ|\mathcal{E}}, \mathrm{P}_{\mathtt{unif}} \times \mathrm{P}_{FZ|\mathcal{E}}\right) \\
&\le 2^{-\frac{\sigma}{2}m-1}, \qquad (17.10)
\end{aligned}$$

which is our bound for "information leakage" to $\mathcal{P}_2$ about $K_{\overline{B}}$.

Note that the only step where information is leaked to $\mathcal{P}_1$ about B is Step 12. But by the first property in Lemma 17.4 the bits received by $\mathcal{P}_1$ are independent of B.

Protocol 17.5 Protocol for computing AND using OT

Input: $\mathcal{P}_1$'s input x_1, $\mathcal{P}_2$'s input x_2, and security parameter n

Output: $x_1 \wedge x_2$ given to both parties

1. Parties use Protocol 17.1 to commit copies of their inputs x_1 and x_2. Further, they use the protocol for committing a value mentioned in Section 17.2 to commit copies of 0 (both commit to two different copies).
2. Parties execute Protocol 17.4 with $\mathcal{P}_1$ using input $(\boxed{K_0}, \boxed{K_1})$ given by $(\boxed{0}, \boxed{x_1})$ for $\mathcal{P}_1$ and the input $\boxed{B}$ given by $\boxed{x_2}$ for $\mathcal{P}_2$. $\mathcal{P}_2$ receives the output $K_B = x_1 \wedge x_2$.
3. Parties execute Protocol 17.4 with $\mathcal{P}_2$ using input $(\boxed{K_0}, \boxed{K_1})$ given by $(\boxed{0}, \boxed{x_2})$ for $\mathcal{P}_2$ and the input $\boxed{B}$ given by $\boxed{x_1}$ for $\mathcal{P}_1$. $\mathcal{P}_1$ receives the output $K_B = x_1 \wedge x_2$.

17.5 Security Analysis of the Protocol for AND

We have all our tools ready for implementing our protocol for AND, depicted in Figure 17.2. The formal description of the combined protocol is given in Protocol 17.5. Before proceeding, we note the following remarks about the protocol.

Remark 17.6 (On committing the output) We note that an alternative strategy is possible. Instead of executing COT two times, we can simply modify COT to commit the output as well and reveal the committed output to $\mathcal{P}_1$ in the next phase. In fact, in our COT protocol, $\mathcal{P}_2$ has already committed to $\boxed{W} = \boxed{U_B}$; $\mathcal{P}_2$ can also commit to the value of $F(\boxed{U_B})$ using Protocol 17.3. Finally, $\mathcal{P}_2$ can commit to the output value $\boxed{Z_2}$ and prove $\boxed{Z_2} = C_B \oplus F(\boxed{U_B})$ as follows: if $C_0 = C_1$, then $\mathcal{P}_2$ proves $\boxed{Z_2} \oplus F(\boxed{U_B}) = C_0$; if $C_0 = 0$ and $C_1 = 1$, then $\mathcal{P}_2$ proves $\boxed{Z_2} \oplus F(\boxed{U_B}) \oplus \boxed{B} = 0$; if $C_0 = 1$ and $C_0 = 0$, then $\mathcal{P}_2$ proves $\boxed{Z_2} \oplus F(\boxed{U_B}) \oplus \boxed{B} = 1$.

We state the main theorem of this chapter below. We reiterate that since we have already seen how to implement BC in a sequentially composably secure manner using OT in Chapter 15, the result below provides a secure protocol for AND using OT alone.

THEOREM 17.10 (Secure protocol for AND) *Set parameters*[12] $m = \sqrt{n}$, $\delta = \frac{1}{2(m+1)}$, $\theta = 1/32$, *and* $\sigma = 1/128$. *Then, there is a constant* $c > 0$ *such that for all* n *sufficiently large, Protocol 17.5 is an* $O(2^{-c\sqrt{n}})$*-sequentially composably secure implementation of* AND *using OT and BC.*

Proof The proof is obtained by combining the security properties of various components that we have established in previous sections. Note that, for the

[12] Note that θ and σ must satisfy $(1 - 2h(\theta)) > 1/2 + 2\sigma$ in COT. The choice of parameters are not optimal.

choice of parameters specified in the theorem, the bounds on the security properties in (17.1)–(17.10) are $\mathcal{O}(2^{-c\sqrt{n}})$ for some $c > 0$. In the following, we will show how these bounds are related to the simulation error.

When both $\mathcal{P}_1$ and $\mathcal{P}_2$ are honest: It can be verified easily that the trivial simulator satisfies $d_{\mathtt{var}}(\mathtt{V}[\pi \circ \pi^f \diamond g], \mathtt{V}[\pi \circ \sigma \circ f]) = 0$ in this case.[13] Namely, Protocol 17.5 computes the AND function with no error.

When $\mathcal{P}_1$ is honest, but $\mathcal{P}_2$ is dishonest: For a given attack by $\mathcal{P}_2$ in the real world, we consider the following simulator σ for the real world. $\mathcal{P}_2$ executes Protocol 17.5 locally, executing only those steps which $\mathcal{P}_2$ can do without the knowledge of $\mathcal{P}_1$'s input and using exactly the same strategy as in the attack for the real world. In fact, most steps of the real protocol can be simulated locally by $\mathcal{P}_2$ in the ideal world – only two steps need to be handled carefully. We highlight them in the remark below.

Remark 17.7 First, the transcripts C_0 and C_1 sent by $\mathcal{P}_1$ in Step 14 in the invocation of COT Protocol 17.4 use $\mathcal{P}_1$'s input, which is not available to an adversarial $\mathcal{P}_2$ in the ideal world. We handle this by introducing a dummy input $\tilde{X}_2'$, by simulating C_B for $B = \tilde{X}_2'$ using the dummy output $\tilde{Z}_2' = X_1 \wedge \tilde{X}_2'$, and by randomly sampling $C_{\overline{B}}$. The validity of randomly sampling $C_{\overline{B}}$ is guaranteed by the leakage analysis of the COT protocol below. This argument is similar to that used in the construction of a simulator for (noncommitted) OT; see the proofs of Proposition 15.4 and Proposition 15.10.

Second, we need to verify if $\mathcal{P}_1$'s output $\tilde{Z}_1 = X_1 \wedge \tilde{X}_2'$ in the ideal world coincides with $\mathcal{P}_1$'s output Z_1 in the real world. Here, "committed" input of COT plays an important role, and $\tilde{Z}_1 = Z_1$ is guaranteed by the fact that $\mathcal{P}_2$ has to use the initially committed bit in the real world.

Specifically, $\mathcal{P}_2$ proceeds as follows.

1. $\mathcal{P}_2$ executes Protocol 17.1 (Committed copy) locally to make copies of $\mathcal{P}_2$'s input (possibly modified due to dishonest behavior). If the output is $\perp$, $\mathcal{P}_2$ sets $\tilde{X}_2' = \perp$; otherwise, $\mathcal{P}_2$ sets the dummy input $\tilde{X}_2'$ to the majority of values stored in copies of X_2.
2. $\mathcal{P}_2$ executes locally Protocol 17.4 (COT) needed in Step 2 of Protocol 17.5 (AND). It proceeds up to Step 13 of Protocol 17.4 (COT) by generating local copies of U_0, U_1, and other random variables needed from $\mathcal{P}_1$. If the output is $\perp$, $\mathcal{P}_2$ sets $\tilde{X}_2' = \perp$; otherwise $\mathcal{P}_2$ retains the previous value of $\tilde{X}_2'$.
3. $\mathcal{P}_2$ uses the dummy input $\tilde{X}_2'$ as input for Phase 1 of the ideal AND function, with $\mathcal{P}_1$ using X_1 as the input. Let $\tilde{Z}_2'$ denote the dummy output obtained.
4. If $\tilde{Z}_2' \neq \perp$, $\mathcal{P}_2$ generates the remaining transcript corresponding to execution of Protocol 17.4 (COT) from Step 13 as follows. $\mathcal{P}_2$ sets $B = \tilde{X}_2'$ and generates $C_B = \tilde{Z}_2' \oplus F(U_B)$ and $C_{\overline{B}}$ is a uniform bit independent of everything else.
5. $\mathcal{P}_2$ executes locally Protocol 17.4 (COT) needed in Step 3 of Protocol 17.5 (AND). Note that this time $\mathcal{P}_2$ is playing the role of $\mathcal{P}_1$ in our description of

[13] By a slight abuse of notation, the oracle g includes both the OT oracle and the BC oracle.

Protocol 17.4 (COT). Also, note that Steps 5–7 seem to require knowledge of $B = X_1$, which $\mathcal{P}_2$ does not possess. However, in our simulator we only execute a modified form of these steps where $\mathcal{P}_2$ simply samples subsets $(\mathcal{I}_0, \mathcal{I}_1)$ of the coordinates of inputs $\mathcal{P}_2$ uses for oracle access to OT and compares them with the corresponding coordinates of U_0 and U_1. Further, $\mathcal{P}_2$ need not execute steps between Step 7 and Step 13, since the goal of these steps is to verify that $\mathcal{P}_1$ has used X_1 as input for B for COT. $\mathcal{P}_2$ proceeds directly to Step 13 and executes it.

If $\perp$ is received as the output at any time, $\mathcal{P}_2$ aborts Phase 2 of the ideal AND function; otherwise $\mathcal{P}_2$ executes Phase 2 without any input. Let $\tilde{Z}_1$ denote the output received by $\mathcal{P}_1$.

6. Finally, $\mathcal{P}_2$ generates the part of the transcript corresponding to the skipped steps of Protocol 17.4 (COT) as follows. For locations j where $U_{0j} = U_{1j}$, $\mathcal{P}_2$ simply reveals this output. For the remaining locations, $\mathcal{P}_2$ emulates the shares that were to be sent by $\mathcal{P}_1$ to prove the linear relation using uniform random bits. $\mathcal{P}_2$ also samples the output $\tilde{Z}_2$ using the same distribution as the real world based on observations simulated so far.

Note that we can define a random variable corresponding to $\tilde{X}_2'$ in the real world as well – it equals the majority value of the copies committed by $\mathcal{P}_2$ in the first step if there is no $\perp$ in the first use of Protocol 17.4 (COT), and $\perp$ otherwise. It can be verified that the random variables $(X_1, X_2, \tilde{X}_2')$ have the same distributions in the real and the ideal worlds. Let $\Pi^{(2)} = (\Pi_1^{(2)}, \Pi_2^{(2)})$ and $\tilde{\Pi}^{(2)} = (\tilde{\Pi}_1^{(2)}, \tilde{\Pi}_2^{(2)})$, respectively, denote the part of the transcript of the protocol seen by $\mathcal{P}_2$ in the real and ideal worlds (the subscript 1 indicates the transcript in Step 1 and Step 2 of Protocol 17.5, and the subscript 2 indicates the transcript in Step 3 of Protocol 17.5). Further, let (Z_1, Z_2) and $(\tilde{Z}_1, \tilde{Z}_2)$ denote the output (passed to the outer protocol) in the real and ideal worlds.[14]

To establish sequential composable security, note that

$$\begin{aligned}
&d_{\mathtt{var}}(\mathtt{V}[\pi \circ \pi^f \diamond g], \mathtt{V}[\pi \circ \sigma \circ f]) \\
&\quad \le d_{\mathtt{var}}(\mathrm{P}_{(X_1,Z_1)(X_2,\tilde{X}_2',\Pi^{(2)},Z_2)}, \mathrm{P}_{(X_1,\tilde{Z}_1)(X_2,\tilde{X}_2',\tilde{\Pi}^{(2)},\tilde{Z}_2)}) \\
&\quad = d_{\mathtt{var}}(\mathrm{P}_{(X_1,Z_1)(X_2,\tilde{X}_2',\Pi^{(2)})}, \mathrm{P}_{(X_1,\tilde{Z}_1)(X_2,\tilde{X}_2',\tilde{\Pi}^{(2)})}) \\
&\quad \le d_{\mathtt{var}}(\mathrm{P}_{\Pi_1^{(2)} X_1 X_2 \tilde{X}_2'}, \mathrm{P}_{\tilde{\Pi}_1^{(2)} X_1 X_2 \tilde{X}_2'}) \\
&\qquad + \mathbb{E}_{\mathrm{P}_{X_1 X_2 \tilde{X}_2' \tilde{\Pi}_1^{(2)}}} \left[d_{\mathtt{var}}(\mathrm{P}_{Z_1 \Pi_2^{(2)} | X_1 X_2 \tilde{X}_2' \Pi_1^{(2)}}, \mathrm{P}_{\tilde{Z}_1 \tilde{\Pi}_2^{(2)} | X_1 X_2 \tilde{X}_2' \tilde{\Pi}_1^{(2)}}) \right], \quad (17.11)
\end{aligned}$$

where we can remove Z_2 and $\tilde{Z}_2$ in the equality since $\tilde{Z}_2$ is sampled by the same distribution as the real world, and the second inequality uses the chain rule for total variation distance (see Chapter 2). Note that the second term on the right-hand side corresponds to the "hybrid world" (see Section 15.6) with expectation over the transcript $\tilde{\Pi}_1^{(2)}$ of the ideal world. We bound each term on

[14] Note that, in the ideal world, $\mathcal{P}_1$ obtains $\tilde{Z}_1$ as the output of the ideal AND function, and $\mathcal{P}_2$ samples $\tilde{Z}_2$ at the end of the simulation.

the right-hand side above separately. At a high level, the first and the second terms correspond to the two concerns discussed in Remark 17.7.

For the first term, denote by V and $\tilde{V}$ the transcripts of the protocol up to Step 13 of Protocol 17.4 (COT) in the real world and ideal world, respectively, which have the same conditional distribution in the real and ideal worlds. Then, we can write

$$d_{\mathtt{var}}(\mathrm{P}_{\Pi_1^{(2)} X_1 X_2 \tilde{X}_2}, \mathrm{P}_{\tilde{\Pi}_1^{(2)} X_1 X_2 \tilde{X}_2}) = d_{\mathtt{var}}(\mathrm{P}_{C_B C_{\overline{B}} V X_1 X_2 \tilde{X}_2'}, \mathrm{P}_{\tilde{C}_B \tilde{C}_{\overline{B}} \tilde{V} X_1 X_2 \tilde{X}_2'}).$$

When $\tilde{Z}_2' = \perp$, which means that the real protocol is aborted in the middle of the first invocation of COT, C_B and $C_{\overline{B}}$ are constants and can be trivially simulated by constant variables $(\tilde{C}_B, \tilde{C}_{\overline{B}})$. When $\tilde{Z}_2' \neq \perp$, for $B = \tilde{X}_2'$, $C_B = F(U_B) \oplus (X_1 \wedge \tilde{X}_2')$ coincides with $\tilde{C}_B = F(U_B) \oplus \tilde{Z}_2'$. Thus, it suffices to bound the error for simulating $C_{\overline{B}}$ with $\tilde{C}_{\overline{B}}$.

This simulation error stems from two cases. The first case is when $\mathcal{P}_2$ invokes more than $(\frac{1}{2} - \frac{\sigma}{2})m$ OTs with input $\overline{B} = \tilde{X}_2' \oplus 1$ in Step 5 of COT, $\mathcal{P}_1$ falsely accepts $\mathcal{P}_2$'s proofs in Step 9 and Step 12 of COT. The probability of this event can be bounded as in (17.1)–(17.4). The second case is when $\mathcal{P}_2$ invokes less than $(\frac{1}{2} - \frac{\sigma}{2})m$ OTs with input $\overline{B}$ in Step 5 of COT. In this case, the simulation error can be evaluated as the leakage analysis in (17.10). Thus, in total, the first term of (17.11) can be bounded by $\mathcal{O}(2^{-c\sqrt{n}})$.

For the second term in (17.11), we note that the expectation is over the ideal world distribution. Note that the transcripts $\Pi_2^{(2)}$ and $\tilde{\Pi}_2^{(2)}$ of the protocol in the real and ideal worlds coincide (have the same conditional distribution given the past). Indeed, the only place they can differ is Step 12, where in the simulator we used random bits instead of the shares of copies of B revealed in proof. But as seen in Property 1 of Lemma 17.4, the bits revealed are independent of everything else. Also, if a $\perp$ is received in the transcript, the outputs coincide with $\perp$. Furthermore, by noting that the output $\tilde{Z}_1$ in the ideal world can only be $\perp$ or $X_1 \wedge \tilde{X}_2'$ and the probabilities of $Z_1 = \perp$ and $\tilde{Z}_1 = \perp$ are the same,[15] we have

$$\begin{aligned}
&\mathbb{E}_{\mathrm{P}_{X_1 X_2 \tilde{X}_2' \tilde{\Pi}_1^{(2)}}} \left[d_{\mathtt{var}}(\mathrm{P}_{Z_1 \Pi_2^{(2)} | X_1 X_2 \tilde{X}_2' \Pi_1^{(2)}}, \mathrm{P}_{\tilde{Z}_1 \tilde{\Pi}_2^{(2)} | X_1 X_2 \tilde{X}_2' \tilde{\Pi}_1^{(2)}}) \right] \\
&\quad = \frac{1}{2} \big[\Pr(\tilde{Z}_1 = \perp) - \Pr(Z_1 = \perp) + \Pr(\tilde{Z}_1 = X_1 \wedge \tilde{X}_2') - \Pr(Z_1 = X_1 \wedge \tilde{X}_2') \\
&\qquad + \Pr(Z_1 = X_1 \wedge \tilde{X}_2' \oplus 1) \big] \\
&\quad = \Pr(Z_1 = X_1 \wedge \tilde{X}_2' \oplus 1).
\end{aligned} \tag{17.12}$$

Note that $Z_1 = X_1 \wedge \tilde{X}_2' \oplus 1$ occurs only if $\mathcal{P}_1$ falsely accept $\mathcal{P}_2$'s proofs in COT (note that $\mathcal{P}_2$ plays the role of sender and $\mathcal{P}_1$ plays the role of receiver in the second invocation of COT). Thus, as we discussed in (17.5)–(17.8), (17.12) can be bounded by $\mathcal{O}(2^{-c\sqrt{n}})$, which gives the desired bound on the second term in (17.11).

[15] Note that, if the protocol is aborted in the real protocol, $\mathcal{P}_2$ aborts Phase 2 of AND.

When $\mathcal{P}_2$ is dishonest, but $\mathcal{P}_1$ is honest: The construction of a simulator for this case and its analysis is similar to the previous case and is omitted. □

17.6 References and Additional Reading

The claim that oblivious transfer is a complete resource even for an active adversary is now widely accepted in the cryptography community. However, to the best of our knowledge, a complete proof is not available in the literature. The first announcement of the claim was made by Kilian in [192] (see also his thesis [193]); it is widely accepted that the credit for the claim is (at least partially) given to [192]. However, it is also recognized that the proof in [192] is not complete. Note that the paper was written before the simulation based security framework was established around 2000. Some time later, an alternative protocol to prove the claim was proposed by Crépeau, Graaf, and Tapp [81] (see also [113]). The ideas of protocol in [81] are very clear, and heuristic reasonings of security are convincing, though a full security proof is not available there. Our exposition in this chapter is mostly based on the protocol in [81]. More recently, another protocol was proposed by Ishai, Prabhakaran, and Sahai [175]. Their protocol is based on the idea of relating the two-party secure computation to multiparty secure computation, and the security analysis is written in the modern language of simulation and universal composability framework. It seems that [175] is the most detailed proof currently available. For computational security, a detailed proof of two-party secure computation is available in [142].

Problems

17.1 Design and analyze a protocol that securely implements the (n, δ)-value copy function described at the end of Section 17.2.

17.2 Prove Lemma 17.7.

HINT Note that any pair of distinct codewords $u, v \in \mathcal{C}$ disagree at least θm coordinates; the probability of not choosing those coordinates is $(1-\theta)^\ell$; use the inequality $1 - t \le e^{-t}$.

17.3 *Sampling with/without replacement*:

1. Let $\mathcal{R} \subset \{1, \ldots, m\}$ be a random subset of size $|\mathcal{R}| = k$ and let σ be a random permutation on $\{1, \ldots, m\}$. For a binary sequence $x^m \in \{0,1\}^m$, verify that
$$\Pr\left(x^m_{\mathcal{R}} = a^k\right) = \Pr\left(\sigma(x^m)_{\mathcal{I}} = a^k\right), \quad \forall a^k \in \{0,1\}^k,$$
where $x^m_{\mathcal{R}} = (x_i : i \in \mathcal{R})$ and $\mathcal{I} = \{1, \ldots, k\}$.
2. For every $x^m, b^m \in \{0,1\}^m$, prove that
$$\Pr\left(\sigma(x^m) = b^m\right) \le (m+1)\bar{\mathrm{P}}^m(b^m),$$
where $\bar{\mathrm{P}} = \mathrm{P}_{x^m}$ is the type of sequence x^m.

HINT Use Corollary 5.11 and Problem 5.7; without using Problem 5.7, we get penalty factor $(m+1)^2$ instead of $(m+1)$.

3. For any $x^m \in \{0,1\}^m$ and a subset $\mathcal{A} \subset \{0,1\}^k$, prove that

$$\Pr\left(\sigma(x^m)_{\mathcal{I}} \in \mathcal{A}\right) \le (m+1)\bar{\mathrm{P}}^k(\mathcal{A}).$$

4. Suppose that the Hamming weight of x^m is δm. Then, prove that

$$\Pr\left(\forall i \in \mathcal{I},\ \sigma(x^m)_i = 0\right) \le (m+1)e^{-\delta k}. \tag{17.13}$$

5. Let $Y^k \sim W^k(\cdot|\sigma(x^m)_{\mathcal{I}})$ for a given channel W from $\{0,1\}$ to $\{0,1\}$. Then, for any $\mathcal{B} \subset \{0,1\}^k \times \{0,1\}^k$, prove that

$$\Pr\left((\sigma(x^m)_{\mathcal{I}}, Y^k) \in \mathcal{B}\right) \le (m+1)(P \times W)^k(\mathcal{B}). \tag{17.14}$$

REMARK The left-hand side of (17.13) is the misdetection probability of the adversary's misbehavior ($x_i = 1$) by sampling without replacement. Even though it is possible to directly evaluate this probability, it is easier to derive an upper bound via sampling with replacement. In the proof of Lemma 17.2, we use a generalized inequality, (17.14), for a more complicated test.

17.4 Let U be a random variable uniformly distributed on $\mathcal{U}$, and let $V = f(U)$ for some function $f\colon \mathcal{U} \to \mathcal{V}$. Then, prove that

$$\sum_{v \in \mathcal{V}} \mathrm{P}_V(v) \max_{u \in f^{-1}(v)} \mathrm{P}_{U|V}(u|v) \le \frac{|\mathcal{V}|}{|\mathcal{U}|}.$$

HINT Use Problem 7.7 and Lemma 7.21.

18 Broadcast, Byzantine Agreement, and Digital Signature

In this chapter and the next, we consider the multiparty secure computation problem for more than two parties. In this chapter, we consider perhaps the most useful multiparty functions for secure computing: *broadcast*, *Byzantine agreement*, and *digital signature*. These functions can be regarded as special cases of the multiparty secure computation problem, but they are also important standalone cryptographic primitives which underlie many recent cryptography applications. In fact, the availability of broadcast is crucial for the multiparty secure computation of other functions, as we will see in the next chapter.

Consider a communication network among n parties $\{\mathcal{P}_1, \ldots, \mathcal{P}_n\}$. Throughout this chapter and the next, we assume that every pair $(\mathcal{P}_i, \mathcal{P}_j)$ of parties is connected by a private authenticated channel. This means that the communication between $\mathcal{P}_i$ and $\mathcal{P}_j$ is concealed from other parties. Moreover, $\mathcal{P}_i$ is confident that the received message is sent from $\mathcal{P}_j$ and vice versa. Alternatively, we can assume that every pair of parties shares pairwise secret keys. Then, every pair of parties can establish a private authenticated channel using secret key encryption and authentication codes seen in Chapters 3 and 8. Furthermore, we assume that the network is synchronous. A network that satisfies these assumptions will be called a *standard network* in this chapter.

At a first glance, standard network seems to be a very strong assumption, and broadcast may seem to be easily realizable if such a strong assumption is made. However, this is not the case. For instance, as we will see later in this chapter, implementing broadcast is not possible even over a standard network if the number of dishonest parties is larger than a certain threshold. Interestingly, we will see that broadcast can be realized from scratch if and only if honest parties form a "supermajority," namely the adversary can influence less than one-third of the parties. Moreover, we will see how to realize broadcast without requiring an honest supermajority when additional resources are available. We also discuss connections between the three fundamental multiparty functions: broadcast, Byzantine agreement, and digital signature.

18.1 Multiparty Communication Model

Formally, a standard network is specified as follows. Consider n parties $\mathcal{P}_1, \ldots, \mathcal{P}_n$ observing inputs $x_i \in \mathcal{X}_i$, $i = 1, \ldots, n$, respectively. Some of the input sets may be singleton, i.e., some parties may not have inputs. The parties seek to evaluate some functions based on their inputs. More specifically, we consider asymmetric functions (cf. Section 12.2), i.e., for $i = 1, \ldots, n$, $\mathcal{P}_i$ seek to output $z_i = f_i(x_1, \ldots, x_n)$ for some $f_i \colon \mathcal{X}_1 \times \cdots \times \mathcal{X}_n \to \mathcal{Z}_i$. To that end, the parties communicate over pairwise private authenticated channels. As we have mentioned above, we assume that the network is synchronous. This means that each party can recognize who should communicate next based on a common clock of the network. For simplicity of description, we assume that the parties communicate in a prescribed order, say the lexicographic order of the set $\{(i, j) : i \neq j \in \{1, \ldots, n\}\}$, i.e., $\mathcal{P}_1$ communicates to $\mathcal{P}_2$ in the first communication slot, $\mathcal{P}_1$ communicates to $\mathcal{P}_3$ in the second communication slot, ..., $\mathcal{P}_n$ communicates to $\mathcal{P}_{n-1}$ in the $\binom{n}{2}$th communication slot, and so on. Since a party is allowed to skip a communication turn by sending a constant message, any protocol with synchronous communication order can be described under this assumption. We also assume that a transcript of each communication slot takes values in a fixed, prescribed set. This means that a dishonest party cannot deviate from the communication format, and the transcript is received by the intended receiver by the end of each communication slot without any timeout, or other protocol disruptions.

A transcript $\Pi_{i \to j}^{[\gamma]}$ sent from $\mathcal{P}_i$ to $\mathcal{P}_j$ in the γth communication slot is communicated over the private authenticated channel between $\mathcal{P}_i$ and $\mathcal{P}_j$. This means that the transcript cannot be seen by other parties. In turn, $\mathcal{P}_i$ can only observe transcripts sent to $\mathcal{P}_i$. Thus, the transcript $\Pi_{i \to j}^{[\gamma]}$ in the γth communication slot is computed by $\mathcal{P}_i$ based on the observation available to $\mathcal{P}_i$: $\mathcal{P}_i$'s input x_i, $\mathcal{P}_i$'s private-coin U_i, the common-coin U_0, and the transcripts that are sent to $\mathcal{P}_i$ up to the γth communication slot. We assume that the common-coin and the public-coins are mutually independent.

At the end of the protocol, based on the observations available, $\mathcal{P}_i$ outputs $\hat{Z}_i$ as an estimate of the function to be computed. In a similar manner to the two-party setting, when U_0 is fixed to constant, a protocol is said to be a private-coin protocol; when $(U_0, U_1, \ldots, U_n)$ are all fixed to constant, a protocol is said to be a deterministic protocol. A protocol with six communication slots is described in Figure 18.1.

In later sections of this chapter, we also consider protocols with resource. For such protocols, in addition to the private- and common-coins, the parties observe correlated randomness $(Y_1, \ldots, Y_n)$, and $\mathcal{P}_i$'s communication and output may depend on Y_i as well.

Unlike the two-party setting, there may be multiple dishonest parties among the n parties. Furthermore, those dishonest parties may collude to increase their benefit as much as possible. Throughout this chapter and the next, we assume that there are at most t dishonest parties, where $1 \leq t \leq n - 1$. Since dishonest parties may collude, we regard dishonest parties as a single entity; we assume that, in each communication slot, all the observations of the dishonest parties are shared among them. In the same manner as the two-party setting, when

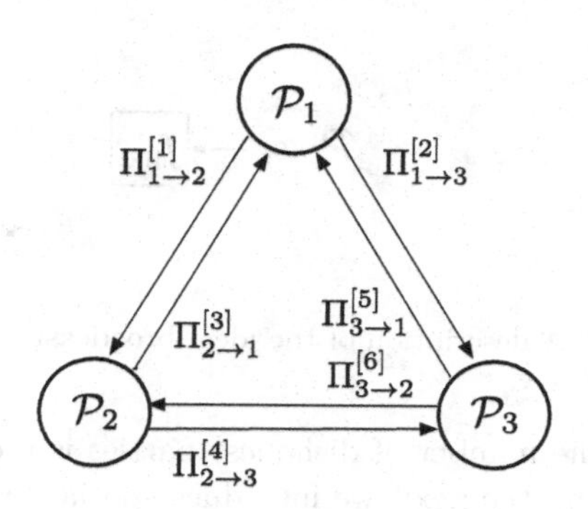

Figure 18.1 An example of a three-party protocol with six communication slots.

we say passively dishonest parties, we mean those dishonest parties which try to gain information but follow the protocol description. On the other hand, actively dishonest parties may not follow the protocol description. For simplicity of exposition, we also assume that the set of dishonest parties is arbitrarily fixed at the beginning of the protocol. This assumption is termed the *static adversary* assumption.[1]

Recall from Chapter 15 that a protocol is secure against an active adversary if dishonest parties can only conduct admissible attacks. This is formally described as before, requiring that the view of the parties in the actual protocol can be simulated in the ideal model. For a subset $\mathcal{A} \subset \mathcal{N} = \{1, \ldots, n\}$ with $|\mathcal{A}| \le t$, suppose that the parties in $\mathcal{A}$ are dishonest and the parties in $\mathcal{H} = \mathcal{N}\backslash\mathcal{A}$ are honest. During a protocol π, each party $\mathcal{P}_i$ observes the transcripts $\Pi_{*\to i}$ that are sent to $\mathcal{P}_i$ via the private channel, and computes the output $\hat{Z}_i$. The view of an honest $\mathcal{P}_i$ is $(X_i, \hat{Z}_i)$; on the other hand, a dishonest $\mathcal{P}_i$'s view is $(X_i, \hat{Z}_i, \Pi_{*\to i})$. For notational convenience, denote $X_{\mathcal{A}} = (X_i : i \in \mathcal{A})$, $X_{\mathcal{H}} = (X_i : i \in \mathcal{H})$, $\hat{Z}_{\mathcal{A}} = (\hat{Z}_i : i \in \mathcal{A})$, $\hat{Z}_{\mathcal{H}} = (\hat{Z}_i : i \in \mathcal{H})$, and $\Pi_{\mathcal{A}} = (\Pi_{*\to i} : i \in \mathcal{A})$. Then, the view of the actual protocol π is given by $\mathtt{V}[\pi] = ((X_{\mathcal{H}}, \hat{Z}_{\mathcal{H}}), (X_{\mathcal{A}}, \hat{Z}_{\mathcal{A}}, \Pi_{\mathcal{A}}))$. Note that the view depends on the set $\mathcal{A}$ of dishonest parties, but we just denote it by $\mathtt{V}[\pi]$ to simplify the notation. We need to guarantee that this view $\mathtt{V}[\pi]$ can be simulated in the ideal model in which each party $\mathcal{P}_i$ receives Z_i from the ideal function $f = (f_1, \ldots, f_n)$ as a response to the inputs $(X_1, \ldots, X_n)$.

DEFINITION 18.1 (Multiparty secure computation for an active adversary) For a given function $f = (f_1, \ldots, f_n)$ and an integer $1 \le t \le n-1$, a communication protocol π in the standard network is an ε-secure implementation of f against a t-active adversary if, for any $\mathcal{A}$ with $|\mathcal{A}| \le t$ and any attack by the parties in $\mathcal{A}$, there exists a simulator σ such that[2]

$$d_{\mathtt{var}}(\mathtt{V}[\pi], \mathtt{V}[\sigma \circ f]) \le \varepsilon.$$

In particular, it is a secure implementation if $\varepsilon = 0$.

[1] The set of dishonest parties may be adaptively chosen based on the transcripts as the protocol proceeds. This type of adversary is termed an *adaptive adversary*.

[2] The simulator may depend on the set $\mathcal{A}$ of dishonest parties and also the attack.

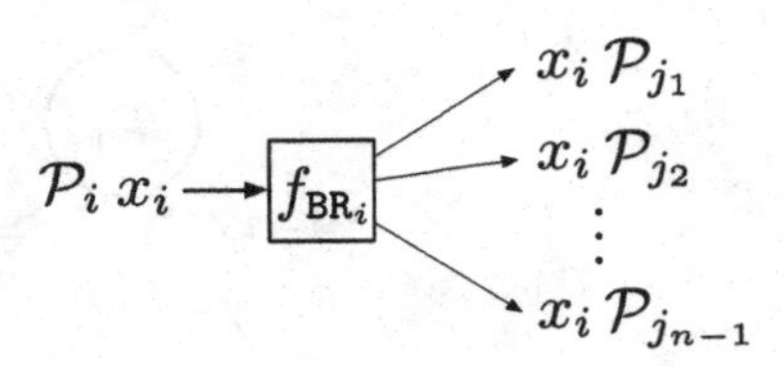

Figure 18.2 A description of the ideal broadcast.

Since the number of dishonest parties is a crucial parameter throughout this chapter and the next, we introduce special terminologies for the two most commonly used thresholds.

DEFINITION 18.2 (Honest majority/supermajority) When the maximum number t of dishonest parties in a protocol satisfies $t < \frac{n}{2}$, it is said to satisfy the *honest majority* assumption; on the other hand, if $t < \frac{n}{3}$ is satisfied, it is said to satisfy the *honest supermajority* assumption.

18.2 Broadcast and Byzantine Agreement

The broadcast function f_{BR_i} is a function with an input x_i by a prescribed party $\mathcal{P}_i$, termed the *sender*. Other parties, termed *receivers*, are supposed to output x_i (see Figure 18.2). Of course, if the sender is honest or passively dishonest, it is trivial to realize broadcast in standard networks: the sender $\mathcal{P}_i$ can broadcast x_i to all the receivers by sending $\Pi_{i \to j} = x_i$ to each receiver $\mathcal{P}_j$. However, when the sender is actively dishonest, since the sender may not send the same value to all the receivers, it is nontrivial to realize a broadcast. As we will see below, even if the sender transmits different values to receivers, honest receivers can cooperate to detect the sender's attack and agree on some common value. However, that common value may not be the same as the sender's original input. In fact, agreeing on some common value is the best thing that can be done by honest parties in this case; as we discussed in Chapter 15, changing the input is an admissible attack when the sender is dishonest. Based on this observation, we can derive the following alternative form of the security requirement for broadcast.

LEMMA 18.3 (Security of broadcast – alternative form) *A communication protocol π is an ε-secure implementation of broadcast f_{BR_i} against a t-active adversary (in the sense of Definition 18.1) if and only if the following hold for any $\mathcal{A} \subset \mathcal{N}$ with $|\mathcal{A}| \le t$ and any attack by the parties in $\mathcal{A}$.*

- *Correctness. When the sender $\mathcal{P}_i$ is honest with input x_i, then all the honest receivers output x_i with probability at least $1 - \varepsilon$.*
- *Agreement. All the honest receivers output the same value z with probability at least $1 - \varepsilon$.*

Proof Fix an arbitrary $\mathcal{A} \subset \mathcal{N}$ with $|\mathcal{A}| \le t$. We first consider the "if" part. When the sender $\mathcal{P}_i$ is honest, i.e., $i \in \mathcal{H} = \mathcal{N}\backslash\mathcal{A}$, by invoking $f_{\mathtt{BR}_i}$ with input x_i, all the parties (both honest and dishonest) obtain x_i in the ideal model. Then, by running protocol π with input x_i among the dishonest parties,[3] the dishonest parties can simulate the transcript $\Pi_{\mathcal{A}}$ and the output $\hat{Z}_{\mathcal{A}}$ of protocol π. Now, we shall compare the distribution $\mathrm{P}_{X_i \hat{Z}_{\mathcal{H}} \hat{Z}_{\mathcal{A}} \Pi_{\mathcal{A}}}$ of the view in the actual protocol and the distribution $\mathrm{Q}_{X_i Z_{\mathcal{H}} \hat{Z}_{\mathcal{A}} \Pi_{\mathcal{A}}}$ of the simulated protocol, where $Z_{\mathcal{H}} = (Z_j : j \in \mathcal{H})$ satisfies $Z_j = X_i$ for every $j \in \mathcal{H}$. Note that $\mathrm{P}_{X_i \hat{Z}_{\mathcal{A}} \Pi_{\mathcal{A}}} = \mathrm{Q}_{X_i \hat{Z}_{\mathcal{A}} \Pi_{\mathcal{A}}}$. Thus, by the correctness requirement

$$\Pr\left(\hat{Z}_j = X_i,\ \forall j \in \mathcal{H}\right) \ge 1 - \varepsilon,$$

we have

$$\begin{aligned}
&d_{\mathrm{var}}\left(\mathrm{P}_{X_i \hat{Z}_{\mathcal{H}} \hat{Z}_{\mathcal{A}} \Pi_{\mathcal{A}}}, \mathrm{Q}_{X_i Z_{\mathcal{H}} \hat{Z}_{\mathcal{A}} \Pi_{\mathcal{A}}}\right) \\
&\quad = \mathbb{E}\left[d_{\mathrm{var}}\left(\mathrm{P}_{\hat{Z}_{\mathcal{H}}|X_i \hat{Z}_{\mathcal{A}} \Pi_{\mathcal{A}}}\left(\cdot|X_i, \hat{Z}_{\mathcal{A}}, \Pi_{\mathcal{A}}\right), \mathrm{Q}_{Z_{\mathcal{H}}|X_i \hat{Z}_{\mathcal{A}} \Pi_{\mathcal{A}}}(\cdot|X_i, \hat{Z}_{\mathcal{A}}, \Pi_{\mathcal{A}})\right)\right] \\
&\quad = 1 - \Pr\left(\hat{Z}_j = X_i,\ \forall j \in \mathcal{H}\right) \\
&\quad \le \varepsilon,
\end{aligned}$$

where the expectation in the second formula is taken over $(X_i, \hat{Z}_{\mathcal{A}}, \Pi_{\mathcal{A}}) \sim \mathrm{P}_{X_i \hat{Z}_{\mathcal{A}} \Pi_{\mathcal{A}}}$ and the second identity follows from the fact that $Z_j = X_i$ with probability 1 in the ideal model (see the maximal coupling lemma in Section 2.9).

When the sender $\mathcal{P}_i$ is dishonest, i.e., $i \in \mathcal{A}$, before invoking $f_{\mathtt{BR}_i}$, the dishonest parties (including $\mathcal{P}_i$) run protocol π among them. Then, they learn what the honest parties would output in the actual protocol, i.e., $\hat{Z}_{j_1}, \ldots, \hat{Z}_{j_{|\mathcal{H}|}}$ with $\mathcal{H} = \{j_1, \ldots, j_{|\mathcal{H}|}\}$. Next, they invoke the ideal function $f_{\mathtt{BR}_i}$ invoked with input $\hat{Z}_{j_1}$,[4] and all the honest parties in the ideal model obtain $Z_j = \hat{Z}_{j_1}, j \in \mathcal{H}$. Now, we shall compare the distribution $\mathrm{P}_{X_i \hat{Z}_{\mathcal{H}} \hat{Z}_{\mathcal{A}} \Pi_{\mathcal{A}}}$ of the view in the actual protocol and the distribution $\mathrm{Q}_{X_i Z_{\mathcal{H}} \hat{Z}_{\mathcal{A}} \Pi_{\mathcal{A}}}$ of the simulated protocol. Note that $\mathrm{P}_{X_i \hat{Z}_{j_1} \hat{Z}_{\mathcal{A}} \Pi_{\mathcal{A}}} = \mathrm{Q}_{X_i Z_{j_1} \hat{Z}_{\mathcal{A}} \Pi_{\mathcal{A}}}$. Thus, by the agreement requirement

$$\Pr\left(\hat{Z}_{j_1} = \cdots = \hat{Z}_{j_{|\mathcal{H}|}}\right) \ge 1 - \varepsilon,$$

we have for $\bar{\mathcal{H}} = \mathcal{H}\backslash\{j_1\}$ that

$$\begin{aligned}
&d_{\mathrm{var}}\left(\mathrm{P}_{X_i \hat{Z}_{\mathcal{H}} \hat{Z}_{\mathcal{A}} \Pi_{\mathcal{A}}}, \mathrm{Q}_{X_i Z_{\mathcal{H}} \hat{Z}_{\mathcal{A}} \Pi_{\mathcal{A}}}\right) \\
&\quad = \mathbb{E}\Big[d_{\mathrm{var}}\Big(\mathrm{P}_{\hat{Z}_{\bar{\mathcal{H}}}|X_i \hat{Z}_{j_1} \hat{Z}_{\mathcal{A}} \Pi_{\mathcal{A}}}\left(\cdot|X_i, \hat{Z}_{j_1}, \hat{Z}_{\mathcal{A}}, \Pi_{\mathcal{A}}\right), \\
&\qquad\qquad \mathrm{Q}_{Z_{\bar{\mathcal{H}}}|X_i Z_{j_1} \hat{Z}_{\mathcal{A}} \Pi_{\mathcal{A}}}(\cdot|X_i, Z_{j_1}, \hat{Z}_{\mathcal{A}}, \Pi_{\mathcal{A}})\Big)\Big] \\
&\quad = 1 - \Pr\left(\hat{Z}_{j_1} = \cdots = \hat{Z}_{j_{|\mathcal{H}|}}\right) \\
&\quad \le \varepsilon.
\end{aligned}$$

[3] As mentioned in Section 18.1, dishonest parties are regarded as a single entity, and they cooperate to run protocol π.

[4] Any one of $\hat{Z}_{j_1}, \ldots, \hat{Z}_{j_{|\mathcal{H}|}}$ will do.

Moving to the "only if" part, note that ε-secure computability of $f_{\mathtt{BR}_i}$ implies the existence of a simulator σ satisfying

$$d_{\mathtt{var}}(\mathtt{V}[\pi], \mathtt{V}[\sigma \circ f_{\mathtt{BR}_i}]) \le \varepsilon.$$

Let $\mathrm{Q}_{X_i Z_{\mathcal{H}}}$ be the marginal of $\mathrm{P}_{\mathtt{V}[\sigma \circ f_{\mathtt{BR}_i}]}$ corresponding to the input by $\mathcal{P}_i$ and honest parties' outputs. Then, since $X_i = Z_j$ for every $j \in \mathcal{H}$ under $\mathrm{Q}_{X_i Z_{\mathcal{H}}}$, correctness follows from

$$\begin{aligned}1 - \Pr\big(\hat{Z}_j = X_i,\ \forall j \in \mathcal{H}\big) &= d_{\mathtt{var}}\big(\mathrm{P}_{X_i \hat{Z}_{\mathcal{H}}}, \mathrm{Q}_{X_i Z_{\mathcal{H}}}\big)\\ &\le d_{\mathtt{var}}\big(\mathtt{V}[\pi], \mathtt{V}[\sigma \circ f_{\mathtt{BR}_i}]\big)\\ &\le \varepsilon.\end{aligned}$$

To derive the agreement requirement, consider the marginal $\mathrm{Q}_{Z_{\mathcal{H}}}$ of $\mathrm{P}_{\mathtt{V}[\sigma \circ f_{\mathtt{BR}_i}]}$. Further, consider the maximal coupling (see Section 2.9) of $\mathrm{P}_{\hat{Z}_{\mathcal{H}}}$ and $\mathrm{Q}_{Z_{\mathcal{H}}}$ satisfying

$$\Pr\big(\hat{Z}_{\mathcal{H}} = Z_{\mathcal{H}}\big) = 1 - d_{\mathtt{var}}\big(\mathrm{P}_{\hat{Z}_{\mathcal{H}}}, \mathrm{Q}_{Z_{\mathcal{H}}}\big).$$

Since $Z_{j_1} = \cdots = Z_{j_{|\mathcal{H}|}}$ with probability 1 under $\mathrm{Q}_{Z_{\mathcal{H}}}$, we have

$$\begin{aligned}\Pr\big(\hat{Z}_{j_1} = \cdots = \hat{Z}_{j_{|\mathcal{H}|}}\big) &\ge \Pr\big(\hat{Z}_{j_1} = Z_{j_1}, \ldots, \hat{Z}_{j_{|\mathcal{H}|}} = Z_{j_{|\mathcal{H}|}}\big)\\ &= 1 - d_{\mathtt{var}}\big(\mathrm{P}_{\hat{Z}_{\mathcal{H}}}, \mathrm{Q}_{Z_{\mathcal{H}}}\big),\end{aligned}$$

which implies that

$$\begin{aligned}1 - \Pr\big(\hat{Z}_{j_1} = \cdots = \hat{Z}_{j_{|\mathcal{H}|}}\big) &\le d_{\mathtt{var}}\big(\mathrm{P}_{\hat{Z}_{\mathcal{H}}}, \mathrm{Q}_{Z_{\mathcal{H}}}\big)\\ &\le d_{\mathtt{var}}\big(\mathtt{V}[\pi], \mathtt{V}[\sigma \circ f_{\mathtt{BR}_i}]\big)\\ &\le \varepsilon.\end{aligned}$$

□

Note that, when the sender is honest, correctness implies agreement in Lemma 18.3; the latter requires that the honest receivers agree on the same value even if the sender is dishonest. Further, when the sender is dishonest, the agreement condition only requires honest parties to output "some" common value that is not necessarily the same as the sender's input.

Next, consider Byzantine agreement, which is closely related to broadcast. To fix ideas, we present an anecdote to clarify the Byzantine context. Once the divisions of the Byzantine armies encircled an enemy city. The generals of the divisions needed to agree on whether to attack the city or to retreat. However, they could communicate only via the messengers between pairs of generals, and some of generals were traitors. The generals wanted to reach an agreement in such a manner that

- all loyal generals could agree on the same plan of action,
- if there is a plan supported by the majority of loyal generals, then all loyal generals could execute that plan.

The Byzantine agreement primitive is what these generals needed!

Formally introducing the ideal function of the Byzantine agreement requires a bit of care. A naive definition, with the same output $Z_i = f_{\mathtt{BA}}(x_1, \ldots, x_n)$ for all parties, may be[5]

$$\tilde{f}_{\mathtt{BA}}(x_1, \ldots, x_n) = \begin{cases} 1, & \text{if } \exists \mathcal{S} \subseteq \mathcal{N},\ |\mathcal{S}| > \frac{n}{2} \text{ such that } x_i = 1\ \forall i \in \mathcal{S}, \\ 0, & \text{else} \end{cases} \tag{18.1}$$

for inputs $x_1, \ldots, x_n \in \{0, 1\}$.[6]

However, this function does not accurately capture Byzantine agreement. In fact, Byzantine agreement need not conceal honest parties' inputs from dishonest parties (see also Remark 18.1). For this reason, we define the ideal function $f_{\mathtt{BA}}$ of the Byzantine agreement as follows. Let $\mathcal{A} \subset \mathcal{N}$ be the set of dishonest parties, and let $\mathcal{H} = \mathcal{N} \backslash \mathcal{A}$ be the set of honest parties. When the function $f_{\mathtt{BA}}$ is invoked with inputs $x_{\mathcal{N}} = (x_1, \ldots, x_n)$, the function first provides honest parties' inputs $x_{\mathcal{H}}$ to dishonest parties. Then, they can change their inputs $x_{\mathcal{A}}$ to $x'_{\mathcal{A}}$. Finally, the function provides $\tilde{f}_{\mathtt{BA}}(x_{\mathcal{H}}, x'_{\mathcal{A}})$ to all the parties, where $\tilde{f}_{\mathtt{BA}}$ is given by (18.1). In words, we define the ideal Byzantine agreement so that "rushing," i.e., dishonest parties changing their inputs after observing honest parties' inputs, is an admissible attack.

The simulation based security with the above defined ideal function $f_{\mathtt{BA}}$ is tantamount to the following property based definition of Byzantine agreement.

LEMMA 18.4 (Security of Byzantine agreement – alternative form) *A communication protocol π is an ε-secure implementation of Byzantine agreement $f_{\mathtt{BA}}$ against a t-active adversary (in the sense of Definition 18.1) if and only if the following hold for any $\mathcal{A} \subset \mathcal{N}$ with $|\mathcal{A}| \le t$ and any attack by the parties in $\mathcal{A}$.*

- *Correctness. If the number N_1 of honest parties with input 1 satisfies $N_1 > \frac{n}{2}$, then all the honest parties output 1 with probability $1 - \varepsilon$; if the number N_0 of honest parties with input 0 satisfies $N_0 \ge \frac{n}{2}$, then all the honest parties output 0 with probability $1 - \varepsilon$.*
- *Agreement. All the honest parties output the same value $z \in \{0, 1\}$ with probability at least $1 - \varepsilon$.*

Proof Fix an arbitrary $\mathcal{A} \subset \mathcal{N}$ with $|\mathcal{A}| \le t$. For the "if" part, by invoking the ideal function $f_{\mathtt{BA}}$, dishonest parties first learn [7] all the inputs $x_{\mathcal{N}} = (x_1, \ldots, x_n)$. Then, by running protocol π with inputs $x_{\mathcal{N}}$ among the dishonest parties, they simulate the transcript $\Pi_{\mathcal{A}}$ and all the outputs $\hat{Z}_{\mathcal{N}}$ of protocol π; in particular, the dishonest parties learn honest parties' outputs $\hat{Z}_{\mathcal{H}} = (\hat{Z}_{j_1}, \ldots, \hat{Z}_{j_{|\mathcal{H}|}})$. Then, the dishonest parties set their dummy inputs as $X'_j = \hat{Z}_{j_1}$ for every $j \in \mathcal{A}$.[8] Finally, the ideal function provides $Z_j = \tilde{f}_{\mathtt{BA}}(X_{\mathcal{H}}, X'_{\mathcal{A}})$ to each party.

[5] For simplicity of exposition, we only consider the binary Byzantine agreement; the nonbinary case can be handled similarly if the tie-break is addressed carefully.

[6] Note that when the numbers of 0 and 1 are a tie, the function outputs 0.

[7] This is because the modified ideal function allows information leakage.

[8] Any one of $\hat{Z}_{j_1}, \ldots, \hat{Z}_{j_{|\mathcal{H}|}}$ will do.

Now, we shall compare the distribution $P_{X_\mathcal{N}\hat{Z}_\mathcal{H}\hat{Z}_\mathcal{A}\Pi_\mathcal{A}}$ of the view in the real protocol and the distribution $Q_{X_\mathcal{N}Z_\mathcal{H}\hat{Z}_\mathcal{A}\Pi_\mathcal{A}}$ of the view in the simulated protocol. In fact, since the analysis can be done almost in the same manner as the proof of Lemma 18.3, we only provide an outline. First, note that $P_{X_\mathcal{N}\hat{Z}_\mathcal{A}\Pi_\mathcal{A}} = Q_{X_\mathcal{N}\hat{Z}_\mathcal{A}\Pi_\mathcal{A}}$. When $N_0 \geq \frac{n}{2}$ or $N_1 > \frac{n}{2}$, the correctness condition guarantees

$$\Pr\left(\hat{Z}_{j_1} = \cdots = \hat{Z}_{j_{|\mathcal{H}|}} = b\right) \geq 1 - \varepsilon$$

for some $b \in \{0, 1\}$. On the other hand, the ideal function $f_{\mathtt{BA}}$ always provides $\hat{Z}_\mathcal{H} = (1, \ldots, 1)$ in the former case, and $\hat{Z}_\mathcal{H} = (0, \ldots, 0)$ in the latter case. By noting these facts, we can derive

$$d_{\mathtt{var}}\big(P_{X_\mathcal{N}\hat{Z}_\mathcal{H}\hat{Z}_\mathcal{A}\Pi_\mathcal{A}}, Q_{X_\mathcal{N}Z_\mathcal{H}\hat{Z}_\mathcal{A}\Pi_\mathcal{A}}\big) \leq \varepsilon.$$

When $N_1 \leq \frac{n}{2}$ and $N_0 < \frac{n}{2}$, by noting that $N_1 + |\mathcal{A}| = n - N_0 > \frac{n}{2}$ and $N_0 + |\mathcal{A}| = n - N_1 \geq \frac{n}{2}$, we can verify that the value $\tilde{f}_{\mathtt{BA}}(X_\mathcal{H}, X'_\mathcal{A})$ coincides with $\hat{Z}_{j_1}$ since $X'_\mathcal{A} = (\hat{Z}_{j_1}, \ldots, \hat{Z}_{j_1})$. Using this observation we can derive

$$d_{\mathtt{var}}\big(P_{X_\mathcal{N}\hat{Z}_\mathcal{H}\hat{Z}_\mathcal{A}\Pi_\mathcal{A}}, Q_{X_\mathcal{N}Z_\mathcal{H}\hat{Z}_\mathcal{A}\Pi_\mathcal{A}}\big) \leq \varepsilon.$$

Next, we consider the "only if" part. Note that ε-secure computability of $f_{\mathtt{BA}}$ implies the existence of a simulator satisfying

$$d_{\mathtt{var}}\big(\mathtt{V}[\pi], \mathtt{V}[\sigma \circ f_{\mathtt{BA}}]\big) \leq \varepsilon.$$

Let $Q_{Z_\mathcal{H}}$ be the marginal of $P_{\mathtt{V}[\sigma\circ f_{\mathtt{BA}}]}$ corresponding to honest parties' outputs. When $N_1 > \frac{n}{2}$ or $N_0 \geq \frac{n}{2}$, by noting that $Z_j = \tilde{f}_{\mathtt{BA}}(X_\mathcal{N})$ for $j \in \mathcal{H}$, the correctness requirement follows from

$$\begin{aligned} 1 - \Pr\left(\hat{Z}_j = \tilde{f}_{\mathtt{BA}}(X_\mathcal{N})\ \forall j \in \mathcal{H}\right) &= d_{\mathtt{var}}\big(P_{\hat{Z}_\mathcal{H}}, Q_{Z_\mathcal{H}}\big) \\ &\leq d_{\mathtt{var}}\big(\mathtt{V}[\pi], \mathtt{V}[\sigma \circ f_{\mathtt{BA}}]\big) \\ &\leq \varepsilon. \end{aligned}$$

On the other hand, agreement can be verified exactly in the same manner as the proof of agreement in Lemma 18.3. □

As we mentioned above, Byzantine agreement is closely related to broadcast. In fact, the former can be constructed from the latter.

LEMMA 18.5 (Byzantine agreement from broadcast) *Suppose that $\pi_{\mathtt{BR}}$ is an ε-secure implementation of broadcast against a t-active adversary.*[9] *Then, by invoking $\pi_{\mathtt{BR}}$, we can construct a protocol $\pi_{\mathtt{BA}}$ that is an $n\varepsilon$-secure implementation of the Byzantine agreement against a t-active adversary.*

Proof A protocol $\pi_{\mathtt{BA}}$ is described in Protocol 18.1. Since $\pi_{\mathtt{BR}}$ is ε-secure, by the union bound, all the honest parties have the same observation with probability $1 - n\varepsilon$ in the second step. In particular, the message $z_{i\to j}$ from honest $\mathcal{P}_i$ coincides with x_i. Thus, the correctness and the agreement of $\pi_{\mathtt{BA}}$ in Lemma 18.4 are satisfied with error $n\varepsilon$. □

[9] We assume that broadcast from every sender is available.

Protocol 18.1 Byzantine agreement from broadcast

Input: $x_1, \ldots, x_n$

Output: $z_1, \ldots, z_n$

1. For $i \in \mathcal{N}$, the parties invoke $\pi_{\mathtt{BR}}$ with sender $\mathcal{P}_i$'s input x_i, and $\mathcal{P}_j$ receives $z_{i \to j}$ for $j \in \mathcal{N}$, where $z_{i \to i} = x_i$.
2. For $j \in \mathcal{N}$, $\mathcal{P}_j$ outputs $z_j = 1$ if there exists $\mathcal{S} \subseteq \mathcal{N}$ with $|\mathcal{S}| > \frac{n}{2}$ such that $z_{i \to j} = 1$ for every $i \in \mathcal{S}$; otherwise, $\mathcal{P}_j$ outputs $z_j = 0$.

Protocol 18.2 Broadcast from Byzantine agreement

Input: Sender $\mathcal{P}_i$'s input $x_i \in \{0, 1\}$

Output: z_j for $j \in \mathcal{N} \backslash \{i\}$

1. For $j \in \mathcal{N}$, the sender $\mathcal{P}_i$ sends $\Pi_{i \to j} = x_i$ to $\mathcal{P}_j$ via the private channel.
2. The parties invoke $\pi_{\mathtt{BA}}$ with inputs $(\Pi_{i \to 1}, \ldots, \Pi_{i \to n})$, where $\Pi_{i \to i} = x_i$, and $\mathcal{P}_j$ receives $z_{i \to j}$ for $j \in \mathcal{N}$.
3. For $j \in \mathcal{N}$, $\mathcal{P}_j$ outputs $z_{i \to j}$.

Remark 18.1 Although Protocol 18.1 is an $n\varepsilon$-secure implementation of the ideal Byzantine agreement $f_{\mathtt{BA}}$, it is not an $n\varepsilon$-secure implementation of $\tilde{f}_{\mathtt{BA}}$ defined in (18.1).

When the honest majority assumption is satisfied, the converse to Lemma 18.5 also holds, i.e., broadcast can be constructed from the Byzantine agreement.

LEMMA 18.6 (Broadcast from Byzantine agreement) *Suppose that $t < \frac{n}{2}$ and $\pi_{\mathtt{BA}}$ is an ε-secure implementation of the Byzantine agreement against a t-active adversary. Then, by invoking $\pi_{\mathtt{BA}}$, we can construct a protocol $\pi_{\mathtt{BR}}$ that is an ε-secure implementation of the (binary) broadcast against a t-active adversary.*

Proof The required protocol $\pi_{\mathtt{BR}}$, when i is the sender, is described in Protocol 18.2. Since $\pi_{\mathtt{BA}}$ is ε-secure, the agreement of $\pi_{\mathtt{BR}}$ is guaranteed. For the correctness of $\pi_{\mathtt{BR}}$, when the sender $\mathcal{P}_i$ is honest, all the honest parties invoke $\pi_{\mathtt{BA}}$ with input x_i. Thus, since $t < \frac{n}{2}$, all the honest parties output $z_{i \to j} = x_i$ with probability $1 - \varepsilon$. □

18.3 Broadcast from Scratch

In order to grasp the difficulty of implementing broadcast in the standard network, consider the following naive attempt to implement broadcast with one dishonest party among three parties, where $\mathcal{P}_1$ is the sender with input $x_1 \in \{0, 1\}$. First, $\mathcal{P}_1$ sends the transcripts $\Pi_{1 \to 2} = x_1$ and $\Pi_{1 \to 3} = x_1$ to $\mathcal{P}_2$ and $\mathcal{P}_3$, respectively. Then, in order to confirm what they received from $\mathcal{P}_1$, $\mathcal{P}_2$ and $\mathcal{P}_3$ forward $\Pi_{2 \to 3} = \Pi_{1 \to 2}$ and $\Pi_{3 \to 2} = \Pi_{1 \to 3}$ to each other. Suppose that $\mathcal{P}_2$ received

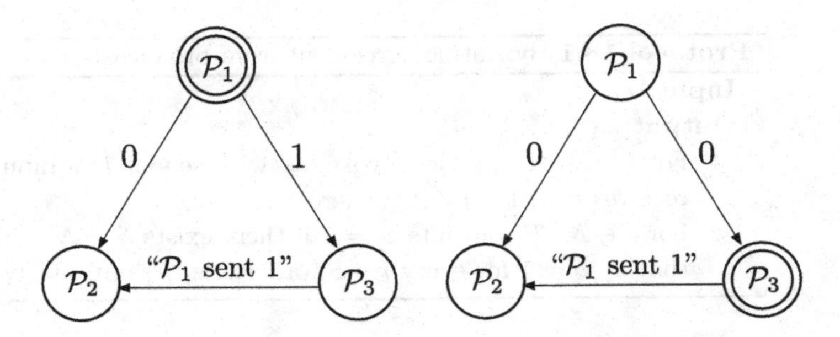

Figure 18.3 Two possible attacks in the naive protocol for broadcast with one dishonest party (double circle) among three parties. $\mathcal{P}_2$ cannot discriminate the situation with dishonest $\mathcal{P}_1$ (left) and the situation with dishonest $\mathcal{P}_3$ (right).

Protocol 18.3 An implementation of four-party broadcast

Input: $\mathcal{P}_1$'s input $x_1 \in \{0, 1\}$
Output: $\mathcal{P}_i$'s output z_i for $i = 2, 3, 4$

1. For $i = 2, 3, 4$, $\mathcal{P}_1$ sends $\Pi_{1\to i} = x_1$ to $\mathcal{P}_i$.
2. For $i = 2, 3, 4$, $\mathcal{P}_i$ sends $\Pi_{i\to j} = \Pi_{1\to i}$ to $\mathcal{P}_j$, where $j \in \{2, 3, 4\}\backslash\{i\}$.
3. For $i = 2, 3, 4$, $\mathcal{P}_i$ sets $z_i = \mathtt{Maj}(\Pi_{j\to i} : j \in \{1, 2, 3, 4\}\backslash\{i\})$.

$\Pi_{1\to 2} = 0$ and $\Pi_{3\to 2} = 1$. What should $\mathcal{P}_2$ output? If $\mathcal{P}_1$ is dishonest and $\mathcal{P}_3$ is honest, $\mathcal{P}_1$ could have sent $\Pi_{1\to 2} = 0$ and $\Pi_{1\to 3} = 1$, and $\mathcal{P}_3$ could have forwarded $\Pi_{3\to 2} = \Pi_{1\to 3} = 1$; alternatively, if $\mathcal{P}_1$ is honest and $\mathcal{P}_3$ is dishonest, even though $\mathcal{P}_1$ sent the same transcript $\Pi_{1\to 2} = \Pi_{1\to 3} = 0$, $\mathcal{P}_3$ could have forwarded the substituted transcript $\Pi_{3\to 2} = 1$. In this situation, there is no way for $\mathcal{P}_2$ to distinguish two possible attacks (cf. Figure 18.3). The problem is that, in the case of three parties, there is no majority among the honest parties. As we will see in Section 18.4, there is no secure implementation of broadcast with one dishonest party among three parties.

When there is only one dishonest party among four parties, the above mentioned difficulty can be overcome. In fact, the above mentioned "send-forward" scheme works well. The key idea is taking the majority of received transcripts. For a vector $(v_1, \ldots, v_n) \in \{0, 1\}^n$, let

$$\mathtt{Maj}(v_1, \ldots, v_n) := \begin{cases} 0, & \text{if } |\{j : v_j = 0\}| \geq \frac{n}{2}, \\ 1, & \text{if } |\{j : v_j = 1\}| > \frac{n}{2} \end{cases}$$

be the majority function. A four-party protocol for broadcast using the majority function is described in Protocol 18.3.

PROPOSITION 18.7 *Protocol 18.3 is a perfect, i.e., $\varepsilon = 0$, implementation of broadcast with one dishonest party among four parties.*

Proof When the sender $\mathcal{P}_1$ is honest, every honest $\mathcal{P}_i$ receives $\Pi_{1\to i} = x_1$. Furthermore, every honest $\mathcal{P}_j$ forwards $\Pi_{j\to i} = \Pi_{1\to j} = x_1$ to $\mathcal{P}_i$. Since there is at

Protocol 18.4 An implementation $\pi_{\mathtt{BR}}[r; \mathcal{S}; \mathcal{P}_i]$ of broadcast for $r = 0$

Input: $\mathcal{P}_i$'s input $x_i \in \{0, 1\}$
Output: $\mathcal{P}_j$'s output z_j for $j \in \mathcal{S}\backslash\{i\}$

1. For $j \in \mathcal{S}\backslash\{i\}$, $\mathcal{P}_i$ sends $\Pi_{i\to j} = x_i$ to $\mathcal{P}_j$.
2. For $j \in \mathcal{S}\backslash\{i\}$, $\mathcal{P}_j$ outputs $z_j = \Pi_{i\to j}$.

Protocol 18.5 An implementation $\pi_{\mathtt{BR}}[r; \mathcal{S}; \mathcal{P}_i]$ of broadcast for $r > 0$

Input: $\mathcal{P}_i$'s input $x_i \in \{0, 1\}$
Output: $\mathcal{P}_\ell$'s output z_ℓ for $\ell \in \mathcal{S}\backslash\{i\}$

1. For $j \in \mathcal{S}\backslash\{i\}$, $\mathcal{P}_i$ sends $\Pi_{i\to j} = x_i$ to $\mathcal{P}_j$.
2. For $j \in \mathcal{S}\backslash\{i\}$, $\mathcal{P}_j$ forwards $\Pi_{i\to j}$ to other parties in $\mathcal{S}\backslash\{i\}$ by invoking $\pi_{\mathtt{BR}}[r-1; \mathcal{S}\backslash\{i\}; \mathcal{P}_j]$ with input $\Pi_{i\to j}$. Denote the output of $\mathcal{P}_\ell$ by $z^{r-1}_{j\to\ell}$.
3. For $\ell \in \mathcal{S}\backslash\{i\}$, using the output $z^{r-1}_{j\to\ell}$ of $\pi_{\mathtt{BR}}[r-1; \mathcal{S}\backslash\{i\}; \mathcal{P}_j]$ for $j \in \mathcal{S}\backslash\{i, \ell\}$ in the previous step and $z^{r-1}_{i\to\ell} := \Pi_{i\to\ell}$ in the first step, $\mathcal{P}_\ell$ outputs $z_\ell = \mathtt{Maj}(z^{r-1}_{j\to\ell} : j \in \mathcal{S}\backslash\{\ell\})$.

most one dishonest party by assumption, the majority function outputs $z_i = x_1$. Thus, the correctness condition is satisfied.

Next, we verify the agreement condition. When $\mathcal{P}_1$ is honest, the agreement follows from the correctness condition. Consider the case when $\mathcal{P}_1$ is dishonest, wherein, since there is at most one dishonest party by assumption, $\mathcal{P}_2$, $\mathcal{P}_3$, and $\mathcal{P}_4$ are honest. Since each $\mathcal{P}_i$ forwards the transcript $\Pi_{i\to j} = \Pi_{1\to i}$ honestly, we have

$$
\begin{aligned}
z_2 &= \mathtt{Maj}(\Pi_{1\to2}, \Pi_{3\to2}, \Pi_{4\to2}) = \mathtt{Maj}(\Pi_{1\to2}, \Pi_{1\to3}, \Pi_{1\to4}),\\
z_3 &= \mathtt{Maj}(\Pi_{1\to3}, \Pi_{2\to3}, \Pi_{4\to3}) = \mathtt{Maj}(\Pi_{1\to3}, \Pi_{1\to2}, \Pi_{1\to4}),\\
z_4 &= \mathtt{Maj}(\Pi_{1\to4}, \Pi_{2\to4}, \Pi_{3\to4}) = \mathtt{Maj}(\Pi_{1\to4}, \Pi_{1\to2}, \Pi_{1\to3}),
\end{aligned}
$$

which implies $z_2 = z_3 = z_4$. □

Now, we present an implementation of broadcast with t dishonest parties among n parties. As we can see from Protocol 18.3, the transcript received from the sender cannot be trusted immediately since the sender may be dishonest; a receiver needs to confirm whether other parties received the same transcript from the sender. In Protocol 18.3, it suffices to do this confirmation cycle just once since there is only one dishonest party by assumption. In general, we iterate the confirmation cycle in multiple rounds depending on the maximum number of dishonest parties. For a set $\mathcal{S} \subseteq \{1, \ldots, n\}$ of parties and a sender $\mathcal{P}_i$ in $\mathcal{S}$, a protocol with r-rounds confirmation cycle is denoted by $\pi_{\mathtt{BR}}[r; \mathcal{S}; \mathcal{P}_i]$. The protocol for $r = 0$ is described in Protocol 18.4 and the protocol for $r > 0$ is described in Protocol 18.5. Our protocol proceeds recursively: a party $\mathcal{P}_j$ forwards a transcript from the sender $\mathcal{P}_i$ in $\pi_{\mathtt{BR}}[r; \mathcal{S}; \mathcal{P}_i]$ by invoking $\pi_{\mathtt{BR}}[r-1; \mathcal{S}\backslash\{i\}; \mathcal{P}_j]$, and $\mathcal{P}_j$ in turn plays the role of the sender in the inner protocol.

It turns out that the protocol $\pi_{\mathtt{BR}}[t; \mathcal{S}; \mathcal{P}_i]$ satisfies the correctness and the agreement conditions as long as the honest supermajority is satisfied. To that end, we first verify the correctness condition.

LEMMA 18.8 *For every $r \geq 0$, the protocol $\pi_{\mathtt{BR}}[r; \mathcal{S}; \mathcal{P}_i]$ satisfies correctness perfectly if $|\mathcal{S}| > 2t + r$ and the number of dishonest parties among $\mathcal{S}$ is at most t.*

Proof For correctness of the protocol, we assume that the sender $\mathcal{P}_i$ is honest. We prove the claim by induction on r. When $r = 0$, an honest $\mathcal{P}_j$ outputs $z_j = \Pi_{i \to j} = x_i$. For $r > 0$, suppose that the protocol $\pi_{\mathtt{BR}}[r-1; \mathcal{S}'; \mathcal{P}_i]$ satisfies the correctness for any set $\mathcal{S}'$ such that $|\mathcal{S}'| > 2t + (r-1)$ and the number of dishonest parties among $\mathcal{S}'$ is at most t. In Step 2 of protocol $\pi_{\mathtt{BR}}[r; \mathcal{S}; \mathcal{P}_i]$, $\mathcal{P}_j$ invokes $\pi_{\mathtt{BR}}[r-1; \mathcal{S}\backslash\{i\}; \mathcal{P}_j]$ to forward $\Pi_{i \to j} = x_i$ to other parties in $\mathcal{S}\backslash\{i\}$. If $\mathcal{P}_j$ is honest, since $|\mathcal{S}| - 1 > 2t + (r-1)$, the induction hypothesis guarantees that an honest $\mathcal{P}_\ell$ outputs $z_{j \to \ell}^{r-1} = \Pi_{i \to j} = x_i$ in Step 2 of protocol $\pi_{\mathtt{BR}}[r; \mathcal{S}; \mathcal{P}_i]$.[10] Thus, in Step 3 of protocol $\pi_{\mathtt{BR}}[r; \mathcal{S}; \mathcal{P}_i]$, since the number of parties satisfies $|\mathcal{S}| - 1 > 2t + (r-1) \geq 2t$, i.e., the honest parties are in a majority, the majority function will output $z_\ell = x_i$. □

Now, we verify the requirements for agreement.

THEOREM 18.9 *For every integer $t \geq 0$, the protocol $\pi_{\mathtt{BR}}[t; \mathcal{S}; \mathcal{P}_i]$ satisfies both correctness and agreement perfectly if $|\mathcal{S}| > 3t$ and the number of dishonest parties among $\mathcal{S}$ is at most t, i.e., the honest supermajority holds.*

Proof We shall prove by induction on t. When all the parties are honest, i.e., $t = 0$, protocol $\pi_{\mathtt{BR}}[0; \mathcal{S}'; \mathcal{P}_i]$ for any set $\mathcal{S}'$ satisfies the correctness and the agreement requirements. For $t > 0$, suppose that $\pi_{\mathtt{BR}}[t-1; \mathcal{S}'; \mathcal{P}_i]$ satisfies the correctness and the agreement requirements when $|\mathcal{S}'| > 3(t-1)$, and that there are at most $(t-1)$ dishonest parties among $\mathcal{S}'$.

When the sender $\mathcal{P}_i$ of protocol $\pi_{\mathtt{BR}}[t; \mathcal{S}; \mathcal{P}_i]$ is honest, by applying Lemma 18.8 with $r = t$, we can show the correctness of $\pi_{\mathtt{BR}}[t; \mathcal{S}; \mathcal{P}_i]$. Furthermore, the agreement also follows from the correctness requirement when the sender is honest.

When the sender $\mathcal{P}_i$ of protocol $\pi_{\mathtt{BR}}[t; \mathcal{S}; \mathcal{P}_i]$ is dishonest, there are at most $t-1$ dishonest parties among $\mathcal{S}\backslash\{i\}$. Since $|\mathcal{S}| - 1 > 3t - 1 > 3(t-1)$, the induction hypothesis guarantees that protocol $\pi_{\mathtt{BR}}[t-1; \mathcal{S}\backslash\{i\}; \mathcal{P}_j]$ invoked in Step 2 of $\pi_{\mathtt{BR}}[t; \mathcal{S}; \mathcal{P}_i]$ satisfies both the correctness and the agreement requirements. Thus, for two honest parties $\mathcal{P}_\ell$ and $\mathcal{P}_{\ell'}$, the outputs $z_{j \to \ell}^{r-1}$ and $z_{j \to \ell'}^{r-1}$ received from $\mathcal{P}_j$ for $j \in \mathcal{S}\backslash\{i, \ell, \ell'\}$ in Step 2 are the same. Furthermore, since $\mathcal{P}_\ell$ and $\mathcal{P}_{\ell'}$ are honest, the output $z_{\ell \to \ell'}^{r-1}$ transmitted from $\mathcal{P}_\ell$ to $\mathcal{P}_{\ell'}$ coincides with $\Pi_{i \to \ell}$ and the output $z_{\ell' \to \ell}^{r-1}$ transmitted from $\mathcal{P}_{\ell'}$ to $\mathcal{P}_\ell$ coincides with $\Pi_{i \to \ell'}$. Therefore, all the

[10] We need not care what $\mathcal{P}_\ell$ receives from a dishonest party.

inputs to the majority function by $\mathcal{P}_\ell$ and $\mathcal{P}_{\ell'}$ are the same – the honest parties get the same output from all the other parties except the sender i, and they all share the value they receive directly from i verbatim with each other. This implies $z_\ell = z_{\ell'}$, whereby agreement follows for the protocol $\pi_{\mathtt{BR}}[t; \mathcal{S}; \mathcal{P}_i]$. □

We illustrate the protocol using an example.

EXAMPLE 18.10 Consider the case with $(n, t) = (7, 2)$, and $\mathcal{P}_1$ is the sender. First, suppose that $\mathcal{P}_1$ is honest, and $\mathcal{P}_6$ and $\mathcal{P}_7$ are dishonest. In $\pi_{\mathtt{BR}}[2; \{1, \ldots, 7\}; \mathcal{P}_1]$, $\mathcal{P}_1$ sends $\Pi_{1 \to j} = x_1$ to $\mathcal{P}_j$ for $j = 2, \ldots, 7$. Since $\mathcal{P}_1$ is honest, the transcripts received by $\mathcal{P}_2, \ldots, \mathcal{P}_7$ are the same. Then, for $j = 2, \ldots, 7$, $\mathcal{P}_j$ forwards $\Pi_{1 \to j}$ to other parties in $\{2, \ldots, 7\}$ by invoking $\pi_{\mathtt{BR}}[1; \{2, \ldots, 7\}; \mathcal{P}_j]$. For the honest sender $\mathcal{P}_j$ and the honest receiver $\mathcal{P}_\ell$ of the inner protocol, since the number of parties 6 is larger than $2t + r = 5$, Lemma 18.8 guarantees that honest $\mathcal{P}_\ell$ receives x_1 from honest $\mathcal{P}_j$.[11] Thus, the majority functions by honest parties in the outer protocol output x_1. Here, note that the correctness of the outer protocol $\pi_{\mathtt{BR}}[2; \{1, \ldots, 7\}; \mathcal{P}_1]$ is guaranteed even though the agreement of the inner protocols $\pi_{\mathtt{BR}}[1; \{2, \ldots, 7\}; \mathcal{P}_6]$ and $\pi_{\mathtt{BR}}[1; \{2, \ldots, 7\}; \mathcal{P}_7]$ are not guaranteed.

Next, suppose that $\mathcal{P}_1$ and $\mathcal{P}_7$ are dishonest. In this case, the transcripts sent by $\mathcal{P}_1$ in $\pi_{\mathtt{BR}}[2; \{1, \ldots, 7\}; \mathcal{P}_1]$ may be different. However, since there is only one dishonest party among $\{2, \ldots, 7\}$ in the inner protocols, both the correctness and the agreement of the inner protocols $\pi_{\mathtt{BR}}[1; \{2, \ldots, 7\}; \mathcal{P}_j]$ for $j = 2, \ldots, 7$ are guaranteed. Thus, all the honest parties have the same inputs to the majority function in the outer protocol, and they output the same value in $\pi_{\mathtt{BR}}[2; \{1, \ldots, 7\}; \mathcal{P}_1]$.

18.4 Impossibility of Broadcast from Scratch without a Supermajority

In Section 18.3, we constructed a secure implementation of broadcast with an honest supermajority. It turns out that an honest supermajority is also necessary to construct a secure implementation of broadcast. In fact, at the beginning of Section 18.3, we discussed the difficulty of implementing broadcast with one dishonest party among three parties. We first prove the impossibility result for the three-party case, from which the impossibility result for the general case follows.

THEOREM 18.11 *For a three-party network with one dishonest party, an ε-secure implementation of broadcast from scratch exists only if $\varepsilon \geq 1/3$.*

Proof Suppose that there exists an ε-secure implementation π of broadcast with sender $\mathcal{P}_1$ and receivers $\mathcal{P}_2$ and $\mathcal{P}_3$. For simplicity of presentation, we first consider the case with $\varepsilon = 0$.

[11] In fact, note that this inner protocol is just one round of a "send-forward" scheme as in Protocol 18.3.

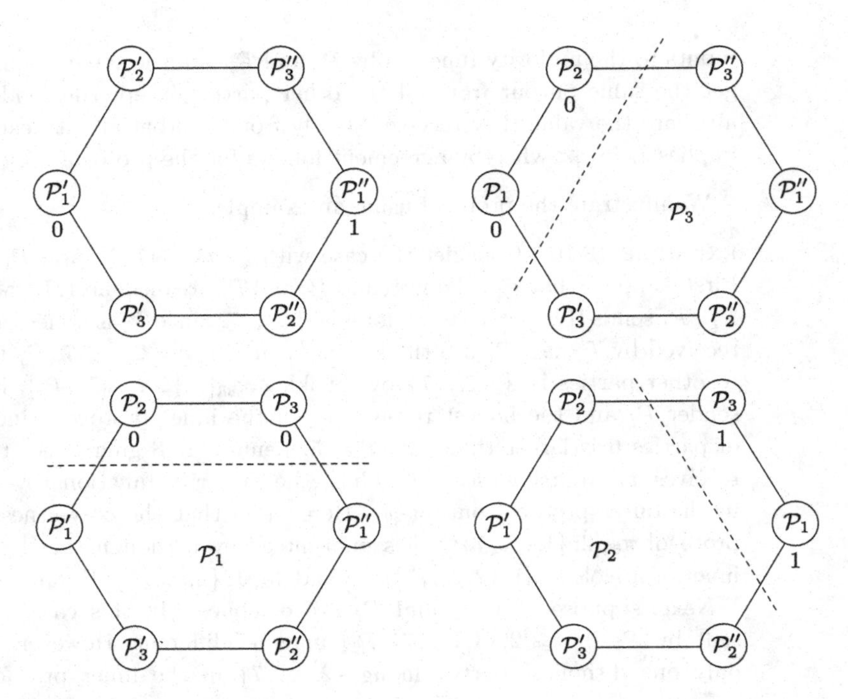

Figure 18.4 A description of the virtual network (top left) considered in the proof of Theorem 18.11 and three possible scenarios in the original protocol: the first scenario in which $\mathcal{P}_3$ is dishonest (top right); the second scenario in which $\mathcal{P}_1$ is dishonest (bottom left); and the third scenario in which $\mathcal{P}_2$ is dishonest (bottom right).

The key idea of the proof is to consider a virtual network that consists of six parties; see the top left of Figure 18.4. This virtual network will be used to design an interesting "coupling" of the distributions induced between the view of the parties across different adversarial behaviors, and finally to conclude through a contradiction that the implementation cannot be secure. Out of these six parties, two will correspond to the honest parties and all the remaining parties will be simulated by the dishonest party in π. In more detail, in one such adversarial setting in the virtual network, we run an alternative protocol $\tilde{\pi}$ instead of π in which $(\mathcal{P}_1', \mathcal{P}_2', \mathcal{P}_3')$ and $(\mathcal{P}_1'', \mathcal{P}_2'', \mathcal{P}_3'')$ follow the protocol description of π with $\mathcal{P}_1'$'s input 0 and $\mathcal{P}_1''$'s input 1, but $\mathcal{P}_2'$ communicates with $\mathcal{P}_3''$ (and not $\mathcal{P}_3'$), $\mathcal{P}_3''$ communicates with $\mathcal{P}_2'$ (and not $\mathcal{P}_2''$). Namely, if $\mathcal{P}_1$ communicates to $\mathcal{P}_2$ in the γth communication slot of protocol π, then $\mathcal{P}_1'$ communicates to $\mathcal{P}_2'$ in the $(2\gamma - 1)$th communication slot and $\mathcal{P}_1''$ communicates to $\mathcal{P}_2''$ in the 2γth communication slot of $\tilde{\pi}$. For the private-coin U_i used in protocol π, we create independent copies U_i' and U_i'' with the same distribution and provide them to $\mathcal{P}_i'$ and $\mathcal{P}_i''$, respectively. For the common-coin U_0 used in π, we provide U_0 to all the parties in $\tilde{\pi}$.

In this manner, the dependency of the transcript on each party's observation in protocol π can be mimicked in protocol $\tilde{\pi}$. Here, it should be noted that there is no dishonest party among the six parties, and all of them follow the protocol $\tilde{\pi}$. The attack is on protocol π. The local behaviors of two adjacent parties in $\tilde{\pi}$ can be regarded as behaviors of the corresponding honest parties in the original protocol π, and the other four parties in $\tilde{\pi}$ are simulated by a dishonest party in π. Based on the parties' behaviors in $\tilde{\pi}$, we can consider three attacks for the original protocol π, and show that one of them must succeed.

Specifically, consider first the scenario in which $\mathcal{P}_3$ is dishonest, where $\mathcal{P}_1$'s input is 0; see the top right of Figure 18.4. In this scenario, by locally simulating $(\mathcal{P}_3', \mathcal{P}_2'', \mathcal{P}_1'', \mathcal{P}_3'')$ of protocol $\tilde{\pi}$, $\mathcal{P}_3$ mimics the behavior of $\mathcal{P}_3'$ when communicating with $\mathcal{P}_1$ and the behavior of $\mathcal{P}_3''$ when communicating with $\mathcal{P}_2$. Then, the transcripts observed by $\mathcal{P}_1$ and $\mathcal{P}_2$ in this scenario are exactly the same as that of $\mathcal{P}_1'$ and $\mathcal{P}_2'$ in protocol $\tilde{\pi}$. Note that this scenario is a valid attack in protocol π with dishonest $\mathcal{P}_3$. Thus, since $\mathcal{P}_1$ is honest, the correctness of protocol π implies that $\mathcal{P}_2$ outputs $\mathcal{P}_1$'s input 0. Since the transcript observed by $\mathcal{P}_2'$ in $\tilde{\pi}$ is the same as the transcript observed by $\mathcal{P}_2$ in this scenario, $\mathcal{P}_2'$ outputs $Z_2' = 0$ as well.

Next, consider the scenario in which $\mathcal{P}_1$ is dishonest; see the bottom left of Figure 18.4. In this scenario, by locally simulating $(\mathcal{P}_1', \mathcal{P}_3', \mathcal{P}_2'', \mathcal{P}_1'')$ of protocol $\tilde{\pi}$, $\mathcal{P}_1$ mimics the behavior of $\mathcal{P}_1'$ when communicating with $\mathcal{P}_2$ and the behavior of $\mathcal{P}_1''$ when communicating with $\mathcal{P}_3$. Then, the transcripts observed by $\mathcal{P}_2$ and $\mathcal{P}_3$ in this scenario are exactly the same as that of $\mathcal{P}_2'$ and $\mathcal{P}_3''$ in protocol $\tilde{\pi}$. Again, note that this scenario is a valid attack in π with dishonest sender $\mathcal{P}_1$. Since the transcript observed by $\mathcal{P}_2$ in this scenario is the same as the transcript observed by $\mathcal{P}_2'$ in $\tilde{\pi}$ and $\mathcal{P}_2'$ outputs $Z_2' = 0$ in $\tilde{\pi}$ (cf. the first scenario), $\mathcal{P}_2$ outputs 0 as well. Then, the agreement of protocol π implies that $\mathcal{P}_3$ outputs the same value 0. Thus, since the transcript observed by $\mathcal{P}_3''$ in $\tilde{\pi}$ is the same as the transcript observed by $\mathcal{P}_3$ in this scenario, $\mathcal{P}_3''$ outputs $Z_3'' = 0$ as well.

Finally, consider the scenario in which $\mathcal{P}_2$ is dishonest, where $\mathcal{P}_1$'s input is 1; see the bottom right of Figure 18.4. In this scenario, by locally simulating $(\mathcal{P}_2', \mathcal{P}_1', \mathcal{P}_3', \mathcal{P}_2'')$ in protocol $\tilde{\pi}$, $\mathcal{P}_2$ mimics the behavior of $\mathcal{P}_2'$ when communicating with $\mathcal{P}_3$ and the behavior of $\mathcal{P}_2''$ when communicating with $\mathcal{P}_1$. Then, the transcripts observed by $\mathcal{P}_1$ and $\mathcal{P}_3$ in this scenario are exactly the same as that of $\mathcal{P}_1''$ and $\mathcal{P}_3''$ in protocol $\tilde{\pi}$. Once again, note that this scenario is a valid attack in π with dishonest $\mathcal{P}_2$. Thus, since $\mathcal{P}_1$ is honest, the correctness of protocol π implies that $\mathcal{P}_3$ outputs $\mathcal{P}_1$'s input 1. Since the transcript observed by $\mathcal{P}_3''$ in $\tilde{\pi}$ is the same as the transcript observed by $\mathcal{P}_3$ in this scenario, $\mathcal{P}_3''$ outputs $Z_3'' = 1$ as well. However, this contradicts the fact that $\mathcal{P}_3''$ outputs $Z_3'' = 0$ (cf. the second scenario). Thus, one of the attacks in the three scenarios must succeed, and there does not exist a secure protocol π for broadcast.

To prove the claim of the theorem for positive ε, we keep track of the probability that $\mathcal{P}_3''$ outputs $Z_3'' = 0$ and $Z_3'' = 1$ assuming the existence of an ε-secure protocol π. By applying the argument above to the first scenario, the correctness of protocol π implies

$$\Pr(Z_2' = 0) = \Pr(Z_2 = 0) \geq 1 - \varepsilon.$$

Then, by applying the argument to the second scenario, the agreement of protocol π implies

$$\begin{aligned}\Pr(Z_2' = 0, Z_3'' = 0) &= \Pr(Z_2' = 0) - \Pr(Z_2' = 0, Z_3'' = 1) \\ &= \Pr(Z_2' = 0) - \Pr(Z_2 = 0, Z_3 = 1) \\ &\geq 1 - 2\varepsilon,\end{aligned}$$

which further implies

$$\Pr(Z_3'' = 0) \geq \Pr(Z_2' = 0, Z_3'' = 0) \geq 1 - 2\varepsilon.$$

By applying the argument to the third scenario, the correctness of protocol π implies

$$\Pr(Z_3'' = 1) = \Pr(Z_3 = 1) \geq 1 - \varepsilon.$$

Thus, we have

$$1 \geq \Pr(Z_3'' = 0) + \Pr(Z_3'' = 1) \geq 2 - 3\varepsilon,$$

which implies $\varepsilon \geq 1/3$. □

The argument in the proof of Theorem 18.11 may be termed "attack-coupling." In summary, it proceeds as follows.

1. Construct the six-party protocol by creating two copies of each party in the original protocol.
2. Construct three attacks, one for each cut of adjacent two parties.
3. Prove that one of these attacks must succeed.

This argument has been used widely to prove impossibility results for distributed computing; see Section 18.7 for more details.

Now, we prove the necessity of an honest supermajority for the case with t dishonest parties among n parties. This general result follows from the three-party result, Theorem 18.11, by partitioning n parties into three groups of parties.

THEOREM 18.12 *For an n-party network with $t \geq n/3$ dishonest parties, an ε-secure implementation of broadcast from scratch exists only if $\varepsilon \geq 1/3$.*

Proof For $\varepsilon < 1/3$, suppose that there exists an ε-secure implementation π of broadcast when the number of dishonest parties is $t \geq n/3$. Let $\mathcal{G}_1, \mathcal{G}_2, \mathcal{G}_3 \subset \{1, \ldots, n\}$ be a partition such that $|\mathcal{G}_i| \leq t$ for every $i = 1, 2, 3$; such a partition exists since t is an integer satisfying $t \geq n/3$. Let $\tilde{\pi}$ be a three-party protocol in which party $\tilde{\mathcal{P}}_i$ emulates all the parties in $\mathcal{G}_i$. Note that the original protocol π is secure whenever the number of dishonest parties is less than or equal to t. When party $\tilde{\mathcal{P}}_i$ is dishonest in protocol $\tilde{\pi}$, $\tilde{\mathcal{P}}_i$ can control $|\mathcal{G}_i|$ parties included in $\mathcal{G}_i$. But, since $|\mathcal{G}_i| \leq t$, the security of protocol π implies that protocol $\tilde{\pi}$ is an ε-secure implementation of broadcast with one dishonest party among three parties, which contradicts the statement of Theorem 18.11. □

We summarize the results in the previous section and this section as the following corollary.

COROLLARY 18.13 (Honest supermajority bound) *For $\varepsilon < 1/3$, an ε-secure implementation of broadcast from scratch exists if and only if the number t of dishonest parties among n parties satisfies $t < n/3$, i.e., the honest supermajority is satisfied.*

A careful inspection of the proof of Theorem 18.11 reveals that the same argument goes through even if every pair of parties shares pairwise correlation. That is, we need not restrict to private-coins U_i in the description of the communication model in Section 18.1, but can even allow every pair of parties $\mathcal{P}_i$ and $\mathcal{P}_j$ to use shared correlation U_{ij}. Even if we allow pairwise correlation, secure implementation of broadcast is possible only if the honest supermajority is satisfied (Problem 18.2). Further, note that Theorem 18.11 is valid for protocols with common-coin. Thus, in order to overcome the honest supermajority bottleneck, we need additional resources to be shared by the parties that are stronger than pairwise correlated random variables and common-coin. Such resources are introduced in the next section.

Remark 18.2 In the proofs of Theorems 18.11 and 18.12, the computational power of the parties never appears. Thus, these impossibility theorems are valid even if we consider the computationally bounded adversary. In other words, the computationally bounded adversary assumption is not helpful to overcome the honest supermajority bottleneck. This is in contrast to the fact that two-party primitives, such as oblivious transfer or bit commitment, can be realized from scratch for computationally bounded adversaries.

18.5 Broadcast without a Supermajority

As we have discussed at the end of Section 18.4, a pairwise correlated resource is not enough for secure implementation of broadcast beyond the honest supermajority bound. In this section, we introduce a cryptographic primitive called the digital signature and discuss the relation between digital signatures and broadcast. In particular, we will see how the digital signature can enable broadcast beyond the honest supermajority.

18.5.1 Digital Signature

The digital signature is closely related to authentication, but there are some significant differences. In fact, the digital signature plays an important role even when all the parties are connected via pairwise authenticated channels. To fix ideas, consider the following concrete situation for three parties.

EXAMPLE 18.14 (Signature on paycheck) $\mathcal{P}_1$ is an employer, $\mathcal{P}_2$ is an employee who is working for $\mathcal{P}_1$, and $\mathcal{P}_3$ is a bank. Suppose that $\mathcal{P}_1$ sends a check to $\mathcal{P}_2$

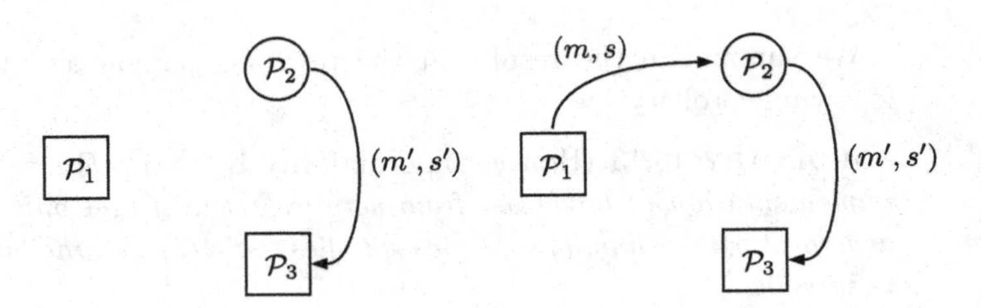

Figure 18.5 A description of forgery situations, where $\mathcal{P}_2$ is the deceiving party.

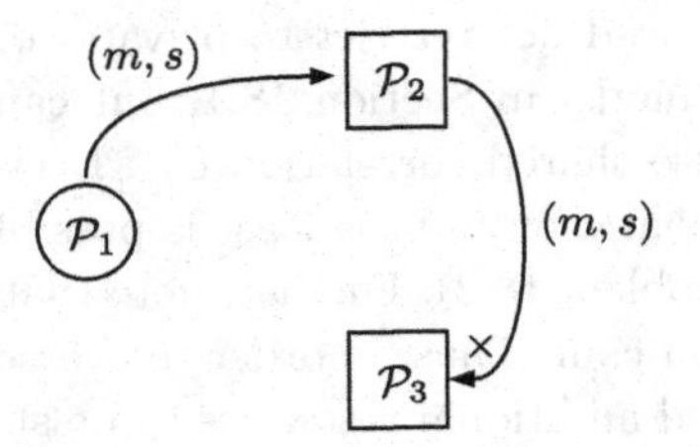

Figure 18.6 A description of transfer situations, where $\mathcal{P}_1$ is the deceiving party.

every month. This check is for the amount of salary m and signature s that guarantees authenticity of the check. Then, $\mathcal{P}_2$ goes to $\mathcal{P}_3$ to cash the check. There are two requirement for this signature scheme to work securely.

The first requirement is *unforgeability*, which is needed to prevent $\mathcal{P}_2$'s attack. For instance, without receiving the check, $\mathcal{P}_2$ might go to $\mathcal{P}_3$ to cash a forged check (m', s') created by $\mathcal{P}_2$ on its own (see Figure 18.5 left). Alternatively, after receiving the authentic check (m, s) from $\mathcal{P}_1$, $\mathcal{P}_2$ might go to $\mathcal{P}_3$ with a falsified check (m', s') so that $\mathcal{P}_2$ can receive a higher salary (see Figure 18.5 right).

The second requirement is *transferability*, which is needed to prevent $\mathcal{P}_1$'s attack. For instance, if $\mathcal{P}_1$ does not want to pay the salary, $\mathcal{P}_1$ might send a check to $\mathcal{P}_2$ so that it will be accepted by $\mathcal{P}_2$ but will be rejected by $\mathcal{P}_3$ (see Figure 18.6).

Unlike other cryptographic primitives, the digital signature was first introduced in the context of public key cryptography; see Sections 18.6 and 18.7. For this reason, commonly used definitions of digital signature in the literature assume that it will be implemented by public key cryptography. In order to discuss an information-theoretically secure digital signature, we introduce the digital signature as a special case of secure function computation. Like bit commitment (cf. Chapter 14), digital signature is a multiphase function with one specific party termed the *signer*. For simplicity of exposition, we restrict our attention to the three-party digital signature. For the three-party case, the ideal digital signature $f_{\text{DS}} = (f_{\text{DS1}}, f_{\text{DS2}})$ consists of two phases (see Figure 18.7). In the first phase of the digital signature f_{DS1}, termed the signing phase, the signer $\mathcal{P}_1$ has input $x_1 \in \mathcal{X} \cup \{\perp\}$ and the intermediate party $\mathcal{P}_2$ receives x_1; here

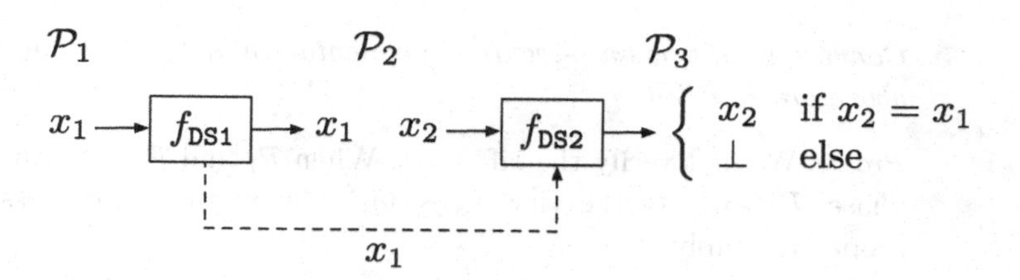

Figure 18.7 A description of the ideal digital signature $f_{\text{DS}} = (f_{\text{DS1}}, f_{\text{DS2}})$.

$x_1 \in \mathcal{X}$ describes the situation that the message is sent with a valid signature while $x_1 = \perp$ describes the situation that the message is sent with an invalid signature. The second phase is only invoked when $x_1 \neq \perp$. In the second phase of the digital signature f_{DS2}, termed the transfer phase, $\mathcal{P}_2$ has input $x_2 \in \mathcal{X}$, and the destination party $\mathcal{P}_3$ receives x_2 if $x_2 = x_1$ and $\perp$ otherwise.

As we can see, $\mathcal{P}_2$'s substitution is detected by $\mathcal{P}_3$ (unforgeability); if $\mathcal{P}_1$ sends a message with a valid signature, i.e., $x_1 \in \mathcal{X}$, then the same message forwarded by honest $\mathcal{P}_2$ will be accepted by $\mathcal{P}_3$.

A protocol π implementing the digital signature is described as follows. To implement the signing phase with $\mathcal{P}_1$'s input $x_1 \in \mathcal{X}$, the parties communicate with each other, and $\mathcal{P}_2$ outputs $z_2 \in \mathcal{X} \cup \{\perp\}$, where $z_2 = \perp$ indicates that $\mathcal{P}_2$ does not accept $\mathcal{P}_1$'s input. To implement the transfer phase, the parties communicate again, and $\mathcal{P}_3$ outputs $z_3 \in \mathcal{X} \cup \{\perp\}$. The connection between simulation based security and an alternative definition, which can be expressed using a few readily verifiable properties of the protocol, is described in the following lemma.

LEMMA 18.15 (Security of digital signature – alternative form) *A communication protocol π is a 2ε-secure implementation of the digital signature f_{DS} against one active adversary (in the sense of Definition 18.1) if the following conditions are satisfied for any attack by the dishonest party.*

- *Correctness. If $\mathcal{P}_1$ and $\mathcal{P}_2$ are honest, then*

$$\Pr\left(z_2 = x_1\right) \geq 1 - \varepsilon.$$

- *Unforgeability. If $\mathcal{P}_1$ and $\mathcal{P}_3$ are honest, then*

$$\Pr\left(z_3 \neq x_1, z_3 \neq \perp\right) \leq \varepsilon.$$

- *Transferability. If $\mathcal{P}_2$ and $\mathcal{P}_3$ are honest, then*

$$\Pr\left(z_2 \neq z_3, z_2 \neq \perp\right) \leq \varepsilon.$$

- *Secrecy. If $\mathcal{P}_1$ and $\mathcal{P}_2$ are honest, then the view $V_3(x_1)$ of $\mathcal{P}_3$ in the signing phase is almost independent of $\mathcal{P}_1$'s input x_1, i.e., there exists Q_{V_3} such that*

$$d_{\mathtt{var}}\left(\mathrm{P}_{V_3(x_1)}, \mathrm{Q}_{V_3}\right) \leq \varepsilon$$

for every x_1.

Conversely, if π is an ε-secure implementation of $f_{\mathtt{DS}}$, then the four requirements above are satisfied.

Proof We first verify the "if" part. When $\mathcal{P}_1$ and $\mathcal{P}_2$ are honest, in the signing phase, $\mathcal{P}_3$ samples the view $V_3 \sim \mathrm{Q}_{V_3}$. Then, the correctness and the secrecy properties imply

$$\begin{aligned}
&d_{\mathtt{var}}\big(\mathrm{P}_{X_1Z_2V_3}, \mathrm{Q}_{X_1X_1} \times \mathrm{Q}_{V_3}\big) \\
&\quad \le d_{\mathtt{var}}\big(\mathrm{P}_{X_1Z_2V_3}, \mathrm{P}_{X_1X_1V_3}\big) + d_{\mathtt{var}}\big(\mathrm{P}_{X_1X_1V_3}, \mathrm{Q}_{X_1X_1} \times \mathrm{Q}_{V_3}\big) \\
&\quad = \mathbb{E}\Big[d_{\mathtt{var}}\big(\mathrm{P}_{Z_2|X_1V_3}(\cdot|X_1,V_3), \mathrm{P}_{X_1|X_1V_3}(\cdot|X_1,V_3)\big)\Big] + d_{\mathtt{var}}\big(\mathrm{P}_{X_1V_3}, \mathrm{Q}_{X_1} \times \mathrm{Q}_{V_3}\big) \\
&\quad \le 2\varepsilon,
\end{aligned}$$

where $\mathrm{Q}_{X_1X_1}$ is the distribution induced by $f_{\mathtt{DS1}}$ with $\mathcal{P}_1$'s input $X_1 \sim \mathrm{P}_{X_1}$. In the transfer phase, since $\mathcal{P}_1$ and $\mathcal{P}_2$ are honest, $\mathcal{P}_3$ learns X_1 in the ideal model. Thus, $\mathcal{P}_3$ can simulate the view V_3' in the transfer phase, and so

$$d_{\mathtt{var}}\big(\mathrm{P}_{X_1Z_2V_3}\mathrm{P}_{V_3'|X_1Z_2V_3}, \mathrm{Q}_{X_1X_1} \times \mathrm{Q}_{V_3}\mathrm{P}_{V_3'|X_1Z_2V_3}\big) \le 2\varepsilon;$$

i.e., there exists a simulator for $\mathcal{P}_3$ with error 2ε.

When $\mathcal{P}_1$ and $\mathcal{P}_3$ are honest, since $\mathcal{P}_2$ receives X_1 in the signing phase $f_{\mathtt{DS1}}$ of the ideal model, $\mathcal{P}_2$ can simulate the view V_2 and $\mathcal{P}_3$'s output Z_3 in the actual protocol π.[12] If $Z_3 = \perp$, then $\mathcal{P}_2$ inputs $X_2' \ne X_1$ to $f_{\mathtt{DS2}}$; otherwise, $\mathcal{P}_2$ inputs $X_2' = X_1$ to $f_{\mathtt{DS2}}$. Then, by unforgeability, the simulated distribution $\mathrm{Q}_{X_1V_2Z_3}$ satisfies[13]

$$\begin{aligned}
&d_{\mathtt{var}}\big(\mathrm{P}_{X_1V_2Z_3}, \mathrm{Q}_{X_1V_2Z_3}\big) \\
&\quad = \mathbb{E}\Big[d_{\mathtt{var}}\big(\mathrm{P}_{Z_3|X_1V_2}(\cdot|X_1,V_2), \mathrm{Q}_{Z_3|X_1V_2}(\cdot|X_1,V_2)\big)\Big] \\
&\quad = \Pr\big(Z_3 \ne X_1, Z_3 \ne \perp\big) \\
&\quad \le \varepsilon.
\end{aligned}$$

Thus, there exists a simulator for $\mathcal{P}_2$ with error ε.

When $\mathcal{P}_2$ and $\mathcal{P}_3$ are honest, $\mathcal{P}_1$ first run π to simulate $\mathcal{P}_2$'s output Z_2. If $Z_2 = \perp$, then $\mathcal{P}_1$ inputs $X_1' = \perp$ to $f_{\mathtt{DS1}}$; otherwise, $\mathcal{P}_1$ inputs $X_1' = Z_2$ to $f_{\mathtt{DS1}}$. Then, by transferability, the simulated distribution $\mathrm{Q}_{X_1V_1Z_2Z_3}$ satisfies[14]

$$\begin{aligned}
&d_{\mathtt{var}}\big(\mathrm{P}_{X_1V_1Z_2Z_3}, \mathrm{Q}_{X_1V_1Z_2Z_3}\big) \\
&\quad = \mathbb{E}\Big[d_{\mathtt{var}}\big(\mathrm{P}_{Z_3|X_1V_1Z_2}(\cdot|X_1,V_1,Z_2), \mathrm{Q}_{Z_3|X_1V_1Z_2}(\cdot|X_1,V_1,Z_2)\big)\Big] \\
&\quad = \Pr\big(Z_2 \ne Z_3, Z_2 \ne \perp\big) \\
&\quad \le \varepsilon.
\end{aligned}$$

[12] Note that $\mathcal{P}_3$ has no input in protocol π, which enables $\mathcal{P}_2$ to play the role of $\mathcal{P}_3$ as well.
[13] Note that $X_2' \ne X_1$ implies $Z_3 = \perp$ in the ideal model.
[14] Note that the transfer phase is not invoked when $Z_2 = \perp$.

Next, we verify the "only if" part. When $\mathcal{P}_1$ and $\mathcal{P}_2$ are honest, the correctness follows from

$$\Pr\left(Z_2 \neq X_1\right) = d_{\mathtt{var}}(\mathrm{P}_{X_1 Z_1}, \mathrm{Q}_{X_1 X_1}).$$

When $\mathcal{P}_1$ and $\mathcal{P}_3$ are honest, consider the maximal coupling of $\mathrm{P}_{X_1 Z_3}$ and $\mathrm{Q}_{X_1 Z_3}$, denoted $\mathrm{R}_{X_1 Z_3 \tilde{X}_1 \tilde{Z}_3}$, where $\mathrm{Q}_{X_1 Z_3}$ is the (X_1, Z_3) marginal of a simulated distribution in the ideal model. Then, since the event $\{\tilde{Z}_3 \neq \tilde{X}_1, \tilde{Z}_3 \neq \perp\}$ never occurs in the ideal model, we have

$$\begin{aligned}\Pr\left(Z_3 \neq X_1, Z_3 \neq \perp\right) &\leq \Pr\left((X_1, Z_3) \neq (\tilde{X}_1, \tilde{Z}_3)\right) \\ &= d_{\mathtt{var}}(\mathrm{P}_{X_1 Z_3}, \mathrm{Q}_{X_1 Z_3}),\end{aligned}$$

from which unforgeability follows. When $\mathcal{P}_2$ and $\mathcal{P}_3$ are honest, by a similar coupling argument as the previous case, we have

$$\Pr\left(Z_2 \neq Z_3, Z_2 \neq \perp\right) \leq d_{\mathtt{var}}(\mathrm{P}_{Z_2 Z_3}, \mathrm{Q}_{Z_2 Z_3}),$$

from which transferability follows. Finally, when $\mathcal{P}_1$ and $\mathcal{P}_2$ are honest, since there exists a simulator generating V_3 without observing X_1, the secrecy is satisfied. □

For digital signatures with more than three parties, it is desired that the message can be transferred multiple times to different parties. For an ordered set $\mathsf{p} = (\mathcal{P}_{i_0}, \ldots, \mathcal{P}_{i_{n-1}})$, a digital signature with transfer path p can be described as follows. The sender $\mathcal{P}_{i_0}$ has input $x_{i_0} \in \mathcal{X} \cup \{\perp\}$, and the first receiver $\mathcal{P}_{i_1}$ receives x_{i_0}; as before, $x_{i_0} \in \mathcal{X}$ describes the situation that the message is sent with a valid signature while $x_{i_0} = \perp$ describes the situation that the message is sent with an invalid signature. The transfer phase is recursively described as follows. For $1 \leq j \leq n-2$, the jth transfer phase is invoked only when $\mathcal{P}_{i_j}$ receives $x_{i_{j-1}} \in \mathcal{X}$; in that case, $\mathcal{P}_{i_j}$ has input $x_{i_j} \in \mathcal{X}$, and the $(j+1)$th receiver $\mathcal{P}_{i_{j+1}}$ receives x_{i_j} if $x_{i_j} = x_{i_{j-1}}$ and $\perp$ otherwise. Note that if the jth transfer phase is invoked, then $\mathcal{P}_{i_{j+1}}$ receives x_{i_0} or $\perp$. Once the structure of the three-party digital signature is well understood, it is relatively easier to extend three-party results to the general case. Interested readers can see the references provided in Section 18.7.

Remark 18.3 The digital signature f_{DS} considered here only provides a part of the properties possessed by the standard digital signature. In the literature, f_{DS} is sometimes referred to as an *information checking signature.*

Remark 18.4 One might wonder why the digital signature should have the secrecy property. In fact, most implementations of the digital signature automatically have the secrecy property. Furthermore, the secrecy property is explicitly needed in some applications, for instance, when the digital signature is used to distribute the shares of secret sharing so that authentic shares will be submitted in the reconstruction phase.

Protocol 18.6 A construction of the three-party broadcast from digital signature

Input: $\mathcal{P}_1$'s input x_1

Output: $\mathcal{P}_j$'s output z_j for $j = 2, 3$

1. $\mathcal{P}_1$ sends $\Pi_{1\to 2} = x_1$ to $\mathcal{P}_2$ by invoking the signing phase of the digital signature, and $\Pi_{1\to 3} = x_1$ to $\mathcal{P}_3$ without signature via the corresponding private channels.
2. $\mathcal{P}_2$ executes the following steps.
   - If $\Pi_{1\to 2} \neq \bot$, then $\mathcal{P}_2$ sets $z_2 = \Pi_{1\to 2}$ and forwards $\Pi_{1\to 2}$ by invoking the transfer phase of the digital signature; $\mathcal{P}_3$ receives $\Pi_{2\to 3}$.
   - If $\Pi_{1\to 2} = \bot$, then $\mathcal{P}_2$ sends $\Pi_{2\to 3} = \bot$ to $\mathcal{P}_3$ via the private channel.
3. $\mathcal{P}_3$ conducts the following.
   - If $\Pi_{2\to 3} \neq \bot$, then $\mathcal{P}_3$ sets $z_3 = \Pi_{2\to 3}$.
   - If $\Pi_{2\to 3} = \bot$, then $\mathcal{P}_3$ sets $z_3 = \Pi_{1\to 3}$, and sends $\Pi_{3\to 2} = \Pi_{1\to 3}$ to $\mathcal{P}_2$ via the private channel.
4. If $\Pi_{1\to 2} = \bot$, then $\mathcal{P}_2$ sets $z_2 = \Pi_{3\to 2}$.

18.5.2 Broadcast and Digital Signature

Broadcast and digital signature are closely related. In the following, we present a couple of methods to construct broadcast from the digital signature and vice versa.

The first method is a construction of three-party broadcast from the digital signature.

THEOREM 18.16 (Three-party broadcast from digital signature) *Suppose that $\pi_{\mathtt{DS}}$ is an ε-secure implementation of the three-party digital signature from $\mathcal{P}_1$ to $\mathcal{P}_3$ via $\mathcal{P}_2$ against one active adversary.*[15] *Then, by invoking $\pi_{\mathtt{DS}}$, we can construct a protocol $\pi_{\mathtt{BR}}$ that is an ε-secure computation of broadcast.*

Proof The desired protocol is described in Protocol 18.6. When $\mathcal{P}_1$ is dishonest, we need to verify whether the agreement is satisfied, i.e., $\mathcal{P}_2$ and $\mathcal{P}_3$ output the same value. If $\Pi_{1\to 2} = \Pi_{2\to 3} \neq \bot$, then both $\mathcal{P}_2$ and $\mathcal{P}_3$ output $z_2 = z_3 = \Pi_{2\to 3}$. On the other hand, if $\Pi_{1\to 2} = \bot$, both $\mathcal{P}_2$ and $\mathcal{P}_3$ output $z_2 = z_3 = \Pi_{1\to 3}$. Thus, ε-transferability of the digital signature implies that $\mathcal{P}_2$ and $\mathcal{P}_3$ output the same value with probability $1 - \varepsilon$.

When $\mathcal{P}_2$ is dishonest, we need to verify whether $\mathcal{P}_3$ outputs $\mathcal{P}_1$'s input x_1. Since $\mathcal{P}_1$ is honest, note that $\mathcal{P}_1$ sends $\Pi_{1\to 2} = x_1$ to $\mathcal{P}_2$ and $\Pi_{1\to 3} = x_1$ to $\mathcal{P}_3$. If $\Pi_{2\to 3} \neq \bot$, then $\mathcal{P}_3$ outputs $z_3 = \Pi_{2\to 3}$; on the other hand, if $\Pi_{2\to 3} = \bot$, then

[15] By symmetry, the assumption can be replaced to the availability of the three-party digital signature from $\mathcal{P}_1$ to $\mathcal{P}_2$ via $\mathcal{P}_3$ against one active adversary.

$\mathcal{P}_3$ outputs $z_3 = \Pi_{1\to3}$. Thus, ε-unforgeability of the digital signature implies that $\mathcal{P}_3$ outputs $\mathcal{P}_1$'s input with probability $1 - \varepsilon$.

When $\mathcal{P}_3$ is dishonest, we need to verify whether $\mathcal{P}_2$ outputs $\mathcal{P}_1$'s input x_1. Since $\mathcal{P}_1$ and $\mathcal{P}_2$ are honest, $\mathcal{P}_1$ sends x_1 to $\mathcal{P}_2$, and $\mathcal{P}_2$ outputs $z_2 = \Pi_{1\to2}$. Thus, ε-correctness of the digital signature implies that $\mathcal{P}_2$ outputs $z_2 = x_1$ with probability $1 - \varepsilon$. □

Remark 18.5 As we can find from the proof of Theorem 18.16, the secrecy property of the digital signature is not necessary to realize secure broadcast.

Remark 18.6 When there are two dishonest parties among three parties, note that the correctness and the agreement requirements of broadcast become trivial. More generally, the requirements of broadcast only make sense when the number t of dishonest parties satisfies $t < n - 1$.

As we have seen in Theorem 18.11, the three-party broadcast from scratch is not possible. Thus, Theorem 18.16 together with Theorem 18.11 implies the following impossibility result.

COROLLARY 18.17 *For a three-party network with one dishonest party, an ε-secure implementation of digital signature from scratch exists only if $\varepsilon \geq 1/3$.*

The next method is a construction of the digital signature from a secure three-party broadcast. The key idea is to use the authentication code introduced in Chapter 8. Recall that, for a message $x \in \mathbb{F}_q$ and key $(k_1, k_2) \in \mathbb{F}_q^2$, if we compute the tag by $\sigma = k_1 \cdot x + k_2$, then the success probability of substitution is $1/q$ (cf. Construction 8.1). Consider the following naive protocol. In the signing phase, the signer $\mathcal{P}_1$ picks (k_1, k_2), sends the key (k_1, k_2) to $\mathcal{P}_3$ via the private channel between $\mathcal{P}_1$ and $\mathcal{P}_3$, and sends (x, σ) to $\mathcal{P}_2$ via the private channel between $\mathcal{P}_1$ and $\mathcal{P}_2$. In the transfer phase, $\mathcal{P}_2$ sends (x, σ) to $\mathcal{P}_3$, and $\mathcal{P}_3$ accepts the message if $\sigma = k_1 \cdot x + k_2$ holds. We can find that $\mathcal{P}_2$'s forgery is prevented by the security of the authentication code used. Secrecy is also satisfied since x is not revealed to $\mathcal{P}_3$ in the signing phase. However, transferability is not guaranteed since $\mathcal{P}_1$ may send an inconsistent message-tag pair $(\tilde{x}, \tilde{\sigma})$ so that it will be rejected by $\mathcal{P}_3$ in the transfer phase. Thus, we need to introduce an additional procedure so that $\mathcal{P}_2$ can verify if (x, σ) is consistent with the key (k_1, k_2) possessed by $\mathcal{P}_3$. Of course, this must be done without disclosing the values of x to $\mathcal{P}_3$ nor (k_1, k_2) to $\mathcal{P}_2$. A protocol incorporating such a procedure is described in Protocol 18.7.

THEOREM 18.18 *Suppose that the ideal broadcast from every party as a sender is available. Then, Protocol 18.7 is an implementation of digital signature with perfect correctness, $1/q$-unforgeability, $1/q$-transferability, and perfect secrecy.*

Proof *Correctness*: Suppose that $\mathcal{P}_1$ and $\mathcal{P}_2$ are honest. Note that $\mathcal{P}_2$ always outputs $z_2 \in \mathcal{X} = \mathbb{F}_q$ in this protocol. Since $\mathcal{P}_1$ is honest, $\mathcal{P}_1$ sends the true value of x to $\mathcal{P}_2$, and it is never updated. In fact, $\mathcal{P}_1$ never says "$\mathcal{P}_2$ is corrupt" in Step 3, and $\mathcal{P}_1$ does not say "accept" when $\mathcal{P}_3$ says "reject" in Step 4.

Protocol 18.7 A construction of the three-party digital signature from broadcast

Input: $\mathcal{P}_1$'s input $x \in \mathbb{F}_q$

Output: $\mathcal{P}_2$'s output $z_2 \in \mathbb{F}_q$ and $\mathcal{P}_3$'s output $z_3 \in \mathbb{F}_q \cup \{\perp\}$

1. **Signing phase** $\mathcal{P}_1$ randomly picks $k_1, k_2, x', k_2' \in \mathbb{F}_q$, sends x, x', $\sigma := k_1 \cdot x + k_2$, and $\sigma' = k_1 \cdot x' + k_2'$ to $\mathcal{P}_2$ via the private channel, and sends (k_1, k_2, k_2') to $\mathcal{P}_3$ via the private channel.
2. $\mathcal{P}_2$ randomly picks $e \in \mathbb{F}_q$, and broadcasts e, $x_e := x' + e \cdot x$, and $\sigma_e := \sigma' + e \cdot \sigma$.
3. $\mathcal{P}_1$ checks if $x_e = x' + e \cdot x$ and $\sigma_e = \sigma' + e \cdot \sigma$ hold and broadcasts "accept" or "$\mathcal{P}_2$ is corrupt" accordingly. In the former case, the parties proceed to the next step of the signing phase. In the latter case, $\mathcal{P}_1$ also broadcasts (x, σ). $\mathcal{P}_2$ updates the value of (x, σ) to the broadcasted one, and $\mathcal{P}_3$ updates the value of k_2 so that $\sigma = k_1 \cdot x + k_2$ holds. Then, the parties terminate the signing phase, where $\mathcal{P}_2$ outputs $z_2 = x$.
4. $\mathcal{P}_3$ checks if $\sigma_e = k_1 \cdot x_e + k_2' + e \cdot k_2$ holds and broadcasts "accept" or "reject" accordingly. $\mathcal{P}_1$ checks if $\mathcal{P}_3$ acted correctly and broadcasts "accept" or "$\mathcal{P}_3$ is corrupt" accordingly.

   - If both $\mathcal{P}_3$ and $\mathcal{P}_1$ said "accept," then the parties terminate the signing phase, where $\mathcal{P}_2$ outputs $z_2 = x$.
   - If $\mathcal{P}_3$ said "reject" and $\mathcal{P}_1$ said "accept," then $\mathcal{P}_1$ broadcasts (x, σ); $\mathcal{P}_2$ updates the value of (x, σ) to the broadcasted one, and $\mathcal{P}_3$ updates the value of k_2 so that $\sigma = k_1 \cdot x + k_2$ holds. The parties terminate the signing phase, where $\mathcal{P}_2$ outputs $z_2 = x$.
   - If $\mathcal{P}_1$ said "$\mathcal{P}_3$ is corrupt," then $\mathcal{P}_1$ broadcasts (k_1, k_2); $\mathcal{P}_3$ updates the value of (k_1, k_2) to the broadcasted one, and $\mathcal{P}_2$ updates the value of σ so that $\sigma = k_1 \cdot x + k_2$ holds. The parties terminate the signing phase, where $\mathcal{P}_2$ outputs $z_2 = x$.

5. **Transfer phase**
6. $\mathcal{P}_2$ sends (x, σ) to $\mathcal{P}_3$ via the private channel.
7. $\mathcal{P}_3$ checks if $\sigma = k_1 \cdot x + k_2$ holds; if so, $\mathcal{P}_3$ outputs $z_3 = x$; otherwise, $\mathcal{P}_3$ outputs $z_3 = \perp$.

Secrecy: Suppose that $\mathcal{P}_1$ and $\mathcal{P}_2$ are honest. The only information $\mathcal{P}_3$ obtains in the signing phase is $x_e = x' + e \cdot x$. Note that σ_e is a function of (x_e, e, k_1, k_2, k_2'). Since x' is picked independently of x, x_e is independent of x.

Unforgeability: Suppose that $\mathcal{P}_1$ and $\mathcal{P}_3$ are honest. If $\mathcal{P}_2$ acts correctly in Step 2, then $\mathcal{P}_1$ says "accept"; if $\mathcal{P}_2$ acts incorrectly in Step 2, then $\mathcal{P}_1$ says "$\mathcal{P}_2$ is corrupt." In the former case, $\mathcal{P}_2$ knows that both $\mathcal{P}_1$ and $\mathcal{P}_3$ will say "accept" in Step 4. Thus, $\mathcal{P}_2$ gets no information other than (x, σ). If $\mathcal{P}_2$ sends $(\tilde{x}, \tilde{\sigma})$ with $\tilde{x} \neq x$ in Step 6, then it will be accepted by $\mathcal{P}_3$ only if $\tilde{\sigma} = k_1 \cdot \tilde{x} + k_2$ is satisfied.

However, by the security of the authentication code used, the probability of acceptance is $1/q$.

Transferability: Suppose that $\mathcal{P}_2$ and $\mathcal{P}_3$ are honest. When $\mathcal{P}_1$ broadcasts (x, σ) or (k_1, k_2), $\mathcal{P}_2$ and $\mathcal{P}_3$ update their values so that (x, σ) will be accepted by $\mathcal{P}_3$ in the transfer phase. Thus, it suffices to consider the case where $\mathcal{P}_1$ does not broadcast either (x, σ) or (k_1, k_2). Suppose that (x, σ) and (k_1, k_2) sent by $\mathcal{P}_1$ are inconsistent, i.e., $\sigma \neq k_1 \cdot x + k_2$. The check by $\mathcal{P}_3$ in Step 4 is accepted only if

$$\sigma' + e \cdot \sigma = k_1 \cdot (x' + e \cdot x) + k_2' + e \cdot k_2.$$

By arranging the terms, we have

$$e = \frac{k_1 \cdot x' + k_2' - \sigma'}{\sigma - k_1 \cdot x - k_2}.$$

Since $e \in \mathbb{F}_q$ is picked independent of other variables, the probability of acceptance is $1/q$. □

18.6 Computationally Secure Digital Signature

In this section, we briefly mention the difference between an information-theoretically secure digital signature and a computationally secure digital signature. For more formal treatment of the computationally secure digital signature, see the references in Section 18.7.

Typically, a computationally secure digital signature is constructed using public key cryptography. Even though the following argument applies for any computationally secure digital signature, consider the digital signature based on RSA cryptography as a specific example. For two prime numbers p and q, let $K_{\mathrm{p}} = (N, e)$ be the public key and $K_{\mathrm{s}} = d$ be the secret key, where $N = pq$ and e and d satisfy $ed \equiv 1 \pmod{(p-1)(q-1)}$.

Suppose that $(K_{\mathrm{s}}, K_{\mathrm{p}})$ are possessed by $\mathcal{P}_1$ and K_{p} is available to $\mathcal{P}_2$ and $\mathcal{P}_3$.[16] Let $h \colon \mathcal{X} \to \{0, \ldots, N-1\}$ be a collision-resistant hash function, where $\mathcal{X}$ is the (finite) message space. In the RSA digital signature, upon observing a message $x \in \mathcal{X}$, $\mathcal{P}_1$ sends (x, σ), where

$$\sigma = D(K_{\mathrm{s}}, K_{\mathrm{p}}, h(x)) = h(x)^d \pmod{N}.$$

Then, $\mathcal{P}_2$ accepts the message if

$$h(x) = E(K_{\mathrm{p}}, \sigma) = \sigma^e \pmod{N}. \tag{18.2}$$

When $\mathcal{P}_1$ is honest, because of the commutativity of the encryption and the decryption of the RSA cryptography, (18.2) holds. When $\mathcal{P}_2$ transfers the message to $\mathcal{P}_3$, $\mathcal{P}_2$ forwards (x, σ) to $\mathcal{P}_3$, and $\mathcal{P}_3$ accepts the message if (18.2) is satisfied.

[16] The number of parties being three is not essential.

In contrast to the information-theoretically secure digital signature, $\mathcal{P}_2$ and $\mathcal{P}_3$ use the same key and the same verification algorithm. Thus, transferability is automatically guaranteed. Roughly speaking, unforgeability is guaranteed based on the security of the RSA cryptography, i.e., $\mathcal{P}_2$ cannot compute the decryption D without having the secret key K_{s} and the collision resistance property of the hash function used.

It is also known that a digital signature scheme can be constructed from a one-way function, which is known as the Lamport signature scheme. Roughly speaking, a one-way function $f: \mathcal{Y} \to \mathcal{Z}$ is a function such that it is possible to efficiently compute $z = f(y)$ for a given $y \in \mathcal{Y}$ but it is computationally not feasible to find y satisfying $z = f(y)$ for a given $z \in \mathcal{Z}$. For randomly generated $Y[0], Y[1] \in \mathcal{Y}$, let $K_{\mathsf{s}} = (Y[0], Y[1])$ be the secret key and $K_{\mathsf{p}} = (Z[0], Z[1]) = (f(Y[0]), f(Y[1]))$ be the public key.

Suppose that $(K_{\mathsf{s}}, K_{\mathsf{p}})$ are possessed by $\mathcal{P}_1$ and K_{p} is available to $\mathcal{P}_2$ and $\mathcal{P}_3$. In the Lamport signature scheme, upon observing $x \in \{0, 1\}$, $\mathcal{P}_1$ sends (x, σ), where $\sigma = Y[x]$. Then, $\mathcal{P}_2$ accepts the message if $f(\sigma) = Z[x]$. When $\mathcal{P}_2$ transfers the message to $\mathcal{P}_3$, $\mathcal{P}_2$ forwards (x, σ) to $\mathcal{P}_3$ and $\mathcal{P}_3$ accepts the message if $f(\sigma) = Z[x]$.

By the same reasoning as for the RSA signature, i.e., since $\mathcal{P}_2$ and $\mathcal{P}_3$ use the same key and the same verification algorithm, transferability is automatically satisfied. The unforgeability requirement is guaranteed based on the fact that f is a one-way function, i.e., it is not feasible for $\mathcal{P}_2$ to find Y satisfying $f(Y) = Z[x \oplus 1]$.

The idea of embedding the message x into the coordinate $Y[x]$ is quite versatile. In fact, it is interesting to observe the resemblance between the Lamport signature scheme and the construction of the digital signature from correlation; see Problem 18.8.

As a final remark of this section, it should be noted that the same public key must be distributed to $\mathcal{P}_2$ and $\mathcal{P}_3$ in advance by a secure protocol; otherwise, transferability is not guaranteed. Thus, even the computationally secure digital signature cannot be realized from scratch. This is in contrast with secure communication or the two-party secure computation.

18.7 References and Additional Reading

The broadcast and the Byzantine agreement problems over standard networks were introduced by Pease, Shostak, and Lamport in [216, 269], where both the scheme and the impossibility result for honest supermajority were proved. The broadcast protocol from scratch in Section 18.3 is from [216, 269]. The attack-coupling argument in the proof of Theorem 18.11 was introduced in [121]. This argument has been widely used for proving various impossibility results in distributed computing; for instance, see [73, 124] and Problems 18.1 and 18.17. For a thorough exposition of the distributed computation over networks problem, including the broadcast and the Byzantine agreement problems, see [227].

The concept of digital signatures was first proposed by Diffie and Hellman in their landmark paper [99]. Along with the proposal of public key cryptography based on integer factorization, Rivest, Shamir, and Adleman proposed to use their public key cryptography for digital signatures as well in [293]. Since their pioneering work, the digital signature under computational assumption has been actively studied. A general construction of digital signature from the one-way function was proposed by Lamport in [215]. For a more detailed exposition on the computationally secure digital signature, see [185].

On the other hand, the study of information-theoretically secure digital signature was initiated by Chaum and Roijakkers [64] (see also [272]); they proposed a digital signature scheme for a one-bit message. More efficient construction based on bivariate polynomials was introduced in [156] (see also [319]). See Protocol 18.8 in Problem 18.3 for a three-party instance of the general construction introduced in [156]. The information-theoretic digital signature was introduced after its computational counterpart. In a similar spirit to Chapter 13 and Chapter 14, it is of interest to develop a theory for constructing broadcast and digital signatures from correlation. However, such a theory is rather immature; only feasibility results have been presented in the literature and the question of efficiency has been barely studied. The necessary condition for the existence of noninteractive digital signature protocols and its sufficiency has been studied in [127]; see Problems 18.4–18.9. In fact, it was claimed in [127] that the same condition is also the necessary condition for the existence of any digital signature protocol.[17] However, as is demonstrated in Problems 18.10 and 18.11, this claim is not true. It is an open problem to derive the necessary and sufficient conditions for the existence of digital signature protocols other than the noninteractive class. The construction of broadcast from digital signature in Protocol 18.6 is from [126]; the construction of digital signature from broadcast in Protocol 18.7, also known as the information checking protocol, is from [76] (see also [77]).

The impossibility results of broadcast, Byzantine agreement, and digital signature in the standard network are based on the various security requirements of each task in a delicate manner and do not apply if one of the requirements is dropped or relaxed. For instance, if the correctness requirement is dropped for the digital signature, then it is possible to construct a digital signature protocol over a standard network [3, 343]; if the agreement requirement is relaxed in broadcast, then it is possible to construct broadcast over a standard network without honest supermajority [126].

Beyond the honest supermajority, it was shown in [216, 269] that broadcast is possible for any number of dishonest parties if a digital signature for arbitrary transfer path is available (see also [109]). The minimum requirement for realizing

[17] More precisely, it was claimed that the violation of either (18.5) or (18.6) is the necessary condition for the existence of the digital signature for either transfer path $\mathcal{P}_1 \to \mathcal{P}_2 \to \mathcal{P}_3$ or $\mathcal{P}_1 \to \mathcal{P}_3 \to \mathcal{P}_2$.

broadcast was studied in [124]. In fact, the n-party global broadcast can be constructed from the three-party broadcast via the graded broadcast as long as the honest majority is satisfied; see Problems 18.12–18.17.

Although the definition of ideal broadcast is rather straightforward, it is subtle how to define ideal functions for Byzantine agreement and digital signature, and there seems to be no consensus. For instance, an ideal function for Byzantine agreement can be found in [72] and one for digital signature can be found in [12]. Our definitions are similar in spirit, though the details differ.

The broadcast protocol in Section 18.3 is a recursive protocol, and it requires $t+1$ rounds of communications. In fact, for deterministic protocols, it has been shown that $t+1$ rounds are necessary [109, 120]. To overcome this lower bound, the idea of using randomized and variable round protocols was introduced in [23, 279]. For honest supermajority, a constant expected round protocol for broadcast and Byzantine agreement was proposed in [118]. For honest majority with digital signature, a constant expected round protocol was proposed in [186].

Another important topic that is not covered in this chapter is broadcast and Byzantine agreement over an asynchronous network. In fact, it was shown in [122] that any deterministic protocol cannot enable Byzantine agreement even when there is one dishonest party. To overcome this impossibility result, the idea of using randomized and variable round protocols was introduced in [23, 279]. For instance, see [8] and the references therein.

Problems

18.1 *Broadcast over incomplete graph:* Consider a communication network $\mathcal{G} = (\mathcal{V}, \mathcal{E})$ (undirected graph) such that $\mathcal{P}_i$ and $\mathcal{P}_j$ are connected by the private authenticated channel if and only if $(i,j) \in \mathcal{E}$. The connectivity $c(\mathcal{G})$ of the graph $\mathcal{G}$ is the minimum number k such that, if k nodes are removed from $\mathcal{G}$, then the resulting graph becomes disconnected. When there are at most t-active adversaries, prove that broadcast over the network $\mathcal{G}$ from scratch is possible if and only if $t < c(\mathcal{G})/2$.

HINT The impossibility can be proved by a similar argument as in the proofs of Theorem 18.11 and Theorem 18.12 [121]: first, prove the impossibility for the four-party network with $\mathcal{E} = \{(1,2),(2,3),(3,4),(4,1),(2,4)\}$, which has connectivity 2; then the general case can be proved by a simulation argument. For the construction, see [106].

18.2 By modifying the proof of Theorem 18.11, prove that an ε-secure implementation of broadcast using pairwise correlation exists only if $\varepsilon \geq 1/3$.

18.3 Consider the following digital signature protocol from a correlation [156]. Let

$$\Phi(\eta,\xi) = K_{00} + K_{10}\eta + K_{01}\xi + K_{11}\eta\xi$$

be a bivariate polynomial on $\mathbb{F}_q$ for randomly chosen coefficients

$$\vec{K} = (K_{00}, K_{01}, K_{10}, K_{11}) \in \mathbb{F}_q^4.$$

Protocol 18.8 Signature scheme with bivariate polynomial

Signing by $\mathcal{P}_1$

When $\mathcal{P}_1$ sends message x, $\mathcal{P}_1$ creates the signature $\sigma(\eta) = \Phi(\eta, x)$ and sends $(x, \sigma(\eta))$ to $\mathcal{P}_2$.

Verification by $\mathcal{P}_2$

When $(x, \sigma(\eta))$ is received, $\mathcal{P}_2$ accepts it if $\sigma(V_2) = g_2(x)$.

Verification by $\mathcal{P}_3$

When $(x, \sigma(\eta))$ is transferred, $\mathcal{P}_3$ accepts it if $\sigma(V_3) = g_3(x)$.

Pick $(V_2, V_3) \in \{(v_2, v_3) \in \mathbb{F}_q^2 : v_2 \neq v_3\}$ at random independently of $\vec{K}$. Then, let $g_2(\xi) = \Phi(V_2, \xi)$ and $g_3(\xi) = \Phi(V_3, \xi)$. We set $Y_1 = \Phi(\eta, \xi)$, $Y_2 = (V_2, g_2(\xi))$, and $Y_3 = (V_3, g_3(\xi))$. Suppose that $\mathcal{P}_i$ observes Y_i for $i = 1, 2, 3$. Prove that Protocol 18.8 is 0-correct, $\frac{1}{q}$-unforgeable, $\frac{1}{q}$-transferable, and 0-secret digital signature.

HINT Note that

$$g_i(x) = (K_{01} + K_{11}V_i)x + (K_{00} + K_{10}V_i)$$

can be regarded as an authentication tag for a message x. Also note that the value of the key $(K_{01}+K_{11}V_i, K_{00}+K_{10}V_i)$ is only known to $\mathcal{P}_i$. By using the bivariate polynomial $\Phi(\eta, \xi)$, $\mathcal{P}_1$ can create the signature polynomial $\sigma(\eta) = \Phi(\eta, x)$. Then, we can consider the decision rule such that the message x will be accepted by $\mathcal{P}_i$ if and only if $\sigma(V_i) = g_i(x)$. Heuristically, unforgeability is guaranteed because $\mathcal{P}_2$ does not know $\mathcal{P}_3$'s key. Furthermore, using the fact that the two authentication codes $g_2(\xi)$ and $g_3(\xi)$ are coupled together via the bivariate polynomial, we can show transferability.

18.4 *Simulatability condition*: Prove that, if $\mathtt{mss}(Y_2|Y_1) \multimap Y_2 \multimap Y_3$, then there exist channels $\mathrm{P}_{\bar{Y}_1|Y_1}$ and $\mathrm{P}_{\hat{Y}_1|Y_2}$ such that $\mathrm{P}_{\bar{Y}_1 Y_2} = \mathrm{P}_{Y_1 Y_2}$ and $\mathrm{P}_{\bar{Y}_1 Y_2 Y_3} = \mathrm{P}_{\hat{Y}_1 Y_2 Y_3}$ [127].

18.5 *Necessary condition of noninteractive digital signature*: Suppose that the condition $\mathtt{mss}(Y_2|Y_1) \multimap Y_2 \multimap Y_3$ holds. Prove that an ε_{c}-correct, ε_{u}-unforgeable, and ε_{t}-transferable noninteractive[18] digital signature protocol exists only if $\varepsilon_{\mathrm{c}} + \varepsilon_{\mathrm{u}} + \varepsilon_{\mathrm{t}} \geq 1$ [127].

HINT As we have seen in Section 14.4.2, $\mathtt{mss}(Y_2|Y_1)$ is the part of Y_1 such that $\mathcal{P}_2$ can detect the substitution by $\mathcal{P}_1$. Thus, $\mathcal{P}_1$'s transcript in the signing phase must be essentially based on $\mathtt{mss}(Y_2|Y_1)$. When $\mathtt{mss}(Y_2|Y_1) \multimap Y_2 \multimap Y_3$ holds, $\mathcal{P}_2$ can locally create a copy of $\mathtt{mss}(Y_2|Y_1)$ such that $\mathcal{P}_3$ cannot distinguish the copy from the authentic one. The claim of the problem follows by formalizing these observations via Problem 18.4.

[18] In the class of noninteractive protocols, only $\mathcal{P}_1$ communicates to $\mathcal{P}_2$ in the signing phase and only $\mathcal{P}_2$ communicates to $\mathcal{P}_3$ in the transfer phase. Note that Protocol 18.8 is categorized into this class. Note also that the secrecy requirement is automatically satisfied for this class of protocols since $\mathcal{P}_3$ does not participate in the signing phase.

18.6 For the correlation (Y_1, Y_2, Y_3) induced by the bivariate polynomial in Problem 18.3, verify that the condition $\mathtt{mss}(Y_2|Y_1) -\!\!\circ\!\!- Y_2 -\!\!\circ\!\!- Y_3$ in Problem 18.5 does not hold.

18.7 *Q-Flip model*: Let $P_{Y_1Y_2Y_3}$ be the uniform distribution on

$$\{(y_1, y_2, y_3) \in \{0, 1, 2\}^3 : y_1 \neq y_2 \neq y_3\}.$$

Verify that the condition $\mathtt{mss}(Y_2|Y_1) -\!\!\circ\!\!- Y_2 -\!\!\circ\!\!- Y_3$ in Problem 18.5 does not hold for this correlation.

COMMENT The correlation in this problem is from [126], which may occur as the measurement outcome of a tripartite quantum entanglement [125].

18.8 For a sequence of i.i.d. random variables $(Y_1^{2n}, Y_2^{2n}, Y_3^{2n})$ distributed according to $P_{Y_1Y_2Y_3}$ in Problem 18.7, consider the following digital signature protocol. Denote $Y_i^{2n}[x] = (Y_{i,x\cdot n+1}, \ldots, Y_{i,x\cdot n+n})$ for $x \in \{0, 1\}$. When $\mathcal{P}_1$ sends $x \in \{0, 1\}$ in the signing phase, $\mathcal{P}_1$ sends (x, Π_1) to $\mathcal{P}_2$, where $\Pi_1 = Y_1^{2n}[x]$. Then, $\mathcal{P}_2$ accepts and outputs $Z_2 = x$ if

$$Y_{2,x\cdot n+j} \neq \Pi_{1,j}, \ \forall 1 \leq j \leq n; \tag{18.3}$$

otherwise, $\mathcal{P}_2$ outputs $Z_2 = \perp$. When $\mathcal{P}_2$ sends x in the transfer phase, $\mathcal{P}_2$ sends (x, Π_1, Π_2), where $\Pi_2 = Y_2^{2n}[x]$. Then, $\mathcal{P}_3$ accepts and outputs $Z_3 = x$ if

$$\big|\{j \in \{1, \ldots, n\} : Y_{3,x\cdot n+j} \notin \{\Pi_{1,j}, \Pi_{2,j}\}\}\big| \geq n(1 - \delta) \tag{18.4}$$

for a prescribed $0 < \delta < 1/2$. Prove that this protocol is 0-correct, 0-secret, ε_{u}-unforgeable, and ε_{t}-transferable, where $\varepsilon_{\mathrm{u}}, \varepsilon_{\mathrm{t}} \to 0$ as $n \to \infty$.

18.9 Suppose that $\mathcal{P}_1$, $\mathcal{P}_2$, and $\mathcal{P}_3$ observe i.i.d. random variables $(Y_1^{2n}, Y_2^{2n}, Y_3^{2n})$ such that $\mathtt{mss}(Y_2|Y_1) -\!\!\circ\!\!- Y_2 -\!\!\circ\!\!- Y_3$ does not hold. Then, prove that we can construct a ε_{c}-correct, ε_{u}-unforgeable, and ε_{t}-transferable noninteractive digital signature protocol such that $\varepsilon_{\mathrm{c}}, \varepsilon_{\mathrm{u}}, \varepsilon_{\mathrm{t}} \to 0$ as $n \to \infty$ [127].

HINT Use a similar idea as the protocol in Problem 18.8; but $\mathcal{P}_1$ uses $K_1 = \mathtt{mss}(Y_2|Y_1)$ instead of Y_1, and the tests in (18.3) and (18.4) are replaced by tests based on joint typicality.

18.10 Let (U, V) be the random variables uniformly distributed on

$$\{(u, v) \in \{0, 1, 2\}^2 : u \neq v\},$$

and let $Y_1 = U$ and $Y_2 = Y_3 = V$. Verify that $\mathtt{mss}(Y_2|Y_1) = \mathtt{mss}(Y_3|Y_1) = Y_1$ and both the simulatability conditions

$$\mathtt{mss}(Y_2|Y_1) -\!\!\circ\!\!- Y_2 -\!\!\circ\!\!- Y_3, \tag{18.5}$$

$$\mathtt{mss}(Y_3|Y_1) -\!\!\circ\!\!- Y_3 -\!\!\circ\!\!- Y_2 \tag{18.6}$$

hold.

COMMENT This means that a noninteractive digital signature protocol cannot be implemented from this correlation.

Protocol 18.9 n-party graded broadcast (n-GBR)

Input: $\mathcal{P}_i$'s input $x_i \in \{0,1\}$

Output: $\mathcal{P}_j$'s output z_j for $j \in \mathcal{N}\backslash\{i\}$

1. For every subset $\mathcal{S} \subset \mathcal{N}\backslash\{i\}$ with $|\mathcal{S}| = 2$, by invoking three-party broadcast, $\mathcal{P}_i$ sends $\Pi_{i\to\mathcal{S}} = x_i$ to the parties in $\mathcal{S}$.
2. For $j \in \mathcal{N}\backslash\{i\}$, $\mathcal{P}_j$ outputs

$$z_j = \begin{cases} 0, & \text{if } \Pi_{i\to\mathcal{S}} = 0 \ \forall \mathcal{S} \text{ such that } j \in \mathcal{S}, \\ 1, & \text{if } \Pi_{i\to\mathcal{S}} = 1 \ \forall \mathcal{S} \text{ such that } j \in \mathcal{S}, \\ *, & \text{else.} \end{cases}$$

18.11 For a sequence of i.i.d. random variables $(Y_1^{2n}, Y_2^{2n}, Y_3^{2n})$ distributed according to $P_{Y_1Y_2Y_3}$ in Problem 18.10, consider the following digital signature protocol, where we use the same notation as Problem 18.8. When $\mathcal{P}_1$ sends $x \in \{0,1\}$ in the signing phase, $\mathcal{P}_1$ sends (x, Π_1) to $\mathcal{P}_2$, where $\Pi_1 = Y_1^{2n}[x]$, and $(\Pi_{3,0}, \Pi_{3,1}) = (Y_1^{2n}[0], Y_1^{2n}[1])$ to $\mathcal{P}_3$. Then, $\mathcal{P}_2$ accepts and outputs $Z_2 = x$ if (18.3) holds; otherwise, $\mathcal{P}_2$ outputs $Z_2 = \perp$. When $\mathcal{P}_2$ sends x in the transfer phase, $\mathcal{P}_2$ sends (x, Π_1) to $\mathcal{P}_3$. Then, $\mathcal{P}_3$ accepts and outputs $Z_3 = x$ if one of the following conditions holds:

1. $Y_{3,j} = \Pi_{3,0,j}$ for some $1 \le j \le n$;
2. $Y_{3,n+j} = \Pi_{3,1,j}$ for some $1 \le j \le n$;
3. $|\{j \in \{1,\ldots,n\} : \Pi_{3,x,j} = \Pi_{1,j}\}| \ge (1-\delta)n$ for a prescribed $0 < \delta < 1/2$.

Otherwise, $\mathcal{P}_3$ outputs $Z_3 = \perp$. Prove that this protocol is 0-correct, 0-secret, ε_u-unforgeable, and ε_t-transferable, where $\varepsilon_u, \varepsilon_t \to 0$ as $n \to \infty$.

18.12 *Graded broadcast*: A communication protocol realizes graded broadcast (GBR) against a t-active adversary if the following hold for any $\mathcal{A} \subset \mathcal{N}$ with $|\mathcal{A}| \le t$ and any attack by the parties in $\mathcal{A}$.

- *Correctness*: When the sender $\mathcal{P}_i$ is honest with input $x_i \in \{0,1\}$, then all honest parties output x_i.
- *Agreement*: Any honest parties $\mathcal{P}_j$ and $\mathcal{P}_\ell$ output $z_j, z_\ell \in \{0,*,1\}$ such that $z_j \ne z_\ell \oplus 1$.

Prove that Protocol 18.9 realizes n-GBR against any number of dishonest parties.

COMMENT The graded broadcast (GBR) is a relaxed version of broadcast [118]: when the sender is honest, the GBR satisfies exactly the same correctness requirement as broadcast; when the sender is dishonest, any pair of honest parties output values that are not opposite, i.e., 0 and 1. Even though the agreement requirement of the GBR is weaker than that of the broadcast, an honest party receiving 0 (or 1) is confident that other honest parties did not receive 1 (or 0). Protocol 18.9 realizes n-party GBR (n-GBR) by 3-BR [73].

18.13 *Two-threshold broadcast (TTBR)*: For $t_c \ge t_a$, a communication protocol realizes (t_c, t_a)-TTBR if the following hold.

Protocol 18.10 (2, 1)-TTBR with four parties

Input: $\mathcal{P}_1$'s input $x_1 \in \{0, 1\}$
Output: $\mathcal{P}_i$'s output z_i for $i \in [2:4]$

1. For $i \in [2:4]$, by invoking four-party graded broadcast, $\mathcal{P}_1$ sends $\Pi_{1\to i} = x_1$ to $\mathcal{P}_i$.
2. For $i \in [2:4]$ and $j \in [2:4]\backslash\{i\}$, by invoking three-party broadcast, $\mathcal{P}_i$ sends $\Pi_{i\to j} = \Pi_{1\to i}$ to $\mathcal{P}_j$.
3. For $j \in [2:4]$, let $L_j[a] = |\{i \in [1:4]\backslash\{j\} : \Pi_{i\to j} = a\}|$, where $a \in \{0, *, 1\}$; then, $\mathcal{P}_j$ decides the following.
   - When $\Pi_{1\to j} = 0$, $\mathcal{P}_j$ outputs $z_j = 0$.
   - When $\Pi_{1\to j} = *$, if $L_j[0] \geq 1$ and $L_j[0] + L_j[*] = 3$, then $\mathcal{P}_j$ outputs $z_j = 0$; otherwise, $\mathcal{P}_j$ outputs $z_j = 1$.
   - When $\Pi_{1\to j} = 1$, $\mathcal{P}_j$ outputs $z_j = 1$.

- *Correctness*: When the sender $\mathcal{P}_i$ is honest with input x_i and there are at most t_c dishonest parties, then all honest parties output x_i.
- *Agreement*: When there are at most t_a dishonest parties, all honest parties output the same value z.[19]

For $t_c \geq t_a$, prove that (t_c, t_a)-TTBR from scratch is possible if and only if $n > 2t_c + t_a$.

COMMENT In Protocol 18.5, we invoke the same protocol for smaller parameter inductively. If the sender of the outer protocol is dishonest, then the number of dishonest parties in the inner protocol decreases from that in the outer protocol since the sender of the outer protocol does not participate in the inner protocol; on the other hand, if the sender of the outer protocol is honest, the number of dishonest parties in the inner protocol remains the same. For this reason, we can expect more flexible construction of protocols if we introduce a variant of the broadcast in which the correctness and the agreement are satisfied for different numbers t_c and t_a of dishonest parties, where $t_c \geq t_a$. For $t_c = t_a = t$, (t_c, t_a)-TTBR is just the standard broadcast that tolerates t dishonest parties.

18.14 Prove that Protocol 18.10 realizes (2, 1)-TTBR for four parties when graded broadcast is available.

18.15 Prove that Protocol 18.11 realizes (t_c, t_a)-TTBR if $|\mathcal{S}| > t_c + t_a$ and $t_c \geq t_a$.

COMMENT In this protocol, we invoke (t_c, t_a)-TTBR inductively to send $\{0, *, 1\}$. However, the input of (t_c, t_a)-TTBR is binary. Thus, we assume that (t_c, t_a)-TTBR is invoked twice with conversion $0 \to (0, 0)$, $1 \to (1, 1)$, and $* \to (0, 1)$;

[19] When the sender is honest, the agreement is subsumed by the correctness since $t_c \geq t_a$. Thus, the agreement essentially requires that all honest receivers output the same value when the sender is dishonest and there are at most $t_a - 1$ dishonest receivers.

Protocol 18.11 (t_c, t_a)-TTBR $\pi_{\mathtt{TTBR}}[t_c, t_a; \mathcal{S}; \mathcal{P}_i]$

Input: $\mathcal{P}_i$'s input $x_i \in \{0,1\}$

Output: $\mathcal{P}_\ell$'s output z_ℓ for $\ell \in \mathcal{S}\backslash\{i\}$

1. If $|\mathcal{S}| = 3$, by invoking three-party broadcast, $\mathcal{P}_i$ sends $\Pi_{i\to j} = m$ to $\mathcal{P}_j$, $j \in \mathcal{S}\backslash\{i\}$; if $|\mathcal{S}| > 3$, by invoking $|\mathcal{S}|$-graded broadcast, $\mathcal{P}_i$ sends $\Pi_{i\to j} = m$ to $\mathcal{P}_j$, $j \in \mathcal{S}\backslash\{i\}$.
2. If $t_a = 0$ or $|\mathcal{S}| = 3$, then $\mathcal{P}_j$, $j \in \mathcal{S}\backslash\{i\}$ outputs $z_j = \Pi_{i\to j}$.
3. For every $j \in \mathcal{S}\backslash\{i\}$, by invoking $\pi_{\mathtt{TTBR}}[t_c, t_a - 1; \mathcal{S}\backslash\{i\}; \mathcal{P}_j]$ twice, $\mathcal{P}_j$ sends $\Pi_{i\to j}$ to other parties in $\mathcal{S}\backslash\{i\}$.
4. For $\ell \in \mathcal{S}\backslash\{i\}$, upon receiving $\Pi_{j\to\ell}$ from $j \in \mathcal{S}\backslash\{i,\ell\}$ in Step 3 and $\Pi_{i\to\ell}$
   in Step 1, let $L_\ell[a] = |\{j \in \mathcal{S}\backslash\{\ell\} : \Pi_{j\to\ell} = a\}|$, where $a \in \{0,*,1\}$; then $\mathcal{P}_\ell$ decides the following.
   - When $\Pi_{i\to\ell} = 0$, if $L_\ell[0] \geq |\mathcal{S}| - t_c - 1$, then $\mathcal{P}_l$ outputs $z_\ell = 0$; otherwise, $\mathcal{P}_l$ outputs $z_\ell = 1$.
   - When $\Pi_{i\to\ell} = *$, if $L_\ell[0] \geq |\mathcal{S}| - t_c - 1$ and $L_\ell[0] + L_\ell[*] \geq |\mathcal{S}| - t_a$, then $\mathcal{P}_l$ outputs $z_\ell = 0$; otherwise, $\mathcal{P}_l$ outputs $z_\ell = 1$.
   - When $\Pi_{i\to\ell} = 1$, if $L_\ell[0] \geq |\mathcal{S}| - t_c - 1$ and $L_\ell[0] + L_\ell[*] \geq |\mathcal{S}| - t_a$, then $\mathcal{P}_l$ outputs $z_\ell = 0$; otherwise, $\mathcal{P}_l$ outputs $z_\ell = 1$.

then, the output (b_1, b_2) is converted as $(0,0) \to 0$, $(1,1) \to 1$, and $(b_1, b_2) \to *$ if $(b_1, b_2) \neq (0,0), (1,1)$.

18.16 When there are at most t dishonest parties among n parties, Prove that Protocol 18.11 with $t_c = t_a = t$ realizes the broadcast if $t < n/2$.

18.17 Under the assumption that three-party broadcast is available, prove that the broadcast is possible if and only if $t < n/2$ [73].

19 Multiparty Secure Computation

In this chapter, we consider multiparty secure computation, a multiparty version of the problem considered in Chapter 12 and 17. In Chapter 12, we have seen that nontrivial functions are not securely computable from scratch in the two-party model. A key feature of multiparty secure computation is that nontrivial functions are securely computable from scratch if the number of dishonest parties is less than certain thresholds. These thresholds depend on the types of adversary, i.e., passive or active, and the availability of the broadcast discussed in Chapter 18.

19.1 Multiparty Computation Formulation

In multiparty secure computation, n parties $\mathcal{P}_1, \ldots, \mathcal{P}_n$ with inputs $x_1, \ldots, x_n$ seek to compute a function $f\colon \mathcal{X}_1 \times \cdots \times \mathcal{X}_n \to \mathcal{Z}$ using a communication protocol. For simplicity of exposition, we confine ourselves to the deterministic symmetric function (cf. Section 12.2 for types of functions).

Unless otherwise stated, we assume that the broadcast is available throughout this chapter. That is, in addition to communication between pairs of parties via private channels (cf. Section 18.1), parties can broadcast messages to all the other parties. In fact, we will see that both the pairwise private communication and the broadcast play important roles in multiparty secure computation. When we consider passive dishonest parties, then the broadcast can be trivially realized by pairwise private communication. Thus, the availability of the broadcast is crucial only when we consider actively dishonest parties. As we discussed in Section 18.3, the broadcast can be realized from scratch if the honest supermajority is satisfied; in this case, the assumption of the availability of broadcast is just for simplicity of exposition. On the other hand, when the honest supermajority is not satisfied, then the availability of the broadcast is crucial; however, it may be substituted with other assumptions, such as the availability of the digital signature (cf. Section 18.5).

For the threat with a passive adversary, we need to guarantee that dishonest parties cannot learn more information about honest parties' inputs than can be learned from the function value. Formally, after running a protocol π, suppose that $\mathcal{P}_i$ observes (X_i, V_i), where V_i consists of $\mathcal{P}_i$'s local randomness U_i, public randomness U, and the transcript $\Pi_{i\leftarrow *}$ that has been sent to $\mathcal{P}_i$ via pairwise private communication and the broadcast. At the end of the protocol, based on the observation (X_i, V_i), $\mathcal{P}_i$ computes an estimate $\hat{Z}_i = \hat{Z}_i(X_i, V_i)$ of $Z = f(X_{\mathcal{N}})$.

DEFINITION 19.1 (Multiparty secure computation for a passive adversary) For a given function f and an integer $1 \le t \le n-1$, a communication protocol π is an (ε, δ)-secure computation protocol against t passively dishonest parties if the reliability condition

$$\Pr\left(\hat{Z}_i = Z,\ \forall i \in \mathcal{N}\right) \ge 1 - \varepsilon$$

is satisfied and the secrecy condition

$$d_{\mathtt{var}}\left(\mathrm{P}_{X_{\mathcal{N}} V_{\mathcal{A}} Z}, \mathrm{P}_{X_{\mathcal{N}} Z} \mathrm{P}_{V_{\mathcal{A}} | X_{\mathcal{A}} Z}\right) \le \delta \tag{19.1}$$

is satisfied for every subset $\mathcal{A} \subset \mathcal{N}$ with $|\mathcal{A}| \le t$. In particular, it is just said to be a secure computation protocol if $\varepsilon = \delta = 0$.

When $\delta = 0$, the secrecy requirement (19.1) is tantamount to saying that the Markov chain $X_{\mathcal{A}^c} -\!\!\circ\!\!- (X_{\mathcal{A}}, Z) -\!\!\circ\!\!- V_{\mathcal{A}}$ holds. In other words, the extra views $V_{\mathcal{A}}$ by the dishonest parties $\mathcal{A}$ can be simulated from the inputs $X_{\mathcal{A}}$ of the dishonest parties and the function output Z.

For an active adversary, as we discussed in Chapter 15, we need to guarantee that dishonest parties cannot do other than admissible misbehaviors; recall that admissible misbehaviors are behaviors that can be simulated in the ideal model where the function to be computed is available. Formally, this can be defined as follows. In the real world, for a subset $\mathcal{A}$ of dishonest parties, the dishonest parties observe the input $X_{\mathcal{A}}$, transcripts $(\Pi_{i\leftarrow *} : i \in \mathcal{A})$, the output $Z_{\mathcal{A}}$, public randomness U, and local randomness $U_{\mathcal{A}}$; on the other hand, honest parties observe input X_i and output Z_i. Using the same convention as in Chapter 15, we denote the entire view of a protocol π by $\mathtt{V}[\pi]$, which depends on the set $\mathcal{A}$ of dishonest parties. In the ideal world, the dishonest parties seek to simulate the view $\mathtt{V}[\pi]$ by invoking a simulator σ and the ideal function f; the simulated view is denoted by $\mathtt{V}[\sigma \circ f]$.

DEFINITION 19.2 (Multiparty secure computation for an active adversary) For a given function f and an integer $1 \le t \le n-1$, a communication protocol π is an ε-secure computation protocol against t actively dishonest parties if, for any misbehaviors by arbitrary t dishonest parties, there exists a simulator σ such that

$$d_{\mathtt{var}}\left(\mathtt{V}[\pi], \mathtt{V}[\sigma \circ f]\right) \le \varepsilon.$$

In particular, it is just said to be a secure computation protocol if $\varepsilon = 0$.

19.2 Secure Computation from Scratch for a Passive Adversary

In the case of two parties, an honest party cannot become a majority since there are only two parties. When there are more than two parties, honest parties can become a majority. It turns out that the honest majority suffices to conduct secure computation for any function against passively dishonest parties. The goal of this section is to prove the following fundamental result.

THEOREM 19.3 *When the honest majority is satisfied, i.e., the number of dishonest parties satisfies $t < \frac{n}{2}$, then there exists a secure computation protocol for any function against passively dishonest parties.*

Note that the above statement holds for perfect reliability and perfect secrecy. The idea of constructing a secure computation protocol is similar to the two-party counterpart discussed in Section 12.6. However, instead of using the NAND gate as a basic building block to construct an arbitrary Boolean circuit, we consider the addition and multiplication over a finite field $\mathbb{F}_q$ as basic building blocks. In the following, admitting the fact that any given function can be described as an arithmetic circuit over $\mathbb{F}_q$, we focus on how to construct a secure computation protocol for arithmetic circuits over $\mathbb{F}_q$. Another main difference is that, while the implementation of the AND, which can be regarded as the multiplication over $\mathbb{F}_2$, required oblivious transfer in the two-party case, the multiplication can be implemented from scratch if the honest majority is satisfied.

Computation of an arithmetic circuit is conducted using a (t, n)-threshold secret sharing scheme; for notation for secret sharing, see Chapter 11. In the initial phase, parties distribute shares of their inputs to each other. Then, the addition and multiplication are applied to those shares directly without reconstructing the secret data from the shares in the middle of the computation. In the final phase, the parties collect shares of the output to reconstruct the output of the circuit. The details of each step are described in the following.

Preprocessing

Let $\alpha_1, \ldots, \alpha_n$ be distinct elements in $\mathbb{F}_q$. Upon observing input $S_i \in \mathbb{F}_q$, $\mathcal{P}_i$ randomly picks a polynomial $f(x) \in \mathbb{P}_t(S_i)$, and distributes share $W_{i,j} = f(\alpha_j)$ to $\mathcal{P}_j$ for $j \in \mathcal{N}$,[1] where $\mathbb{P}_t(S_i)$ is the set of all polynomials $f(x)$ of degree less than or equal to t such that $f(0) = S_i$. If $\mathcal{P}_i$ has more than one input, $\mathcal{P}_i$ does the same for the other inputs.

Interpolation revisited

Before presenting the schemes for the addition and multiplication, let us revisit Lagrange interpolation, which plays a key role in both the addition and multiplication. For an arbitrary polynomial $f(x)$ of degree less than or equal to $n-1$, let

[1] The ith share $W_{i,i} = f(\alpha_i)$ is kept by $\mathcal{P}_i$.

$S = f(0)$ and $W_i = f(\alpha_i)$ for $i \in \mathcal{N} = \{1, \ldots, n\}$. By the Lagrange interpolation formula, we can write[2]

$$f(x) = \sum_{i=1}^{n} \prod_{j \neq i} \frac{(x - \alpha_j)}{(\alpha_i - \alpha_j)} W_i.$$

By substituting $x = 0$, we have

$$\begin{aligned} S &= f(0) \\ &= \sum_{i=1}^{n} \prod_{j \neq i} \frac{-\alpha_j}{(\alpha_i - \alpha_j)} W_i \\ &= \sum_{i=1}^{n} \beta_i W_i, \end{aligned} \tag{19.2}$$

where $\beta_i = \prod_{j \neq i} \frac{-\alpha_j}{(\alpha_i - \alpha_j)}$. Here, note that the coefficients $\{\beta_i\}_{i=1}^n$ do not depend on the polynomial f. In other words, once the values $f(\alpha_1), \ldots, f(\alpha_n)$ of the polynomial evaluated at $\alpha_1, \ldots, \alpha_n$ are given, the constant term $f(0)$ can be described as a linear combination of those values and coefficients that do not depend on the polynomial.

EXAMPLE 19.4 For $n = 3$ and $q = 5$, let $\alpha_1 = 1$, $\alpha_2 = 2$, and $\alpha_3 = 3$. Then, we can verify that $\beta_1 = 3$, $\beta_2 = 2$, and $\beta_3 = 1$. For the polynomial $f(x) = 2 + x + 3x^2$ on $\mathbb{F}_5$, if we are given three values $f(1) = 1$, $f(2) = 1$, and $f(3) = 2$, then the constant term $f(0) = 2$ can be described by a linear combination as $2 = 3 \cdot 1 + 2 \cdot 1 + 1 \cdot 2$.

Addition

For $i \in \mathcal{N}$, suppose that $\mathcal{P}_i$ has shares $(W_{j,i} : j \in \mathcal{N})$, where $W_{j,i}$ is the ith share of S_j. The parties would like to compute the addition of $S_1, \ldots, S_n$ without reconstructing them from the shares.[3] In fact, we consider a slightly more general task of computing a linear combination

$$S = \sum_{i=1}^{n} \gamma_i S_i$$

for some $(\gamma_1, \ldots, \gamma_n) \in \mathbb{F}_q^n$.

Let $f_j(x) \in \mathbb{P}_t(S_j)$ be the polynomial used to create shares $W_{j,1}, \ldots, W_{j,n}$. A key observation of the computation of the linear combination is that

$$W_i = \sum_{j=1}^{n} \gamma_i W_{j,i}$$

[2] Although the interpolated formula appears to have degree $n - 1$, it has degree strictly less than $n - 1$ if $W_1, \ldots, W_n$ are created from a polynomial of degree strictly less than $n - 1$.

[3] Here, S_i need not be the input by $\mathcal{P}_i$ if the addition is used as a building block of an arithmetic circuit.

Protocol 19.1 Secret shared linear combination

Input: Shares $(W_{j,i} : j \in \mathcal{N})$ owned by $\mathcal{P}_i$ for $i \in \mathcal{N}$, where
$S_j = \sum_{i=1}^n \beta_i W_{j,i}$

Output: Share W_i owned by $\mathcal{P}_i$ for $i \in \mathcal{N}$, where $S = \sum_{i=1}^n \beta_i W_i$

1. For $i \in \mathcal{N}$, $\mathcal{P}_i$ locally computes $W_i = \sum_{j=1}^n \gamma_j W_{j,i}$.

is the value of polynomial

$$f(x) = \sum_{j=1}^n \gamma_j f_j(x)$$

evaluated at $x = \alpha_i$. In fact, we can verify that

$$\begin{aligned} f(\alpha_i) &= \sum_{j=1}^n \gamma_j f_j(\alpha_i) \\ &= \sum_{j=1}^n \gamma_j W_{j,i} \\ &= W_i. \end{aligned}$$

Furthermore, we also note that $f(0) = S$ and the degree of $f(x)$ is less than or equal to t. In fact, we can also verify that $f(x)$ is uniformly distributed on $\mathbb{P}_t(S)$ if $f_j(x)$ is uniformly distributed on $\mathbb{P}_t(S_j)$ for $j \in \mathcal{N}$. By using this observation, a protocol for computing the linear combination is described in Protocol 19.1.

Multiplication

For $i \in \mathcal{N}$, suppose that $\mathcal{P}_i$ has shares $(W_{1,i}, W_{2,i})$, where $W_{1,i}$ and $W_{2,i}$ are ith shares of S_1 and S_2, respectively. The parties would like to compute the multiplication $S = S_1 \cdot S_2$ without reconstructing the secret data from the shares.

The first key observation is that, when shares $(W_{1,i} : i \in \mathcal{N})$ and $(W_{2,i} : i \in \mathcal{N})$ are created by polynomials $f_1(x) \in \mathbb{P}_t(S_1)$ and $f_2(x) \in \mathbb{P}_t(S_2)$, then $W_i = W_{1,i} \cdot W_{2,i}$ is the value of polynomial

$$f(x) = f_1(x) f_2(x)$$

evaluated at $x = \alpha_i$. Furthermore, the constant term of $f(x)$ is $S_1 \cdot S_2$. However, $(W_1, \ldots, W_n)$ cannot be regarded as shares of $S_1 \cdot S_2$ by a (t, n)-threshold scheme since the degree of $f(x)$ may be larger than t.

The second key observation is that, if $2t < n$, then $S_1 \cdot S_2 = f(0)$ can be written as (cf. (19.2))

$$S_1 \cdot S_2 = \sum_{i=1}^n \beta_i W_i.$$

Thus, computing $S_1 \cdot S_2$ is now reduced to computing the linear combination, where $\mathcal{P}_i$ observes W_i.

Protocol 19.2 Secret shared multiplication

Input: Shares $(W_{1,i}, W_{2,i})$ owned by $\mathcal{P}_i$ for $i \in \mathcal{N}$, where $S_j = \sum_{i=1}^n \beta_i W_{j,i}$
Output: Share $\bar{W}_i$ owned by $\mathcal{P}_i$ for $i \in \mathcal{N}$, where $S_1 \cdot S_2 = \sum_{i=1}^n \beta_i \bar{W}_i$

1. For $i \in \mathcal{N}$, $\mathcal{P}_i$ locally computes $W_i = W_{1,i} \cdot W_{2,i}$.
2. For $j \in \mathcal{N}$, $\mathcal{P}_j$ randomly picks $\bar{f}_j(x) \in \mathbb{P}_t(W_j)$, and distributes $\bar{W}_{j,i} = \bar{f}_j(\alpha_i)$ to $\mathcal{P}_i$.
3. For $i \in \mathcal{N}$, $\mathcal{P}_i$ locally computes $\bar{W}_i = \sum_{j=1}^n \beta_j \bar{W}_{j,i}$.

Let $\bar{f}_j(x) \in \mathbb{P}_t(W_j)$, and let $\bar{W}_{j,i} = \bar{f}_j(\alpha_i)$ for $i \in \mathcal{N}$. Then,

$$\bar{W}_i = \sum_{j=1}^n \beta_j \bar{W}_{j,i}$$

is the value of polynomial

$$\bar{f}(x) = \sum_{j=1}^n \beta_j \bar{f}_j(x)$$

evaluated at $x = \alpha_i$. Furthermore, since

$$\begin{aligned}\bar{f}(0) &= \sum_{j=1}^n \beta_j \bar{f}_j(0) \\ &= \sum_{j=1}^n \beta_j W_j \\ &= S_1 \cdot S_2,\end{aligned}$$

we have $\bar{f} \in \mathbb{P}_t(S_1 \cdot S_2)$. By using these observations, a protocol for computing the multiplication is described in Protocol 19.2. Note that the outputs of the protocol satisfy

$$\begin{aligned}\sum_{i=1}^n \beta_i \bar{W}_i &= \sum_{i=1}^n \beta_i \sum_{j=1}^n \beta_j \bar{W}_{j,i} \\ &= \sum_{j=1}^n \beta_j \sum_{i=1}^n \beta_i \bar{W}_{j,i} \\ &= \sum_{j=1}^n \beta_j W_j \\ &= S_1 \cdot S_2.\end{aligned}$$

EXAMPLE 19.5 Let us consider the same parameters as in Example 19.4, i.e., $n = 3$, $q = 5$, $(\alpha_1, \alpha_2, \alpha_3) = (1, 2, 3)$ and $(\beta_1, \beta_2, \beta_3) = (3, 2, 1)$. In order to compute the multiplication of $S_1 = 2$ and $S_2 = 3$, suppose that $f_1(x) = 2 + 2x$ and $f_2(x) = 3 + 3x$ are picked. Then, shares of the parties are given as

$$\begin{array}{lll} W_{1,1} = 4, & W_{1,2} = 1, & W_{1,3} = 3, \\ W_{2,1} = 1, & W_{2,2} = 4, & W_{2,3} = 2, \\ W_1 = 4, & W_2 = 4, & W_3 = 1. \end{array}$$

If we redistribute the shares (W_1, W_2, W_3) using polynomials $\bar{f}_1(x) = 4 + 2x$, $\bar{f}_2(x) = 4 = 3x$, and $\bar{f}(x) = 1 + x$, then we obtain shares

$$\begin{array}{lll} \bar{W}_{1,1} = 1, & \bar{W}_{1,2} = 3, & \bar{W}_{1,3} = 0, \\ \bar{W}_{2,1} = 2, & \bar{W}_{2,2} = 0, & \bar{W}_{2,3} = 3, \\ \bar{W}_{3,1} = 2, & \bar{W}_{3,2} = 3, & \bar{W}_{3,3} = 4, \\ \bar{W}_1 = 4, & \bar{W}_2 = 2, & \bar{W}_3 = 0. \end{array}$$

We can verify that $S_1 \cdot S_2 = 1$ is given by $3 \cdot 4 + 2 \cdot 2 + 1 \cdot 0$.

Reconstruction

In order to securely compute a given arithmetic circuit, the parties first distribute their inputs by preprocessing, and then sequentially apply the linear combination and the multiplication. When applications of these operations are done, $\mathcal{P}_i$ has a share Z_i of the output Z of the arithmetic circuit. Then, the parties exchange their shares, and each party reconstructs the output by

$$Z = \sum_{i=1}^{n} \beta_i Z_i.$$

EXAMPLE 19.6 For $n = 3$, suppose that the arithmetic circuit to be computed is $3 \cdot S_1 \cdot S_2 + 2 \cdot S_3$, where S_i is $\mathcal{P}_i$'s input. In the preprocessing phase, $\mathcal{P}_i$ distributes shares $(W_{i,1}, W_{i,2}, W_{i,3})$ of S_i. Then, the multiplication is first applied with $\mathcal{P}_i$'s input $(W_{1,i}, W_{2,i})$. Then, using the output $\bar{W}_i$ of the multiplication, each party computes $Z_i = 3 \cdot \bar{W}_i + 2 \cdot W_{3,i}$. By exchanging each other's Z_i, the parties reconstruct $\beta_1 Z_1 + \beta_2 Z_2 + \beta_3 Z_3$, which coincides with $Z = 3 \cdot S_1 \cdot S_2 + 2 \cdot S_3$.

Security analysis

Since the outputs of the parties in each basic operation are the shares of the output of that basic operation, it is not difficult to see that the final shares are the shares of the final output of the arithmetic circuit. In order to verify the secrecy of the computation protocol, for a given subset $\mathcal{A} \subset \mathcal{N}$ with $|\mathcal{A}| \le t$, we need to verify that the extra view $V_{\mathcal{A}}$ obtained by dishonest parties $(\mathcal{P}_i : i \in \mathcal{A})$ during the protocol can be simulated by themselves in the ideal model. Note that the extra view $V_{\mathcal{A}}$ consists of shares observed by $(\mathcal{P}_i : i \in \mathcal{A})$ in the preprocessing and intermediate operations, and the final shares $(Z_i : i \in \mathcal{N})$ exchanged to reconstruct the output Z of the arithmetic circuit.

In the preprocessing phase, the shares $(W_{j,i} : i \in \mathcal{N})$ of S_j for $j \in \mathcal{A}$ can be simulated since $(S_j : j \in \mathcal{A})$ are available to $(\mathcal{P}_j : j \in \mathcal{A})$. On the other hand, since $|\mathcal{A}| \le t$, by the security of the (t, n)-threshold scheme, $(W_{j,i} : i \in \mathcal{A})$ are uniformly distributed and independent of S_j for $j \in \mathcal{A}^c$. Thus, $(W_{j,i} : i \in \mathcal{A})$ for $j \in \mathcal{A}^c$ can be simulated as well.

In the linear combination operation, no new information is added to the dishonest parties. In the multiplication operation, there is no new information in the first step. In the second step, the shares $(\bar{W}_{j,i} : i \in \mathcal{N})$ for $j \in \mathcal{A}$ can be simulated since W_j for $j \in \mathcal{A}$ are available; note that the simulation is conducted sequentially along the same flow as the application of the basic operations, and thus W_j has been simulated at this point. Also, since $|\mathcal{A}| \leq t$, again by the security of the (t, n)-threshold scheme, the shares $(\bar{W}_{j,i} : i \in \mathcal{A})$ for $j \in \mathcal{A}^c$ are uniformly distributed and independent of W_j for $j \in \mathcal{A}^c$, and can be simulated.

In the final reconstruction phase, the dishonest parties additionally obtain shares $(Z_i : i \in \mathcal{A}^c)$. For simplicity, first suppose that $|\mathcal{A}| = t$. In the ideal model, the dishonest parties observe the output Z and $(Z_i : i \in \mathcal{A})$, where the latter is simulated by themselves in this reconstruction phase. Then, they can simulate $(Z_i : i \in \mathcal{A}^c)$ as follows. First, the dishonest parties reproduce the polynomial $f(x)$ of degree less than or equal to t such that $f(\alpha_i) = Z_i$ and $f(0) = Z$. Then, they compute $(f(\alpha_i) : i \in \mathcal{A}^c)$. Since $(Z_i : i \in \mathcal{A}^c)$ and $(Z_i : i \in \mathcal{A})$ constitute shares of Z by the (t, n)-threshold scheme in the real model, the observations simulated in this manner have the desired distribution. When $|\mathcal{A}| < t$, then, for a subset $\tilde{\mathcal{A}}$ with $|\mathcal{A} \cup \tilde{\mathcal{A}}| = t$, the dishonest parties first randomly sample Z_i for $i \in \tilde{\mathcal{A}}$; again by the security of the (t, n)-threshold scheme, the sampled observations have the same distribution as those in the real model. Then, the dishonest parties do the same simulation as above using $(Z_i : i \in \mathcal{A} \cup \tilde{\mathcal{A}})$.

The reader may have noticed that the honest majority condition $t < \frac{n}{2}$ is only used in the multiplication operation, and the linear combination operation works securely as long as $t \leq n - 1$. For later use, we state this observation separately.

PROPOSITION 19.7 *When the number of dishonest parties satisfies $t \leq n - 1$, there exists a secure computation protocol for any linear combination function.*

Note that Proposition 19.7 is valid even for $n = 2$. In fact, we have seen in Example 12.3 that the modulo-sum function is securely computable in a trivial manner.

19.3 Verifiable Secret Sharing

The next problem we would like to tackle is multiparty secure computation for an active adversary. The main obstacle to that end is how to conduct the secret sharing in the presence of an active adversary. In Section 11.4, we considered secret sharing with dishonest share holders; in such a case, dishonest share holders may submit forged shares. In this section, we consider the case such that even the dealer may be dishonest. Such secret sharing is termed *verifiable secret sharing*.

19.3.1 Dishonest Dealer

Since protecting against the dealer's attack requires new ingredients, let us first consider the case in which only the dealer may be dishonest and all the share holders are honest.

When the dealer is dishonest, what kind of threat should we protect against? To fix ideas, let us consider the $(2, 4)$-threshold scheme using a polynomial (cf. Protocol 11.1). The dealer is supposed to create four shares $W_i = f(\alpha_i)$ using a degree 1 polynomial $f(x)$ with $f(0) = S$. If the dealer is dishonest, the dealer may create shares $W_1 = f(\alpha_1)$, $W_2 = f(\alpha_2)$, $W_3 = g(\alpha_3)$, and $W_4 = g(\alpha_4)$ using another polynomial $g(x)$ with $g(0) \neq S$ so that $\mathcal{P}_3$ and $\mathcal{P}_4$ together will reproduce something different from $\mathcal{P}_1$ and $\mathcal{P}_2$. More generally, the dishonest dealer may create "inconsistent" shares so that different subsets of share holders will reproduce different values. Thus, we need to introduce a mechanism that prevents the dealer from distributing inconsistent shares without revealing the value of secret data in the distribution phase. The following example presents a naive way to protect against inconsistent shares.

EXAMPLE 19.8 (Random sampling) Let us consider the $(2, 4)$-threshold scheme with secret data $S \in \mathbb{F}_q$. The dealer first randomly samples $S_1, \ldots, S_m \in \mathbb{F}_q$, and creates shares $(V_{1,j}, \ldots, V_{4,j})$ of S_j for $j = 1, \ldots, m$ using the distribution phase of Protocol 11.1. Then, the shares $(V_{i,j} : 1 \leq j \leq m)$ are provided to ith share holder $\mathcal{P}_i$. In order to check consistency of the shares, the share holders randomly sample $\mathcal{J} \subset \{1, \ldots, m\}$ of size $|\mathcal{J}| = m - 1$, and, by revealing $(V_{1,j}, \ldots, V_{4,j})$ for $j \in \mathcal{J}$ to each other, they verify whether there exists a degree 1 polynomial $f_j(x)$ satisfying $f_j(\alpha_i) = V_{i,j}$ for $i = 1, \ldots, 4$ and $j \in \mathcal{J}$. Since the dealer does not know which subset $\mathcal{J}$ will be sampled in advance, if the above verification test is passed and m is sufficiently large, then it is highly likely that the remaining share $(V_{1,\ell}, \ldots, V_{4,\ell})$, where $\ell \notin \mathcal{J}$, is also consistent. Now, the dealer provides $\Pi = S + S_\ell$ to all the share holders. Then, in the reconstruction phase, each size 2 subset of share holders can recover the data by first recovering S_ℓ and then subtracting it from Π. Here, the recovered data may not be S if the dealer is dishonest; however, different subsets of share holders will recover the same value with high probability.

Now, let us consider a more sophisticated scheme using the bivariate polynomial. For a given $s \in \mathbb{F}_q$, let

$$\mathbb{B}_{k-1}(s) := \left\{ \sum_{i,j=0}^{k-1} a_{i,j} x^i y^j : a_{i,j} \in \mathbb{F}_q, a_{0,0} = s \right\}$$

be the set of all bivariate polynomials of degree less than or equal to $k - 1$ such that the constant term is s. A threshold scheme for dishonest dealer is described in Protocol 19.3 (see also Figure 19.1).

Protocol 19.3 (k, n)-threshold scheme with dishonest dealer

1. **Distribution phase**
2. The dealer $\mathcal{P}_d$ randomly chooses $F(x, y) \in \mathbb{B}_{k-1}(S)$, and sends $f_i(x) = F(x, \alpha_i)$ and $g_i(y) = F(\alpha_i, y)$ to the ith share holder $\mathcal{P}_i$ via the private channel.
3. For $i \in \mathcal{N}$, if $f_i(\alpha_i) \neq g_i(\alpha_i)$ or the degree of $f_i(x)$ or $g_i(y)$ is larger than $k-1$, then $\mathcal{P}_i$ broadcasts "accuse the dealer."
4. For $i, j \in \mathcal{N}$, $\mathcal{P}_i$ sends $\Pi_{i \to j} = f_i(\alpha_j)$ to $\mathcal{P}_j$ via the private channel.
5. For $i, j \in \mathcal{N}$, if $g_j(\alpha_i) \neq \Pi_{i \to j}$, then $\mathcal{P}_j$ broadcasts "accuse the dealer."
6. If the dealer is not accused by any share holder, then $\mathcal{P}_i$ keeps $W_i = f_i(0)$ as ith share for $i \in \mathcal{N}$.
7. **Reconstruction phase**
8. When a subset $(\mathcal{P}_i : i \in \mathcal{I})$ of share holders with $|\mathcal{I}| \geq k$ submit their shares, they reconstruct the polynomial $g(y)$ from $(W_i : i \in \mathcal{I})$ using the Lagrange interpolation formula, (11.3), and output $g(0)$.

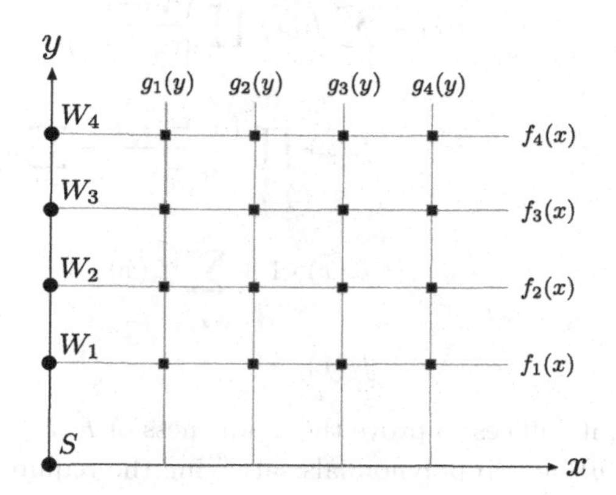

Figure 19.1 A description of the bivariate polynomial based secret sharing with four shares.

THEOREM 19.9 *The (k, n)-threshold scheme in Protocol 19.3 is secure in the following sense.*

1. *When the dealer is honest, then, for every subset $\mathcal{I} \subset \mathcal{N}$ with $|\mathcal{I}| \leq k-1$, the secret data S is independent of $(f_i(x), g_i(y), \Pi_{j \to i} : i \in \mathcal{I}, j \in \mathcal{N})$.*
2. *For any attack by the dealer, if the share holders output shares $(W_1, \ldots, W_n)$ without any accusation, then there exists a unique polynomial $g(y)$ of degree less than or equal to $k-1$ such that $g(\alpha_i) = W_i$ for every $i \in \mathcal{N}$.*

Before proving Theorem 19.9, let us present a few technical results concerning the bivariate polynomial. The first result is a bivariate extension of the polynomial interpolation.

LEMMA 19.10 *Let $\alpha_1, \ldots, \alpha_k$ be distinct elements of $\mathbb{F}_q$, and let $f_1(x), \ldots, f_k(x)$ be polynomials on $\mathbb{F}_q$ of degree less than or equal to $k-1$. Then, there exists a unique bivariate polynomial $F(x,y)$ on $\mathbb{F}_q$ with degree less than or equal to $k-1$ such that*

$$F(x, \alpha_i) = f_i(x) \tag{19.3}$$

for every $i = 1, \ldots, k$.

Proof Let

$$F(x,y) = \sum_{i=1}^{k} f_i(x) \prod_{\substack{j=1:\\ j\neq i}}^{k} \frac{(y-\alpha_j)}{(\alpha_i - \alpha_j)}$$

be the bivariate version of the Lagrange interpolation. In fact, we can readily verify that it satisfies (19.3) as follows:

$$\begin{aligned} F(x,\alpha_\ell) &= \sum_{i=1}^{k} f_i(x) \prod_{\substack{j=1:\\ j\neq i}}^{k} \frac{(\alpha_\ell-\alpha_j)}{(\alpha_i - \alpha_j)} \\ &= f_\ell(x) \prod_{\substack{j=1:\\ j\neq \ell}}^{k} \frac{(\alpha_\ell-\alpha_j)}{(\alpha_\ell - \alpha_j)} + \sum_{\substack{i=1:\\ i\neq \ell}}^{k} f_i(x) \prod_{\substack{j=1:\\ j\neq i}}^{k} \frac{(\alpha_\ell-\alpha_j)}{(\alpha_i - \alpha_j)} \\ &= f_\ell(x)\cdot 1 + \sum_{\substack{i=1:\\ i\neq \ell}}^{k} f_i(x)\cdot 0 \\ &= f_\ell(x). \end{aligned}$$

Thus, it suffices to prove the uniqueness of $F(x,y)$. To that end, let $F_1(x,y)$ and $F_2(x,y)$ be two polynomials satisfying the requirement, and let

$$\begin{aligned} G(x,y) &= F_1(x,y) - F_2(x,y) \\ &= \sum_{i,j=0}^{k-1} b_{i,j} x^i y^j \\ &= \sum_{i=0}^{k-1} \left(\sum_{j=0}^{k-1} b_{i,j} y^j \right) x^i. \end{aligned} \tag{19.4}$$

By substituting $y = \alpha_\ell$ for $\ell = 1, \ldots, k$, we have

$$\begin{aligned} G(x,\alpha_\ell) &= F_1(x,\alpha_\ell) - F_2(x,\alpha_\ell) \\ &= f_\ell(x) - f_\ell(x) \\ &= 0, \end{aligned}$$

where the second equality follows from the assumption that $F_1(x,y)$ and $F_2(x,y)$ satisfy (19.3). From (19.4), we can also write

$$\sum_{i=0}^{k-1}\left(\sum_{j=0}^{k-1} b_{i,j}\alpha_\ell^j\right)x^i = 0.$$

This means that

$$\sum_{j=0}^{k-1} b_{i,j}\alpha_\ell^j = 0 \tag{19.5}$$

for every $\ell = 1, \ldots, k$. Furthermore, (19.5) means that the univariate polynomial

$$h_i(y) = \sum_{j=0}^{k-1} y^j$$

for $i = 0, \ldots, k-1$ satisfies $h_i(\alpha_\ell) = 0$ for every $\ell = 1, \ldots, k$. Since $h_i(y)$ has k roots and the degree is less than or equal to $k-1$, it must be the 0 polynomial, i.e., $b_{i,j} = 0$ for every $j = 0, \ldots, k-1$. Consequently, we have $b_{i,j} = 0$ for every $i, j = 0, \ldots, k-1$, and thus $F_1(x,y) = F_2(x,y)$. □

Even though Lemma 19.10 guarantees the existence of a unique bivariate polynomial of degree less than or equal to $k-1$ that is compatible with given k univariate polynomials, such a bivariate polynomial may not exist in general if more than k univariate polynomials are given. However, if additional constraints, $f_i(\alpha_j) = g_j(\alpha_i)$ for $i, j \in \mathcal{I}$, are satisfied, then the existence of a unique bivariate polynomial is guaranteed as follows.

LEMMA 19.11 *For a subset $\mathcal{I} \subseteq \mathcal{N}$ with $|\mathcal{I}| \geq k$ and given polynomials $(f_i(x), g_i(y) : i \in \mathcal{I})$ of degree less than or equal to $k-1$ such that $f_i(\alpha_j) = g_j(\alpha_i)$ for every $i, j \in \mathcal{I}$, there exists a unique bivariate polynomial $F(x,y)$ of degree less than or equal to $k-1$ such that*

$$F(x, \alpha_i) = f_i(x), \ \forall i \in \mathcal{I}, \tag{19.6}$$

$$F(\alpha_i, y) = g_i(y), \ \forall i \in \mathcal{I}. \tag{19.7}$$

Proof Let $\tilde{\mathcal{I}} \subseteq \mathcal{I}$ be an arbitrary subset with $|\tilde{\mathcal{I}}| = k$. By Lemma 19.10, there exists a unique bivariate polynomial $F(x,y)$ of degree less than or equal to $k-1$ such that

$$F(x, \alpha_i) = f_i(x), \ \forall i \in \tilde{\mathcal{I}}.$$

Thus, it suffices to show that this $F(x,y)$ also satisfies other constraints in (19.6) and (19.7). Let $\tilde{f}_i(x) = F(x, \alpha_i)$ for $i \in \mathcal{I}\backslash\tilde{\mathcal{I}}$ and $\tilde{g}_i(y) = F(\alpha_i, y)$ for $i \in \mathcal{I}$. For $i \in \tilde{\mathcal{I}}$ and $j \in \mathcal{I}$, since

$$\tilde{g}_j(\alpha_i) = F(\alpha_j, \alpha_i) = f_i(\alpha_j) = g_j(\alpha_i),$$

$\tilde{g}_j(y)$ and $g_j(y)$ share $|\tilde{\mathcal{I}}| = k$ common points, which implies $\tilde{g}_j(y) = g_j(y)$ by the Lagrange interpolation formula for the univariate polynomial. Similarly, for $i \in \mathcal{I}\backslash\tilde{\mathcal{I}}$ and $j \in \mathcal{I}$, since

$$\begin{aligned}\tilde{f}_i(\alpha_j) &= F(\alpha_j, \alpha_i)\\ &= \tilde{g}_j(\alpha_i)\\ &= g_j(\alpha_i)\\ &= f_i(\alpha_j),\end{aligned}$$

we have $\tilde{f}_i(x) = f_i(x)$. Thus, $F(x,y)$ satisfies all the constraints in (19.6) and (19.7). □

The following corollary immediately follows from Lemma 19.11, which is used to prove security against a dishonest dealer.

COROLLARY 19.12 *For a subset $\mathcal{I} \subseteq \mathcal{N}$ with $|\mathcal{I}| \geq k$ and given polynomials $(f_i(x), g_i(y) : i \in \mathcal{I})$ of degree less than or equal to $k-1$ such that $f_i(\alpha_j) = g_j(\alpha_i)$ for every $i, j \in \mathcal{I}$, there exists a unique polynomial $g(y)$ of degree less than or equal to $k-1$ such that*

$$g(\alpha_i) = w_i, \ \forall i \in \mathcal{I},$$

where $w_i = f_i(0)$ for $i \in \mathcal{I}$.

Proof By Lemma 19.11, there exists a unique $F(x,y)$ of degree less than or equal to $k-1$ such that (19.6) and (19.7) are satisfied. Then, $g(y) = F(0,y)$ apparently satisfies the requirement of the corollary; the uniqueness follows since $|\mathcal{I}| \geq k$. □

When the number of given univariate polynomial is less than k, a compatible bivariate polynomial is not unique, which means that the secret data cannot be determined uniquely. The following result states that the number of candidate bivariate polynomials does not depend on given univariate polynomials, which is a key property of guaranteeing the secrecy.

LEMMA 19.13 *For a subset $\mathcal{I} \subset \mathcal{N}$ with $|\mathcal{I}| \leq k-1$ and given polynomials $g(y)$ and $(f_i(x), g_i(y) : i \in \mathcal{I})$ of degree less than or equal to $k-1$ such that $f_i(\alpha_j) = g_j(\alpha_i)$ for $i, j \in \mathcal{I}$ and $f_i(0) = g(\alpha_i)$ for $i \in \mathcal{I}$, let*

$$\begin{aligned}\mathcal{F}\big(g(y), (f_i(x), g_i(y) : i \in \mathcal{I})\big) := \big\{&F(x,y) : \mathit{deg}F(x,y) \leq k-1, F(0,y) = g(y),\\ &(F(x,\alpha_i), F(\alpha_i, y)) = (f_i(x), g_i(y)) \ \forall i \in \mathcal{I}\big\}\end{aligned}$$

be the set of all bivariate polynomials of degree less than or equal to $k-1$ that are compatible with the given univariate polynomials. Then, we have

$$\big|\mathcal{F}\big(g(y), (f_i(x), g_i(y) : i \in \mathcal{I})\big)\big| = \big(q^{k-1-|\mathcal{I}|}\big)^{k-|\mathcal{I}|}$$

irrespective of given univariate polynomials.

Proof Without loss of generality, let $\mathcal{I} = \{1, \ldots, \kappa\}$, where $\kappa \leq k-1$. For given $(g(y), (f_i(x), g_i(y) : i \in \mathcal{I}))$, Lemma 19.10 implies that additional $(k-\kappa)$ polynomials $f_{\kappa+1}(x), \ldots, f_k(x)$ together with $f_1(x), \ldots, f_\kappa(x)$ uniquely determine the bivariate polynomial $F(x,y)$ of degree less than or equal to $k-1$ such that

$$F(x, \alpha_i) = f_i(x)$$

for $i = 1, \ldots, k$. Thus, the number of compatible $F(x, y)$ is equal to the degree of freedom of $f_{\kappa+1}(x), \ldots, f_k(x)$. Without any constraint, each $f_i(x)$ has the degree of freedom q^k since it is a polynomial of degree less than or equal to $k-1$. However, since $f_i(\alpha_j) = F(\alpha_j, \alpha_i) = g_j(\alpha_i)$ for $j = 1, \ldots, \kappa$ and $f_i(0) = F(0, \alpha_i) = g(\alpha_i)$ must be satisfied, the effective degree of freedom of each polynomial is $q^{k-1-\kappa}$; thus, in total, we have $\left(q^{k-1-\kappa}\right)^{k-\kappa}$. □

Remark 19.1 Note that, without the constraint $f_i(\alpha_j) = g_j(\alpha_i)$ for $i, j \in \mathcal{I}$ and $f_i(0) = g(\alpha_i)$ for $i \in \mathcal{I}$, the set $\mathcal{F}\big(g(y), (f_i(x), g_i(y) : i \in \mathcal{I})\big)$ is the empty set.

Proof of Theorem 19.9
To prove Claim 1, when the dealer is honest, note that $\Pi_{j \to i} = g_i(\alpha_j)$ for every $i \in \mathcal{I}$ and $j \in \mathcal{N}$. Thus, it suffices to show that $(f_i(x), g_i(y) : i \in \mathcal{I})$ are independent of S. In fact, we can prove a slightly stronger claim that $(f_i(x), g_i(y) : i \in \mathcal{I})$ are independent of $g(y) = F(0, y)$. This stronger claim follows from Lemma 19.13 since $F(x, y) \in \mathbb{B}_{k-1}(S)$ is chosen at random and the number of bivariate polynomials that are compatible with $(f_i(x), g_i(y) : i \in \mathcal{I})$ and $g(y)$ is constant.

To prove Claim 2, note that, if there is no accusation, $(f_i(x), g_i(y) : i \in \mathcal{N})$ are of degree less than or equal to $k-1$ and satisfy $f_i(\alpha_j) = g_j(\alpha_i)$ for every $i, j \in \mathcal{N}$. Thus, Corollary 19.12 with $\mathcal{I} = \mathcal{N}$ guarantees the existence of a unique polynomial $g(y)$ satisfying the requirement. □

19.3.2 Scheme for an Honest Supermajority

Now, let us consider the situation in which the dealer may be dishonest and at most t share holders may be dishonest. Since some share holders are dishonest, Protocol 19.3 does not work. For instance, the message $\Pi_{i \to j}$ sent in Step 4 may be falsified; in that case, the person who should be blamed is $\mathcal{P}_i$ instead of the dealer. To overcome these new challenges, let us consider a "two phase complaint procedure" as in Protocol 19.4.

THEOREM 19.14 *When the number of dishonest share holders satisfies $t \le \frac{n-k}{2}$, then Protocol 19.4 is secure in the following sense.*

1. *When the dealer is honest, the protocol is not aborted.*
2. *For every subset $\mathcal{I} \subset \mathcal{N}$ with $|\mathcal{I}| \le k-1$, if the dealer and share holders in $\mathcal{N} \backslash \mathcal{I}$ are honest,*[4] *then the secret data S is independent of the view $\mathtt{V}_{\mathcal{I}}$ by $\mathcal{I}$, where $\mathtt{V}_{\mathcal{I}}$ consists of $(f_i(x), g_i(y), \Pi_{j \to i} : i \in \mathcal{I}, j \in \mathcal{N})$, $D_{i,j}$ if $\mathcal{P}_j$ has broadcast "(i, j) is suspicious" in Step 5, and $\tilde{f}(x)$ if $\mathcal{P}_i$ has broadcast "accuse the dealer" in Step 3 or Step 7.*
3. *For any attack by the dealer and dishonest share holders, if the distribution phase terminates with less than or equal to t accusations, then there exists a unique polynomial $g(y)$ of degree less than or equal to $k-1$ such that $g(\alpha_i) =$*

[4] This means that t dishonest share holders are included in the set $\mathcal{I}$.

Protocol 19.4 Verifiable (k, n)-threshold scheme for honest supermajority

1. **Distribution phase**
2. The dealer $\mathcal{P}_d$ randomly chooses $F(x, y) \in \mathbb{B}_{k-1}(S)$, and sends $f_i(x) = F(x, \alpha_i)$ and $g_i(y) = F(\alpha_i, y)$ to the ith share holder $\mathcal{P}_i$ via the private channel.
3. For $i \in \mathcal{N}$, if $f_i(\alpha_i) \neq g_i(\alpha_i)$ or the degree of $f_i(x)$ or $g_i(y)$ is larger than $k - 1$, then $\mathcal{P}_i$ broadcasts "accuse the dealer."
4. For $i, j \in \mathcal{N}$, $\mathcal{P}_i$ sends $\Pi_{i \to j} = f_i(\alpha_j)$ to $\mathcal{P}_j$ via the private channel.
5. For $i, j \in \mathcal{N}$, if $g_j(\alpha_i) \neq \Pi_{i \to j}$, then $\mathcal{P}_j$ broadcasts "(i, j) is suspicious."
6. If $\mathcal{P}_j$ broadcast "(i, j) is suspicious" in the previous step, then the dealer broadcasts $D_{i,j} = F(\alpha_j, \alpha_i)$.
7. If $f_i(\alpha_j) \neq D_{i,j}$, then $\mathcal{P}_i$ broadcasts "accuse the dealer"; if $g_j(\alpha_i) \neq D_{i,j}$, then $\mathcal{P}_j$ broadcasts "accuse the dealer."
8. For $i \in \mathcal{N}$, if $\mathcal{P}_i$ has broadcast "accuse the dealer" so far, then the dealer broadcasts $\tilde{f}_i(x) = F(x, \alpha_i)$.
9. If the degree of $\tilde{f}_i(x)$ is larger than $k - 1$, then the protocol is aborted.
10. For $j \in \mathcal{N}$, if $g_j(\alpha_i) \neq \tilde{f}_i(\alpha_j)$, then $\mathcal{P}_j$ broadcasts "accuse the dealer."
11. If the number of accusing share holders is less than or equal to t, then $\mathcal{P}_i$ keeps $W_i = f_i(0)$ as ith share for $i \in \mathcal{N}$, where $f_i(x)$ is replaced by $\tilde{f}_i(x)$ if $\tilde{f}_i(x)$ is broadcast; otherwise, the protocol is aborted.
12. **Reconstruction phase**
13. All share holders exchange their shares with each other via the private channels.
14. Let $(\tilde{W}_{j \to i} : j \in \mathcal{N})$ be the shares collected by $\mathcal{P}_i$. Then, $\mathcal{P}_i$ reconstructs the polynomial $\hat{g}(y)$ from $(\tilde{W}_{j \to i} : j \in \mathcal{N})$ such that $\hat{g}(\alpha_j) = \tilde{W}_{j \to i}$, $\forall j \in \mathcal{I}$ holds for some $\mathcal{I} \subset \mathcal{N}$ with $|\mathcal{I}| \geq n - t$, and outputs $\hat{g}(0)$.

W_i for every honest share holder $\mathcal{P}_i$. In particular, if the dealer is honest, $g(0) = S$.

Note that honest share holders exchange authentic shares with each other in the reconstruction phase. Thus, Claim 3 of Theorem 19.14 implies that, by the same reasoning as in Theorem 11.17, all the honest share holders output $g(0)$ in the reconstruction phase.

Proof of Theorem 19.14
When the dealer is honest, honest share holders never accuse the dealer. Since the number of dishonest share holders is less than or equal to t, the protocol is never aborted (Claim 1).

Since the dealer and the share holders in $\mathcal{N} \backslash \mathcal{I}$ are assumed to be honest in Claim 2, $\Pi_{j \to i} = g_i(\alpha_j)$ can be computed by the share holders in $\mathcal{I}$. Furthermore, $D_{i,j}$ is broadcast by the dealer only when either $i \in \mathcal{I}$ or $j \in \mathcal{I}$. In either case, $D_{i,j} = f_i(\alpha_j)$ or $D_{i,j} = g_j(\alpha_i)$ can be computed by the share holders in

$\mathcal{I}$. Similarly, $\tilde{f}_i(x)$ is broadcast by the dealer only when $i \in \mathcal{I}$. in this case, $\tilde{f}_i(x) = f_i(x)$ is not extra information for the share holders in $\mathcal{I}$. Consequently, it suffices to show that S is independent of $(f_i(x), g_i(y) : i \in \mathcal{I})$, which follows from Claim 19.9 in Theorem 19.9.

To prove Claim 3, first note that there are at least k honest share holders that did not accuse the dealer during the protocol; this follows from the assumption that the number of accusing share holders is less than t, the number of dishonest share holders is less than t, and the assumption $n - 2t \geq k$. For those share holders, which we denote $\mathcal{I}_0$, note that $f_i(\alpha_j) = g_j(\alpha_i)$ holds for every $i, j \in \mathcal{I}_0$; otherwise, at least one of $\mathcal{P}_i$ or $\mathcal{P}_j$ must have accused the dealer in Step 3 or Step 5. Thus, by Lemma 19.11, there exists a unique bivariate polynomial $F(x, y)$ of degree less than or equal to $k - 1$ such that

$$F(x, \alpha_i) = f_i(x), \ \forall i \in \mathcal{I}_0,$$
$$F(\alpha_i, y) = g_i(y), \ \forall i \in \mathcal{I}_0.$$

By letting $g(y) = F(0, y)$, it also satisfies $g(\alpha_i) = W_i$ for $i \in \mathcal{I}_0$.

Let $\mathcal{I}_1$ be the set of honest share holders other than $\mathcal{I}_0$. The remaining task is to show that $g(y)$ above also satisfies $g(\alpha_i) = W_i$ for $i \in \mathcal{I}_1$. To that end, note that $g_j(\alpha_i) = f_i(\alpha_j)$ is satisfied for $j \in \mathcal{I}_0$ and $i \in \mathcal{I}_1$; otherwise, $\mathcal{P}_j$ must have accused the dealer in Step 7 or Step 10. Since $|\mathcal{I}_0| \geq k$ and

$$F(\alpha_j, \alpha_i) = g_j(\alpha_i) = f_i(\alpha_j), \ \forall j \in \mathcal{I}_0,$$

$F(x, \alpha_i) = f_i(x)$ must hold by the Lagrange interpolation formula for the univariate polynomial. Thus, we also have

$$g(\alpha_i) = F(0, \alpha_i) = f_i(0) = W_i$$

for $i \in \mathcal{I}_1$. Finally, when the dealer is honest, the bivariate polynomial reconstructed above must be the same as the bivariate polynomial chosen by the dealer in Step 2; thus $g(0) = S$ must hold. □

19.4 Secure Computation for an Active Adversary

The goal of this section is to provide high-level ideas of how to prove the following basic result on secure computation for an active adversary.

THEOREM 19.15 *When the honest supermajority assumption is satisfied, i.e., the number of dishonest parties satisfies* $t < \frac{n}{3}$, *then there exists a (perfectly) secure implementation of any function against an active adversary in the standard network setting. Furthermore, when the honest majority assumption is satisfied and broadcast is available, there exists a statistically secure implementation of any function against an active adversary.*

At a high level, in a similar manner as in Section 19.2, all parties distribute shares of their inputs, and subsequently process those shares. However, the main difficulty in the presence of an active adversary is that dishonest parties may not follow the protocol. We combat an adversary using verifiable secret sharing (VSS) seen in the previous section.

In the following sections, we explain how the VSS in Section 19.3.2 is used in the honest supermajority case. The full proof requires a lot more work; we only provide an overview. For a detailed account, the reader can see the references listed in Section 19.6.

It should be noted that the honest supermajority assumption in Theorem 19.15 is tight in the sense that, otherwise, secure computation for an active adversary is not possible in the standard network. This is because the broadcast (or the Byzantine agreement) in Chapter 18 can be regarded as an instance of secure computation (see Problem 19.3); as we have seen in Chapter 18, the broadcast cannot be realized in the standard network unless the honest supermajority is not satisfied.

When the broadcast is available as an assumption, then secure computation is possible under an honest majority; however, a protocol under that assumption requires additional machineries, which is beyond the scope of this book. See the references in Section 19.6.

19.4.1 Commitment Using Verifiable Secret Sharing

As we have seen in Chapter 17, the key idea for enabling secure computation in the presence of an active adversary is that the parties commit to their inputs at the beginning, and then they process those committed values throughout the protocol. In the case of two parties, since bit commitment cannot be realized from scratch, we realized it from oblivious transfer. When the honest supermajority assumption is satisfied, we can construct bit commitment from scratch as follows.

Suppose that $\mathcal{P}_i$ would like to commit to a value $S \in \mathbb{F}_q$. To that end, the parties invoke the VSS scheme in Protocol 19.4 with parameters $(k, n) = (t + 1, n)$, where $\mathcal{P}_i$ plays the role of the dealer as well as one of the share holders (for the inputs of other parties, who themselves act as a dealer for their inputs). Here, note that $\mathcal{P}_i$ may not be an honest party. After running the VSS, honest parties $(\mathcal{P}_j : j \in \mathcal{H})$ obtain shares $(W_j : j \in \mathcal{H})$. Since $t < n/3$, by the security of Protocol 19.4, the following properties are guaranteed.

1. There exists a unique polynomial $f(x)$ of degree less than or equal to t such that $f(\alpha_j) = W_j$ for every $j \in \mathcal{H}$. This means that $\mathcal{P}_i$ is committed to the value $f(0)$, which may not coincide with S if $\mathcal{P}_i$ is dishonest. The committed value $f(0)$ can be revealed when honest parties exchange their shares. Note that, since $t < n/3$, honest parties can reconstruct the value $f(0)$ irrespective of the shares submitted by dishonest parties.
2. If $\mathcal{P}_i$ is honest, then $f(0) = S$ and the observations of dishonest parties are independent of S. In other words, the value S is concealed from the dishonest parties until the honest parties come together to reveal the value $f(0) = S$.

19.4.2 Committed Addition

For $i \in \mathcal{N}$, suppose that $\mathcal{P}_i$ has shares $(W_{j,i} : j \in \mathcal{M})$ of value S_j for $j \in \mathcal{M} = \{1, \ldots, m\}$, where each S_j is committed by distributing shares in the manner explained in Section 19.4.1; this means that, for each $j \in \mathcal{M}$, there exists a polynomial $f_j(x) \in \mathbb{P}_t(S_j)$ such that $f_j(\alpha_i) = W_{j,i}$ for every $i \in \mathcal{N}$.

The parties would like to compute the linear combination

$$S = \sum_{j=1}^{m} \gamma_j S_j$$

for some $(\gamma_1, \ldots, \gamma_m) \in \mathbb{F}_q^m$. To that end, let

$$f(x) = \sum_{j=1}^{m} \gamma_j f_j(x). \tag{19.8}$$

Then, we can verify that

$$\begin{aligned} f(\alpha_i) &= \sum_{j=1}^{m} \gamma_j f_j(\alpha_i) \\ &= \sum_{j=1}^{m} \gamma_j W_{j,i} \\ &=: W_i, \end{aligned}$$

i.e., W_i is the value of polynomial $f(x)$ evaluated at $x = \alpha_i$. Furthermore, we have

$$\begin{aligned} f(0) &= \sum_{j=1}^{m} \gamma_j f_j(0) \\ &= \sum_{j=1}^{m} \gamma_j S_j \\ &= S. \end{aligned}$$

Thus, W_i is a share of S distributed by the polynomial $f(x)$. By using this observation, a protocol for computing the linear combination is described in Protocol 19.5. In fact, this protocol is the same as the protocol for computing the linear combination against a passive adversary (cf. Protocol 19.1). Of course, there is no guarantee that a dishonest party $\mathcal{P}_i$ computes W_i correctly. However, since the polynomial $f(x)$ in (19.8) is of degree less than or equal to t and $t < n/3$, honest parties $(\mathcal{P}_i : i \in \mathcal{H})$ can uniquely reconstruct $f(x)$ from their shares $(W_i : i \in \mathcal{H})$ irrespective of the shares submitted by dishonest parties.

Protocol 19.5 Committed linear combination

Input: Shares $(W_{j,i} : j \in \mathcal{M})$ owned by $\mathcal{P}_i$ for $i \in \mathcal{N}$, where $W_{j,i} = f_j(\alpha_i)$ for a polynomial $f_j(x) \in \mathbb{P}_t(S_j)$

Output: Share W_i owned by $\mathcal{P}_i$ for $i \in \mathcal{N}$, where $W_i = f(\alpha_i)$ for a polynomial $f(x) \in \mathbb{P}_t(S)$ and $S = \sum_{j=1}^{m} \gamma_j S_j$

1. For $i \in \mathcal{N}$, $\mathcal{P}_i$ locally computes $W_i = \sum_{j=1}^{m} \gamma_j W_{j,i}$.

19.4.3 Committed Multiplication

For $i \in \mathcal{N}$, suppose that $\mathcal{P}_i$ has shares $(W_{1,i}, W_{2,i})$, where $W_{1,i}$ and $W_{2,i}$ are shares of S_1 and S_2. Here, each S_a for $a = 1, 2$ is committed, i.e., there exists a polynomial $f_a(x) \in \mathbb{P}_t(S_a)$ such that $f_a(\alpha_i) = W_{a,i}$ for every $i \in \mathcal{N}$.

The parties would like to compute the multiplication $S = S_1 \cdot S_2$. For simplicity of exposition, we assume that $t < n/4$; at the end of this section, we will show which part of the protocol needs to be modified in order to handle the case with $t < n/3$.

First, we review secure implementation of multiplication for a passive adversary. Our secure implementation of multiplication for an active adversary will build on that for a passive adversary seen in Section 19.2. Let $f(x) := f_1(x)f_2(x)$, $W_{1,i} := f_1(\alpha_i)$, $W_{2,i} := f_2(\alpha_i)$, for $1 \le i \le n$, and $W_i := W_{1,i}W_{2,i}$. As we have seen in Section 19.2, since the degree of $f(x)$ is $2t < n$, $S = f(0) = f_1(0)f_2(0) = S_1 S_2$ can be reconstructed as the linear combination

$$S = \sum_{i=1}^{n} \beta_i W_i.$$

In the protocol for a passive adversary (cf. Protocol 19.2), each party further distributed subshares $(\bar{W}_{j,1}, \ldots, \bar{W}_{j,n})$ of W_j by computing $\bar{W}_{j,i} = \bar{f}_j(\alpha_i)$ for a randomly choosen polynomial $\bar{f}_j(x) \in \mathbb{P}_t(W_j)$. Then,

$$\bar{W}_i = \sum_{j=1}^{n} \beta_j \bar{W}_{j,i}$$

computed locally by $\mathcal{P}_i$ is a share of $S = S_1 \cdot S_2$ computed using the polynomial

$$\bar{f}(x) = \sum_{j=1}^{n} \beta_j \bar{f}_j(x).$$

In fact, we have

$$\begin{aligned} \bar{f}(\alpha_i) &= \sum_{j=1}^{n} \beta_j \bar{f}_j(\alpha_j) \\ &= \sum_{j=1}^{n} \beta_j \bar{W}_{j,i} \\ &= \bar{W}_i. \end{aligned} \tag{19.9}$$

In the case of an active adversary, there is no guarantee that a dishonest $\mathcal{P}_j$ correctly distributes the subshare $\bar{W}_{j,i}$ of W_j to honest $\mathcal{P}_i$. Then, the final share $\bar{W}_i$ created by honest $\mathcal{P}_i$ may not be a share of S. To overcome this problem, we make the following modification to the protocol proposed earlier for a passive adversary.

1. The distribution of subshares is done using VSS.
2. After distribution of subshares, the parties identify dishonest parties by using an error correcting code; specifically, we use the Reed–Solomon code of Section 11.5.
3. Subshares of dishonest parties are redistributed using constant polynomials. Namely, they are given fixed values W_j which we view as evaluations of the constant polynomial $f(x) = W_j$ to simplify our analysis.

Formally, we present the steps in Protocol 19.6, and provide elaboration on each step below.

Note that even if VSS is used in the distribution of subshares, dishonest parties may not use correct inputs. To indicate clearly that these shares may have been adversarially modified, in Step 2, we denote these shares by $\tilde{W}_j$ instead of W_j. Clearly, $\tilde{W}_j = W_j$ when $\mathcal{P}_j$ is honest.

Note that we apply two kinds of error correction procedures, one in Step 4 and the other in Step 5. First, in order to compute the syndrome $\mathbf{s} = \tilde{W}_{\mathcal{N}} H^T$ using Protocol 19.5, the error correction procedure is conducted $(n-k) = n - 2t - 1$ times; this error correction is needed since dishonest parties may submit forged versions of $C_{i,j}$ when s_i is reconstructed in Step 4. Second, in order to identify the error vector $\mathbf{e}$ from the syndrome $\mathbf{s}$, the error correction procedure is used once more; here, only the error vector is identified. Furthermore, it should be noted that the error correction capabilities of the first and second procedures are different. Since the tuple $(\tilde{W}_{j,1}, \ldots, \tilde{W}_{j,n})$ is distributed in Step 2 using the VSS with parameters $(t+1, n)$, the sequence $(C_{i,1}, \ldots, C_{i,n})$ of shares of s_i in Step 3 is a codeword of the Reed–Solomon code with parameters $(t+1, n)$. Thus, the first error correction procedure works as long as $t < n/3$. On the other hand, the tuple $(W_1, \ldots, W_n)$ comprised the evaluations of polynomial $f(x) = f_1(x) f_2(x)$ of degree less than or equal to $2t$ at different points. Thus, $(W_1, \ldots, W_n)$ is a codeword of the Reed–Solomon code with parameters $(k, n) = (2t+1, n)$. Since we assumed $t < n/4$, the number of correctable errors is

$$\frac{n-k}{2} = \frac{n-2t-1}{2} \geq t. \tag{19.10}$$

In other words, the success of the second error correction procedure relies on the condition $t < n/4$.

In Step 7, for j with $e_j \neq 0$, the parties set $\bar{W}_{j,i} = \tilde{W}_j - e_j$, which coincides with W_j. This can be interpreted as W_j being distributed by the constant polynomial[5] $\bar{f}_j(x) = W_j$, which satisfies $\bar{f}_j(x) \in \mathbb{P}_t(W_j)$. On the other hand, since an honest

[5] This notation is for convenience of analysis.

Protocol 19.6 Committed multiplication

Input: For $i \in \mathcal{N}$, shares $(W_{1,i}, W_{2,i})$ owned by $\mathcal{P}_i$, where, for $a = 1, 2$, there exists a polynomial $f_a(x) \in \mathbb{P}_t(S_a)$ such that $f_a(\alpha_i) = W_{a,i}$ for every $i \in \mathcal{N}$

Output: For $i \in \mathcal{N}$, share $\bar{W}_i$ owned by $\mathcal{P}_i$, where there exists a polynomial $\bar{f}(x) \in \mathbb{P}_t(S_1 \cdot S_2)$ such that $\bar{f}(\alpha_i) = \bar{W}_i$ for every $i \in \mathcal{N}$

1. For $i \in \mathcal{N}$, $\mathcal{P}_i$ locally computes $W_i = W_{1,i} \cdot W_{2,i}$.
2. For $j \in \mathcal{N}$, $\mathcal{P}_j$ distributes share $\tilde{W}_{j,i}$ of $\tilde{W}_j$ to $\mathcal{P}_i$ using the VSS (Protocol 19.4).
3. In order to compute the syndrome

$$\mathbf{s} = (s_1, \ldots, s_{n-k}) = \tilde{W}_{\mathcal{N}} H^T,$$

   where H is the parity check matrix of the Reed–Solomon code with parameter $(k, n) = (2t + 1, n)$, the parties invoke the committed linear combination (Protocol 19.5) $n - k$ times so that $\mathcal{P}_j$ will obtain a share $C_{i,j}$ of s_i for $1 \leq i \leq n - k$.
4. The parties exchange the shares obtained in the previous step, and reconstruct s_i by using the reconstruction phase of Protocol 11.3 for each $1 \leq i \leq n - k$.
5. By using the syndrome decoding (cf. Section 11.5), the parties identify the error vector $\mathbf{e} = (\tilde{W}_{\mathcal{N}} - W_{\mathcal{N}})$ from the syndrome reconstructed in the previous step.
6. For $i, j \in \mathcal{N}$, if $e_j \neq 0$, $\mathcal{P}_i$ sends $\tilde{W}_{j,i}$ to the other parties.
7. For $i \in \mathcal{N}$, upon receiving $(\tilde{W}_{j,1}, \ldots, \tilde{W}_{j,n})$ for j with $e_j \neq 0$, $\mathcal{P}_i$ reconstructs $\tilde{W}_j$ and sets $\bar{W}_{j,i} = \tilde{W}_j - e_j$; for j with $e_j = 0$, $\mathcal{P}_i$ sets $\bar{W}_{j,i} = \tilde{W}_{j,i}$.
8. For $i \in \mathcal{N}$, $\mathcal{P}_i$ locally computes $\bar{W}_i = \sum_{j=1}^{n} \beta_j \bar{W}_{j,i}$.

$\mathcal{P}_j$ distributes $\tilde{W}_j = W_j$ correctly in Step 2, there exists a polynomial $\bar{f}_j(x) \in \mathbb{P}_t(W_j)$ such that $\bar{W}_{j,i} = \bar{f}_j(\alpha_i)$ for honest $\mathcal{P}_i$. Thus, by (19.9), we can verify that the output $\bar{W}_i$ of honest $\mathcal{P}_i$ satisfies the desired property.

Regarding the security, we can verify that additional information obtained by dishonest parties in each step can be simulated locally by themselves; here, the ideal committed multiplication function is such that it has the same inputs and outputs as Protocol 19.6 without any exchange of information among the parties. In Step 1, there is no additional information. As we have discussed in the proof of Theorem 19.14, information exchanged during the execution of VSS can be simulated by dishonest parties locally by themselves; furthermore, shares distributed to dishonest parties from honest parties are independent of the W_i owned by honest parties. Thus, the information obtained by dishonest parties in Step 2 can be simulated. In Step 3, the committed linear combination is invoked. Since computing the shares of $\mathbf{s} = \tilde{W}_{\mathcal{N}} H^T$ is a local operation, it can be simulated. Regarding the exchange of the shares to reconstruct $\mathbf{s} = \tilde{W}_{\mathcal{N}} H^T$, we can verify that it can be simulated as follows. Note that, since $\mathbf{s} = \mathbf{e}H^T$ and $\mathbf{e}$ is known to dishonest parties, they can compute $\mathbf{s}$ by themselves; then, using $\mathbf{s}$ and t shares of s, other shares of $\mathbf{s}$ can be reconstructed. In Step 6, since $e_j \neq 0$ implies $\mathcal{P}_j$ is dishonest and $\tilde{W}_{j,i}$ is created by $\mathcal{P}_j$, it can be simulated. Finally, since Steps 7 and 8 are local operations, there is no additional information leaked.

We conclude this section with a remark on the assumption $t < n/4$. As we have discussed above, the assumption $t < n/4$ is used to derive (19.10), the error correcting capability of the procedure in Step 5. Under the weaker assumption $t < n/3$, the redundancy of the Reed–Solomon code is not enough, and the error vector $\mathbf{e}$ cannot be identified in Step 5. To avoid this problem, we resort to a procedure termed "degree reduction." Since this procedure is rather involved and requires a few technical preparations, we do not provide it here; interested readers can see the references provided in Section 19.6.

19.5 Beyond an Honest Majority

19.5.1 Securely Computable Functions from Scratch

For a passive adversary, we have seen in Theorem 19.3 that any function can be securely computed from scratch as long as the honest majority assumption holds. When the honest majority assumption is not satisfied, i.e., the number of dishonest parties is $t \geq n/2$, the possibility of secure computation depends on the functions to be computed.

Let $\mathcal{N} = \mathcal{A}_1 \cup \mathcal{A}_2$ be a partition of the set of parties such that $\max\{|\mathcal{A}_1|, |\mathcal{A}_2|\} \leq t$; since $t \geq n/2$, such a partition exists. Then, we can regard the n-party function $f(x_1, \ldots, x_n)$ as the two-party function $f(x_{\mathcal{A}_1}, x_{\mathcal{A}_2})$. Suppose that, for n-party computation with $\mathcal{P}_1, \ldots, \mathcal{P}_n$, there exists a protocol that securely computes $f(x_1, \ldots, x_n)$ for a t-passive adversary. Then, for two-party computation with parties $\mathcal{P}_1'$ and $\mathcal{P}_2'$, where $\mathcal{P}_1'$ plays the roles of $(\mathcal{P}_i : i \in \mathcal{A}_1)$ and $\mathcal{P}_2'$ plays the roles of $(\mathcal{P}_i : i \in \mathcal{A}_2)$, we can construct a two-party protocol that securely computes the two-party function $f(x_{\mathcal{A}_1}, x_{\mathcal{A}_2}) = f(x_1, \ldots, x_n)$. However, as we have seen in Section 12.3, a function is not securely computable in the two-party setting if its refinement is not an interactive function. Thus, if there exists a partition $\mathcal{N} = \mathcal{A}_1 \cup \mathcal{A}_2$ with $\max\{|\mathcal{A}_1|, |\mathcal{A}_2|\} \leq t$ such that the refinement $f_{\mathtt{ref}}(x_{\mathcal{A}_1}, x_{\mathcal{A}_2})$ is not interactive,[6] then we have a contradiction to the assumption that f is securely computable in the n-party computation. By this observation, we have the following necessary condition for securely computable functions.

THEOREM 19.16 *When the honest majority assumption is not satisfied, i.e., $t \geq n/2$, a function $f(x_1, \ldots, x_n)$ is securely computable for a t-passive adversary only if $f_{\mathtt{ref}}(x_{\mathcal{A}_1}, x_{\mathcal{A}_2})$ is an interactive function for every partition $\mathcal{N} = \mathcal{A}_1 \cup \mathcal{A}_2$ with* $\max\{|\mathcal{A}_1|, |\mathcal{A}_2|\} \leq t$.

[6] Note that the refinement of this function may depend on the partition $\{\mathcal{A}_1, \mathcal{A}_2\}$.

Table 19.1 $f(x_1, x_2, x_3)$ in Example 19.17

x_1	x_2	x_3	$f(x_1, x_2, x_3)$
$*$	1	1	1
2	$*$	2	2
3	3	$*$	3
others			$x_{\mathcal{N}}$

Table 19.2 $f(x_{\mathcal{A}_1}, x_{\mathcal{A}_2})$ in Example 19.17

x_1 \ x_2 / x_3	1 / 1	1 / 2	1 / 3	2 / 1	2 / 2	2 / 3	3 / 1	3 / 2	3 / 3
1	1	$x_{\mathcal{N}}$	$x_{\mathcal{N}}$	$x_{\mathcal{N}}$	$x_{\mathcal{N}}$	$x_{\mathcal{N}}$	$x_{\mathcal{N}}$	$x_{\mathcal{N}}$	$x_{\mathcal{N}}$
2	1	2	$x_{\mathcal{N}}$	$x_{\mathcal{N}}$	2	$x_{\mathcal{N}}$	$x_{\mathcal{N}}$	2	$x_{\mathcal{N}}$
3	1	$x_{\mathcal{N}}$	$x_{\mathcal{N}}$	$x_{\mathcal{N}}$	$x_{\mathcal{N}}$	$x_{\mathcal{N}}$	3	3	3

One might be curious whether the necessary condition in Theorem 19.16 is also sufficient. In fact, the condition is not sufficient as is illustrated in the following example.

EXAMPLE 19.17 Consider $n = 3$ party computation with $t = 2$, and let $f(x_1, x_2, x_3)$ be the function on $\{1, 2, 3\}^3$ described in Table 19.1, where $*$ indicates an arbitrary symbol in $\{1, 2, 3\}$ and $x_{\mathcal{N}} = (x_1, x_2, x_3)$ denotes that the output is the same as the input tuple. For the partition $(\mathcal{A}_1, \mathcal{A}_2) = (\{1\}, \{2, 3\})$, the function table of $f(x_{\mathcal{A}_1}, x_{\mathcal{A}_2})$ is described in Table 19.2. We can easily verify that this is an interactive function. By symmetry, $f(x_{\mathcal{A}_1}, x_{\mathcal{A}_2})$ is an interactive function for other partitions $(\mathcal{A}_1, \mathcal{A}_2) = (\{2\}, \{1, 3\})$ and $(\mathcal{A}_1, \mathcal{A}_2) = (\{3\}, \{1, 2\})$ as well. Thus, this function satisfies the necessary condition in Theorem 19.16. However, the function f is not securely computable.

To see this, consider a transcript τ^i produced by a protocol after i rounds. In a similar manner as for the proof of Theorem 12.21, we can prove inductively that, if the protocol is secure, then $\Pr(\tau^i | x_1, x_2, x_3)$ takes the same value for every (x_1, x_2, x_3) if the protocol constitutes a (perfectly) secure implementation of f. Specifically, by symmetry, assume that τ_1 is sent by $\mathcal{P}_1$. Then, for $(x_1, x_2, x_3) \neq (x_1', x_2', x_3')$,

$$\begin{aligned}\Pr(\tau_1 | x_1, x_2, x_3) &= \Pr(\tau_1 | x_1, 1, 1) \\ &= \Pr(\tau_1 | x_1', 1, 1) \\ &= \Pr(\tau_1 | x_1', x_2', x_3'),\end{aligned}$$

where the first and the third identity hold since τ_1 is computed by $\mathcal{P}_1$ locally and the second identity holds since information about $\mathcal{P}_1$'s observation is leaked to $\mathcal{P}_2$ and $\mathcal{P}_3$ otherwise. Similarly, assuming that $\Pr(\tau^{i-1} | x_1, x_2, x_3)$ does not depend on (x_1, x_2, x_3), we can use the steps above to show that $\Pr(\tau_i | x_1, x_2, x_3, \tau^{i-1})$ does not depend on (x_1, x_2, x_3) either. Hence, $\Pr(\tau^i | x_1, x_2, x_3)$ is the same for every (x_1, x_2, x_3), whereby the protocol will have the same output distribution for every input. Thus, this protocol cannot compute the function f correctly.

Even though the necessary condition in Theorem 19.16 is not sufficient in general, for Boolean functions, we can show that the condition is sufficient as well. In fact, using Corollary 12.22, we can derive a more explicit necessary and sufficient condition.

THEOREM 19.18 *When $t \geq n/2$, a Boolean function $f(x_1, \ldots, x_n)$ is securely computable for a t-passive adversary if and only if there exist Boolean functions $f_i \colon \mathcal{X}_i \to \{0,1\}$ for $i = 1, \ldots, n$ such that*

$$f(x_1, \ldots, x_n) = f_1(x_1) \oplus \cdots \oplus f_n(x_n).$$

Proof The "if" part follows from Proposition 19.7. To prove the "only if" part, note that for any partition $\mathcal{N} = \mathcal{A}_1 \cup \mathcal{A}_2$ with $\max\{|\mathcal{A}_1|, |\mathcal{A}_2|\} \leq t$, Corollary 12.22 implies that

$$f(x_1, \ldots, x_n) = f_{\mathcal{A}_1}(x_{\mathcal{A}_1}) \oplus f_{\mathcal{A}_2}(x_{\mathcal{A}_2}) \tag{19.11}$$

for some Boolean functions $f_{\mathcal{A}_1}$ and $f_{\mathcal{A}_2}$. By using (19.11), we first prove that, for each $i \in \mathcal{N}$, there exists a partition $\mathcal{X}_i = \mathcal{B}_i \cup \mathcal{C}_i$ satisfying the following: for every $b_i \neq b_i' \in \mathcal{B}_i$, $c_i \in \mathcal{C}_i$, and $x_j \in \mathcal{X}_j$ for $j \in \mathcal{N} \backslash \{i\}$,

$$f(b_i, x_{\mathcal{N}\backslash\{i\}}) = f(b_i', x_{\mathcal{N}\backslash\{i\}}), \tag{19.12}$$
$$f(b_i, x_{\mathcal{N}\backslash\{i\}}) \neq f(c_i, x_{\mathcal{N}\backslash\{i\}}). \tag{19.13}$$

For notational convenience, we prove the case with $i = 1$, but other cases can be proved similarly. Let $\mathcal{S} = \{2, 3, \ldots, \lceil n/2 \rceil\}$ and $\mathcal{T} = \{\lceil n/2 \rceil + 1, \ldots, n-1\}$. If the function f does not depend on x_1, then we can set $\mathcal{B}_1 = \mathcal{X}_1$ and $\mathcal{C}_1 = \emptyset$. Otherwise, there must exist $y_1 \neq y_1' \in \mathcal{X}_1$, $y_{\mathcal{S}} \in \mathcal{X}_{\mathcal{S}}$, $y_{\mathcal{T}} \in \mathcal{X}_{\mathcal{T}}$, and $y_n \in \mathcal{X}_n$ such that

$$f(y_1, y_{\mathcal{S}}, y_{\mathcal{T}}, y_n) \neq f(y_1', y_{\mathcal{S}}, y_{\mathcal{T}}, y_n).$$

Let

$$\mathcal{B}_1 = \big\{b_1 \in \mathcal{X}_1 : f(b_1, y_{\mathcal{S}}, y_{\mathcal{T}}, y_n) = f(y_1, y_{\mathcal{S}}, y_{\mathcal{T}}, y_n)\big\},$$
$$\mathcal{C}_1 = \big\{c_1 \in \mathcal{X}_1 : f(c_1, y_{\mathcal{S}}, y_{\mathcal{T}}, y_n) = f(y_1', y_{\mathcal{S}}, y_{\mathcal{T}}, y_n)\big\}.$$

Since f is Boolean, it is a partition $\mathcal{X}_1 = \mathcal{B}_1 \cup \mathcal{C}_1$. From the definition of $\mathcal{B}_1$ and $\mathcal{C}_1$, it holds that

$$f(b_1, y_{\mathcal{S}}, y_{\mathcal{T}}, y_n) \neq f(c_1, y_{\mathcal{S}}, y_{\mathcal{T}}, y_n) \tag{19.14}$$

for every $b_1 \in \mathcal{B}_1$ and $c_1 \in \mathcal{C}_1$. We claim that, for every $(x_{\mathcal{S}}, x_{\mathcal{T}}, x_n) \in \mathcal{X}_{\mathcal{N}\backslash\{1\}}$,

$$f(b_1, x_{\mathcal{S}}, x_{\mathcal{T}}, x_n) \neq f(c_1, x_{\mathcal{S}}, x_{\mathcal{T}}, x_n). \tag{19.15}$$

To prove by contradiction, assume that (19.15) does not hold for some $(x_{\mathcal{S}}, x_{\mathcal{T}}, x_n)$.

Then, by applying (19.11) for partition $\mathcal{A}_1 = \{1\} \cup \mathcal{S}$ and $\mathcal{A}_2 = \mathcal{T} \cup \{n\}$, we have

$$\begin{aligned} f_{\{1\}\cup\mathcal{S}}(b_1, x_\mathcal{S}) \oplus f_{\mathcal{T}\cup\{n\}}(x_\mathcal{T}, x_n) &= f(b_1, x_\mathcal{S}, x_\mathcal{T}, x_n) \\ &= f(c_1, x_\mathcal{S}, x_\mathcal{T}, x_n) \\ &= f_{\{1\}\cup\mathcal{S}}(c_1, x_\mathcal{S}) \oplus f_{\mathcal{T}\cup\{n\}}(x_\mathcal{T}, x_n), \end{aligned}$$

which implies

$$f_{\{1\}\cup\mathcal{S}}(b_1, x_\mathcal{S}) = f_{\{1\}\cup\mathcal{S}}(c_1, x_\mathcal{S}).$$

Furthermore, by applying (19.11) for the same partition, we have

$$\begin{aligned} f(b_1, x_\mathcal{S}, y_\mathcal{T}, y_n) &= f_{\{1\}\cup\mathcal{S}}(b_1, x_\mathcal{S}) \oplus f_{\mathcal{T}\cup\{n\}}(y_\mathcal{T}, y_n) \\ &= f_{\{1\}\cup\mathcal{S}}(c_1, x_\mathcal{S}) \oplus f_{\mathcal{T}\cup\{n\}}(y_\mathcal{T}, y_n) \\ &= f(c_1, x_\mathcal{S}, y_\mathcal{T}, y_n). \end{aligned}$$

Next, by applying (19.11) for partition $\mathcal{A}_1 = \{1\} \cup \mathcal{T}$ and $\mathcal{S} \cup \{n\}$, we have

$$\begin{aligned} f_{\{1\}\cup\mathcal{T}}(b_1, y_\mathcal{T}) \oplus f_{\mathcal{S}\cup\{n\}}(x_\mathcal{S}, y_n) &= f(b_1, x_\mathcal{S}, y_\mathcal{T}, y_n) \\ &= f(c_1, x_\mathcal{S}, y_\mathcal{T}, y_n) \\ &= f_{\{1\}\cup\mathcal{T}}(c_1, y_\mathcal{T}) \oplus f_{\mathcal{S}\cup\{n\}}(x_\mathcal{S}, y_n), \end{aligned}$$

which implies

$$f_{\{1\}\cup\mathcal{T}}(b_1, y_\mathcal{T}) = f_{\{1\}\cup\mathcal{T}}(c_1, y_\mathcal{T}).$$

Furthermore, by applying (19.11) for the same partition, we have

$$\begin{aligned} f(b_1, y_\mathcal{S}, y_\mathcal{T}, y_n) &= f_{\{1\}\cup\mathcal{T}}(b_1, y_\mathcal{T}) \oplus f_{\mathcal{S}\cup\{n\}}(y_\mathcal{S}, y_n) \\ &= f_{\{1\}\cup\mathcal{T}}(c_1, y_\mathcal{T}) \oplus f_{\mathcal{S}\cup\{n\}}(y_\mathcal{S}, y_n) \\ &= f(c_1, y_\mathcal{S}, y_\mathcal{T}, y_n), \end{aligned}$$

which contradicts (19.14). Thus, (19.15) must hold for every $(x_\mathcal{S}, x_\mathcal{T}, x_n)$. Now, for every $b_1 \neq b_1' \in \mathcal{B}_1$, $c_1 \in \mathcal{C}_1$, and $x_j \in \mathcal{X}_j$ for $j \in \mathcal{N}\backslash\{1\}$, we have

$$\begin{aligned} f(b_1, x_{\mathcal{N}\backslash\{1\}}) &\neq f(c_1, x_{\mathcal{N}\backslash\{1\}}), \\ f(b_1', x_{\mathcal{N}\backslash\{1\}}) &\neq f(c_1, x_{\mathcal{N}\backslash\{1\}}). \end{aligned}$$

Since f is Boolean, we have (19.12) and (19.13).

By using the partitions constructed above, let

$$f_i(x_i) = \begin{cases} 0, & \text{if } x_i \in \mathcal{B}_i, \\ 1, & \text{if } x_i \in \mathcal{C}_i. \end{cases}$$

Fix arbitrary $(b_1, \ldots, b_n) \in \mathcal{B}_1 \times \cdots \times \mathcal{B}_n$. Without loss of generality, assume that[7]

$$f(b_1, \ldots, b_n) = 0.$$

[7] Otherwise, we can flip one of the functions $f_1, \ldots, f_n$ in their definition.

For any $x_{\mathcal{N}} = (x_1, \ldots, x_n)$, let $\mathcal{J} = \mathcal{J}(x_{\mathcal{N}}) = \{i : x_i \in \mathcal{B}_i\}$. Then, by applying (19.12) for each coordinate in $\mathcal{J}$ and (19.13) for each coordinate in $\mathcal{J}^c$ one by one, respectively, we have

$$\begin{aligned} f(x_{\mathcal{J}}, x_{\mathcal{J}^c}) &= f(b_{\mathcal{J}}, x_{\mathcal{J}^c}) \\ &= f(b_{\mathcal{J}}, b_{\mathcal{J}^c}) \oplus \left(\bigoplus_{j \in \mathcal{J}^c} 1 \right) \\ &= \bigoplus_{j \in \mathcal{J}^c} f_j(x_j) \\ &= f_1(x_1) \oplus \cdots \oplus f_n(x_n). \end{aligned}$$

□

19.5.2 Secure Computation Using OT

When the honest majority assumption is not satisfied, since secure computation from scratch is not possible in general, we consider secure computation using some additional resource. As we have seen in Section 12.6, any two-party function can be securely computed if oblivious transfer (OT) is available. In this section, we show that any multiparty function can be securely computed against a t-passive adversary for $t \le n-1$ if OTs are available between every pair of parties.

For simplicity, we restrict to binary vectors as inputs and Boolean functions. The protocol proceeds along the same line as that in Section 19.2, i.e., it consists of the preprocessing by secret sharing, addition (modulo-sum), multiplication (AND function), and finally reconstruction.

Preprocessing

Upon observing $S_i \in \mathbb{F}_2$, $\mathcal{P}_i$ randomly generates $R_j \in \mathbb{F}_2$ for $j \in \mathcal{N}\backslash\{i\}$, distributes share $W_{i,j} = R_j$ to $\mathcal{P}_j$ for $j \in \mathcal{N}\backslash\{i\}$, and keeps

$$W_{i,i} = S_i \oplus \bigoplus_{j \in \mathcal{N}\backslash\{i\}} R_j$$

so that

$$\bigoplus_{j \in \mathcal{N}} W_{i,j} = S_i.$$

For more than one input, $\mathcal{P}_i$ repeats the same procedure for the other inputs.

Addition

For $i \in \mathcal{N}$, suppose that $\mathcal{P}_i$ has shares $(W_{i,j} : 1 \le j \le m)$, where the shares satisfy $\oplus_{i=1}^n W_{j,i} = S_j$. The parties would like to compute the addition

$$S = \bigoplus_{j=1}^m S_j.$$

To that end, each party computes

$$W_i = \bigoplus_{j=1}^{m} W_{j,i}.$$

Then, we have

$$S = \bigoplus_{i=1}^{n} W_i.$$

Multiplication

For $i \in \mathcal{N}$, suppose that $\mathcal{P}_i$ has shares $(W_{1,i}, W_{2,i})$ of S_1 and S_2, where $\oplus_{i=1}^{n} W_{j,i} = S_j$ for $j = 1, 2$. The parties would like to compute the product $S = S_1 \cdot S_2$. Note that

$$S_1 \cdot S_2 = \left(\bigoplus_{i=1}^{n} W_{1,i} \cdot W_{2,i} \right) \oplus \left(\bigoplus_{i \neq k} W_{1,i} \cdot W_{2,k} \right). \tag{19.16}$$

Each party $\mathcal{P}_i$ can locally compute $W_{1,i} \cdot W_{2,i}$. On the other hand, we use OT between $\mathcal{P}_i$ and $\mathcal{P}_k$ to compute $V_{i,k}$ and $V'_{i,k}$ satisfying

$$V_{i,k} \oplus V'_{i,k} = W_{1,i} \cdot W_{2,k}. \tag{19.17}$$

Specifically, $\mathcal{P}_i$ randomly generates $V_{i,k} \in \mathbb{F}_2$. Then, $\mathcal{P}_i$ and $\mathcal{P}_k$ invoke the OT with $\mathcal{P}_i$'s inputs $(V_{i,k}, V_{i,k} \oplus W_{1,i})$ and $\mathcal{P}_k$'s input $W_{2,k}$. Finally, $\mathcal{P}_k$ sets $V'_{i,k}$ as the output of the OT. We can verify that $V_{i,k}$ and $V'_{i,k}$ created in this manner satisfy (19.17).

Finally, for $i \in \mathcal{N}$, $\mathcal{P}_i$ sets

$$W_i = W_{1,i} \cdot W_{2,i} \oplus \bigoplus_{k \neq i} V_{i,k} \oplus \bigoplus_{k \neq i} V'_{k,i}.$$

From (19.16) and (19.17), we can verify that

$$S_1 \cdot S_2 = \bigoplus_{i=1}^{n} W_i.$$

Reconstruction

In order to compute a function securely using a decomposition of the function into a binary arithmetic circuit, the parties apply the addition and multiplication operations sequentially. When computation of all the operations is done, $\mathcal{P}_i$ has a share Z_i of the output Z.[8] Then, the parties exchange their shares, and each party reconstructs the output by

$$Z = \bigoplus_{i=1}^{n} Z_i.$$

[8] When the output of the function is not binary, the parties have sequences of shares.

Security analysis

Since the outputs of the parties in each basic operation are the shares of the output of that basic operation, it is not difficult to see that the final shares are the shares the output of the circuit.

In order to verify the security of the protocol, for a given subset $\mathcal{A} \subset \mathcal{N}$ with $|\mathcal{A}| \leq n-1$, we need to verify that the extra view $V_{\mathcal{A}}$ obtained by dishonest parties $(\mathcal{P}_i : i \in \mathcal{A})$ during the protocol can be simulated by the parties themselves in the ideal model. Note that the extra view $V_{\mathcal{A}}$ consists of shares observed by $(\mathcal{P}_i : i \in \mathcal{A})$ in the preprocessing phase, the output of OTs, and the final shares exchanged in the reconstruction phase.

In the preprocessing phase, since all shares are created by an (n, n)-threshold secret sharing scheme, the shares observed by dishonest parties are uniformly distributed irrespective of honest parties' inputs.

In the addition operation, no new information is provided to dishonest parties. In the multiplication operation, when $i, k \in \mathcal{A}$, then $V_{i,k}$ and $V'_{i,k}$ can be generated by themselves from what they have observed so far. When $i \in \mathcal{A}$ and $k \notin \mathcal{A}$, $V_{i,k}$ can be generated by $\mathcal{P}_i$. When $i \notin \mathcal{A}$ and $k \in \mathcal{A}$, since $V_{i,k}$ is randomly generated and is not available to dishonest parties, $V'_{i,k}$ is uniformly distributed irrespective of other variables created so far in the protocol. Thus, the information obtained in the multiplication phases can be simulated.

In the final reconstruction phase, when $|\mathcal{A}| = n - 1$, Z_i for $i \notin \mathcal{A}$ can be simulated as

$$Z_i = Z \oplus \bigoplus_{j \in \mathcal{A}} Z_j. \tag{19.18}$$

When $|\mathcal{A}| < n-1$, we add a subset $\tilde{\mathcal{A}}$ with $|\mathcal{A} \cup \tilde{\mathcal{A}}| = n-1$; since $(Z_i : i \in \mathcal{N})$ are shares of Z obtained using an (n, n)-threshold scheme, $(Z_i : i \in \tilde{\mathcal{A}})$ are uniformly distributed, and the remaining share Z_i for $i \notin \mathcal{A} \cup \tilde{\mathcal{A}}$ can be simulated as in (19.18).

In summary, we have the following result.

THEOREM 19.19 *When OT is available between every pair of parties, any function can be securely computed for a t-passive adversary when* $t \leq n - 1$.

19.6 References and Additional Reading

The concept of secure computation was introduced by Yao in [360]. A general construction for computationally secure multiparty computation was given in [144]. General constructions for information-theoretically secure multiparty computation were given in [25, 63].

The concept of verifiable secret sharing was introduced in [66], where a computationally secure scheme was proposed. An information-theoretically secure verifiable secret sharing scheme using bivariate polynomial was introduced in [25].

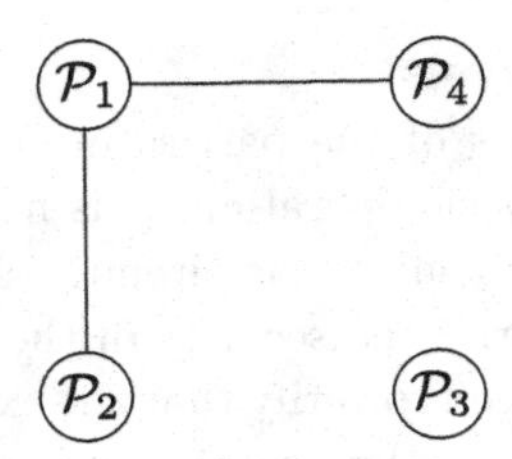

Figure 19.2 A description of the network in Problem 19.6. The OT is available between connected vertices.

The multiplication protocol described in Section 19.2 is from [77], which is somewhat different from the protocol in [25]; see [77, Section 5.6] for details. Other useful expositions of multiparty secure computation are provided in [6, 114].

For an active adversary, in order to simplify the exposition, we have presented a protocol that is secure only for $t < n/4$ dishonest parties. In order to handle the threshold $t < n/3$, we need to incorporate an additional procedure of reducing the degree of the polynomial in the middle of the protocol. For full treatment of secure computation against $t < n/3$ dishonest parties, see [77] or [6]. For $t < n/3$, it is possible to construct a protocol with perfect security. Furthermore, even the availability of broadcast can be waived since, as we have seen in Chapter 18, the broadcast protocol can be constructed from scratch for $t < n/3$ dishonest parties. When the number of dishonest parties is $t \geq n/3$, it is necessary to assume the availability of broadcast. By considering approximate security, it is still possible to realize secure computation against $t < n/2$ dishonest parties [281].

Beyond the honest majority assumption (Section 19.5), the complete characterization of securely computable Boolean functions (Theorem 19.18) was obtained in [68]. More recently, the complete characterization of securely computable ternary functions was obtained in [206]. Characterization for general functions is an open problem; the counterexample in Example 19.17 for the sufficiency of the partition condition in Theorem 19.16 is from [67]. In general, it is not possible to realize secure computation if the dishonest parties constitute a majority. However, if the parties in the network share pairwise OT, then it is possible to realize secure computation for any function (Theorem 19.19); this result on the OT network is from [157]. In fact, it is not necessary that every pair of parties share OT; some pairs of parities can create OT by cooperation of other parties, i.e., a full OT network can be created from certain partial OT networks (see also Problem 19.6). A graph-theoretic characterization of construct on of a full OT network was obtained in [208].

Problems

19.1 Prove that the honest majority requirement in Theorem 19.3 is also a necessary condition for every function to be securely computable.

HINT Consider a partition $\mathcal{N} = \mathcal{A}_1 \cup \mathcal{A}_2$ such that $|\mathcal{A}_i| \leq \frac{n}{2}$ for $i = 1, 2$. If the number of dishonest parties t satisfies $t \geq \frac{n}{2}$, a protocol is secure even if all the parties in either $\mathcal{A}_1$ or $\mathcal{A}_2$ are dishonest. In order to derive a contradiction to the result in Section 12.3, consider a two-party protocol where each party simulates the role of the parties in $\mathcal{A}_1$ and $\mathcal{A}_2$.

19.2 In order to compute the summation function once, consider the following protocol: $\mathcal{P}_1$ randomly generates $U \in \mathbb{F}_q$ and sends $\Pi_1 = X_1 + U$ to $\mathcal{P}_2$; $\mathcal{P}_i$ sends $\Pi_i = \Pi_{i-1} + X_i$ to $\mathcal{P}_{i+1}$ for $2 \leq i \leq n$, where $\mathcal{P}_{n+1} = \mathcal{P}_1$; $\mathcal{P}_1$ computes $\Pi_n - U$ and broadcasts it to the other parties. Verify that this protocol is a secure implementation of the summation function for $t = 1$ passive adversary.

COMMENT Compared to the secret sharing based protocol, this type of simple protocol saves on the amount of randomness and communication used in the protocol [97, 219].

19.3 Prove that the honest supermajority requirement in Theorem 19.15 is also a necessary condition for every function to be securely computable in the standard network.

HINT Note that the Byzantine agreement of Chapter 18 can be regarded as a special case of multiparty secure computation.

19.4 Verify that Protocol 19.6 works even if dishonest parties do not cooperate to send $\tilde{W}_{j,i}$ in Step 6.

HINT After "time-out," the messages from dishonest parties can be set as a prescribed constant.

19.5 Fill in the detail of the proof of the claim in Example 19.17.

19.6 In a four-party network, suppose that OT is available between $\mathcal{P}_1$ and $\mathcal{P}_2$ and between $\mathcal{P}_1$ and $\mathcal{P}_4$; cf. Figure 19.2.[9] When $t = 2$ parties may be (passively) dishonest, construct a four-party protocol to implement OT from $\mathcal{P}_3$ to $\mathcal{P}_1$.

HINT Consider the following protocol: $\mathcal{P}_3$ creates shares of K_0 and K_1 by using a $(3, 3)$-threshold scheme, respectively, and sends each share to $\mathcal{P}_1$, $\mathcal{P}_2$, and $\mathcal{P}_4$; then, $\mathcal{P}_1$ asks $\mathcal{P}_2$ and $\mathcal{P}_4$ to send shares of K_B via OT.

[9] Since the direction of OT can be flipped (see Problem 13.3), it suffices to assume availability of OT in one direction for each pair.

Appendix Solutions to Selected Problems

Solutions to Problems in Chapter 8

Problem 8.6

1. The proof proceeds along the line of the proof of Theorem 8.5. For fixed message m, let $\mathcal{D}_m$ be the set of tag-observation pairs (t, y) for which the receiver accepts the message m. Since the MAC is ε-sound, we must have $\mathrm{P}_{TY}(\mathcal{D}_m) \geq 1 - \varepsilon$. Let $\overline{X}$ be an independent copy of X, and consider an impersonation attack that sends $(m, \overline{T})$, $\overline{T} = \varphi(m, \overline{X})$. Then, we must have $\mathrm{P}_{\overline{T}Y}(\mathcal{D}_m) \leq P_{\mathrm{I}}$. Here, let us consider hypothesis testing between P_{TY} and $\mathrm{P}_{\overline{T}Y} = \mathrm{P}_{\overline{T}} \times \mathrm{P}_Y$. Then, the above bounds imply

$$\begin{aligned}
-\log P_{\mathrm{I}} &\leq -\log \beta_\varepsilon(\mathrm{P}_{TY}, \mathrm{P}_{\overline{T}Y}) \\
&\leq \frac{D(\mathrm{P}_{TY} \| \mathrm{P}_{\overline{T}Y}) + h(\varepsilon)}{1 - \varepsilon} \\
&= \frac{I(T \wedge Y) + h(\varepsilon)}{1 - \varepsilon}.
\end{aligned} \tag{A.1}$$

Next, for arbitrarily fixed $m \neq \tilde{m}$, let $\overline{X}$ be distributed according to $\mathrm{P}_{X|T}(\cdot|T)$, where $T = \varphi(m, X)$. Consider a substitution attack that replaces (m, T) with $(\tilde{m}, \overline{T})$, where $\overline{T} = \varphi(\tilde{m}, \overline{X})$. Then, the success probability of this substitution attack must satisfy $\mathrm{P}_{\overline{T}Y}(\mathcal{D}_{\tilde{m}}) \leq P_{\mathrm{S}}$.

On the other hand, hypothetically consider the legitimate tag $\tilde{T} = \varphi(\tilde{m}, X)$ for $\tilde{m}$. Then, since the MAC is ε-sound, we must have $\mathrm{P}_{\tilde{T}Y}(\mathcal{D}_{\tilde{m}}) \geq 1 - \varepsilon$. Here, let us consider hypothesis testing between $\mathrm{P}_{\tilde{T}Y}$ and $\mathrm{P}_{\overline{T}Y}$. Then, the above bounds imply

$$\begin{aligned}
-\log P_{\mathrm{S}} &\leq -\log \beta_\varepsilon(\mathrm{P}_{\tilde{T}Y}, \mathrm{P}_{\overline{T}Y}) \\
&\leq \frac{D(\mathrm{P}_{\tilde{T}Y} \| \mathrm{P}_{\overline{T}Y}) + h(\varepsilon)}{1 - \varepsilon} \\
&\leq \frac{D(\mathrm{P}_{XYT} \| \mathrm{P}_{\overline{X}YT}) + h(\varepsilon)}{1 - \varepsilon} \\
&= \frac{I(X \wedge Y|T) + h(\varepsilon)}{1 - \varepsilon},
\end{aligned} \tag{A.2}$$

where the identity follows because $\mathrm{P}_{\overline{X}YT}$ is of the form $\mathrm{P}_{X|T}\mathrm{P}_{YT}$. Finally, by noting $T \multimap X \multimap Y$, which is true since the tag is computed only from X and

m, we have $I(X \wedge Y|T) = I(X \wedge Y) - I(T \wedge Y)$; this together with (A.1) and (A.2) imply (8.17).

2. From the definition of λ_ε, it is apparent that the scheme is ε-sound. Note that $\mathrm{P}_Y(y) \le \mathrm{P}_{Y|X}(y|x)\, 2^{-\lambda_\varepsilon}$ for every $(x, y) \in \mathcal{S}(\lambda_\varepsilon)$. For any $\overline{X}$ that is independent of the legitimate parties' observation $((X_0, X_1), (Y_0, Y_1))$, we have

$$\begin{aligned} \Pr\left((\overline{X}, Y_m) \in \mathcal{S}(\lambda_\varepsilon)\right) &= \sum_{x,y} \mathrm{P}_{\overline{X}}(x)\, \mathrm{P}_Y(y)\, \mathbf{1}[(x,y) \in \mathcal{S}(\lambda_\varepsilon)] \\ &\le \sum_{x,y} \mathrm{P}_{\overline{X}}(x)\, \mathrm{P}_{Y|X}(y|x)\, 2^{-\lambda_\varepsilon} \\ &= 2^{-\lambda_\varepsilon}. \end{aligned}$$

Thus, we have $P_{\mathrm{I}} \le 2^{-\lambda_\varepsilon}$.

For $\tilde{m} \neq m$, let $\tilde{X}$ be any random variable generated from X_m. Since $\tilde{X}$ is independent of $(X_{\tilde{m}}, Y_{\tilde{m}})$, in a similar manner as above, we can derive

$$\Pr\left((\tilde{X}, Y_{\tilde{M}}) \in \mathcal{S}(\lambda_\varepsilon), \tilde{M} \neq m\right) \le 2^{-\lambda_\varepsilon},$$

which implies $P_{\mathrm{S}} \le 2^{-\lambda_\varepsilon}$.

Solutions to Problems in Chapter 12

Problem 12.4

Since Z is a function of $Z_{\mathtt{ref}}$, we have

$$\mathtt{mcf}((X_1, Z_{\mathtt{ref}}), (X_2, Z_{\mathtt{ref}})) \equiv \mathtt{mcf}((X_1, Z_{\mathtt{ref}}, Z), (X_2, Z_{\mathtt{ref}}, Z)).$$

Since Z is a common function of (X_1, Z) and (X_2, Z), we have

$$\begin{aligned} \mathtt{mcf}((X_1, Z_{\mathtt{ref}}, Z), (X_2, Z_{\mathtt{ref}}, Z)) &\equiv \mathtt{mcf}((X_1, Z), (X_2, Z)) \\ &\equiv Z_{\mathtt{ref}}. \end{aligned}$$

Solutions to Problems in Chapter 13

Problem 13.4

Let $V_1 = g_1(X_1)$, where $g_1(x_1) = [x_1]$. Since

$$\mathrm{P}_{X_2|V_1X_1}(x_2|v_1, x_1) = \mathrm{P}_{X_2|V_1X_1}(x_2|v_1, x_1')$$

whenever $g_1(x_1) = g(x_1') = v_1$, we have $X_1 \multimap V_1 \multimap X_2$, i.e., V_1 is a sufficient statistic. On the other hand, for any sufficient statistic $U = g(X_1)$, since

$$\mathrm{P}_{X_2|X_1}(x_2|x_1) = \mathrm{P}_{X_2|UX_1}(x_2|u, x_1) = \mathrm{P}_{X_2|UX_1}(x_2|u, x_1') = \mathrm{P}_{X_2|X_1}(x_2|x_1')$$

whenever $g(x_1) = g(x_1') = u$, there is a function g' such that $g(x_1) = g'(g(x_1))$.

Problem 13.5

Let $(\mathcal{X}_1 \cup \mathcal{X}_2, \mathcal{E})$ be the bipartite graph such that $(x_1, x_2) \in \mathcal{E}$ if and only if. $\mathrm{P}_{X_1X_2}(x_1, x_2) > 0$, and let $\mathcal{C}$ be the set of all connected components. By noting $X_2 -\!\!\circ\!\!- X_1 -\!\!\circ\!\!- V$, we have

$$\mathrm{P}_{V|X_1X_2}(v|x_1, x_2) = \mathrm{P}_{V|X_1}(v|x_1) = \mathrm{P}_{V|X_1X_2}(v|x_1, x_2')$$

whenever $\mathrm{P}_{X_1X_2}(x_1, x_2)\,\mathrm{P}_{X_1X_2}(x_1, x_2') > 0$. Similarly, $X_1 -\!\!\circ\!\!- X_2 -\!\!\circ\!\!- V$ implies $\mathrm{P}_{V|X_1X_2}(v|x_1, x_2) = \mathrm{P}_{V|X_1X_2}(v|x_1', x_2)$ whenever $\mathrm{P}_{X_1X_2}(x_1, x_2)\,\mathrm{P}_{X_1X_2}(x_1', x_2) > 0$. Thus, we have $\mathrm{P}_{V|X_1X_2}(v|x_1, x_2) = \mathrm{P}_{V|X_1X_2}(v|x_1', x_2')$ whenever (x_1, x_2) and (x_1', x_2') are in the same connected component of the graph. By noting the explicit characterizations of the maximum common function (cf. Lemma 12.11) and the minimum sufficient statistic (cf. Problem 13.4), it is not difficult to see that $\mathtt{mss}(V|X_1, X_2)$ is a function of $\mathtt{mcf}(X_1, X_2)$, and also $\mathtt{mcf}(X_1, X_2)$ is a sufficient statistic of V given (X_1, X_2).

Problem 13.9

1. By the property of interactive communication (cf. Lemma 10.7), we have

$$\begin{aligned} I(Y_1 \wedge Y_2|\mathtt{mcf}(Y_1, Y_2)) &\geq I(Y_1, \Pi \wedge Y_2, \Pi|\mathtt{mcf}(Y_1, Y_2), \Pi) \\ &\geq I(Y_1, \Pi \wedge Y_2, \Pi|\mathtt{mcf}((Y_1, \Pi), (Y_2, \Pi))), \end{aligned}$$

where the second inequality follows from the fact that $(\mathtt{mcf}(Y_1, Y_2), \Pi)$ is a common function of $((Y_1, \Pi), (Y_2, \Pi))$ and Lemma 13.25.

To prove the second inequality, let $V_1 = \mathtt{mss}(Y_2|Y_1)$. Since $Y_1 -\!\!\circ\!\!- V_1 -\!\!\circ\!\!- Y_2$, we have $(Y_1, \Pi) -\!\!\circ\!\!- (V_1, \Pi) -\!\!\circ\!\!- (Y_2, \Pi)$ by the property of interactive communication. This means that (V_1, Π) is a sufficient statistic for (Y_2, Π) given (Y_1, Π). Thus, $\mathtt{mss}(Y_2, \Pi|Y_1, \Pi)$ is a function of (V_1, Π), and we have

$$\begin{aligned} H(V_1|Y_2) &\geq H(V_1|Y_2, \Pi) \\ &\geq H(\mathtt{mss}(Y_2, \Pi|Y_1, \Pi)|Y_2, \Pi). \end{aligned}$$

We can prove the third inequality similarly.

2. Denoting $V_0 = \mathtt{mcf}((Y_1', Z_1), (Y_2', Z_2))$, the conditions $Y_1' -\!\!\circ\!\!- Z_1 -\!\!\circ\!\!- Z_2$ and $Z_1 -\!\!\circ\!\!- Z_2 -\!\!\circ\!\!- Y_2'$ imply $V_0 -\!\!\circ\!\!- Z_1 -\!\!\circ\!\!- Z_2$ and $Z_1 -\!\!\circ\!\!- Z_2 -\!\!\circ\!\!- V_0$, i.e., the double Markov chain. Thus, the maximum common function $\tilde{V}_0 = \mathtt{mcf}(Z_1, Z_2)$ satisfies the Markov chain $(Z_1, Z_2) -\!\!\circ\!\!- \tilde{V}_0 -\!\!\circ\!\!- V_0$ (cf. Problem 13.5), and we have

$$\begin{aligned} I(Y_1', Z_1 \wedge Y_2', Z_2|V_0) &\geq I(Z_1 \wedge Z_2|V_0) \\ &= I(Z_1, \tilde{V}_0 \wedge Z_2, \tilde{V}_0|V_0) \\ &\geq I(Z_1 \wedge Z_2|\tilde{V}_0, V_0) \\ &= I(Z_1 \wedge Z_2|\tilde{V}_0). \end{aligned}$$

To prove the second inequality, note that $V_1 = \mathtt{mss}(Y_2', Z_2|Y_1', Z_1)$ satisfies $(Y_1', Z_1) -\!\!\circ\!\!- V_1 -\!\!\circ\!\!- (Y_2', Z_2)$, which implies $(Y_1', Z_1) -\!\!\circ\!\!- V_1 -\!\!\circ\!\!- Z_2$. Thus, V_1 is a sufficient statistic of Z_2 given (Y_1', Z_1). We also note that $\mathtt{mss}(Z_2|Y_1', Z_2) = \mathtt{mss}(Z_2|Z_1)$ when $Y_1' \multimap Z_1 \multimap Z_2$ (cf. Problem 13.4). By these observations, we have

$$\begin{aligned} H(V_1|Y_2', Z_2) &\geq H(\mathtt{mss}(Z_2|Y_1', Z_1)|Y_2', Z_2) \\ &= H(\mathtt{mss}(Z_2|Z_1)|Y_2', Z_2) \\ &= H(\mathtt{mss}(Z_2|Z_1)|Z_2), \end{aligned}$$

where the last identity follows from $Z_1 \multimap Z_2 \multimap Y_2'$ and the fact that $\mathtt{mss}(Z_2|Z_1)$ is a function of Z_1.

3. We have already proved the first inequality in Lemma 13.27. Let us use the same notation as in Lemma 13.27. Let $V_1 = \mathtt{mss}(Y_2|Y_1)$. Since $Y_1 \multimap V_1 \multimap Y_2$, by noting that V_1 is a function of Y_1 and by applying Lemma 13.25 and Lemma 13.27, we have

$$\begin{aligned} 0 &= I(Y_1, V_1 \wedge Y_2, V_1|V_1) \\ &\geq I(Y_1, V_1 \wedge Y_2, V_1|\mathtt{mcf}((Y_1, V_1), (Y_2, V_1))) \\ &\geq I(Y_1', V_1 \wedge Y_2', V_1|\mathtt{mcf}((Y_1', V_1), (Y_2', V_1))), \end{aligned}$$

which implies $Y_1' \multimap \mathtt{mcf}((Y_1', V_1), (Y_2', V_1)) \multimap Y_2'$. This means that $\mathtt{mcf}$ $((Y_1', V_1), (Y_2', V_1))$ is a sufficient statistics of Y_2' given Y_1'.[1] Thus, we have

$$\begin{aligned} H(V_1|Y_2) &\geq H(V_1|Y_2') \\ &\geq H(\mathtt{mcf}((Y_1', V_1), (Y_2', V_1))|Y_2') \\ &\geq H(\mathtt{mss}(Y_2'|Y_1')|Y_2'). \end{aligned}$$

The third inequality can be proved similarly.

Problem 13.13

We have already established the equivalence of Claim 1 and Claim 2 in Theorem 12.37. Claim 5 follows from Problem 13.12, and Claim 3 and Claim 4 are special cases of Claim 5 (Claim 4 is a special case of Lemma 13.27 as well). Conversely, Claim 4 implies that, for the uniform distribution on $\mathcal{X}_1 \times \mathcal{X}_2$, $X_1 \multimap Z_{\mathtt{ref}} \multimap X_2$ holds, where $Z_{\mathtt{ref}} = f_{\mathtt{ref}}(X_1, X_2)$. Thus, the conditional distribution $\mathrm{P}_{X_1X_2|Z_{\mathtt{ref}}}$ factorizes as $\mathrm{P}_{X_1|Z_{\mathtt{ref}}} \times \mathrm{P}_{X_2|Z_{\mathtt{ref}}}$, which means that $f_{\mathtt{ref}}$ is rectangular.

Solutions to Problems in Chapter 15

Problem 15.3

Set the dummy input as $X_1' = 1$ if $\Pi_1 =$ yes and $X_1' = 3$ if $\Pi_1 =$ no. Then, it is not difficult to check that $\mathcal{P}_2$'s output Z_2 in the real protocol and the ideal protocol coincide. When $\Pi_1 =$ no, Π_2 can be computed from $Z_1' = f(X_1', X_2)$.

[1] Note that V_1 is a function of Y_1'.

Problem 15.5

Set the dummy input $X_1' = \Pi_1$.

Problem 15.6

1. For the attack in which dishonest $\mathcal{P}_2$ sends $\Pi_2 = 0$, the output Z_1 of $\mathcal{P}_1$ is 0 irrespective of the value of X_1. Then, $\mathcal{P}_2$'s dummy input must be $X_2' = 0$. However, $\Pi_1 = X_1$ cannot be simulated unless $X_2' = 1$.
2. For the attack in which dishonest $\mathcal{P}_2$ inputs 1 to the bit OT and sends $\Pi_2 = 0$, the output Z_1 of $\mathcal{P}_1$ is 0 irrespective of the value of X_1. Then, $\mathcal{P}_2$'s dummy input must be $X_2' = 0$. However, the output of the bit OT, X_1, cannot be simulated unless $X_2' = 1$.

Problem 15.7

1. For the attack in which dishonest $\mathcal{P}_2$ sends $\mathcal{P}_2 = \Pi_1 = X_1$, the output Z_1 of $\mathcal{P}_1$ is 0 irrespective of the value of X_1. But, unless X_1 is constant, $f(X_1, X_2')$ cannot be constant 0 for any dummy input X_2' satisfying $X_2' -\!\!\circ\!\!- X_2 -\!\!\circ\!\!- X_1$.
2. When $\mathcal{P}_1$ is dishonest, set the dummy input as $X_1' = \Pi_1$, which satisfies $X_1' -\!\!\circ\!\!- X_1 -\!\!\circ\!\!- X_2$ by the hiding property of the bit commitment. Then, $\mathcal{P}_2$'s output Z_2 in the real protocol and the ideal protocol coincide. The output $\Pi_2 = X_2$ of the commitment can be simulated from $Z_1' = X_1' \oplus X_2$ and X_1'.

   When $\mathcal{P}_2$ is dishonest, set the dummy input as $X_2' = \Pi_2$, which satisfies $X_2' -\!\!\circ\!\!- X_2 -\!\!\circ\!\!- X_1$ by the binding property of the bit commitment. Then, $\mathcal{P}_1$'s output Z_1 in the real protocol and the ideal protocol coincide. The transcript $\Pi_1 = X_1$ can be simulated from $Z_2' = X_1 \oplus X_2'$ and X_2'.

Problem 15.9

When $\mathcal{P}_1$ is dishonest, set the dummy input as $A_0' = K_{1,0}$, $A_1' = K_{1,1} \oplus K_{2,0}$, $A_3' = K_{1,1} \oplus K_{2,1} \oplus K_{3,0}$, and $A_4' = K_{1,1} \oplus K_{2,1} \oplus K_{3,1}$.

When $\mathcal{P}_2$ is dishonest, set the dummy input B' and simulate $(\hat{K}_1, \hat{K}_2, \hat{K}_3)$ using the dummy output Z_2' of the $\binom{4}{1}$-OT as follows.

- If $C_1 = 0$, set $B' = 0$ and $\hat{K}_1 = Z_2'$, and sample $\hat{K}_2$ and $\hat{K}_3$ uniformly.
- If $(C_1, C_2) = (1, 0)$, set $B' = 1$, sample $\hat{K}_1$ and $\hat{K}_3$ uniformly, and set $\hat{K}_2 = Z_2' \oplus \hat{K}_1$.
- If $(C_1, C_2, C_3) = (1, 1, 0)$, set $B' = 2$, sample $\hat{K}_1$ and $\hat{K}_2$ uniformly, and set $\hat{K}_3 = Z_2' \oplus \hat{K}_1 \oplus \hat{K}_2$.
- If $(C_1, C_2, C_3) = (1, 1, 1)$, set $B' = 3$, sample $\hat{K}_1$ and $\hat{K}_2$ uniformly, and set $\hat{K}_3 = Z_2' \oplus \hat{K}_1 \oplus \hat{K}_2$.

Solutions to Problems in Chapter 18

Problem 18.3

When $\mathcal{P}_1$ and $\mathcal{P}_2$ are honest, it is apparent that Protocol 18.8 is 0-correct and 0-secret since $\mathcal{P}_3$ is not involved in the signing phase. To analyze transferability, note that dishonest $\mathcal{P}_1$ may send an arbitrary polynomial $\tilde{\sigma}(\eta) = e_0 + e_1\eta$ as a signature for message x. If $\tilde{\sigma}(\eta) = \sigma(\eta) := \Phi(\eta, x)$, then $(x, \tilde{\sigma}(\eta))$ is accepted by both $\mathcal{P}_2$ and $\mathcal{P}_3$; thus this event is not counted as a failure of the transfer. We claim that, if $\tilde{\sigma}(\eta) \neq \sigma(\eta)$, then the probability that $(x, \tilde{\sigma}(\eta))$ is accepted by $\mathcal{P}_2$ is less than $\frac{1}{q}$, which implies that Protocol 18.8 is $\frac{1}{q}$-transferable. To verify the claim, note that $(x, \tilde{\sigma}(\eta))$ is accepted only if

$$e_0 + e_1 V_2 = (K_{00} + K_{01}x) + (K_{10} + K_{11}x)V_2.$$

Since $\tilde{\sigma}(\eta) \neq \sigma(\eta)$ implies $(e_0, e_1) \neq (K_{00}+K_{01}x, K_{10}+K_{11}x)$ and V_2 is uniformly distributed on $\mathbb{F}_q$ and is independent of (e_0, e_1) and $(K_{00} + K_{01}x, K_{10} + K_{11}x)$, by the same reasoning as the construction of the UHF (cf. Lemma 4.9), we have[2]

$$\Pr\left(e_0 + e_1 V_2 = (K_{00} + K_{01}x) + (K_{10} + K_{11}x)V_2\right) \leq \frac{1}{q}.$$

Now, we analyze the success probability of forging by $\mathcal{P}_2$. Upon observing $(V_2, g_2(\xi))$ and $(x, \sigma(\eta))$, $\mathcal{P}_2$ forges $(\tilde{x}, \tilde{\sigma}(\eta))$ with $\tilde{x} \neq x$. It will be accepted by $\mathcal{P}_3$ if $\tilde{\sigma}(V_3) = g_3(\tilde{x}) = \Phi(V_3, \tilde{x})$. Fix a realization $(V_2, V_3) = (v_2, v_3)$; note that $v_2 \neq v_3$. For given $g_2(\xi)$, $\sigma(\eta)$, and $\tilde{\sigma}(\eta)$, the probability of acceptance by $\mathcal{P}_3$, i.e., $\tilde{\sigma}(v_3) = \Phi(v_3, \tilde{x})$, is given by

$$\frac{|\{\Phi(\eta,\xi) : \Phi(v_2,\xi) = g_2(\xi), \Phi(\eta,x) = \sigma(\eta), \Phi(v_3,\tilde{x}) = \tilde{\sigma}(v_3)\}|}{|\{\Phi(\eta,\xi) : \Phi(v_2,\xi) = g_2(\xi), \Phi(\eta,x) = \sigma(\eta)\}|}. \tag{A.3}$$

Since $v_2 \neq v_3$, by Lemma 19.10, $\Phi(\eta, \xi)$ is uniquely determined if $g_2(\xi)$ and $g_3(\xi)$ are given. Note that $g_3(x) = \sigma(v_3)$ and $g_3(\tilde{x}) = \tilde{\sigma}(v_3)$ for $x \neq \tilde{x}$; since $g_3(\xi)$ is degree 1, it is uniquely determined given these constraints. Thus, the number of $\Phi(\eta, \xi)$ satisfying the constraints in the numerator of (A.3) is 1. On the other hand, in the denominator, $g_3(\xi)$ has degree of freedom 1, and thus $\Phi(\eta, \xi)$ has degree of freedom 1 as well. Thus, (A.3) is given by $\frac{1}{q}$, which implies that Protocol 18.8 is $\frac{1}{q}$-unforgeable.

Problem 18.4

By the definition of the minimum sufficient statistic (cf. Definition 13.13), $K_1 = \mathtt{mss}(Y_2|Y_1)$ satisfies $Y_1 \multimap K_1 \multimap Y_2$. Note also that $K_1 = g(Y_1)$ is a function of Y_1. Thus, by letting

$$\mathrm{P}_{\bar{Y}_1|Y_1}(\bar{y}_1|y_1) := \mathrm{P}_{Y_1|K_1}(\bar{y}_1|g(y_1)),$$

[2] As we can find from the analysis, it is not crucial that V_3 is concealed from $\mathcal{P}_1$; if V_3 is known to $\mathcal{P}_1$, then the upper bound on the success probability of $\mathcal{P}_1$ increases to $1/(q-1)$.

we have

$$\begin{aligned} \mathrm{P}_{\bar{Y}_1 Y_2}(\bar{y}_1, y_2) &= \sum_{y_1} \mathrm{P}_{Y_1|K_1}(\bar{y}_1|g(y_1))\, \mathrm{P}_{Y_1 Y_2}(y_1, y_2) \\ &= \sum_{k_1} \mathrm{P}_{Y_1|K}(\bar{y}_1|k_1)\, \mathrm{P}_{K_1 Y_2}(k_1, y_2) \\ &= \mathrm{P}_{Y_1 Y_2}(y_1, y_2). \end{aligned}$$

Now, by noting that $K_1 \multimap Y_2 \multimap Y_3$, let

$$\mathrm{P}_{\hat{Y}_1|Y_2}(\hat{y}_1|y_2) := \sum_{k_1} \mathrm{P}_{Y_1|K_1}(\hat{y}_1|k_1)\, \mathrm{P}_{K_1|Y_2}(k_1|y_2).$$

Then, we have

$$\begin{aligned} \mathrm{P}_{\hat{Y}_1 Y_2 Y_3}(\hat{y}_1, y_2, y_3) &= \sum_{k_1} \mathrm{P}_{Y_1|K_1}(\hat{y}_1|k_1)\, \mathrm{P}_{K_1|Y_2}(k_1|y_2)\, \mathrm{P}_{Y_2 Y_3}(y_2, y_3) \\ &= \sum_{k_1} \mathrm{P}_{Y_1|K_1}(\hat{y}_1|k_1)\, \mathrm{P}_{K_1 Y_2 Y_3}(k_1, y_2, y_3) \\ &= \sum_{y_1} \mathrm{P}_{Y_1|K_1}(\hat{y}_1|g(y_1))\, \mathrm{P}_{Y_1 Y_2 Y_3}(y_1, y_2, y_3) \\ &= \sum_{y_1} \mathrm{P}_{\bar{Y}_1|Y_1}(\hat{y}_1|y_1)\, \mathrm{P}_{Y_1 Y_2 Y_3}(y_1, y_2, y_3) \\ &= \mathrm{P}_{\bar{Y}_1 Y_2 Y_3}(\hat{y}_1, y_2, y_3). \end{aligned}$$

Problem 18.5

Let us consider an arbitrary noninteractive protocol as follows: for a given message $x \in \mathcal{X}$, $\mathcal{P}_1$ sends $\Pi_1 = \Pi_1(x, Y_1)$ to $\mathcal{P}_2$, and $\mathcal{P}_2$ outputs $Z_2(\Pi_1, Y_2) \in \mathcal{X} \cup \{\perp\}$ in the signing phase. Then, $\mathcal{P}_2$ sends $\Pi_2 = \Pi_2(\Pi_1, Y_2)$ to $\mathcal{P}_3$, and $\mathcal{P}_3$ outputs $Z_3(\Pi_2, Y_3) \in \mathcal{X} \cup \{\perp\}$ in the transfer phase. By Problem 18.4, there exist $\mathrm{P}_{\bar{Y}_1|Y_1}$ and $\mathrm{P}_{\hat{Y}_1|Y_2}$ such that $\mathrm{P}_{\bar{Y}_1 Y_2} = \mathrm{P}_{Y_1 Y_2}$ and $\mathrm{P}_{\bar{Y}_1 Y_2 Y_3} = \mathrm{P}_{\hat{Y}_1 Y_2 Y_3}$. Let $\bar{\Pi}_1 = \Pi_1(x, \bar{Y}_1)$. Since $\mathrm{P}_{\bar{Y}_1 Y_2} = \mathrm{P}_{Y_1 Y_2}$, we have $\mathrm{P}_{\bar{\Pi}_1 Y_2} = \mathrm{P}_{\Pi_1 Y_2}$. Thus, ε_{c}-correctable implies

$$\begin{aligned} \Pr\left(Z_2(\bar{\Pi}_1, Y_2) = x\right) &= \Pr\left(Z_2(\Pi_1, Y_2) = x\right) \\ &\geq 1 - \varepsilon_{\mathrm{c}}. \end{aligned} \tag{A.4}$$

Since $\bar{Y}_1$ can be locally generated from Y_1, sending $\bar{\Pi}_1$ is a valid attack by $\mathcal{P}_1$. Thus, ε_{t}-transferability says that, if $Z_2(\bar{\Pi}_1, Y_2) \neq \perp$ and $\mathcal{P}_2$ transfers $\bar{\Pi}_2 = \Pi_2(\bar{\Pi}_1, Y_2)$ to $\mathcal{P}_3$, then we have

$$\Pr\left(Z_3(\bar{\Pi}_2, Y_3) \neq Z_2(\bar{\Pi}_1, Y_2), Z_2(\bar{\Pi}_1, Y_2) \neq \perp\right) \leq \varepsilon_{\mathrm{t}},$$

which implies

$$1 - \varepsilon_{\mathrm{t}} \leq \sum_{x' \in \mathcal{X}} \Pr\left(Z_3(\bar{\Pi}_2, Y_3) = Z_2(\bar{\Pi}_1, Y_2) = x'\right) + \Pr\left(Z_2(\bar{\Pi}_1, Y_2) = \perp\right)$$

$$\leq \Pr\left(Z_3(\bar{\Pi}_2, Y_3) = x\right) + \sum_{\substack{x' \in \mathcal{X} \cup \{\bot\}: \\ x' \neq x}} \Pr\left(Z_2(\bar{\Pi}_1, Y_2) = x'\right)$$

$$= \Pr\left(Z_3(\bar{\Pi}_2, Y_3) = x\right) + \Pr\left(Z_2(\bar{\Pi}_1, Y_2) \neq x\right)$$

$$\leq \Pr\left(Z_3(\bar{\Pi}_2, Y_3) = x\right) + \varepsilon_c, \tag{A.5}$$

where the last inequality follows from (A.4). Since $\hat{Y}_1$ can be locally generated from Y_2, when $\mathcal{P}_1$'s message is $x' \neq x$, sending $\hat{\Pi}_2 = \Pi_2(\Pi_1(x, \hat{Y}_1), Y_2)$ is a valid attack by $\mathcal{P}_2$; furthermore, since $\mathrm{P}_{\bar{Y}_1 Y_2 Y_3} = \mathrm{P}_{\hat{Y}_1 Y_2 Y_3}$, we have $\mathrm{P}_{\bar{\Pi}_2 Y_3} = \mathrm{P}_{\hat{\Pi}_2 Y_3}$. Thus, ε_u-unforgeability implies

$$\Pr\left(Z_3(\bar{\Pi}_2, Y_3) = x\right) = \Pr\left(Z_3(\hat{\Pi}_2, Y_3) = x\right) \leq \varepsilon_u. \tag{A.6}$$

By combining (A.5) and (A.6), we have the desired claim.

Problem 18.8

Since $Y_{1,x\cdot n+j} \neq Y_{2,x\cdot n+j}$, this protocol is 0-correct. It is 0-secret since it is noninteractive protocol. To verify the transferability, note that, if $\mathcal{P}_1$ wants to make $\mathcal{P}_3$ reject x, then $\mathcal{P}_1$ needs to send Π_1 such that $\Pi_{1,j} \neq Y_{1,x\cdot n+j}$ for at least δ fraction of $1 \leq j \leq n$; otherwise, (18.4) is satisfied. However, the probability that $\mathcal{P}_2$ will accept x with such Π_1 is bounded above by $2^{-n\delta}$. To verify the unforgeability, $(\tilde{\Pi}_1, \tilde{\Pi}_2)$ forged by $\mathcal{P}_2$ satisfies $(\tilde{\Pi}_1, \tilde{\Pi}_2) \multimap Y_2^{2n}[\bar{x}] \multimap (Y_1^{2n}[\bar{x}], Y_3^{2n}[\bar{x}])$, where $\bar{x} = x \oplus 1$. Thus, for each $1 \leq j \leq n$, the probability that $Y_{3,\bar{x}\cdot n+j} \notin \{\tilde{\Pi}_{1,j}, \tilde{\Pi}_{2,j}\}$ is bounded above by $1/2$, which implies that (18.4) is satisfied with exponentially small probability as $n \to \infty$.

Problem 18.11

Apparently, the protocol is 0-correct and 0-secret. To verify transferability, let us first explain the meaning of the three conditions in $\mathcal{P}_3$'s test. The conditions 1 and 2 check whether $\mathcal{P}_1$ is believable or not; if either 1 or 2 holds, $\mathcal{P}_3$ recognizes $\mathcal{P}_1$ as dishonest, and $\mathcal{P}_3$ accepts x irrespective of the value of Π_1; if both 1 and 2 hold, then $\mathcal{P}_1$ is not suspicious at this point, and $\mathcal{P}_3$ accepts x depending on the value of Π_1 and $\Pi_{3,x}$. When $\mathcal{P}_1$ is dishonest and would like to make $\mathcal{P}_3$ reject x, $\mathcal{P}_1$ needs to send Π_1 and $\Pi_{3,x}$ such that

$$\left|\{j \in \{1, \ldots, n\} : \Pi_{3,x,j} \neq Y_{1,x\cdot n+j} \text{ or } \Pi_{1,j} \neq Y_{1,x\cdot n+j}\}\right| > \delta n. \tag{A.7}$$

Otherwise, the condition 3 is satisfied. On the other hand, the event $\{Z_2 \neq \bot, Z_3 = \bot\}$ occurs only when $\Pi_{1,j} \neq Y_{2,x\cdot n+j}$ for every $1 \leq j \leq n$ and $\Pi_{3,x,j} \neq Y_{3,x\cdot n+j}$ for every $1 \leq j \leq n$. Since $Y_{2,x\cdot n+j} = Y_{3,x\cdot n+j}$ are uniformly distributed on $\{0,1,2\} \backslash \{Y_{1,x\cdot n+j}\}$, for Π_1 and $\Pi_{3,x}$ satisfying (A.7), the probability of the event $\{Z_2 \neq \bot, Z_3 = \bot\}$ is bounded above by $2^{-n\delta}$, which implies transferability.

When $\mathcal{P}_2$ is dishonest, $(\bar{x}, \tilde{\Pi}_1)$ forged by $\mathcal{P}_2$ can be accepted by $\mathcal{P}_3$ only by the condition 3. In this case, since $\Pi_{3,\bar{x},j} = Y_{1,\bar{x}\cdot n+j}$ is uniformly distributed on

$\{0,1,2\}\backslash\{Y_{2,\bar{x}\cdot n+j}\}$ for $1 \le j \le n$, the probability that $\tilde{\Pi}_{1,j} = \Pi_{3,\bar{x},j}$ for at least $1-\delta$ fraction of $1 \le j \le n$ converges to 0 as $n \to \infty$, which implies unforgeability.

Problem 18.12

The correctness follows since $\Pi_{i\to\mathcal{S}} = x_i$ for every $\mathcal{S}$. For the agreement, suppose that $\mathcal{P}_j$ and $\mathcal{P}_\ell$ are honest. If $z_j = a \in \{0,1\}$, then $\Pi_{i\to\mathcal{S}} = a$ for every $\mathcal{S}$ with $j \in \mathcal{S}$. In particular, this is true for $\mathcal{S} = \{j,\ell\}$. Thus, z_ℓ must be a or $*$, which implies the agreement.

Problem 18.14

First, let us verify the correctness. When the sender $\mathcal{P}_1$ is honest, an honest $\mathcal{P}_i$ receives $\Pi_{1\to i} = x_i$ in Step 1. Thus, by the decision rule, $\mathcal{P}_i$ outputs $z_i = x_i$.

Next, let us verify the agreement. Since $t_{\mathsf{c}} = 2 \ge 1 = t_{\mathsf{a}}$, the agreement follows from the correctness when $\mathcal{P}_1$ is honest. So, suppose that $\mathcal{P}_1$ is dishonest. Then, since $t_{\mathsf{a}} = 1$, all the receivers are honest. Thus, any two (honest) receivers $\mathcal{P}_j$ and $\mathcal{P}_\ell$ receive consistent messages $\Pi_{i\to j} = \Pi_{i\to\ell}$ in Step 2,[3] which implies $L_j[a] = L_\ell[a]$ for every $a \in \{0,*,1\}$. Thus, if $\Pi_{1\to j} = \Pi_{1\to\ell}$, $\mathcal{P}_j$ and $\mathcal{P}_\ell$ output the same value $z_j = z_\ell$.

Suppose that $\Pi_{1\to j} \ne \Pi_{1\to\ell}$. By the property of 4-GBR, $\Pi_{1\to j} \ne \Pi_{1\to\ell} \oplus 1$. Without loss of generality, assume $\Pi_{1\to j} \in \{0,1\}$.

- When $\Pi_{1\to j} = 0$, $\mathcal{P}_j$ outputs $z_j = 0$. In this case, $\Pi_{1\to\ell} = *$. By the property of GBR and since all the receivers are honest, $L_\ell[0] + L_\ell[*] = 3$. Thus, $\mathcal{P}_\ell$ outputs $z_\ell = 0$.
- When $\Pi_{1\to j} = 1$, $\mathcal{P}_j$ outputs $z_j = 1$. In this case, $\Pi_{1\to\ell} = *$. Since $L_\ell[1] \ge 1$ implies $L_\ell[0] + L_\ell[*] \le 2$, $\mathcal{P}_\ell$ outputs $z_\ell = 1$.

Consequently, the agreement is satisfied.

Problem 18.15

First, let us verify the correctness. We shall prove by the induction on the size of the set $\mathcal{S}$ and t_{a}. When $t_{\mathsf{a}} = 0$ or $|\mathcal{S}| = 3$, then the correctness is satisfied since each honest receiver just outputs the received message and the sender is honest. Assume that $\pi_{\mathsf{TTBR}}[t_{\mathsf{c}}, t_{\mathsf{a}}-1; \mathcal{S}\backslash\{i\}; \mathcal{P}_j]$ satisfies the correctness when there are at most t_{c} dishonest parties.[4] When the sender $\mathcal{P}_i$ is honest, an honest receiver $\mathcal{P}_\ell$ receives $\Pi_{i\to\ell} = m$ in Step 1. By the inductive hypothesis, $\mathcal{P}_\ell$ receives $\Pi_{j\to\ell} = \Pi_{i\to j} = m$ from an honest $\mathcal{P}_j$ in Step 3. Since there are $|\mathcal{S}| - t_{\mathsf{c}}$ honest parties, $L_\ell[m] \ge |\mathcal{S}| - t_{\mathsf{c}} - 1$ is satisfied. When $m = 0$, the decision rule immediately

[3] In fact, since 3-BR is invoked in Step 2, honest receivers get a consistent message anyway.
[4] When we prove the correctness, we only need the induction hypothesis for the correctness.

implies that $\mathcal{P}_\ell$ outputs $z_\ell = 0$. When $m = 1$, $L_\ell[1] \geq |\mathcal{S}| - t_c - 1$ together with $|\mathcal{S}| > t_c + t_a$ imply

$$\begin{aligned} L_\ell[0] + L_\ell[*] &= |\mathcal{S}| - L_\ell[1] - 1 \\ &\leq t_c \\ &< |\mathcal{S}| - t_a. \end{aligned}$$

Thus, $\mathcal{P}_\ell$ outputs $z_\ell = 1$.

Next, let us verify the agreement. Again, we shall prove by induction on the size of the set $\mathcal{S}$ and t_a. When $t_a = 0$ or $|\mathcal{S}| = 3$, the agreement is trivially satisfied since there is no dishonest party in the former case and the 3-BR is used in Step 1 in the latter case. When the sender is honest, the agreement follows from the correctness since $t_c \geq t_a$. So, assume that $t_a > 0$, $|\mathcal{S}| > 3$, and the sender $\mathcal{P}_i$ is dishonest.

As the inductive hypothesis, assume that $\pi_{\text{TTBR}}[t_c, t_a - 1; \mathcal{S}\backslash\{i\}; \mathcal{P}_j]$ satisfies the agreement (as we have already proved, it also satisfies the correctness). Since the sender $\mathcal{P}_i$ is dishonest, there are at most $t_a - 1$ dishonest parties among $\mathcal{S}\backslash\{i\}$. Thus, the inductive hypothesis guarantees that any two honest parties $\mathcal{P}_\ell$ and $\mathcal{P}_{\ell'}$ receive a consistent message $\Pi_{j\to\ell} = \Pi_{j\to\ell'}$ in Step 3, which implies $L_\ell[a] = L_{\ell'}[a]$ for every $a \in \{0, *, 1\}$. Thus, if $\Pi_{i\to\ell} = \Pi_{i\to\ell'}$, $\mathcal{P}_\ell$ and $\mathcal{P}_{\ell'}$ follow the same decision rule and output a consistent value $z_\ell = z_{\ell'}$. Suppose that $\Pi_{i\to\ell} \neq \Pi_{i\to\ell'}$. By the property of the GBR, $\Pi_{i\to\ell} \neq \Pi_{i\to\ell'} \oplus 1$. Without loss of generality, assume $\Pi_{i\to\ell} \in \{0, 1\}$.

- When $(\Pi_{i\to\ell}, \Pi_{i\to\ell'}) = (0, *)$, suppose that $\mathcal{P}_\ell$ outputs $z_\ell = 0$. In this case, $L_{\ell'}[0] = L_\ell[0] \geq |\mathcal{S}| - t_c - 1$. Furthermore, note that there are at least $|\mathcal{S}| - t_a$ honest parties among $\mathcal{S}\backslash\{i\}$, and $\mathcal{P}_{\ell'}$ receives correct $\Pi_{j\to\ell'} = \Pi_{i\to j}$ from honest $\mathcal{P}_j$ by the correctness of $\pi_{\text{TTBR}}[t_c, t_a - 1; \mathcal{S}\backslash\{i\}; \mathcal{P}_i]$. By the property of the GBR, $\Pi_{i\to j} \in \{0, *\}$. Thus, $L_{\ell'}[0] + L_{\ell'}[*] \geq |\mathcal{S}| - t_a$, which implies that $\mathcal{P}_{\ell'}$ outputs $z_{\ell'} = 0$.
- When $(\Pi_{i\to\ell}, \Pi_{i\to\ell'}) = (0, *)$, suppose that $\mathcal{P}_\ell$ outputs $z_\ell = 1$. In this case, $L_{\ell'}[0] = L_\ell[0] < |\mathcal{S}| - t_c - 1$. Thus, $\mathcal{P}_{\ell'}$ outputs $z_{\ell'} = 1$.
- When $(\Pi_{i\to\ell}, \Pi_{i\to\ell'}) = (1, *)$, $\mathcal{P}_\ell$ and $\mathcal{P}_{\ell'}$ output the same value since the decision rules are the same for $*$ and 1 and $L_\ell[a] = L_{\ell'}[a]$ for every $a \in \{0, *, 1\}$.

Problem 18.17

The "if" is Problem 18.16. To prove the "only if" part, we use a similar technique as for the proof of Theorem 18.11, the attack-coupling technique. Since the details are the same as the proof of Theorem 18.11, we only provide an outline of how to construct attack-coupling. Let us consider the scenarios described in Figure A.1. In the figure, the directed edges indicate private channels between parties, and the triangles indicate 3-BR among those parties. Assume that $\mathcal{P}_1'$ and $\mathcal{P}_1''$ have input 0 and 1, respectively. We prove a slightly stronger claim, i.e., $(n, t) = (4, 2)$

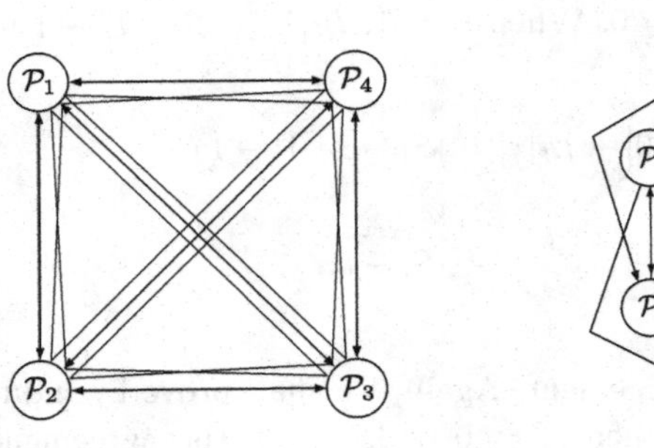

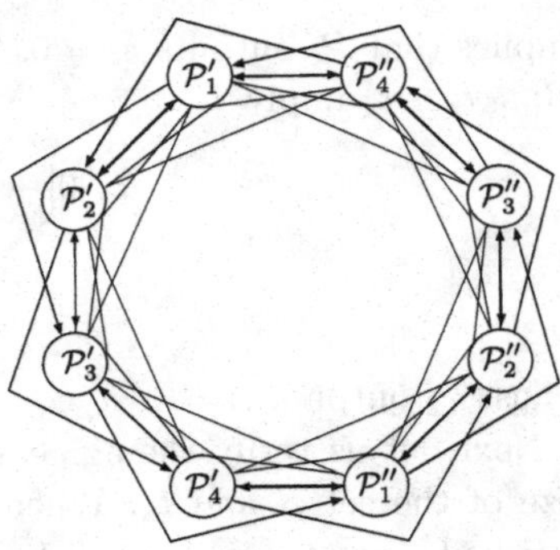

Figure A.1 Scenarios to prove the "only if" part of Problem 18.17.

is not possible if two adjacent parties are dishonest. We consider four scenarios in the eight-party network.

1. $\mathcal{P}_1'$ and $\mathcal{P}_2'$ are honest. Then, $\mathcal{P}_2'$ outputs 0 by the correctness.
2. $\mathcal{P}_2'$ and $\mathcal{P}_3'$ are honest. Then, $\mathcal{P}_3'$ outputs 0 by the agreement.
3. $\mathcal{P}_3'$ and $\mathcal{P}_4'$ are honest. Then, $\mathcal{P}_4'$ outputs 0 by the agreement.
4. $\mathcal{P}_1''$ and $\mathcal{P}_4'$ are honest. Then, $\mathcal{P}_4'$ outputs 1 by the correctness, which is a contradiction.

For general n and t with $t \geq n/2$, let us partition $\{1, \ldots, n\}$ into four sets $\mathcal{G}_1, \mathcal{G}_2, \mathcal{G}_3, \mathcal{G}_4$ so that $|\mathcal{G}_1 \cup \mathcal{G}_2| \leq t$, $|\mathcal{G}_2 \cup \mathcal{G}_3| \leq t$, $|\mathcal{G}_3 \cup \mathcal{G}_4| \leq t$, and $|\mathcal{G}_4 \cup \mathcal{G}_1| \leq t$. Such a partition is possible since $t \geq n/2$ and we only consider unions of two adjacent sets. Thus, by the simulation argument, this contradicts the fact that $(n, t) = (4, 2)$ is not possible when two adjacent parties are dishonest.

References

[1] R. Ahlswede and I. Csiszár, "Common randomness in information theory and cryptography – part I: Secret sharing," *IEEE Transactions on Information Theory,* vol. 39, no. 4, pp. 1121–1132, July 1993.

[2] R. Ahlswede and I. Csiszár, "On oblivious transfer capacity," in *Information Theory, Combinatorics, and Search Theory,* Lecture Notes in Computer Science, vol. 7777. Springer, 2013, pp. 145–166.

[3] R. Amiri, A. Abidin, P. Wallden, and E. Andersson, "Efficient unconditionally secure signatures using universal hashing," in *Proc. Applied Cryptography and Network Security,* Lecture Notes in Computer Science, vol. 10892. Springer, 2018, pp. 143–162.

[4] R. Arnon-Friedman, R. Renner, and T. Vidick, "Non-signaling parallel repetition using de Finetti reductions," *IEEE Transactions on Information Theory,* vol. 62, no. 3, pp. 1440–1457, November 2016.

[5] S. Arora and B. Barak, *Computational Complexity.* Cambridge University Press, 2009.

[6] G. Asharov and Y. Lindell, "A full proof of the BGW protocol for perfectly secure multiparty computation," *Journal of Cryptology,* vol. 30, no. 1, pp. 58–151, 2017.

[7] S. Asoodeh, M. Diaz, F. Alajaji, and T. Linder, "Information extraction under privacy constraints," *Information,* vol. 7, no. 1, March 2016.

[8] J. Aspnes, "Randomized protocols for asynchronous consensus," *Distributed Computing,* vol. 16, pp. 165–175, 2003.

[9] M. Atici and D. R. Stinson, "Universal hashing and multiple authentication," in *Advances in Cryptology – CRYPTO '96,* Lecture Notes in Computer Science, vol. 1109. Springer, 1996, pp. 16–30.

[10] K. Audenaert, "A sharp Fannes-type inequality for the von Neumann entropy," *Physical Review A,* vol. 40, pp. 8127–8136, 2007, arXiv: quant-ph/0610146.

[11] M. Backes, J. Müller-Quade, and D. Unruh, "On the necessity of rewinding in secure multiparty computation," in *Theory of Cryptography Conf. – TCC,* Lecture Notes in Computer Science, vol. 4392. Springer, 2007, pp. 157–173.

[12] C. Badertscher, U. Maurer, and B. Tackmann, "On composable security for digital signatures," in *Public-Key Cryptography – PKC 2018,* Lecture Notes in Computer Science, vol. 10769. Springer, 2018, pp. 494–523.

[13] L. A. Bassalygo and V. Burnashev, "Estimate for the maximal number of messages for a given probability of successful deception," *Problems of Information Transmission,* vol. 30, no. 2, pp. 129–134, 1994.

[14] D. Beaver, "Perfect privacy for two party protocols," *Technical Report TR-11-89,* Harvard University, 1989.

[15] D. Beaver, "Foundations of secure interactive computing," in *Advances in Cryptology – CRYPTO '91*, Lecture Notes in Computer Science, vol. 576. Springer, 1991, pp. 377–391.

[16] D. Beaver, "Precomputing oblivious transfer," in *Advances in Cryptology – CRYPTO '95*, Lecture Notes in Computer Science, vol. 963. Springer, 1995, pp. 97–109.

[17] D. Beaver, "Commodity-based cryptography," in *Proc. ACM Symposium on Theory of Computing – STOC '97.* ACM, 1997, pp. 446–455.

[18] A. Beimel and T. Malkin, "A quantitative approach to reductions in secure computation," in *Theory of Cryptography Conf. – TCC*, Lecture Notes in Computer Science, vol. 2951. Springer, 2004, pp. 238–257.

[19] J. Beirbrauer, T. Johansson, G. Kabatianskii, and B. Smeets, "On families of hash functions via geometric codes and concatenation," in *Advances in Cryptology – CRYPTO '93*, Lecture Notes in Computer Science, vol. 773. Springer, 1994, pp. 331–342.

[20] M. Bellare and P. Rogaway, "Random oracles are practical: a paradigm for designing efficient protocols," in *Proc. 1st ACM Conference on Computer and Communications Security – CCS '93.* ACM, 1993, pp. 62–73.

[21] M. Bellare and A. Sahai, "Non-malleable encryption: equivalence between two notions, and an indistinguishability-based characterization," in *Advances in Cryptology – CRYPTO '99*, Lecture Notes in Computer Science, vol. 1666. Springer, 1999, pp. 519–536.

[22] M. Bellare, S. Tessaro, and A. Vardy, "Semantic security for the wiretap channel," in *Advances in Cryptology – CRYPTO 2012*, Lecture Notes in Computer Science, vol. 7417. Springer, 2012, pp. 294–311.

[23] M. Ben-Or, "Another advantage of free choice: completely asynchronous protocols," in *Proc. Annual ACM Symposium on Principles of Distributed Computing.* ACM, 1983, pp. 27–30.

[24] M. Ben-Or, S. Goldwasser, J. Kilian, and A. Wigderson, "Multi-prover interactive proofs: how to remove intractability assumptions," in *Proc. ACM Symposium on Theory of Computing – STOC '88.* ACM, 1988, pp. 113–131.

[25] M. Ben-Or, S. Goldwasser, and A. Wigderson, "Completeness theorems for non-cryptographic fault-tolerant distributed computation," in *Proc. ACM Symposium on Theory of Computing – STOC '88.* ACM, 1988, pp. 1–10.

[26] J. Benaloh and J. Leichter, "Generalized secret sharing and monotone functions," in *Advances in Cryptology – CRYPTO '88*, Lecture Notes in Computer Science, vol. 403. Springer, 1990, pp. 27–35.

[27] C. H. Bennett, G. Brassard, C. Crépeau, and U. M. Maurer, "Generalized privacy amplification," *IEEE Transactions on Information Theory*, vol. 41, no. 6, pp. 1915–1923, November 1995.

[28] C. H. Bennett, G. Brassard, and J.-M. Robert, "Privacy amplification by public discussion," *SIAM Journal on Computing*, vol. 17, no. 2, pp. 210–229, 1988.

[29] T. Berger and R. Yeung, "Multiterminal source encoding with encoder breakdown," *IEEE Transactions on Information Theory*, vol. 35, no. 2, pp. 237–244, March 1989.

[30] D. J. Bernstein, "The Poly1305-AES message-authentication code," in *Fast Software Encryption*, H. Gilbert and H. Handschuh, Eds. Springer, 2005, pp. 32–49.

[31] J. Bierbrauer, "Construction of orthogonal arrays," *Journal of Statistical Planning and Inference*, vol. 56, pp. 39–47, 1996.

[32] J. Black, S. Halevi, H. Krawczyk, T. Krovetz, and P. Rogaway, "UMAC: fast and secure message authentication," in *Advances in Cryptology – CRYPTO '99*, Lecture Notes in Computer Science, vol. 1666. Springer, 1999, pp. 216–233.

[33] R. E. Blahut, "Hypothesis testing and information theory," *IEEE Transactions on Information Theory*, vol. 20, no. 4, pp. 405–417, July 1974.

[34] R. E. Blahut, *Cryptography and Secure Communication*. Cambridge University Press, 2014.

[35] G. R. Blakley, "Safeguarding cryptographic keys," in *International Workshop on Managing Requirements Knowledge*. IEEE, 1979, pp. 313–317.

[36] G. R. Blakley and C. Meadows, "Security of ramp schemes," in *Advances in Cryptology – CRYPTO '84*, Lecture Notes in Computer Science, vol. 196. Springer, 1985, pp. 242–268.

[37] M. Bloch and J. Barros, *Physical-Layer Security*. Cambridge University Press, 2011.

[38] M. Blum, "Coin flipping by telephone a protocol for solving impossible problems," *SIGACT News*, vol. 15, no. 1, pp. 23–27, January 1983.

[39] M. Blum and S. Micali, "How to generate cryptographically strong sequences of pseudo-random bits," *SIAM Journal on Computing*, vol. 13, no. 4, pp. 850–864, 1984.

[40] S. Boucheron, G. Lugosi, and P. Massart, *Concentration Inequalities – A Nonasymptotic Theory of Independence*. Oxford University Press, 2013.

[41] G. Brassard, C. Crépeau, and J.-M. Robert, "Information theoretic reduction among disclosure problems," in *27th IEEE Annual Symposium on Foundations of Computer Science, FOCS*. IEEE, 1986, pp. 168–173.

[42] G. Brassard, C. Crépeau, and M. Sántha, "Oblivious transfers and intersecting codes," *IEEE Transactions on Information Theory*, vol. 42, no. 6, pp. 1769–1780, November 1996.

[43] G. Brassard, C. Crépeau, and S. Wolf, "Oblivious transfer and privacy amplification," *Journal of Cryptology*, vol. 16, no. 4, pp. 219–237, 2003.

[44] M. Braverman and A. Rao, "Information equals amortized communication," in *52nd IEEE Annual Symposium on Foundations of Computer Science, FOCS*. IEEE, 2011, pp. 748–757.

[45] E. F. Brickell and D. Davenport, "On the classification of ideal secret sharing schemes," *Journal of Cryptology*, vol. 4, no. 2, pp. 123–134, 1991.

[46] C. Brzuska, M. Fischlin, H. Schröder, and S. Katzenbeisser, "Physically unclonable functions in the universal composition framework," in *Advances in Cryptology – CRYPTO 2011*, Lecture Notes in Computer Science, vol. 6841. Springer, 2011, pp. 51–70.

[47] H. Buhrman, S. Fehr, and C. Schaffner, "On the parallel repetition of multiplayer games: the no-signaling case," in *9th Conference on the Theory of Quantum Computation, Communication and Cryptography, TQC 2014*, Leibniz International Proceedings in Informatics, vol. 27. Leibniz Centrum für Informatik, 2014, pp. 24–35.

[48] C. Cachin, C. Crépeau, and J. Marcil, "Oblivious transfer with a memory-bounded receiver," in *39th IEEE Annual Symposium on Foundations of Computer Science, FOCS.* IEEE, 1998, pp. 493–502.

[49] C. Cachin, C. Crépeau, J. Marcil, and G. Savvides, "Information-theoretic interactive hashing and oblivious transfer to a storage-bounded receiver," *IEEE Transactions on Information Theory,* vol. 61, no. 10, pp. 5623–5635, 2015.

[50] F. P. Calmon, A. Makhdoumi, M. Médard, M. Varia, and K. R. Duffy, "Principal inertia components and applications," *IEEE Transactions on Information Theory,* vol. 63, no. 8, pp. 5011–5038, August 2017.

[51] F. P. Calmon, M. Varia, M. Médard, M. M. Christiansen, K. R. Duffy, and S. Tessaro, "Bounds on inference," in *51st Annual Allerton Conference on Communication, Control, and Computation (Allerton).* IEEE, 2013, pp. 567–574.

[52] R. Canetti, "Security and composition of multiparty cryptographic protocols," *Journal of Cryptography,* vol. 3, no. 1, pp. 143–202, 2000.

[53] R. Canetti, "Universally composable security: a new paradigm for cryptographic protocols," in *42nd Annual Symposium on Foundations of Computer Science, FOCS.* IEEE, 2001, pp. 136–145, Cryptology ePrint Archive, Report 2000/067.

[54] R. M. Capocelli, A. D. Santis, L. Gargano, and U. Vaccaro, "On the size of shares for secret sharing schemes," *Journal of Cryptography,* vol. 6, no. 3, pp. 157–167, 1993.

[55] J. L. Carter and M. N. Wegman, "Universal classes of hash functions," *Journal of Computer and System Sciences,* vol. 18, no. 2, pp. 143–154, 1979.

[56] N. Cerf, S. Massar, and S. Schneider, "Multipartite classical and quantum secrecy monotones," *Physical Review A,* vol. 66, no. 4, p. 042309, October 2002.

[57] C. Chan, "On tightness of mutual dependence upperbound for secret-key capacity of multiple terminals," arXiv: 0805.3200, 2008.

[58] C. Chan, "Multiterminal secure source coding for a common secret source," in *49th Annual Allerton Conference on Communication, Control, and Computing (Allerton).* IEEE, September 2011, pp. 188–195.

[59] C. Chan, "Agreement of a restricted secret key," in *Proc. IEEE International Symposium on Information Theory (ISIT).* IEEE, 2012, pp. 1782–1786.

[60] C. Chan, A. Al-Bashabsheh, J. B. Ebrahimi, T. Kaced, and T. Liu, "Multivariate mutual information inspired by secret-key agreement," *Proceedings of the IEEE,* vol. 103, no. 10, pp. 1883–1913, 2015.

[61] C. Chan, A. Al-Bashabsheh, Q. Zhou, N. Ding, T. Liu, and A. Sprintson, "Successive omniscience," *IEEE Transactions on Information Theory,* vol. 62, no. 6, pp. 3270–3289, 2016.

[62] C. Chan, M. Mukherjee, N. Kashyap, and Q. Zhou, "On the optimality of secret key agreement via omniscience," *IEEE Transactions on Information Theory,* vol. 64, no. 4, pp. 2371–2389, 2018.

[63] D. Chaum, C. Crépeau, and I. Damgård, "Multi-party unconditionally secure protocols," in *Proc. ACM Symposium on Theory of Computing – STOC '88.* ACM, 1988, pp. 11– 19.

[64] D. Chaum and S. Roijakkers, "Unconditionally-secure digital signatures," in *Advances in Cryptology – CRYPTO '90,* Lecture Notes in Computer Science, vol. 537. Springer, 1991, pp. 206–214.

[65] B. Chor and O. Goldreich, "Unbiased bits from sources of weak randomness and probabilistic communication complexity," *SIAM Journal on Computing*, vol. 17, no. 2, pp. 230–261, 1988.

[66] B. Chor, S. Goldwasser, S. Micali, and B. Awerbuch, "Verifiable secret sharing and achieving simultaneity in the presence of faults," in *26th IEEE Annual Symposium on Foundations of Computer Science, FOCS*. IEEE, 1985, pp. 383–395.

[67] B. Chor and Y. Ishai, "On privacy and partition arguments," *Information and Computation*, vol. 167, no. 1, pp. 2–9, May 2001.

[68] B. Chor and E. Kushilevitz, "A zero-one law for boolean privacy," *SIAM Journal on Discrete Mathematics*, vol. 4, no. 1, pp. 36–47, 1991.

[69] R. A. Chou and M. R. Bloch, "Separation of reliability and secrecy in rate-limited secret-key generation," *IEEE Transactions on Information Theory*, vol. 60, no. 8, pp. 4941–4957, 2014.

[70] M. Christandl, A. Ekert, M. Horodecki, P. Horodecki, J. Oppenheim, and R. Renner, "Unifying classical and quantum key distillation," in *Theory of Cryptography Conf. – TCC*, Lecture Notes in Computer Science, vol. 4392. Springer, 2007, pp. 456–478.

[71] J. F. Clauser, M. A. Horne, A. Shimony, and R. A. Holt, "Proposed experiment to test local hidden-variable theories," *Physical Review Letters*, vol. 23, no. 15, pp. 880–884, October 1969.

[72] R. Cohen, S. Coretti, J. Garay, and V. Zikas, "Probabilistic termination and composability of cryptographic protocols," *Journal of Cryptology*, vol. 32, pp. 690–741, 2019.

[73] J. Considine, M. Fitzi, M. Franklin, L. A. Levin, U. Maurer, and D. Metcalf, "Byzantine agreement given partial broadcast," *Journal of Cryptology*, vol. 18, pp. 191–217, 2005.

[74] T. Cover, "A proof of the data compression theorem of Slepian and Wolf for ergodic sources," *IEEE Transactions on Information Theory*, vol. 22, no. 2, pp. 226–228, March 1975.

[75] T. M. Cover and J. A. Thomas, *Elements of Information Theory*. Wiley-Interscience, 2006.

[76] R. Cramer, I. Damgård, S. Dziembowski, M. Hirt, and T. Rabin, "Efficient multiparty computations secure against an adaptive adversary," in *Advances in Cryptology – EUROCRYPT '99*, Lecture Notes in Computer Science, vol. 1592. Springer, 1999, pp. 311–326.

[77] R. Cramer, I. Damgård, and J. Nielsen, *Secure Multiparty Computation and Secret Sharing*. Cambridge University Press, 2015.

[78] C. Crépeau, "Equivalence between two flavours of oblivious transfer," in *Advances in Cryptology – CRYPTO '87*, Lecture Notes in Computer Science, vol. 293. Springer, 1988, pp. 350–354.

[79] C. Crépeau, "Efficient cryptographic protocols based on noisy channels," in *Advances in Cryptology – EUROCRYPT '97*, Lecture Notes in Computer Science, vol. 1233. Springer, 1997, pp. 306–317.

[80] C. Crépeau, R. Dowsley, and C. A. Nascimento, "On the commitment capacity of unfair noisy channel," *IEEE Transactions on Information Theory*, vol. 66, no. 6, pp. 3745–3752, 2020.

[81] C. Crépeau, J. Graaf, and A. Tapp, "Committed oblivious transfer and private multi-party computation," in *Advances in Cryptology – CRYPTO '95*, Lecture Notes in Computer Science, vol. 963. Springer, 1995, pp. 110–123.

[82] C. Crépeau and J. Kilian, "Weakening security assumptions and oblivious transfer," in *Advances in Cryptology – CRYPTO '88*, Lecture Notes in Computer Science, vol. 403. Springer, 1990, pp. 2–7.

[83] C. Crépeau, K. Morozov, and S. Wolf, "Efficient unconditionally oblivious transfer from almost any noisy channel," in *Proc. 4th International Conference on Security in Communication Networks, SCN 2004*, Lecture Notes in Computer Science, vol. 3352. Springer, 2004, pp. 47–59.

[84] C. Crépeau and G. Savvides, "Optimal reductions between oblivious transfer using interactive hashing," in *Advances in Cryptology – EUROCRYPT 2006*, Lecture Notes in Computer Science, vol. 4004. Springer, 2006, pp. 201–221.

[85] C. Crépeau, G. Savvides, C. Schaffner, and J. Wullschleger, "Information-theoretic conditions for two-party secure function evaluation," in *Advances in Cryptology -EUROCRYPT 2006*, Lecture Notes in Computer Science, vol. 4004. Springer, 2006, pp. 538–554.

[86] C. Crépeau and J. Wullschleger, "Statistical security conditions for two-party secure function evaluation," in *International Conference on Information Theoretic Security, ICITS 2008*, Lecture Notes in Computer Science, vol. 5155. Springer, 2008, pp. 86–99.

[87] I. Csiszár, "The method of types," *IEEE Transactions on Information Theory*, vol. 44, no. 6, pp. 2505–2523, October 1998.

[88] I. Csiszár and J. Körner, *Information Theory: Coding Theorems for Discrete Memoryless Channels*, 2nd edition. Cambridge University Press, 2011.

[89] I. Csiszár and P. Narayan, "Common randomness and secret key generation with a helper," *IEEE Transactions on Information Theory*, vol. 46, no. 2, pp. 344–366, March 2000.

[90] I. Csiszár and P. Narayan, "Secrecy capacities for multiple terminals," *IEEE Transactions on Information Theory*, vol. 50, no. 12, pp. 3047–3061, December 2004.

[91] I. Csiszár and P. Narayan, "Secrecy capacities for multiterminal channel models," *IEEE Transactions on Information Theory*, vol. 54, no. 6, pp. 2437–2452, June 2008.

[92] I. Csiszár and P. Narayan, "Secrecy generation for multiaccess channel models," *IEEE Transactions on Information Theory*, vol. 59, no. 1, pp. 17–31, 2013.

[93] I. Damgård, "On σ protocols," available online, 2014.

[94] I. Damgård, S. Fehr, L. Salvail, and C. Schaffner, "Cryptography in the bounded-quantum-storage model," *SIAM Journal on Computing*, vol. 37, no. 6, pp. 1865–1890, 2008.

[95] I. Damgård, J. Kilian, and L. Salvail, "On the (im)possibility of basing oblivious transfer and bit commitment on weakened security assumption," in *Advances in Cryptology – EUROCRYPT '99*, Lecture Notes in Computer Science, vol. 1592. Springer, 1999, pp. 56–73.

[96] I. Damgård, T. P. Pedersen, and B. Pfitzmann, "Statistical secrecy and multibit commitments," *IEEE Transactions on Information Theory*, vol. 44, no. 3, pp. 1143–1151, May 1998.

[97] D. Data, V. M. Prabhakaran, and M. M. Prabhakaran, "Communication and randomness lower bounds for secure computation," *IEEE Transactions on Information Theory*, vol. 62, no. 7, pp. 3901–3929, July 2016.

[98] N. Datta and R. Renner, "Smooth entropies and the quantum information spectrum," *IEEE Transactions on Information Theory*, vol. 55, no. 6, pp. 2807–2815, June 2009.

[99] W. Diffie and M. Hellman, "New directions in cryptography," *IEEE Transactions on Information Theory*, vol. 22, no. 6, pp. 644–654, 1976.

[100] Y. Z. Ding, D. Harnik, A. Rosen, and R. Shaltiel, "Constant-round oblivious transfer in the bounded storage model," *Journal of Cryptology*, vol. 20, no. 2, pp. 165–202, 2007.

[101] Y. Dodis, "Shannon impossibility, revisited," in *International Conference on Information Theoretic Security, ICITS 2011*, Lecture Notes in Computer Science, vol. 6673. Springer, 2011, pp. 100–110.

[102] Y. Dodis, "Randomness in cryptography," *Lecture Notes*, New York University, 2014. Available at cs.nyu.edu/~dodis/randomness-in-crypto

[103] Y. Dodis and S. Micali, "Lower bounds for oblivious transfer reductions," in *Advances in Cryptology – EUROCRYPT '99*, Lecture Notes in Computer Science, vol. 1592. Springer, 1999, pp. 42–55.

[104] Y. Dodis and A. Smith, "Entropic security and the encryption of high entropy messages," in *Theory of Cryptography Conf. – TCC*, Lecture Notes in Computer Science, vol. 3378. Springer, 2005, pp. 556–577.

[105] Y. Dodis and D. Wichs, "Non-malleable extractors and symmetric key cryptography from weak secrets," in *Proc. ACM Symposium on Theory of Computing – STOC '09*. ACM, 2009, pp. 601–610.

[106] D. Dolev, "The Byzantine generals strikes again," *Journal of Algorithms*, vol. 3, pp. 14–30, 1982.

[107] D. Dolev, C. Dwork, and M. Naor, "Non-malleable cryptography," *SIAM Journal on Computing*, vol. 30, no. 2, pp. 391–437, 2000.

[108] D. Dolev, C. Dwork, O. Waarts, and M. Yung, "Perfectly secure message transmission," *Journal of the ACM*, vol. 40, no. 1, pp. 17–47, 1993.

[109] D. Dolev and R. Strong, "Authenticated algorithms for Byzantine agreement," *SIAM Journal on Computing*, vol. 12, no. 4, pp. 656–666, 1979.

[110] M. J. Donald, M. Horodecki, and O. Rudolph, "The uniqueness theorem for entanglement measure," *Journal of Mathematical Physics*, vol. 43, no. 9, pp. 4252–4272, 2002.

[111] R. Dowsley, F. Lacerda, and C. A. Nascimento, "Commitment and oblivious transfer in the bounded storage model with errors," *IEEE Transactions on Information Theory*, vol. 64, no. 8, pp. 5970–5984, 2018.

[112] R. Dowsley and A. C. Nascimento, "On the oblivious transfer capacity of generalized erasure channels against malicious adversaries: the case of low erasure probability," *IEEE Transactions on Information Theory*, vol. 63, no. 10, pp. 6819–6826, October 2017.

[113] G. Estren, "Universally composable committed oblivious transfer and multi-party computation assuming only black-box primitives," Master's Thesis, McGill University, 2004.

[114] D. Evans, V. Kolesnikov, and M. Rosulek, *A Pragmatic Introduction to Secure Multi-Party Computation.* Hanover, MA: Now Publishers, 2018.

[115] S. Even, O. Goldreich, and A. Lempel, "A randomized protocol for signing contracts," *Communications of the ACM*, vol. 28, no. 6, pp. 637–647, June 1985.

[116] S. Fehr and S. Berens, "On the conditional Rényi entropy," *IEEE Transactions on Information Theory*, vol. 60, no. 11, pp. 6801–6810, 2014.

[117] U. Feige, A. Fiat, and A. Shamir, "Zero-knowledge proofs of identity," *Journal of Cryptology*, vol. 1, pp. 77–94, 1988.

[118] P. Feldman and S. Micali, "Optimal probabilistic protocol for synchronous Byzantine agreement," *SIAM Journal on Computing*, vol. 26, no. 4, pp. 873–933, 1997.

[119] A. Fiat and A. Shamir, "How to prove yourself: practical solutions to identification and signature problems," in *Advances in Cryptology – CRYPTO '86*, Lecture Notes in Computer Science, vol. 263. Springer, 1987, pp. 186–194.

[120] M. J. Fischer and N. A. Lynch, "A lower bound for the time to assure interactive consistency," *Information Processing Letters*, vol. 14, no. 4, pp. 183–186, June 1982.

[121] M. J. Fischer, N. A. Lynch, and M. Merritt, "Easy impossibility proofs for distributed consensus problems," *Distributed Computing*, vol. 1, no. 1, pp. 26–39, 1986.

[122] M. J. Fischer, N. A. Lynch, and M. S. Paterson, "Impossibility of distributed consensus with one faulty process," *Journal of the ACM*, vol. 32, no. 2, pp. 374–382, April 1985.

[123] M. J. Fischer, M. S. Paterson, and C. Rackoff, "Secret bit transmission using a random deal of cards," in *Distributed Computing and Cryptography*, AMS DIMACS Series in Discrete Mathematics and Theoretical Computer Science, vol. 2. AMS, 1991, pp. 173–181.

[124] M. Fitzi, J. A. Garay, U. Maurer, and R. Ostrovsky, "Minimal complete primitive for secure multi-party computation," *Journal of Cryptology*, vol. 18, pp. 37–61, 2005.

[125] M. Fitzi, N. Gisin, and U. Maurer, "Quantum solution to Byzantine agreement problem," *Physical Review Letters*, vol. 87, no. 21, p. 217901, November 2001.

[126] M. Fitzi, N. Gisin, U. Maurer, and O. V. Rotz, "Unconditional Byzantine agreement and multi-party computation secure against dishonest minorities from scratch," in *Advances in Cryptology – EUROCRYPT 2002*, Lecture Notes in Computer Science, vol. 2332. Springer, 2002, pp. 482–501.

[127] M. Fitzi, S. Wolf, and J. Wullschleger, "Pseudo-signatures, broadcast, and multiparty computation from correlated randomness," in *Advances in Cryptology – CRYPTO 2004*, Lecture Notes in Computer Science, vol. 3152. Springer, 2004, pp. 562–578.

[128] L. Fortnow, "Complexity-theoretic aspects of interactive proof systems," Ph.D. Dissertation, Massachusetts Institute of Technology, 1989.

[129] M. Franklin and M. Yung, "Communication complexity of secure computation," in *Proc. ACM Symposium on Theory of Computing – STOC '92.* ACM, 1992, pp. 699–710.

[130] P. Gács and J. Körner, "Common information is far less than mutual information," *Problems of Control and Information Theory*, vol. 2, no. 2, pp. 149–162, 1973.

[131] A. El Gamal and Y.-H. Kim, *Network Information Theory.* Cambridge University Press, 2011.

[132] C. Gehrmann, "Cryptanalysis of the Gemmell and Naor multiround authentication protocol," in *Advances in Cryptology – CRYPTO '94*, Lecture Notes in Computer Science, vol. 839. Springer, 1994, pp. 121–128.

[133] C. Gehrmann, "Multiround unconditionally secure authentication," *Journal of Designs, Codes and Cryptography*, vol. 15, no. 1, pp. 67–86, 1998.

[134] P. Gemmell and M. Naor, "Codes for interactive authentication," in *Advances in Cryptology – CRYPTO '93*, Lecture Notes in Computer Science, vol. 773. Springer, 1994, pp. 355–367.

[135] E. N. Gilbert, F. J. MacWilliams, and N. J. Sloane, "Codes which detect deception," *Bell System Technical Journal*, vol. 53, pp. 405–424, 1974.

[136] A. A. Gohari and V. Anantharam, "Information-theoretic key agreement of multiple terminals: Part i," *IEEE Transactions on Information Theory*, vol. 56, no. 8, pp. 3973–3996, August 2010.

[137] A. A. Gohari and V. Anantharam, "Information-theoretic key agreement of multiple terminals: Part ii," *IEEE Transactions on Information Theory*, vol. 56, no. 8, pp. 3997–3996, August 2010.

[138] A. A. Gohari and V. Anantharam, "Comments on 'information-theoretic key agreement of multiple terminals: Part i'," *IEEE Transactions on Information Theory*, vol. 63, no. 8, pp. 5440–5442, August 2017.

[139] A. A. Gohari, O. Günlü, and G. Kramer, "Coding for positive rate in the source model key agreement problem," to appear in *IEEE Transactions on Information Theory.*

[140] O. Goldreich, *Modern Cryptography, Probabilistic Proofs and Pseudorandomness.* Springer, 1998.

[141] O. Goldreich, *Foundations of Cryptography: Basic Tools.* Cambridge University Press, 2001.

[142] O. Goldreich, *Foundations of Cryptography: II Basic Applications.* Cambridge University Press, 2004.

[143] O. Goldreich, S. Goldwasser, and S. Micali, "How to construct random functions," *Journal of the ACM*, vol. 33, no. 4, pp. 792–807, October 1986.

[144] O. Goldreich, S. Micali, and A. Wigderson, "How to play any mental game – a completeness theorem for protocols with honest majority," in *Proc. ACM Symposium on Theory of Computing – STOC '87.* ACM, 1987, pp. 218–229.

[145] O. Goldreich, S. Micali, and A. Wigderson, "Proofs that yield nothing but their validity, or all language in NP have zero-knowledge proof systems," *Journal of the ACM*, vol. 38, no. 3, pp. 690–728, 1991.

[146] O. Goldreich and R. Vainish, "How to solve any protocol problem – an efficiency improvement," in *Advances in Cryptology – CRYPTO '87*, Lecture Notes in Computer Science, vol. 293. Springer, 1988, pp. 73–86.

[147] S. Goldwasser and S. Micali, "Probabilistic encryption," *Journal of Computer and System Sciences*, vol. 28, no. 2, pp. 270–299, 1984.

[148] S. Goldwasser, S. Micali, and C. Rackoff, "The knowledge complexity of interactive proof systems," *SIAM Journal on Computing*, vol. 18, no. 1, pp. 186–208, 1989.

[149] S. D. Gordon, C. Hazay, J. Katz, and Y. Lindell, "Complete fairness in secure two-party computation," *Journal of the ACM*, vol. 58, no. 6, pp. 24:1–37, 2011.

[150] A. Guntuboyina, "Lower bounds for the minimax risk using f-divergences, and applications," *IEEE Transactions on Information Theory*, vol. 57, no. 4, pp. 2386–2399, 2011.

[151] T. S. Han, *Information-Spectrum Methods in Information Theory* [English translation], Stochastic Modelling and Applied Probability, vol. 50. Springer, 2003.

[152] T. S. Han and S. Verdú, "Approximation theory of output statistics," *IEEE Transactions on Information Theory*, vol. 39, no. 3, pp. 752–772, May 1993.

[153] G. Hanaoka, "Some information theoretic arguments for encryption: nonmalleability and chosen-ciphertext security (invited talk)," in *International Conference on Information Theoretic Security, ICITS 2008*, Lecture Notes in Computer Science, vol. 5155. Springer, 2008, pp. 223–231.

[154] G. Hanaoka, Y. Hanaoka, M. Hagiwara, H. Watanabe, and H. Imai, "Unconditionally secure chaffing-and-winnowing: a relationship between encryption and authentication," in *International Symposium on Applied Algebra, Algebraic Algorithms and Error-Correcting Codes (AAECC)*, Lecture Notes in Computer Science, vol. 3857. Springer, 2006, pp. 154–162.

[155] G. Hanaoka, J. Shikata, Y. Hanaoka, and H. Imai, "Unconditionally secure anonymous encryption and group authentication," *The Computer Journal*, vol. 49, no. 3, pp. 310–321, May 2006.

[156] G. Hanaoka, J. Shikata, Y. Zheng, and H. Imai, "Unconditionally secure digital signature schemes admitting transferability," in *Advances in Cryptology – ASIACRYPT 2000*, Lecture Notes in Computer Science, vol. 1976. Springer, 2000, pp. 130–142.

[157] D. Harnik, Y. Ishai, and E. Kushilevitz, "How many oblivious transfers are needed for secure multiparty computation," in *Advances in Cryptology – CRYPTO 2007*, Lecture Notes in Computer Science, vol. 4622. Springer, 2007, pp. 284–302.

[158] D. Harnik, Y. Ishai, E. Kushilevitz, and J. B. Nielsen, "OT-combiners via secure computation," in *Theory of Cryptography Conf. – TCC*, Lecture Notes in Computer Science, vol. 4948. Springer, 2008, pp. 393–411.

[159] M. Hayashi, "Information spectrum approach to second-order coding rate in channel coding," *IEEE Transactions on Information Theory*, vol. 55, no. 11, pp. 4947–4966, November 2009.

[160] M. Hayashi, "Exponential decreasing rate of leaked information in universal random privacy amplification," *IEEE Transactions on Information Theory*, vol. 57, no. 6, pp. 3989–4001, June 2011.

[161] M. Hayashi, *Quantum Information Theory: Mathematical Foundation*, 2nd edition. Springer, 2016.

[162] M. Hayashi, "Security analysis of ε-almost dual universal$_2$ hash functions: smoothing of min entropy versus smoothing of Rényi entropy of order 2," *IEEE Transactions on Information Theory*, vol. 62, no. 6, pp. 3451–3476, June 2016.

[163] M. Hayashi and T. Tsurumaru, "More efficient privacy amplification with less random seeds via dual universal hash function," *IEEE Transactions on Information Theory*, vol. 62, no. 4, pp. 2213–2232, April 2016.

[164] M. Hayashi, H. Tyagi, and S. Watanabe, "Secret key agreement: general capacity and second-order asymptotics," *IEEE Transactions on Information Theory*, vol. 62, no. 7, May 2016.

[165] Y. Hayashi and H. Yamamoto, "Coding theorems for the Shannon cipher system with a guessing wiretapper and correlated source outputs," *IEEE Transactions on Information Theory*, vol. 54, no. 6, pp. 2808–2817, June 2008.

[166] C. Hazay and Y. Lindell, *Efficient Secure Two-Party Protocols: Techniques and Constructions*. Springer, 2010.

[167] T. Holenstein, "Parallel repetition: simplifications and the no-signaling case," *Theory of Computing*, vol. 5, pp. 141–172, 2009.

[168] J. Håstad, R. Impagliazzo, L. A. Levin, and M. Luby, "A pseudorandom generator from any one-way function," *SIAM Journal on Computing*, vol. 28, no. 4, pp. 1364–1396, 1999.

[169] H. Imai, J. Müller-Quade, A. Nascimento, and A. Winter, "Rates for bit commitment and coin tossing from noisy correlation," in *Proc. IEEE International Symposium on Information Theory (ISIT)*. IEEE, 2004, p. 47.

[170] H. Imai, K. Morozov, A. Nascimento, and A. Winter, "On the oblivious transfer capacity of the erasure channel," in *Proc. IEEE International Symposium on Information Theory (ISIT)*. IEEE, 2006, pp. 1428–1431.

[171] H. Imai, K. Morozov, A. C. Nascimento, and A. Winter, "Efficient protocols achieving the commitment capacity of noisy correlations," in *Proc. IEEE International Symposium on Information Theory (ISIT)*. IEEE, 2006, pp. 1432–1436.

[172] R. Impagliazzo, L. A. Levin, and M. Luby, "Pseudo-random generation from one-way functions," in *Proc. ACM Symposium on Theory of Computing – STOC '89*. ACM, 1989, pp. 12–24.

[173] R. Impagliazzo and D. Zuckerman, "How to recycle random bits," in *30th Annual Symposium on Foundations of Computer Science, FOCS*. IEEE, 1989, pp. 248–253.

[174] Y. Ishai, E. Kushilevitz, R. Ostrovsky, M. Prabhakaran, A. Sahai, and J. Wullschleger, "Constant-rate oblivious transfer from noisy channels," in *Advances in Cryptology – CRYPTO 2011*, Lecture Notes in Computer Science, vol. 6841. Springer, 2011, pp. 667–684.

[175] Y. Ishai, M. Prabhakaran, and A. Sahai, "Founding cryptography on oblivious transfer efficiently," in *Advances in Cryptology – CRYPTO 2008*, Lecture Notes in Computer Science, vol. 5157. Springer, 2008, pp. 572–591.

[176] I. Issa, A. B. Wagner, and S. Kamath, "An operational approach to information leakage," *IEEE Transactions on Information Theory*, vol. 66, no. 3, pp. 1625–1657, March 2020.

[177] M. Itoh, A. Saito, and T. Nishizeki, "Secret sharing scheme realizing general access structure," in *Proc. IEEE Globecom*. IEEE, 1987, pp. 99–102.

[178] M. Iwamoto, H. Koga, and H. Yamamoto, "Coding theorems for a (2, 2)-threshold scheme with detectability of impersonation attacks," *IEEE Transactions on Information Theory*, vol. 58, no. 9, pp. 6194–6206, September 2012.

[179] M. Iwamoto and K. Ohta, "Security notions for information theoretically secure encryptions," in *Proc. IEEE International Symposium on Information Theory (ISIT)*. IEEE, 2011, pp. 1777–1781.

[180] M. Iwamoto, K. Ohta, and J. Shikata, "Security formalization and their relationships for encryption and key agreement in information-theoretic cryptography," *IEEE Transactions on Information Theory*, vol. 64, no. 1, pp. 654–685, January 2018.

[181] M. Iwamoto and J. Shikata, "Information theoretic security for encryption based on conditional Rényi entropies," in *International Conference on Information Theoretic Security, ICITS 2013*, Lecture Notes in Computer Science, vol. 8317. Springer, 2014, pp. 103–121.

[182] M. Iwamoto and H. Yamamoto, "Strongly secure ramp secret sharing schemes for general access structure," *Information Processing Letters*, vol. 97, no. 2, pp. 52–57, January 2006.

[183] M. Iwamoto, H. Yamamoto, and H. Ogawa, "Optimal multiple assignment based on integer programming in secret sharing schemes with general access structures," *IEICE Transactions on Fundamentals of Electronics, Communications and Computer Sciences*, vol. E90-A, no. 1, pp. 101–112, January 2007.

[184] J. Justesen and T. Høholdt, *A Course in Error-Correcting Codes*, 2nd edition. European Mathematical Society, 2017.

[185] J. Katz, *Digital Signature*. Springer, 2010.

[186] J. Katz and C.-Y. Koo, "On expected constant-round protocols for Byzantine agreement," *Journal of Computer and System Sciences*, vol. 75, no. 2, pp. 91–112, 2009.

[187] J. Katz and Y. Lindell, *Introduction to Modern Cryptography*, 3rd edition. Chapman & Hall/CRC, 2020.

[188] J. Katz and M. Yung, "Characterization of security notions for probabilistic private-key encryption," *Journal of Cryptology*, vol. 19, pp. 67–95, 2006.

[189] A. Kawachi, C. Portman, and K. Tanaka, "Characterization of the relations between information-theoretic non-malleability, secrecy, and authenticity," in *International Conference on Information Theoretic Security, ICITS 2011*, Lecture Notes in Computer Science, vol. 6673. Springer, 2011, pp. 6–24.

[190] D. Khurana, D. Kraschewski, H. K. Maji, M. Prabhakaran, and A. Sahai, "All complete functionalities are reversible," in *Advances in Cryptology – EUROCRYPT 2016*, Lecture Notes in Computer Science, vol. 9665. Springer, 2016, pp. 213–242.

[191] D. Khurana, H. K. Maji, and A. Sahai, "Secure computation from elastic noisy channel," in *Advances in Cryptology – EUROCRYPT 2016*, Lecture Notes in Computer Science, vol. 9665. Springer, 2016, pp. 184–212.

[192] J. Kilian, "Founding cryptography on oblivious transfer," in *Proc. ACM Symposium on Theory of Computing – STOC '88*. ACM, 1988, pp. 20–31.

[193] J. Kilian, "Uses of randomness in algorithms and protocols," Ph.D. Dissertation, Massachusetts Institute of Technology, 1989.

[194] J. Kilian, "A general completeness theorem for two-party games," in *Proc. ACM Symposium on Theory of Computing – STOC '91*. ACM, 1991, pp. 553–560.

[195] J. Kilian, "More general completeness theorems for secure two-party computation," in *Proc. ACM Symposium on Theory of Computing – STOC 2000*. ACM, 2000, pp. 316–324.

[196] D. Knuth and A. Yao, "The complexity of nonuniform random number generation," in *Algorithms and Complexity, New Directions and Results.* Academic Press, 1976, pp. 357–428.

[197] H. Koga, "Coding theorems on the threshold scheme for a general source," *IEEE Transactions on Information Theory,* vol. 54, no. 6, pp. 2658–2677, June 2008.

[198] H. Koga, "New coding theorems for fixed-length source coding and Shannon's cipher system with a general source," in *Proc. International Symposium on Information Theory and its Applications (ISITA).* IEEE, 2008, pp. 1–6.

[199] H. Koga, "Characterization of the smooth Rényi entropy using majorization," in *2013 IEEE Information Theory Workshop (ITW).* IEEE, 2013, pp. 604–608.

[200] H. Koga and H. Yamamoto, "Coding theorem for secret-key authentication systems," *IEICE Transactions on Fundamentals of Electronics, Communications and Computer Sciences,* vol. E83-A, no. 8, pp. 1691–1703, August 2000.

[201] R. König, R. Renner, and C. Schaffner, "The operational meaning of min- and max-entropy," *IEEE Transactions on Information Theory,* vol. 55, no. 9, pp. 4337–4347, September 2009.

[202] R. König, S. Wehner, and J. Wullschleger, "Unconditional security from noisy quantum storage," *IEEE Transactions on Information Theory,* vol. 58, no. 3, pp. 1962–1984, March 2012.

[203] J. Körner and K. Marton, "Comparison of two noisy channels," in *Colloquia Mathematica Societatis, János Bolyai, 16, Topics in Information Theory.* North-Holland, 1977, pp. 411–424.

[204] D. Kraschewski and J. Müller-Quade, "Completeness theorems with constructive proofs for finite deterministic 2-party functions," in *Theory of Cryptography Conf. – TCC,* Lecture Notes in Computer Science, vol. 6597. Springer, 2011, pp. 364–381.

[205] D. Kraschewski, H. K. Maji, M. Prabhakaran, and A. Sahai, "A full characterization of completeness for two-party randomized function evaluation," in *Advances in Cryptology – EUROCRYPT 2014,* Lecture Notes in Computer Science, vol. 8441. Springer, 2014, pp. 659–676.

[206] G. Kreitz, "A zero-one law for secure multi-party computation with ternary outputs," in *Theory of Cryptography Conf. – TCC,* Lecture Notes in Computer Science, vol. 6597. Springer, 2011, pp. 382–399.

[207] S. Kullback, *Information Theory and Statistics.* Dover Publications, 1968.

[208] R. Kumaresan, S. Raghuraman, and A. Sealfon, "Network oblivious transfer," in *Advances in Cryptology – CRYPTO 2016,* Lecture Notes in Computer Science, vol. 9814. Springer, 2016, pp. 366–396.

[209] J. Kurihara, T. Uyematsu, and R. Matsumoto, "Secret sharing schemes based on linear codes can be precisely characterized by the relative generalized Hamming weight," *IEICE Transactions on Fundamentals of Electronics, Communications and Computer Sciences,* vol. E95-A, no. 11, pp. 2067–2075, November 2012.

[210] K. Kurosawa, W. Kishimoto, and T. Koshiba, "A combinatorial approach to deriving lower bounds for perfectly secure oblivious transfer reductions," *IEEE Transactions on Information Theory,* vol. 54, no. 6, pp. 2566–2571, June 2008.

[211] K. Kurosawa, S. Obana, and W. Ogata, "t-cheater identifiable (k, n) threshold secret sharing schemes," in *Advances in Cryptology – CRYPTO '95,* Lecture Notes in Computer Science, vol. 963. Springer, 1995, pp. 410–423.

[212] E. Kushilevitz, "Privacy and communication complexity," *SIAM Journal on Discrete Mathematics*, vol. 5, no. 2, pp. 273–284, 1992.

[213] E. Kushilevitz, Y. Lindell, and T. Rabin, "Information-theoretically secure protocols and security under composition," *SIAM Journal on Computing*, vol. 39, no. 5, pp. 2090–2112, 2010.

[214] L. Lai, H. E. Gamal, and H. V. Poor, "Authentication over noisy channels," *IEEE Transactions on Information Theory*, vol. 55, no. 2, pp. 906–916, February 2009.

[215] L. Lamport, "Constructing digital signatures from a one-way function," *SRI International Technical Report*, vol. CSL-98, 1979.

[216] L. Lamport, R. Shostak, and M. Pease, "The Byzantine general problem," *ACM Transactions on Programming Languages and Systems*, vol. 4, no. 3, pp. 382–401, July 1982.

[217] C. Lancien and A. Winter, "Parallel repetition and concentration for (sub-)nosignalling games via a flexible constrained de Finetti reduction," *Chicago Journal of Theoretical Computer Science*, no. 11, pp. 1–22, 2016.

[218] S. Lang, *Undergraduate Algebra*. Springer, 2005.

[219] E. J. Lee and E. Abbe, "A Shannon approach to secure multi-party computations," arXiv: 1401.7360, 2014.

[220] E. L. Lehmann and J. P. Romano, *Testing Statistical Hypotheses*, 3rd edition. Springer, 2005.

[221] S. K. Leung-Yan-Cheong, "Multi-user and wiretap channels including feedback," Ph.D. Dissertation, Stanford University, 1976.

[222] D. A. Levin and Y. Peres, *Markov Chains and Mixing Times*, 2nd edition. American Mathematical Society, 2017.

[223] C. T. Li and A. E. Gamal, "Maximal correlation secrecy," *IEEE Transactions on Information Theory*, vol. 64, no. 5, pp. 3916–3926, May 2018.

[224] Y. Lindell, "General composition and universal composability in secure multi-party computation," in *44th Annual Symposium on Foundations of Computer Science, FOCS*. IEEE, 2003, pp. 394–403.

[225] J. Liu, P. Cuff, and S. Verdú, "Key capacity for product sources with application to stationary Gaussian processes," *IEEE Transactions on Information Theory*, vol. 62, no. 2, pp. 984–1005, 2017.

[226] H.-K. Lo, "Insecurity of quantum secure computation," *Physical Review A*, vol. 97, no. 2, p. 1154, August 1997.

[227] N. A. Lynch, *Distributed Algorithms*. Morgan Kaufmann, 1996.

[228] M. Madiman and P. Tetali, "Information inequalities for joint distributions, with interpretations and applications," *IEEE Transactions on Information Theory*, vol. 56, no. 6, pp. 2699–2713, June 2010.

[229] H. K. Maji, M. Prabhakaran, and M. Rosulek, "Complexity of multi-party computation problems: the case of 2-party symmetric secure function evaluation," in *Theory of Cryptography Conf. – TCC*, Lecture Notes in Computer Science, vol. 5444. Springer, 2009, pp. 256–273.

[230] H. K. Maji, M. Prabhakaran, and M. Rosulek, "A unified characterization of completeness and triviality for secure function evaluation," in *International Conference on Cryptology in India – INDOCRYPT 2012*, Lecture Notes in Computer Science, vol. 7668. Springer, 2012, pp. 40–59.

[231] H. K. Maji, M. Prabhakaran, and M. Rosulek, "Complexity of multi-party computation functionalities," in *Secure Multi-Party Computation*, Cryptography and Information Security Series, vol. 10. IOS Press, 2013, pp. 249–283.

[232] Y. Mansour, N. Nisan, and P. Tiwari, "The computational complexity of universal hashing," *Theoretical Computer Science*, vol. 107, no. 1, pp. 121–133, 1993.

[233] J. Martí-Farré and C. Padró, "On secret sharing schemes, matroids and polymatroids," in *Theory of Cryptography Conf. – TCC*, Lecture Notes in Computer Science, vol. 4392. Springer, 2007, pp. 253–272.

[234] U. Martínez-Peñas, "On the similarities between generalized rank and Hamming weights and their applications to network coding," *IEEE Transactions on Information Theory*, vol. 62, no. 7, pp. 4081–4095, July 2016.

[235] J. L. Massey, "An introduction to contemporary cryptology," *Proceedings of the IEEE*, vol. 76, no. 5, pp. 533–549, May 1988.

[236] U. Maurer, "The role of information theory in cryptography," in *Proc. IMA International Conference on Cryptography and Coding (IMACC)*. IMA, 1993, pp. 49–71.

[237] U. Maurer, "Secret key agreement by public discussion from common information," *IEEE Transactions on Information Theory*, vol. 39, no. 3, pp. 733–742, May 1993.

[238] U. Maurer, "Information-theoretic cryptography," in *Advances in Cryptology – CRYPTO '99*, Lecture Notes in Computer Science, vol. 1666. Springer, 1999, pp. 47–55.

[239] U. Maurer, "Authentication theory and hypothesis testing," *IEEE Transactions on Information Theory*, vol. 46, no. 4, pp. 1350–1356, July 2000.

[240] U. Maurer, "Constructive cryptography: a new paradigm for security definitions and proofs," in *Theory of Security and Applications, TOSCA 2011*, Lecture Notes in Computer Science, vol. 6993. Springer, 2011, pp. 33–56.

[241] U. Maurer and S. Wolf, "Unconditionally secure key agreement and the intrinsic conditional information," *IEEE Transactions on Information Theory*, vol. 45, no. 2, pp. 499–514, March 1999.

[242] U. Maurer and S. Wolf, "Information-theoretic key agreement: from weak to strong secrecy for free," in *Advances in Cryptology – EUROCRYPT 2000*, Lecture Notes in Computer Science, vol. 1807. Springer, 2000, pp. 351–368.

[243] U. Maurer and S. Wolf, "Secret-key agreement over unauthenticated public channels – part I: Definitions and a completeness result," *IEEE Transactions on Information Theory*, vol. 49, no. 4, pp. 822–831, April 2003.

[244] U. Maurer and S. Wolf, "Secret-key agreement over unauthenticated public channels – part III: Privacy amplification," *IEEE Transactions on Information Theory*, vol. 49, no. 4, pp. 839–851, April 2003.

[245] U. Maurer and S. Wolf, "Secret-key agreement over unauthenticated public channels – part II: The simulatability condition," *IEEE Transactions on Information Theory*, vol. 49, no. 4, pp. 832–838, April 2003.

[246] L. McAven, R. Safavi-Naini, and M. Yung, "Unconditionally secure encryption under strong attacks," in *Proc. Australasian Conference on Information Security and Privacy*, Lecture Notes in Computer Science, vol. 3108. Springer, 2004, pp. 427–439.

[247] R. J. McEliece and D. V. Sarwate, "On sharing secrets and Reed–Solomon codes," *Communications of the ACM*, vol. 24, no. 9, pp. 583–584, September 1981.

[248] N. Merhav, "Universal coding with minimum probability of codeword length overflow," *IEEE Transactions on Information Theory*, vol. 37, no. 3, pp. 556–563, May 1991.

[249] N. Merhav and E. Arikan, "The Shannon cipher system with a guessing wiretapper," *IEEE Transactions on Information Theory*, vol. 45, no. 6, pp. 1860–1866, September 1999.

[250] S. Micali and P. Rogaway, "Secure computation," in *Advances in Cryptology – CRYPTO '91*, Lecture Notes in Computer Science, vol. 576. Springer, 1991, pp. 391–404.

[251] M. Mishra, B. K. Dey, V. M. Prabhakaran, and S. N. Diggavi, "Wiretapped oblivious transfer," *IEEE Transactions on Information Theory*, vol. 63, no. 4, pp. 2560–2595, April 2017.

[252] S. Miyake and F. Kanaya, "Coding theorems on correlated general sources," *IEICE Transactions on Fundamentals of Electronics, Communications and Computer Sciences*, vol. E78-A, no. 9, pp. 1063–1070, September 1995.

[253] J. Muramatsu, K. Yoshimura, and P. Davis, "Secret key capacity and advantage distillation capacity," *IEICE Transactions on Fundamentals of Electronics, Communications and Computer Sciences*, vol. E89-A, no. 10, pp. 2589–2596, October 2006.

[254] H. Nagaoka and M. Hayashi, "An information-spectrum approach to classical and quantum hypothesis testing for simple hypotheses," *IEEE Transactions on Information Theory*, vol. 53, no. 2, pp. 534–549, February 2007.

[255] M. Naor, "Bit commitment using pseudorandomness," *Journal of Cryptology*, vol. 4, no. 2, pp. 151–158, 1991.

[256] M. Naor, R. Ostrovsky, R. Venkatesan, and M. Yung, "Perfect zero-knowledge arguments for NP can be based on general complexity assumptions," in *Advances in Cryptology – CRYPTO '92*, Lecture Notes in Computer Science, vol. 740. Springer, 1993, pp. 196–214.

[257] M. Naor, G. Segev, and A. Smith, "Tight bounds for unconditional authentication protocols in the manual channel and shared key models," *IEEE Transactions on Information Theory*, vol. 54, no. 6, pp. 2408–2425, June 2008.

[258] P. Narayan and H. Tyagi, *Multiterminal Secrecy by Public Discussion.* Hanover, MA: Now Publishers, 2016.

[259] P. Narayan, H. Tyagi, and S. Watanabe, "Common randomness for secure computing," in *Proc. IEEE International Symposium on Information Theory (ISIT)*. IEEE, 2015, pp. 949– 953.

[260] A. C. A. Nascimento and A. Winter, "On the oblivious-transfer capacity of noisy resources," *IEEE Transactions on Information Theory*, vol. 54, no. 6, pp. 2572–2581, 2008.

[261] M. A. Nielsen and I. L. Chuang, *Quantum Computation and Quantum Information.* Cambridge University Press, 2010.

[262] M. Nishiara and K. Takizawa, "Strongly secure secret sharing scheme with ramp threshold based on Shamir's polynomial interpolation scheme," *IEICE Transactions on Fundamentals of Electronics, Communications and Computer Sciences*, vol. J92-A, no. 12, pp. 1009–1013, December 2009 (in Japanese).

[263] T. Ogawa and H. Nagaoka, "Strong converse and Stein's lemma in quantum hypothesis testing," *IEEE Transactions on Information Theory*, vol. 46, no. 7, pp. 2428–2433, November 2000.

[264] D. Ostrev, "Composable, unconditionally secure message authentication without any secret key," in *Proc. IEEE International Symposium on Information Theory (ISIT)*. IEEE, 2019, pp. 622–626.

[265] D. Ostrev, "An introduction to the theory of unconditionally secure message authentication using the constructive cryptography framework." Available at http://hdl.handle.net/10993/39497

[266] R. Ostrovsky, A. Scafuro, I. Visconti, and A. Wadia, "Universally composable secure computation with (malicious) physically unclonable functions," in *Advances in Cryptology - EUROCRYPT 2013*, Lecture Notes in Computer Science, vol. 7881. Springer, 2013, pp. 702–718.

[267] R. Ostrovsky, R. Venkatesan, and M. Yung, "Fair games against an all-powerful adversary," in *Advances in Computational Complexity Theory*, AMS DIMACS Series in Discrete Mathematics and Theoretical Computer Science, vol. 13. AMS, 1993, pp. 155–169.

[268] O. Ozel and S. Ulukus, "Wiretap channels: implications of more capable condition and cyclic shift symmetry," *IEEE Transactions on Information Theory*, vol. 59, no. 4, pp. 2153–2164, April 2013.

[269] M. Pease, R. Shostak, and L. Lamport, "Reaching agreement in the presence of faults," *Journal of the ACM*, vol. 27, no. 2, pp. 228–234, April 1980.

[270] T. P. Pedersen, "Non-interactive and information-theoretic secure verifiable secret sharing," in *Advances in Cryptology - CRYPTO '91*, Lecture Notes in Computer Science, vol. 576. Springer, 1992, pp. 129–140.

[271] B. Pfitzmann, M. Schunter, and M. Waidner, "Secure reactive system," *Technical Report RZ 3206*, IBM Zurich, 2000.

[272] B. Pfitzmann and M. Waidner, "Information-theoretic pseudosignatures and Byzantine agreement for $t \geq n/3$" *IBM Research Technical Report*, vol. RZ 2882 (#90830), 1996.

[273] A. C. B. Pinto, R. Dowsley, A. Morozov, and. C. Nascimento, "Achieving oblivious transfer capacity of generalized erasure channels in the malicious model," *IEEE Transactions on Information Theory*, vol. 57, no. 8, pp. 5566–5571, August 2011.

[274] Y. Polyanskiy, H. V. Poor, and S. Verdú, "Channel coding rate in the finite blocklength regime," *IEEE Transactions on Information Theory*, vol. 56, no. 5, pp. 2307–2359, May 2010.

[275] Y. Polyanskiy and Y. Wu, "Lecture notes on information theory," Massachusetts Institute of Technology, 2014.

[276] G. Pope, "Distinguishing advantage lower bounds for encryption and authentication protocols," *Technical Report*, Department of Computer Science, ETH Zurich, 2008.

[277] C. Portmann, "Key recycling in authentication," *IEEE Transactions on Information Theory*, vol. 60, no. 7, pp. 4383–4396, July 2014.

[278] V. Prabhakaran and M. Prabhakaran, "Assisted common information with an application to secure two-party sampling," *IEEE Transactions on Information Theory*, vol. 60, no. 6, pp. 3413–3434, June 2014.

[279] M. O. Rabin, "Randomized Byzantine generals," in *24th Annual Symposium on Foundations of Computer Science, FOCS*. IEEE, 1983, pp. 403–409.

[280] M. O. Rabin, "How to exchange secrets with oblivious transfer," Cryptology ePrint Archive, Report 2005/187, 2005, http://eprint.iacr.org/

[281] T. Rabin and M. Ben-Or, "Verifiable secret sharing and multiparty protocols with honest majority," in *Proc. ACM Symposium on Theory of Computing – STOC '89*. ACM, 1989, pp. 73–85.

[282] S. Ranellucci, A. Tapp, S. Winkler, and J. Wullschleger, "On the efficiency of bit commitment reductions," in *Advances in Cryptology – ASIACRYPT 2011*, Lecture Notes in Computer Science, vol. 7073. Springer, 2011, pp. 520–537.

[283] A. Rao, "Parallel repetition in projection games and a concentration bound," *SIAM Journal on Computing*, vol. 40, no. 6, pp. 1871–1891, 2011.

[284] K. S. Rao and V. M. Prabhakaran, "A new upperbound for the oblivious transfer capacity of discrete memoryless channel," in *2014 IEEE Information Theory Workshop (ITW)*. IEEE, 2014, pp. 35–39.

[285] R. Raz, "A parallel repetition theorem," *SIAM Journal on Computing*, vol. 27, no. 3, pp. 763–803, 1998.

[286] R. Renner, "Security of quantum key distribution," Ph.D. Dissertation, ETH Zurich, 2005.

[287] R. Renner, N. Gisin, and B. Kraus, "Information-theoretic security proof for quantum key distribution protocols," *Physical Review A*, vol. 72, no. 1, p. 012332, July 2005.

[288] R. Renner and S. Wolf, "New bounds in secret-key agreement: the gap between formation and secrecy extraction," in *Advances in Cryptology – EUROCRYPT 2003*, Lecture Notes in Computer Science, vol. 2656. Springer, 2003, pp. 562–577.

[289] R. Renner and S. Wolf, "Unconditional authentication and privacy from an arbitrary weak secret," in *Advances in Cryptology – CRYPTO 2003*, Lecture Notes in Computer Science, vol. 2729. Springer, 2003, pp. 78–95.

[290] R. Renner and S. Wolf, "Simple and tight bounds for information reconciliation and privacy amplification," in *Advances in Cryptology – ASIACRYPT 2005*, Lecture Notes in Computer Science, vol. 3788. Springer, 2005, pp. 199–216.

[291] A. Rényi, "On measures of entropy and information," in *Proc. Fourth Berkeley Symposium on Mathematical Statistics and Probability*, vol. 1. University of California Press, 1961, pp. 547–561.

[292] R. Rivest, "Unconditionally secure commitment and oblivious transfer schemes using private channels and trusted initializer," unpublished manuscript (available online), 1999.

[293] R. L. Rivest, A. Shamir, and L. Adleman, "A method for obtaining digital signatures and public-key cryptosystems," *Communications of the ACM*, vol. 21, no. 2, pp. 120–126, 1978.

[294] U. Rührmair and M. van Dijk, "On the practical use of physical unclonable functions in oblivious transfer and bit commitment protocols," *Journal of Cryptographic Engineering*, vol. 3, pp. 17–28, 2013.

[295] A. Russell and H. Wang, "How to fool an unbounded adversary with a short key," *IEEE Transactions on Information Theory*, vol. 52, no. 3, pp. 1130–1140, March 2006.

[296] D. Salsburg, *The Lady Tasting Tea: How Statistics Revolutionized Science in the Twentieth Century*. Henry Holt, 2002.

[297] M. Santha and U. Vazirani, "Generating quasi-random sequence from slightly-random sources," in *25th IEEE Annual Symposium on Foundations of Computer Science, FOCS*. IEEE, 1984, pp. 434–440.

[298] D. V. Sarwate, "A note on universal classes of hash functions," *Information Processing Letters*, vol. 19, pp. 41–45, 1980.

[299] G. Savvides, "Interactive hashing and reduction between oblivious transfer variants," Ph.D. Dissertation, McGill University, 2007.

[300] C. Schieler and P. Cuff, "Rate-distortion theory for secrecy systems," *IEEE Transactions on Information Theory*, vol. 60, no. 12, pp. 7584–7605, December 2014.

[301] C. Schieler and P. Cuff, "The henchman problem: measuring secrecy by the minimum distortion in a list," *IEEE Transactions on Information Theory*, vol. 62, no. 6, pp. 3426–3450, June 2016.

[302] C. P. Schnorr, "Efficient signature generation by smart cards," *Journal of Cryptology*, vol. 4, pp. 161–174, 1991.

[303] A. Shamir, "How to share a secret," *Communications of the ACM*, vol. 22, no. 11, pp. 612–613, 1979.

[304] C. E. Shannon, "A mathematical theory of communication," *Bell System Technical Journal*, vol. 27, pp. 379–423, 1948.

[305] C. E. Shannon, "Communication theory of secrecy systems," *Bell System Technical Journal*, vol. 28, pp. 656–715, 1949.

[306] J. Shikata, "Tighter bounds on entropy of secret keys in authentication codes," in *2017 IEEE Information Theory Workshop (ITW)*. IEEE, 2017, pp. 259–263.

[307] J. Shikata and D. Yamanaka, "Bit commitment in the bounded storage model: tight bound and simple optimal construction," in *Proc. IMA International Conference on Cryptography and Coding (IMACC)*, Lecture Notes in Computer Science, vol. 7089. Springer, 2011, pp. 112–131.

[308] G. J. Simmons, "Authentication theory/coding theory," in *Advances in Cryptology – CRYPTO '84*, Lecture Notes in Computer Science, vol. 196. Springer, 1985, pp. 411–431.

[309] G. J. Simmons, "A survey of information authentication," *Proceedings of the IEEE*, vol. 76, no. 5, pp. 603–620, May 1988.

[310] S. Singh, *The Code Book: The Secrets Behind Codebreaking*. Ember, 2003.

[311] M. Sipser, *Introduction to the Theory of Computation*, 3rd edition. Cengage Learning, 2013.

[312] D. Slepian and J. Wolf, "Noiseless coding of correlated information source," *IEEE Transactions on Information Theory*, vol. 19, no. 4, pp. 471–480, July 1973.

[313] J. Soni and R. Goodman, *A Mind at Play: How Claude Shannon Invented the Information Age*. Simon & Schuster, 2017.

[314] D. R. Stinson, "Combinatorial techniques for universal hashing," *Journal of Computer and System Sciences*, vol. 48, pp. 337–346, 1994.

[315] D. R. Stinson, "Decomposition constructions for secret sharing," *IEEE Transactions on Information Theory*, vol. 40, no. 1, pp. 118–125, January 1994.

[316] D. R. Stinson, "Universal hashing and authentication codes," *Designs, Codes and Cryptography*, vol. 4, pp. 369–380, 1994.

[317] D. R. Stinson, "On the connections between universal hashing, combinatorial designs and error-correcting codes," *Congressus Numerantium*, vol. 114, pp. 7–27, 1996.

[318] D. R. Stinson, *Cryptography: Theory and Practice*, 3rd edition. Chapman & Hall/CRC, 2006.

[319] C. M. Swanson and D. R. Stinson, "Unconditionally secure signature schemes revisited," *Journal of Mathematical Cryptology*, vol. 10, pp. 35–67, 2016.

[320] V. Y. F. Tan, *Asymptotic Estimates in Information Theory with Non-Vanishing Error Probabilities.* Hanover, MA: Now Publishers, 2014.

[321] K. Tochikubo, "Efficient secret sharing schemes realizing general access structures," *IEICE Transactions on Fundamentals of Electronics, Communications and Computer Sciences*, vol. E87-A, no. 7, pp. 1788–1797, July 2004.

[322] K. Tochikubo, "Efficient secret sharing schemes based on unauthorized subsets," *IEICE Transactions on Fundamentals of Electronics, Communications and Computer Sciences*, vol. E91-A, no. 10, pp. 2860–2867, October 2008.

[323] K. Tochikubo, "New secret sharing schemes realizing general access structure," *Journal of Information Processing*, vol. 23, no. 5, pp. 570–578, September 2015.

[324] K. Tochikubo, T. Uyematsu, and R. Matsumoto, "Efficient secret sharing schemes based on authorized subsets," *IEICE Transactions on Fundamentals of Electronics, Communications and Computer Sciences*, vol. E88-A, no. 1, pp. 322–326, January 2005.

[325] M. Tomamichel and M. Hayashi, "A hierarchy of information quantities for finite block length analysis of quantum tasks," *IEEE Transactions on Information Theory*, vol. 59, no. 11, pp. 7693–7710, November 2013.

[326] M. Tompa and H. Woll, "How to share a secret with cheaters," *Journal of Cryptology*, vol. 1, no. 3, pp. 133–138, 1988.

[327] R. Tonicelli, A. C. A. Nascimento, R. Dowsley, J. Müller-Quade, H. Imai, G. Hanaoka, and A. Otsuka, "Information-theoretically secure oblivious polynomial evaluation in the commodity-based model," *International Journal of Information Security*, vol. 14, pp. 73–84, 2015.

[328] L. Trevisan, "Cryptography," Lecture Notes, Stanford University, 2009.

[329] W. Tu and L. Lai, "Keyless authentication and authenticated capacity," *IEEE Transactions on Information Theory*, vol. 64, no. 5, pp. 3696–3714, May 2018.

[330] P. Tuyls, B. Škorić, and T. Kevenaar (Eds), *Security with Noisy Data.* Springer, 2007.

[331] H. Tyagi, "Common information and secret key capacity," *IEEE Transactions on Information Theory*, vol. 59, no. 9, pp. 5627–5640, 2013.

[332] H. Tyagi and S. Watanabe, "A bound for multiparty secret key agreement and implications for a problem of secure computing," in *Advances in Cryptology – EUROCRYPT 2014*, Lecture Notes in Computer Science, vol. 8441. Springer, 2014, pp. 369–386.

[333] H. Tyagi and S. Watanabe, "Converses for secret key agreement and secure computing," *IEEE Transactions on Information Theory*, vol. 61, pp. 4809–4827, 2015.

[334] H. Tyagi and S. Watanabe, "Universal multiparty data exchange and secret key agreement," *IEEE Transactions on Information Theory*, vol. 63, pp. 4057–4074, April 2017.

[335] H. Tyagi and S. Watanabe, "A new proof of nonsignaling multiprover parallel repetition theorem," in *Proc. IEEE International Symposium on Information Theory (ISIT)*. IEEE, 2019, pp. 967–971.

[336] O. Uchida and T. S. Han, "The optimal overflow and underflow probabilities of variable-length coding for the general sources," *IEICE Transactions on Fundamentals of Electronics, Communications and Computer Sciences*, vol. E84-A, no. 10, pp. 2457–2465, October 2001.

[337] T. Uyematsu, "A new unified method for fixed-length source coding problems of general sources," *IEICE Transactions on Fundamentals of Electronics, Communications and Computer Sciences*, vol. E93-A, no. 11, pp. 1868–1877, November 2010.

[338] S. P. Vadhan, "A study of statistical zero-knowledge proofs," Ph.D. Dissertation, Harvard University, 1999.

[339] M. van Dijk, "On a special class of broadcast channels with confidential messages," *IEEE Transactions on Information Theory*, vol. 43, no. 2, pp. 712–714, March 1997.

[340] S. Vembu and S. Verdú, "Generating random bits from an arbitrary source: fundamental limits," *IEEE Transactions on Information Theory*, vol. 41, no. 5, pp. 1322–1332, 1995.

[341] J. von Neumann, "Various techniques used in connection with random digits," *National Bureau of Standards, Applied Math Series*, vol. 12, pp. 36–38, 1951.

[342] N. Walenta, A. Burg, D. Caselunghe, J. Constantin, N. Gisin, O. Guinnard, R. Houlmann, P. Junod, B. Korzh, N. Kulesza, M. Legré, C. W. Lim, T. Lunghi, L. Monat, C. Portman, M. Soucarros, R. T. Thew, P. Trinkler, G. Trolliet, F. Vannel, and H. Zbinden, "A fast and versatile quantum key distribution system with hardware key distribution and wavelength multiplexing," *New Journal of Physics*, vol. 16, p. 013047, January 2014.

[343] P. Wallden, V. Dunjko, A. Kent, and E. Andersson, "Quantum digital signature with quantum-key-distribution components," *Physical Review A*, vol. 91, no. 4, p. 042304, April 2015.

[344] Y. Wang, P. Ishwar, and S. Rane, "An elementary completeness proof for secure two-party computation primitives," in *2014 IEEE Information Theory Workshop (ITW)*. IEEE, 2014, pp. 521–525.

[345] S. Watanabe, R. Matsumoto, T. Uyematsu, and Y. Kawano, "Key rate of quantum key distribution with hashed two-way classical communication," *Physical Review A*, vol. 76, no. 3, p. 032312, September 2007.

[346] S. Watanabe and Y. Oohama, "Secret key agreement from vector Gaussian sources by rate limited public communication," *IEEE Transactions on Information Forensics and Security*, vol. 6, no. 3, pp. 541–550, 2011.

[347] M. N. Wegman and J. L. Carter, "New hash functions and their use in authentication and set equality," *Journal of Computer and System Sciences*, vol. 22, pp. 265–279, 1981.

[348] M. M. Wilde, *Quantum Information Theory*. Cambridge University Press, 2017.

[349] S. Winkler and J. Wullschleger, "On the efficiency of classical and quantum secure function evaluation," *IEEE Transactions on Information Theory*, vol. 60, no. 6, pp. 3123–3143, June 2014.

[350] A. Winter, A. C. A. Nascimento, and H. Imai, "Commitment capacity of discrete memoryless channels," in *Proc. IMA International Conference on Cryptography and Coding (IMACC)*, Lecture Notes in Computer Science, vol. 2898. Springer, 2003, pp. 35–51.

[351] S. Wolf, "Unconditional security in cryptography," in *Lectures on Data Security*, Lecture Notes in Computer Science, vol. 1561. Springer, 1999, pp. 217–250.

[352] S. Wolf and J. Wullschleger, "Oblivious transfer is symmetric," in *Advances in Cryptology – EUROCRYPT 2006*, Lecture Notes in Computer Science, vol. 4004. Springer, 2006, pp. 222–232.

[353] S. Wolf and J. Wullschleger, "New monotones and lower bounds in unconditional two-party computation," *IEEE Transactions on Information Theory*, vol. 54, no. 6, pp. 2792–2797, June 2008.

[354] A. D. Wyner, "The common information of two dependent random variables," *IEEE Transactions on Information Theory*, vol. 21, no. 2, pp. 163–179, March 1975.

[355] H. Yamamoto, "Secret sharing system using (k, l, n) threshold scheme," *Electronics and Communications in Japan*, vol. 69, no. 9, pp. 945–952, 1986. English translation of Japanese version published in 1985.

[356] H. Yamamoto, "Information theory in cryptography," *IEICE Transactions on Fundamentals of Electronics, Communications and Computer Sciences*, vol. E74-A, no. 9, pp. 2456–2464, September 1991.

[357] H. Yamamoto, "Coding theorems for Shannon's cipher system with correlated source outputs, and common information," *IEEE Transactions on Information Theory*, vol. 40, no. 1, pp. 85–95, January 1994.

[358] H. Yamamoto, "Rate-distortion theory for the Shannon cipher system," *IEEE Transactions on Information Theory*, vol. 43, no. 3, pp. 827–835, May 1997.

[359] W. Yang, R. F. Schaefer, and H. V. Poor, "Wiretap channels: nonasymptotic fundamental limits," *IEEE Transactions on Information Theory*, vol. 65, no. 7, pp. 4069–4093, July 2019.

[360] A. C. Yao, "Protocols for secure computations," in *23rd Annual Symposium on Foundations of Computer Science, FOCS*. IEEE, 1982, pp. 160–164.

[361] A. C. Yao, "Theory and applications of trapdoor functions," in *23rd Annual Symposium on Foundations of Computer Science, FOCS*. IEEE, 1982, pp. 80–91.

[362] Z. Zhang, "Estimating mutual information via Kolmogorov distance," *IEEE Transactions on Information Theory*, vol. 53, no. 9, pp. 3280–3282, September 2007.

Symbol Index

Index